High-Strain Zones: Structure and Physical Properties

Special Publication reviewing procedures

The Society makes every effort to ensure that the scientific and production quality of its books matches that of its journals. Since 1997, all book proposals have been refereed by specialist reviewers as well as by the Society's Books Editorial Committee. If the referees identify weaknesses in the proposal, these must be addressed before the proposal is accepted.

Once the book is accepted, the Society has a team of Book Editors (listed above) who ensure that the volume editors follow strict guidelines on refereeing and quality control. We insist that individual papers can only be accepted after satisfactory review by two independent referees. The questions on the review forms are similar to those for *Journal of the Geological Society*. The referees' forms and comments must be available to the Society's Book Editors on request.

Although many of the books result from meetings, the editors are expected to commission papers that were not presented at the meeting to ensure that the book provides a balanced coverage of the subject. Being accepted for presentation at the meeting does not guarantee inclusion in the book.

Geological Society Special Publications are included in the ISI Index of Scientific Book Contents, but they do not have an impact factor, the latter being applicable only to journals.

More information about submitting a proposal and producing a Special Publication can be found on the Society's web site: www.geolsoc.org.uk.

It is recommended that reference to all or part of this book should be made in one of the following ways:

Bruhn, D. & Burlini, L. (eds) 2005. *High-Strain Zones: Structure and Physical Properties*. Geological Society, London, Special Publications, **245**.

Resor, P.G. & Snoke, A.W. 2005. Laramie Peak shear system, central Laramie Mountains, Wyoming, USA: regeneration of the Archean Wyoming province during Palaeoproterozoic accretion. *In*: Bruhn, D. & Burlini, L. (eds). *High-Strain Zones: Structure and Physical Properties*. Geological Society, London, Special Publications, **245**, 81–107.

GEOLOGICAL SOCIETY SPECIAL PUBLICATION NO. 245

High-Strain Zones: Structure and Physical Properties

EDITED BY

DAVID BRUHN

GFZ Potsdam, Germany

and

LUIGI BURLINI

Geological Institute, ETH-Z, Zurich, Switzerland

2005
Published by
The Geological Society
London

THE GEOLOGICAL SOCIETY

The Geological Society of London (GSL) was founded in 1807. It is the oldest national geological society in the world and the largest in Europe. It was incorporated under Royal Charter in 1825 and is Registered Charity 210161.

The Society is the UK national learned and professional society for geology with a worldwide Fellowship (FGS) of 9000. The Society has the power to confer Chartered status on suitably qualified Fellows, and about 2000 of the Fellowship carry the title (CGeol). Chartered Geologists may also obtain the equivalent European title, European Geologist (EurGeol). One fifth of the Society's fellowship resides outside the UK. To find out more about the Society, log on to www.geolsoc.org.uk.

The Geological Society Publishing House (Bath, UK) produces the Society's international journals and books, and acts as European distributor for selected publications of the American Association of Petroleum Geologists (AAPG), the American Geological Institute (AGI), the Indonesian Petroleum Association (IPA), the Geological Society of America (GSA), the Society for Sedimentary Geology (SEPM) and the Geologists' Association (GA). Joint marketing agreements ensure that GSL Fellows may purchase these societies' publications at a discount. The Society's online bookshop (accessible from www.geolsoc.org.uk) offers secure book purchasing with your credit or debit card.

To find out about joining the Society and benefiting from substantial discounts on publications of GSL and other societies worldwide, consult www.geolsoc.org.uk, or contact the Fellowship Department at: The Geological Society, Burlington House, Piccadilly, London W1J 0BG: Tel. +44 (0)20 7434 9944; Fax +44 (0)20 7439 8975; E-mail: enquiries@geolsoc.org.uk.

For information about the Society's meetings, consult *Events* on www.geolsoc.org.uk. To find out more about the Society's Corporate Affiliates Scheme, write to enquiries@geolsoc.org.uk.

Published by The Geological Society from:
The Geological Society Publishing House
Unit 7, Brassmill Enterprise Centre
Brassmill Lane
Bath BA1 3JN, UK

Orders: Tel. +44 (0)1225 445046
          Fax +44 (0)1225 442836
Online bookshop: www.geolsoc.org.uk/bookshop

The publishers make no representation, express or implied, with regard to the accuracy of the information contained in this book and cannot accept any legal responsibility for any errors or omissions that may be made.

British Library Cataloguing in Publication Data

A catalogue record for this book is available from the British Library.

ISBN 1-86239-178-5

Typeset by Techset Composition, Salisbury, UK
Printed by The Alden Press, Oxford, UK

Distributors

USA
  AAPG Bookstore
  PO Box 979
  Tulsa
  OK 74101-0979
  USA
  Orders: Tel. +1 918 584-2555
            Fax +1 918 560-2652
            E-mail bookstore@aapg.org

India
  Affiliated East-West Press Private Ltd
  Marketing Division
  G-1/16 Ansari Road, Darya Ganj
  New Delhi 110 002
  India
  Orders: Tel. +91 11 2327-9113/2326-4180
            Fax +91 11 2326-0538
            E-mail affiliat@vsnl.com

Japan
  Kanda Book Trading Company
  Cityhouse Tama 204
  Tsurumaki 1-3-10
  Tama-shi, Tokyo 206-0034
  Japan
  Orders: Tel. +81 (0)423 57-7650
            Fax +81 (0)423 57-7651
            E-mail geokanda@ma.kcom.ne.jp

Contents

Preface

Most of the deformation on Earth is concentrated in relatively narrow high-strain zones (e.g. plate boundaries). The purpose of this volume was to address different aspects dealing with high-strain zones, from the map scale to the processes active in high-strain zones to the physical properties of highly strained rocks. Several of the contributions were originally presented in a special session entitled 'High-Strain Zones' at the EGS-AGU-EUG meeting 2003 in Nice, France, which inspired the compilation of this book.

An introduction to the subject of deformation in high-strain zones is given by **Burlini & Bruhn**. They discuss recent developments in laboratory studies on high strain with an emphasis on torsion and the consequences of the results of these studies on our understanding of strain-softening mechanisms. The other contributions are grouped by common themes. To some extent the grouping of papers seems arbitrary, as some papers address several of the aspects chosen as headers for the different thematic groups. In the first group, on high-strain zones in nature, studies often include investigations on physical properties or the interaction of deformation and metamorphism. In their contribution on the Kohistan palaeo-island arc, in Pakistan, **Burg et al.** present the result of many years of fieldwork in the area. Their detailed investigations show that the shear zones in the area are Pre-Himalayan and synmagmatic, probably consistent with the bulk flow direction of the subduction zone and may be mistaken for structures documenting continental collision. In their contribution they include a complete cross section (foldout) which can be used as an excellent example of the mantle crustal transition of an exhumed island arc. In the next contribution, **Teyssier et al.** discuss the collapse, partial melting and detachment in parts of the Canadian cordillera in a combined study of anisotropy of magnetic susceptibility (AMS), microstructure and thermochronology. The following paper is a geometric analysis of a shear zone, which was repeatedly intruded by granitic veins during deformation, presented by **Passchier et al.** Based on this analysis, they discuss potential constraints on the stress field and the mode of deformation during synkinematic vein intrusion. In the next paper, **Resor & Snoke** present a detailed study of a Precambrian shear zone in the Laramie Mountains of Wyoming, USA. The last paper in this group is by **Edwards & Ratschbacher**, who investigated a Himalayan fault zone that exposes a range of exhumed fault rock assemblages, allowing the study of deeper crustal deformation processes, fault zone rheology and possible weakening mechanisms.

The next group of papers addresses the physical properties of high-strain zones, both in field studies and through laboratory investigations. **Sidman et al.** describe the development of magnetic fabrics across a mylonitic shear zone and present a rolling-hinge model to explain the rotation of fabrics forming a wide arch when approaching the shear zone. In the next contribution, **Ritter et al.** compare field electrical resistivity models of fault zones in diverse environments. The subsequent contribution is a laboratory study on the seismic properties at pressure and temperature of progressively deformed rocks by **Burlini et al.**, with an emphasis on the effect of temperature. Then **Schubnel et al.** present results of laboratory measurements of compressional and shear wave velocities in samples of limestone as they are stressed to brittle failure. Finally, the laboratory study by **Baud et al.** addresses the issue of anisotropy induced by bedding or foliation and its effect on rock physical properties, on brittle strength, and on the brittle–ductile transition in porous rocks.

In the next group of papers about the rheology of high-strain zones, **Mainprice & Paterson** contribute results on laboratory deformation of flint. Their study presents a comprehensive set of approaches, including constant strain rate triaxial compression and load relaxation tests, microstructural and fabric analysis, IR spectroscopy, and application of the results to test micromechanical models of dislocation and diffusion creep. The next paper is a study on the rheology of gypsum deformed to high strains by **Barberini et al.**, who present new experimental data that show evolution of deformational behaviour with increasing strain and during ongoing dehydration. **Bruhn et al.** then address the reciprocal effect of grain size and melt distribution on the rheology of partially molten peridotites. In the next paper **Niemeijer & Spiers** present experimental results and modelling of phyllonite generation and its effects on fault strength in the brittle–ductile transition. The last contribution in this group is by **Evans**, who reviews potential ways to define mechanical laws for rocks undergoing plastic deformation to high strains. The emphasis of this review paper is on the mechanical equation of state approach to studying deformation of rocks with evolving microstructures.

An area that has received increasing attention, mainly in the field but also by some laboratory studies is the interaction of deformation and metamorphism. Although several of the papers in the first group also address the subject, it is the central issue of the two contributions in this group. One of the major questions in this context is that of reaction weakening, which has preoccupied structural geologists for almost 25 years now, but its mechanical effects on large-scale tectonic systems such as foreland thrusts, as presented here by **Wibberley**, have only recently come under scrutiny. The other contribution in this group is by **Rossi et al.**, who address the chemical changes in the host rock induced by the formation of a shear zone in the Mont Blanc Massif in the French–Italian Alps.

New methods in the investigation of phenomena related to high strain in rocks are the subject of two papers in the next group. **Muto et al.** combined cathodoluminescence observations with IR-spectroscopy measurements in deformed quartzite samples to reveal hydrogen distribution patterns, **Fernández et al.** developed a new approach to analyse shape fabrics and determine finite strain using a Geographical Information System (GIS).

The last group in this compilation includes two contributions with a numerical modelling approach to high-strain zones. **Schmid** investigates rotation and pressure perturbations in and around rigid polygons in shear zones using a combination of finite element method calculations with Muskhelishvili-type analytical solutions instead of the commonly employed ellipsoid-based shape approximation. Then **Issen & Challa** present a comprehensive explanation and summary of the bifurcation approach for analysis of strain localization in porous rocks. Using this approach they clarify they conditions under which compaction and dilation bands may develop under axisymmetric compression and extension loading.

In conclusion, we would like to express our thanks to the many authors and reviewers who provided manuscripts and comments to this special volume. We are also indebted to those who helped by discussing scientific and organisational issues, first of all our scientific series editor Bob Holdsworth, and all others who are too numerous to name. Our special thanks go to Angharad Hills, Sarah Joomun and Sally Oberst at Geological Society Publishing House who so patiently put up with all the delays and shortcomings during the editorial and publication processes.

David Bruhn
Luigi Burlini

High-strain zones: laboratory perspectives on strain softening during ductile deformation

L. BURLINI[1] & D. BRUHN[2]

[1]ETH-Zürich, Geologisches Institut, 8092 Zürich, Switzerland
(e-mail: burlini@erdw.ethz.ch)

[2]GeoForschungsZentrum Potsdam, Telegrafenberg, 14473-Potsdam,
Germany (e-mail: dbruhn@gfz-potsdam.de)

Abstract: Deformation in the Earth's outer shell is mostly localized into narrow high-strain zones. Because they can have displacements up to several hundreds or thousands of kilometres, they can affect the entire lithosphere. The properties of high-strain zones control the kinematics and dynamics of our planet, and are therefore of key importance for an understanding of plate tectonics, stress accumulation and release (e.g. earthquakes), mountain building, etc.

One of the requirements of shear zone formation in ductile rocks is localized strain softening (Hobbs *et al.* 1990). In this paper we review the strain softening mechanisms that were identified and proposed 25 years ago and analyse their relevance in light of recent experimental results conducted to large strains. For this purpose, some of the newer developments in experimental deformation techniques that permit high strain in torsion are summarized and recent results are reviewed. Using these results we discuss mechanisms, processes and conditions that lead to localization.

Since the first deployment of GPS (Global positioning system) to monitor plate motion in the early 1980s (e.g. Engelis *et al.* 1984), a major effort has been made by the international scientific community to achieve a complete view of the Earth's surface kinematics. For example the ILP Project (II-8 A Global strain rate map) resulted in a map of plate motions (Kreemer *et al.* 2000) and the second strain rate invariant (Kreemer *et al.* 2003); (Fig. 1). Blank areas are intraplate areas where the strain rates are known to be very low and that have been considered as rigid spherical caps (for details see Kreemer *et al.* 2003). In contrast, the local strain rate at plate boundaries is several orders of magnitude higher and is restricted to narrow zones, even within more diffuse plate boundary zones such as central Asia. Since displacements along these high strain rate zones are of the order of hundreds of kilometres or more, they must affect the entire lithosphere. Paterson (2001) pointed out that to model the rheology of the crust, one should consider it to be divided into blocks separated by faults (upper crust, where deformation is time-independent) and shear zones (lower crust or even mantle, where the deformation is time-dependent). The bulk rheology depends predominantly upon the behaviour of these shear zones.

In trying to summarize and define the status quo of our knowledge about the formation and mechanical properties of shear zones it is helpful to take a look at a prescient summary published a quarter of a century ago by White *et al.* (1980). This and other publications we refer to were the result of a conference on shear zones in rocks held in Barcelona, Spain in 1979. We will follow the arguments of White *et al.* (1980) and include other publications which attempted to define the status quo at the time, for example the Penrose conference report on mylonites by Tullis *et al.* (1982). Although the general topic is broad, we will focus on one area, the formation of ductile shear zones due to localized relative strain softening. The questions posed at that time are discussed in the light of new results. Other important issues related to shear zones include reactivation, brittle faulting, and the role of high-strain zones as conduits for fluids, to name but a few, but these will only be addressed in so far as they affect the question of deformation mechanisms in ductile high-strain zones.

Deformation and deformation mechanisms at depth in high strain zones

Shear zones are generally divided into brittle fault rocks (including cataclasite and pseudotachylite)

From: BRUHN, D. & BURLINI, L. (eds) 2005. *High-Strain Zones: Structure and Physical Properties.*
Geological Society, London, Special Publications, **245**, 1–24.
0305-8719/05/$15.00 © The Geological Society of London 2005.

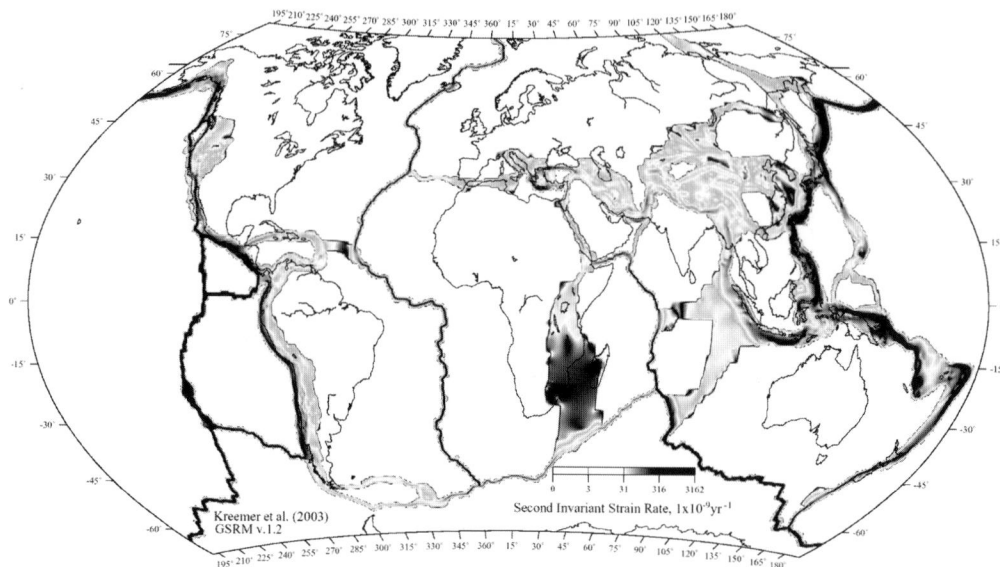

Fig. 1. Second invariant of crustal strain rate associated with plate motion and plate boundary deformation (C. Kreemer pers. comm., 2005). Note that high-strain rate is concentrated in narrow zones which, apart from oceanic ridges, are faults and subduction zones, i.e. narrow area where the deformation is localized. (Updated from Kreemer, C., Holt, W.E. & Haines, A.J. 2003. An integrated global model of present-day plate motions and plate boundary deformation. *Geophysical Journal International*, **154**, 8–34.)

and ductile fault rocks (including mylonites and ultramylonites). The ductile fault rocks are thought to deform primarily by intracrystalline plastic mechanisms, with brittle deformation playing a subordinate role. In contrast, brittle fault rocks are the product of comminution (cataclasis) and frictional melting (pseudotachylite generation) in the brittle regime. The presence of pseudotachylite is generally accepted to represent direct evidence of seismic activity of a fault zone, as it is thought to form by local melting of the rock due to heat generated by rapid frictional sliding (e.g. Sibson 1975; Spray 1987, 1992).

In general, the dominant deformation mechanism in the shear deformation of rocks changes with depth due to the competing effects of temperature, which enhances thermally activated deformation mechanisms (intracrystalline plasticity and diffusion creep) and stress, which enhances frictional sliding. Since the latter occurs typically with an increase of volume (dilatancy), it is suppressed by the confining pressure due to depth of burial. Therefore at depth, deformation is macroscopically plastic whereas it is cataclastic close to the surface. Cataclastic deformation may also occur at great depth, provided that there is enough pore pressure caused by the presence of fluids (vapour, melt, etc), according to Terzaghi's effective stress rule which

applies to rock failure. The influence of pressure, temperature, differential stress (and strain rate) on the different deformation mechanisms is illustrated in Figure 2.

However, this distinction of dominant deformation mechanisms with increasing depth is only applicable for monomineralic materials. In polymineralic rocks, plastic deformation in one phase often occurs in parallel with cataclastic flow in another. Evidence for such behaviour is found frequently, for example in microstructures of quartz–feldpathic mylonites, where feldspar clasts are broken in a fine-grained quartz-rich matrix that shows evidence of plastic deformation. Whichever the deformation mechanism, deformation may either be localized or diffuse: in the brittle field one may observe localized faulting or cataclastic flow or, in the ductile regime, pervasive deformation or formation of localized high-strain zones.

Strain softening

A wealth of observations demonstrates that ductile intracrystalline plastic deformation is often localized, and that localization occurs on a wide range of scales. The question how strain becomes localized in rocks deforming at nominally constant volume and what factors serve to keep it concentrated in relatively narrow zones,

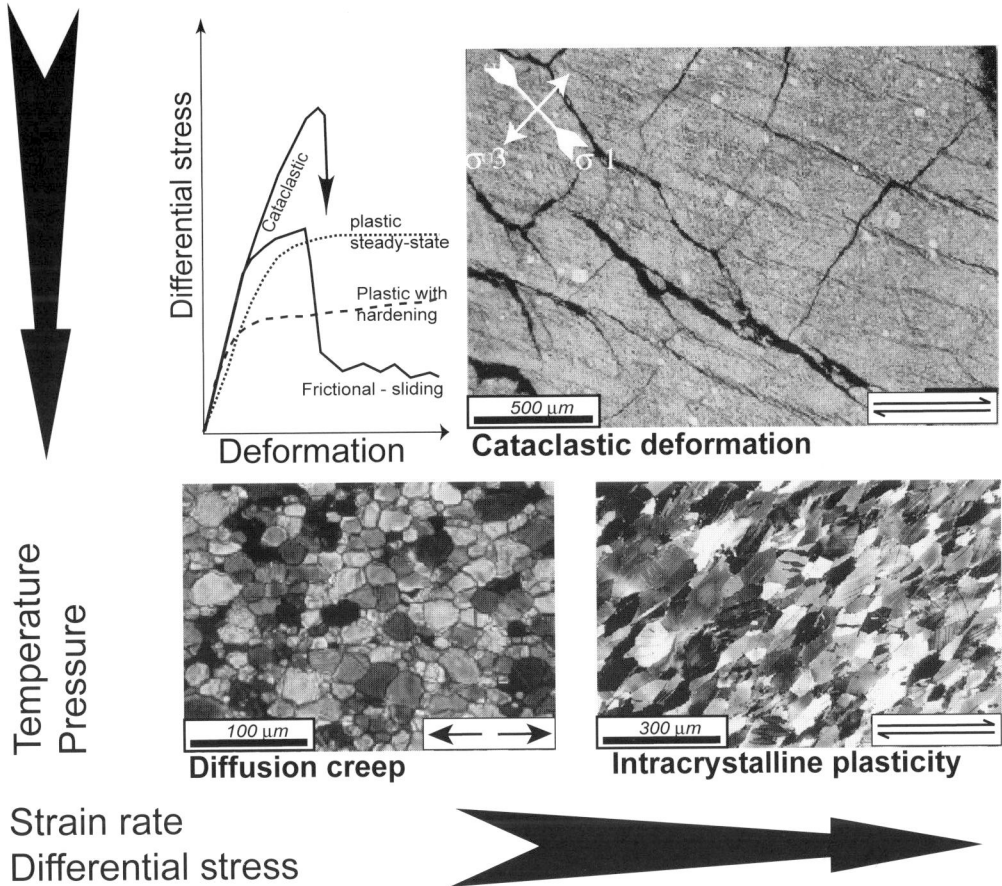

Fig. 2. Relationships between deformation mechanisms with pressure, temperature, stress and strain rate. The three main microstructures and relative schematic stress–strain curves are shown. The microphotograph illustrating the cataclastic deformation is a specimen of quartz aggregate deformed in torsion, which developed cracks oriented with their normal parallel to the least stress; the illustration of the dislocation creep is a specimen of Carrara marble deformed in torsion at about $\gamma = 1$ at 727 °C and strain rate of $3 \times 10^{-4}\,\mathrm{s}^{-1}$. The illustration of diffusion creep is a specimen of synthetic calcite (much finer grain size with respect to the Carrara marble) deformed in extension at 700 °C and strain rate of about $10^{-6}\,\mathrm{s}^{-1}$.

even in homogeneous rocks, was defined as a key problem at the Penrose conference on mylonites in San Diego, California in 1981 (Tullis *et al.* 1982). Maintenance of localized flow requires local weakening relative to the host rock. Several strain softening mechanisms have been proposed, all of which could work separately or in combination. Many of the arguments summarized at the Barcelona meeting were based on field observations and on results of laboratory experiments on the deformation of metals and rocks (Poirier 1980; White *et al.* 1980).

Following the arguments by White *et al.* (1980) and Poirier (1980) plastic strain softening can occur by a number of processes which will be discussed below. We have added the last process

in the list (partial melting), even though it was not explicitly mentioned in the papers resulting from the Barcelona conference, but it is a process that has received increased attention since the late 1980s.

Suggested strain softening mechanisms include:

1 Change in deformation mechanism from dislocation creep to grain size sensitive flow as a consequence of grain size reduction by recrystallization or cataclasis.
2 Continual or cyclic dynamic recrystallization produces new grains free of internal strain.

3   Geometric, or fabric, softening through the
     development of a crystallographic preferred
     orientation (CPO), where easy slip systems
     rotate into parallelism with stretching linea-
     tion and align favourably to the macro-
     scopic shear with progressive deformation.
4   Reaction softening would aid softening by
     producing small, strain-free grains promot-
     ing a change in deformation mechanism.
     In addition, the formation of new, soft reac-
     tion products may further reduce the
     strength of the rock.
5   Chemical softening occurs by changing the
     trace element content or point defect con-
     centration of a mineral, for example in
     contact with water.
6   Pore fluid effects reduce the strength of a
     rock if pore pressure is increased by lower-
     ing the fracture strength and favour
     embrittlement due to the lower effective
     stress.
7   Shear heating or viscous dissipation can
     lead to an increased temperature due to
     deformation in a narrow zone, which
     enhances plastic deformation.
8   Structural softening can be caused in poly-
     phase rocks by the evolution with strain of
     the aggregate structure towards an intercon-
     nected weak layer configuration.
9   Thermal softening/mineral phase changes
     occur when minerals undergo a phase tran-
     sition and possibly induce a catastrophic
     softening by transformation plasticity or
     fluid weakening.
10   Partial melting can lead to considerable
     weakening, as melts usually have a much
     lower viscosity than solid rocks.

These mechanisms and their relevance in the formation of shear zones will be discussed in the
context of latest experimental and theoretical
progress.

High-strain deformation in the laboratory

Our knowledge of the mechanical behaviour,
microstructural evolution and physical properties
of rocks from high-strain zones is mostly derived
from laboratory experiments on rock specimens,
rock physics and measurements of microstructure and texture by electron microscopy, texture
goniometry and electron backscattered diffraction (EBSD). In general, the studies that generated descriptions of the flow behaviour of rocks
at high temperatures have assumed that a
steady-state distributed flow develops after an
initial transient creep period under constant
stress, temperature and strain rate, and that this

steady state continues indefinitely (e.g. Carter
1976). Studies of naturally deformed rocks,
however, show that at high strains, flow commonly tends to localize into shear zones, which
are microstructurally very different from their
protoliths. Strength characteristics by implication must also evolve progressively during
large plastic strain increments (e.g. Rutter 1999).

To address the question of localization and
strain softening in the laboratory, new experimental techniques and devices that allow deformation to high strains have been developed. For
simple shear strain paths these include the so-
called diagonal cut cylinder (or direct shear
test) and torsion testing (Fig. 3). For axisymmetric strain paths the limitations of compression
testing have been mentioned, but high strains can
be obtained in axisymmetric extension (up to
logarithmic strains of 2.5) by taking advantage
of necking instabilities (Rutter 1998). In the
diagonal cut assembly the axial movement of
the piston is converted to lateral shear displacement due to the geometrical arrangement (e.g.
Zhang & Karato 1995; Dell'Angelo & Tullis
1989). This has some advantages, since the technique is relatively simple and secure, and can be
employed in gas and solid medium apparatus to
very high confining pressures. The disadvantages
are that amount of lateral displacement of the
forcing blocks is limited, and to reach large
strains the thickness of the deforming sample is
limited (e.g. few hundreds of micrometres),
which puts tight constraints on the sample grain
size that can be used. Moreover, the tendency
of the column to buckle can produce apparent
hardening, which can preclude its use of the
results for rheological determination. With progressive strain, the surface area of the specimen
changes, therefore the load must be continuously
corrected to determine the shear stress. In
addition, the thickness of the sheared zone may
change during deformation, which poses problems for the determination of the real state of
stress and strain within the resulting shear zone
(simple shear and pure shear components
cannot be separated directly, even if this
becomes less of a problem the thinner the specimen). With the use of strain markers in the
sample, it has become possible to separate the
two strain components, but only the initial state
and final state can be defined. Some of these
limitations can be overcome with the use of the
torsion technique (Paterson & Olgaard 2000).

The first studies on torsion deformation at
room temperature on natural rocks were by
Handin et al. (1960). The first high temperature
(up to 1600 K), high pressure (up to 500 MPa)
deformation apparatus developed for torsion

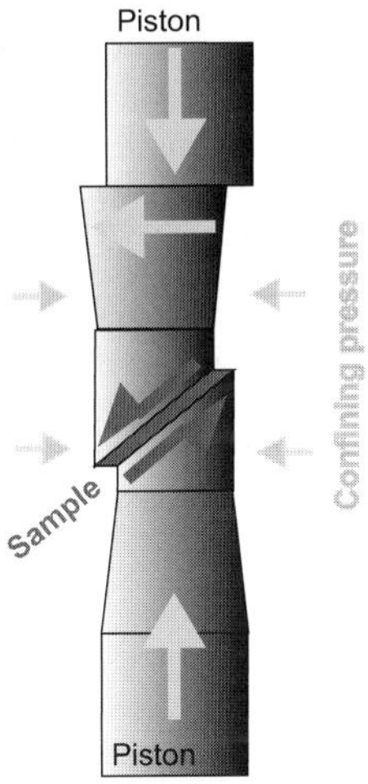

	Diagonal cut cylinder		Torsion
	Axial movement of piston	Displacement induced by	Rotation of the piston
	Axial load. It is converted to shear stress by the geometry. An axial component is always present	Force transmitted to the sample by	Torque. It is transmitted through interfaces by the friction generated by the confining pressure.
	Often heterogeneous, expecially at the edge. Variable with strain. Difficult to separate pure from simple shear components.	Stress distribution	Max at the rim and 0 at the cylinder axis. Varies according to the rheology of the sample (increases c.a. with the cube of radius). Approximate simple shear quite well.
	Often heterogeneous, expecially at the edge. Strain markers are used to separate pure from simple shear deformation.	Strain distribution	Linearly variable from the centre (zero) to the rim of the cylinder (max).
	About $\gamma = 7$	Max. shear strain obtained so far	About $\gamma = 50$

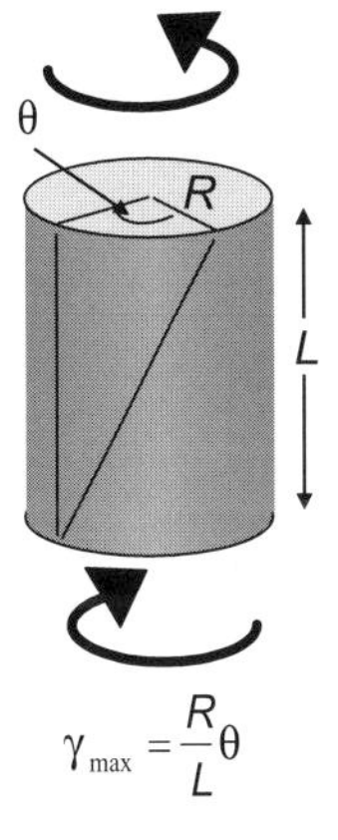

$$\gamma_{max} = \frac{R}{L}\theta$$

Fig. 3. Comparison between the diagonal cut cylinder and the torsion techniques in rock deformation. These are the actual techniques used to reach large non coaxial strains.

testing of geological materials (Paterson & Olgaard 2000) was installed in 1995 at the Rock Deformation Laboratory, ETH Zürich. Since then, similar apparatus has been installed at several other laboratories around the world.

This apparatus represents a breakthrough in research into high strain and its effects on microstructures and physical properties of rocks. It then became possible to reach strains comparable to those of natural shear zones (e.g. $\gamma = 50$, i.e. one order of magnitude higher than in the diagonal cut assembly). Moreover, the torque variation with strain gave the first complete picture of the mechanical property evolution occurring during the dynamic recrystallization of the deformed rock. The disputed concept of steady-state deformation was answered with steady state being reached only after the rock microstructure was entirely refreshed (Barnhoorn *et al.* 2004). The main limitation of torsion today is given by the maximum torque that can be accessed. The torque is transmitted from the motor to the specimen through friction occurring at the interfaces between pistons and sample, which is enhanced by the confining pressure. This limits the use of fluids (vapour, melts etc.), since the pore pressure reduces the normal stress across the area between sample and piston. Another point of contention is that stress and strain vary across the sample from the rim to the centre. Strain varies linearly, but the stress (and the torque) varies in a manner that depends on the rheology (Paterson &

Olgaard 2000). If the rheology is constant throughout the sample, two samples deformed at the same angular displacement rate are sufficient to determine the distribution of stress and torque within the probing cylinder. Otherwise, if an abrupt change in deformation mechanism occurs, five samples are needed. In case of more complicated variations it is not possible to determine local stress and torque. Such stress calibrations showed that the higher the stress exponent in a flow law, the smaller the influence of the part of the sample which is towards the centre (Barnhoorn 2004). In other words, the part of the sample located inside two thirds of the radius affects the total torque by only about 20% if the stress exponent is equal to or higher than 3. In this case the variation of the shear stress from the exterior surface to the two thirds of the radius is of only a few percent. The central part can therefore be ignored for rheological determinations. Moreover it must be noted that the stress exponent is generally determined only on two or three orders of magnitude of strain rate. Strain rate itself was relatively fast (generally from 10^{-3} to 10^{-5} s^{-1}) in most of the studies, due to the large amount of strain aimed in a laboratory convenient timescale.

Extensive studies in torsion were conducted on natural and synthetic monomineralic aggregates of anhydrite (Heidelbach *et al.* 2001), calcite (Casey *et al.* 1998; Pieri *et al.* 2001a, b; Barnhoorn *et al.* 2004), olivine (Bystricky *et al.* 2000), quartz (Schmocker *et al.* 2003),

plagioclase (Bystricky pers. com.), magnesiowüstite (Heidelbach et al. 2003) and two-phase
aggregates under stable conditions (calcite and
quartz, Rybacki et al. 2003; calcite and anhydrite
Barnhoorn et al. 2003; quartz and anorthite, Ji
et al. 2004), or on rocks during metamorphic
reactions, such as dehydration of serpentinite
(Hirose et al. 2004) and gypsum (Barberini
et al. this volume).

Most of the monomineralic aggregates
deformed homogeneously (Fig. 4), and reached
a mechanical steady-state flow stress only after
shear strain >5, after initial hardening was
followed by a weakening phase (Fig. 5).

Weakening was important in anhydrite (50%)
(Fig. 6), probably induced by extensive recrystallization and grain refinement which led to a
switch in deformation mechanism from dislocation to diffusion creep (Heidelbach et al.
2001). Weakening was less pronounced for
calcite and olivine, which deformed mainly in
the dislocation creep regime (Bystricky et al.
2000; Barnhoorn et al. 2004). In both
minerals, the microstructure changed remarkably, from a deformation microstructure at low
strains (where grains were essentially stretched,
maintaining their volume almost entirely) to a
recrystallization microstructure (Pieri et al.
2001b), where new grains developed, beginning
at the grain boundaries and then replacing all
the old grains, as shown in the series of microstructural pictures taken at progressive strains
(Fig. 7).

Recrystallization occurred both by progressive
subgrain rotation and nucleation and grain
boundary migration (Bystricky et al. 2000;
Pieri et al. 2001b). Lattice preferred orientation
(LPO) increased progressively with strain, commonly with changes of pattern at low strains,
reflecting the difference between deformation
and recrystallization microstructures (Barnhoorn
2004). The LPO was still evolving after $\gamma = 30$
(Fig. 7), but at a slower rate then before $\gamma = 10$.
This finding poses a serious question about the
concept of steady state. Commonly, steady state
is thought to occur when stress, microstructure
and fabric do not vary with strain anymore.
With the studies of Barnhoorn et al. (2004),
even if stress stabilized at about $\gamma = 10$ and the
microstructure apparently remained constant,
the LPO intensity continued to strengthen till
the end of the highest strain experiment. It is not
yet possible to determine when this tendency
will stabilize, even if the uppermost limit is
given by a single crystal orientation for the
whole aggregate (which has never been observed
in nature or in experiments). Unlike calcite and
olivine, fine-grained quartz super-saturated with
water (flint), which deformed predominantly by
granular flow, displayed a more complex hardening behaviour (Schmocker et al. 2003). The
mineral grains did not track the imposed bulk
strain, even at very low deformation levels. As
an example, a micrograph of the microstructure
of quartz and calcite are compared for a deformation of about $\gamma = 4$ in Figure 8.

In all of these cases, deformation was macroscopically homogeneous. If aggregates contain
excess water or melt or more than one mineral
phase, deformation is more complex, and often
becomes localized (shear bands, shear zones,
Riedel shears etc. Fig. 9).

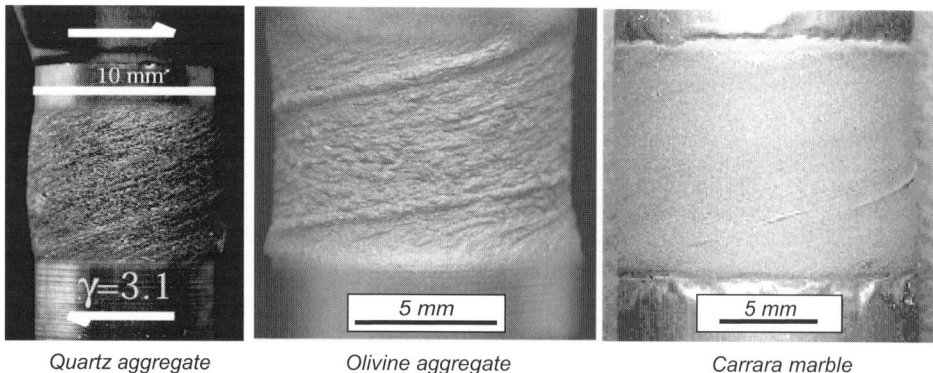

Quartz aggregate   Olivine aggregate   Carrara marble

Fig. 4. Pictures of monomineralic samples deformed in torsion to large strains, showing homogeneous deformation.
All the samples are still in their iron jacket. The little wrinkles on the outer surfaces where originally vertical, and
outline the amount of shear deformation attained. Sense of movement: top to the right. From left to right: sample of
quartz aggregate (flint) deformed to $\gamma = 3$ at 1027 °C; sample of olivine aggregate deformed to $\gamma = 5$ at temperature of
1200 °C; sample of Carrara marble deformed to $\gamma = 5$ at temperature of 727 °C.

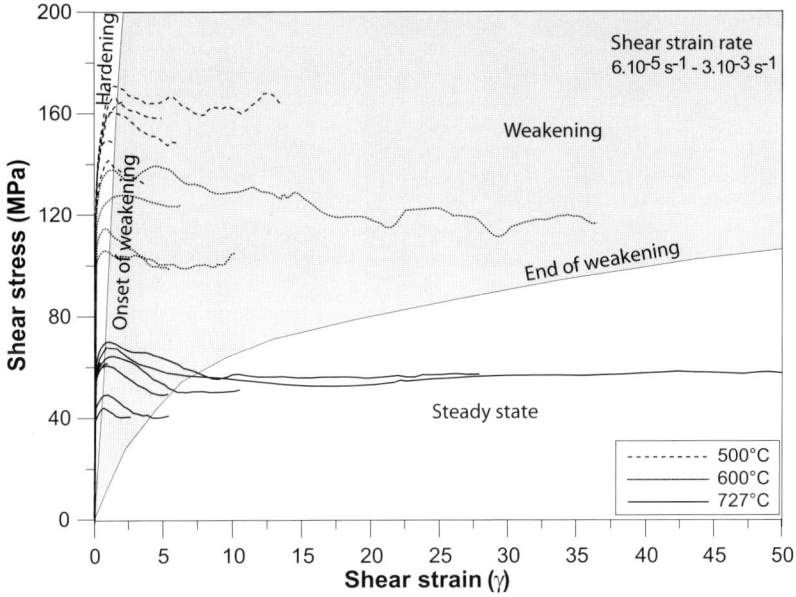

Fig. 5. Compilation of shear stress vs. shear strain curves of Carrara marble deformed in torsion at different temperatures and strain rates (Pieri *et al.* 2001; Barnhoorn *et al.* 2004). For each temperature the curve at higher stress is for the faster strain rate and that at lower stress is for the slower strain rate. The grey area outlines the strain weakening field. Note that the end of the strain weakening has probably never been reached at 600 and 500 °C.

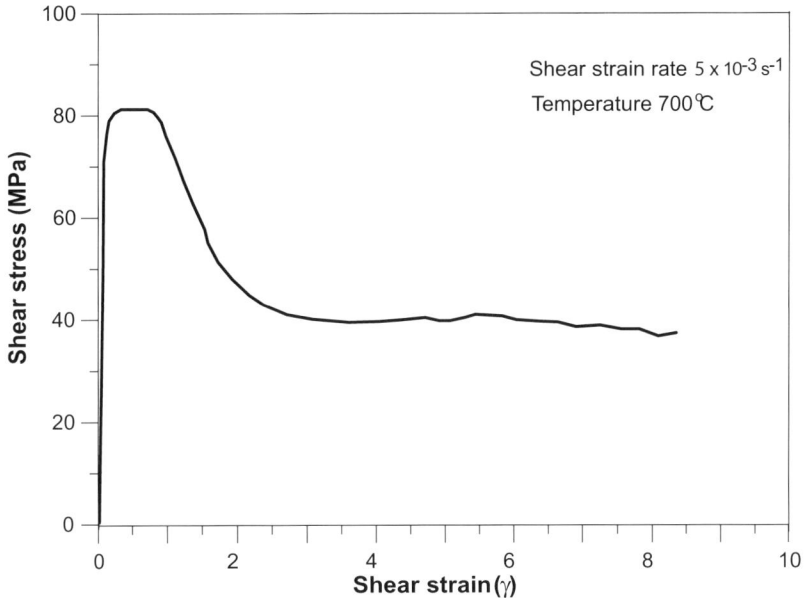

Fig. 6. An example of shear stress vs. shear strain curve for an anhydrite aggregate deformed in torsion (Heidelbach *et al.* 2001, redrawn). Note the substantial strain weakening occurred between $\gamma = 1$ and 3.

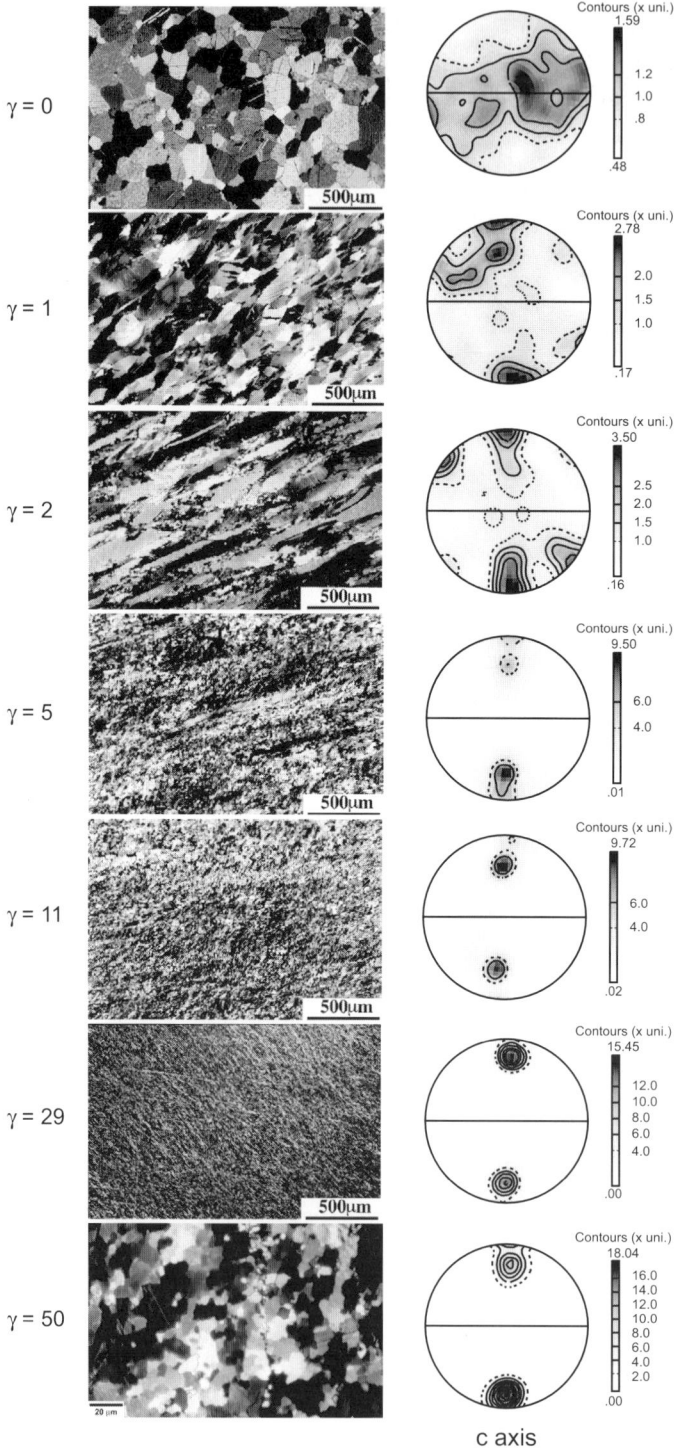

Fig. 7. Microphotographs under cross polar of Carrara marble deformed in torsion to increasing strain up to $\gamma = 50$. The stereoplots of the C-axis distribution of calcite is also shown (contours in multiple of uniform distribution). Note that the intensity of the maximum of concentration of C-axis increases even from $\gamma = 29$ to $\gamma = 50$.

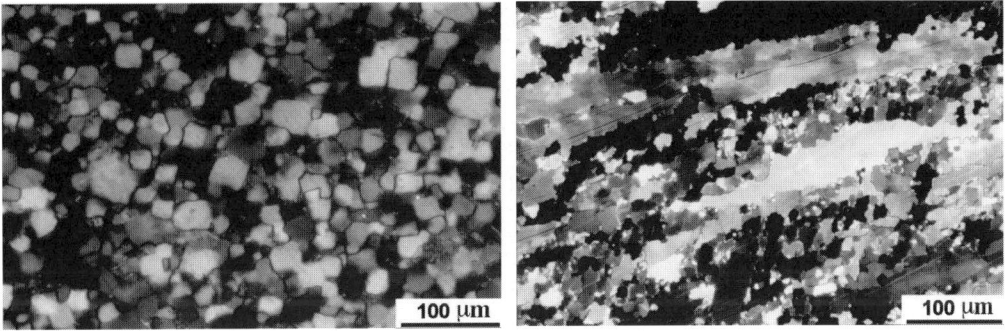

Fig. 8. Microphotograph (cross polars) of quartz (left) and calcite (right) after a shear strain of about $\gamma = 4$. Note that quartz grains are isometric, whilst the large calcite grains are tracking the imposed bulk strain.

The first extensive study on two phase aggregate deformation in torsion was that of Rybacki *et al.* (2003) on calcite and quartz. They reported that the aggregate was stronger than predicted by averaging between the strengths of the two component phases and that deformation was inhomogeneous. They attributed inhomogeneity to variation in sample dimensions, porosity and thermal gradient occurring during experiments. More recently, Barnhoorn *et al.* (2003) deformed calcite and anhydrite aggregates with varying volume proportions of the phases. They also reported inhomogeneous deformation, which was interpreted as being due to temperature gradients or variation in diameter originated by pore collapse. On the other hand, Barberini

et al. (this volume) reported the formation of shear bands in gypsum deformed in torsion at temperatures near the dehydration reaction gypsum = bassanite + water. The heterogeneous deformation was partially interpreted as driven by the reaction (Fig. 9). Deformation of serpentinite under dehydration conditions is the subject of ongoing studies (Hirose *et al.* 2004). Similar to gypsum at comparable conditions, deformation is not homogeneous and leads to strain localization.

The effect of partial melting on the rheology was investigated preliminarily by Mecklenburgh *et al.* (1999), who deformed an aggregate of quartz and granitic melt in torsion. Deformation localized in the centre of the sample (Fig. 9).

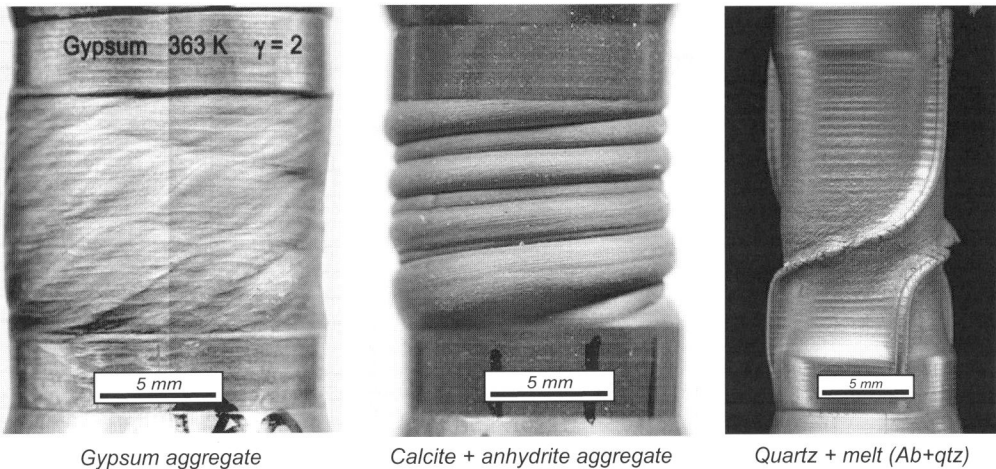

Gypsum aggregate   Calcite + anhydrite aggregate   Quartz + melt (Ab+qtz)

Fig. 9. Pictures of polyphase samples deformed in torsion to large strains, showing heterogeneous deformation, in contrast to the monomineralic samples shown in Figure 4. All the samples are still in their jackets. Sense of movement: top to the right. From left to right: gypsum deformed at 90 °C, aggregate of calcite and anhydrite deformed at 600 °C and finally the aggregate of quartz and melt of quartz + albite at 1100 °C. Sense of movement: top to the right. Note the localization of deformation.

Discussion

The results of the torsion experiments presented above and of some other recent studies will be used to discuss the 10 strain softening mechanisms suggested in the introduction as to their validity and relevance in light of recent scientific progress, mainly in laboratory studies.

Change in deformation mechanism

A change in deformation mechanism from dislocation creep to grain size sensitive flow has been suggested to follow grain size reduction by recrystallization. Such a change in mechanism has been proposed by numerous authors, for example by Rutter (1995) in calcite, and by Karato et al. (1986) for olivine. Early discussions of localization centred on the formation of strain-free grains recrystallized during dislocation flow, followed by switch to diffusion creep (e.g. Poirier 1980). Based on observations of natural microstructure, several workers have suggested such a switch (e.g. Pfiffner 1982; Behrmann 1983; Burkhard 1990; Busch & van der Pluijm 1995; Ulrich et al. 2002). However, in laboratory experiments using conventional compression tests, strains are usually not large enough to demonstrate a drop in strength associated with a change in deformation mechanism unequivocally.

All torsion experiments on monomineralic materials resulted in eventual strain softening, but only anhydrite displayed signs of a change in deformation mechanism (Stretton & Olgaard 1997). All other materials suggested no switch in deformation mechanism. Most materials show extensive grain refinement due to recrystallization after high strains, but rheological parameters do not indicate a change in deformation mechanism towards grain-size sensitive flow. These findings prompted a comment by Mackwell & Rubie (2000), who emphasized that 'The new results on torsion deformation, albeit collected over a limited range of experimental conditions, show no such changes in deformation mechanism for olivine- and calcite-dominated rocks despite substantial grain-size reduction. In high-strain axisymmetric extension tests on calcite marble (Rutter 1995) also found no change in deformation mechanism despite total dynamic recrystallization to a finer grain size and evidence of strain weakening under some conditions. Ultimate resolution of the question as to whether grain-size reduction by dynamic recrystallization changes the deformation mechanism in the upper mantle must await further experimental data covering a broader range of conditions. Until then, we should be careful not to infer deformation

mechanism changes, for example from microstructural observations alone of dynamic grain size reduction in naturally deformed rocks, in the absence of other lines of evidence'.

The general mechanism of strain weakening by grain size reduction and a change in deformation mechanism was questioned by de Bresser et al. (1998) on the basis of experimental observations on a metallic rock analogue and for theoretical reasons. The authors argue that grain growth would slow deformation in the grain size sensitive deformation regime until grain sizes were large enough for deformation by dislocation creep to be more efficient, which in turn would lead to grain refinement by recrystallization. The competition between these two processes would prevent significant strain softening. The authors therefore conclude that 'the widely accepted idea that dynamic recrystallization can lead to major rheological weakening in the Earth may not hold'. The process is illustrated in Figure 10 which shows the deformation mechanism map of calcite rocks extrapolated to a constant strain rate typical of geological deformation. If a rock with a starting grain size of few millimetres is deformed at a temperature of 250 °C (Point A), it will recrystallize according to the palaeopiezometric relations (e.g. Schmid et al. 1980; Rutter 1995) with a grain-size defined by point B. The corresponding shear stress for this grain size will be defined by point C, as the strain rate and temperature do not change. At this point the microstructure will evolve towards the recrystallization line again, following the isotherm of 250 °C to point D. At point D the microstructure will be in equilibrium. This point, and all the points along the recrystallization line, is between the diffusion and dislocation creep fields.

This theoretical argument that dynamic recrystallization would lead to a balance between grain-size reduction and grain growth is obviously in conflict with the observation on strain softening in anhydrite (Stretton & Olgaard 1997), as anhydrite seems to stay weak towards high strain. If the hypothesis by de Bresser et al. (1998) applies, then the peak stress in Stretton & Olgaard's experiments at low strain represents deformation by pure grain size insensitive (dislocation/GSI) creep, but then dynamic recrystallization sets in and reduces the average grain size, resulting in a deformation from pure dislocation creep to combined dislocation creep and grain size sensitive (diffusion/GSS) creep. The material weakens as a consequence, but there is no complete switch in deformation mechanism, just a relative increase in GSS component. There is no strain hardening since after sufficient

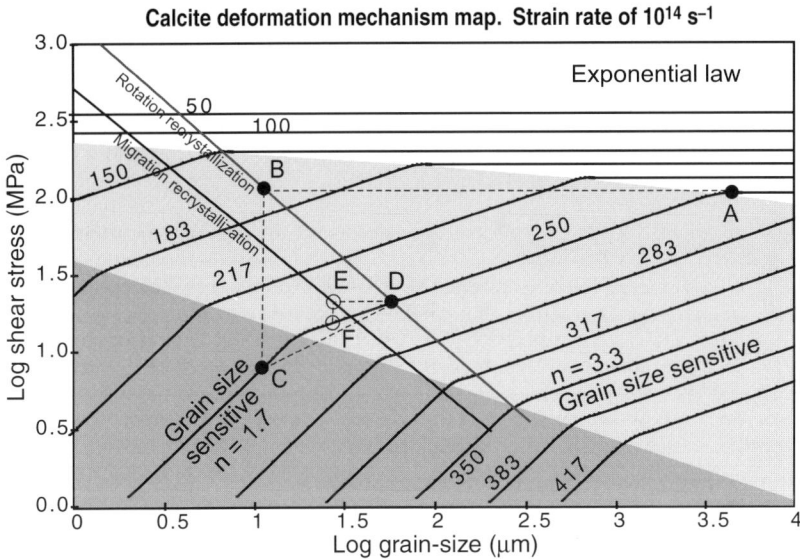

Fig. 10. Map of deformation mechanisms of calcite aggregates for a strain rate of natural deformation (10^{-14} s^{-1}). For the explanation of the points A to F see text. The three shadows represent the grain size sensitive flow with stress exponent 1.7 (dark), the grain size sensitive flow with stress exponent 3.3 (light grey) and the white area the exponential form of the flow law. Contours are temperatures in °C. The flow laws used are those summarized in Schmidt *et al.* (1980) and Rutter (1995). The two lines labelled rotation and migration recrystallization are after Schmidt *et al.* (1980) and Rutter (1995).

reduction in grain size, a balance would develop between reduction and grain growth: small grains grow and large grains are broken up, effectively keeping the average grain size constant and sample strength at the same level. The strong textures measured by Heidelbach *et al.* (2001) in these samples indicate a significant contribution of GSI dislocation creep and thus corroborate such an explanation.

In an elaboration on their hypothesis, de Bresser *et al.* (2001) further specify that significant weakening by grain-size reduction in localized shear zones is possible only if caused by a process other than dynamic recrystallization (such as syntectonic reaction or cataclasis) or if grain growth is inhibited. The effect of grain growth was already pointed out by Olgaard (1990), who emphasized the role of second phases in stabilizing grain size. Impurities or second phases inhibit grain growth and can keep deformation in the grain-size sensitive regime that way. However, small recrystallized grains share large parts of their grain boundary area with grains of the same phase such that grain growth is still possible.

The theoretical argument on the competition between recrystallization and grain growth assumes isothermal conditions. However, if the loading conditions change (Montesi & Zuber 2002) or if syndeformational cooling occurs (Braun *et al.* 1999), then theory suggests that the onset of softening might depend on a competition of grain-growth kinetics (during diffusion creep) and dynamic recrystallization (during dislocation creep). In nature, there may be a background rise in temperature or change of strain rate so that grain refinement produced at a low temperature may lead to a deformation mechanism switch with rising temperatures. Also, in nature a hot wall of a shear zone may be displaced against a cooler wall, heating the shear zone contents more effectively than conductive temperature rise (e.g. Brodie & Rutter 1987).

Continual or cyclic dynamic recrystallization

Waves of recrystallization replacing strained crystals with dislocation free grains would lead to softening by ensuring that there are always some newly formed strain-free grains. The extent of the softening will depend upon the volume fraction of these at any given instant. This mechanism has been described for experimentally deformed feldspar aggregates by Tullis & Yund (1985). In high strain torsion experiments on calcite (e.g. Pieri *et al.* 2001*a*) weakening begins and ends parallel to the beginning and completion of dynamic recrystallization, as described

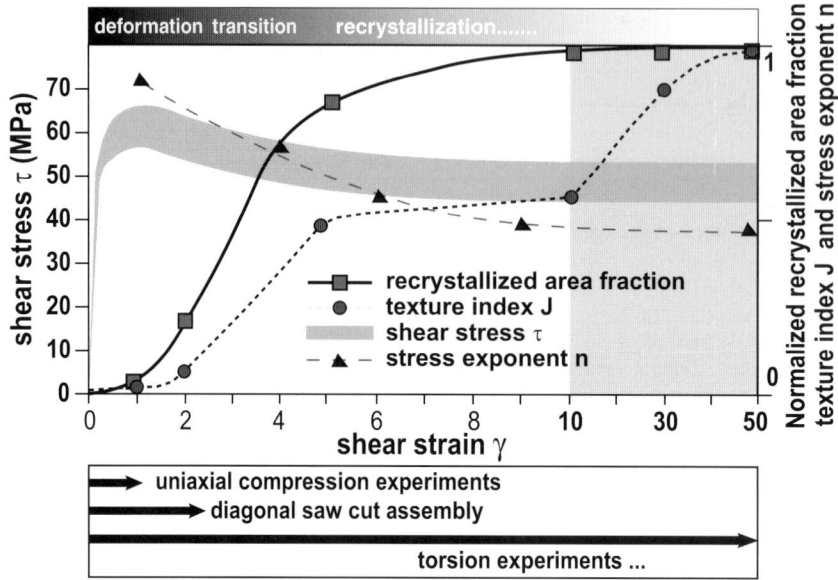

Fig. 11. Summary of the evolution of shear stress, amount of recrystallization, texture index and stress exponent with shear strain for calcite rocks. The shaded area indicates the change in horizontal scale. Note the relative amount of deformation that can be investigated with the three more common deformation techniques.

above (Fig. 11). There is no indication of a switch in mechanism, as there is no significant change in the stress exponent. The direct interpretation is that grain size reduction and refreshing of the microstructure reduces the strength of the rocks by replacing large strain-hardened grains with new dislocation-free grains. Similarly, the high strains in olivine did not result in a switch in deformation mechanism but nonetheless in considerable weakening. However, in all of these samples, it is hard to separate this softening mechanism from the one induced by the alignment of favourable crystallographic slip systems discussed below.

Geometrical, or fabric, softening

Geometrical softening has been proposed to occur where easy slip systems rotate into parallelism with stretching lineation and align favourably to the macroscopic shear with continuing deformation (e.g. Etchecopar 1974, 1977; Urai et al. 1986; Schmid et al. 1987). A strong lattice preferred orientation (LPO) develops and the flow stress for deformation by dislocation glide processes is at a minimum.

The effect of the LPO development on strength is probably less pronounced and it is not possible to discriminate on calcite rocks deformed in torsion. This observation is different from the interpretation of strain softening

observed in experiments on Carrara marble in a simple-shear geometry (diagonal cut cylinders) by Schmid et al. (1987). There the softening was attributed to the alignment of favourable slip systems parallel to the shear plane. In torsion, LPO seems to intensify continuously with strain even for the largest strains reported above, but the samples do not continue to weaken after a steady-state microstructure developed (Barnhoorn et al. 2004). One may therefore be tempted to discard fabric softening as a viable softening mechanism and attribute the softening to the recovery processes discussed above. On the other hand, olivine aggregates continue to weaken (Bystricky et al. 2000), albeit at a slower rate, even when the microstructures show complete recrystallization. The slower weakening may be caused by the LPO build up, where grains are preferentially oriented for easy slip and therefore render the aggregate softer. Additional support for the hypothesis of strain softening by this mechanism comes from a recent field study by Michibayashi & Mainprice (2004). They report that pre-existing fabric in olivines influenced development of a shear zone in the Oman ophiolite.

Reaction softening

Metamorphic reactions may be inferred to aid softening by producing small grains that are likely to promote a change in deformation mechanism and also by producing soft, strain-free grains (e.g. Rutter & Brodie 1988). In addition, the reaction of hard phases such as feldspars to form mechanically weaker assemblages of quartz and sericite would be expected to reduce the strength of the rock further.

Most of the constraints for this mechanism are inferred from field studies, as there are not many laboratory studies on the interaction of deformation and metamorphic reactions (e.g. Holyoke & Tullis 2001, 2003; de Ronde et al. 2004). The reactions that can lead to strain softening depend heavily on the temperature, fluid pressure and composition of the rocks. In upper crustal rocks the syntectonic alteration of feldspars to clays or to micas during fluid–rock interaction is often associated with cataclastic deformation (e.g. Evans 1990). Such an alteration may lead to a transition from feldspar and quartz-dominated rheologies to phyllosilicate-dominated rheologies due to fluid-assisted alteration of the feldspars to mica and other fine-grained reaction products, mainly quartz (e.g. Evans 1990; Mitra 1992). The newly formed matrix of small strain-free grains promotes dislocation and diffusion creep deformation ('reaction-enhanced ductility'; White & Knipe 1978). This switch in mechanism from cataclastic deformation to grain-size sensitive flow potentially causes significant weakening and shifts the position of the brittle plastic transition in the crust (e.g. Schmid & Handy 1991; Stewart et al. 2000; Imber et al. 2001). At higher metamorphic grades, feldspatization of mica or albitization of orthoclase may occur (e.g. Wintsch & Knipe 1985), potentially leading to reaction hardening.

Strain softening due to a solid-solid state reaction between olivine and plagioclase was shown experimentally by de Ronde et al. (2004). The reaction depended on the Ca-content of the plagioclase and produced fine-grained opx–cpx–spinel aggregates, which accommodate a large fraction of the finite strain. Deformation and reaction were localized within tiny shear zones. Another metamorphic mechanism leading to strain softening was observed in deformation experiments on dehydrating serpentinite by Rutter & Brodie (1988). Strain was accommodated by tiny shear zones of ultra-fine-grained olivine which formed by the breakdown of serpentine. Deformation occurred by grain-size sensitive diffusion assisted grain boundary sliding.

In a natural example of a metamorphic dehydration reaction leading to transient strain softening, Urai & Feenstra (2001) found strongly deformed metabauxite lenses surrounded by marbles. The metabauxite deformation is explained by the dehydration from diasporite, which is also present in the area. Although the diasporites are essentially undeformed, the corundum-rich rocks are strongly deformed, even though both diasporites and corundum-rich rocks are much stronger than the surrounding intensely deformed marbles. The observed structures are explained as an effect of high fluid pressures during the prograde diaspore–corundum dehydration reaction, which causes dramatic temporary weakening of the metabauxites, allowing the dehydrating bauxite mass to deform together with the surrounding marbles.

A summary of the mechanisms that can lead to reaction enhanced strain softening is given, for example, by Rubie (1990). In most of these mechanisms, fluids play an important role. These fluids are commonly supplied by metamorphic reactions involving dehydration of hydrous minerals such as gypsum, amphiboles or serpentine. Very often metamorphic reactions lead to strain softening by one or several of the other mechanisms listed here, such as a switch in deformation mechanism due to grain refinement, cataclastic flow due to elevated pore pressure, enhanced grain boundary diffusion rates due to the presence of water or transformation plasticity.

Chemical softening

The scientific interest in strain softening caused by changes in trace elements has been largely restricted to the effect of water. The weakening due to the chemical effects of water was originally described for single crystals of quartz (Griggs & Blacic 1964), where it is referred to as 'hydrolytic weakening', but water also decreases the strength of polycrystalline quartzites (e.g. Kronenberg & Tullis 1984). The exact mechanism by which water reduces the creep strength of quartz is still unknown, but it is presumed to depend on the concentration of a water-related crystal defect, which is controlled by the water activity (and thus by the water fugacity). Detailed reviews on hydrolytic weakening in single crystals and the water-related defects observed in quartz are given by Paterson (1989) and Kronenberg (1994).

Water was also found to weaken olivine dramatically, both single crystals (Mackwell et al. 1985) and natural and synthetic polycrystals (Carter & Avé Lallement 1970; Chopra & Paterson 1984; Karato et al. 1986). In their more recent deformation experiments on water-saturated synthetic olivine aggregates, Mei & Kohlstedt (2000a, b) found a systematic dependence of sample strength on water fugacity. On the basis of their experimental results, the authors argue that water influences creep strength through its effect on the concentration of intrinsic point defects, which in turn affects ionic diffusion and dislocation climb.

The water effect can interact with several other strain softening mechanisms discussed here. In a study on the crystallographic preferred orientations developed in experimentally deformed olivines with water added, Jung & Karato (2001) found that fabrics can change with water content. This change suggests that the dominant slip systems in olivine may change by the addition of water.

Beyond the effect of water, there is evidence for the weakening effect of other trace elements such as Mn in calcite (Wang et al. 1996; Freund et al. 2004). Increasing Mn-concentration from 10–670 ppm was found to enhance creep rate by up to an order of magnitude in the diffusion creep regime. This enhancement of creep rate by Mn is attributed to the substitution of Ca by Mn, which may increase grain-boundary diffusivity of calcite. In the dislocation regime, the effect is less pronounced and is interpreted as a combination of dislocation climb and glide with the exchange of Ca by aliovalent Mn, which is thought to affect the point defect concentration. One of the remaining questions with respect to the relevance of this mechanism to processes occurring in nature is how such an exchange of trace elements would occur and lead to strain softening.

Pore fluid effects

Pore fluids may affect the strength of rocks in several ways. One is by chemical interaction, which has been addressed above. Fluids may also accelerate metamorphic reactions and weaken the rocks as discussed below. A purely mechanical effect is that of pore pressure reducing the effective pressure and with it the resistance against brittle deformation, as discussed for example in the classic paper by Hubbert & Rubey (1959). The effect was demonstrated for serpentinite by Raleigh & Paterson (1965) and for gypsum by Heard & Rubey (1966), for example. It is widely held to be responsible for enabling intermediate depth earthquakes in subduction zones (e.g. Kirby 1996a). It has not yet been demonstrated experimentally that a progressive rise in pore fluid pressure with rising temperature whilst under a progressively relaxing differential stress can lead to a brittle faulting instability; however a rise in pore water pressure through a dehydration reaction can also promote chemical softening because water fugacity will increase (Rutter & Brodie 1995). In turn grain boundary migration recrystallization can facilitate equilibration of the internal crystal defect concentration with the ambient pore pressure (Rutter & Brodie 2004), thus several of these different strain rate-enhancing processes may interact. A second effect of pore fluid is an increase in ductility by enhancing pressure solution (e.g. Rutter 1976; Bos et al. 2000; Hellmann et al. 2002).

Shear heating

Shear heating (viscous dissipation) is the increase in temperature that can accompany deformation, especially if it is concentrated into narrow zones and the rate of strain is high enough to overcome the effects of conductive heat loss. In a review on theoretical work on shear heating, Brun & Cobbold (1980) present models which indicate that temperature rises of a few hundred degrees can be expected in major shear zones. They conclude that 'the resulting temperature gradients should be detectable geologically, but evidence is scanty. The resulting thermal softening is sufficient to concentrate most of the deformation in narrow zones. Thus strain heating is an important crustal phenomenon which should be incorporated in models of large-scale tectonic processes. It may even contribute to local partial melting in some shear zones.' In contrast, White et al. (1980) concluded that 'significant shear heating can result from high stresses and fast strain rates. However, it is unlikely that high stresses are maintained after softening has occurred. Shear heating is not regarded as a significant process during shearing and is unlikely to be a major softening process in most shear zones.' Only in very narrow zones and by shearing across stratified layers (Fleitout & Froidevaux 1980), heating was found to be significant and lead to strain softening. However, there is no experimental evidence for this mechanism.

Recent theoretical advances mainly in the group of D. Yuen at the University of Minnesota have proposed shear heating during plastic deformation (e.g. Kameyama et al. 1997; van den Berg & Yuen 1997; Regenauer-Lieb & Yuen 1998; Schott et al. 2000). Viscous heating might drive localization of deformation below a critical temperature (which is material dependent, and therefore can occur at different depths in the lithosphere), as discussed in the classical works of Hobbs & Ord (1988) and Hobbs et al. (1990). Experimental results on shear heating are only available for the brittle field (e.g. Mair & Marone 2000; Hirose & Shimamoto 2003), we are not aware of experiments on shear heating in the plastic regime in the published literature. Such experiments would require a modification from traditional deformation experiments for which temperature is kept constant by the control system. So the possible influence of shear heating is excluded from experimental stress–strain curves. Shear heating during plastic deformation should enhance weakening during deformation. Such shear weakening is conceivable for materials with high activation enthalpies (e.g. olivine, pyroxenes, etc.) and low thermal conductivity (olivine, mica, etc.), so that the increase in temperature generated by shear heating will be sustained within the high-strain zone. Heat is produced both from the release in energy due to the motion of dislocations and vacancies and from the effect of recrystallization, which eliminates grains containing dense populations of dislocations (high strain energy).

Structural softening

Poirier (1980) proposed the arrangement of second phases in a polyphase rock to be a source of structural softening, that is the evolution with strain of the phase structure towards an interconnected weak layer microstructure and rheology (e.g. Handy 1990 and references therein). Such behaviour has long been proposed for natural shear zones based on observations on mylonites (e.g. Hippertt et al. 2001; Vernon & Flood 1988; Boullier & Bouchez 1978; Wakefield 1977) and cataclasites (e.g. Imber et al. 1997, 2001). Experimental studies on the deformation of two-phase aggregates have also noted strain softening that occurs when progressive strain leads to an increasing interconnectedness of the weaker phase (e.g. Jordan 1987; Ross & Bauer 1987; Tullis & Wenk 1994; Bons & Urai 1994; Dell'Angelo & Tullis 1996). From these studies, it is now accepted that even small amounts (<15%; Dell'Angelo & Tullis 1996) of a weak phase can significantly reduce the strength of an aggregate; the weaker phase generally becomes more interconnected with progressive strain, and eventually controls the deformation of the aggregate. A theoretical model for the development of the strength of polyphase rocks with strain and structural evolution was developed, for example, by Handy (1990).

The contiguous volume of the phases (Gurland 1958) rather than the volume fraction was proposed to control the strength of coarse-grained calcite–halite aggregates (Bloomfield & Covey-Crump 1993). In that case, the ratio of the number of grain boundaries between like phases to the number of phase boundaries between grains of unlike phases would determine the sample strength if deformation was fully crystal plastic (no contribution of grain boundary sliding).

The ratio of grain boundaries to phase boundaries may also affect sample strength if diffusion and grain boundary sliding do occur, as proposed by Wheeler (1992). In a model calculation for forsterite–enstatite rocks he suggested that diffusion along grain boundaries between unlike phases may be faster than along boundaries between like phases. Grain boundary sliding would be enhanced and materials could be weaker than the pure phases. This mechanism has been proposed to occur in other two-phase rocks based on experimental observations by Bruhn et al. (1999). Hitchings et al. (1989) also reported a relative weakness for experimentally deformed olivine–enstatite samples, which would support the Wheeler (1992) model. However, their observation was not confirmed by similar experiments reported by Daines & Kohlstedt (1996). Enhanced diffusion along phase boundaries was measured on quartz–feldspar aggregates by Farver & Yund (1995). Their results imply that deformation could be localized in polyphase regions of a shear zone rather than in an interconnected layer of the weakest phase, if interphase boundary diffusion is significantly faster than grain boundary diffusion. And indeed, such a localization has been described for natural shear zones (e.g. Stünitz & FitzGerald 1993; Fliervoet 1997). In torsion experiments reported so far on polyphase aggregates, deformation was always localized. In these experiments we favour the interpretation that the grain boundaries of unlike phases can act as preferential pathways for ions, enhancing diffusion on local scale. Moreover, with progressive deformation the grain boundaries between unlike phases, which are less mobile than those with like-phase due to the chemical barrier at the interface, will rotate their normal with the sense of shear, enhancing the contribution of grain boundary sliding to accommodate the imposed deformation. This can stabilize the localization of the deformation.

Thermal softening/Mineral phase changes

A softening mechanism that has long been proposed for Earth's crust and mantle is weakening due to solid state transformation of mineral phases, often referred to as 'transformation plasticity' (e.g. Gordon 1971; Sammis & Dein 1974;

Poirier 1982). If straining occurs at temperatures not far below the temperature for a phase transformation or a chemical reaction between minerals, it is conceivable that even a slight temperature increase might be enough to drive the system through the transition and possibly induce a catastrophic softening by transformation plasticity or fluid weakening (e.g. Poirier 1982; Kirby 1996*b*), hence making shear instabilities possible. Such a phase change has often been proposed to cause faulting at great depths such as in the mantle transition zone (e.g. Burnley *et al.* 1991; Kirby *et al.* 1996*b*). Although phase transformations have been shown to affect elastic properties of rocks (e.g. Coe & Paterson 1969; Carpenter *et al.* 1998) and to cause faulting (e.g. Burnley *et al.* 1991; Brodie & Rutter 2000), unequivocal experimental evidence of enhanced plasticity in geologically relevant materials is still scarce. Significant mechanical weakening due to transformation plasticity has been reported for other materials, such as several metallic systems (e.g. Greenwood & Johnson 1965; Taleb *et al.* 2001), some ionic crystals (e.g. McLaren & Meike 1996) and oxides including ice (e.g. Dunand *et al.* 2001). However, the transformation plasticity effect is often said to be transient because it cannot produce much strain unless a rock is cycled through the transformation (e.g. Poirier 1980).

Apart from the dehydration reactions in rocks such as serpentinite or diasporite, which were discussed under 'reaction softening' above, the only phase transformation in a silicate mineral for which we have some experimental evidence of weakening and enhanced deformation is the $\alpha-\beta$ transition of quartz. The first study to show an effect of the phase transition on plastic deformation was by Chaklader (1963). In three-point bending tests of quartz single crystals he observed some irreversible deformation, albeit very small, by cycling across the $\alpha-\beta$ transition temperature. More recently, Schmidt *et al.* (2003) studied the deformation of fluid inclusions in a quartz crystal using a hydrothermal diamond-anvil cell. Pressure differences required to deform the quartz around a fluid inclusion were significantly lower at the phase transition than in either the stability fields of α- or β-quartz. These results indicate that transformation plasticity of silicates can indeed cause a considerable localized reduction in the strength of Earth materials.

Partial melting

Partial melting can affect the strength of rocks in several ways. One possibility is enhanced diffusion along grain boundaries wetted by melts. Such a mechanism has been proposed for partially molten granitic aggregates by Dell'Angelo *et al.* (1987) and for olivine aggregates containing basaltic melt by, for example Hirth & Kohlstedt (1995).

Another possibility is a network of melt in a solid rock, which can also weaken the material simply because the viscosity of the melt is much lower than that of the minerals. Such a mechanism would be similar to the one discussed under 'structural softening' above. With increasing melt fraction, the strength of partially molten rock decreases from a value characteristic of a solid to a value nearly equal to that of the melt. The dependence of aggregate strength on melt fraction is subdivided into three regions: A where aggregate strength is controlled by the solid framework; B where it is transitional; and C where it is controlled by the melt viscosity, respectively. The melt fraction at the transition between region A and C has been called the 'rheologically critical melt percentage' (RCMP). Early experimental examples of such behaviour were published by Arzi (1978) for aplite aggregates and by van der Molen & Paterson (1979) for granitic samples. In both of these studies, a strength drop by several orders of magnitude was observed for melt volumes between 12% and 25% melt in the samples, while smaller amounts of melt reduced the strength by less than an order of magnitude. In contrast to these studies, Rutter & Neumann (1995) report no sudden strength drop for partially molten granite samples and thus no evidence of a 'rheologically critical melt percentage', but rather a continuous weakening with increasing melt fraction. As an explanation for these observed differences, Renner *et al.* (2000) point out that the viscosities of the melts varied considerably between the different sample materials and melt viscosity strongly affects the transport properties and strength of rocks. Unfortunately, the melt viscosity in the experiments by Rutter & Neumann (1995) was incorrectly calculated and is reported two orders of magnitude too high, which makes the considerations by Renner *et al.* (2000) invalid.

In a study on melt viscosity with increasing fraction of solid crystals, Lejeune & Richet (1995) found that sample viscosity started to become non-Newtonian, suggesting that material viscosity was not controlled by the melt alone, with the solid volume larger than 40%. When the solid volume exceeded 60%, the material viscosity was controlled by the solid framework. These results suggest that the RCMP is between 40 and 60% melt volume. Most studies on rock deformation never investigated the rheology of materials with melt fractions in that range (Rutter & Neumann (1995) reached 45%), as sample strengths are usually below the resolution limits of triaxial testing apparatus.

The usefulness of the concept of the RCMP is questioned in a recent review of experimental work on the deformation of partially molten rocks by Rosenberg & Handy (2005). By looking at the absolute drops (on a linear scale) in material strength with increasing melt fraction Rosenberg & Handy (2005) argue that the most significant decrease occurs for relatively small melt volumes (<7%). The changes observed for higher melt contents span more orders of magnitude but are much smaller in absolute numbers. This drop in strength at low melt volumes is considered the far more important mechanism for strain softening than the effect observed for the RCMP. In summary, there is still a lot of controversy about the significance of specific mechanisms that lead to strain softening in partially molten rocks. However, from what we know so far the conclusion seems to emerge that the factor that has by far the most dramatic effect on strength is melt fraction, whereas melt viscosity and composition are relatively secondary, even though still significant.

The melt fraction in shear zones is often higher than in surrounding rocks. Melts (and other fluids) are focused into shear zones, due to recrystallization induced grain refinement (e.g. Wark & Watson 2000). This melt focusing further weakens the shear zone and helps to stabilize localization. Such an interaction of deformation and focusing of melts into finer-grained regions of experimentally deformed rock samples is described by Bruhn et al. (this volume). In general, percolation of melts is greatly enhanced by deformation, leading to a channelization of melt flow (e.g. Daines & Kohlstedt 1997; Holtzman et al. 2003) and to an anisotropy in the permeability, as shear zones serve as conduits for melt transport. Even melts that are supposed to be immobile due to high interfacial energies were found to percolate and localize in channels in an experimental study on the deformation of olivine with metallic melts by Bruhn et al. (2000).

Partial melting may interact with the other strain softening mechanisms discussed here in many ways. A switch in deformation mechanism by recrystallization and grain refinement can be promoted as grain-boundary diffusion is enhanced along boundaries wetted by melt. In addition, melt may prevent grain growth by pinning grain boundaries, effectively keeping the material weak. On the other hand recrystallization by grain boundary migration is more difficult with melt along grain boundaries, such that

one of the recovery mechanisms leading to weak, strain free grains is inhibited. The lattice-preferred orientation of minerals can have a strong effect on the melt-distribution. For example, basaltic melt in peridotite is found to preferentially wet olivine crystallographic (010) faces because the surface energy of olivine is anisotropic, and melt–grain interfaces become faceted rather than smoothly rounded indicative of isotropic wetting (Waff & Faul 1992). However, such LPO-controlled melt-distribution was only found for a static equilibrium. In an experimental study on the deformation of partially molten peridotite in the diagonal-cut cylinder design, Holtzman et al. (2003) showed that the presence of melt weakens the alignment of olivine a-axes and when melt segregates and forms networks of weak shear zones, strain partitions between weak and strong zones, resulting in an alignment of olivine a-axes $90°$ from the shear direction in three-dimensional deformation. Other authors (e.g. Daines & Kohlstedt 1997; Zimmerman et al. 1999) showed that deformation localized melts in channels controlled by the stress field rather than by olivine crystallography. A similar stress-controlled melt distribution observation in naturally deformed granites was described by Rosenberg & Riller (2000).

The combination of water and partial melting also has a strong effect on rock strength in several ways. First of all, water reduces the melting temperature and increases the melt fraction compared to dry samples at the same temperature. This water-rich melt usually has a lower viscosity than the 'dry' melt, which has a strong effect on the strain softening mechanism, as discussed by Renner et al. (2000). The effect of water on the rheology of partially molten olivine-basalt samples was investigated by Mei et al. (2002). Their samples deformed at least a factor of ten faster under water-saturated conditions than samples with the same melt fractions deformed at the same temperature under unhydrous conditions. The authors emphasize that under both hydrous and anhydrous conditions, the creep rate of olivine-basalt samples depends exponentially (or even more strongly) on melt fraction.

Partial melting may also lead to a strengthening of rocks containing some water, as water is preferentially partitioned into silicate melts and thus 'sucked' out of the rocks as the melt segregates. This drying out can effectively makes the rock stronger, as discussed by Hirth & Kohlstedt (1996).

Concluding remarks

One of the more frequent causes of localization in ductile shear zones is initial faulting followed by some of the softening mechanisms discussed above. For example, faulting can lead to comminution of particles followed by a switch in mechanism to grain boundary sliding and diffusion creep (at elevated temperatures). Fracture zones serve as conduits for fluids, which in turn can lower the strength of the zone mechanically by increasing pore pressure (e.g. Sibson 1983; Scholz 1988). Fluids in fracture zones can lead to enhanced pressure solution and to metamorphic alteration and the formation of phyllosilicate minerals which facilitate semibrittle or even plastic flow in the shear zone. This process has been proposed by theoretical studies (e.g. Wintsch et al. 1995), was described for natural shear zones (e.g. Imber et al. 1997, 2001; Stewart et al. 2000) and has been addressed by deformation experiments including clay-rich aggregates (Bos & Spiers 2002). It is likely to be widespread during fault reactivation.

If there is no brittle precursor to plastic deformation, it is more difficult to explain why any of the strain softening mechanisms in the viscous field would lead to localization rather than affect the whole rock. Such cases are extensively illustrated in the classical work of Hobbs & Ord (1988) or Hobbs et al. (1990). In most torsion experiments, softening was not associated with strain localization. From the comparison between experiments performed on monomineralic rocks and those on multiphase, we suggest that localization occurs preferentially in materials that contain two or more mechanically different phases. Homogeneous monophase aggregates do not produce macroscopic localization in torsion, even if a change of deformation mechanism occurred during the experiment (e.g. on anhydrite; Heidelbach et al. 2001).

The results of torsion experiments that did show localization contained more than one mechanical phase either because they were polymineralic to begin with or because mineral reactions such as gypsum dehydration or partial melting introduced a new mechanically different phase which provided the heterogeneity required for localization to be initiated. Similarly, natural shear zones often originate in conditions similar to those in polyphase rocks (e.g. Fliervoet et al. 1997), by dehydration of serpentinite (e.g. Barnes et al. 2004) or in fine-grained, partially molten rocks (e.g. Dijkstra et al. 2002). Because natural rocks are essentially multiphase aggregates, we contend that localization is the rule rather than the exception. The question should probably be inverted: why does deformation not localize in some cases, both in the field and in nature? Means & Williams (1972) explaining the pervasive deformation of mica

and salt in compression experiments reached the conclusion that friction between different grain boundaries may control localization: since the friction between mica and mica is much higher than between mica and salt, as soon as a shear band is formed, it was deactivated since more micas come in contact with other micas. Deformation therefore proceeds with the formation of another shear band next to the previous one. This mechanism may explain the gneissic texture, which can be controlled by mica and quartzofeldspathic grains, but not all the other cases (e.g. the homogeneous deformation of mantle peridotites such as that of Balmuccia in the Ivrea zone or some crustal gabbros).

In summary, the technical progress made with the introduction of torsion testing at high temperatures and pressures has provided new insights into the development of the mechanical behaviour of rocks with increasing strain. But even though this has been an important first step to addressing issues such as the relationship of mechanical and microstructural/textural steady-state and the nature of strain softening processes, many questions remain open. We have pointed out that the 'how and why' of localization is still poorly understood. In this context, several of the strain softening mechanisms discussed above await experimental verification or further specification, e.g. viscous shear heating and transformation plasticity. Other effects that have been addressed open up new questions, such as the mechanical and chemical interaction with deformation, why localization occurs in homogeneous polyphase rocks.

Progress in our understanding of mechanical behaviour with strain also requires progress in scientific theory. A potential way for theoretical calculations to proceed with high strain experiments is proposed by Evans (this volume) who pointed out the need to formulate constitutive laws for rocks as their structure (i.e. fabric, texture, and mineralogy) changes. For this purpose, structural variables need to be identified and to be quantified by experimental observation.

Some of the variables may not be quantifiable by laboratory testing. For example, it will never be possible to access the slow strain rates of geological processes in the laboratory and, in order to achieve plastic deformation at human time-scales, often temperature has to be kept unreasonably high (e.g. Paterson 1987). For this reason polyphase systems can be investigated experimentally in only limited fields of P and T conditions. However, the development of computer models may permit the fast modelling of complex systems such as rocks, allowing the variation of several parameters. We think therefore that the combined effort of laboratory and numerical modelling investigation should be the focus of future research in order to better understand the problem of localization of plastic rocks. However, when using mechanical data resulting from torsion experiments, we have to keep in mind that the strain (and therefore strain rate) decreases from the mantle to the cylinder axis; therefore the deformation mechanism may change inside the samples. We do not know if a switch in deformation mechanism inside the specimen would lead to a stabilization of the deformation. Thus before any attempt is made to conduct numerical modelling experiments we need to ascertain whether torsion experiments are significantly different from natural deformation. This question awaits further investigation. Last but not least, all of the models and explanations we propose from laboratory experiments and/or numerical models have to be verified by observation in nature, such that future progress depends critically on the interaction between experts in these different fields.

This paper benefited from discussions and suggestions by J.-P. Burg, J.H.P. de Bresser and M. Bystricky and from the comments of the reviewers E. Rutter and C. Spiers as well as of the editor R. Holdsworth. The torsion experimental research at ETH was supported by the ETH grant n. 0-42073-93/3392 and NF project # 2000-58809.99.

References

ARZI, A.A. 1978. Critical phenomena in the rheology of partially molten rocks. *Tectonophysics*, **44**, 173–184.

BARNES, J.D., SELVERSTONE, J. & SHARP, Z.D. 2004. Interactions between serpentinite devolatization, metasomatism and strike–slip strain localization during deep-crustal shearing in the Eastern Alps. *Journal of Metamorphic Geology*, **22**, 283–300.

BARNHOORN, A. 2004. *Rheological and microstructural evolution of carbonate rocks during large strain torsion experiments*. PhD thesis, ETH Zürich, No. 15309.

BARNHOORN, A., BYSTRICKY, M., KUNZE, K. & BURLINI, L. 2003. High strain deformation of calcite-anhydrite aggregates. *Geophysical Research Abstracts*, **5**, 07537.

BARNHOORN, A., BYSTRICKY, M., KUNZE, K. & BURLINI, L. 2004. The role of recrystallisation on the deformation behaviour of calcite rocks: large strain torsion experiments on Carrara marble. *Journal of Structural Geology*, **26**, 885–903.

BEHRMANN, J.H. 1983. Microstructure and fabric transitions in calcite tectonites from the Sierra Alhamilla (Spain). *Geologisches Rundschau*, **72**, 605–618.

BLOOMFIELD, J.P. & COVEY-CRUMP, S.J. 1993. Correlating mechanical data with microstructural observations in deformation experiments on synthetic two-phase aggregates. *Journal of Structural Geology*, **15**, 1007–1019.

BONS, P.D. & URAI, J.L. 1994. Experimental deformation of two-phase rock analogues. *Materials Science Engineering*, **A 175**, 221–224.

BOS, B. & SPIERS, C.J. 2002. Frictional-viscous flow of phyllosilicate-bearing fault rock: Microphysical model and implications for crustal strength profiles. *Journal of Geophysical Research*, **107**, B2, 10.1029/2001JB000301.

BOS, B., PEACH, C.J. & SPIERS, C.J. 2000. Slip behaviour of simulated gouge-bearing faults under conditions favoring pressure solution. *Journal of Geophysical Research*, **105**, 16699–16717.

BOULLIER, A.-M. & BOUCHEZ, J.-L. 1978. Le quartz en rubans dans les mylonites. *Bulletin of the Geological Society, France* **7**, 253–262.

BRAUN, J., CHERY, J., POLIAKOV, A., MAINPRICE, D., VAUCHEZ, A., TOMASSI, A. & DAIGNIERES, M. 1999. A simple parameterization of strain localization in the ductile regime due to grain-size reduction: a case study for olivine. *Journal of Geophysical Research – Solid Earth*, **104**, 25167–25181.

BRODIE, K.H & RUTTER, E.H. 1987. Deep crustal extensional faulting in the Ivrea Zone of Northern Italy. *Tectonophysics*, **140**, 193–212.

BRODIE, K.H. & RUTTER, E.H. 2000. Rapid stress release caused by polymorphic transformation during the experimental deformation of quartz. *Geophysical Research Letters*, **27**, 3089–3092.

BRUHN, D., OLGAARD, D.L. & DELL'ANGELO, L.N. 1999. Evidence for enhanced deformation on two-phase rocks: Experiments on the rheology of calcite-anhydrite aggregates. *Journal of Geophysical Research*, **104**, 707–724.

BRUHN, D., GROEBNER, N. & KOHLSTEDT, D.L. 2000. An interconnected network of core-forming melts produced by shear deformation. *Nature*, **403**, 883–886.

BRUN, J.P. & COBBOLD, P.R. 1980. Strain heating and thermal softening in continental shear zones: a review. *Journal of Structural Geology*, **2**, 149–158.

BURKHARD, M. 1990. Ductile deformation mechanisms in micritic limestones naturally deformed at low temperatures (150–350°C). *In*: KNIPE, R.J. & RUTTER, E.H. (eds) *Deformation mechanisms, Rheology and Tectonics*. Geological Society, London, Special Publications, **54**, 241–257.

BURNLEY, P.C., GREEN, H.W. & PRIOR, D.J. 1991. Faulting associated with the olivine to spinel transformation in Mg (sub 2) GeO (sub 4) and its implications for deep-focus earthquakes. *Journal of Geophysical Research*, **96**, 425–443.

BUSCH, J.P. & VAN DER PLUIJM, B.A. 1995. Calcite textures, microstructures and rheological properties of marble mylonites in the Bancroft shear zone, Ontario, Canada. *Journal of Structural Geology*, **17**, 677–688.

BYSTRICKY, M., KUNZE, K., BURLINI, L. & BURG, J.-P. 2000. High shear strain of olivine aggregates: rheological, textural and seismic consequences. *Science*, **290**, 1564–1567.

CARPENTER, M.A., SALJE, E.K.H., GRAEME-BARBER, A., WRUCK, B., DOVE, M.T. & KNIGHT, K.S. 1998. Calibration of excess thermodynamic properties and elastic constant variations associated with the $\alpha \leftrightarrow \beta$ transition in quartz. *American Mineralogist*, **83**, 2–22.

CARTER, N. 1976. Steady-state flow of rock. *Reviews of Geophysics and Space Physics*, **14**, 301–360.

CARTER, N.L. & AVÉ LALLEMANT, H.G. 1970. High temperature flow of dunite and peridotite. *Geological Society of America Bulletin*, **81**, 2181–2202.

CASEY, M., KUNZE, K. & OLGAARD, D.L. 1998. Texture of Solnhofen limestone deformed to high strains in torsion. *Journal of Structural Geology*, **20**, 255–267.

CHAKLADER, A.C.D. 1963. Deformation of quartz crystals at the transformation temperature. *Nature*, **197**, 791–792.

CHOPRA, P.N. & PATERSON, M.S. 1984. The role of water in the deformation of dunite. *Journal of Geophysical Research*, **89**, 7861–7876.

COE, R.S. & PATERSON, M.S. 1969. The $\alpha-\beta$ inversion in quartz: a coherent phase transition under nonhydrostatic stress. *Journal of Geophysical Research*, **74**, 4921–4948.

DAINES, M.J. & KOHLSTEDT, D.L. 1996. Rheology of olivine-pyroxene aggregates. *Eos, Transactions of the American Geophysical Union*, **77**, 46, Supplement, 711.

DAINES, M.J. & KOHLSTEDT, D.L. 1997. Influence of deformation on melt topology in peridotites. *Journal of Geophysical Research*, **107**, 10257–10271.

DE BRESSER, J.H.P., PEACH, C.J., REIJS, J.P.J. & SPIERS, C.J. 1998. On dynamic recrystallisation during solid state flow; effects of stress and temperature. *Geophysical Research Letters*, **25**, 3457–3460.

DE BRESSER, J., TER HEEGE, J. & SPIERS, C. 2001. Grain size reduction by dynamic recrystallisation: can it result in major rheological weakening? *International Journal of Earth Sciences*, **90**, 28–45.

DELL'ANGELO, L.N. & TULLIS, J. 1989. Fabric development in experimentally sheared quartzites. *Tectonophysics*, **169**, 1–21.

DELL'ANGELO, L.N. & TULLIS, J. 1996. Textural and mechanical evolution with progressive strain in experimentally deformed aplite. *Tectonophysics*, **256**, 57–82.

DELL'ANGELO, L.N., TULLIS, J. & YUND, R.A. 1987. Transition from dislocation creep to melt-enhanced diffusion creep in fine-grained granitic aggregates. *Tectonophysics*, **139**, 325–332.

DE RONDE, A.A., HEILBRONNER, R., STÜNITZ, H. & TULLIS, J. 2004. Spatial correlation of deformation and mineral reaction in experimentally deformed plagioclase–olivine aggregates. *Tectonophysics*, **389**, 93–109.

DIJKSTRA, A.H., DRURY, M.R. & FRIJHOFF, R.M. 2002. Microstructures and lattice fabrics in the Hilti mantle section (Oman Ophiolite): evidence for shear localization and melt weakening in the crust–mantle transition zone? *Journal of Geophysical Research*, **107**, 2270–2287.

DUNAND, D.C., SCHUH, C. & GOLDSBY, D.L. 2001. Pressure-induced transformation plasticity of H_2O ice. *Physical Reviews Letters*, **86/4**, 668–671.

ENGELIS, T., RAPP, R. & TSCHERNING, C. 1984. The precise computation of geoid undulation differences with comparison to results obtained from the global positioning system. *Geophysical Research Letters*, **11**, 821–824.

ETCHECOPAR, A. 1974. *Simulation par ordinateur de la deformation progressive d'un aggregat polycristallin. Etude du développement de structures orientées par écrasement et cisaillement*. PhD thesis, Université de Nantes, France, 135 pp.

ETCHECOPAR, A. 1977. A plane kinematic model of progressive deformation in a polycrystalline aggregate. *Tectonophysics*, **39**, 121–139.

EVANS, J.P. 1990. Textures, deformation mechanisms, and the role of fluids in the cataclastic deformation of granitic rocks. *In*: KNIPE, R.J. & RUTTER, E.H. (eds) *Deformation Mechanisms, Rheology and Tectonics*. Geological Society, London, Special Publications, **54**, 29–39.

FARVER, J.R. & YUND, R.A. 1995. Interphase boundary diffusion of oxygen and potassium in K-feldspar/quartz aggregates. *Geochimica et Cosmochimica Acta*, **59**, 3697–3705.

FLEITOUT, L. & FROIDEVAUX, C. 1980. Thermal and mechanical evolution of shear zones. *Journal of Structural Geology*, **2**, 159–164.

FLIERVOET, T.F., WHITE, S.H. & DRURY, M.R. 1997. Evidence for dominant grain-boundary sliding deformation in greenschist- and amphibolite-grade polymineralic ultramylonites from the Redbank deformed zone, central Australia. *Journal of Structural Geology*, **19**, 1495–1520.

FREUND, D., WANG, Z., RYBACKI, E. & DRESEN, G. 2004. High-temperature creep of synthetic calcite aggregates: influence of Mn content. *Earth and Planetary Science Letters*, **226**, 433–448.

GORDON, R.B. 1971. Observation of crystal plasticity under high pressure with applications to the earth's mantle, *Journal of Geophysical Research*, **76**, 1248–1254.

GREENWOOD, G.W. & JOHNSON, R.H. 1965. The deformation of metals under small stresses during phase transformations. *Proceedings of the Royal Society of London*, Series A, **283**, 403–422.

GRIGGS, D.T. & BLACIC, J.D. 1964. The strength of quartz in the ductile regime. *Transactions of the Americal Geophysical Union*, **45**, 102–103.

GURLAND, J. 1958. The measurement of grain contiguity in two-phase alloys. *Transactions of the Metallurgical Society AIME*, **212**, 452–455.

HANDIN, J., HIGGS, D.V. & O'BRIEN, J.K. 1960. Torsion of Yule marble under confining pressure. In: GRIGGS, D. & HANDIN, J. (eds) *Rock deformation*. Geological Society of America Memoir, **79**, 245–274.

HANDY, M.R. 1990. The solid-state flow of poly-mineralic rocks. *Journal of Geophysical Research*, **95**, 8647–8661.

HEARD, H.C. & RUBEY, W.W. 1966. Tectonic implications of gypsum dehydration. *Geological Society of America Bulletin*, **77**, 741–760.

HEIDELBACH, F., STRETTON, I.C. & KUNZE, K. 2001. Texture development of polycrystalline anhydrite deformed in torsion. *International Journal of Earth Sciences (Geologische Rundschau)*, **90**, 118–126.

HEIDELBACH, F., STRETTON, I., LANGENHORST, F. & MACKWELL, S. 2003. Fabric evolution during high shear strain deformation of magnesiowüstite (Mg0.8Fe0.2O). *Journal of Geophysical Research*, **108**, 2154.

HELLMANN, R., RENDERS, P.J.N., GRATIER, J.-P. & GUIGUET, R. 2002. Experimental pressure solution compaction of chalk in aqueous solutions. *In*: HELLMANN, R. & WOOD, S.A. (eds) *Water–Rock Interactions, Ore Deposits, and Environmental Geochemistry; a tribute to David A. Crerar*. Geochemical Society, Special Publications, **7**, 129–152.

HIPPERTT, J., ROCHA, A., LANA, C., EGYDIO-SILVA, M. & TAKESHITA, T. 2001. Quartz plastic segregation and ribbon development in high-grade striped gneisses. *Journal of Structural Geology*, **23**, 67–80.

HIROSE, T. & SHIMAMOTO, T. 2003. Fractal dimension of molten surfaces as a possible parameter to infer the slip-weakening distance of faults from natural pseudotachylytes. *Journal of Structural Geology*, **25**, 1569–1574.

HIROSE, T., BYSTRICKY, M., STÜNITZ, H. & KUNZE, K. 2004. Large strain deformation of serpentinite during dehydration reaction. IGC Florence 2004, 309–311.

HIRTH, G. & KOHLSTEDT, D.L. 1995. Experimental constraints on the dynamics of the partially molten upper mantle; deformation in the diffusion creep regime. *Journal of Geophysical Research*, **100**, 1981–2001.

HIRTH, G. & KOHLSTEDT, D.L. 1996. Water in the oceanic upper mantle; implications for rheology, melt extraction and the evolution of the lithosphere. *Earth and Planetary Science Letters*, **144**, 93–108.

HITCHINGS, R.S., PATERSON, M.S. & BITMEAD, J. 1989. Effects of iron and magnetite additions in olivine-pyroxene rheology. *Physics of the Earth and Planetary Interior*, **55**, 277–291.

HOBBS, B.E. & ORD, A. 1988. Plastic instabilities: implications for the origin of intermediate and deep focus earthquakes. *Journal of Geophysical Research*, **93**-B9, 10521–10540.

HOBBS, B.E., MÜHLHAUS, H.-B. & ORD, A. 1990. Instability, softening and localization of deformation. *In*: KNIPE, R.J. & RUTTER, E.H. (eds) *Deformation Mechanisms, Rheology and Tectonics*. Geological Society, London, Special Publications, **54**, 143–165.

HOLTZMAN, B.K., KOHLSTEDT, D.L., ZIMMERMAN, M.E., HEIDELBACH, F., HIRAGA, T., HUSTOFT, J. 2003. Melt segregation and strain partitioning: implications for seismic anisotropy and mantle flow. *Science*, **301**, 1227–1230.

HOLYOKE, C.W. & TULLIS, J. 2001. Initiation of ductile shear zones. *Abstracts with Programs – Geological Society of America*, **33**, 324–325.

HOLYOKE, C.W. & TULLIS, J. 2003. Processes of strain weakening; comparing syndeformational reaction and weak phase interconnection. *Abstracts with Programs – Geological Society of America*, **35**, 604.

HUBBERT, M.K. & RUBEY, W.W. 1959. Role of fluid pressure in mechanics of overthrust faulting. *Geological Society America, Bulletin*, **70**, 115–166.

IMBER, J., HOLDSWORTH, R.E., BUTLER, C.A. & LLOYD, G.E. 1997. Fault-zone weakening processes along the reactivated Outer Hebrides fault zone, Scotland. *Journal of the Geological Society of London*, **154**, 105–109.

IMBER, J., HOLDSWORTH, R.E., BUTLER, C. A. & STRACHAN, R.A. 2001. A reappraisal of the Sibson-Scholz fault zone model; the nature of the frictional to viscous ('brittle-ductile') transition along a long-lived, crustal-scale fault, Outer Hebrides, Scotland. *Tectonics*, **20**, 601–624.

JI, S.C., JIANG, Z.T., RYBACKI, E., WIRTH, R., PRIOR, D. & XIA, B. 2004. Strain softening and microstructural evolution of anorthite aggregates and quartz–anorthite layered composites deformed in torsion. *Earth and Planetary Science Letters*, **222**, 377–390.

JORDAN, P. 1987.The deformational behaviour of bimineralic limestone–halite aggregates. *Tectonophysics*, **135**, 185–197.

JUNG, H. & KARATO, S.-I. 2001. Water-induced fabric transitions in olivine. *Science*, **293**, 1460–1462.

KAMEYAMA, M., YUEN, D.A. & FUJIMOTO, H. 1997. The interaction of viscous heating with grain-size dependent rheology in the formation of localised slip zones. *Geophysical Research Letters*, **24**, 2523–2526.

KARATO, S., PATERSON, M.S. & FITZGERALD, J.D. 1986. Rheology of synthetic olivine aggregates: influence of grain size and water. *Journal of Geophysical Research*, **91**, 8151–8176.

KIRBY, S.H., ENGDAHL, E.R. & DENLINGER, R. 1996a. Intermediate-depth intraslab earthquakes and arc volcanism as physical expressions of crustal and uppermost mantle metamorphism in subducting slabs. *In*: BEBOUT, G.E. SCHOLL, D. & KIRBY, S. (eds) *Subduction Top to Bottom, Geophysical Monograph*. American Geophysical Union, Washington, D.C., 195–214.

KIRBY, S.H., STEIN, S., OKAL, E.A. & RUBIE, D.C. 1996b. Metastable mantle phase transformations and deep earthquakes in subducting oceanic lithosphere. *Reviews of Geophysics*, **34**, 261–306.

KREEMER, C., HAINES, A.J., HOLT, W.E., BLEWITT, G. & LAVALÉE, D. 2000. On the determination of a global strain rate model. *Earth Planets and Space*, **52**, 765–770.

KREEMER, C., HOLT, W.E. & HAINES, A.J. 2003. An integrated global model of present-day plate motions and plate boundary deformation. *Geophysical Journal International*, **154**, 8–34.

KRONENBERG, A.K. 1994. Hydrogen speciation and chemical weakening of quartz. *In*: HEANEY, P.J., PREWITT, C.T. & GIBBS, G.V. (eds) Silica: Physical Behaviour, Geochemistry, and Materials Applications. *Mineralogical Society of America, Reviews of Mineralogy*, **29**, 123–176.

KRONENBERG, A.K. & TULLIS, J. 1984. Flow strengths of quartz aggregates: grain size and pressure effects due to hydrolytic weakening. *Journal of Geophysical Research*, **89**, 4281–4297.

LEJEUNE, A.-M. & RICHET, P. 1995. Rheology of crystal-bearing silicate melts; an experimental study at high viscosities. *Journal of Geophysical Research*, **100**, 4215–4229.

MACKWELL, S. & RUBIE, D. 2000. Earth under strain. *Science*, **290**, 1514–1515.

MACKWELL, S.J., KOHLSTEDT, D.L. & PATERSON, M.S. 1985. The role of water in the deformation of olivine single crystals. *Journal of Geophysical Research*, **90**, 11319–11333.

MAIR, K. & MARONE, C. 2000. Shear heating in granular layers. *In*: MORA, P., MADARIAGA, R, MATSU-URA, M. & MINSTER, J.B. (eds) Microscopic and macroscopic simulation; towards predictive modelling of the earthquake process. *Pure and Applied Geophysics*, **157**, 1847–1866.

McLAREN, A.C. & MEIKE, A. 1996. Transformation plasticity in single and two-component ionic polycrystals in which only one component transforms. *Physics and Chemistry of Minerals*, **23**, 439–451.

MEANS, W.D. & WILLIAMS, P.F. 1972. Crenulation cleavage and faulting in an artificial salt–mica schist. *Journal of Geology*, **80**, 569–591.

MECKLENBURGH, J., RUTTER, E., BURLINI, L. & BYSTRICKY, M. 1999. Deformation of partially molten synthetic granite: implications for crustal rheology and granite pluton formation. AGU Fall Meeting. T51G-04.

MEI, S. & KOHLSTEDT, D.L. 2000a. Influence of water on plastic deformation of olivine aggregates 1. Diffusion creep regime. *Journal of Geophysical Research*, **105**, 21457–21469.

MEI, S. & KOHLSTEDT, D.L. 2000b. Influence of water on plastic deformation of olivine aggregates 2. Dislocation creep regime. *Journal of Geophysical Research*, **105**, 21471–21481.

MEI, S., BAI, W., HIRAGA, T. & KOHLSTEDT, D.L. 2002. Influence of melt on the creep behaviour of olivine-basalt aggregates under hydrous conditions. *Earth and Planetary Science Letters*, **201**, 491–507.

MICHIBAYASHI, K. & MAINPRICE, D. 2004. The role of pre-existing mechanical anisotropy on shear zone development within oceanic mantle lithosphere: An example from the Oman ophiolite. *Journal of Petrology*, **45**, 405–414.

MITRA, G. 1992. Deformation of granitic basement rocks along fault zones at shallow to intermediate crustal levels. *In*: MITRA, S. & FISHER, G.W. (eds) *Structural Geology of Fold and Thrust Belts*. The John Hopkins University Press, Baltimore, 123–144.

MONTESI, L. & ZUBER, M. 2002. A unified description of localization for application to large-scale tectonics. *Journal of Geophysical Research — Solid Earth*, **107**, 2045, doi: 10.1029/2001JB000465.

OLGAARD, D.L. 1990. The role of second phase in localizing deformation. *In*: KNIPE, R.J & RUTTER, E.H. (eds) *Deformation Mechanisms, Rheology and Tectonics*. Geological Society, London, Special Publications, **54**, 175–181.

PATERSON, M.S. 1987. Problems in the extrapolation of laboratory rheological data. *Tectonophysics*, **133**, 33–43.

PATERSON, M.S. 1989. The interaction of water with quartz and its influence in dislocation flow – an overview. *In*: KARATO, S.-I. & TORIUMI, M. (eds) *Rheology of Solids and of the Earth*. Oxford University Press, Oxford, 107–142.

PATERSON, M.S. 2001. Relating experimental and geological rheology. *International Journal of Earth Sciences* (Geologische Rundschau), **90**, 157–167.

PATERSON, M.S. & OLGAARD, D.L. 2000. Rock deformation tests to large shear strains in torsion. *Journal of Structural Geology*, **22**, 1341–1358.

PFIFFNER, A.O. 1982. Deformation mechanisms and flow regimes in limestone from the Helvetic zone of the Swiss Alps. *Journal of Structural Geology*, **4**, 429–444.

PIERI, M., BURLINI, L., KUNZE, K., OLGAARD, D. & STRETTON, I.C. 2001a. Dynamic recrystallisation of Carrara marble during high temperature torsion experiments. *Journal of Structural Geology*, **23**, 1393–1413.

PIERI, M., KUNZE, K., BURLINI, L., STRETTON, I., OLGAARD, D.L., BURG, J.-P. & WENK, H.R. 2001b. Texture development of calcite by deformation and dynamic recrystallisation at 1000 K during torsion experiments of marble to large strains. *Tectonophysics*, **330**, 119–140.

POIRIER, J.P. 1980. Shear localization and shear instability in materials in the ductile field. *Journal of Structural Geology*, **2**, 135–142.

POIRIER, J.P. 1982. On transfomation plasticity. *Journal of Geophysical Research*, **87**, 6791–6797.

RALEIGH, C.B. & PATERSON, M.S. 1965. Experimental deformation of serpentinite and its tectonic implications. *Journal of Geophysical Research*, **70**, 3965–3985.

REGENAUER-LIEB, K. & YUEN, D.A. 1998. Rapid conversion of elastic energy into plastic shear heating during incipient necking of the lithosphere. *Geophysical Research Letters*, **25**, 2737–2740.

RENNER, J., EVANS, B. & HIRTH, G. 2000. On the rheologically critical melt fraction. *Earth and Planetary Science Letters*, **181**, 585–594.

ROSENBERG, C.L. & RILLER, U. 2000. Partial-melt topology in statically and dynamically recrystallised granite. *Geology*, **28**, 7–10.

ROSENBERG, C.L. & HANDY, M. 2005. Experimental deformation of partially-melted granite revisited: implications for the continental crust. *Journal of Metamorphic Geology*, **23**, 19–28.

ROSS, J.V., BAUER, S.J. & HANSEN, F.D. 1987. Textural evolution of synthetic anhydrite–halite mylonites. *Tectonophysics*, **140**, 307–326.

RUBIE, D. 1990. Mechanisms of reaction-enhanced deformability in minerals and rocks. *In*: BARBER, D.J. & MEREDITH, P.G. (eds) *Deformation mechanisms in minerals, ceramics and rocks*. The Mineralogical Society Series, Unwin Hyman, London, 262–295.

RUTTER, E.H. 1976. The kinetics of rock deformation by pressure solution. *Philosophical Transactions of the Royal Society of London, Series A*, **283**, 203–219.

RUTTER, E.H. 1995. Experimental study of the influence of stress, temperature, and strain on the dynamic recrystallisation of Carrara marble. *Journal of Geophysical Research*, **100**, 24651–24663.

RUTTER, E.H. 1998. The use of extension testing to investigate the high strain rheological behaviour of marble. *Journal of Structural Geology*, **20**, 243–254.

RUTTER, E.H. 1999. On the relationship between the formation of shear zones and the form of the flow law for rocks undergoing dynamic recrystallisation. *Tectonophysics*, **303**, 147–158.

RUTTER, E.H. & BRODIE, K.H. 1988. Experimental "syntectonic" dehydration of serpentinite under conditions of controlled pore water pressure. *Journal of Geophysical Research*, **93**, 4907–4932.

RUTTER, E.H. & BRODIE, K.H. 1995. Mechanistic interactions between deformation and metamorphism. *Geological Journal*, **30**, 27–240.

RUTTER, E.H. & BRODIE, K.H. 2004. Experimental intracrystalline plastic deformation of hot-pressed aggregates of Brazilian quartz. *Journal of Structural Geology*, **26**, 259–270.

RUTTER, E.H. & NEUMANN, D.H. 1995. Experimental deformation of partially molten Westerly granite under fluid-absent conditions, with implications for the extraction of granitic magmas. *Journal of Geophysical Research*, **100**, 15697–15715.

RYBACKI, E., PATERSON, M.S., WIRTH, R., & DRESEN, G. 2003. Rheology of calcite–quartz aggregates deformed to large strain in torsion. *Journal of Geophysical Research*, **108**, art. no. 2089.

SAMMIS, C.G. & DEIN, J.L. 1974. On the possibility of transformational superplasticity in the Earth's mantle. *Journal of Geophysical Research*, **79**, 2961–2965.

SCHMID, S.M. & HANDY, M.R. 1991. Towards a genetic classification of fault rocks: geological usage and tectonophysical implications. *In*: MULLER, D.W., MCKENZIE, J.A. & WEISSERT, H. (eds) *Controversies in Modern Geology*. Academic Press, London, 339–361.

SCHMID, S.M., PATERSON, M.S. & BOLAND, J.N. 1980. High temperature flow and dynamic recrystallization in Carrara marble. *Tectonophysics*, **65**, 245–280.

SCHMID, S.M., PANOZZO, R. & BAUER, S. 1987. Simple shear experiments on calcite rocks: Rheology and microfabric. *Journal of Structural Geology*, **9**, 747–778.

SCHMIDT, C., BRUHN, D. & WIRTH, R. 2003. Experimental evidence of transformation plasticity in silicates: minimum of creep strength in quartz. *Earth and Planetary Science Letters*, **205**, 273–280.

SCHMOCKER, M., BYSTRICKY, M., BURLINI, L., KUNZE, K., STUENIZ, O. & BURG, J.-P. 2003. Experimental deformation to large shear strains of natural quartz aggregates. *Journal of Geophysical Research*, **108**, 2242.

SCHOLZ, C.H. 1988. The brittle–plastic transition and the depth of seismic faulting. *In*: ZANKL, H., BELLIERE, J. & PRASHNOWSKY, A. (eds) Detachment and shear. *Geologische Rundschau*, **77**, 319–328.

SCHOTT, B., YUEN, D.A. & SCHMELING, H. 2000. The significance of shear heating in continental delamination. *Physics of the Earth and Planetary Interiors*, **118**, 273–290.

SIBSON, R.H. 1975. Generation of pseudotachylyte by ancient seismic faulting. *The Geophysical Journal of the Royal Astronomical Society*, **43**, 775–794.

SIBSON, R.H. 1983. Continental fault structure and the shallow earthquake source. *Journal of the Geological Society, London*, **140**, 741–767.

SPRAY, J.G. 1987. Artificial generation of pseudotachylyte using friction welding apparatus; simulation of melting on a fault plane. *Journal of Structural Geology*, **9**, 49–60.

SPRAY, J.G. 1992. Physical basis for the frictional melting of some rock-forming minerals. *In*: MAGLOUGHLIN, J.F. & SPRAY, J.G. (eds) Frictional melting processes and products in geological materials. *Tectonophysics*, **204**, 205–221.

STEWART, M.A., HOLDSWORTH, R.E. & STRACHAN, R.A. 2000. Deformation processes and weakening mechanisms within the frictional–viscous transition zone of major crustal faults: insights from the Great Glen Fault Zone, Scotland. *Journal of Stuctural Geology*, **22**, 543–560.

STRETTON, I.C. & OLGAARD, D.L. 1997. A transition in deformation mechanism through dynamic recrystallisation; evidence from high strain, high temperature torsion experiments. *Eos, Transactions of the American Geophysical Union*, **78**, 723.

STÜNITZ, H. & FITZGERALD, J.D. 1993. Deformation of granitoids at low metamorphic grade II: Granular flow in albite-rich mylonites. *Tectonophysics*, **221**, 299–324.

TALEB, L., CAVALLO, N. & WAECKEL, F. 2001. Experimental analysis of transformation plasticity. *International Journal of Plasticity*, **17/I**, 1–20.

TULLIS, J., SNOKE, A.W. & TODD, V. 1982. Significance and petrogenesis of mylonitic rocks. Penrose conference report. *Geology*, **10**, 227–230.

TULLIS, J. & WENK, H.-R. 1994. The effect of muscovite on the strength and lattice preferred orientations of experimentally deformed quartz aggregates. *Materials Science Engineering*, **A175**, 209–220.

TULLIS, J. & YUND, R.A. 1985. Dynamic recrystallisation of feldspar: a mechanism for ductile shear zone formation. *Geology*, **13**, 238–241.

ULRICH, S., SCHULMANN, K. & CASEY, M. 2002. Microstructural evolution and rheological behaviour of marbles deformed at different crustal levels. *Journal of Structural Geology*, **24**, 979–995.

URAI, J.L., MEANS, W.D. & LISTER, G.S. 1986. Dynamic recrystallisation of minerals. *In*: HOBBS, B.E.H. & HEARD, H.C. (eds) *Mineral and Rock Deformation: Laboratory Studies, The Paterson Volume*. American Geophysical Union, Washington, D.C., 161–200.

URAI, J.L. & FEENSTRA, A. 2001. Weakening associated with the diaspore-corundum dehydration reaction in metabauxites; an example from Naxos (Greece). *Journal of Structural Geology*, **23**, 941–950.

VAN DEN BERG, A.P. & YUEN, D.A. 1997. The role of shear heating in lubricating mantle flow. *Earth and Planetary Science Letters*, **151**, 33–42.

VAN DER MOLEN, I. & PATERSON, M.S. 1979. Experimental deformation of partially melted granite. *Contributions to Mineralogy and Petrology*, **70**, 299–318.

VERNON, R.H. & FLOOD, R.H. 1988. Contrasting deformation of S- and I-type granitoids in the Lachlan Fold Belt, eastern Australia. *Tectonophysics*, **147**, 127–143.

WAFF, H.S. & FAUL, U.H. 1992. Effects of crystalline anisotropy on fluid distribution in ultramafic partial melts. *Journal of Geophysical Research*, **97**, 9003–9014.

WAKEFIELD, J. 1977. Mylonitization in the Lethakane shear zone, eastern Botswana. *Journal of the Geological Society, London*, **133**, 263–275.

WANG, Z., BAI, Q., DRESEN, G., WIRTH, R. & EVANS, B. 1996. High temperature deformation of calcite single crystals. *Journal of Geophysical Research*, **101**, 20377– 20390.

WARK, D.A. & WATSON, B.E. 2000. Effect of grain size on the distribution and transport of deep-seated fluids and melts. *Geophysical Research Letters*, **27**, 2029–2032.

WHEELER, J. 1992. Importance of pressure solution and coble creep in the deformation of polymineralic rocks. *Journal of Geophysical Research*, **97**, 4579–4586.

WHITE, S.H. & KNIPE, R.J. 1978. Transformation and reaction enhanced ductility in rocks. *Journal of the Geological Society, London*, **135**, 513–516.

WHITE, S.H., BURROWS, S.E., CARRERAS, J., SHAW, N.D. & HYMES, F.J. 1980. On mylonites in ductile shear zones. *Journal of Structural Geology*, **2**, 175–187.

WINTSCH, R.P. & KNIPE, R.J. 1985. The possible effects of deformation on chemical processes in metamorphic fault zones. *In*: THOMPSON, A.B. & RUBIE, D.C. (eds) *Advances in Physical Chemistry* 4, Springer-Verlag, New York, 251–268.

WINTSCH, R.P., CHRISTOFFERSEN, R. & KRONENBERG, A.K. 1995. Fluid–rock reaction weakening of fault zones. *Journal of Geophysical Research*, **100**, 13021–13032.

ZHANG, S. & KARATO, S.-I. 1995. Lattice preferred orientation of olivine aggregates deformed in simple shear. *Nature*, **375**, 774–777.

ZIMMERMAN, M.E., ZHANG, S., KOHLSTEDT, D.L. & KARATO, S. 1999. Melt distribution in mantle rocks deformed in shear. *Geophysical Research Letters*, **26**, 1505–1508.

Shear strain localization from the upper mantle to the middle crust of the Kohistan Arc (Pakistan)

J.-P. BURG[1], L. ARBARET[2], N. M. CHAUDHRY[3], H. DAWOOD[4],
S. HUSSAIN[4] & G. ZEILINGER[5]

[1]Geologisches Institut, ETH-Zentrum and Zurich University, Sonneggstrasse 5, CH-8092,
Zürich, Switzerland (e-mail: jean-pierre.burg@erdw.ethz.ch)
[2]Institut des Sciences de la Terre, UMR 6113 – CNRS/Université d'Orléans,
1A rue de la Férollerie, 45071 Orléans cédex 2, France
[3]Institute of Geology, Punjab University, Quaid-e-Azam Campus,
Lahore 54590, Pakistan
[4]Museum of Natural History, Garden Avenue, Shakar Parrian,
Islamabad 44000, Pakistan
[5]Institut of Geological Sciences, University of Bern, Baltzerstrasse 1–3,
CH-3012, Bern, Switzerland

Abstract: Shear structures from mantle to middle crust levels of the Kohistan palaeo-island arc, in Pakistan, are described. Pre-Himalayan ductile shear zones show a wide variety in size and shape, and developed from gabbro subsolidus to amphibolite facies conditions. Their lithological context and geological history give insights into mechanisms that initiate shear strain localization, factors that control stabilization of deformation in shear zones and flow properties at the mantle–crust transition. Shear strain localization began within compositional gradients. Gabbros were more prone to localization into anastomosing patterns than diorites and granites, which show more homogeneous strain. Shear strain localization during cooling led to less numerous but longer and thicker shear zones. Viscous heating within shear zones resulted in melt production and segregation in deformation structures, and seems to have taken part in the plutonic history of the arc. Using Kohistan as an example, we suggest that the plutonic, lower crust of arcs is strongly affected by subhorizontal, synmagmatic shear zones, probably consistent with the bulk flow direction of the subduction zone. These features can obviously be preserved in collision orogens and may be mistaken for structures documenting the continental collision.

The nature of deformation structures and the structural relationships between major lithological units are a subject of major importance in understanding the dynamics of crustal evolution. For this purpose, seismic reflection and refraction profiling are nowadays the main tools for deep structural mapping on a regional scale (e.g. Bois et al. 1990; Nelson 1991; Ludden & Hynes 2000). How lithological units are geometrically assembled is the first indispensable knowledge required to really understand the processes of crustal evolution. The increasing demand for higher resolution from seismic information places emphasis on the need for detailed geological and, in particular, structural documentation of regions where mantle–crust transitions and deep crustal levels are exposed. The geological details may in turn be used to generate seismic models (e.g. Fountain 1986; Rutter et al. 1999) and new constraints for seismic interpretations (e.g. Fountain et al. 1990; Khazanehdari et al. 2000). In the frame of this interdisciplinary effort, we describe the style of shear deformation across the transition from mantle to middle, plutonic crust of the Kohistan palaeo-island arc, in Pakistan (Fig. 1). Numerous time markers in the form of intrusive bodies make the Kohistan Arc an excellent place to study the role of magmatic structures at the mantle–crust transition and in the deep crust of an arc system, in this particular case with respect to younger Himalayan tectonics.

From: BRUHN, D. & BURLINI, L. (eds) 2005. High-Strain Zones: Structure and Physical Properties.
Geological Society, London, Special Publications, **245**, 25–38.
0305-8719/05/$15.00 © The Geological Society of London 2005.

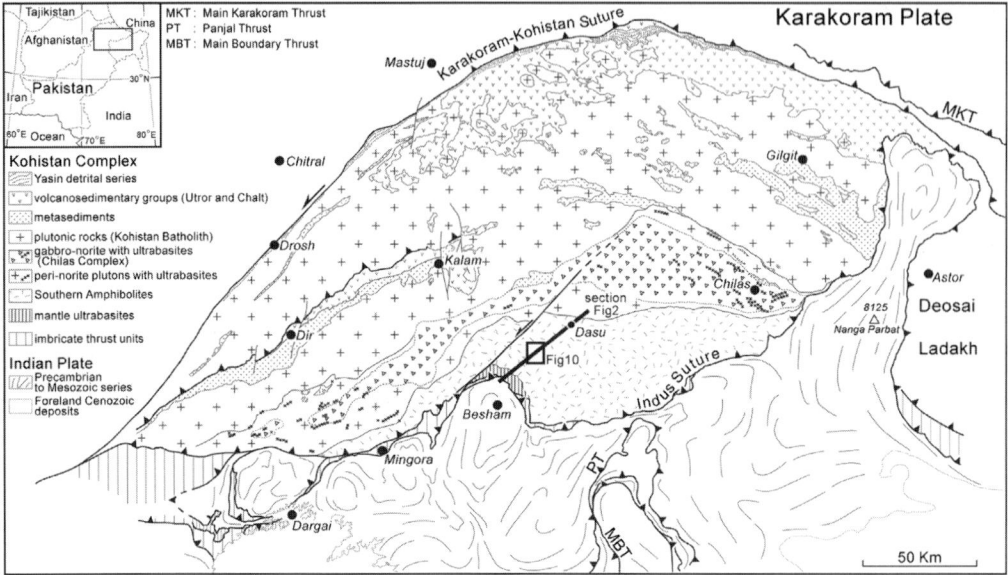

Fig. 1. Sketch map of the Kohistan Complex (modified from Bard 1983 and Burg *et al.* 1998).

Pre-Himalayan shear zones show a wide variety in size and shape at the base of the plutonic crust. Evaluating their lithological context and geological history provides insights into mechanisms that initiate shear strain localization, factors that control stabilization of deformation in shear zones and flow properties at the mantle–crust transition. Our case history is the Kohistan–Karakoram Highway (KKH) section, which we mapped in more structural detail than previous work with the intention of providing a geometrical and structural framework in which petrological, geochemical and geophysical work on island arcs and, possibly, other deep crustal regions can be placed.

In the section studied, gabbros were more prone to the development of anastomosing patterns than diorites and granites, which show more homogeneous strain. This demonstrates a lithological control on shear strain localization. Localization occurred at compositional gradients rather than within domains rich in the weak mineral phases because deformation prefers to follow rheological contrasts rather than weak rheologies (Arbaret & Burg 2003). Therefore, ductile shear zones easily propagated along lithological boundaries and anastomosing patterns of shear zones formed in a nearly 10 km-thick layer of the crust. Owing to shear-zone propagation, magmatic structures are only preserved in the thick plutonic bodies, which demonstrates a geometrical control on strain distribution. Shear

zones developed from super-solidus conditions, but above the rheologically critical melt fraction, i.e. at advanced crystallization of the magmas, through subsolidus to amphibolite facies conditions, with strain localization during cooling leading to less numerous but longer and thicker active shear zones. Magmatic fabrics have significantly smaller intensities than solid-state fabrics, which suggests that seismic reflections and anisotropy reflect solid-state structures (e.g. Ji *et al.* 1993). Viscous heating within shear zones resulted in melt production and segregation. We suggest that this deformation-assisted process may take part in the plutonic history of the arcs.

Geological setting

In NW Pakistan, the Indian and Asian plates are separated by the dominantly basic–intermediate rocks of the Kohistan Arc Complex, which was developed as an island arc above the northward subduction zone of the Tethys Ocean during the late Mesozoic (Tahirkheli *et al.* 1979; Bard *et al.* 1980; Bard 1983; Coward *et al.* 1986). The tectonic history involves the 102–75 Ma old accretion to the Asian plate, to the north (Le Fort *et al.* 1983; Petterson & Windley 1985; Treloar *et al.* 1989), intra-arc rifting and magma upwelling that produced the 85 Ma old calc-alkaline noritic gabbros of the Chilas Complex (Treloar *et al.* 1996; Burg *et al.* 1998) and closure of the Tethys Ocean, to the south, at *c.* 55 Ma, followed by obduction of the Kohistan Complex onto India (Coward *et al.* 1987).

The southern, lower boundary of the Kohistan Arc Complex is a N-dipping fault zone, the Indus Suture (Fig. 1), along which ultrabasic rocks, possibly derived from the Tethys oceanic lithosphere, are strongly serpentinized and on which the Kohistan Arc Complex has been obducted over continental India (e.g. Tahirkheli *et al.* 1979; Coward *et al.* 1986). The 'magmatostratigraphic' build-up of the Kohistan Arc Complex as exposed today is a primary feature that predates the India–Asia collision (e.g. Treloar *et al.* 1996). Just north of the Indus Suture, the Jijal ultramafic complex is composed of peridotites, pyroxenites, garnetites and hornblendites (e.g. Miller *et al.* 1991). The Jijal Complex includes the arc crust–mantle boundary with mantle rocks overlain by garnet-bearing granulite facies gabbros. The Jijal granulite facies metagabbro is in turn overlain by the so-called Southern (Kamila) Amphibolites, which are mostly strongly deformed metagabbros and metadiorites, intruded, to the north, by the 300 km-long Chilas gabbronorite (Fig. 1). This work does not document the Chilas gabbronorite and the Kohistan Batholith, further to the north. We focus on the structural characteristics of the intensely sheared Jijal Complex and overlying Southern Amphibolites with the aim of documenting the structural expression of shear deformation into the deepest levels of an arc section. We show that strain localization takes place from the magmatic to the solid-state stages of magma emplacement.

Indus river section

The deep parts of the Kohistan Arc Complex are remarkably exposed along the Indus valley, where relationships between exposed rock types can be studied in details. We used numerous outcrop and landscape photographs to construct the section nearly orthogonal to the regional trends while respecting shapes and structural styles identified on the photographs and in the field. Minor offsets due to late faulting were removed. First, we briefly describe the lithological section from the bottom to the top, i.e. from south to north (Fig. 2).

Mantle

The lower part of the Jijal section ('the mantle section', Fig. 2) is dominated by N-dipping dunites and wehrlites interlayered with subordinate pyroxenites (Jan & Howie 1981; Jan &

Windley 1990; Miller *et al.* 1991; Burg *et al.* 1998) (Fig. 3). The presence of harzburgites, with a high-temperature tectonite microstructure, indicates that peridotites were part of a residual 'arc mantle'. Flames of dunite in websterites indicate that these rocks have replaced peridotites through melt–rock reactions, at pressure–temperature (P–T) conditions close to the peridotite solidus (Burg *et al.* 1998). The proportion of mafic rocks (websterites and minor hornblendites and garnetites) increases up-section, up to an intrusive sharp contact between ultramafic rocks and overlying garnet granulites. Trails of chromite and rotated clinopyroxenite dykes underline the compositional layering parallel to the strong mineral foliation, which reflects mantle deformation during magmatic emplacement as observed in other, near Moho level active environments (Nicolas 1992; Boudier *et al.* 1996). Post-magmatic structures are kink bands in olivine, exsolution lamellae in pyroxene (Kausar *et al.* 1997) and asymmetric folds that suggest large shear solid-state deformation.

Dykes of greenish Cr-rich clinopyroxenites cut the dunitic rocks and locally build an angular framework (Fig. 4). The breccia-like fragments of the welded peridotites point to brittle behaviour of the mantle, as similar Cr-clinopyroxenites are known only in association with refractory mantle rocks and are ascribed to the segregation and transport of boninitic melts (Bodinier & Godard 2003) or to the high-pressure crystallization of basaltic magma (Müntener *et al.* 2001) at supra-subduction mantle depths.

Fig. 3. Interlayered dunites and clinopyroxenite dykes. Arrow: intersection between clinopyroxenite 'layers' that have been sheared and rotated towards near-parallelism, with peridotite foliation marked by chromite. For location see Figure 2.

Fig. 4. Dykes of greenish Cr-clinopyroxenites that cut ultrabasites, hence demonstrating brittle behaviour at mantle depth. For location see Figure 2.

Lower crust

The granulite facies, garnet-rich metagabbro forms the 'crustal' unit of the Jijal Complex (Yamamoto 1993; Ringuette *et al.* 1998). Its lower contact dips gently to the north, parallel to the mantle 'layering' (Fig. 2). It is intrusive into the hornblendites and garnetites mentioned and attributed above to the mantle; this boundary is interpreted as the exhumed petrological arc-'Moho' (Burg *et al.* 1998). The garnet-gabbro is intrusive into, and has been intruded by, hornblendites (Fig. 5). Xenoliths of peridotite found up to Patan (Fig. 2) support the interpretation that this body has intruded mantle rocks.

The upper contact of the granulite facies garnet-gabbro marks the boundary with the Southern Amphibolites (Treloar *et al.* 1990; Khan *et al.* 1993), which we interpret as the 'deep- to mid-crustal' section of the arc in the light of thermobarometric information: deepest magmas crystallized at 800 °C and 0.8–1.1 GPa, and metamorphosed at 550–650 °C under pressures

Fig. 5. Angular hornblendite enclaves (dark) welded by leucocratic veins derived from the granulite facies, Jijal Complex garnet-gabbro. For location see Figure 2.

of 0.9–1.0 GPa prior to 83 ± 1 Ma (Bard 1983; Treloar *et al.* 1990; Yoshino *et al.* 1998).

Lower–middle crust

From the south to the north, the Southern Amphibolites are divided into three stratum levels.

The southern, lower stratum corresponds to the Patan Complex of Miller *et al.* (1991). The bottom Sarangar metagabbro (Arbaret *et al.* 2000) cuts the magmatic fabric of the garnet-gabbro, which, along with a few granulitic enclaves near the irregular, locally digitized contact, demonstrates that it has intruded into the granulite facies, Jijal Complex metagabbro. Preserved plagioclase–clinopyroxene assemblages indicate that the Sarangar gabbro crystallized at 800 °C and 0.8–1.1 GPa (Yoshino *et al.* 1998) at 98.9 ± 0.4 Ma (^{206}Pb/^{238}U zircon age: Schaltegger *et al.* 2002). The emplacement age of the Jijal Complex garnet-gabbro has been estimated from 91 ± 6.3 to 95.7 ± 2.7 Ma (Sm–Nd cooling ages based on pyroxene–garnet assemblages: Yamamoto & Nakamura 1996; Anczkiewicz & Vance 2000; Schaltegger *et al.* 2002). Structural relationships prove the Sarangar gabbro to have intruded a fully crystallized, hence cooler than solidus garnet-gabbro (Arbaret *et al.* 2000). The Jijal Complex garnet-gabbro is therefore older than the Sarangar gabbro. Sm–Nd ages are likely to represent system closure during cooling at 750–650 °C (Anczkiewicz & Vance 2000).

The Sarangar gabbro has also recorded granulite facies conditions (Yoshino *et al.* 1998) and displays structural features similar to those found in the underlying Jijal Complex garnet-gabbro. Therefore, there is no structural or lithological ground to separate the Jijal Complex from the overlying Southern Amphibolites.

The overlying sequence is mainly composed of amphibolite-facies metagabbros, hornblende-gabbros, diorites and subordinate tonalites (Fig. 2). They have locally preserved igneous layering, have intruded each other without any identified logic, and have been intruded by small volumes of hornblendite pegmatoids and plagioclase–quartz ± amphibole pegmatite veins. Rare calc-silicate enclaves imply that these rocks have intruded sediments; their presence further suggests that some amphibolite xenoliths found in any plutonic body may derive from basalts. The plagioclase–quartz ± amphibole veins comprise synmagmatic differentiation veins containing the same mineralogical components as the bulk rock. The petrography and geochemistry of the plutonic rocks have been described by Jan & Howie (1981), Treloar *et al.* (1990), Miller *et al.* (1991), Yamamoto (1993) and Yoshino *et al.* (1998). All authors agree that they represent calc-alkaline magmas emplaced during the arc activity from a partially molten mantle source with mid-ocean ridge basalt (MORB)-type isotopic characteristics (Schaltegger *et al.* 2002). A magmatic breccia characterizes the bottom of this sequence (Miller *et al.* 1991) (Fig. 6).

To the north, the Kiru Amphibolites (Fig. 2) form the intermediate level of the Southern Amphibolites. They are comprised of more than 2500 m-thick interlayered and magmatically imbricate sills and/or dykes of intensely deformed metagabbros and metadiorites (Fig. 7). The rocks are mainly composed of amphibole and plagioclase with irregular occurrence of garnet that depends on the bulk composition (Treloar *et al.* 1990). The marked shape preferred orientation (SPO) of amphiboles delineates the mineral and stretching lineation. Rocks of the Kiru Amphibolites intrude the underlying gabbros and diorites of the Patan Complex. They are in turn intruded by a coarse-grained hornblende diorite dated at 91.8 ± 1.4 Ma (U/Pb zircon age: Schaltegger *et al.* 2002).

The top level, the Kamila Amphibolites, *sensu stricto* (Fig. 2) is a sequence of various plutonic bodies with gabbroic–tonalitic compositions, which have intruded into a crustal sequence of metamorphosed pelites, carbonates, volcano-detritic and volcanic rocks (Fig. 8). Granitic and pegmatitic veins are variously foliated and folded but cross-cut the banding of the country Kamila Amphibolites (Fig. 9). Granites are composed of quartz, plagioclase, muscovite, garnet, euhedral epidote–clinozoisite intergrown with

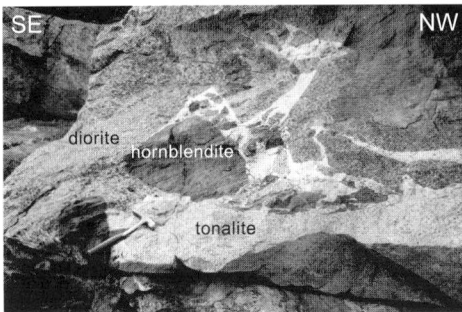

Fig. 7. Imbricate sills and/or dykes of intensely deformed metagabbros and metadiorites, on a typical outcrop of Kiru Amphibolites. Note the undeformed, coarse-grained gabbro boudins within intensely sheared diorites. For location see Figure 2.

quartz, chlorite, opaque phases, apatite and zircon. One of the thickest granites is dated at 97.1 ± 0.2 Ma (U/Pb zircon age: Schaltegger *et al.* 2002). The Kamila Amphibolites were strongly deformed during contact metamorphism along their northern boundary with the intrusive

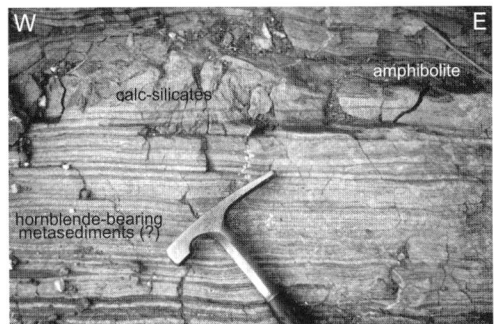

Fig. 6. Magmatic breccia: a leucocratic tonalite has welded angular fragments of coarse grained diorite containing hornblendite enclaves with smooth contours, and the whole has been intruded in turn by a white, quartz-felsic rich magma. For location see Figure 2.

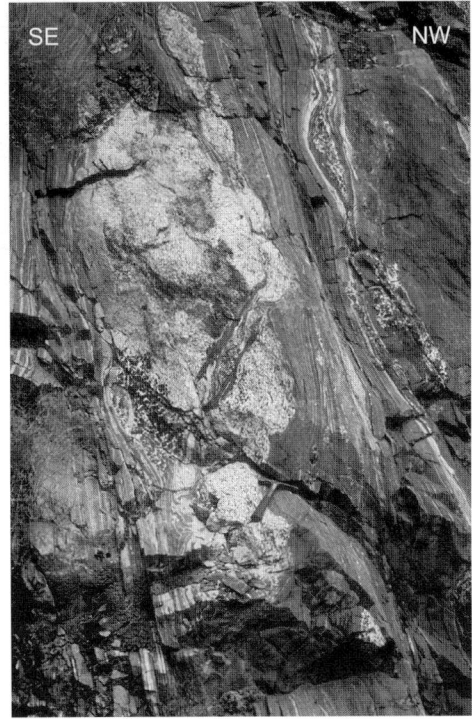

Fig. 8. Volcano-sedimentary sequence of the Kamila Amphibolite *sensu stricto*. For location see Figure 2.

Fig. 9. Variably deformed granite veins intruding the volcano-sedimentary sequence of the Kamila Amphibolite *sensu stricto*. Note the boudinaged, calc-silicate layer in the large enclave. For location see Figure 2.

85 Ma old gabbronoritic Chilas Complex (Schaltegger *et al.* 2002) and associated gabbros and tonalites. The lack of a strong and pronounced mineral or rodding lineation, and the absence of systematically asymmetric structures in the metasediments and associated metavolcanites (Figs 8 and 9, except for localized shear zones), points to a high flattening component during deformation. Intense vertical crenulation in places marks the transition to constriction. The alternation of strongly deformed volcano-sediments and undeformed plutonic units with preserved intrusive boundaries cutting the fabric of country rocks suggests that regional deformation is essentially due to emplacement of different intrusions into the island-arc system rather than strain localization in volcano-sedimentary sequences. This is also probably true for the strain of the deeper crustal levels on which we focused in this work.

Magmatic structures

The shape-preferred orientation (SPO or shape fabric) of magmatic minerals defines magmatic foliations and lineations nearly everywhere throughout the plutonic section (Arbaret *et al.* 2000). The magmatic fabric was acquired by alignment of euhedral pyroxene, amphibole, biotite and plagioclase eventually organized in cumulate textures. The subsequent development of fine-grained and granoblastic, metamorphic textures with crystallization of nearly equant grains reduced the fabric intensity, in particular in granulite facies rocks (Yoshino *et al.* 1998). Therefore, metamorphic facies were excluded from the fabric analysis. The intensity of the shape fabric corresponds to the aspect ratio of the fabric ellipse measured on conventional [*XZ*] sections of the rocks by using the intercept method (Launeau & Robin 1996), with *X* parallel to the lineation and *Z* orthogonal to the foliation. It is accepted that magmatic shape fabrics developed during magma emplacement by rotation and/or oriented crystallization of igneous minerals into the flow directions (e.g. Blumenfeld & Bouchez 1988; Ildefonse *et al.* 1992; Nicolas 1992). Aspect ratios of the fabric ellipses (long axis/short axis) correspond to strain ratios (March 1932; Ramsay & Graham 1970), which usually are weak in magmatic rocks. Igneous textures are thus easily separated from later, high-strain ratios produced during solid-state deformation of weaker minerals, such as plagioclase, in the studied rocks.

Lower crust

Weak intensity (1.05–1.2) magmatic fabrics are preserved in the core of the garnet-gabbro (Arbaret *et al.* 2000). The cumulate assemblage is composed of hypidiomorphic pyroxene, garnet, plagioclase and amphibole in textural equilibrium (Ringuette *et al.* 1998). Fabrics of this assemblage defines the *c.* 40° NW-dipping foliation and the *c.* 35° NW-plunging lineation (Figs 2 and 10). The magmatic foliation is sub-parallel to the modal layering, with cyclic plagioclase-rich layers grading into garnet-rich levels. Asymmetric cross-bedding features of igneous origin locally disrupt the compositional layering (Miller *et al.* 1991). Towards the basal contact with mantle rocks, higher intensity fabrics (1.2–>1.6; Arbaret *et al.* 2000) reflect plastic deformation of plagioclase and rigid rotation of pyroxene and other mineral phases in the border zone of the gabbro. S–C structures and the angular relationship between the mineral fabric and magmatic layering indicate bulk SW-ward shear.

Low–middle crust

In the overlying Sarangar gabbro, on a maximum thickness of about 2000 m, high-intensity fabrics

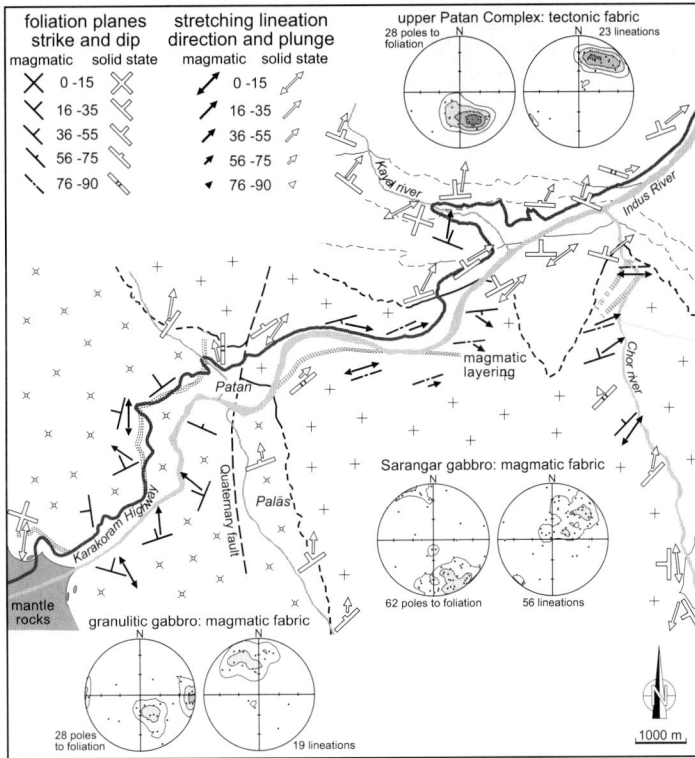

Fig. 10. Map (located in Fig. 1) of the magmatic and solid-state foliations and lineations in the upper Jijal, Patan and Kiru complexes. Lower-hemisphere equal-area projections. Starkey density contours: 2, 4, 6 and 8%.

(1.2–1.4) correspond to plastically deformed plagioclase and rotated pyroxene at the intrusive contact. The fabric intensity decreases upwards, toward the inner parts of the pluton. Smallest intensities (<1.08) were measured in coarse-grained magmatic cumulates with a garnet-free assemblage of hypidioblastic orthopyroxene, clinopyroxene, plagioclase, amphibole, epidote and accessory quartz (Yoshino *et al.* 1998). The weak fabric delineates an E–W-trending, subvertical magmatic foliation that bears a W-plunging lineation (Fig. 10). Where present, graded compositional layers parallel to the foliation suggest flow-sorting processes during magmatic flow parallel to the foliation. The hornblende-gabbro and diorite laccoliths that have intruded the Sarangar gabbro, at its top, have a 35–40° N-dipping foliation with a NE-striking lineation. Strong orientation intensity fabrics (1.3–1.4) are associated with elongated enclaves and deformed felsic veins. The magmatic fabric is rarely preserved in small boudins and the deformation of these rocks is fundamentally solid state with rotated porphyroclasts indicating a SW-ward

sense of shear, as recognized everywhere in the Southern Amphibolites (Treloar *et al.* 1990).

Late magmatic structures

Late magmatic structures are identified as such because they are consistent with the magmatic fabrics, but intracrystalline plastic deformation is pervasive. Related strain documents late solid-state emplacement of magmas at near-solidus conditions (e.g. Nicolas 1992).

Deepest and deep crust

In both the garnet-gabbros and the Sarangar gabbros, shear zones less than 1 m long and a few centimetres wide (set 1 of Arbaret *et al.* 2000) (Fig. 11) already pertain to the generation of late magmatic structures. They typically bend the magmatic foliation and layering without noticeable change in grain size; the magmatic paragenesis remains stable in the most deformed zones and discrete, feldspar-rich magmatic veins cut these zones (Fig. 12). Therefore, these shear

Fig. 11. Magmatic shape fabric and layering cut by a late magmatic shear zone in the Sarangar gabbro. For location see Figure 2.

zones formed at temperature conditions above the solidus. They define two NW–SE-striking orientation populations. One dips $40° \pm 10°$ NE, the other dips $20° \pm 10°$ SW. Both fit Riedel orientations described in brittle movement zones (Fig. 13) (Riedel 1929), in support of the interpretation that arrays of shear zones respect slip-system orientations (Cobbold & Gapais 1986). Normal SW-verging shear zones may be compared to Riedel shears (R), while reverse NE-dipping shear zones are consistent

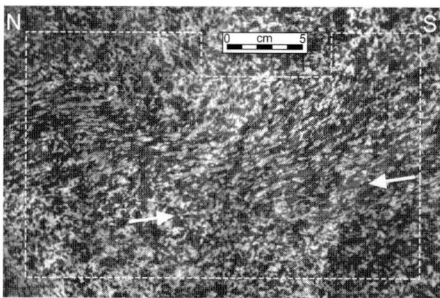

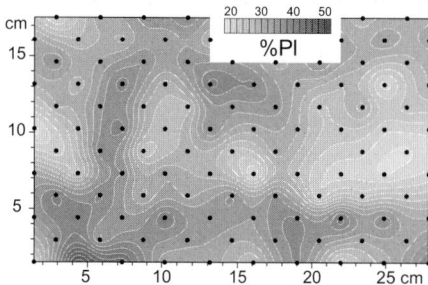

Fig. 12. *XZ* section of a lengthwise terminating shear zone cross-cut by a plagioclase-rich magmatic joint (arrows). Below: map of plagioclase content. For location see Figure 2.

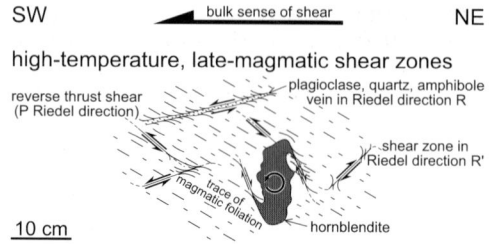

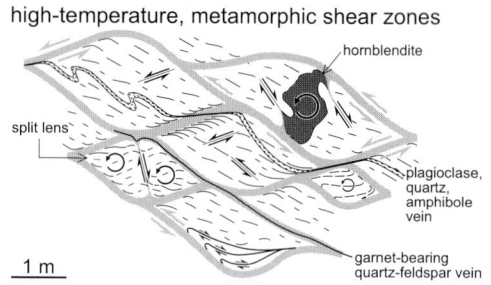

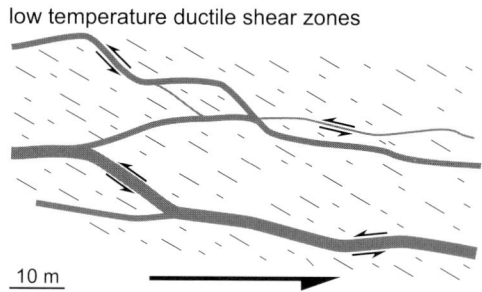

Fig. 13. Synthetic sketch of anastomosing ductile shear zones. Partial melting in the mylonite yielded garnet-bearing quartzo-feldspathic veins. NE-ward normal shear zones developed essentially in response to rotation of hornblendite bodies and split lenses. Modified after Arbaret *et al.* (2000).

with complementary thrust shears (P) (Tchalenko 1968) (Fig. 13). Both orientation populations developed contemporaneously in a general, nearly horizontal SW-verging shear regime before tilting of the Kohistan Complex during the Himalayan collision (Tahirkheli *et al.* 1979; Coward *et al.* 1986). Variations in per cent area of plagioclase and strain ratios were calculated together from image analysis on [*XZ*] sections of a lengthwise-terminated, SW-verging Riedel shear zone (Fig. 12). Low strain ratios (<1.2) and homogeneous surface plagioclase content ($34.27 \pm 5.33\%$) characterize the undeformed protolith (Fig. 14). Rock domains with smallest plagioclase content (19%) do not match the maximum strain ratio

(1.68 indicating shear strain $\gamma < 5$; March 1932; Ramsay 1980) along the shear zone, on a domain where the plagioclase content is the same as in the undeformed protolith (Figs 11 and 14). Strain localization took place within a primarily steep gradient in relative plagioclase content, which is the measured weak phase (Fig. 12). Therefore, gradients in the distribution of the weakest mineralogical phase seem to play a more vital role in strain localization than relative proportions of weak and strong mineral phases (Goodwin & Tikoff 2002).

Middle crust

Magmatic and late magmatic fabrics are diversely preserved in the overlying Southern Amphibolites. In the cores of several meter-thick metagabbro bodies, centimetre-wide, low shear-strain shear zones looking very similar to those observed in the Sarangar gabbro (Fig. 13), developed along Riedel-like directions. They represent late magmatic structures. In the metadiorites, independently from the intrusion thickness, these discrete, subsolidus shear zones are rare and disperse, even possibly absent; an homogeneous and often strong shape fabric, difficult to discriminate from subsequent lower amphibolite facies deformation, represents the late magmatic structures. In centimetre–metre-thick gabbro and diorite intrusions,

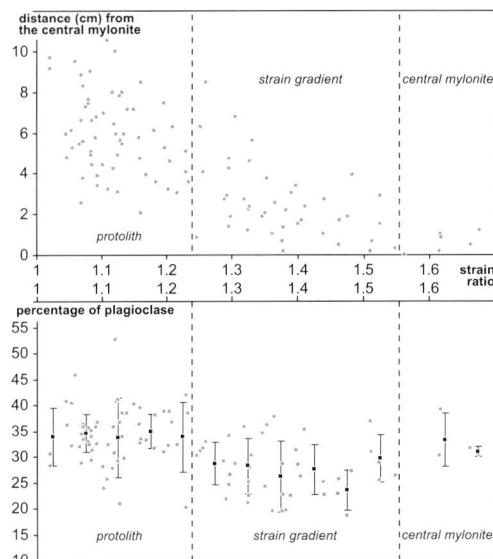

Fig. 14. Comparison of the strain ratio across the Riedel shear zone of Figure 12 with distance from the shear plane (top) and relative plagioclase content (bottom).

without regard to the rock composition, amphibolite-facies shear zones whose assemblages are quartz, plagioclase, hornblende and accessory garnet (Yoshino *et al.* 1998) are pervasive and have penetratively erased earlier magmatic features. Hornblende long axes define a SW-striking mineral and stretching lineation developed during solid-state shear deformation (Fig. 10).

High-temperature metamorphic shear zones

Mylonitic shear zones form a pervasive pattern of N-dipping to flat-lying, SW-verging reverse shear zones bearing a *c.* 30° NE-plunging lineation and branching into horizontal to W-dipping, SW-verging normal shear zones bearing a *c.* 15° SW-plunging lineation (Arbaret *et al.* 2000). Consistently asymmetric lenses at all scales indicate that enclosing shear zones developed during distributed non-coaxial deformation (Gapais *et al.* 1987). Earliest, subsolidus shear zones are small, spaced by a few centimetres (Figs 12 and 13), and shear strain in the most deformed parts do not exceed $\gamma = 5$. These earliest shear zones remain inconspicuous. The prominent pattern is built of high-temperature shear zones that formed once the rock had fully crystallized, hence at lower temperatures than inconspicuous, subsolidus shear zones. The anastomosing pattern begins about 1000 m above the 'palaeo-Moho', at a level where the granulitic gabbro contains many hornblendite boudins (Fig. 2). Anastomosing becomes denser upwards, with an average spacing of 8–10 m in the Sarangar gabbro (Fig. 13). Higher up, in the hornblende-gabbros and diorite laccoliths of the Southern Amphibolites, spacing between shear zones is about 50 m, but varies according to lithologies.

In the Sarangar gabbro, less than 1% water content conditions governed the progressive replacement of diopside and hypersthene by ferroan-pargasitic to tschermakitic hornblendes in the shear strain gradients, and the crystallization of the quartz, plagioclase, pargasitic to tschermakitic hornblende, garnet-bearing assemblages within the subsolidus shear zones (Treloar *et al.* 1990; Arbaret & Burg 2003). These nearly dry conditions are consistent with the presence of pyroxene porphyroclasts preserved from full retrogression in the mylonite. Thermo-barometric conditions were estimated at 550–650 °C under pressures of 0.9–1.0 GPa (Bard 1983; Treloar *et al.* 1990).

Continuous strain gradients include progressively bent foliations along which fabric strain ratios increase up to 1.9 (Arbaret *et al.* 2000). Strain gradients show a variety of morphology, and, contrary to classical knowledge, they are rarely symmetric. Strain gradients on both sides of mylonite zones often have different shapes and widths (from 20 to <1 cm), a gradient on one side even being absent in extreme cases (Fig. 15). Differences in width and foliation curvature in strain gradients are regarded as expressing differences in the stress exponent of the power-law rheology (e.g. Wilkinson 1960; Turcotte & Schubert 2002) and/or in the weakening processes, such as variation in grain size and water content (see Talbot 1999; Mulchrone 2001; and discussion in Sonder 2001; Talbot 2001). In the anastomosing pattern studied (set 2 of Arbaret *et al.* 2000), differences in shear strain gradients are explained by variations in finite strain and in thinning components due to complex flow forced to wrap lenses of nearly undeformed rock. The structural asymmetry reflects differences in strain propagation along the mylonite border, which are induced by compositional heterogeneities such as soft, plagioclase-rich and syndeformational veins on one side (Arbaret & Burg 2003). The angular

relationship between the mylonite and the protolith foliation also plays an important role on the asymmetry of strain gradient (Ghosh & Sengupta 1987).

Partial melting in mylonites generated quartz–plagioclase–garnet-bearing segregation veins, most voluminous in tensional zones between lenses (Fig. 16). Incipient melting, possibly triggered/enhanced by dehydration of hornblendite, implies temperatures exceeding 650–700 °C under pressures of >0.8 GPa (Burnham 1979), in agreement with the upper amphibolite facies crystal plasticity identified by (Treloar *et al.* 1990). The fact that melt products are found in shear zones only meets two interpretations: (1) fluids that facilitate melting are concentrated and permeate mylonites; however, geochemical data demonstrate isochemical deformation (Arbaret & Burg 2003), which contradicts the pervasive role of incoming fluids; or (2) viscous heating (Gruntfest 1963; Melosh 1976; Regenauer-Lieb & Yuen 2000) exceeds the ability of heat to escape from the rock volume undergoing shear deformation, hence reaching melting temperatures (>750–800 °C) in the mylonites only. Shear melting would further

Fig. 15. Asymmetric, high-temperature shear zone. Note the absence of a strain gradient on the left-hand side of the mylonite. For location see Figure 2.

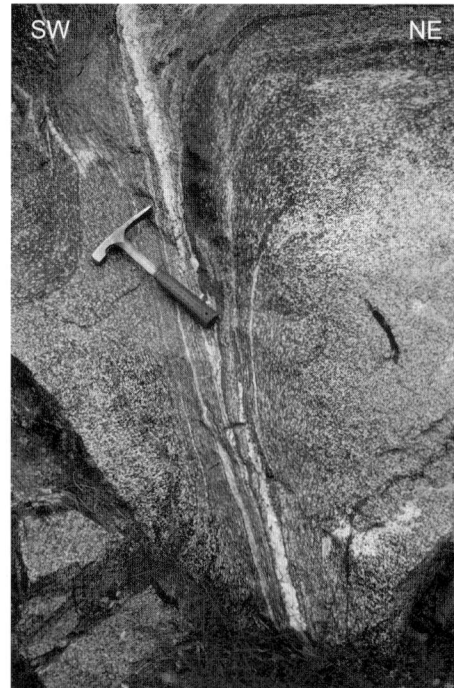

Fig. 16. Mylonites with quartz–plagioclase–garnet-bearing segregation veins. For location see Figure 2.

localize deformation into the shear zones as melt has a very low viscosity (e.g. Shaw 1972; Scaillet *et al.* 1997). This hypothesis fits observation better than the intervention of fluids; we conclude that viscous heating and shear melting were critical strain localization processes in the deep crust of the Kohistan Arc, as they are in other crustal faults (Brun & Cobbold 1980; West & Hubbard 1997; Jin *et al.* 1998).

We wonder whether this mechanism may lead to pervasive upward-migration of a small melt fraction, such as those that have fed the granite and pegmatite veins intruding the Southern Amphibolites, and in particular the top Kamila Amphibolites (Fig. 2). The granite veins bear variously intense shear fabrics parallel to the regional foliation of the host amphibolites, both with rotated porphyroblasts and asymmetrically folded granite veins denoting intense SW-ward shearing (Treloar *et al.* 1990) during emplacement, at 97.1 ± 0.2 Ma (Schaltegger *et al.* 2002), a short time after emplacement of the deep gabbros (Sarangar at 98.9 ± 0.4 Ma, Schaltegger *et al.* 2002). This near-contemporaneousness indicates that melt generation, ascent and emplacement of differentiated magmas into the middle crust took place immediately after emplacement of parental magmas in the deep crust; that is, while the high temperature, subsolidus–late magmatic solid-state shear zones were active. Partial melting in active shear zones may be an underestimated, deformation-assisted mechanism producing and transferring granitic components from the bottom to shallower crustal levels of island arcs.

Low-temperature ductile shear zones

The low-temperature ductile shear zones are about 3 m thick and several tens of metres long in the same lithologies as previously described (e.g. the Sarangar gabbro). They dip shallowly towards the south and north, with a wavelength >50 m (Fig. 13). Spacing is more than 100 m. The mylonite assemblage comprises porphyroclastic garnet, amphibole and plagioclase derived from stretched pegmatite veins embedded in a strongly foliated and fine-grained matrix of quartz, plagioclase, calcic amphibole and epidote. The matrix assemblage points to epidote amphibolite facies, which is probably 83–80 Ma old, the regional $^{39}Ar/^{40}Ar$ cooling age measured on amphiboles (Treloar *et al.* 1989; Wartho *et al.* 1996).

Drastic grain size reduction and large porphyroclast rotations indicate shear strains of over $\gamma = 10$. Such high strains were rarely measured in the high-temperature shear zones. Having noted the larger spacing of low-temperature, high shear strain zones, we conclude that shear strains concentrates into fewer but more important movement zones, a clear shear strain localization during cooling of the deep–middle plutonic crust of the arc. It is probable that part of the shear strain is related to the stretching and rotation of shear zones during the development of the anastomosing pattern, as noted in analogue experiments of high-grade shear zones (Grujic & Mancktelow 1998).

Conclusions

In summary, the following points have been argued.

- Magmatic and solid-state shear structures are kinematically consistent from mantle to middle crust rocks of the Kohistan palaeo-arc. The magmatic fabrics and the subsequent late magmatic shear zones record the emplacement of mafic, calc-alkaline magmas into the base of the crust of the arc and document a continuum of SW-ward shear-dominated flow that influenced the intrusion modes and deformation styles of the nascent crust.

- Strain localization took place within steep gradients between zones with different mineral contents as early as magmatic stages in large gabbro intrusions, somewhat more favourably than in dioritic intrusions. Therefore, shear strain localization is likely to occur from the earliest stages of arc building, along with crystallization of the first gabbroic magma patches. In the Kohistan deep crust, lengthwise propagation and coalescence of discrete shear zones during subsequent solid-state deformation resulted in the pervasive asymmetric pattern of mylonitic shear zones. There is an obvious relationship between spacing and wavelength of anastomosing shear zones and the temperature conditions at which they evolved.

- Penetrative deformation of relatively soft and/or thin sill-in-sill (or dyke-in-dyke) intrusions is responsible for thick, regular and melt-absent shear zones. Partial melting in high-temperature shear zones, enhanced by viscous shear heating, is apparently a constructive factor of arc formation, favouring generation and mobilization of differentiated magma and transferring granite components from bottom to shallower crustal levels. These observations indicate the importance of deformation-assisted processes in the

creation of arc lithospheres, which are seeds
to continental lithosphere. Pervasive deformation and intracrustal melting necessary to
transform arc crust into mature continental
crust begin to occur during arc building and
are not diagnostic of any subsequent continental collision.

• Morphologically, ductile shear zones appear
thicker and longer with decreasing temperatures. Individual shear zones are probably
too small to be imaged by reflection profiling.
However, the anastomosing system occupies
a thick part of the lower crust and intensifies
heterogeneity both in terms of lithology and
fabric anisotropy. Therefore, anastomosing
shear zones should bear a significant role in
the seismic signature of deep crustal levels.

The Swiss National Science Foundation, grants 20-
49372.96 and 20-61465.00, supported this work. The
Pakistan Museum of Natural History supports S. Hussain
and H. Dawood, and the University of Punjab in Lahore
supports N. Chaudhry. We thank the editors, an anonymous reviewer and P. Ulmer for pertinent reviews that
helped to clarify this description.

References

ANCZKIEWICZ, R. & VANCE, D. 2000. Isotopic constraints on the evolution of metamorphic conditions
in the Jijal–Patan complex and the Kamila Belt of
the Kohistan arc, Pakistan Himalaya. *In*: KHAN,
M.A., TRELOAR, P.J., SEARLE, M.P. & JAN, M.Q.
(eds) *Tectonics of the Nagna Parbat Syntaxis and
the Western Himalaya*. Geological Society,
London, Special Publications, **170**, 321–331.

ARBARET, L. & BURG, J.-P. 2003. Complex flow in
lowest crustal, anastomosing mylonites: Strain gradients in a Kohistan gabbro, northern Pakistan.
Journal of Geophysical Research, **108**, 2467, doi:
10.1029/2002JB002295.

ARBARET, L., BURG, J.-P., ZEILINGER, G., CHAUDHRY,
N., HUSSAIN, S. & DAWOOD, H. 2000. Pre-
collisional anastomosing shear zones in the
Kohistan arc, NW Pakistan. *In*: KHAN, M.A.,
TRELOAR, P.J., SEARLE, M.P. & JAN, M.Q. (eds)
*Tectonics of the Nanga Parbat Syntaxis and the
Western Himalaya*. Geological Society, London,
Special Publications, **170**, 295–311.

BARD, J.-P. 1983. Metamorphism of an obducted
island arc: Example of the Kohistan sequence
(Pakistan) in the Himalayan collided range. *Earth
and Planetary Science Letters*, **65**, 133–144.

BARD, J.-P., MALUSKI, H., MATTE, P. & PROUST,
F. 1980. The Kohistan sequence: crust and mantle
of an obducted island arc. *Geological Bulletin of
the University of Peshawar*, **11**, 87–94.

BLUMENFELD, P. & BOUCHEZ, J.-L. 1988. Shear criteria in granite and migmatite deformed in the

magmatic and solid states. *Journal of Structural
Geology*, **10**, 361–372.

BODINIER, J.-L. & GODARD, M. 2003. Orogenic,
ophiolitic, and abyssal peridotites. *In*: TUREKIAN,
K., HOLLAND, H.E. & CARLSON, R.W.v.E. (eds)
*Treatise on Geochemistry Volume 2: Geochemistry
of the Mantle and Core*. Elsevier Science,
Amsterdam, 103–170.

BOIS, C., LEFORT, J.-P., LE GALL, B., SIBUET, J.-C.,
GARIEL, O., PINET, B. & CAZES, M. 1990.
Superimposed Variscan, Caledonian and Proterozoic features inferred from deep seismic profiles
recorded between southern Ireland, southwestern
Britain ans western France. *Tectonophysics*, **177**,
15–37.

BOUDIER, F., NICOLAS, A. & ILDEFONSE, B. 1996.
Magma chambers in the Oman ophiolite: fed
from the top and the bottom. *Earth and Planetary
Science Letters*, **144**, 239–250.

BRUN, J.-P. & COBBOLD, P.R. 1980. Strain heating
and thermal softening in continental shear zones:
a review. *Journal of Structural Geology*, **2**,
149–158.

BURG, J.-P., BODINIER, J.-L., CHAUDHRY, M.N.,
HUSSAIN, S. & DAWOOD, H. 1998. Infra-arc
mantle–crust transition and intra-arc mantle
diapirs in the Kohistan Complex (Pakistani Himalaya): petro-structural evidence. *Terra Nova*, **10**,
74–80.

BURNHAM, C.W. 1979. Magmas and hydrothermal
fluids. *In*: BARNES, H.L. (ed.) *Geochemistry of
Hydrothermal Ore Deposits*. Wiley-Interscience,
New York, 71–136.

COBBOLD, P.R. & GAPAIS, D. 1986. Slip-system
domains. I. Plane-strain kinematics of arrays of
coherent bands with twinned fibre orientations.
Tectonophysics, **131**, 113–132.

COWARD, M.P., BUTLER, R.W.H., KHAN, M.A. &
KNIPE, R.J. 1987. The tectonic history of Kohistan
and its implications for Himalayan structure.
Journal of the Geological Society, London, **144**,
377–391.

COWARD, M.P., WINDLEY, B.F. *ET AL.* 1986. Collision
tectonics in the NW Himalayas. *In*: COWARD, M.P.
& RIES, A.C. (eds) *Collision Tectonics*. Geological
Society, London, Special Publications, **19**, 203–
219.

FOUNTAIN, D.M. 1986. Implications of deep crustal
evolution for seismic reflection interpretation. *In*:
BARAZANGI, M. & BROWN, L. (eds) *Reflection
Seismology: The Continental Crust*. Geodynamic
Series, **14**. American Geophysical Union,
Washington, D.C., 1–7.

FOUNTAIN, D.M., SALISBURY, M.H. & PERCIVAL,
J. 1990. Seismic structure of the continental crust
based on rock velocity measurements from the
Kapuskasing uplift. *Journal of Geophysical
Research*, **95**, 1167–1186.

GAPAIS, D., BALÉ, P., CHOUKROUNE, P., COBBOLD, P.,
MAHJOUB, Y. & MARQUER, D. 1987. Bulk
kinematics from shear zone patterns: some field
examples. *Journal of Structural Geology*, **9**,
635–646.

GHOSH, S.K. & SENGUPTA, S. 1987. Progressive development of structures in a ductile shear zone. *Journal of Structural Geology*, **9**, 277–287.

GOODWIN, L.B. & TIKOFF, B. 2002. Competency contrast, kinematics, and the development of foliations and lineations in the crust. *Journal of Structural Geology*, **24**, 1065–1085.

GRUJIC, D. & MANCKTELOW, N.S. 1998. Melt-bearing shear zones: analogue experiments and comparison with examples from southern Madagascar. *Journal of Structural Geology*, **20**, 673–680.

GRUNTFEST, I.J. 1963. Thermal feedback in liquid flow: plane shear at constant stress. *Transactions of the Society of Rheology*, **7**, 195–207.

ILDEFONSE, B., LAUNEAU, P., FERNANDEZ, A. & BOUCHEZ, J.-L. 1992. Effect of mechanical interactions on development of shape preferred orientations: a two-dimensional experimental approach. *Journal of Structural Geology*, **14**, 73–83.

JAN, M.Q. & HOWIE, R.A. 1981. The mineralogy and geochemistry of the metamorphosed basic and ultrabasic rocks of the Jijal complex, Kohistan, NW Pakistan. *Journal of Petrology*, **22**, 85–126.

JAN, M.Q. & WINDLEY, B.F. 1990. Chromian spinel-silicate chemistry in ultramafic rocks of the Jijal Complex, Northwest Pakistan. *Journal of Petrology*, **31**, 667–715.

JI, S., SALISBURY, M.H. & HANMER, S. 1993. Petrofabric, P-wave anisotropy and seismic reflectivity of high-grade tectonites. *Tectonophysics*, **222**, 195–226.

JIN, D.H., KARATO, S.I. & OBATA, M. 1998. Mechanisms of shear localization in the continental lithosphere: inference from the deformation microstructures of peridotites from the Ivrea zone, northwestern Italy. *Journal of Structural Geology*, **20**, 195–209.

KAUSAR, A.B., PICARD, C., KELLER, F. & ZAFAR, M. 1997. Jijal mafic-ultramafic Complex: Crystallisation story and implcations for crustal evolution in the Kohistan Arc, Pakistan. *Terra Nova* **9**, Abstract Supplement 1, 383.

KHAN, M.A., JAN, M.Q. & WEAVER, B.L. 1993. Evolution of the lower arc crust in Kohistan, N. Pakistan: temporal arc magmatism through early, mature and intra-arc rift stages. *In*: TRELOAR, P.J. & SEARLE, M.P. (eds) *Himalayan Tectonics*. Geological Society, London, Special Publications, **74**, 123–138.

KHAZANEHDARI, J., RUTTER, E.H. & BRODIE, K.H. 2000. High-pressure-high-temperature seismic velocity structure of the midcrustal and lower crustal rocks of the Ivrea–Verbano zone and Serie dei Laghi, NW Italy. *Journal of Geophysical Research*, **105**, 13 843–13 858.

LAUNEAU, P. & ROBIN, P.-Y.F. 1996. Fabric analysis using the intercept method. *Tectonophysics*, **267**, 91–119.

LE FORT, P., MICHARD, A., SONET, J. & ZIMMERMANN, J.-L. 1983. Petrography, geochemistry and geochronology of some samples from the Karakorum batholith (Northern Pakistan). *In*: SHAMS, F.A. (ed.) *Granites of Himalayas,*

Karakoram and Hindu Kush. Punjab University, Lahore, 377–387.

LUDDEN, J. & HYNES, A. 2000. The Lithoprobe Abitibi-Grenville transect: two billion years of crust formation and recycling in the Precambrian Shield of Canada. *Canadian Journal of Earth Sciences*, **37**, 459–476.

MARCH, A. 1932. Mathematische Theorie der Regelung nach der Korngestalt bei affiner Deformation. *Zeitschrift für Kristallographie, Mineralogie und Petrographie*, **81**, 285–298.

MELOSH, H.J. 1976. Plate motion and thermal instability in the asthenosphere. *Tectonophysics*, **35**, 363–390.

MILLER, D.J., LOUCKS, R.R. & ASHRAF, M. 1991. Platinum-group element mineralization in the Jijal layered ultramafic–mafic complex, Pakistani Himalayas. *Economic Geology*, **86**, 1093–1102.

MULCHRONE, K.F. 2001. Quantitative estimation of exponents of power-low flow with confidence intervals in ductile shear zones. *Journal of Structural Geology*, **23**, 803–806.

MÜNTENER, O., KELEMEN, P.B. & GROVE, T.L. 2001. The role of H_2O during crystallization of primitive arc magmas under uppermost mantle conditions and genesis of igneous pyroxenites: an experimental study. *Contributions to Mineralogy and Petrology*, **141**, 643–658.

NELSON, K.D. 1991. A unified view of craton evolution motivatedby recent deep seismic reflection and refraction results. *Geophysical Journal International*, **105**, 25–35.

NICOLAS, A. 1992. Kinematics in magmatic rocks with special reference to gabbros. *Journal of Petrology*, **33**, 891–915.

PETTERSON, M.G. & WINDLEY, B.F. 1985. Rb–Sr dating of the Kohistan arc-batholith in the Trans-Himalaya of north Pakistan, and tectonic implications. *Earth and Planetary Science Letters*, **74**, 45–57.

RAMSAY, J.G. 1980. Shear zone geometry: a review. *Journal of Structural Geology*, **2**, 83–99.

RAMSAY, J.G. & GRAHAM, R.H. 1970. Strain variation in shear belts. *Canadian Journal of Earth Sciences*, **7**, 786–813.

REGENAUER-LIEB, K. & YUEN, D.A. 2000. Quasi-adiabatic instabilities associated with necking processes of an elasto-viscoplastic lithosphere. *Physics of the Earth and Planetary Interiors*, **118**, 89–102.

RIEDEL, W. 1929. Zur Mechanik geologischer Brucherscheinungen. *Zentralblatt für Mineralogie, Geologie und Paläontologie*, **Abteilung B, Geologie und Paläontologie**, 354–368.

RINGUETTE, L., MARTIGNOLE, J. & WINDLEY, B.F. 1998. Pressure–Temperature evolution of garnet-bearing rocks from the Jijal complex (western Himalayas, northern Pakistan): from high-pressure cooling to decompression and hydration of a magmatic arc. *Geological Bulletin, University of Peshawar*, **31**, 167–168.

RUTTER, E.H., KHAZANEHDARI, J., BRODIE, K.H., BLUNDELL, D.J. & WALTHAM, D.A. 1999. Synthetic seismic reflection profile through the

Ivrea zone–Serie dei Laghi continental crustal section, northwestern Italy. *Geology*, **27**, 79–82.

SCAILLET, B., HOLTZ, F. & PICHAVANT, M. 1997. Rheological properties of granitic magmas in their crystallization range. *In*: BOUCHEZ, J.-L., HUTTON, D. & STEPHEN, W.E. (eds) *Granite: From Segregation of Melt to Emplacement Fabrics*. Kluwer, Dordrecht, 11–29.

SCHALTEGGER, U., ZEILINGER, G., FRANK, M. & BURG, J.-P. 2002. Multiple mantle sources during island arc magmatism: U–Pb and Hf isotopic evidence from the Kohistan arc complex, Pakistan. *Terra Nova*, **14**, 461–468.

SHAW, H.R. 1972. Viscosities of magmatic silicate liquid: an empirical method of prediction. *American Journal of Science*, **272**, 870–893.

SONDER, L.J. 2001. Ductile shear zones as counterflow boundaries in pseudoplastic fluids: Discussion and theory. *Journal of Structural Geology*, **23**, 149–153.

TAHIRKHELI, R.A.K., MATTAUER, M., PROUST, F. & TAPPONNIER, P. 1979. The India Eurasia Suture Zone in Northern Pakistan: Synthesis and interpretation of recent data at plate scale. *In*: FARAH, A. & DE JONG, K.A. (eds) *Geodynamics of Pakistan*. Geological Survey of Pakistan, Quetta, 125–130.

TALBOT, C.J. 1999. Ductile shear zones as counterflow boundaries in pseudoplastic fluids. *Journal of Structural Geology*, **21**, 1535–1551.

TALBOT, C.J. 2001. Ductile shear zones as counterflow boundaries in pseudoplastic fluids: Reply. *Journal of Structural Geology*, **23**, 157–159.

TCHALENKO, J.S. 1968. The evolution of kink-bands and the development of compression textures in sheared clays. *Tectonophysics*, **6**, 159–174.

TRELOAR, P.J., BRODIE, K.H. ET AL. 1990. The evolution of the Kamila Shear Zone, Kohistan, Pakistan. *In*: SALLISBURY, M.H. & FOUNTAIN, D.M. (eds) *Exposed Cross-Sections of the Continental Crust*. Kluwer Academic Press, Amsterdam, 175–214.

TRELOAR, P.J., PETTERSON, M.G., QASIM JAN, M. & SULLIVAN, M.A. 1996. A re-evaluation of the stratigraphy and evolution of the Kohistan arc sequence, Pakistan Himalaya: implications for magmatic and tectonic arc-building processes. *Journal of the Geological Society, London*, **153**, 681–693.

TRELOAR, P.J., REX, D.C. ET AL. 1989. K/Ar and Ar/Ar geochronology of the Himalayan collision in NW Pakistan: constraints on the timing of suturing, deformation, metamorphism and uplift. *Tectonics*, **8**, 881–909.

TURCOTTE, D.L. & SCHUBERT, G. 2002. *Geodynamics*. Cambridge University Press, Cambridge.

WARTHO, J.-A., REX, D.C. & GUISE, P.G. 1996. Excess argon in amphiboles linked to greenschist facies alterations in the Kamila Amphibolite Belt, Kohistan island arc system, northern Pakistan: insights from $^{40}Ar/^{39}Ar$ step-heating and acid leaching experiments. *Geological Magazine*, **133**, 595–606.

WEST, D.P. & HUBBARD, M.S. 1997. Progressive localization of deformation during exhumation of a major strike-slip shear zone: Norumbega fault zone, south-central Maine, USA. *Tectonophysics*, **273**, 185–201.

WILKINSON, W.L. 1960. *Non-Newtonian Fluids: Fluid Mechanics, Mixing and Heat Transfer*. Pergamon Press, London.

YAMAMOTO, H. 1993. Contrasting metamorphic P–T–time paths of the Kohistan granulites and tectonics of the western Himalayas. *Journal of the Geological Society, London*, **150**, 843–856.

YAMAMOTO, H. & NAKAMURA, E. 1996. Sm–Nd dating of garnet granulites from the Kohistan complex, northern Pakistan. *Journal of the Geological Society, London*, **153**, 965–969.

YOSHINO, T., YAMAMOTO, H., OKUDAIRA, T. & TORIUMI, M. 1998. Crustal thickening of the lower crust of the Kohistan arc (N. Pakistan) deduced from Al zoning in clinopyroxene and plagioclase. *Journal of Metamorphic Geology*, **16**, 729–748.

Flow of partially molten crust and origin of detachments during collapse of the Cordilleran Orogen

C. TEYSSIER[1], E. C. FERRÉ[2], D. L. WHITNEY[1], B. NORLANDER[1], O. VANDERHAEGHE[3] & D. PARKINSON[4,5]

[1]Department of Geology and Geophysics, University of Minnesota, Minneapolis, MN 55455, USA

[2]Department of Geology, Southern Illinois University, Mailcode 4324, Carbondale, IL 62901, USA

[3]Université de Nancy 1, UMR 7566 G2R, BP 239, 54506 Vandoeuvre-les-Nancy, France

[4]Department of Geological Sciences, University of California, Santa Barbara, CA 93106, USA

[5]Present address: 429 N 190th Street, Seattle, WA 98133, USA

Abstract: In metamorphic core complexes two types of detachments develop, coupled by flow of partially molten crust: a channel detachment and a rolling-hinge detachment. The channel detachment, on the hinterland side of the orogen, represents the long-lived interface that separates the partially molten crust flowing in a channel from the rigid upper crustal lid. On the foreland side of the core complex, a rolling-hinge detachment develops. This detachment dips toward the foreland, probably affects the whole crust, and its geometry is governed by strain localization at the critical interface between cold foreland and hot hinterland. Activation of the rolling-hinge detachment drives rapid decompression and melting, leading to the diapiric rise of migmatite domes in the footwall of the detachment. A kinematic hinge (switch in sense of shear) separates the two types of detachments. Structural, metamorphic and geo/thermochronological studies in the Shuswap core complex (North American Cordillera), combined with an anisotropy of magnetic susceptibility study of leucogranites concentrated in the detachments, suggest that this orogen collapsed rapidly through the development of channel and rolling-hinge detachments in the early Eocene. The kinematic hinge is currently located approximately 40 km west of the footwall in which it originated, corresponding to a mean exhumation rate of >5 km Ma^{-1}, which explains the near-isothermal decompression recorded within the migmatite dome.

A new paradigm for orogeny involves the development of a topographic plateau and a thermally and rheologically layered crust, with a relatively thin, rigid lid of upper crust overlying a partially molten layer that makes up a significant fraction of the crust (Nelson *et al.* 1996; Schilling & Partzsch 2001). This low-viscosity layer may flow in a channel (Royden 1996; Royden *et al.* 1997) that decouples the upper crust from the lower crust–upper mantle (Nelson *et al.* 1996; McKenzie *et al.* 2000; Beaumont *et al.* 2001; Rey *et al.* 2001; Vanderhaeghe & Teyssier 2001*a*, *b*; Vanderhaeghe *et al.* 2003*a*). The behaviour of partially molten crust has significant implications for the mechanics of orogens, and in particular for the origin and evolution of the interface between the partially molten layer and the rigid upper crust.

In exhumed orogens, anatectic migmatite domes are commonly exposed, suggesting upward flow of partially molten crust, and the interface between upper crust and migmatites is commonly represented by a shallowly dipping detachment. Structural, metamorphic and geo/thermochronological data show that melt crystallization in migmatite domes and leucogranites, activation of detachments, and cooling/exhumation of the footwall rocks are largely coeval (Vanderhaeghe & Teyssier 2001*b*). Therefore, the formation and evolution of detachments may be fundamentally tied to the dynamics of partially molten crust, including both lateral

From: Bruhn, D. & Burlini, L. (eds) 2005. *High-Strain Zones: Structure and Physical Properties*. Geological Society, London, Special Publications, **245**, 39–64.
0305-8719/05/$15.00 © The Geological Society of London 2005.

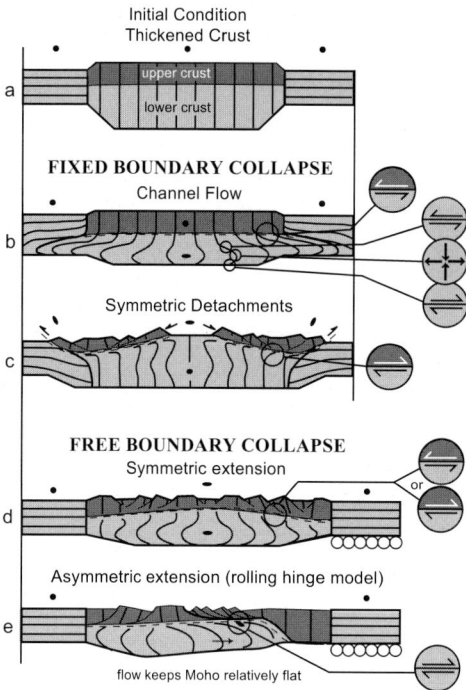

Fig. 1. Fixed-boundary and free-boundary modes of collapse (modified from Rey *et al.* 2001) with expected kinematics at critical localities of these systems, as discussed in the text.

(channel) and vertical (diapiric) flow, as the orogen evolves from an orogenic plateau phase to a collapse phase (Fig. 1) (Rey *et al.* 2001). The kinematic relationship between upper crustal extension and lower crustal flow has been debated vigorously with regard to the origin of metamorphic core complexes (MCC) (Crittenden *et al.* 1980; Lister & Davis 1989; Malavieille 1993; Wills & Buck 1997; Zheng *et al.* 2004). The prevailing view is that the ductile crust flows passively to accommodate extension of the upper crust (Bird 1991; Axen *et al.* 1998). The lower crust flows toward the thinned zone (Buck 1988, 1991; Brun *et al.* 1994) to fill the gap beneath the stretched upper crust and to spread the thinning (Bertotti *et al.* 2000). The progressive rotation of early formed normal faults in the footwall of active high-angle normal faults, as in a 'rolling-hinge' model, explains the existence of shallowly dipping detachments.

In the Shuswap MCC, North American Cordillera, the interface between upper crust and partially molten crust has been exhumed. We present new metamorphic and geochronological

data, and synthesize existing structural, metamorphic and age data from this complex to evaluate the evolution of high strain detachment zones. In addition, we use the anisotropy of magnetic susceptibility to study the deformation of well-dated leucogranite laccoliths concentrated in the detachments.

Conceptual models for flow of partially molten crust

Two types of models describe the relative contribution of the upper crust and lower crust to the collapse process: one in which collapse of thickened crust occurs without extension at the boundaries (fixed-boundary collapse); and another in which the boundaries of the thickened crust extend (free-boundary collapse) (Rey *et al.* 2001) (Fig. 1). Channel flow (Royden 1996) beneath a rigid upper crustal lid (blind collapse of Rey *et al.* 2001) (Fig. 1b) displays a sense of flow from the centre of the orogen towards the foreland, driven by the gravitational potential of the thick crust. Three elements of this model are important for kinematics: (1) the lower–middle crust flows relative to the rigid lid, imposing centripetal (top to the hinterland) sense of shear relative to the orogen at the interface between the channel and the lid; (2) the velocity gradient in the channel imposes a reversal of sense of shear across the channel from centripetal at the top to centrifugal at the bottom (Fig. 1b); and (3) a gradient in the component of simple shear occurs from coaxial flow in the centre of the orogen to non-coaxial flow toward the margin. In the centre of the channel, velocity gradients are minimal and coaxial flow dominates. An alternative model is one in which sliding of upper crust away from the kinematic axis of the orogen translates into thrusting in the foreland (Fig. 1c) (Axen *et al.* 1998). In this case, the sense of shear at the base of crust is centrifugal relative to the orogen and, significantly, opposite to the case of channel flow. Therefore, in the history of the orogen, if detachment tectonics followed a period of channel flow, the sense of shear would reverse.

If the orogen extends at the boundaries according to the free-boundary collapse of Rey *et al.* (2001), the compatibility of deformation between upper and lower crust and within the lower crust dictates the kinematics of the system. During symmetric extension (Fig. 1d), sense of shear beneath the upper crust is either centripetal or centrifugal dependent on local differential velocities between the upper and lower crust. The lower crust is characterized by channel flow with a component of coaxial deformation associated with the far-field extension and crustal thinning. During asymmetric extension (Lister & Davis 1989) by a rolling hinge (Buck 1988, 1991), sense of shear in the detachment is centrifugal. Low-viscosity crust flows into the localized extensional region and maintains a flat Moho (Fig. 1e). Depending on the rate of thinning of the lower crust and extension of the upper crust, and whether this thinning is uniform or localized, these models predict characteristic pressure–temperature (P–T) paths for the deep crust (Teyssier & Whitney 2002).

Geological setting

The tectonic evolution of the Canadian Cordillera (Fig. 2) involved Mesozoic terrane accretion and crustal thickening, Paleocene–Eocene orogenic collapse involving flow of partially molten crust and generation of leucogranite, and mid-Eocene and later E–W extension along crustal-scale high-angle faults (Crittenden et al. 1980; Armstrong 1982; Brown et al. 1986; Price 1986; Parrish et al. 1988; Carr 1992; Johnson & Brown 1996; Vanderhaeghe & Teyssier 1997; Crowley et al. 2001).

The Shuswap MCC is located within the N–S-trending Omineca Belt, SE Canadian Cordillera in British Columbia, where a series of metamorphic core complexes are exposed. One view is that the Omineca Belt represents a discrete welt of thickened crust in the hinterland of the Rocky Mountains (Brown et al. 1986). Another view is that the belt represents the eastern edge of a collapsed Paleocene–Eocene continental plateau that extended west to the Coast Mountains (Whitney et al. 2004). This argument is supported by palaeobotanical evidence (Wolfe et al. 1998) for high palaeoelevations in southern British Columbia and NE Washington in the early Eocene. In addition, Mulch et al. (2004) recovered the isotopic composition of 49–48 Ma meteoric water from recrystallized white mica in quartzite mylonites of the eastern Shuswap detachment, and calculated >4000 m elevation for the catchment area located west of the present trace of the detachment. In this plateau model, orogenic crust flowed to the east and west, and was exhumed as metamorphic belts in the Eocene.

Structural history

Structural studies have documented the deformation of gneiss and migmatite in the Shuswap MCC, the mylonitic deformation in the detachment zones and the high-angle normal faults (Reesor & Moore 1971; Read & Brown 1981; Mattauer et al. 1983; Okulitch 1984; Brown & Journeay 1987; Carr 1991, 1992). At the latitude of the Thor–Odin Dome, the MCC is divided into three structural units (Fig. 2): an upper unit of Mesozoic crystalline rocks overlain by low-grade–unmetamorphosed Early Tertiary sedimentary and volcanic strata (Mathews 1981; Archibald et al. 1983; Colpron et al. 1996); a middle unit of Tertiary migmatitic rocks (Proterozoic–Palaeozoic protoliths) and 60–55 Ma leucogranites (Sevigny et al. 1989; Carr 1991, 1992; Parkinson 1992); and a lower unit of diatexite-dominated migmatite (Thor–Odin Dome; Reesor & Moore 1971) whose age has been debated to be either Proterozoic or Tertiary (Vanderhaeghe & Teyssier 1997; Vanderhaeghe et al. 1999). The detachments juxtapose the upper and middle structural units, and therefore separate low-grade rocks from high-grade (upper amphibolite facies) gneiss. The Thor–Odin Dome is part of a belt of N–S-trending migmatite domes located at the eastern edge of the core complex. The belt of domes continues south into Washington, USA (Fig. 2a).

The Shuswap MCC is delimited by outward-dipping detachment zones: the Okanagan detachment to the west (Tempelman-Kluit & Parkinson 1986), and the Columbia River detachment to the east (Read & Brown 1981) (Fig. 2). These zones are characterized by progressive overprinting of the high-grade fabric of the middle unit by a greenschist facies mylonitic fabric, with the local occurrence of pseudotachylite and cataclasite zones (Read & Brown 1981; Vanderhaeghe & Teyssier 1997). The detachments also concentrated leucogranite laccoliths up to several hundred metres thick that developed magmatic–solid-state fabrics. These laccoliths are typically homogeneously deformed and may have intruded synkinematically (Gapais 1989).

The southern Canadian Cordillera (Fig. 2) is affected by an array of brittle normal faults linked by large strike-slip faults (Ewing 1981; Struik 1993). The Shuswap MCC is cross-cut at its northern tip by the dextral strike-slip Rocky Mountain trench, which connects with the Wolverine core complex a few hundred kilometres north (Struik 1993). Near Revelstoke (Fig. 2), normal faults affecting all units are dominantly N–S-trending and associated with dip-slip or slightly oblique-slip movement. The W-dipping Victor Lake normal fault and the E-dipping Columbia River normal fault define a horst that extends south to the Thor–Odin Dome (Fig. 2). The Columbia River fault cross-cuts the Columbia River detachment (Lane 1984;

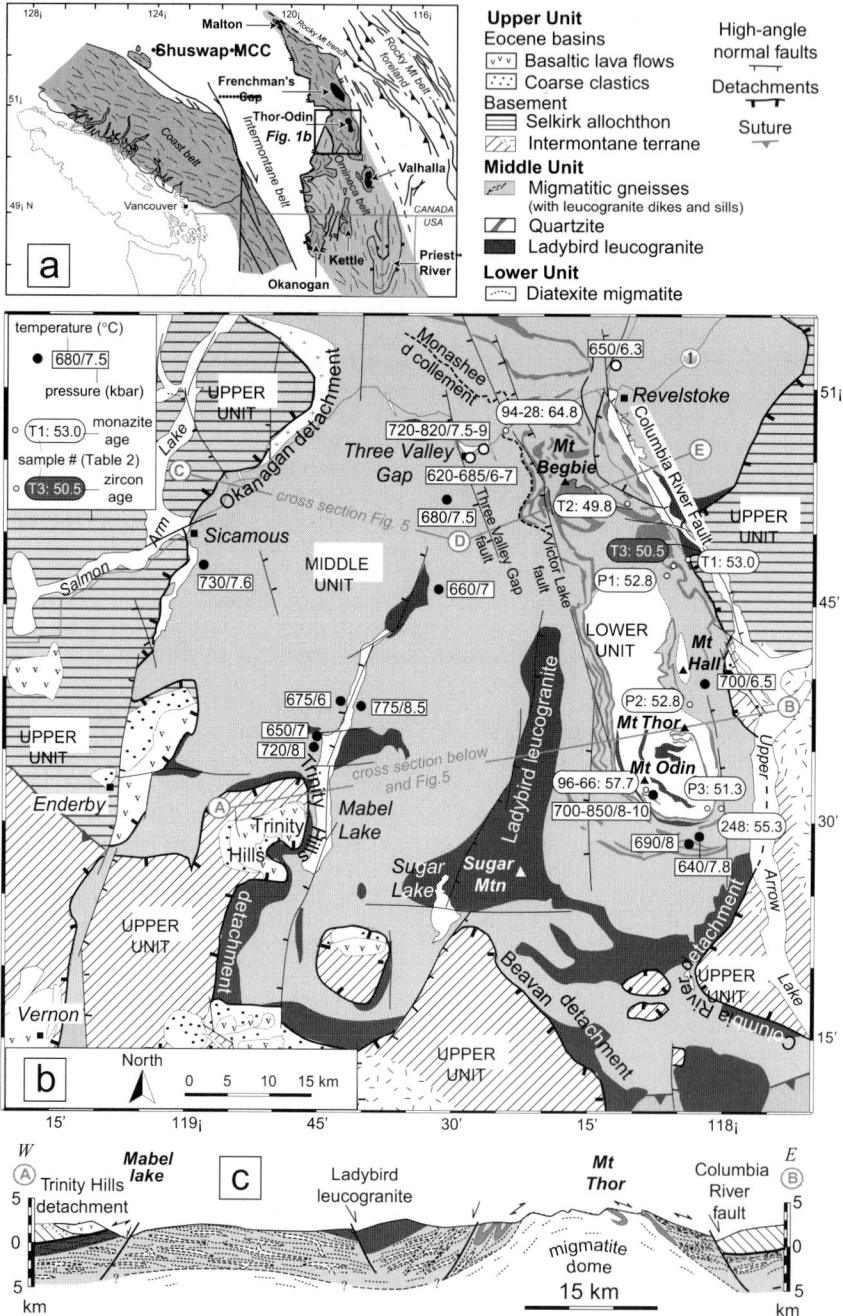

Fig. 2. (**a**) Regional geological map of the Shuswap MCC showing dome structures along the eastern margin (Frenchman Cap dome and Valhalla complex to the north and south of the Thor–Odin Dome, respectively). (**b**) Simplified geological map of the Thor–Odin Dome region of the Shuswap MCC (after Vanderhaeghe *et al.* 1999); following Carr (1992) the Ladybird leucogranite is shown where it comprises >40% of the total volume of outcrops; the map shows the location of samples in which pressure and temperature were estimated, as well as location of geochronology samples and ages obtained on monazite and zircon (Table 2). (**c**) The cross-section across the Thor–Odin Dome shows the relationship of the Ladybird leucogranite suite to upper, middle and lower units, and low-angle detachment faults.

Vanderhaeghe *et al.* 1999, 2003*b*), causing tilting of the hanging wall. The Columbia River fault is a prominent geomorphic feature and is associated with the formation of a small graben basin of unknown age east of Mt Hall (Fig. 2); this fault probably controls the trend of the Columbia River and the alignment of the rugged Columbia Mountains. In the southern continuation of the fault, the Slocan Lake fault is identified on a Lithoprobe seismic profile, where it is shown to affect at least the upper crust (Cook *et al.* 1992).

Metamorphic history

The transition between mantling gneiss (middle unit) and dome (lower unit) is characterized by an increase in leucosome abundance and granitic intrusions, and has been mapped as a metatexite–diatexite transition (Vanderhaeghe *et al.* 1999). The gneiss dome core is comprised of migmatitic metasedimentary and meta-igneous rocks, granite and amphibolite (Reesor & Moore 1971; Norlander *et al.* 2002). Thermobarometric data and analysis of petrogenetic grids for assemblages in Mg–Al-rich gedrite amphibolites document $T = 750-800\ °C$ at $P \geq 8-10$ kbar. The replacement of garnet by cordierite and anorthite $\pm$ amphibole $\pm$ biotite, and the replacement of kyanite by sillimanite accompanied by reaction of kyanite and gedrite ($\pm$ biotite) to cordierite $+$ spinel $\pm$ corundum $\pm$ sapphirine indicates decompression to $3-5$ kbar at $T > 700\ °C$ (Fig. 3).

The metamorphic rocks that mantle the dome consist of metapelite, amphibolite, calc-silicate, marble and quartzite. An extensive network of granitic sills and dykes intruded these rocks and can be traced into larger leucogranitic laccoliths (hundreds of metres thick) in the detachment zones. In some regions that are mapped as metamorphic rock (Fig. 2) granitic material comprises 10–40% of the total rock volume. The metamorphic history of the mantling sequence has been studied in several localities near the Thor–Odin Dome. Orthopyroxene-bearing amphibolite boudins from an outcrop on Highway 1 (west of Three Valley Gap, Fig. 2) record $T = 620-685\ °C$ at $6-7$ kbar (Ghent *et al.* 1977). Migmatitic metapelitic rocks (garnet–sillimanite–K-feldspar gneiss) on Highway 1, east of Three Valley Gap, record $720-820\ °C$ at approximately $8-9$ kbar (Nyman *et al.* 1995). In metapelitic gneiss, the assemblage biotite $+$ garnet $+$ plagioclase $+$ quartz $+$ sillimanite $+$ K-feldspar occurs from the Thor–Odin dome to the detachment faults, indicating that the entire sequence at current exposure levels records conditions above the second sillimanite isograd. Representative analyses of garnet, biotite and plagioclase are given in Table 1.

Petrography and mineral compositions

The metapelitic rocks of the mantling gneiss are migmatitic, with mesosomes comprised of biotite $+$ garnet $+$ sillimanite (prismatic and fibrous) $+$ plagioclase $+$ quartz $+$ K-feldspar $+$ ilmenite $\pm$ pyrite (Fig. 3). Plagioclase is typically unzoned, with compositional variation of An_{15-30}. Matrix biotite is homogeneous in a variety of textural settings ($X_{Mg} = 0.52$), but biotite inclusions in garnet are more Mg-rich than matrix biotite. Melanosomes are dominated by biotite, garnet and sillimanite. Leucosomes are concordant with the mesosome foliation, $1-5$ cm thick, and contain quartz $+$ K-feldspar $+$ plagioclase $\pm$ garnet $\pm$ biotite. These phases are typically coarser than the same phases in adjacent mesosome–melanosome. Leucosome garnet is typically smaller, Mn-enriched and inclusion-free compared to mesosome–melanosome garnet (<2 mm, 12 mol% spessartine). In some outcrops, centimetre-scale leucosomes coalesce into larger dykes and sills ($0.5-1$ m). Mesosome garnets are Fe-rich ($Alm_{80}Sps_2Prp_{14}Grs_9$), homogeneous or slightly zoned, and vary in size from 2 mm to 1 cm. Some garnets have a thin ($<50\ \mu m$) retrograde rim zone that is slightly enriched in Fe and Mn relative to the adjacent interior regions of the garnet. Common inclusions are quartz, plagioclase, ilmenite, biotite, prismatic and fibrous sillimanite, and K-feldspar. In some garnets inclusion suites are the same as the matrix assemblage. Garnets in the transition zone near the Thor–Odin Dome have partially decomposed to a corona of symplectitic plagioclase $+$ biotite $+$ quartz.

Garnet amphibolite is interlayered with migmatitic metapelitic gneiss and contains centimetre- to metre-scale granitic dykes and veins, many of them pegmatitic. The primary assemblage is hornblende $+$ garnet $+$ plagioclase $+$ quartz $+$ biotite $+$ ilmenite. Garnets are large ($10-12$ mm) and, in samples located near the Thor–Odin Dome (e.g. Mt Symonds locality), partially to completely replaced by coronas of symplectitic plagioclase $+$ hornblende $+$ quartz $+$ ilmenite $\pm$ magnetite (Fig. 3d & e). In deformed samples, garnets also have pressure shadows of plagioclase $+$ quartz. X-ray maps of decomposed garnet show that the relict grains are homogeneous ($Alm_{74}Sps_9Prp_{12}Grs_5$). Isolated garnets within the symplectitic coronas are less Fe-rich ($X_{Alm} = 0.66$). The matrix is

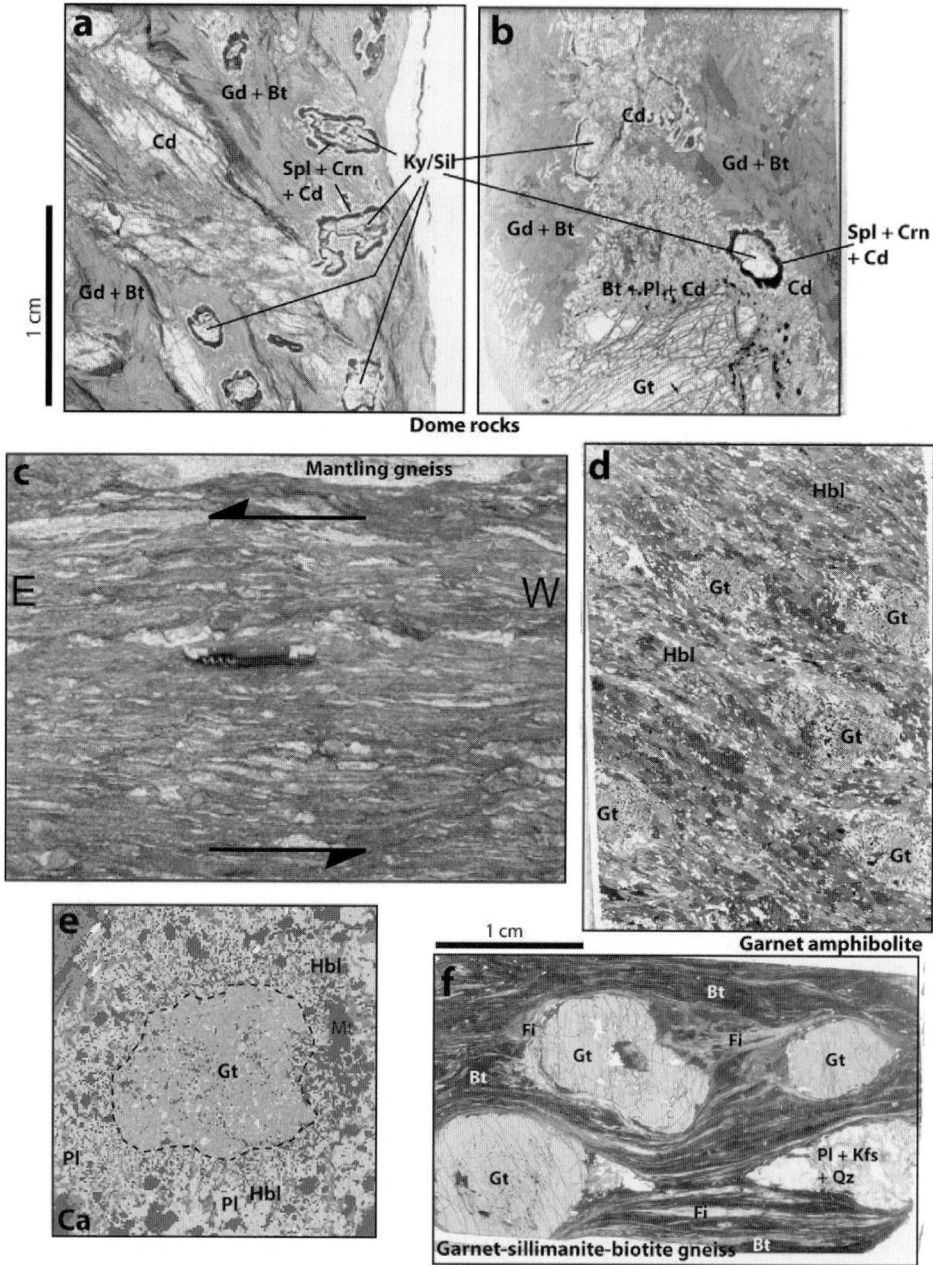

Fig. 3. (**a & b**) Scanned images of thin sections from Thor–Odin Dome gedrite-cordierite rocks: kyanite has been pseudomorphed by sillimanite and is surrounded by symplectitic spinel + cordierite + corundum (and in some samples, sapphirine). The matrix is dominated by gedrite + biotite + cordierite. The thin section shown in (**b**) also contains a large garnet that has partially decomposed to biotite + gedrite + plagioclase + cordierite. (**c**) Outcrop photograph of a migmatitic metapelitic gneiss from Mt Symonds, on the southern rim of the Thor–Odin Dome; shear bands indicate top-to-the-east sense of shear. (**d**) Scanned image of a thin section of garnet amphibolite from Mt Symonds (southern margin of the Thor–Odin Dome), in a transitional region between the dome and the mantling gneiss. Garnets exist as relict cores surrounded by coronas of symplectitic hornblende + plagioclase, as seen in the Ca X-ray map in (**e**). The relict garnet has been outlined. The rest of the field of view of the X-ray map is symplectite, with coarse-grained magnetite. (**f**) Thin section of a metapelitic gneiss from the mantling gneiss (Mable Lake locality).

Table 1. *Representative mineral compositions from metapelitic rocks: mantling gneiss (MG) and dome**

	MG			Dome		
	Grt	Bt	Pl	Grt	Bt	Pl
SiO_2	37.51	35.17	63.40	37.62	33.94	55.55
TiO_2	<d.l.	3.38		<d.l.	3.43	
Al_2O_3	21.39	19.98	23.68	21.33	19.57	28.72
FeO	35.70	21.20	0.34	34.50	21.31	0.15
MnO	1.02	0.00		1.78	0.14	
MgO	3.48	7.72		3.98	7.59	
CaO	1.45	<d.l.	4.22	1.69	0.08	10.11
Na_2O		0.23	9.17		0.19	5.60
K_2O		9.81	0.09		9.65	0.26
Total	100.55	97.48	100.89	100.90	95.91	100.38
Cations						
Si	2.99	5.28	2.78	2.99	5.21	2.49
Ti		0.38			0.40	
Al	2.01	2.72	1.22	2.00	2.79	1.52
Fe^{2+}	2.38	2.66	0.01	2.29	2.74	0.01
Mn	0.07	0.00		0.12	0.02	
Mg	0.41	1.73		0.47	1.74	
Ca	0.12		0.20	0.14	0.01	0.49
Na		0.07	0.78		0.06	0.49
K		1.88	0.01		1.89	0.26
X_{Alm}	0.80			0.76		
X_{Sps}	0.02			0.04		
X_{Prp}	0.14			0.16		
X_{Grs}	0.04			0.05		
X_{Mg}		0.39			0.39	
X_{An}			0.20			0.49

*Mineral compositions and major element distribution maps were obtained using a JEOL JXA-8900 electron microprobe at the University of Minnesota. Operating conditions for quantitative (WDS) analysis were 15 kV accelerating voltage, 20–25 nA beam current, and a range of beam diameters (focused for garnet, defocused to 5 μm for biotite and plagioclase). X-ray maps were determined using a beam current of 100 nA, 50 ms dwelltime and 1–9 μm beam diameters. Natural mineral standards and the ZAF matrix correction routine were used.
<d.l., less than detection limit of the microprobe.
Cation normalization: garnet = 12 oxygen; biotite = 22 oxygen; plagioclase = 8 oxygen.
X_{Mg} for biotite calculated as $Mg/(Mg + Fe + Mn + Ti + Al^{VI})$.

composed mainly of large, pleochroic-green ferro-tschermakitic ($X_{Fe} = 0.65$) hornblende. Small (200–400 μm) hornblende occurs in symplectitic regions around garnet, and is less Fe-rich ($X_{Fe} = 0.59$) than matrix hornblende. Ca-rich plagioclase occurs both in the matrix and in the symplectitic corona ($X_{An} = 0.89-0.94$).

Pressure–temperature conditions

Peak metamorphic conditions were determined for garnet–sillimanite gneiss and garnet amphibolite using mineral compositions and the internally consistent thermodynamic database of Berman (1991; TWQ version 2.02, including the garnet solution model of Berman 1990), except for equilibria involving hornblende as indicated below.

Metapelitic gneiss records peak temperatures of 600–800 °C, calculated using matrix biotite compositions and garnet compositions interior to the thin retrograde rim zoning; most samples record 600–700 °C. Garnet rim-adjacent biotite compositions give lower temperatures (500–550 °C). High temperatures are consistent with conditions expected for sillimanite + K-feldspar-bearing rocks in which these phases formed from the breakdown of muscovite + quartz (Fig. 4). The *P*-sensitive equilibrium garnet–sillimanite–plagioclase–quartz records 7–9 kbar at temperatures of 600–700 °C.

Garnet amphibolites record $T = 675-725$ °C using both the garnet–hornblende Fe–Mg exchange thermometer (Graham & Powell 1984) and hornblende–plagioclase thermometer (Holland & Blundy 1994). Garnets are partially

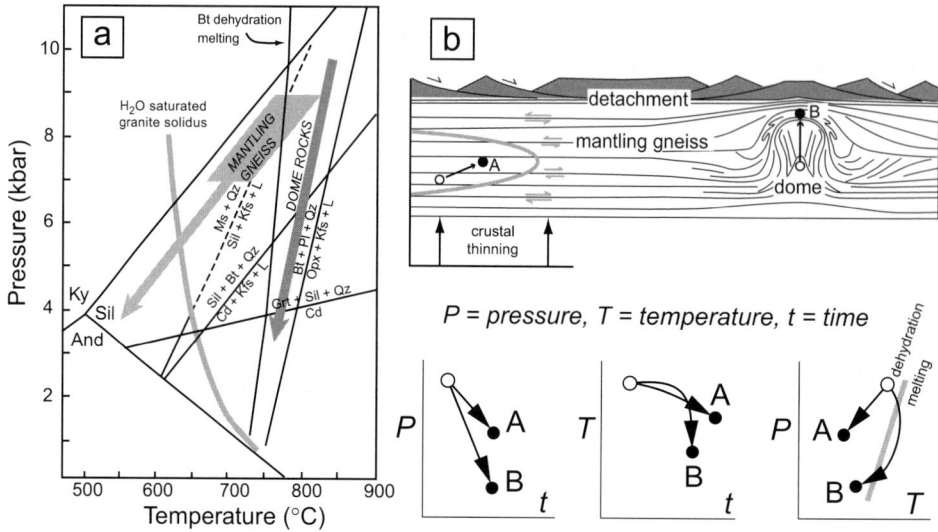

Fig. 4. (a) $P-T$ paths of middle unit and lower unit in the Thor–Odin region; (b) conceptual sketch of lateral and vertical flow in channel and diapir, respectively, with predicted $P-T$ paths.

decomposed, but similar temperatures were calculated for relatively intact garnets (and neighbouring hornblende + plagioclase) as well as relict garnets (and symplectitic hornblende + plagioclase). The P-sensitive equilibria garnet + H_2O = hornblende + plagioclase + quartz accounts for the breakdown of garnet to the corona/symplectite of hornblende + plagioclase, and records $P = 6-7$ kbar at approximately 700 °C.

Granitic leucosomes that probably derived by melting of the metasedimentary gneiss are an additional source of information on metamorphic conditions. Above about 6 kbar, the breakdown of muscovite (+quartz + plagioclase) may involve partial melting to produce sillimanite and K-feldspar + melt (Fig. 4). Garnet-bearing leucosomes may have formed by involvement of biotite in a melting reaction: e.g. biotite + sillimanite + plagioclase + quartz = garnet + K-feldspar + melt (Le Breton & Thompson 1988; Vielzeuf & Clemens 1992).

Temperatures and pressures are relatively consistent (6–9 kbar, 600–800 °C; Fig. 4) throughout the mantling gneiss, independent of sample location or rock type, including samples collected immediately beneath the western detachment. The consistency of temperatures in middle unit rocks might be explained by thermal buffering if most of the region was partially molten at the same time.

The mantling gneisses record pressures that are not correlated to structural position within the approximately 5–10 km-thick unit. Rocks

from the structurally deepest locality (e.g. Mt Symonds) do not record significantly higher pressures than rocks located immediately beneath the detachment faults (Fig. 2b). Variation in calculated pressure may reflect uncertainty in the thermodynamic properties of the minerals, in the application of thermobarometric techniques and/or in inferences about structural position of the samples.

The mantling gneisses record only slightly lower $P-T$ conditions than rocks in the dome. Assemblages and textures in the dome suggest $P > 9$ kbar at $T > 800$ °C. Ubiquitous corona/symplectite textures on garnet and Al_2SiO_5 (including decomposition of garnet + kyanite to a cordierite-bearing assemblage) indicate that the dome rocks were exhumed from greater depths – at elevated temperatures – compared to the mantling rocks (Fig. 4). It is difficult to reconstruct the post-peak (decompression) path for the mantling rocks, particularly those away from the dome, because they lack textures and assemblages that provide this information. The lack of cordierite in mantling gneisses may provide some limits on the decompression path, but whole-rock (inductively coupled plasma-mass spectrometry – ICP-MS) analyses of two samples of garnet–sillimanite–biotite gneiss show that cordierite is not predicted for rocks of these low-Mg bulk compositions (2–3 wt% MgO, compared to cordierite-bearing dome rocks with >9 wt% MgO). Nevertheless, the lack of corona/symplectite textures and the presence of abundant sillimanite in rocks with no widespread evidence for kyanite suggest that the mantling gneiss did not experience either the high pressure or near-isothermal decompression of the dome (Fig. 4).

Thermochronology

Hanging wall of detachments

U–Pb ages on zircon and monazite from calc-alkaline batholiths in the hanging wall of the detachments range from 174 to 161 Ma. These ages are interpreted as mid-Jurassic emplacement of the plutons in an island arc (Parrish & Wheeler 1983; Parrish & Armstrong 1987; Carr 1991). Mesozoic cooling from c. 500 to c. 300 °C is recorded by K–Ar and $^{40}Ar/^{39}Ar$ dating of hornblende and micas (Mathews 1981; Archibald et al. 1983; Colpron et al. 1996). Thermal modelling of K-feldspar from the Selkirk allochthon in the hanging wall of the Columbia River fault indicates that an episode of mid-Jurassic cooling at about 170 Ma was followed by $1-3$ °C Ma^{-1} cooling

until Eocene time (Colpron et al. 1996). Sedimentary sequences deposited in extensional basins in the hanging wall of the western Okanagan detachment are intruded and capped by 49–47 Ma rhyolitic–basaltic volcanic rocks (Mathews 1981). The sediments contain clasts from the underlying calc-alkaline batholith, greenschist units, and the leucogranite and gneiss exhumed in the footwall of the detachments, indicating that these rocks were exposed to erosion before about 49 Ma (Matthews 1981; Lorencak et al. 2001).

Footwall of detachments

The U–Pb method has been used to identify the protolith of the migmatitic gneiss in the core of the Thor–Odin Dome. Zircons yield slightly discordant Precambrian ages, indicating an affinity with North American basement (Wanless & Reesor 1975; Armstrong et al. 1991; Parkinson 1991, 1992; Wheeler & McFeely 1991; Parrish 1995; Vanderhaeghe et al. 1999) (Fig. 5). Leucogranites of the Ladybird suite (Figs 2 and 5), intruded in the middle unit and concentrated in

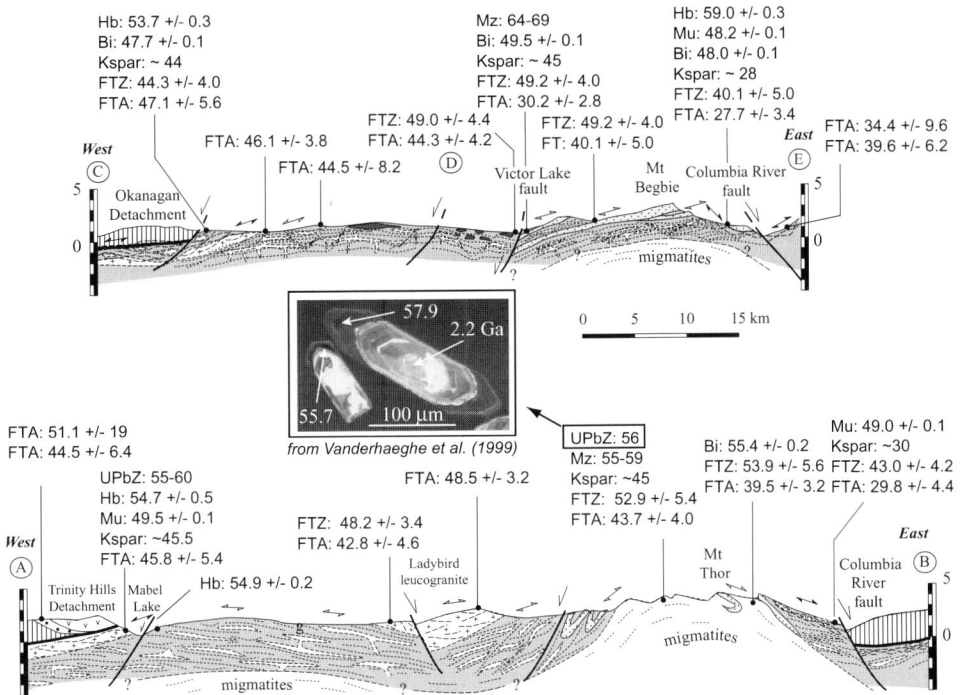

Fig. 5. Summary of thermochronologic study (Vanderhaeghe et al. 1999, 2003b; Lorencak et al. 2001) displayed on two cross-sections located in Figure 2 across the Shuswap MCC (ages in Ma). At most sites, results show rapid cooling between 55 and 45 Ma.

the detachment zone, exhibit U–Pb monazite and zircon ages ranging from *c.* 100 to *c.* 50 Ma, with a dominant population at 60–55 Ma (Carr 1991, 1992; Parkinson 1992; Johnston *et al.* 2000). SHRIMP (Sensitive High Resolution Ion Microprobe) analysis of a leucogranite west of Mabel Lake (Vanderhaeghe *et al.* 1999) documents the main phase of zircon growth at 60 Ma, and shows the existence of old zircon cores with discordant ages as old as 1600 Ma. Carr (1992) reported monazite ages from the same samples in which zircon was dated; the monazite ages are typically younger than the zircon ages and cluster around 55–56 Ma. The crystallization of the Ladybird leucogranite is clearly within the 60–55 Ma bracket, with a tendency towards the younger end of this range. The presence of inherited Precambrian zircons and the geochemistry of the Ladybird leucogranite suite are consistent with an anatectic crustal source for the leucogranite (Sevigny *et al.* 1989; Carr 1992).

Monazite ages from deformed leucosome and pegmatite at the southern and eastern margins of the Thor–Odin Dome cluster at 55–53 Ma, with ages as young as 50 Ma from zircons in an undeformed cross-cutting pegmatite close to the Columbia River fault (Table 2; Appendix 1). All monazite analyses plot above concordia due to excess radiogenic ^{206}Pb* interpreted to be the product of the decay of excess ^{230}Th incorporated in monazite at the time of crystallization (Parrish 1990). Therefore, only the ^{207}Pb*/^{235}U age has been used for age interpretations. The 50 ± 0.5 Ma zircon age from the undeformed pegmatite indicates that high-temperature deformation had ceased by this time.

Vanderhaeghe *et al.* (1999) conducted a U–Pb SHRIMP study of zircon from two granitic leucosomes in the Thor–Odin Dome. Cathodoluminescence imaging of zircon from migmatite (Fig. 5) displays inherited cores with Palaeoproterozoic ages, and U-rich rims that show magmatic oscillatory zoning and yield ages of 56.4 ± 1.4 and 55.9 ± 3.1 Ma. The similarity between leucogranite and leucosome crystallization demonstrates that early Tertiary crustal melting was prevalent in the region. In addition, the shared inheritance history of zircons in both leucogranite and migmatite suggests a genetic link between them.

Rb–Sr K-feldspar/muscovite mineral isochrons, and K–Ar dating of hornblende, white mica and biotite, yield a range from 64 to 45 Ma (Mathews 1981; Parrish *et al.* 1988). More recently, the cooling history of the Shuswap MCC has been deciphered by ^{40}Ar/^{39}Ar thermochronology (Vanderhaeghe *et al.* 2003*b*). With the exception of a few samples contaminated by excess argon, the analytical results indicate a consistent range of early Tertiary ^{40}Ar/^{39}Ar ages throughout the area. Hornblende yields ages ranging from 59 to 54 Ma (with evidence of some excess argon), muscovite and biotite show flat argon-release spectra with plateau ages clustered between 49.5 and 47 Ma, and K-feldspar shows progressive closure ranging from 50 to 43 Ma, except for samples in the immediate footwall of the Columbia River fault that yield ages as young as 28 Ma. Mulch (2004) performed a detailed ^{40}Ar/^{39}Ar study of white mica on a mylonitic quartzite section within the eastern detachment located just north of Mt Hall and along the Columbia River (Fig. 2), and found plateau ages ranging from 49.0 to 47.9 Ma in the upper part of the detachment zone. In addition, Mulch (2004) conducted a laser-based ^{40}Ar/^{39}Ar spot analysis on white micas and concluded from the absence of diffusion profiles that the ages obtained are probable crystallization ages. Therefore, Mulch (2004) probably dated the cessation of ductile deformation and recrystallization in the mylonitic level of the east detachment at approximately 49–48 Ma.

Apatite and zircon fission-track data from Lorencak *et al.* (2001) show that the time at which rocks cooled from $T > 300$ to $< 110\,^{\circ}$C varies throughout the Shuswap MCC. In most regions, zircon and apatite fission-track ages follow the mica argon ages, with zircon ages clustered at 49–48 Ma and apatite at 45–43 Ma, respectively (Fig. 5). However, on the eastern side, in the vicinity of the eastern detachment and around the dome, and in particular near normal faults, significantly younger apatite ages occur (Fig. 5). This region is transected by late normal faults along which hydrothermal fluid circulation may have perturbed the thermal history of the core complex after its main period of exhumation.

These fission-track and ^{40}Ar/^{39}Ar ages, combined with U–Pb ages on zircon and monazite, define the cooling history for several localities in the Shuswap MCC (Fig. 5). The migmatitic core of the complex was affected by rapid cooling (from *c.* 700 to *c.* 300 °C) between approximately 56 and 48 Ma. Based on thermochronological and structural analysis, Vanderhaeghe *et al.* (2003*b*) proposed that this rapid cooling followed an event of rapid exhumation associated with the formation of the MCC by ductile thinning of a previously thickened and partially molten crust, activation of low-angle detachments and formation of migmatite domes.

Table 2. *U–Pb analyses of monazite from the Thor–Odin region (see Appendix 1 for sample description and locations; preferred ages in box)*

Sample number	Wt (mg)	$^{206}Pb^*$ (ppm)	U (ppm)	$^{206}Pb/^{204}Pb$	$^{208}Pb/^{206}Pb$	$^{206}Pb^*/^{238}U$	$^{206}Pb^*/^{238}U$ age	$^{207}Pb^*/^{235}U$	$^{207}Pb^*/^{235}U$ Age (error Ma)	$^{207}Pb^*/^{206}Pb^*$	$^{207}Pb^*/^{206}Pb^*$ age (error Ma)
P-1 (81787-2)											
M1:	0.4	108	14,753	754	0.969	0.00844	54.2 (0.1)	0.05334	52.8 (0.3)	0.04585	<0 (5.7)
P-2 (81089-3)											
M1: +145 μ	0.17	79.5	10,825	4290	1.2827	0.00848	54.5 (0.1)	0.05303	52.5 (0.1)	0.04533	<0 (2.5)
M2: +145 μ	0.1	62.2	8,414	736	1.6849	0.00854	54.8 (0.1)	0.05364	53.1 (0.2)	0.04557	<0 (7.7)
P-3 (82789-2)											
M1: +145 μ	1.6	29.2	4,186	3873	2.054	0.00805	51.7 (0.2)	0.0518	51.3 (0.2)	0.04664	31.0 (1.6)
T-1 (82188-1.2)											
M1: large cld, yelbrn, eu, incls (n = 11)	0.3	39.7	5,291	962	2.306	0.00867	55.6 (0.1)	0.05387	53.3 (0.2)	0.04507	<0 (6.1)
M2: small clr, yel, glassy, no incls (n = 70)	0.2	74.1	9,898	1947	2.173	0.00865	55.5 (0.1)	0.0533	52.7 (0.1)	0.0447	<0 (3.5)
T-2 (82188-8)											
M1: +145 μ, yel, eu (n = 12)	0.4	69	9,576	3178	0.952	0.00832	53.4 (0.2)	0.05028	49.8 (0.2)	0.04381	<0 (0.9)
M2: 100–125 μ (n = 15)	0.2	175	24,268	10450	0.804	0.00833	53.5 (0.2)	0.05028	49.8 (0.2)	0.04377	<0 (0.7)
T-3 (81887-1)											
Z1: M − 80 μ	0.5	65.2	9,662	751	0.053	0.00779	50.0 (0.1)	0.05053	50.1 (0.2)	0.04705	52.0 (10.3)
Z2: M + 125 μ	0.3	267	39,719	1741	0.0252	0.00777	49.9 (0.1)	0.05054	50.1 (0.1)	0.04718	58.3 (2.0)
Z3: NM + 80–125 μ	0.6	17	2,502	250	0.1562	0.00783	50.3 (0.1)	0.05086	50.4 (0.2)	0.04709	54.1 (6.0)
Z4: NM + 125 μ	0.23	40	5,931	321	0.1232	0.00779	50.0 (0.1)	0.05052	50.0 (0.2)	0.04706	52.2 (6.6)
Z5: NM + 80–125 μ, abr	0.3	44.9	6,574	3342	0.0152	0.0079	50.7 (0.1)	0.0525	52.0 (0.1)	0.04822	110.3 (3.6)
248.1											
M1	1	130	3650	3275	3.65658	0.0087061	55.9 (0.1)	0.0559983	55.3 (0.1)	0.046650	31.3 (1.7)
M2	1	177	5294	2292	3.33289	0.0088223	56.6 (0.1)	0.0558575	55.2 (0.1)	0.045919	<0
94-28c											
M1	4.6	132	2498	2397	4.76467	0.0104299	66.9 (0.1)	0.0656177	64.5 (0.1)	0.045533	<0
M2	2.5	126	2081	1612	5.44456	0.0106794	68.5 (0.1)	0.0662473	65.1 (0.1)	0.045847	<0
96-66b											
M1	4.6	163	5190	2838	2.95266	0.0090632	58.2 (0.1)	0.0584962	57.7 (0.2)	0.047118	39.6 (2.9)
M2	2	153	4796	2062	2.95285	0.0092092	59.1 (0.1)	0.0584393	57.7 (0.1)	0.046301	<0

Notes: Z, zircon; M, monazite; NM/M, non-magnetic/magnetic @ 0.5 sidetilt, 1.8 A; euh, euhedral bipyramids; eq, equant; cl, cloudy; clr, clear; st, stubby; ab, abraded; incl, inclusions; μ, size in μm; n, number of crystals; ^{206}Pb, radiogenic ^{206}Pb; $^{207}Pb^*$, radiogenic ^{207}Pb.
$^{207}Pb/^{206}Pb$, $^{208}Pb/^{206}Pb$ errors are 0.03 and 0.06%, respectively. $^{206}Pb/^{204}Pb$ errors are 0.1–4% depending on ratio (all errors 2σ).
Common Pb corrections made using Stacey & Kramers (1975) values.
Ages calculated using following decay constants: $^{238}U = 1.55125 \times 10^{-10}$, and $^{235}U = 9.8485 \times 10^{-10}$.

Deformation of leucogranite – anisotropy of magnetic susceptibility

We have conducted a combined microstructural and anisotropy of magnetic susceptibility (AMS) study of the Ladybird leucogranite suite in the Thor–Odin region. The leucogranites are pervasive throughout the middle unit and are concentrated within the low-angle detachment zones, in a position similar to that mapped in the Himalaya (Murphy & Harrison 1999). It is at this level in the crust that the kinematics of the end-member models (channel flow, symmetric detachments, symmetric extension and asymmetric extension) are expected to differ (Fig. 1). The fabric recorded in the granites varies across the complex. At the level of the detachments a solid-state mylonitic fabric grades downward into a higher temperature solid-state fabric and a magmatic fabric (Fig. 6). Low-temperature solid-state fabrics correspond to mylonitic deformation involving grain size reduction, the formation of quartz ribbon grains, significant deformation of micas and brittle behaviour of feldspar (Snoke *et al.* 1998). The characteristics of high-temperature solid-state fabrics include well-recrystallized quartz ribbons, blocky mica grains and ductile deformation of feldspar (Passchier & Trouw 1996). Magmatic fabrics are characterized by relatively equant quartz grains, blocky micas, and feldspar that has retained its crystal faces and zoning (Vernon 1976).

In order to investigate the fabrics in the granites, we used the anisotropy of magnetic susceptibility (Appendix 2). This highly sensitive technique has proven useful in revealing both solid-state and magmatic fabrics in granitoids, even when the rocks are weakly deformed and lack a macroscopic fabric discernible in the field (e.g. Hrouda 1982; Borradaile 1988; Bouchez *et al.* 1990; Archanjo *et al.* 1994). In addition, this method has yielded valuable results in synanatectic leucogranites similar to the Ladybird suite, for example from the Variscan Massif Central (Jover *et al.* 1989; Talbot *et al.* 2004), Taylor Valley, Antarctica (Allibone & Norris 1992), and in the Himalaya (Guillot *et al.* 1993; Rochette *et al.* 1994; Scaillet *et al.* 1995).

AMS results

Samples of the Ladybird leucogranite suite were collected from 90 locations in the region of the Thor–Odin Dome along a 70 km-traverse from the western to the eastern detachments (Fig. 7). At the level of the western detachment, the granite occurs in the middle unit as a

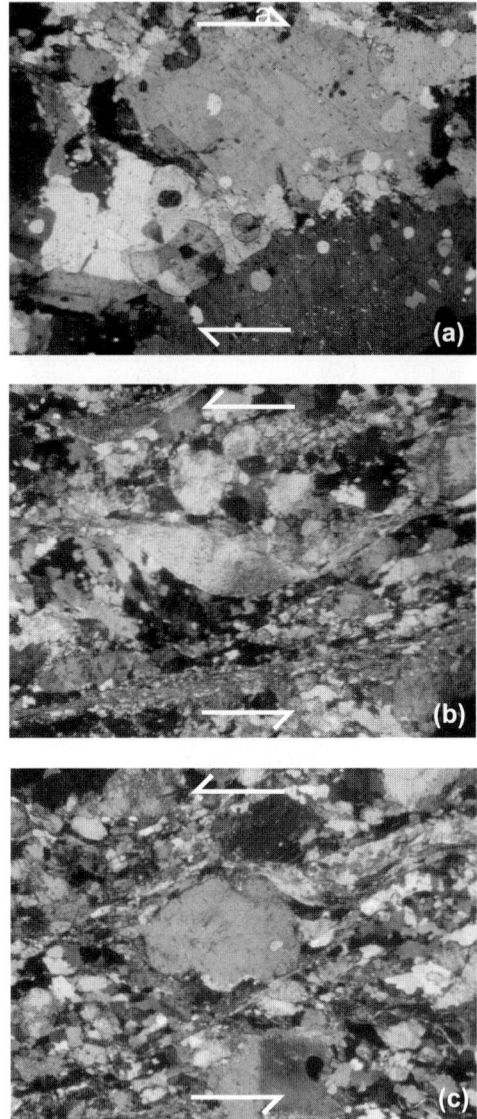

Fig. 6. Photomicrographs of leucogranite samples used to determine conditions of deformation and sense of shear. (**a**) Sample SM-9G: tiling of feldspar grains shows top-to-the-right (west) sense of shear in high-temperature fabric. K_1 is parallel to long axis of photomicrograph (long dimension = 7.5 mm). (**b**) Sample 9629: Mica fish shows top-to-the-left (west) sense of shear and low-temperature fabric (long dimension = 4 mm). (**c**) Sample 9729B: feldspar σ-clast in low-temperature fabric shows top-to-the-left (west) sense of shear (long dimension = 4 mm).

400–500 m-thick sheet with boundaries oriented parallel to the main foliation in the surrounding metamorphic rocks. At other locations, the granite occurs in 1–10 m-thick sills and dykes

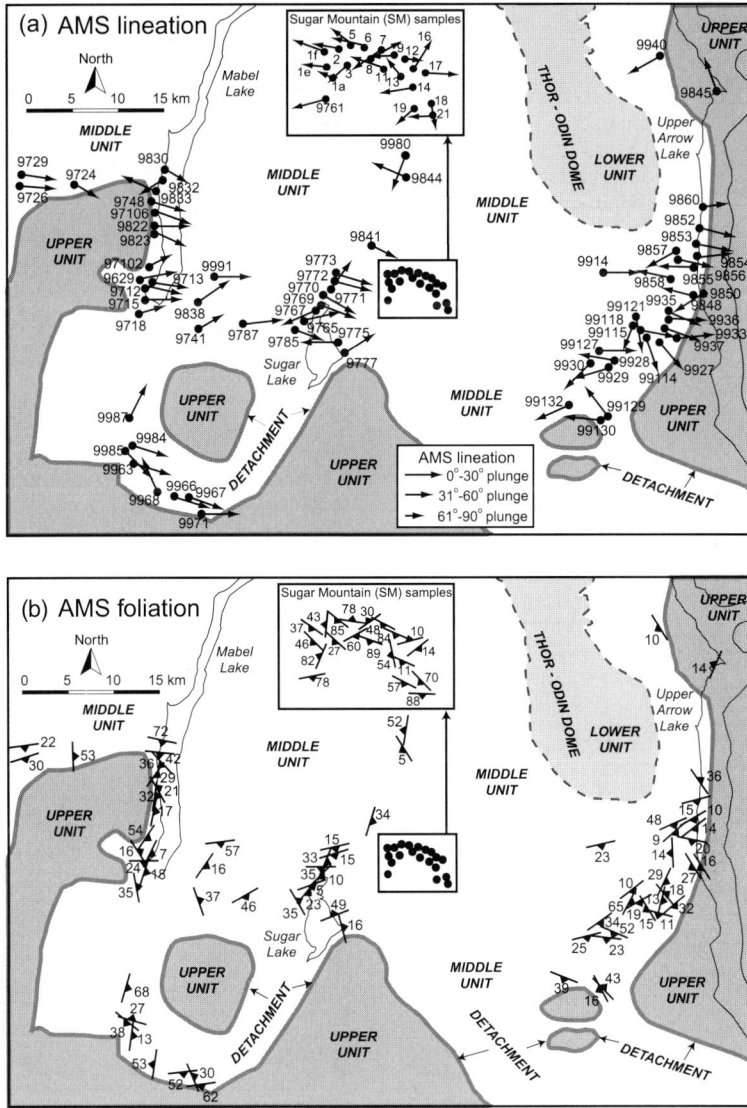

Fig. 7. (**a**) Average AMS lineation (K_1 direction) for each location (the numbers next to stations are locality numbers) (Table 3); the length of the arrows is a function of plunge; most lineations have <30° plunge; (**b**) average AMS foliation (strike and dip) for each location.

with the exception of another relatively large body located at Sugar Mountain (Figs 2 and 7). Samples collected from sills and dykes were chosen from near the centre of the body in order to avoid the possible complication of emplacement fabrics developed along the margins. The sample cores were measured for AMS at low field (10^{-4} T (tesla), 875 Hz) on a Kappabridge KLY-3S instrument at the University of Wisconsin-Madison.

The AMS average data for each station (Table 3) show that the bulk magnetic

susceptibility, K_m, for the leucogranite specimens ranges from 1×10^{-6} to 4×10^{-2} SI. Such magnetic susceptibilities are characteristic of paramagnetic granites (Rochette *et al.* 1992; Bouchez 1997). A few cores that yielded negative magnetic susceptibilities due to the predominance of diamagnetic minerals (quartz and feldspar) were not included in further analyses. The bulk magnetic susceptibility displays no particular spatial distribution. The magnetic susceptibility values of 372 specimens, plotted in a histogram (Fig. 8), show a unimodal distribution

Table 3. *Low-field magnetic data for the Ladybird leucogranites*[*][†]

Number	Anisotropy factors					AMS principal directions						n
	K $(10^{-6}$ SI)	P	F	L	T	K_1		K_2		K_3		
						dec	inc	dec	inc	dec	inc	
SH022	22	0.097	1.072	1.025	0.48	86	32	178	6	272	58	3
SH023	78	1.175	1.122	1.053	0.36	114	17	23	1	287	73	4
SH029	19	1.100	1.073	1.027	0.46	76	14	345	6	234	74	2
SH030	302	1.234	1.129	1.105	0.05	128	47	272	37	16	18	3
SH032	12	1.173	1.079	1.093	−0.08	235	16	129	22	358	48	4
SH033	104	1.089	1.063	1.025	0.41	288	9	196	33	26	54	4
SH038	30	1.162	1.121	1.041	0.48	42	4	134	17	295	74	4
SH041	25	1.161	1.105	1.056	0.29	120	34	219	13	329	56	5
SH044	80	1.163	1.113	1.050	0.36	300	1	210	4	39	85	3
SH045	32	1.157	1.125	1.033	0.57	341	9	249	10	114	76	5
SH048	22	1.024	1.016	1.008	0.36	288	11	187	7	62	63	4
SH050	52	1.016	1.010	1.006	0.21	243	15	335	0	57	74	4
SH052	64	1.117	1.078	1.039	0.32	92	2	185	13	335	75	4
SH053	70	1.107	1.081	1.026	0.49	95	4	186	9	344	80	6
SH054	56	1.103	1.085	1.018	0.64	84	7	176	12	324	76	2
SH055	30	1.085	1.059	1.026	0.38	224	13	225	20	15	70	2
SH056	82	1.179	1.147	1.032	0.62	101	8	192	4	310	81	2
SH057	47	1.061	1.039	1.022	0.28	241	4	147	49	331	42	3
SH058	23	1.061	1.047	1.013	0.55	291	13	199	5	89	76	2
SH060	12	1.55	1.523	1.027	0.77	85	32	346	14	235	54	1
SH112	45	1.143	1.100	1.043	0.38	103	5	195	23	1	66	4
SH113	55	1.129	1.097	1.032	0.48	102	7	192	2	296	83	4
SH115	26	1.136	1.094	1.042	0.38	95	18	4	1	271	72	2
SH118	9	1.023	1.018	1.005	0.56	222	20	119	29	269	55	7
SH124	28200	1.697	1.399	1.298	0.01	116	50	10	12	271	37	2
SH126	43	1.102	1.058	1.045	0.11	86	13	184	27	332	60	2
SH129	73	1.153	1.125	1.029	0.61	103	8	197	21	353	68	3
SH141	32	1.100	1.053	1.047	0.04	60	36	154	6	252	53	3
SH148	87	1.140	1.093	1.047	0.3	107	27	201	8	305	61	3
SH161	20	1.022	1.005	1.018	−0.59	255	11	127	75	349	12	4
SH165	27	1.058	1.043	1.015	0.49	77	13	172	18	312	67	2
SH167	48	1.183	1.132	1.051	0.42	237	3	147	2	22	85	2
SH169	34	1.091	1.061	1.030	0.33	115	10	206	1	304	80	4
SH170	32	1.064	1.031	1.033	−0.03	118	20	218	25	354	57	4
SH171	8	1.061	1.038	1.023	0.27	34	2	125	15	296	75	3
SH172	53	1.152	1.113	1.040	0.46	105	9	197	12	340	75	2
SH173	22	1.089	1.062	1.027	0.38	102	4	192	14	355	75	8
SH175	98	1.274	1.237	1.037	0.70	266	17	13	44	161	41	3
SH177	31	1.154	1.136	1.018	0.76	57	15	149	5	257	74	4
SH185	23	1.054	1.039	1.015	0.43	122	36	14	16	256	55	5
SH187	32	1.127	1.092	1.035	0.43	84	22	193	38	331	44	6
SH202	6	1.031	1.025	1.005	0.66	61	34	182	36	303	36	2
SH206	6897	2.295	1.520	1.775	−0.38	111	13	17	15	239	69	6
SH314	89	1.231	1.168	1.063	0.42	88	8	182	21	337	67	3
SH327	16	1.027	1.018	1.009	0.34	140	5	230	9	18	79	3
SH328	49	1.248	1.202	1.046	0.41	279	18	169	47	24	38	1
SH329	60	1.197	1.174	1.024	0.75	256	8	163	21	6	67	4
SH330	1385	1.218	1.171	1.047	0.53	216	19	121	14	348	65	3
SH333	25	1.113	1.089	1.024	0.57	79	18	178	25	316	58	3
SH335	11	1.156	1.113	1.043	0.45	117	23	17	5	296	61	3
SH336	59	1.245	1.188	1.058	0.50	92	25	5	10	276	72	3
SH337	612	1.521	1.333	1.188	0.18	102	13	193	4	297	77	4
SH340	20	1.379	1.342	1.037	0.65	242	10	152	1	55	80	1
SH363	25	1.015	1.006	1.009	−0.17	111	5	19	11	221	77	3
SH366	478	1.477	1.294	1.183	0.14	108	18	217	46	3	38	3
SH367	12	1.145	1.071	1.074	−0.01	101	23	0	22	235	60	3

(Continued)

Table 3. *Continued*

Number	K $(10^{-6}$ SI)	P	F	L	T	K_1 dec	K_1 inc	K_2 dec	K_2 inc	K_3 dec	K_3 inc	n
SH368	18	1.038	1.011	1.027	−0.43	335	30	220	42	88	37	4
SH371	56	1.168	1.130	1.038	0.53	88	11	198	59	352	28	2
SH380	13	1.061	1.022	1.040	−0.29	200	10	303	50	105	38	2
SH384	57	1.100	1.026	1.075	−0.50	288	1	17	27	197	63	3
SH385	56	1.077	1.026	1.051	−0.33	310	4	217	37	44	52	3
SH387	35	1.218	1.158	1.060	0.42	10	33	146	44	268	22	2
SH391	975	1.288	1.272	1.016	0.88	89	10	195	55	353	33	1
SH414	16	1.048	1.026	1.022	0.09	130	1	220	18	52	75	4
SH415	28	1.096	1.060	1.036	0.23	106	18	197	6	306	71	3
SH418	24	1.045	1.038	1.008	0.65	202	20	76	57	303	25	3
SH421	4	1.029	1.025	1.004	0.74	182	8	91	6	321	80	2
SH427	51	1.213	1.167	1.045	0.55	88	17	188	28	330	56	2
SH429	17	1.292	1.230	1.062	0.35	322	5	57	42	227	48	1
SH430	67	1.193	1.172	1.021	0.77	271	14	179	6	66	74	4
SH432	61	1.164	1.118	1.046	0.42	249	13	153	35	357	51	4
SM001	19	1.033	1.025	1.008	0.50	294	86	17	6	113	8	8
SM004	108	3.533	3.181	1.352	0.56	276	31	166	38	45	44	8
SM005	46	1.100	1.090	1.010	0.80	290	16	222	25	36	53	15
SM006	21	1.048	1.035	1.013	0.46	280	40	187	3	95	47	6
SM007	43	1.069	1.060	1.009	0.72	227	33	126	10	42	63	12
SM009	7	1.044	1.018	1.026	−0.16	307	31	136	59	39	5	5
SM010	17	1.045	1.028	1.017	0.22	283	11	73	76	189	12	5
SM011	28	1.040	1.023	1.017	0.13	235	10	354	42	144	60	13
SM012	22	1.033	1.020	1.012	0.24	61	2	168	55	329	30	7
SM013	40	1.064	1.033	1.031	0.03	268	32	154	32	28	42	44
SM015	27	1.058	1.030	1.028	0.04	289	13	117	76	20	1	3
SM016	49	1.076	1.037	1.039	−0.03	97	70	289	21	23	6	22
SM017	55	1.054	1.024	1.030	−0.12	322	36	205	34	78	36	17
SM018	19	1.045	1.031	1.014	0.38	261	11	174	8	24	79	8
SM020	37	1.100	1.072	1.028	0.43	32	5	298	0	165	80	9
SM021	34	1.074	1.037	1.037	0.00	93	12	186	9	322	76	7
SM022	313	1.135	1.110	1.025	0.43	228	55	124	10	27	33	1
SM023	11	1.111	1.106	1.005	1.00	169	57	306	25	46	20	1
SM025	109	1.047	1.031	1.015	0.34	268	49	87	37	2	2	14
Average	*471*	*1.176*	*1.126*	*1.050*	*0.33*							
SD	*3048*	*0.304*	*0.241*	*0.093*	*0.32*							

*Data from 90 stations.

[†]K, bulk magnetic susceptibility; P, degree of anisotropy; F, planar anisotropy (K_1/K_3); L, linear anisotropy (K_1/K_2); T, shape parameter; K_1, K_2, K_3, maximum, intermediate, and minimum susceptibilities; n, number of cores measured.

which, combined with low magnitudes of K (median $K \approx 34 \times 10^{-6}$ SI), would indicate the contribution of one main paramagnetic phase. The presence of a ferromagnetic contribution was tested on 25 specimens by measuring the hysteresis properties on a Vibrating Sample Magnetometer at the University of Minnesota. All but one specimen host ferromagnetic minerals. Large variations in hysteresis parameters (M_s, saturation magnetization; M_r, remanence magnetization at saturation; H_c, coercive field) point to a complex magnetic mineralogy similar to that observed in leucogranites elsewhere (Jover *et al.* 1989). Magnetite, maghemite and pyrrhotite are the most likely ferromagnetic

carriers and may occur either as magmatic crystals in between other grains or even as micro-inclusions in the lattice of ferromagnesian silicates as suggested for other granites by Borradaile & Werner (1994). Details of the ferromagnetic mineralogy are beyond the scope of this paper.

The relationships among K_m (mean susceptibility), P_j (corrected degree of anisotropy (Jelínek 1978)) and T (shape parameter) are plotted in Figure 9. Most specimens have a relatively low degree of anisotropy with $P_j < 1.5$ (Fig. 9). Magnetic lineations in the leucogranites are relatively consistent across the complex, and are dominantly shallowly plunging and oriented

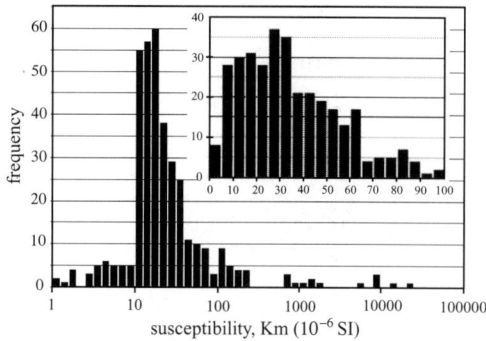

Fig. 8. Histogram of bulk magnetic susceptibilities (K_m) distribution in 372 Shuswap leucogranite core samples. The inset shows distribution of samples with mean susceptibility $< 100 \times 10^{-6}$ SI.

E–W (Figs 7 and 10). The magnetic foliation, defined as perpendicular to K_3, is relatively flat lying (Figs 7 and 10), with the exception of the region around Sugar Mountain where both lineation and foliation are significantly steeper. Magnetic fabrics are fairly consistent at the scale of the station and between neighbouring stations. The magnetic foliations and lineations in the leucogranites conform with the fabrics measured in the field where the fabric is sufficiently strong to be measureable.

Kinematic analysis

In order to evaluate sense of shear in the leucogranite, we cut 60 samples along the plane defined by the AMS K_1–K_3 axes. These samples were reoriented in the laboratory and a laser device projected the K_1–K_3 plane onto the sample. The sample was cut along this plane and the K_1–K_3 axes were located on the cut face. Observation of all cut sections and selected thin sections allowed the characterization of the conditions of fabric development according to magmatic, high-temperature solid state, and low-temperature solid state (Fig. 6). Sense of shear criteria used in this study for solid-state deformation include C–S fabrics and the presence of shear bands (Berthé *et al.* 1979), mica fish, and asymmetric sigma and delta porphyroclasts (Passchier & Simpson 1986) (Fig. 6). Determining the sense of shear in rocks deformed in the magmatic state is more challenging (Hanmer & Passchier 1991). Criteria used in this study are based on particle interactions and include piggyback (Tikoff & Teyssier 1994) and tiled (Blumenfeld & Bouchez 1988) feldspar grains (Fig. 6).

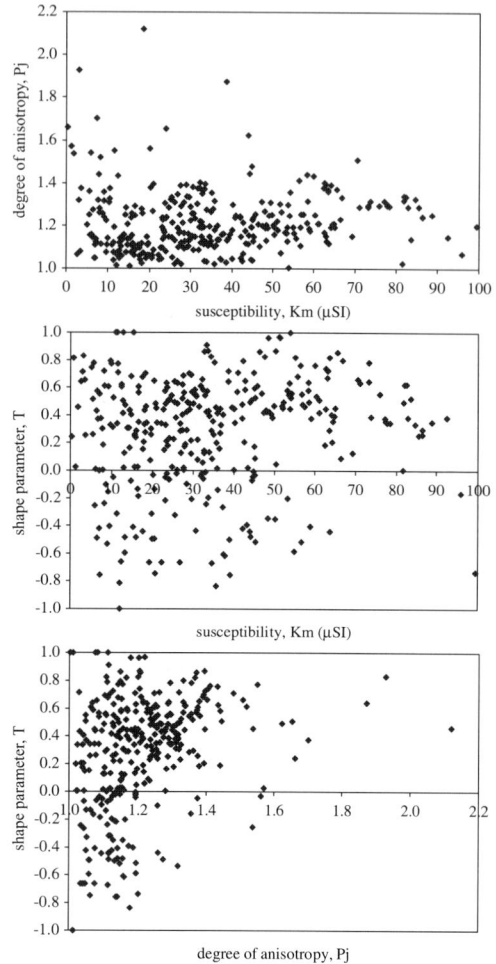

Fig. 9. Anisotropy of magnetic susceptibility data. Graphs of relations between K_m and P_j (top), K_m and T (middle), and P_j and T (bottom). See the text for discussion.

The sense of shear recorded in leucogranites deformed in the low-temperature solid state is symmetric relative to Sugar Mountain with top-to-the-west in the west and top-to-the-east in the east (Fig. 11). Three samples located around Sugar Lake show evidence of coaxial deformation (conjugate shear bands and absence of consistent shear sense). The occurrence of high-temperature fabrics is mostly restricted to the region around Sugar Mountain as well as the SW corner of the field area (Fig. 11). Sense of shear in the SW region is dominantly top-to-the-west. Again, in the region of Sugar Lake, high-temperature fabric indicates coaxial deformation. Sense of shear in

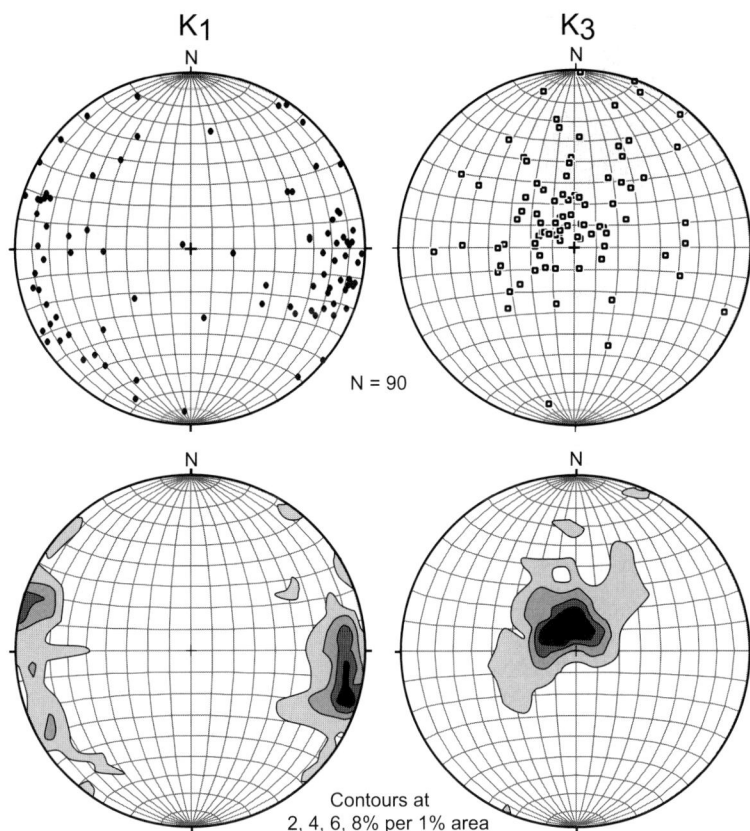

Fig. 10. Equal-area, lower-hemisphere stereonets showing the average AMS K_1 directions (lineation – closed circles) and K_3 (pole to foliation – open boxes), and associated contoured diagrams (below) for the 90 sample locations in the Shuswap MCC (sample locations in Fig. 7).

the leucogranite of Sugar Mountain is somewhat ambiguous but is dominantly top-to-the-east. The variability in both direction and sense of shear in this region may be associated with a larger component of coaxial deformation. Fabrics developed in the magmatic state are preserved only in the Sugar Mountain region (Fig. 11), where the AMS lineation plunges more steeply, and possibly indicate the presence of a feeder zone (Amice & Bouchez 1989; Aranguren *et al.* 2003). At one location (SM-9, Fig. 7a) four samples were analysed for sense of shear and three of these samples indicate clear top-to-the-east relations from feldspar tiling.

These results identify Sugar Mountain as the 'kinematic hinge' of this part of the Shuswap MCC. From the limited observations of fabrics developed under magmatic and high-temperature deformation, this kinematic pattern may have also prevailed under higher temperature

conditions. The steep fabrics at Sugar Mountain may represent a feeder zone for the leucogranite.

Discussion

Channel detachment v. rolling-hinge detachment

If a channel of partially molten crust formed in response to crustal thickening and development of an orogenic plateau (Fig. 2), this channel would flow toward the eastern and western foreland regions. Therefore, along the eastern edge of the plateau (Omineca belt), flow would have been toward the east (Fig. 12a). This process of blind orogenic collapse (Rey *et al.* 2001) has been inferred to explain the growth of the Tibetan plateau (Royden *et al.* 1997). Inception of Eocene extension in the Canadian Cordillera, presumably related to a change in plate

Magmatic and High-Temperature

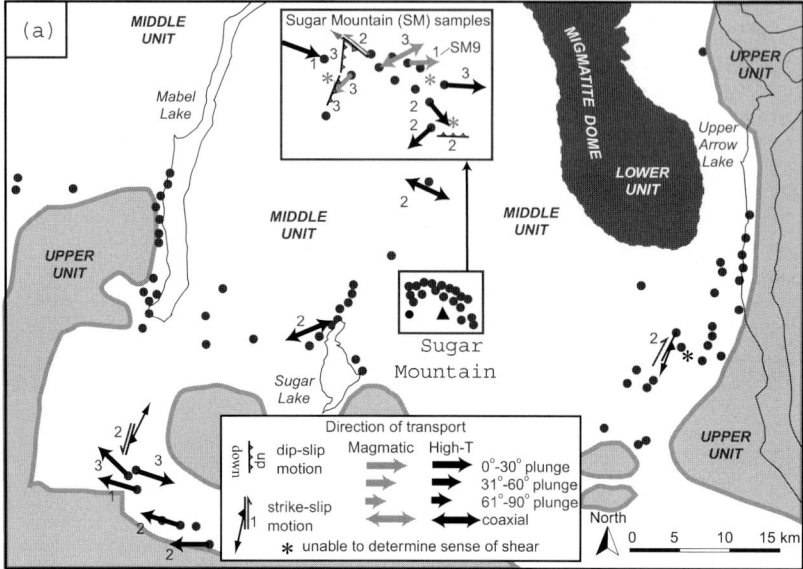

Low-Temperature

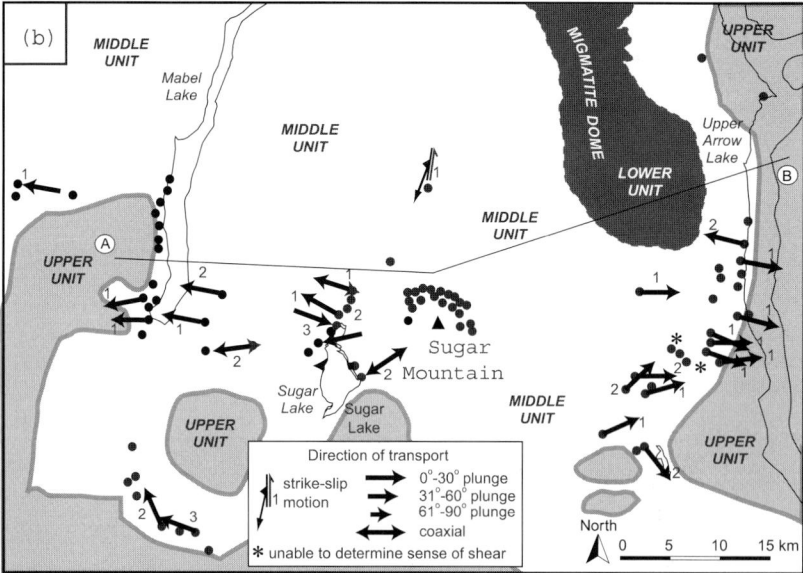

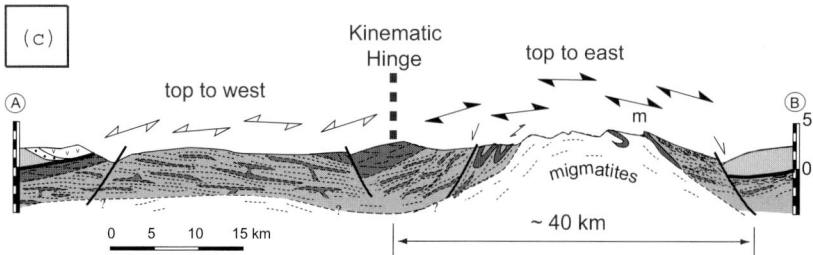

kinematics, determined the transition from fixed-boundary to free-boundary collapse (Fig. 1, after Rey *et al.* 2001). During extension and crustal boudinage, strain localized at the interface between hot crust and cold crust (Fig. 12a & b) and the Columbia River detachment originated. Two types of detachments are identified: a flat ('channel') detachment at the roof of the partially molten layer; and a crustal-scale ('rolling-hinge') detachment where crustal-scale extension localized (Fig. 12b) (Buck 1988, 1991; Brun & Van Den Driessche 1994; Brun *et al.* 1994; Axen *et al.* 1995).

In this interpretation, sense of shear on the channel detachment is top-to-the-west, and sense of shear associated with the rolling-hinge detachment is top-to-the-east. Top-to-the-west shear has been recognized in association with the western detachment, not only in the low-*T* mylonite zones, but also in the high-temperature deformation zones located in the southern region of the west detachment (Fig. 11a). This west detachment has a peculiar geometry: it is structurally very flat and does not appear to cut down-section substantially, at least over the approximately 30 km width where rafts of upper crust are preserved in the southern part of the area (Figs 2 and 11). This flat attitude implies the existence of a subhorizontal shear zone at the base of the upper crust that possibly accommodated the differential motion between the rigid upper crust and the ductile crust flowing in a channel (Glombick *et al.* 2002). Top-to-the-west shearing in the 60–55 Ma leucogranite may be attributable to the rooting of W-directed normal faults that dissected the upper crust in a domino-like fashion (Fig. 12c). This flat detachment may also have recorded substantial shearing related to channel flow prior to leucogranite intrusion.

Toward the east, past the 'kinematic hinge' at Sugar Mountain (Figs 11 and 12c), sense of shear in the leucogranite is consistently top-to-the-east all the way to the Columbia River fault. Top-to-the-east shear criteria are recognized not only in the leucogranite sheets but also in high-grade gneisses around the Thor–Odin Dome, with spectacular examples along the southern

margin of the dome (Vanderhaeghe *et al.* 1999). Therefore, these top-to-the-east kinematics are consistent over an approximately 5 km-thick section of crust, from Sugar Mountain to the edge of the dome (Fig. 11). We propose that this deformation is associated with a rolling-hinge detachment (Buck 1988, 1991; Brun & Van Den Driessche 1994; Brun *et al.* 1994; Axen *et al.* 1995). Unlike the flat western detachment that marks the limit between upper crust and lower crust, the foreland-dipping rolling-hinge detachment cuts deeply into the crust and can exhume the deepest rocks of the orogen. In a rolling hinge, high-angle normal faults form sequentially as they rotate and become inactive, allowing early formed ductile fabrics to be exhumed in the footwall of the detachment (Buck 1988, 1991). These early formed fabrics rotate progressively, explaining the existence of flat shear zones in the footwall of the detachment. The Columbia River fault is the structure that formed during the last increment of extension. In the footwall of the Columbia River fault, mylonites of the Columbia River detachment (Lane 1984; Vanderhaeghe *et al.* 2003b) have been exhumed. Following this logic, all E-directed structures in the footwall of the Columbia River detachment may be related to the same process, including the Monashee décollement (Vanderhaeghe *et al.* 1999). The high-grade tectonites that wrap around the Thor–Odin Dome may represent the deep roots of the Columbia River detachment (as in experiments of Brun *et al.* 1994), and not an E-directed thrust, as previously interpreted (Journeay & Brown 1986; Brown *et al.* 1992).

This interpretation integrates not only the kinematic data, but also the metamorphic and thermochronological evolution of the Shuswap MCC, the production of leucogranites and the emplacement of the Thor–Odin gneiss dome into a single tectonic framework. Activation of the rolling-hinge detachment (Fig. 12b) sets off a process of upward flow of deep crust accompanied by partial melting. The incipient stages of decompression may have generated a pulse of leucogranite that was extracted and emplaced beneath the detachment. The main

Fig. 11. Maps showing shear sense for AMS sample locations that exhibit: (**a**) high-temperature (black arrows) and magmatic (grey arrows) fabric, respectively; and (**b**) low-temperature fabrics. Arrows are drawn along the AMS lineation and in the direction of the motion of the top of the sample. Star depicts samples that appear isotropic (no observable fabric) in (a) or that show K_1–K_3 orientation inconsistent with macroscopic fabric observed in the hand sample. A rating of 1–3 represents the confidence in determination of sense of shear at each location: 1, excellent; 2, good; 3, weak. (**c**) Cross-section A–B (Fig. 2) showing the kinematic hinge at Sugar Mountain located approximately 40 km from the footwall of the Columbia River fault.

feeder zone for the leucogranites coincides with the kinematic hinge of the system, with W-directed shear sense in the west and E-directed shear in the east (Fig. 12b).

With further extension, the orogenic crust in the footwall of the rolling hinge rises rapidly, leading to increased partial melting during isothermal decompression (Norlander *et al.* 2002; Teyssier & Whitney 2002; Whitney *et al.* 2004). At this point, a diapiric instability takes place, with *in mass* upward flow of the crust that forms a diatexite-dominated migmatite dome. The Thor–Odin Dome is now located

only 5–10 km from the trace of the Columbia River fault; therefore, the exhumation process provided by the rolling hinge and the diapiric instability shut off shortly after the emplacement of the migmatite dome.

How does this interpretation fare with the timing of events measured by geochronology? In general, one would expect a younging of cooling ages from west to east on the east side of the kinematic hinge. The leucogranites were generated and emplaced at approximately 60–55 Ma (Carr 1992), the crystallization of migmatites occurred at about 56 Ma

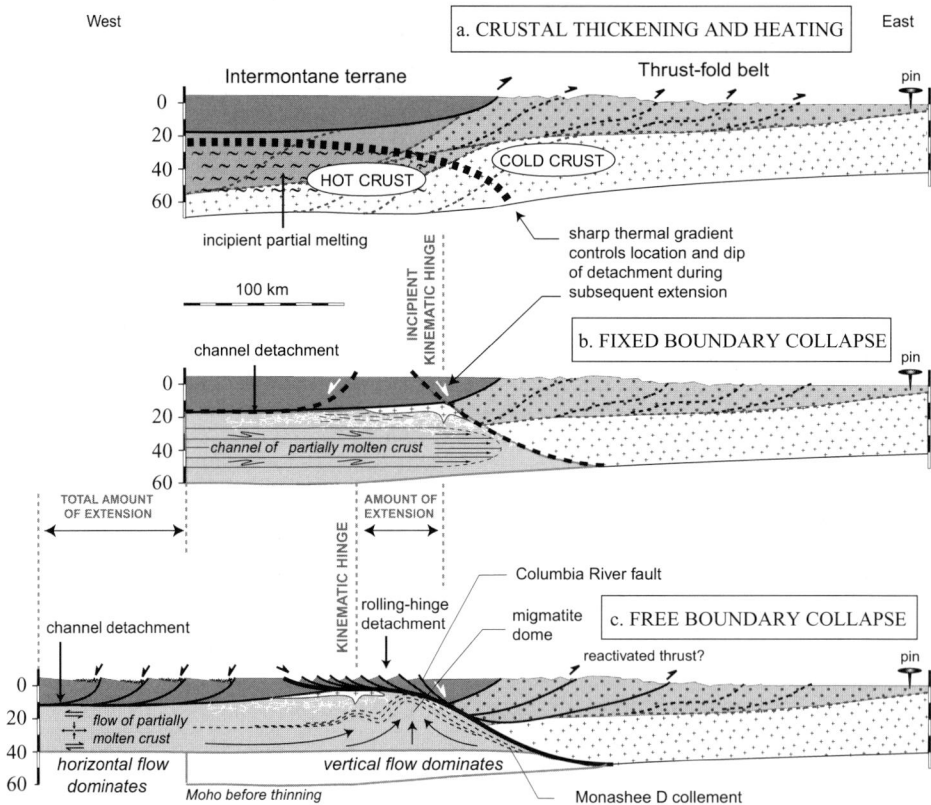

Fig. 12. (**a**) Schematic cross-section of the Canadian Cordillera with a fold–thrust belt foreland, intermontane terrane thrust sheet and crustal duplex that probably characterized the thickening processes. (**b**) Following thickening, the thick hinterland partially melts and the partially molten crust flows in a channel; this evolution determines the geometry of the channel detachment and the foreland-dipping detachment that will become a rolling-hinge detachment, when far-field extension occurs at a later time, during free-boundary collapse; activation of rolling-hinge detachment leads to decompression of deep rocks and production of melt that is extracted to form the leucogranite. A kinematic hinge develops that separates top-to-the-west shear in the channel detachment zone and top-to-the-east shear in the rolling-hinge detachment zone. (**c**) With further extension, the rolling-hinge detachment has moved toward the east relative to the kinematic hinge; this displacement is a direct measure of extension; decompression of deep rocks enhances partial melting and triggers a diapiric instability, forming a migmatite dome; the low-viscosity crust flows both laterally and vertically to accommodate crustal extension (arrows); this coupled system allows the Moho to remain flat (Brun *et al.* 1994; Brun & Van Den Driessche 1994; Bertotti *et al.* 2000).

(Vanderhaeghe *et al.* 1999), and the dome rocks, especially on the eastern margin, show the youngest monazite ages in leucosomes and pegmatites as 55–50 Ma (Parkinson 1992). By 49–48 Ma, the rocks had cooled below temperatures necessary for ductile deformation (Lorencak *et al.* 2001; Vanderhaeghe *et al.* 2003*b*). Therefore, the scenario presented in Figure 12b & c represents a short event, on the order of 5–10 Ma. In addition, this model explains why the high-grade rocks cooled rapidly after their partial exhumation: the rigid lid was removed very efficiently by domino-style extension or some other form of thinning (Fig. 12c). In contrast upper crustal dominoes are still preserved in the western part of the region, to the west of the kinematic hinge, where they are overlain by tilted sedimentary basins capped by volcanic rocks (Enderby and Trinity Hills basins, Fig. 3). This observation further implies an asymmetry in the development of the Shuswap MCC that resulted in a better preservation of the Eocene palaeosurface (nonconformity at base of basins) in the west compared to the east, and therefore a different mode of detachment tectonics.

The present-day distance between the leucogranite feeder zone (Sugar Mountain) and the footwall of the Columbia River detachment is approximately 40 km (Fig. 12). Therefore, if it took 5–10 Ma for the system to develop, the rate of extension was 4–8 km Ma^{-1}, a rather rapid rate for a single exhuming structure. This average rate implies that the rate of ascent of rocks in the steep footwall of the rolling-hinge detachment was also on the order of 5 km Ma^{-1}, with faster rates in the core of the domes as they 'intrude' the mantling gneisses. Rates of approximately 5–10 km Ma^{-1} explain the near-isothermal decompression observed in the dome (Fayon *et al.* 2004).

Therefore, the coupled rolling-hinge–channel-flow model explains: (1) the kinematic hinge observed in the leucogranites with top-to-the-east sense of shear east of Sugar Mountain and top-to-the-west sense of shear to the west; (2) the position of the leucogranite feeder zone recognized by AMS; (3) the position of the migmatite dome located close to the Columbia River detachment today (asymmetric relative to the core complex), and the fact that the dome overprints the top-to-the-east fabrics; (4) the distinct metamorphic paths in the dome and the mantling gneiss; (5) the relative timing of events measured by geochronology, although the whole process took place in a relatively short time (*c.* 5 Ma) such that many ages obtained are within error of each other; and (6) the relatively flat Moho

interpreted on a Lithoprobe profile (Cook *et al.* 1992; Clowes *et al.* 1995).

Conclusions

Metamorphic core complexes such as the Shuswp MCC may form by asymmetric boudinage of the crust generated by the high lateral thermal gradient between cold foreland and hot hinterland in which partially molten crust has developed. During orogenic collapse, lateral and vertical flow of partially molten crust are intimately linked to detachments. Two detachment types develop: (1) a long-lived *channel detachment* separates the rigid upper crust from the flowing orogenic crust; and (2) a short-lived *rolling-hinge detachment* develops during the boudinage instability. The channel detachment probably exists during the period of fixed-boundary or blind collapse (flow of partially molten crust in a channel). As the system evolves into free-boundary collapse associated with far-field extension, a region of high thermal gradient at the front of the flowing channel determines the foreland-dipping geometry of the rolling-hinge detachment. Fast exhumation of deep rocks (near-isothermal decompression) produces leucogranite that becomes a temporal marker for the melting event and a kinematic marker for the syn- and post-emplacement deformation. Further extension and exhumation produce substantial partial melting of the orogenic crust and trigger diapiric instabilities. The effective vertical mass and heat transfer through diapirism produces migmatite domes in which isothermal decompression metamorphic paths are preserved. Such an efficient heat and mass transfer halts the flow of orogenic crust and marks the end of orogeny. Further extension becomes distributed along more brittle normal faults (Basin-and-Range type faulting).

Research supported by NSF grants EAR-9526915, EAR-9814669 and EAR-0106953, the Department of Geology and Geophysics and the Graduate School, University of Minnesota, and GSA and Sigma Xi student grants. D. Parkinson thanks Jim Mattinson and the Geological Sciences Department at UCSB for financial assistance, and B. Tikoff is thanked for kindly providing access to the Kappabridge KLY-3S instrument at the University of Wisconsin-Madison. Reviews by N. Mancktelow, P. Glombick and J. Van Den Driessche helped improve the manuscript.

Appendix 1: U–Pb sample locations and descriptions (Table 2)

Samples were hand-picked in alcohol and some abraded using the procedure outlined by Krogh

(1982). Zircons were dissolved by a method slightly modified from Parrish (1987). Monazites were dissolved using 12N HCl in microcapsules in hydrothermal bombs at 220 °C. Pb and U from both zircon and monazite solutions were separated by method modified after Krogh (1973). Pb blanks 10 pg. Mass spectrometry was performed on a Finnigan Mat 261 multicollector instrument, with fractionation and precision controlled by repeated runs of NBS 981 and 983 Pb standards. Samples 248, 94-28 and 96-66 are located and described in Vanderhaeghe (1997).

81787-2: Elevation: 4500', Lat: 118°06.6', Long: 50°47.25'.
Medium–coarse-grained biotite pegmatite from core gneiss. Concordant pod in foliation, deformed and not cross-cutting. Monazites: yellow clear glassy–cloudy yellow-orange with inclusions, euhedral.

81089-3: Elevation: 4000', Lat: 118°02.5', Long: 50°38.5'.
Coarse-grained biotite pegmatite, with large K-feldspars. Foliation parallel pod within basement supracrustal gneiss. Monazites: 'gem' quality, euhedral, equant, yellow, clear, glassy with no inclusions; or large, cloudy, euhedral, yellow-orange with inclusions.

82789-2: Elevation: 3400', Lat: 118°00.4', Long: 50°30.5'.
Biotite–quartz–feldspar gneiss.

82188-1.2: Elevation: 2200', Lat: 118°04.1', Long: 50°48.7'.
Biotite pegmatite (2 m thick), foliation parallel. From within basement gneisses on Highway 23. Monazites: Large cloudy, some dark inclusions, yellow-brown-orange, cloudy–clear, complete crystals platy and euhedral; to subequant, euhedral, no inclusions, yellow orange, glassy, 'gem' quality.

82188-8: Elevation: 2600', Lat: 118°09.55', Long: 50°52.7'.
Tourmaline pegmatite, slightly deformed. Coarse–medium grained. 1 mm black tourmaline and muscovite books. Located in new logging scar, as a pod in quartzite.

81887-1: Elevation: 3600', Lat: 118°05.8', Long: 50°48.05'.
Undeformed cross-cutting tourmaline–biotite pegmatite, zoned with medium-grained core and coarse-grained margin. Collected medium-grained core for most part. Zircons: orange–orange-brown elongate sharp faceted, cloudy, euhedral grains.

Appendix 2: Anisotropy of magnetic susceptibility

The magnetic susceptibility of a specimen (K) is the ratio of the induced magnetization (M) to the applied magnetic field (H). The magnetic susceptibility is anisotropic and generally measured in low magnetic field (*c.* 10^{-4} tesla (T)) for a number of different directions. Reviews of AMS principles and applications can be found in Hrouda (1982), Rochette (1988), Tarling & Hrouda (1993) and Borradaile & Henry (1997). The anisotropy of magnetic susceptibility (AMS) is approximated by a symmetric second-rank tensor with three principal axes ($K_1 \geq K_2 \geq K_3$). The bulk magnetic susceptibility, K, is the arithmetic mean of the principal susceptibilities [$K_m = 1/3$ $(K_1 + K_2 + K_3)$]. The total anisotropy is given by $P_j = \exp[2\ \Sigma(\ln K_i/K)^2]^{1/2}$ ($i = 1-3$) and the ellipsoid shape is given by the parameter $T = [(2\ln K_2 - \ln K_1 - \ln K_3)/(\ln K_1 - \ln K_3)]$ (Jelínek 1978). The AMS results mainly from lattice-preferred orientation and shape-preferred orientation of mineral grains and therefore, as initially suggested by Graham (1954), can be used as proxy for mineral fabric in rocks. K_1 is the magnetic lineation and K_3 is the pole to the magnetic foliation.

In granitic rocks, diamagnetic minerals such as quartz and feldspars do not contribute significantly to the bulk AMS because of their weak intrinsic susceptibility. In contrast, paramagnetic minerals, such as ferromagnesian silicates, and ferromagnetic minerals, such as iron oxides and sulphides, control the AMS (Rochette *et al.* 1992). When present, even in small amounts, ferromagnetic minerals tend to dominate the AMS (e.g. Borradaile & Henry 1997; Bouchez 1997). In a few cases, the interpretation of AMS is complicated by the presence of minerals that have an inverse magnetic fabric, such as tourmaline, ferran calcite or single domain magnetite (Potter & Stephenson 1988; Rochette *et al.* 1992, 1994). In such cases, when the inverse contribution is larger than 10%, important variations in degree of anisotropy and shape factors occur even when the orientation of the AMS principal axes are not substantially affected (Ferré 2002).

References

ALLIBONE, A.H. & NORRIS, J. 1992. Segregation of leucogranite microplutons during syn-anatectic deformation: an example from the Taylor Valley, Antarctica. *Journal of Metamorphic Geology*, **10**, 589–600.

AMICE, M. & BOUCHEZ, J.-L. 1989. Susceptibilite magnetique et zonation du batholite granitique de Cabeza de Araya (Extremadura, Espagne). *Comptes Rendus de l'Academie des Sciences, Serie II*, **308** (13), 1171–1178.

ARANGUREN, A., CUEVAS, J., TUBIA, J.M., ROMAN-BERDIEL, T., CASAS-SAINZ, A. & CASAS-PONSATI, A. 2003. Granite laccolith emplacement in the Iberian Arc; AMS and gravity study of the La Tojiza Pluton (NW Spain). *Journal of the Geological Society, London*, **160**, 435–445.

ARCHANJO, C.J., BOUCHEZ, J.L., CORSINI, M. & VAUCHEZ, A. 1994. The Pombal granite pluton: magnetic fabric, emplacement and relationships with the Brasiliano strike-slip setting of NE Brazil (Paraiba State). *Journal of Structural Geology*, **16**, 323–335.

ARCHIBALD, D.A., CARMICHAEL, D.M., GLOVER, J.K., PRICE, R.A. & FARRAR, E. 1983. Geochronology and tectonic implications of magmatism and metamorphism, southern Kootenay Arc and neighbouring regions, southeastern British Columbia. Part I: Jurassic to mid-Cretaceous (Canada). *Canadian Journal of Earth Sciences*, **20**, 1891–1913.

ARMSTRONG, R.L. 1982. Cordilleran metamorphic core complexes – From Arizona to southern Canada. *Annual Review of Earth and Planetary Sciences*, **10**, 129–154.

ARMSTRONG, R.L., PARRISH, R.R., VAN DER HEYDEN, P., SCOTT, K., RUNKLE, D. & BROWN, R.L. 1991. Early Proterozoic basement exposures in the southern Canadian Cordillera: core gneiss of Frenchman Cap, Unit I of the Grand Forks Gneiss, and the Vaseaux Formation. *Canadian Journal of Earth Sciences*, **28**, 1169–1201.

AXEN, G.J., BARTLEY, J.M. & SELVERSTONE, J. 1995. Structural expression of a rolling hinge in the footwall of the Brenner Line normal fault, eastern Alps. *Tectonics*, **14**, 1380–1392.

AXEN, G.J., SELVERSTONE, J., BYRNE, T. & FLETCHER, J.M. 1998. If the strong crust leads, will the weak crust follow? *GSA Today*, **8**, 1–8.

BEAUMONT, C., JAMIESON, R.A., NGUYEN, M.H. & LEE, B. 2001. Himalayan tectonics explained by extrusion of a low-viscosity crustal channel coupled to focused surface denudation. *Nature*, **414**, 738–742.

BERMAN, R.G. 1990. Mixing properties of Ca–Mg–Fe–Mn garnets. *American Mineralogist*, **75**, 328–344.

BERMAN, R.G. 1991. Thermobarometry using multi-equilibrium calculations: a new technique, with petrologic applications. *Canadian Mineralogist*, **29**, 833–855.

BERTHÉ, D., CHOUKROUNE, P. & JAGOUZO, P. 1979. Orthogneiss, mylonite and non-coaxial deformation of granites: the example of the South Armorican shear zone. *Journal of Structural Geology*, **1**, 34–42.

BERTOTTI, G., PODLADCHIKOV, Y. & DAEHLER, A. 2000. Dynamic link between the level of ductile crustal flow and style of normal faulting of brittle crust. *Tectonophysics*, **320**, 195–218.

BIRD, P. 1991. Lateral extrusion of lower crust from under high topography, in the isostatic limit. *Journal of Geophysical Research*, **96**, 10 275–10 286.

BLUMENFELD, P. & BOUCHEZ, J.L. 1988. Shear criteria in granite and migmatite deformation in the magmatic and solid states. *Journal of Structural Geology*, **10**, 361–372.

BORRADAILE, G.J. 1988. Magnetic susceptibility, petrofabrics and strain. *Tectonophysics*, **156**, 1–20.

BORRADAILE, G.J. & HENRY, B. 1997. Tectonic applications of magnetic susceptibility and its anisotropy. *Earth-Science Review*, **42**, 49–93.

BORRADAILE, G.J. & WERNER, T. 1994. Magnetic anisotropy of some phyllosilicates. *Tectonophysics*, **235**, 223–248.

BOUCHEZ, J.L. 1997. Granite is never isotropic: an introduction to AMS studies of granitic rocks. *In*: BOUCHEZ, J.L., HUTTON, D.H.W. & STEPHENS, W.E. (eds) *Granite: From Segregation of Melt to Emplacement Fabrics*. Kluwer, Dordrecht, 95–112.

BOUCHEZ, J.L., GLEIZES, G., DJOUADI, T. & ROCHETTE, P. 1990. Microstructure and magnetic susceptibility applied to emplacement kinematics of granites; the example of the Foix Pluton (French Pyrenees). *Tectonophysics*, **184**, 157–171.

BROWN, R.L. & JOURNEAY, J.M. 1987. Tectonic denudation of the Shuswap metamorphic terrane of southeastern British Columbia (Canada). *Geology*, **15**, 142–146.

BROWN, R.L., CARR, S.D., JOHNSON, B.J., COLEMAN, V.J., COOK, F.A. & VARSEK, J.L. 1992. The Monashee décollement of the southern Canadian Cordillera: a crustal-scale shear zone linking the Rocky Mountain Foreland belt to lower crust beneath accreted terranes. *In*: MCCLAY, K.R. (ed.) *Thrust Tectonics*. Chapman & Hall, London, 357–364.

BROWN, R.L., JOURNEAY, J.M., LANE, L.S., MURPHY, D.C. & REES, C.J. 1986. Obduction, backfolding and piggyback thrusting in the metamorphic hinterland of the southeastern Canadian Cordillera. *Journal of Structural Geology*, **8**, 255–268.

BRUN, J.-P., SOKOUTIS, D. & VAN DEN DRIESSCHE, J. 1994. Analogue modeling of detachment fault systems and core complexes. *Geology*, **22**, 319–322.

BRUN, J.-P. & VAN DEN DRIESSCHE, J. 1994. Extensional gneiss domes and detachment fault systems; structure and kinematics. *Bulletin de la Société Géologique de France*, **165**, 519–530.

BUCK, W.R. 1988. Flexural rotation of normal faults. *Tectonics*, **7**, 959–973.

BUCK, W.R. 1991. Modes of continental lithospheric extension. *Journal of Geophysical Research*, **96**, 20 161–20 178.

CARR, S.D. 1991. U–Pb zircon and titanite ages of three Mesozoic igneous rocks south of the Thor–Odin–Pinnacles area, southern Omineca Belt, British Columbia. *Canadian Journal of Earth Sciences*, **28**, 1877–1882.

CARR, S.D. 1992. Tectonic setting and U–Pb geochronology of the Early Tertiary Ladybird leucogranite suite, Thor–Odin–Pinnacles area, southern Omineca Belt, British Columbia. *Tectonics*, **11**, 258–278.

CLOWES, R.M., ZELT, C.A., AMOR, J.R. & ELLIS, R.M. 1995. Lithospheric structure in the southern Canadian Cordillera. *Canadian Journal of Earth Sciences*, **32**, 1485–1513.

COLPRON, M., PRICE, R.A., ARCHIBALD, D.A. & CARMICHAEL, D.M. 1996. Middle Jurassic exhumation along the western flank of the Selkirk fan structure: thermobarometric and thermochronometric constrains from the Illecillewaet synclinorium, southeastern British Columbia. *Geological Society of America Bulletin*, **108**, 1372–1392.

COOK, F.A., Varsek. J.L. ET AL. 1992. Lithoprobe crustal reflection cross section of the southern Canadian Cordillera, 1, Foreland thrust and fold belt to Fraser River Fault. *Tectonics*, **11**, 12–35.

CRITTENDEN, M.D., JR., CONEY, P.J. & DAVIS, G.H. 1980. *Cordilleran Metamorphic Core Complexes*. Memoir of the Geological Society of America, **153**.

CROWLEY, J.L., BROWN, R.L. & PARRISH, R.R. 2001. Diachronous deformation and a strain gradient beneath the Selkirk allochthon, northern Monashee complex, southeastern Canadian Cordillera. *Journal of Structural Geology*, **23**, 1103–1121.

EWING, T.E. 1981. Regional stratigraphy and structural setting of the Kamloops Group, south-central British Columbia. *Canadian Journal of Earth Sciences*, **18**, 1464–1477.

FAYON, A.K., WHITNEY, D.L. & TEYSSIER, C. 2004. Exhumation of orogenic crust: diapiric ascent vs. low-angle normal faulting. *In*: WHITNEY, D.L., TEYSSIER, C. & SIDDOWAY, C.S. (eds) *Gneiss Domes in Orogeny*. Geological Society of America Special Paper, **380**, 129–139.

FERRÉ, E.C. 2002. Theoretical models of intermediate and inverse AMS fabrics. *Geophysical Research Letters*, **29**, 31-1–31-4.

GAPAIS, D. 1989. Shear structures within deformed granites: mechanical and thermal indicators. *Geology*, **17**, 1144–1147.

GHENT, E.D., NICHOLLS, J., STOUT, M.Z. & ROTTENFUSSER, B. 1977. Clinopyroxene amphibolite boudins from Three Valley Gap, British Columbia. *Canadian Mineralogist*, **15**, 269–282.

GLOMBICK, P.M., THOMPSON, R.I. & ERDMER, P. 2002. The role of a melt-rich middle crust layer in core complex formation; evidence from the Shuswap metamorphic complex, south-central British Columbia. *Geological Society of America, Abstracts with Programs*, **34**, 109.

GRAHAM, C.M. & POWELL, R. 1984. A garnet-hornblende geothermometer: calibration, testing, and application to the Pelona Schist, Southern California. *Journal of Metamorphic Geology*, **2**, 13–31.

GRAHAM, J.W. 1954. Magnetic susceptibility anisotropy, an unexploited petrofabric element. *Geological Society of America Bulletin*, **65**, 1257–1258.

GUILLOT, S., PÊCHER, A., ROCHETTE, P. & LE FORT, P. 1993. The emplacement of the Manaslu granite of central Nepal: field and magnetic susceptibility constraints. *In*: TRELOAR, P.J. (ed.) *Himalayan Tectonics*. Geological Society, London, Special Publications, **74**, 413–428.

HANMER, S. & PASSCHIER, C. 1991. *Shear-sense Indicators; A Review*. Geological Survey of Canada Paper, **90-17.**

HOLLAND, T. & BLUNDY, J. 1994. Non-ideal interactions in calcic amphiboles and their bearing on amphibole–plagioclase thermometry. *Contributions to Mineralogy and Petrology*, **116**, 433–447.

HROUDA, F. 1982. Magnetic anisotropy of rocks and its application in geology and geophysics. *Geophysical Surveys*, **5**, 37–82.

JELÍNEK, V. 1978. Statistical processing of anisotropy of magnetic susceptibility measured on groups of specimens. *Studia Geophyzika et Geodetika*, **22**, 50–62.

JOHNSON, B.J. & BROWN, R.L. 1996. Crustal structure and early Tertiary extensional tectonics of the Omineca belt at 51°N latitude, southern Canadian Cordillera. *Canadian Journal of Earth Sciences*, **33**, 1596–1611.

JOHNSTON, D.H., WILLIAMS, P.F., BROWN, R.L., CROWLEY, J.L. & CARR, S.D. 2000. Northeastward extrusion and extensional exhumation of crystalline rocks of the Monashee complex, southeastern Canadian Cordillera. *Journal of Structural Geology*, **22**, 603–625.

JOURNEAY, M. & BROWN, R.L. 1986. Major tectonic boundaries of the Omineca Belt in southern British Columbia: A progress report. *Current Research, Part A, Geological Survey of Canada*, **86-1A**, 81–88.

JOVER, O., ROCHETTE, P., LORAND, J.P., MAEDER, M. & BOUCHEZ, J.L. 1989. Magnetic mineralogy of some granites from the French Massif Central: origin of their low field susceptibility. *Physics of the Earth and Planetary Interiors*, **55**, 79–92.

KROGH, T.E. 1973. A low-contamination method for hydrothermal dissolution of zircons and extraction of U and Pb for isotopic age determinations. *Geochimica et Cosmochimica Acta*, **37**, 485–494.

KROGH, T.E. 1982. Improved accuracy of U–Pb zircon ages by the creation of more concordant systems using an air abrasion technique. *Geochimica et Cosmochimica Acta*, **46**, 637–649.

LANE, L.S. 1984. Brittle deformation in the Columbia River fault zone near Revelstoke, southeastern British Columbia. *Canadian Journal of Earth Sciences*, **21**, 584–598.

LE BRETON, N. & THOMPSON, A.B. 1988. Fluid-absent (dehydration) melting of biotite in metapelites in the early stages of crustal anatexis. *Contributions to Mineralogy and Petrology*, **99**, 226–237.

LISTER, G.A. & DAVIS, G.A. 1989. The origin of metamorphic core complexes and detachment faults formed during Tertiary continental extension in the northern Colorado River region, U.S.A. *Journal of Structural Geology*, **11**, 65–94.

LORENCAK, M., SEWARD, D., VANDERHAEGHE, O., TEYSSIER, C. & BURG, J.-P. 2001. Low temperature cooling history of the Shuswap metamorphic core complex, British Columbia: Constraints from apatite and zircon fission-track analysis. *Canadian Journal of Earth Sciences*, **38**, 1615–1625.

MALAVIEILLE, J. 1993. Late orogenic extension in mountain belts; insights from the Basin and Range and the late Paleozoic Variscan belt. *Tectonics*, **12**, 1115–1130.

MATHEWS, W.H. 1981. Early Cenozoic resetting of potassium–argon dates and geothermal history of North Okanagan area, British Columbia. *Canadian Journal of Earth Sciences*, **18**, 1310–1319.

MATTAUER, M., COLLOT, B. & VAN DEN DRIESSCHE, J. 1983. Alpine model for the internal metamorphic zones of the North American Cordillera. *Geology*, **11**, 11–15.

MCKENZIE, D., MILLER, E.L., NIMMO, F., JACKSON, J.A. & GANS, P.B. 2000. Characteristics and consequences of flow in the lower crust. *Journal of Geophysical Research B: Solid Earth*, **105**, 11 029–11 046.

MULCH, A. 2004. *Integrated high-spatial resolution $^{40}Ar/^{39}Ar$ geochronology, stable isotope geochemistry, and structural analysis of extensional detachment systems: case studies form the Porsgrunn–Kristiansand shear zone (S-Norway) and the Shuswap metamorphic core complex (Canada).* PhD thesis, University of Lausanne.

MULCH, A., TEYSSIER, C., COSCA, M.A., VANDERHAEGHE, O. & VENNEMANN, T.W. 2004. Reconstructing paleoelevation in eroded orogens. *Geology*, **32**, 525–528.

MURPHY, M.A. & HARRISON, T.M. 1999. Relationship between leucogranites and the Qomolangma detachment in the Rongbuk Valley, South Tibet. *Geology*, **27**, 831–834.

NELSON, K.D., ZHAO, W., ET AL. 1996. Partially molten middle crust beneath southern Tibet: Synthesis of Project INDEPTH results. *Science*, **274**, 1684–1688.

NORLANDER, B.H., WHITNEY, D.L., TEYSSIER, C. & VANDERHAEGHE, O. 2002. Partial melting and decompression of the Thor–Odin dome, Shuswap metamorphic core complex, Canadian Cordillera. *Lithos*, **61**, 103–125.

NYMAN, M.W., PATTISON, D.R.M. & GHENT, E.D. 1995. Melt extraction during formation of K-feldspar + sillimanite migmatites, west of Revelstoke, British Columbia. *Journal of Petrology*, **36**, 351–372.

OKULITCH, A.V. 1984. The role of the Shuswap Metamorphic Complex in Cordilleran tectonism: A review. *Canadian Journal of Earth Sciences*, **21**, 1171–1193.

PARKINSON, D.L. 1991. Age and isotopic character of Early Proterozoic basement gneisses in the southern Monashee Complex, southeastern British Columbia. *Canadian Journal of Earth Sciences*, **28**, 1159–1168.

PARKINSON, D.L. 1992. *Age and tectonic evolution of the southern Monashee complex, southeastern British Columbia: a window into the deep crust.* PhD thesis, University of Santa Barbara.

PARRISH, R.R. 1987. An improved micro-capsule for zircon dissolution in U-Pb geochronology, *Isotope Geoscience*, **66**, 99–102.

PARRISH, R.R. 1990. U–Pb dating of monazite and its implication to geological problems, *Canadian Journal of Earth Sciences*, **27**, 1431–1450.

PARRISH, R.R. 1995. Thermal evolution of the southeastern Canadian Cordillera. *Canadian Journal of Earth Sciences*, **32**, 1618–1642.

PARRISH, R.R. & ARMSTRONG, R.L. 1987. The ca. 162 Ma Galena Bay stock and its relationship to the Columbia River fault zone, southeast British Columbia. *In: Radiogenic age and isotopic studies: Report 1*. Geological Survey of Canada Paper, **87-2**, 25–32.

PARRISH, R.R. & WHEELER, J.O. 1983. A U–Pb zircon age from the Kuskanex batholith, southeastern British Columbia (Canada). *Canadian Journal of Earth Sciences*, **20**, 1751–1756.

PARRISH, R.R., CARR, S.D. & PARKINSON, D.L. 1988. Eocene extensional tectonics and geochronology of the southern Omineca Belt, British Columbia and Washington. *Tectonics*, **7**, 181–212.

PASSCHIER, C.W. & SIMPSON, C. 1986. Porphyroclast systems as kinematic indicators. *Journal of Structural Geology*, **8**, 831–843.

PASSCHIER, C.W. & TROUW, R.A.J. 1996. *Microtectonics*. Springer, Berlin.

PRICE, R.A. 1986. The southeastern Canadian Cordillera: thrust faulting, tectonic wedging, and delamination of the lithosphere. *Journal of Structural Geology*, **8**, 239–254.

POTTER, D.K. & STEPHENSON, A. 1988. Single-domain particles in rocks and magnetic fabric analysis. *Geophysical Research Letters*, **15**, 1097–1100.

READ, P.B. & BROWN, R.L. 1981. Columbia River Fault Zone: southeastern margin of the Shuswap and Monashee complexes, southern British Columbia. *Canadian Journal of Earth Sciences*, **18**, 1127–1145.

REESOR, J.E. & MOORE, J.M., JR. 1971. *Petrology and Structure of Thor–Odin Gneiss Dome, Shuswap Metamorphic complex, British Columbia*. Geological Survey of Canada Bulletin, **195**.

REY, P., VANDERHAEGHE, O. & TEYSSIER, C. 2001. Gravitational collapse of the continental crust: definition, regimes and modes. *Tectonophysics*, **342**, 435–449.

ROCHETTE, P. 1988. Inverse magnetic fabric carbonate bearing rocks. *Earth and Planetary Science Letters*, **90**, 229–237.

ROCHETTE, P., JACKSON, M. & AUBOURG, C. 1992. Rock magnetism and the interpretation of anisotropy of magnetic susceptibility. *Reviews in Geophysics*, **30**, 209–226.

ROCHETTE, P., SCAILLET, B., GUILLOT, S., LE FORT, P. & PÊCHER, A. 1994. Magnetic properties of the High Himalayan leucogranites: Structural implications. *Earth and Planetary Science Letters*, **126**, 217–234.

ROYDEN, L.H. 1996. Coupling and decoupling of crust and mantle in convergent orogens: Implications for strain partitioning in the crust. *Journal of Geophysical Research*, **101**, 17 679–17 705.

ROYDEN, L.H., BURCHFIEL, B.C., KING, R.W., WANG, E., CHEN, Z., SHEN, F. & LIU, Y. 1997. Surface deformation and lower crustal flow in eastern Tibet. *Science*, **276**, 788–790.

SCAILLET, B., Pêcher, A., ROCHETTE, P. & CHAMPENOIS, M. 1995. The Gangotri granite (Garhwal Himalaya): laccolithic emplacement in an extending collisional belt. *Journal of Geophysical Research B: Solid Earth*, **100**, 585–607.

SCHILLING, F.R. & PARTZSCH, G.M. 2001. Quantifying partial melt fraction in the crust beneath the central Andes and the Tibetan Plateau. *Physics and Chemistry of the Earth (A)*, **26**, 239–246.

SEVIGNY, J.H., PARRISH, R.R. & GHENT, E.D. 1989. Petrogenesis of peraluminous granites, Monashee Mountains, southeastern Canadian Cordillera. *Journal of Petrology*, **30**, 557–581.

SNOKE, A.W., TULLIS, J. & TODD, V.R. 1998. Fault-related Rocks: A Photographic Atlas. Princeton University Press, Princeton, NJ.

STACEY, J.S. & KRAMERS, J.D. 1975. Approximation of terrestrial lead isotope evolution by a two-stage model. *Earth and Planetary Science Letters*, **26**, 207–221.

STRUIK, L.C. 1993. Intersecting intracontinental Tertiary transform fault systems in the North American Cordillera. *Canadian Journal of Earth Sciences*, **30**, 1262–1274.

TALBOT, J.-Y., FAURE, M., MARTELET, G., COURRIOUX, G. & CHEN, Y. 2004. Emplacement in an extensional setting of the Mont Lozère-Borne granitic complex (SE France) inferred from comprehensive AMS, structural and gravity studies. *Journal of Structural Geology*, **26**, 11–28.

TARLING, D.H. & HROUDA, F. 1993. *The Magnetic Anisotropy of Rocks*. Chapman & Hall, London.

TEMPELMAN-KLUIT, D. & PARKINSON, D. 1986. Extension across the Eocene Okanagan crustal shear in southern British Columbia (Canada). *Geology*, **14**, 318–321.

TEYSSIER, C. & WHITNEY, D.L. 2002. Gneiss domes and orogeny. *Geology*, **30**, 1139–1142.

TIKOFF, B. & TEYSSIER, C. 1994. Strain and fabric analyses based on porphyroclast interaction. *Journal of Structural Geology*, **16**, 477–491.

VANDERHAEGHE, O. 1997. *Role of partial melting during late-orogenic collapse*. PhD. thesis, University of Minnesota.

VANDERHAEGHE, O. & TEYSSIER, C. 1997. Formation of the Shuswap metamorphic core complex during late-orogenic collapse of the Canadian Cordillera: Role

of ductile thinning and partial melting of the mid- to lower crust. *Geodinamica Acta*, **10**, 41–58.

VANDERHAEGHE, O. & TEYSSIER, C. 2001*a*. Crustal-scale rheological transitions during late-orogenic collapse. *Tectonophysics*, **335**, 211–228.

VANDERHAEGHE, O. & TEYSSIER, C. 2001*b*. Partial melting and flow of orogens. *Tectonophysics*, **342**, 451–472.

VANDERHAEGHE, O., TEYSSIER, C. & WYSOCZANSKI, R. 1999. Structural and geochronological constraints on the role of partial melting during the formation of the Shuswap metamorphic core complex at the latitude of the Thor–Odin Dome, British Columbia. *Canadian Journal of Earth Sciences*, **36**, 917–943.

VANDERHAEGHE, O., JAMIESON, R.A., MEDVEDEV, S., FULLSACK, P. & BEAUMONT, C. 2003*a*. Evolution of orogenic wedges and continental plateaux: Insights from crustal thermal–mechanical models overlying subducting mantle lithosphere. *Geophysical Journal International*, **153**, 27–51.

VANDERHAEGHE, O., MCDOUGALL, I., DUNLAP, W.J. & TEYSSIER C. 2003*b*. Cooling and exhumation of the Shuswap metamorphic core complex constrained by $^{40}Ar-^{39}Ar$ thermochronology. *Geological Society of America Bulletin*, **115**, 200–216.

VERNON, R.H. 1976. *Metamorphic Processes*. George Allen and Unwin, London.

VIELZEUF, D. & CLEMENS, J.D. 1992. The fluid-absent melting of phlogopite + quartz: experiments and models. *American Mineralogist*, **77**, 1206–1222.

WANLESS, R.K. & REESOR, J.E. 1975. Precambrian zircon age of orthogneiss in the Shuswap Metamorphic Complex, British Columbia. *Canadian Journal of Earth Sciences*, **12**, 326–332.

WHEELER, J.O. & MCFEELY, P. 1991. *Tectonic Assemblage Map of the Canadian Cordillera and Adjacent Parts of the United States of America*. Geological Survey of Canada, Map 1712A.

WHITNEY, D.L., TEYSSIER, C. & FAYON, A.K. 2004. Isothermal decompression, partial melting, and the exhumation of deep continental crust. *In*: GROCOTT, J., MCCAFFREY, K.J.W., TAYLOR, G. & TIKOFF, B. (eds) *Vertical Coupling and Decoupling in the Lithosphere*. Geological Society, London, Special Publications, **227**, 15–33.

WILLS, S. & BUCK, W.R. 1997. Stress-field rotation and rooted detachment faults: a Coulomb failure analysis. *Journal of Geophysical Research*, **102**, 20 503–20 514.

WOLFE, J.A., FOREST, C.E. & MOLNAR, P. 1998. Paleobotanical evidence of Eocene and Oligocene paleolatitudes in midlatitude western North America. *Geological Society of America Bulletin*, **110**, 664–678.

ZHENG, Y., WANG, T., MINGBO, M. & DAVIS, G.A. 2004. Maximum effective moment criterion and the origin of low-angle normal faults. *Journal of Structural Geology*, **26**, 271–285.

Geometric aspects of synkinematic granite intrusion into a ductile shear zone – an example from the Yunmengshan core complex, northern China

C. W. PASSSCHIER[1], J. S. ZHANG[2] & J. KONOPÁSEK[1]

[1]Institut für Geowissenschaften, University of Mainz, 55099 Mainz, Germany
[2]State Seismological Bureau, China

Abstract: The Cretaceous Yungmengshan core complex in northern China contains a large syntectonic granodiorite batholith that intrudes a slightly older diorite intrusion. A major gently dipping ductile décollement shear zone is developed along the contact of the diorite and granodiorite. The shear zone is invaded by a large volume of granitic and pegmatite veins associated with the main granodiorite batholith during activity of the shear zone under high-grade metamorphic conditions. Progressively older veins are more strongly deformed into tight cylindrical fold structures rotated into parallelism with the lineation and foliation in the shear zone. Parallelism of veins to the foliation is partly due to this rotation, but also to foliation-parallel injection of younger syntectonic pegmatite veins. Several small-scale structures have been recognized that allow distinction of solid-state deformation of veins. Granite veins do not extend much above the ductile shear zone that seems to act as a lid and an effective depository to intruding granite veins from the underlying batholith. There was considerable volume increase in the footwall and lower part of the shear zone by vein intrusion.

Gently dipping ductile shear zones are a common feature of many continental metamorphic core complexes throughout the world, for example in the Basin and Range (Reynolds & Lister 1987; Fletcher & Bartley 1994; Foster & Fanning 1997; Foster et al. 2001), Algeria (Caby et al. 2001), New Guinea (Baldwin et al. 1993) and in the Aegean Sea (Lister et al. 1984; Walcott & White 1998; Pe-Piper et al. 2002). The shear zones are thought to accommodate most of the crustal-scale extension that is typical for the development of such core complexes. Core complexes are also commonly associated with the intrusion of granitoid plutons and associated rhyolitic volcanism (e.g. Caby et al. 2001; Foster et al. 2001; Pe-Piper et al. 2002). Many granitoid intrusions in core complexes can be shown to have intruded during active deformation of the ductile shear zones of the complex (Caby et al. 2001; Foster et al. 2001; Pe-Piper et al. 2002). In such cases, parts of a pluton may cut mylonitic rocks, while other parts are mylonitized with the same orientation of mylonitic structures as in the truncated older parts, and this is used as evidence of syntectonic intrusion.

In the Yunmengshan of northern China, a Cretaceous core complex developed in Archaean basement and Proterozoic metasediments at the northern rim of the North China Craton (Fig. 1) (Davis et al. 1996). The core complex is centred on a granodiorite batholith, which is truncated at the northern, eastern and southern sides by brittle and ductile shear-zone segments. The segments dip gently away from the granodiorite batholith towards the north, east and south (Figs 1 and 2). Mapping by Davis et al. (1996) has shown that, although these shear-zone segments are of different metamorphic grade and presumably of different age in different parts of the pluton, they share a common orientation of stretching lineations. They are therefore inferred to have formed as part of one tectonic shear-zone system that acted as a major décollement of the core complex. Its present variation in dip direction and dip is probably due to folding over the developing core complex, a feature also observed in many other complexes (e.g. Fletcher & Bartley 1994).

Along the NE side of the core complex, the Yunmengshan granodiorite is bordered by metadioritic rocks and the contact is affected by a ductile shear zone, the Sihetang shear zone discussed in this paper (Fig. 2). The metadiorites are intrusive into Proterozoic metasediments (Fig. 2) (Miyun metadiorite; Davis et al. 1996),

From: BRUHN, D. & BURLINI, L. (eds) 2005. High-Strain Zones: Structure and Physical Properties.
Geological Society, London, Special Publications, **245**, 65–80.
0305-8719/05/$15.00 © The Geological Society of London 2005.

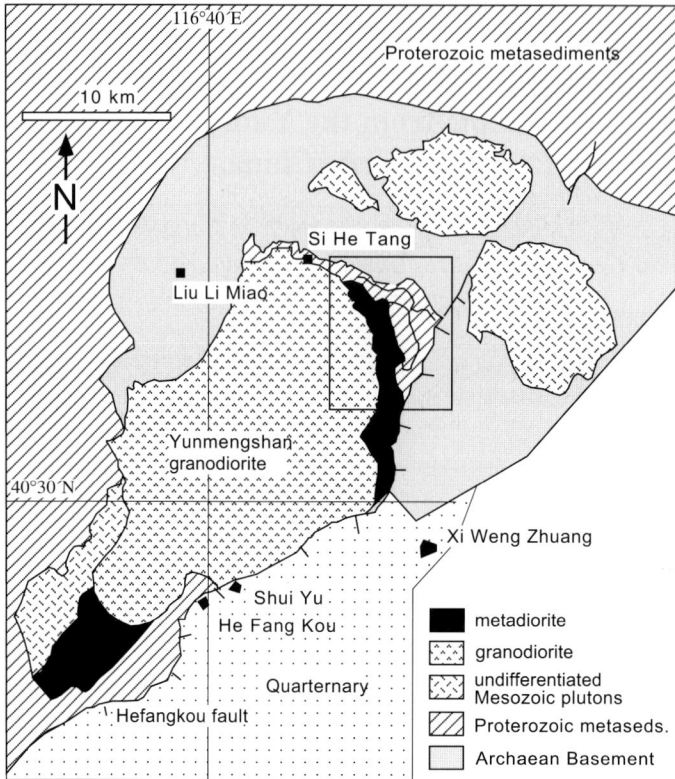

Fig. 1. General map of the Yunmengshan granodiorite core complex, after Davis *et al.* (1996). H, Hefangkou; L, Liulimiao; X, Xiwengzhuang.

but the intrusive contact is rarely exposed and is marked by a Mesozoic brittle thrust fault in most places (Fig. 2). This fault post-dates intrusion of the diorite and granodiorite, and seems to post-date and cut the Sihetang ductile shear zone. The thrust and the Sihetang shear zone are cut in the SE by the brittle Hefangkou normal fault (Fig. 1).

The Sihetang shear zone is an amphibolite facies ductile shear zone with a thickness of 20–50 m (Fig. 2) (Davis *et al.* 1996). The shear zone is centred on the contact between the metadiorite and the granodiorite, and has strongly mylonitized the igneous rocks. Despite the presence of the shear zone, it is clear that the contact between the granite and granodiorite of the Yunmengshan batholith and the Miyun metadiorite was intrusive: xenoliths of metadiorite occur in the granite, and granite veins intrude the metadiorite. The metadiorite therefore predates the granitic phases. Dating of both units has given concordant U–Pb zircon ages of 159 ± 2 Ma for the Miyun metadiorite and

ages between 151 and 127 Ma for the Yunmeng-shan granodiorite (Davis *et al.* 1996).

Although the main body of the granodiorite is deformed by the Sihetang shear zone, there is a large volume of granite dykes that show different stages of deformation, and a decrease in deformation intensity with a decrease in relative age within the shear zone, based on cross-cutting relations. This suggests that the shear zone was active while intrusive material was added from lower levels of the granodiorite pluton. The Sihetang shear zone is not affected by late brittle deformation and forms a flat 'roof' to the batholith. It is only invaded by minor veins, although these represent a considerable volume of material. Some work has been published on synkinematic intrusion of granite veins into ductile shear zones (Zurbriggen *et al.* 1998; Pawley *et al.* 2002), but three-dimensional (3D) control on geometry is limited in these studies. The excellent outcrop condition and the sharp colour contrast between the metadiorite and the leucocratic veins in the Sihetang shear zone

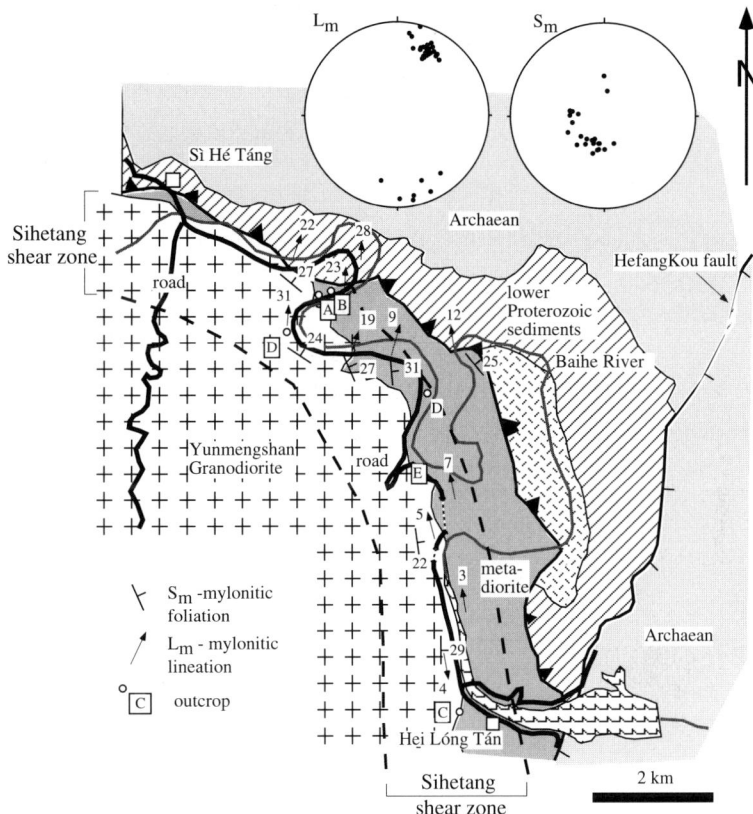

Fig. 2. Schematic map of the NE Yunmengshan granodiorite with the approximate position of the Sihetang shear zone, after Davis *et al.* (1996). Orientation of the mylonitic foliation (S_m) and lineation (L_m) measured in this study in this part of the shear zone is shown as inset stereograms.

allow a closer look at the internal structure of this type of ductile shear zone.

Host rock lithology

The Miyun metadiorite (Davis *et al.* 1996) consists of hornblende, plagioclase, biotite and quartz. It is metaluminous, alkaline to calc-alkaline–tholeiitic in composition with xenoliths rich in hornblende and garnet (Davis *et al.* 1996). Locally, evidence exists for several pulses of dioritic magma with different amount of dark minerals. The grain size is uniform at 0.5–3 mm, and phenocrysts of plagioclase are rare. Where the metadiorite is deformed, it shows weak banding indicating that originally some variation in composition may have been present.

The Yunmengshan batholith away from the shear zone is composed of granodiorite with few granite and pegmatite veins. The granodiorite outside the contact zone is coarse grained,

with plagioclase grains up to 1 cm in diameter and relatively biotite-rich. In the top of the batholith close to the contact with the Miyun metadiorite, the batholith contains several generations of granite veins with xenoliths or blocks of granodiorite and late pegmatite and microgranite veins.

Sihetang shear zone

The Sihetang shear zone at the contact of the granodiorite and metadiorite has a single, penetrative foliation and a well-developed object lineation (terminology as in Piazolo & Passchier 2002) in both the granodiorite and the metadiorite (Fig. 3b). Strain seems to be highest along the contact of both units, and a mylonitic foliation and lineation is visible over at least 15 km along the contact (Fig. 2) (Davis *et al.* 1996). The main foliation consists of a preferred orientation of hornblende, biotite and of xenoliths in

Fig. 3. (**a**) General structure of the Sihetang shear zone in outcrop C of Figure 2, looking west. The shear zone is developed in metadiorite with numerous veins on top of undeformed granodiorite. The contact between the granodiorite and the overlying metadiorite is indicated with an arrow. Deformed granite and pegmatite veins lie at a small angle to the contact in the metadiorite. The rock face lies at a small angle to the mylonitic lineation. The height of outcrop is approximately 50 m. (**b**) View from the core of the Sihetang shear zone in outcrop A of Figure 2, looking towards the south and the undeformed granodiorite. The outcrop in the foreground shows the strong N–S-trending object lineation characteristic of the deformed metadiorite.

the metadiorite, and a shape fabric of aggregates of recrystallized quartz, feldspar and biotite in the granodiorite. No magmatic foliation is visible in the granodiorite or metadiorite and dykes are undeformed away from the Sihetang shear zone.

In thin section, the deformed metadiorite consists of equigranular plagioclase, hornblende, biotite and quartz with minor sphene and zircon. Hornblende and biotite occur as clusters of euhedral or subhedral grains with a strong preferred orientation defining a foliation. Feldspar grains are commonly slightly elongate parallel to this foliation. Biotite, plagioclase and hornblende crystals are usually unzoned and strain-free. Albite twins in most plagioclase grains are straight and transect the whole grain, although some plagioclase grains show weak undulous extinction and tapering deformation albite twins. Quartz crystals show weak undulous extinction and elongate subgrains. Interphase

boundaries and grain boundaries are mostly gently curving, but some boundaries, especially between quartz crystals, are lobate. The lack of small subgrains and new grains around old quartz crystals, and the lobate grain and interphase boundaries indicate the activity of high-temperature grain-boundary migration recrystallization in quartz above 500 °C (Stipp *et al.* 2002). However, chessboard-subgrain structures in quartz (Stipp *et al.* 2002) are lacking and all subgrains are of elongate prismatic type. The euhedral or subhedral shape of biotite and hornblende, and the gently curved boundaries of plagioclase grains indicate grain-boundary migration and grain-boundary area-reduction processes (Passchier & Trouw 1996).

The deformed granodiorite and the granite veins have a similar quartz–feldspar microstructure as the metadiorite and in many cases a strong preferred orientation of euhedral–subhedral biotite. Quartz occurs in ribbons 0.5–1 mm wide and up to 8 mm long in granite veins. Myrmekite growth is common. However, in the less deformed parts of the granodiorite, and in the younger, little deformed granite and pegmatite veins, some plagioclase grains show zoning that is probably a relict of the magmatic fabric. Such plagioclase grains are surrounded by a mantle of smaller grains inferred to have formed by recrystallization. These recrystallized grains have a diameter of 0.05–1 mm.

Relicts of magmatic microstructures are rare in the Sihetang shear zone. Although some of the microstructures observed above are also observed in undeformed igneous rocks, the most typical features such as feldspar synneusis, euhedral faces on plagioclase to quartz, zoning and truncated growth twins in plagioclase, and wedge- or corridor-shaped quartz grains between euhedral feldspar grains (Paterson et al. 1989, 1998) are missing in the oldest veins and in the metadiorite; they are present as relict in the youngest, weakly deformed veins.

The object lineation is a hornblende grain lineation in the metadiorite, and an aggregate lineation of quartz and feldspar in the granodiorite (Fig. 3b). The orientation of the foliation and lineation is remarkably constant within the shear zone. A deflection of the foliation is only visible close to veins or other features: in all other places it is straight (Fig. 3). Although most structures in the shear zone are symmetric in sections parallel to the lineation, local metre-scale ductile shear bands, shear-band boudins (Goscombe & Passchier 2001) and asymmetric σ-shaped feldspar porphyroclasts indicate S-directed transport of the hanging wall for at least part of the deformation history. Some shear bands have tension gashes filled with pegmatite that give the same shear sense. In a marble horizon within the metasediments overlying the metadiorite, asymmetric boudins, some with the geometry of delta objects

and up to 50 cm in diameter, indicate top-to-the-south displacement as well.

The stability of hornblende, biotite, plagioclase and garnet in the samples, and the evidence for high-temperature grain-boundary migration in quartz and of grain-boundary area reduction in other minerals, indicate that deformation took place under amphibolite facies conditions. Migmatic veins are absent in the metadiorite, and no structures indicative of local melting or melt accumulation were seen; all veins are formed by intrusion from outside the observed outcrops.

An attempt was made to use amphibole–plagioclase thermometry on five samples from the metadiorite in the shear zone (Holland & Blundy 1994) to determine the temperature of deformation more precisely. All samples show the quartz-bearing mineral assemblage (amphibole–plagioclase–K-feldspar–quartz–biotite ± titanite) and thus both the edenite–tremolite and edenite–richterite calibrations can be used. The composition of amphiboles in all samples varies around the junction edenite–magnesiohastingsite–magnesiohornblende and the anorthite content of plagioclases is in the range of 17–28%. For all five studied samples, both calibrations gave similar but surprisingly high average temperatures of 662–722 °C (Table 1: standard deviation of c. ±15 °C) in the range of pressures between 6 and 12 kbar. As no partial melt structures were observed in the shear zone, such high temperatures could be due to non-equilibration of magmatic hornblende in the samples.

Main contact of granodiorite batholith and metadiorite

We investigated the geometry of structures in the shear zone in some detail in a number of well-exposed road cuts along the main road from Hei Long Tan to Si He Tang (Fig. 2).

Table 1. *Average temperature estimates (including standard deviations) for studied samples**

Sample	Number of Am–Pl pairs	Average $T\,(P = 6\,\text{kbar})$ ed–tr (°C)	SD (°C)	Average $T\,(P = 12\,\text{kbar})$ ed–tr (°C)	SD (°C)	Average $T\,(P = 6\,\text{kbar})$ ed–ri (°C)	SD (°C)	Average $T\,(P = 12\,\text{kbar})$ ed–ri (°C)	SD (°C)
MY 18	13	694	15	657	16	667	15	692	16
MY 30	29	697	14	662	12	668	13	694	12
MY 38	16	710	16	676	14	682	13	710	13
MY 49	23	717	12	689	9	691	10	722	8
MY 7	17	691	19	663	20	669	15	699	16

*The temperatures were calculated using the amphibole–plagioclase thermometer of Holland & Blundy (1994). Abbreviations: ed–tr, edenite–tremolite calibration; ed–ri, edenite–richterite calibration.

In all localities, the contact between the dark metadiorite and light-coloured granodiorite rocks of the Yunmengshan batholith is sharp and dipping parallel to the main foliation, S_m (Figs 2 and 3). The granodiorite has only a few inclusions of metadiorite, but the latter is invaded by a large number of granite and pegmatite veins in a zone 50–100 m wide parallel to the contact (Fig. 3a). Up to 40% vol. of the rock can be made up of such veins. Higher up in the metadiorite, the veins are uncommon. The Proterozoic marbles and quartzite overlying the metadiorite (Fig. 2) contain occasional veins of metadiorite and granite, and the granite veins in

all cases cut metadiorite veins, even in these metasediments. In the top of the Yunmengshan granodiorite batholith a metre-scale xenolith of alternating marble and quartzite was intruded by a diorite vein; all three units are cut off by the surrounding granite.

Nature of the veins

The veins intruding the metadiorite are richer in SiO_2 than the parent granodiorite and consist of several types of medium-to coarse-grained granite, alkali-granite and pegmatite. The veins range in thickness from 1 mm to several metres,

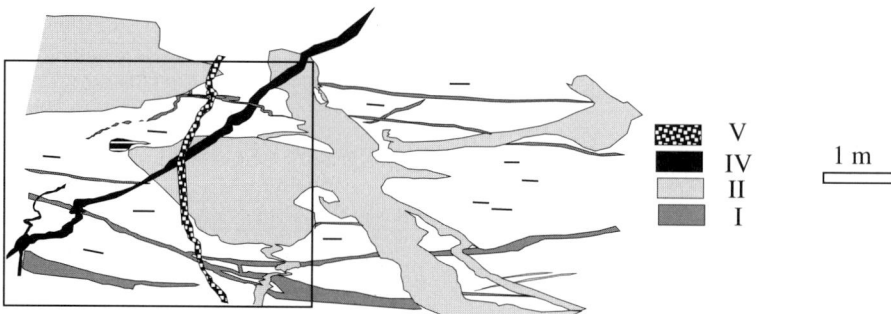

Fig. 4. Part of outcrop A in Figure 2 looking north, seen from the outcrop in Figure 3b. The age relationship of dykes visible in the photograph (I–V) is given in the sketch. Relatively younger veins are less folded and boudinaged, and at a higher angle to the mylonitic foliation than older veins.

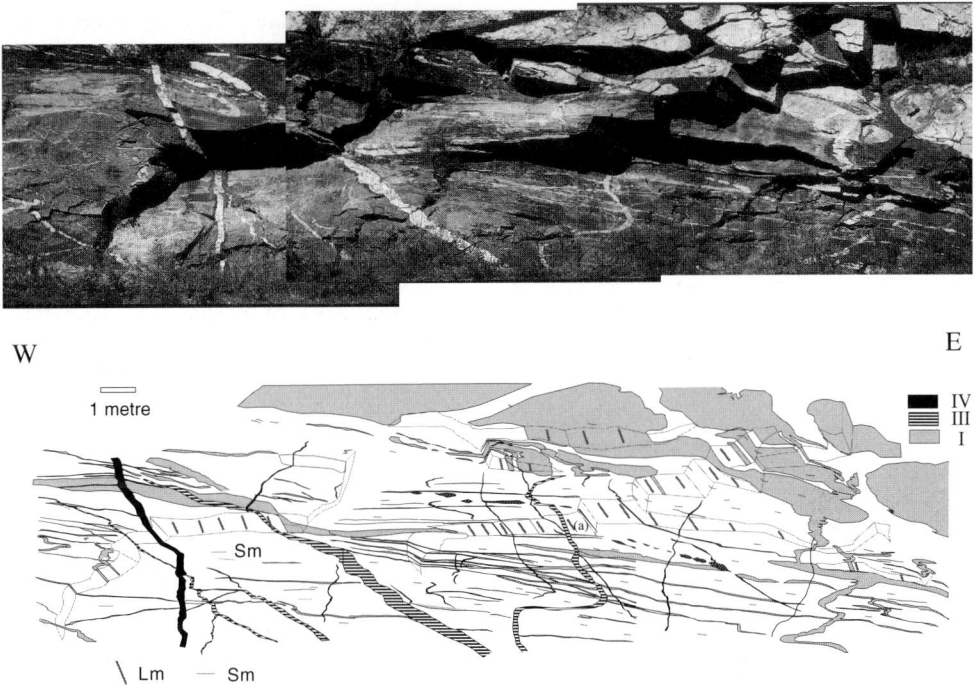

Fig. 5. Composite image of outcrop B in Figure 2 looking north. The age relationship of dykes visible in the photograph is given in the sketch. Relatively younger veins are less folded and boudinaged and at a higher angle to the mylonitic foliation than older veins. In the centre of the drawing (a) shows the site of Figure 6a.

have sharp boundaries and their relative age can be established by cross-cutting relations (Figs 4 and 5). All veins, even the oldest ones, show signs that they intruded the metadiorite after a foliation was established. This can be seen from the fact that parts of the contact or branches of the veins follow S_m (Figs 4–6c & f). Progressively younger veins were systematically subject to less deformation, but even the youngest veins recognized in the shear zone are slightly deformed. All veins in the shear zone are therefore synkinematic.

A sequence of veins can be recognized in every outcrop, suggesting that intrusion was phased, but this is not convertible from outcrop to outcrop, and they are not all separable by their bulk chemistry. There is a weak tendency for earlier veins to be granitic and medium grained, while later veins are more commonly pegmatitic, more alkaline and more coarse grained. However, exceptions to this trend are common. Flow structures have not been preserved in any veins.

Geometry of the veins

The geometry of the deformed veins has been studied in a number of fresh road cuts and

cliffs of different orientation along the main Baihe River, as indicated in Figure 2. The geometry of the deformed veins is dramatically different on outcrop surfaces of different orientation. The outcrops A and B shown in Figures 4 and 5 are the most important data source for the vein network. The general characteristics are as follows.

Rock faces normal to L_m and S_m

On surfaces normal to the lineation, a complex network of veins of different colour, grain size and composition is visible (Figs 4 and 5). Cross-cutting relationships show that the oldest veins (I) are granite and alkali-granite veins subparallel to the foliation, branching at small angles into ladder- or irregular-shaped networks, usually with one set of branches parallel to S_m. These are followed by several generations of grey and red granite, and pegmatite veins (II–V), with increasing steepness towards the foliation in the metadiorite, although some branches of each system tend to be parallel to S_m. The youngest, little or undeformed veins (IV and V) are pegmatites or granite veins at a high angle to

Fig. 6. (**a**) Lower part of an S_m-surface in metadiorite in outcrop B of Figure 5 showing strong object lineations and three types of veins of generations I, III and IV. Younger veins are more oblique to L_m than older veins. The width of the view is 1 m; (**b**) outcrop E normal to L_m looking south. Three sets of deformed veins of different age lie in the metadiorite; the oldest ones are horizontal parallel to S_m; the next generation is tightly folded (at right), the last generation is gently folded (on the left). The width of the view is 3 m. An enlargement of the tightly folded vein is given below (b). (**c**) Detail of a deformed vein of generation I with tight folding and a junction structure (Fig. 8a). Outcrop E of Figure 2. The width of the view is 1 m; (**d**) view of outcrop B in a section parallel to L_m and normal to S_m. All veins are subparallel or at a small angle to S_m. One vein is boudinaged. The width of the view is 4 m; (**e**) cuspate–lobate folding in a vein of generation I in outcrop A (Fig. 4). The width of the view is 1.5 m; (**f**) feeder vein in pegmatite with a large number of minor veins branching off into the metadiorite parallel to S_m. Flanking folds have developed in the branching veinlets in the rim of the feeder vein and show that the angle between the feeder and branching veins has decreased considerably. The width of the view is about 5 m.

the foliation. The hornblende–biotite foliation in the metadiorite continues into the veins as a shape or biotite foliation. Veins that lie at a low angle to S_m are either straight with a variable thickness or show pinch-and-swell or boudinage

(Figs 4–6c & d). The straight nature of some veins may be an original intrusive feature as seen from steps in the contact. Veins in steeper orientation are commonly folded (Figs 4–6b & c). The scale and nature of the folds depends on thickness and composition of the vein. Older veins have tighter folds, with thin limbs and massive hinges (Fig. 6b, inset), more intense internal foliation and better developed boudin structures than younger veins (Figs 4 and 5). Some veins seem to have been shortened without being folded, as seen from adjacent parallel veins of a different composition but with tight folding. Other wide veins show cuspate–lobate folding in the contact, cusps pointing towards the granitic vein in all cases (Fig. 6e). Where cusps are not opposite lobes, these features could also be described as mullions (Fig. 6e). Thin branches from the same vein show open–tight parallel or similar folding. In most cases, the fold axis is at a small angle to the lineation in the rock. There are also rare examples of round- or ring-shaped vein segments in outcrops normal to L_m. These could be parts of sheath-fold-like structures in the veins (Fig. 5, left-hand side centre).

Rock faces parallel to L_m and normal to S_m

On surfaces parallel to the lineation and normal to the foliation (Fig. 6d), the geometry of the veins is dramatically different, and if both surfaces had not been observed as part of one outcrop they could be taken for non-related features (compare Figs 4–6d). In surfaces parallel to L_m, the veins are nearly all parallel (Fig. 6d). If outcrops are large enough, veins can be seen to be cross-cutting at very low angles below $10°$ (Fig. 6d, top). Deformed medium-grained veins are mostly straight and planar, and may have a weak internal shape fabric parallel to the hornblende foliation in the metadiorite. Veins are straight or show minor pinch-and-swell and occasionally boudinage on the metre–10 m scale (Fig. 6d, bottom). There is only one set of late pegmatite veins that are folded, and which lie at a high angle to L_m and S_m. These are the only veins that are visible on faces parallel to L_m which are prominently oblique.

Rock faces parallel to S_m

On surfaces parallel to S_m, the lineation L_m is prominently present (Figs 3b and 6a). On these outcrop surfaces, most veins are straight and subparallel to the lineation, but there is a clear relation between the angle of vein intersection with S_m and L_m, and vein age. Figure 5 shows a road cut normal to L_m on which at least three generations of granite and pegmatite veins can be clearly distinguished by their cross-cutting relations. The younger veins are progressively more oblique to S_m and less intensely folded

and boudinaged. The same veins can be followed in this outcrop to surfaces parallel to S_m, on which the angle between vein intersection and L_m can be observed (Fig. 6a). The oldest veins are subparallel to L_m, the second generation makes an angle of $2–6°$, while the last weakly deformed generation of pegmatite veins is oblique, $30–50°$ to L_m (Figs 5 and 6a).

Three-dimensional vein geometry

The overall 3D geometry of the veins as reconstructed from the outcrops described above is a system of straight cylindrical folds intersecting subparallel to L_m (Fig. 7a). The oldest veins are generally most strongly folded and boudinaged, and the fold axes make the smallest angle with

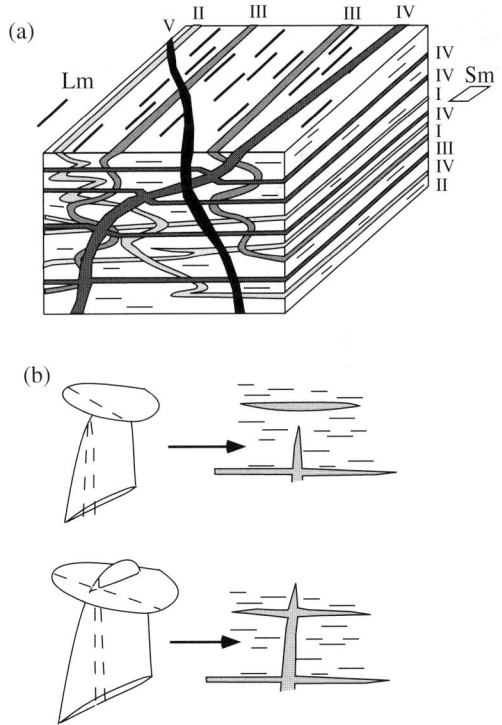

Fig. 7. (a) Schematic block diagram showing the 3D interpretation of vein geometry in the shear zone. Older veins (light colours) are more deformed than young veins (dark colours). Older veins are cylindrical and folded parallel to L_m; younger veins are less folded and more oblique to L_m. The dark vein IV is a feeder dyke with foliation-parallel injection veins. Roman numbers refer to vein generation; **(b)** explanation for foliation-parallel injection veins which branch from a feeder dyke on both sides; originally each foliation-parallel dyke spreads out from a tip of the feeder, but is subsequently cut by the expanding and propagating feeder dyke itself.

L_m. As a result, these veins appear as banded structures in sections parallel to L_m, and are tightly folded and boudinaged on sections normal to L_m. In most types of veins, boudins occur as strips parallel to L_m, and are therefore only visible in sections normal to L_m (Fig. 5). In some outcrops boudins occur that have a disk-shape, and are visible on surfaces normal and parallel to L_m. With decreasing age of veins, the tightness of folds in the veins decreases, and the angle between the fold axis and L_m increases (Figs 4–6b). The youngest pegmatite and granite veins can have orientations oblique or even normal to L_m.

Feeder veins

The parallelism of some of the granite and peg-matite veins to S_m is not only due to deformation, but partly an original intrusive feature. Although this is not always easy to show for the more deformed veins, the younger, less deformed veins clearly show evidence for intrusion parallel to S_m, such as foliation-parallel jogs in the veins (Fig. 8h), and rectangular xenoliths of the wallrock with sides parallel to the foliation in the xenolith. A number of feeder veins have been found that are connected to swarms of foli-ation-parallel veins, mainly of late pegmatite and granite veins (Fig. 6f). Feeder veins are between 0.3 and 2 m wide, and intrude highly oblique to S_m, although sideward jogs parallel to S_m of a few metres long and much thinner vein segments have been observed. In all cases these feeder veins show bristling sets of minor veins, 0.1– 0.5 m thick, branching off the main vein parallel to S_m (Figs 6f and 7a). Owing to deformation, the angle between the feeder veins and the offshoots has in many cases been reduced, in some instances to less than 30°. The original steep orientation of the feeder veins can still be recog-nized by the presence of flanking folds (Passchier 2001) in the rim of the feeder veins (Fig. 6f). The amplitude of these flanking folds is never more than the diameter of the feeder vein. In most cases, the feeder veins are strongly affected by cuspate–lobate folding or mullion formation in the contact, and, in some cases, by 10 m scale open folding of the whole vein. In 3D the feeder veins are most commonly at a small angle to L_m. However, this applies to the most defor-med examples, and other less deformed veins also occur oblique or even orthogonal to L_m. The fold axes in such veins are equally normal or oblique to L_m, the only examples of folds with this orientation in the Sihetang shear zone. An interesting feature of the feeder veins is that there is usually only a single feeder vein

without oblique branches or neighbours of similar nature. No examples were seen of veins parallel to S_m that are connected to more than one feeder. This is doubtlessly partly due to the limited size of the outcrops, but also shows that few feeder dykes are needed to produce the observed dyke swarms.

Many offshoots on the feeder dykes occur exactly opposite each other, i.e. they branch out from the feeder in two directions from the same point (Fig. 6f). This may give information on the nature of vein growth. If the feeder forms first and the branches later, there would be no reason for veins to branch off at the same site, unless there is a specific layer of lithological break that they follow. This does not seem to be the case here. A possible explanation is that the feeders and branching veins grow upwards and sideways at the same time (Fig. 7b).

Structures indicative of deformation in planar veins

Deformation of veins in granitoid rocks leads to folding and boudinage of the veins if these have a rheology significantly different from that of the host rock. However, a specific problem in deformed granitoid rocks is that veins and host rock have in many cases similar composition and grain size. As a result, veins act as passive markers and deformation leads to homogeneous thinning or thickening and change in angle between veins, and shape fabrics are commonly the only visible evidence of deformation. In such cases, some other more subtle structures may be useful to recognize vein deformation in the field. Several of these have been recognized in the Sihetang shear zone and are listed below. They may also be useful in other, similar settings.

Junction structure

Where an older vein is cut by a younger one some veins show significant thickening of the older vein towards the younger one (Figs 6c and 8a). In many cases, the younger vein is also thickened. This structure may form when the younger vein is oblique to the shortening direction and deforms by folding while the older one deforms by homogeneous thinning; where both veins are in contact, the thinning vein is hampered in its deformation by the pre-sence of the younger one, and reveals the minimum original thickness of the older vein.

Vein refraction structure

Where two veins cross-cut obliquely, or at right angles, they may be deformed together so that

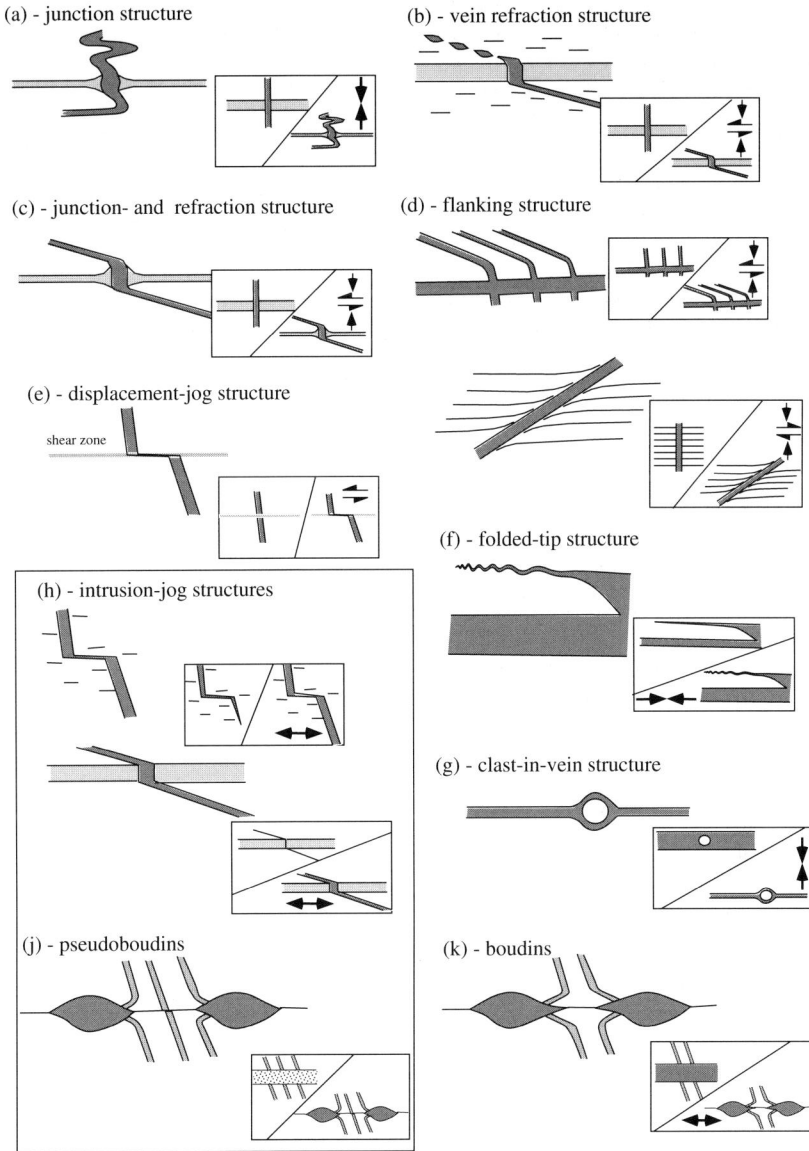

Fig. 8. Several types of mesostructures recognized in the Sihetang shear zone that can be used to recognize solid-state ductile deformation of intrusive dykes. Inset boxes show inferred mode of development, and arrows show whether the structure is associated with shortening, shear or extension. Intrusion-jog structures and pseudoboudins form by intrusion, not solid-state deformation and are therefore shown in a separate box. (**a**) Junction structure – an older, cross-cut vein widens towards the younger vein. As the younger vein is folded, this must be due to thinning of the older vein; (**b**) vein refraction structure – a younger vein has a jog where it cuts the older vein. This is due to relative rotation of both veins where the intersection domain is protected by the more rigid older vein; (**c**) combination of (b) and (c); (**d**) two types of flanking structure with deflection of veins close to a feeder dyke (top) or single dyke (bottom); (**e**) displacement jog structure due to offset; (**f**) folded tip structure. Thin branches of a dyke show folding, but wider parts are shortened more homogeneously. (**g**) Clast-in-vein structure – a large monocrystal of feldspar lies in a thinner vein due to deformation and thinning of the vein during or after intrusion; (**h**) intrusion jog structure due to jogged geometry of the original fracture before opening. This can have similar geometry as (b) or (e), but has undeformed material in the jog; (**j**) pseudoboudins, which can have older veins in the gap between the boudins, running up to the suture separating them. These structures form by collapse or inflation of a vein without vein-parallel extension; (**k**) solid-state boudins formed by extension of a vein; here, older veins cannot touch the suture between the veins but are linked to the boudins.

the intersection region is homogeneously flattened. However, if the older vein deforms less than the host rock, the intersection region preserves the original high angle to some extend. This leads to a sharp jog in the younger vein where it crosses the older one (Fig. 8b). This kind of structure is commonly observed in shear zones if older veins are subparallel to the flow plane. In many cases the younger vein is thicker in the junction area than in the wallrock as it is less deformed. Vein refraction structures can be confused with *intrusion jogs* formed during intrusion of a vein (Fig. 8h), but refraction structures have deformed and in some cases boudinaged limbs, can be oblique to the foliation in the wallrock and have less sharp hinges at the edge of the older vein (Fig. 8b).

Vein refraction structures may combine with junction structures (Fig. 8c). This kind of structure forms if the intersection area or the entire older vein has a higher competency contrast to the matrix than the younger vein. In the area investigated, this kind of structure only occurs around the youngest, least deformed pegmatite veins. This may be due to a decrease in temperature during late stages of the deformation.

Flanking structures

Some veins show rims of deviating orientation of the foliation or veins in the wallrock along the long axis of the vein (Figs 6f and 8d). Such structures were described by Passchier (2001) as flanking structures. Two types have been described, with either steepening or flattening of elements towards the central vein. In the area investigated here, both types have been observed. Veins branching from a feeder dyke are commonly developed as steepening flanking folds (Fig. 6f). Some veins in metadiorite have a rim of biotite–plagioclase where the hornblende has been transformed to biotite. In these biotite rims, the foliation flattens towards the vein (Fig. 8d, lower part). Such structures can form by either dextral or sinistral rotation of the vein with respect to the foliation (Graseman *et al.* 2003).

Displacement jogs

Some veins show sharp, angular jogs in the vein with a short segment parallel or at a low angle to S_m (Fig. 8e) that can form by localized shear flow, displacing the vein. However, this kind of structure is very similar to some intrusion jogs (Fig. 8h). It is not easy to distinguish both types, but presence or absence of deformation features in the jog is decisive. In the studies area, the intrusion jog type is most common.

Folded-tip structures

Some veins oblique to S_m may seem undeformed with straight sharp contacts that cut S_m. However, where thin side veins branch off such veins their tips may show folding where the large veins do not (Fig. 8f). This type of feature has already been recognized by Ramsay & Huber (1983) and is due to the dependence of fold wavelength on vein thickness (Biot 1961).

Solid-state boudins

Although the recognition of solid-state boudinage may seems trivial, Bons *et al.* (2004) demonstrated the existence of pseudoboudin structures in intrusive veins with a very similar geometry to boudins. Pseudoboudin structures (Fig. 8j) form by collapse or inflation of parts of a vein when it is still magma-filled. The resulting structure has a geometry very similar to solid-state boudins, with one major exception; older veins can pass up to a suture line separating pseudoboudins (Fig. 8j), but in solid-state boudins, all older veins must be connected to the boudins themselves and cannot be joined to the suture line (Fig. 8k).

Clast-in-vein structures

Some narrow veins parallel to S_m show no macroscopic signs of deformation but contain local porphyroclasts that consist of a large single grain of feldspar with a diameter exceeding the thickness of the vein and larger than any feldspar grain in the matrix (Fig. 8g). This structure is particularly common in thin branches of pegmatite veins. Such *clast-in-vein structures* are similar in geometry to mantled clasts, but represent a much smaller strain and are not diagnostic of solid-state deformation; they may have formed during vein intrusion or by later deformation. A magma-filled vein with large phenocrysts of feldspar may have collapsed by migration of the magma while few phenocrysts were left behind, which became compressed between the vein walls. Alternatively, during deformation the vein may have been stretched and thinned but, because of small rheology contrast, no other features formed. The large feldspar crystals, however, were not much reduced in size and show that the vein was initially much wider.

Discussion

Deformation described above is not due to normal movement on the Hefangkau fault zone (Fig. 1), as that zone has in general greenschist facies or lower grade fabric. Early deformation in amphibolite facies in the Sihetang shear zone has been by non-coaxial flow induced by S-directed transport of the hanging wall. This is in the same direction as thrusting on the brittle fault in the hanging wall, but thrust geometry in the present orientation may well have been a normal shear zone at an earlier time, tilted by folding of the entire shear zone over the batholith (cf. Passchier 1984).

Intrusion geometry

One of the most conspicuous properties of the Sihetang shear zone is the dominant parallelism of old, strongly deformed veins and L_m, and the gradual increase in obliqueness of veins to L_m with decreasing age (Fig. 6a). The shear-zone fabric indicates strong ductile extension in a N–S direction, a relative transport of the hanging wall to the south and vertical shortening in the shear zone. If the shear zone operated by simple shear or another monoclinic flow type (Passchier 1998) veins should have opened at right angles to the lineation, as shown in Figure 9a. However, even at very high strain it is impossible to rotate deformed veins into parallelism with the lineation in homogeneous monoclinic flow (Fig. 9a). A possible explanation for the observed structure could be orthogonal intrusion followed by *inhomogeneous* monoclinic flow, where some parts of the shear zone move faster than others (Fig. 9b). In fact, flow in the shear zone is locally triclinic in that case (Jiang & Williams 1998). This would rotate veins into

the observed orientation, but it requires very high strain gradients in the mylonite. A third possibility is that veins already intruded oblique or even at a small angle to the lineation and were rotated in some type of monoclinic flow (Fig. 9c). There are strong indications for this scenario as late pegmatite veins are mostly observed in orientations oblique but not orthogonal to lineation, L (Fig. 6a).

In a simple model, a ductile shear zone operates by a stress field that is symmetrically arranged with respect to the resulting fabric elements and remains unchanged in orientation with respect to the wallrock during the development of the shear-zone fabric. An explanation for initial oblique intrusion of veins into such a simple shear zone could be that veins intrude from the solidified top of the underlying batholith upwards. If the stress field in the underlying batholith differs in orientation from that in the shear zone, the upwards intruding vein tips could simply copy the orientation of veins in the batholith into the shear zone. It is more likely, however, that a propagating vein tip would twist to adapt to the stress field in the shear zone.

A second possibility is that veins intrude as a conjugate set oblique to the principal stress directions in the shear zone. In that case they would have a shear component, i.e. they would not open normal to the plane of the vein. No displacement was observed along the veins except on a pegmatite vein but this applies to outcrop faces normal to L_m. On outcrop faces parallel to L_m and normal to S_m (Fig. 6d) strain is too high to make definite statements, and outcrops parallel

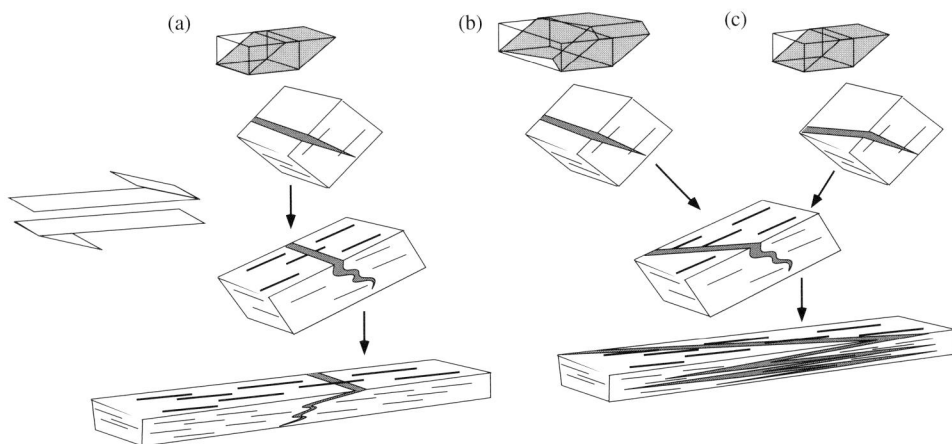

Fig. 9. (a) The present geometry of veins in the shear zone cannot be explained by deformation of veins originally orthogonal to the extension direction in simple shear; they would remain orthogonal. It can be explained if: (**b**) they were orthogonal but deformed in inhomogeneous monoclinic or triclinic flow; or (**c**) they intruded originally oblique to the extension direction. The latter seems to apply in the Sihetang shear zone.

to S_m (Figs 3b and 6a) are small and relatively rare. This type of oblique intrusion of veins would imply that some extension is possible in the plane of the foliation normal to the lineation, and that flow in the shear zone was not by plane strain.

Finally, the assumption of invariable orientation of stress in a developing ductile shear zone may be wrong. Mylonitic deformation is slow, and intrusion of a vein into the zone must be many orders of magnitude faster than the ductile flow. It might be that on the timescale of vein intrusion and on the outcrop-scale stress orientation changes periodically, but that the net result on the ductile fabric gives rise to a 'mean' orientation of stress averaged over the time it takes for the shear zone to develop. During vein intrusion, stress axes could have been significantly oblique to this 'mean' orientation and give rise to oblique vein intrusion, even in tension. In this case, the veins would give a unique insight in the variability of stress orientation throughout the active life of a ductile shear zone.

Shear-zone model

Gently dipping ductile shear zones in core complexes can operate by a variety of flow types. If volume is to be conserved, the simplest possible model is one of simple shear (Fig. 10a). However, if the wallrock is relatively hot, it is also possible that it stretches coaxially while conserving volume in the shear zone (Fig. 10b).

The Sihetang shear zone is also unusual as it cannot have conserved volume. It is clear from field observations that, during the process of deformation, 20–40% vol. of granitic material has been added to the shear zone in the form of dykes. This can simply be established from the surface area of vein material in Figures 4 and 5, and extrapolated to volume due to the cylindrical nature of the folds. The presence of folds, boudins and clast-in-vein structures shows that considerable vertical shortening occurred in the zone during magma emplacement. We have seen above that dykes seem to intrude oblique to S_m and L_m, although numerous branches intruded parallel to S_m. There is in principle no space problem normal to S_m, as the shear zone is gently dipping and the roof of the batholith may simply be uplifted.

The metadioritic rock of the hanging wall away from the Sihetang shear zone has only a weak deformation fabric, and it therefore seems likely that the top of the zone deformed by a regime close to plane strain simple shear. In this case, the rotation of the veins must be entirely due to plane strain non-coaxial progressive deformation, although a gradient of deviation from plane strain

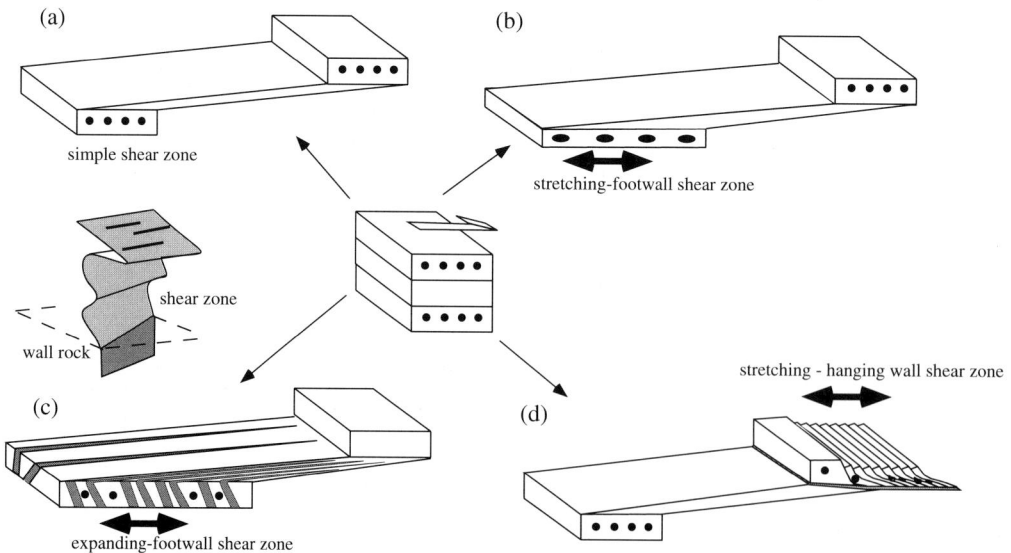

Fig. 10. Possible modes of deformation in a gently dipping ductile shear zone over a core complex: (**a**) simple shear; (**b**) volume-constant homogeneous ductile extension of the footwall and volume constant flow with a vorticity gradient in the shear zone; (**c**) volume increase in the footwall and shear zone by vein intrusion; (**d**) volume-constant brittle extension in the hanging wall, separated from an older simple shear ductile fabric in the footwall by a cataclasite zone. The Sihetang shear zone is of type (c). Inset in (c) shows deformation and rotation of a vein in the shear zone.

towards the footwall cannot be excluded. It may be that veins are injected one by one oblique to L_m and that flow is in reality not plane strain, but has a shortening component parallel to the rotation axis. Such a flow regime would cause a stronger rotation of veins towards parallelism with the lineation than in simple shear.

A strong mylonitic fabric with good planar and linear shape fabric developed in the metadiorite, but also in the top of the granodiorite batholith. Apparently, this part of the batholith was already in a solid state when the shear zone was active. The density of intruded granite veins rapidly decreases upwards from the contact between the metadiorite and the granodiorite. Few veins are found in the shear zone in the meta-diorite at more than 50 m away from the contact, nor in the less deformed and undeformed meta-diorite in the hanging wall. The mylonite zone seems to form an effective lid on the batholith, preventing the escape of magma to higher levels. It could be that the veins have difficulty transecting the highly anisotropic mylonite zone, and that most of the magma travelling up the veins is injected in sheets parallel to the foli-ation. The dominantly folded nature of older veins suggests, however, that they intruded oblique to the foliation in the shear zone. The most likely model for the development of the observed structures is that veins intruded from the granodiorite batholith into the active shear zone in a extensional setting, and were sub-sequently deformed. If the hanging wall is rigid, this can only occur by extension of the footwall to accommodate vein intrusion contemporaneous with relative N–S displacement of footwall and hanging wall along the shear zone (Fig. 10c). Stretching of the footwall rocks would be most effective if it was at a small angle to the shear direction in the shear zone. Such stretching of the footwall could also be accommodated by ductile pure shear in the footwall, and in that case the shear zone would be thinning (Fig. 10b). As there is no ductile deformation fabric in the granodiorite underlying the Sihetang shear zone, we can exclude this possibility.

The picture emerging from the shear zone is therefore one with simple shear in the top, and general non-coaxial flow with volume increase and stretching by vein intrusion towards the foot-wall. This type of shear zone could be referred to as an *expanding-footwall shear zone* (Fig. 10c). Interestingly, this is not the kind of shear zone that is commonly described from metamorphic core complexes. Usually, a brittle fault zone is overlying and reworking older mylonites, with a hanging wall consisting of brittle fault blocks (Fig. 10d) (Armstrong 1982). In many such cases, it is apparently the footwall that was rigid during brittle deformation. If other core complexes have evolved in a similar way to the Yunmengshan, with a stage of abundant vein intrusion, the décollement shear zone of the complex may show a change from ductile and intrusion-related expanding-footwall geometry, to brittle extending hanging wall behaviour, or possibly more complex transitions.

Conclusions

- The Sihetang shear zone operated under amphibolite facies metamorphic conditions in the absence of partial melting.
- A large number of granitic veins, up to 40% of the rock volume, intruded during myloni-tic deformation in the shear zone.
- Veins parallel to the foliation form both by high strain and by foliation-parallel intrusion from feeder dykes.
- Veins seem to have intruded oblique but not orthogonal to the mylonitic lineation L_m.
- Deformed veins form a complex 3D network of late oblique structures and older progress-ively more deformed, folded veins that trend to parallelism with the lineation.
- The shear zone seems to have acted as a repo-sitory for the vein material; few veins breach the roof of the shear zone into overlying host rock material that is of identical composition, but less deformed.
- Deformation in the shear zone must have deviated significantly from simple shear because of the volume increase involved with granite vein intrusion. Most likely is a model with an expanding footwall. This is geometrically the opposite to brittle hanging wall extension in core complexes with a brittle overprint on the ductile shear zone.

This paper is based on fieldwork during short visits to Beijing of C.W. Passchier financed by the Dutch Ministry of Internal Affairs. The project is supported by DFG grants PA578-3 and PA578-5 to CP. J.S. Zhang acknowledges support in project No. 49772153 by the National Natural Science Foundation of China (NSFC). J. Konopásek grate-fully acknowledges support by the Marie Curie Fellowship scheme of the EC (contract No. HPMF-CT-2000-01101). We thank S. Coelho and D. Iacopini for useful sugges-tions, and C. Rosenberg and A. Vauchez for critical and very constructive reviews.

References

ARMSTRONG, R.L. 1982. Cordilleran metamorphic core complexes – from Arizona to southern Canada. *Annual Review of Earth and Planetary Sciences*, **10**, 129–154.

BALDWIN, S.L., LISTER, G.S., HILL, E.J., FOSTER, D.A. & McDOUGALL, I. 1993. Thermochronologic constraints on the tectonic evolution of an active metamorphic core complex, D'Entrecasteaux Islands, Papua New Guinea. *Tectonics*, **12**, 611–628.

BIOT, M.A. 1961. Theory of folding of stratified viscoelastic media and its implications in tectonics and orogenesis. *Geological Society of America Bulletin*, **72**, 1595–1620.

BONS, P.D., DRUGUET, E., HAMAN, I., CARRERAS, J. & PASSCHIER, C.W. 2004. Apparent boudinage in dykes. *Journal of Structural Geology*, **26**, 625–636.

CABY, R., HAMMOR, D. & DELOR C. 2001. Metamorphic evolution, partial melting and Miocene exhumation of lower crust in the Edough metamorphic core complex, west Mediterranean orogen, eastern Algeria. *Tectonophysics*, **342**, 239–273.

DAVIS, G.A., QIAN, X. *ET AL.* 1996. Mesozoic deformation and plutonism in the Yunmeng Shan: a metamorphic core complex north of Beijing, China. *In*: YIN, A. & HARRISON, M. (eds) *Tectonic Evolution of Asia.* Cambridge University Press, New York, 253–280.

FLETCHER, J.M. & BARTLEY, J.M. 1994. Constrictional strain in a noncoaxial shear zone – implications for fold and rock fabric development, central Mojave metamorphic core complex, California. *Journal of Structural Geology*, **16**, 555–570.

FOSTER, D.A. & FANNING, C.M. 1997. Geochronology of the northern Idaho batholith and the Bitterroot metamorphic core complex: magmatism preceding and contemporaneous with extension. *Geological Society of America Bulletin*, **109**, 379–394.

FOSTER, D.A., SCHAFER, C., FANNING, C.M. & HYNDMAN, D.W. 2001. Relationships between crustal partial melting, plutonism, orogeny, and exhumation: Idaho–Bitterroot batholith. *Tectonophysics*, **342**, 313–350.

GOSCOMBE, B.D. & PASSCHIER, C.W. 2003. Asymmetric boudins as shear sense indicators – an assessment from field data. *Journal of Structural Geology*, **25**, 575–589.

GRASEMAN, B., STÜWE, K. & VANNAY, J.-C. 2003. Sense and non-sense of shear in flanking structures. *Journal of Structural Geology*, **25**, 19–34.

HOLLAND, T.J.B. & BLUNDY, J. 1994. Non-ideal interactions in calcic amphiboles and their bearing on amphibole–plagioclase thermometry. *Contributions to Mineralogy and Petrology*, **116**, 433–447.

JIANG, D. & WILLIAMS, P.F. 1998. High strain zones: a unified model. *Journal of Structural Geology*, **21**, 933–937.

LISTER, G.S., BANGA, G. & FEENSTRA, A. 1984. Metamorphic core complexes of the Cordilleran type in the Cyclades, Aegean Sea, Greece. *Geology*, **12**, 221–225.

PASSCHIER, C.W. 1984. Mylonite-dominated footwall geometry in a low-angle fault culmination, Central Pyrenees. *Geological Magazine*, **121**, 429–436.

PASSCHIER, C.W. 1998. Monoclinic model shear zones. *Journal of Structural Geology*, **20**, 1121–1137.

PASSCHIER, C.W. 2001. Flanking structures. *Journal of Structural Geology*, **23**, 951–962.

PASSCHIER, C.W. & TROUW, R.A.J. 1996. *Microtectonics.* Springer, Berlin.

PATERSON, S.R., VERNON, R.H. & TOBISCH, O.T. 1989. A review of criteria for the identification of magmatic and tectonic foliations in granitoids. *Journal of Structural Geology*, **11**, 349–363.

PATERSON, S.R., FOWLER T.K., JR., SCHMIDT, K.L., YOSHINOBU, A.S., YUAN, E.S. & MILLER, R.B. 1998. Interpreting magmatic fabric patterns in plutons. *Lithos*, **44**, 53–82.

PAWLEY, M.J., COLLINS, W.J. & VAN KRANENDONK, M.J. 2002. Origin of fine-scale sheeted granites by incremental injection of magma into active shear zones: examples from the Pilbara Craton, NW Australia. *Lithos*, **61**, 127–139.

PE-PIPER, G., PIPER, D.J.W. & MATARANGAS, D. 2002. Regional implications of geochemistry and style of emplacement of Miocene I-type diorite and granite, Delos, Cyclades, Greece. *Lithos*, **60**, 47–66.

PIAZOLO, S. & PASSCHIER, C.W. 2002. Controls on lineation development in low to medium grade shear zones: a study from the Cap de Creus peninsula, NE Spain. *Journal of Structural Geology*, **24**, 25–44.

RAMSAY, J.G. & HUBER, M.I. 1983. *The Techniques of Modern Structural Geology. Volume 2, Folds and Fractures.* Academic Press, London.

REYNOLDS, S.J. & LISTER, G.S. 1987. Folding of mylonite zones in Cordilleran metamorphic core complexes: evidence from near the mylonitic front. *Geology*, **18**, 21–219.

STIPP, M., STUNITZ, H., HEILBRONNER, R. & SCHMID, S.M. 2002. The eastern Tonale fault zone: a 'natural laboratory' for crystal plastic deformation of quartz over a temperature range from 250 to 700 degrees C. *Journal of Structural Geology*, **24**, 1861–1884.

WALCOTT, C.R. & WHITE, S.H. 1998. Constraints on the kinematics of post-orogenic extension imposed by stretching lineations in the Aegean region. *Tectonophysics*, **298**, 155–175.

ZURBRIGGEN, R., KAMBER, B.S., HANDY, M.R. & NAGLER, T.F. 1998. Dating synmagmatic folds: a case study of Schlingen structures in the Strona-Ceneri Zone (Southern Alps, northern Italy). *Journal of Metamorphic Geology*, **16**, 403–414.

Laramie Peak shear system, central Laramie Mountains, Wyoming, USA: regeneration of the Archean Wyoming province during Palaeoproterozoic accretion

PHILLIP G. RESOR[1,2] & ARTHUR W. SNOKE[1]

[1]*Department of Geology and Geophysics, Department 3006, 1000 East University Avenue, University of Wyoming, Laramie, WY 82071, USA (e-mail: snoke@uwyo.edu)*
[2]*Present address: Department of Earth and Environmental Sciences, Wesleyan University, 265 Church Street, Middletown, CT 06459-0139, USA*

Abstract: The Laramie Peak shear system (LPSS) is a 10 km-thick zone of heterogeneous general shear (non-coaxial) that records significant tectonic regeneration of middle–lower crustal rocks of the Archean Wyoming province. The shear system is related to the 1.78–1.74 Ga Medicine Bow orogeny that involved the collision of an oceanic-arc terrane (Colorado province or Green Mountain block or arc) with the rifted, southern margin of the Wyoming province. The style and character of deformation associated with the LPSS is distinctive: a strong, penetrative (mylonitic) foliation commonly containing a moderately steep, SW-plunging elongation lineation. In mylonitic quartzo-feldspathic gneisses of the Fletcher Park shear zone, shear-sense indicators indicate southside-up, and this interpretation is supported by metamorphic and geochronological studies across the LPSS. We argue that distributed general shear (non-coaxial) involving high-strain zones and multiple folding events yielded a broad, en-masse uplift (Palmer Canyon block) during the late stages of the Medicine Bow orogeny. The LPSS is thus an excellent example of how crystal-plastic strain is distributed in sialic crust during an oceanic arc–continental margin collision. As magmatism (and attendant thermal softening) did not occur in the Wyoming province during its partial subduction beneath the oceanic-arc terranes of the Colorado province, the crystal-plastic strain manifested within the Wyoming province is mechanical in nature and was concurrent with crustal thickening. Strain is localized into discrete shear zones separated by weakly deformed rocks. These high-strain zones are commonly located along contacts between differing rock types and we propose that mechanical and chemical weakening processes may have contributed to strain localization.

The rheological behaviour of the middle and lower crust is commonly cited as playing an important role in the mechanics and geometry of continental plate tectonics (e.g. Molnar 1988; Northrup 1996; Axen *et al.* 1998; Beaumont *et al.* 2001; Jackson 2002; McKenzie & Jackson 2002). In many cases, lower crustal deformation is inferred to involve broad-scale, viscous flowage; e.g. Basin and Range province of western North America (Gans 1987; Block & Royden 1990; Wernicke 1990; Kruse *et al.* 1991). This lower crustal flow appears to compensate for large variations in upper crustal deformation (extension in the case of the Basin and Range province) leading to relatively constant crustal thickness and a flat-lying Moho (Klemperer *et al.* 1986; Hauser *et al.* 1987). However, in other cases, the deep crust does

not appear to flow over large areas and Moho topography is maintained over long periods (e.g. Alps of western Europe: Bois & ECORS Scientific Party 1990). It has been suggested that this variation in the behaviour of the deep crust is associated with the presence or absence of tectonic thickening and/or syntectonic magmatism – processes that raise the temperature of the lower crust and facilitate its viscous flowage (e.g. McKenzie & Jackson 2002).

This general model of lower crustal deformation is based in a large part on experimental studies, mechanical modelling and geophysical observations. Surface exposures of rocks that previously occupied lower crustal depths are relatively scarce. In this paper we present the results of detailed field mapping of one such exposure of deep crustal rocks from the

From: BRUHN, D. & BURLINI, L. (eds) 2005. *High-Strain Zones: Structure and Physical Properties.*
Geological Society, London, Special Publications, **245**, 81–107.
0305-8719/05/$15.00 © The Geological Society of London 2005.

Laramie Mountains of SE Wyoming, USA. The Laramie Mountains expose Archean and Protero-zoic rocks that are interpreted to have occupied middle–lower crustal depths (>25 km) during an inferred Palaeoproterozoic arc–continent collision (i.e. the 1.78–1.74 Ga Medicine Bow orogeny). In contrast to many previous studies of deep crustal exposures (e.g. MacCready et al. 1997; Klepeis et al. 2004; Karlstrom & Williams 2005) these rocks did not experience syntectonic magmatism, and thus the strain developed in a purely mechanical fashion without the role of magmatic softening as a deformational agent.

We argue that field observations demonstrate that deformation in the lower crust was characterized by strain localization with discrete shear zones separated by undeformed–weakly deformed rocks. Furthermore, we suggest that this style of deformation is typical of lower crustal deformation in the absence of syntectonic magmatism and thus presents an opportunity to directly observe lower crustal behaviour where broad-scale flowage is absent. Finally, we discuss the implications of these observations to tectonic models of arc–continent collisions. These results have implications for inferred Phanerozoic arc–continent collisions where a similar geological situation has been proposed (e.g. Dewey & Bird 1970; Speed & Sleep 1982; Suppe 1987; Snyder et al. 1996) as well as for general models of collisional orogenesis (e.g. Beaumont & Quinlan 1994; Snyder 2002).

Geological setting and structural subdivisions

The Precambrian rocks of SE Wyoming are exposed in late Cretaceous–early Eocene (Laramide) basement-involved uplifts (Fig. 1). These uplifts typically are bound on one or more of their flanks by a deep-rooted, reverse and/or thrust fault system that developed during the contractional Laramide orogeny. Three ranges (E–W) expose much of the Precambrian rocks of the area: Laramie Mountains, Medicine Bow Mountains and Sierra Madre. Other less extensive exposures of Precambrian rocks in the region include the Elk, Coad and Pennock mountains, Casper Mountain, Richeau and Cooney hills, and Hartville uplift (Fig. 1). The basement rocks exposed in these ranges record a long and complex Precambrian history that probably began in the Middle Archean(?) and continued into the Mesoproterozoic. Numerous papers summarize parts of this history, but especially useful review articles include Karlstrom &

Houston (1984), Duebendorfer & Houston (1987), Houston et al. (1989, 1993) Houston (1993) and Chamberlain (1998).

A key tectonic event in the Precambrian geological evolution of SE Wyoming, USA, is the Medicine Bow orogeny (1.78–1.74 Ga) interpreted as a collisional orogeny involving the Colorado province (Palaeoproterozoic oceanic arc and associated rocks; also called the Green Mountain block or arc of Karlstrom & the CD-ROM Working Group 2002; Tyson et al. 2002, respectively) and the Archean Wyoming province (rifted continental margin) (see Chamberlain 1998 for a recent summary). This collisional orogeny is commonly interpreted as the initial accretion event along the southern margin of the Wyoming province (however, see Sims 1995; Bauer & Zeman 1997; Hill & Bickford 2001; Chamberlain et al. 2002; Sims & Stein 2003 for comments on a possible earlier collisional history) and is thus considered the harbinger of the eventual addition of approximately 1200 km of juvenile Palaeoproterozoic crust to the Wyoming craton (Condie 1982; Nelson & DePaolo 1985; Reed et al. 1987; Karlstrom & Bowring 1988; Bowring & Karlstrom 1990; Van Schmus et al. 1993).

Based on petrological, structural and geochro-nological studies (e.g. Condie & Shadel 1984; Karlstrom & Houston 1984; Reed et al. 1987), previous workers have suggested a tectonic model in which an inferred early S-dipping sub-duction zone accommodated northward migration (in present geographic co-ordinates) of an oceanic arc (i.e. Colorado province rocks). The Wyoming province was thus drawn into the subduction zone during collision (Houston 1993, fig. 14F). Subsequent uplift and erosion has exposed rocks that occupied deep crustal levels in the Palaeopro-terozoic (Chamberlain et al. 1993).

The Laramie Peak shear system is located in the Laramie Mountains approximately 50 km north of the inferred trace of the Cheyenne belt, i.e. the suture zone along which the Proterozoic Colorado province and Archean Wyoming pro-vince are juxtaposed (Duebendorfer & Houston 1987). Chamberlain et al. (1993) originally coined the name 'Laramie Peak shear zone', although portions of the shear zone (system) had been recognized by previous workers (Condie 1969; Segerstrom et al. 1977; Langstaff 1984; Snyder et al. 1995). Chamberlain et al. (1993) delineated the Laramie Peak shear zone as consisting chiefly of the Garrett–Fletcher Park shear zone but also noted a splay shear zone that they called the Cottonwood Park shear zone (Fig. 1) (Chamberlain et al. 1993, fig. 2). These zones of concentrated ductile

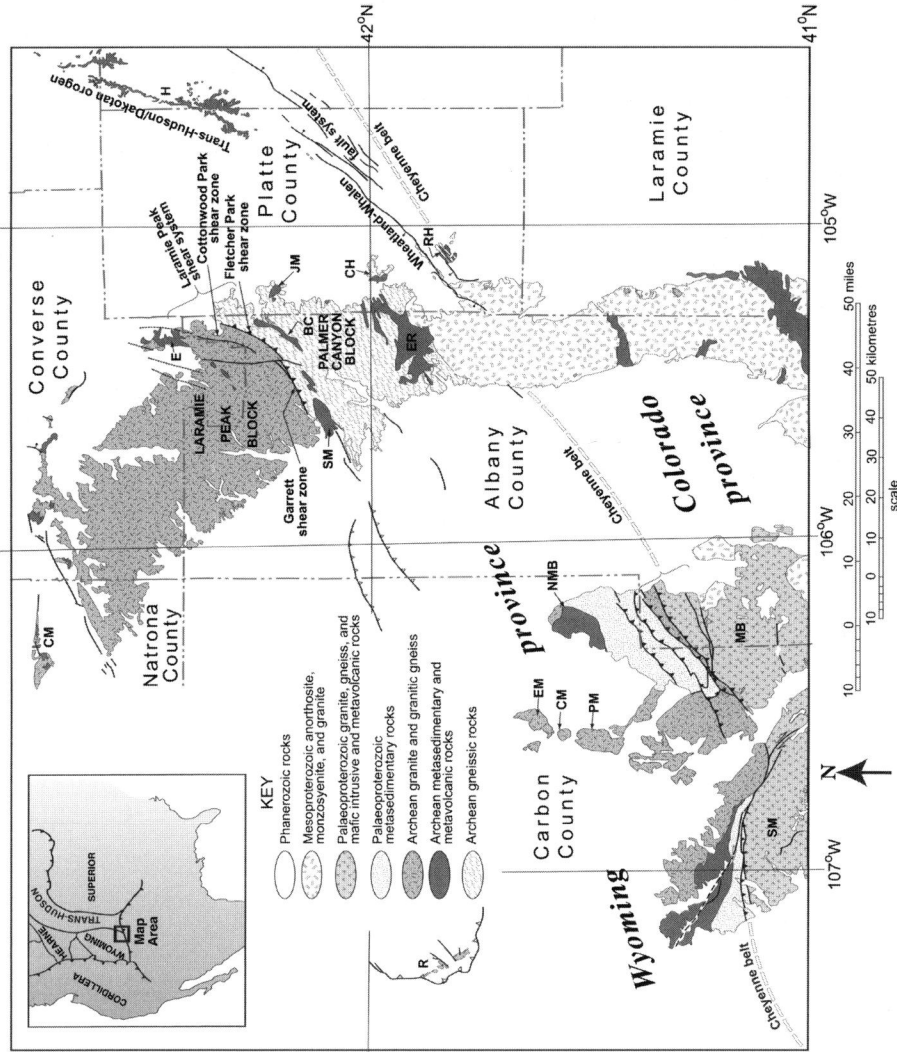

Fig. 1. Geological index map of Precambrian rocks and selected Phanerozoic structural features in SE Wyoming (base derived from Love & Christiansen 1985; Snyder *et al.* 1995). Inset map was derived from Hoffman (1989, fig. 13). Abbreviations: fragmentary supracrustal sequences (metasedimentary and metavolcanic rocks) – ER, Elmers Rock; BC, Brandel Creek; JM, Johnson Mountain; E, Esterbrook; SM, Sellers Mountain; topographic features – RH, Richleau Hills; CH, Cooney Hills; H, Hartville uplift; CM, Casper Mountain; NMB, Northern Medicine Bow Mountains; MB, Medicine Bow Mountains; SM, Sierra Madre; EM, Elk Mountain; CM, Coad Mountain; PM, Pennock Mountain; R, Rawlins uplift.

Fig. 2. Structural geological map of a part of the Laramie Peak shear system, central Laramie Mountains, Wyoming. See Figure 3 for an explanation of rock units and symbols. NLR, North Laramie River. The map area of Figure 3 is delineated. The area described by Resor *et al.* (1996) is delineated, which includes the mafic dyke (Xad) from which syntectonic titanite was extracted and radiometrically dated using U–Pb techniques. Structural domains (subareas 1–4) are also delineated. Base derived from the South Mountain and Fletcher Park, Wyoming, US Geological Survey 7.5-minute quadrangles. Geology mapped by P.G. Resor and A.W. Snoke (1993–1998, non-inclusive). Some geological data have been adapted from Snyder *et al.* (1995).

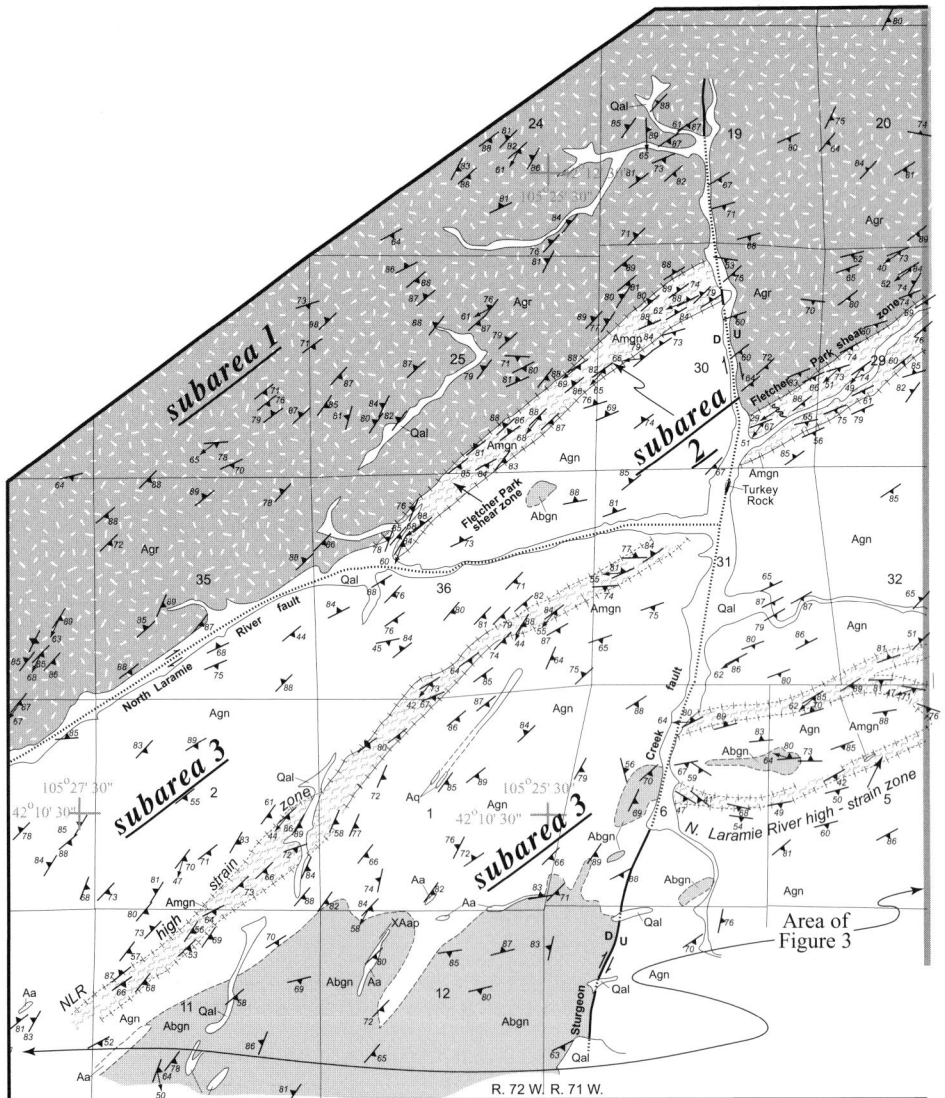

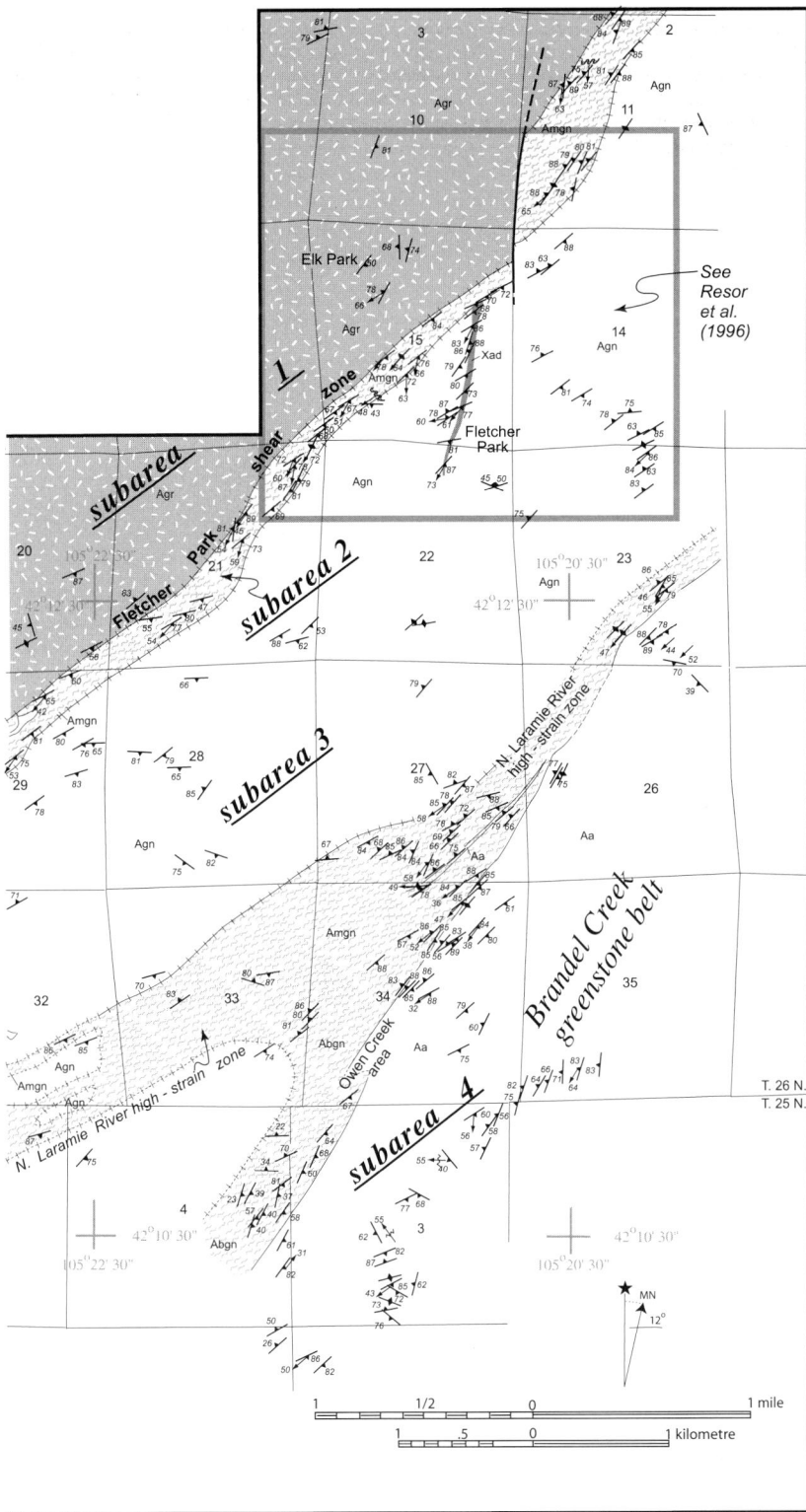

deformation are inferred to be of Palaeoprotero-zoic age, although only the crystal-plastic defor-mation associated with the Fletcher Park shear zone, as exposed in Fletcher Park (Fig. 2), has been directly dated with radiometric age techniques (c. 1.76 Ga, Resor et al. 1996).

In this paper we expand the overall width of the belt of ductile deformation associated with the Laramie Peak shear zone by including another NE–SW-striking zone of concentrated shear strain, the North Laramie River high-strain zone, that occurs SE of the Fletcher Park shear zone (Fig. 2). Recently, Allard (2003) has recog-nized additional evidence of heterogeneous, high strain deformation to the south and east of our map area, which he correlates with the same deformational phase as manifested in the Laramie Peak shear zone of Chamberlain et al. (1993). Furthermore, fabric elements (NE–SW-striking penetrative foliation and SW-trending elongation lineation) characteristic of the shear zone deformation also occur between the deli-neated zones of concentrated shear strain as well as sporadically SE of these zones, suggesting a broad, albeit heterogeneous, zone of ductile deformation associated with the LPSS. Given this significant expansion in distribution of high strain rocks, as well as to avoid future confusion in nomenclature, we have designated all these zones of high strain as the 'Laramie Peak shear system (LPSS)' in this paper. Within this shear system individual shear zones are given local names as delineated in Figures 1 and 2.

The LPSS is a significant structural, meta-morphic and geochronological boundary. The rocks within this broadly defined zone of high strain have experienced upper amphibolite facies dynamic metamorphism and have been locally converted to well-foliated and lineated mylonitic gneisses (typically an L–S tectonite but commonly L > S). Deformation was at a relatively high temperature (>450 °C), and the mylonitic rocks exhibit widespread evidence of recovery and recrystallization.

Rocks to the north of the Garrett–Fletcher Park shear zone are largely undeformed. These rocks have been referred to as the Laramie batho-lith by Condie (1969) or Laramie Peak block by Patel et al. (1999) (Fig. 1). Narrow zones (typi-cally <1 m wide) of concentrated strain are locally present and perhaps are also related to the shear strain deformation associated with the LPSS. Metamorphic grade is low-pressure amphibolite facies with pressure–temperature (P–T) estimated at c. 2.5 kb and c. 575 °C (Patel et al. 1999). The metamorphism north of the Garrett–Fletcher Park shear zone is inter-preted as Archean, based on U–Pb apatite cooling ages that are >2.1 Ga (Chamberlain

et al. 1993; Patel et al. 1999; Chamberlain & Bowring 2000).

Rocks to the south of the Garrett–Fletcher Park shear zone are variably deformed and have been referred to as the 'central meta-morphic complex' (Condie 1969) or the Palmer Canyon block (Patel et al. 1999). Rocks in this block include granitic gneiss–gneissic granite (chiefly Archean in crystallization age but also include subordinate, pretectonic Palaeoprotero-zoic granitic rocks: Snyder et al. 1995), Archean migmatitic banded gneiss, Archean supracrustal rocks, ultramafic rocks, and Archean(?) and Palaeoproterozoic mafic dykes. Locally these rocks are strongly deformed and partitioned by anastomosing zones of high strain. Metamorphic grade is upper amphibolite with P–T estimated at c. 7.5 kb and c. 625 °C (Patel et al. 1999). Apatite cooling ages within and south of the Garrett–Fletcher Park shear zone are less than 1.78 Ga (Chamberlain et al. 1993; Patel et al. 1999; Chamberlain & Bowring 2000). In this light, the LPSS is a zone of Palaeoproterozoic high strain that defines the northern margin of the Palmer Canyon block. Consequently, the LPSS is the 'deformation front' for Palaeoproterozoic strain in the Archean Wyoming province.

Rock units

Although Archean granitic rocks constitute the bulk of the rock types exposed in the central Laramie Mountains, a diverse variety of other rock types provide an unusual spectrum of lithological variation including: an Archean supracrustal sequence, ultramafic rocks, Palaeo-proterozoic (pretectonic) granitic rocks, and Archean(?) and Palaeoproterozoic mafic dykes (Fig. 3). Geochronometric data indicate that these rocks were deposited or intruded from the Middle Archean(?) through to the Palaeoproterozoic.

Granitic rocks

Migmatitic banded gneiss. The migmatitic banded gneiss is considered the oldest rock unit in the area (Johnson & Hills 1976; Snyder et al. 1995) and is most extensively exposed in our study area south of the Fletcher Park shear zone (Figs 2 and 3). Johnson & Hills (1976) reported Rb–Sr whole-rock isochrons of 2759 ± 152 Ma for grey granitic gneiss and 2776 ± 35 Ma for leucogranite from a similar unit in the northern Laramie Mountains. The migmatitic banded gneiss is characterized by compositional banding defined chiefly by vari-ations in biotite content. Foliation is parallel to compositional banding, and mineral lineation is uncommon in these rocks. The compositional banding is commonly folded and cut by leucogranite layers. These layers are interpreted as melt and/or magma derived from *in situ* anatexis of the enclosing gneiss during high-temperature, Late Archean metamorphism (Allard 2003). The modal composition of the leucocratic layers indicates a minimum melt composition consisting of about equal parts plagioclase feldspar, alkali feldspar and quartz; other phases only account for <1% of the modal mineral content. The gneiss, exclusive of the leucocratic layers, is tonalitic with plagioclase feldspar typically accounting for approximately 60% of the mineral content of the rock. The remaining mineral phases are approximately 30% quartz, 2% alkali feldspar and 8% biotite. The migmatitic banded gneiss is intruded by variably deformed gneissic granite (Late Archean and/or Palaeoproterozoic: Allard 2003), and the migmatitic banded gneiss commonly occurs as enclaves (<1 to >100 m in longest dimension) within gneissic granite.

Undeformed–weakly deformed granite. North of the Garrett–Fletcher Park shear zone (i.e. Laramie Peak block of Patel *et al.* 1999) the predominant granitic rock is an undeformed–weakly deformed granite. The most common rock type is a megacrystic, two-feldspar biotite monzogranite previously referred to as the 'Laramie granite' by Johnson & Hills (1976). These authors reported a Rb–Sr whole-rock isochron age of 2567 ± 25 Ma for the granite (Johnson & Hills 1976). Condie (1969) mapped a vast area as the 'Laramie batholith' that included the granitic rocks north of the Garrett–Fletcher Park shear zone as exposed in our map area. These granitic rocks are commonly intruded by mafic dykes (either Archean(?) and/or Palaeoproterozoic in age) and scarce ultramafic bodies (e.g. Elk Park peridotite of Snyder *et al.* 1995), and locally contain supracrustal enclaves (wallrock xenoliths).

Variably deformed gneissic granite. The variably deformed gneissic granite is the most heterogeneous unit in the map area. The gneissic granite occurs south of the Fletcher Park shear zone but north of the migmatitic banded gneiss (Figs 2 and 3). Thus, it is part of the 'central metamorphic complex' of Condie (1969) or 'igneous and metamorphic complex of Laramie River' of Snyder *et al.* (1995, 1998) or Palmer Canyon block of Patel *et al.* (1999). However, at least parts of this unit bear a marked affinity to granitic rocks north of the Fletcher Park shear zone, and thus these variably deformed

granitic rocks may be deeper level (and deformed) equivalents of the Late Archean 'Laramie granite'. On the other hand, some of the granitic rocks that comprise this unit are distinctly different from any granitic rock type in the 'Laramie granite' and may not even be Archean in age. Snyder *et al.* (1995) reported a U–Pb zircon age of 2051 ± 9 Ma for a distinctive granodiorite pluton in the east-central Laramie Mountains (SE of our map area). Thus, we view the 'variably deformed gneissic granite' as a composite unit consisting of deformed Archean granitic rocks but probably also including younger granitic rocks of possible Palaeoproterozoic crystallization age.

The granites of this unit appear to have an intrusive contact relationship with the migmatitic banded gneiss. The contact is highly irregular (e.g. Fig. 3) and enclaves of migmatitic banded gneiss occur in the variably deformed gneissic granite. The gneissic granite can be seen locally cutting across the fabrics of the migmatitic banded gneiss. Compositionally, the granitic rocks are chiefly monzogranitic containing approximately equal proportions of plagioclase feldspar and alkali feldspar in addition to 25–40% quartz; biotite is the typical varietal mineral and garnet is an occasional accessory phase. Virtually all the granitic rocks south of the Fletcher Park shear zone exhibit some evidence of solid-state deformation. The fabric in these granitic rocks is predominantly defined by elongated quartz and feldspar grains, and in this light it is not uncommon for the deformed granitic rocks to be L > S tectonites.

Mylonitic granitic gneiss. Mylonitic granitic gneiss is found in shear zones throughout the map area, but it was only mapped as a discrete unit in the two largest shear zones – Fletcher Park shear zone and North Laramie River high-strain zone (described in the section 'Structural features and rock fabrics'). The mylonitic granitic gneiss was derived from the other granitic units during high-temperature (>450 °C) crystal-plastic strain. Some outcrops of mylonitic granitic gneiss are characterized by a distinctive pinstripe layering defined by mineralogical and grain-size variations (Fig. 4a). The mylonitic gneisses are typically L–S tectonites with strong mylonitic foliation (fluxion structure) that invariably contains a moderately steep-plunging elongation lineation. Recovery and recrystallization outpaced or kept up with deformation, thus, at the grain scale, evidence of crystal-plastic strain commonly is not well developed. However, at the outcrop scale there is abundant evidence that these rocks experienced substantial strain during their deformational

Geologic map of the
South Mountain area,
Albany County,
Wyoming, U.S.A.

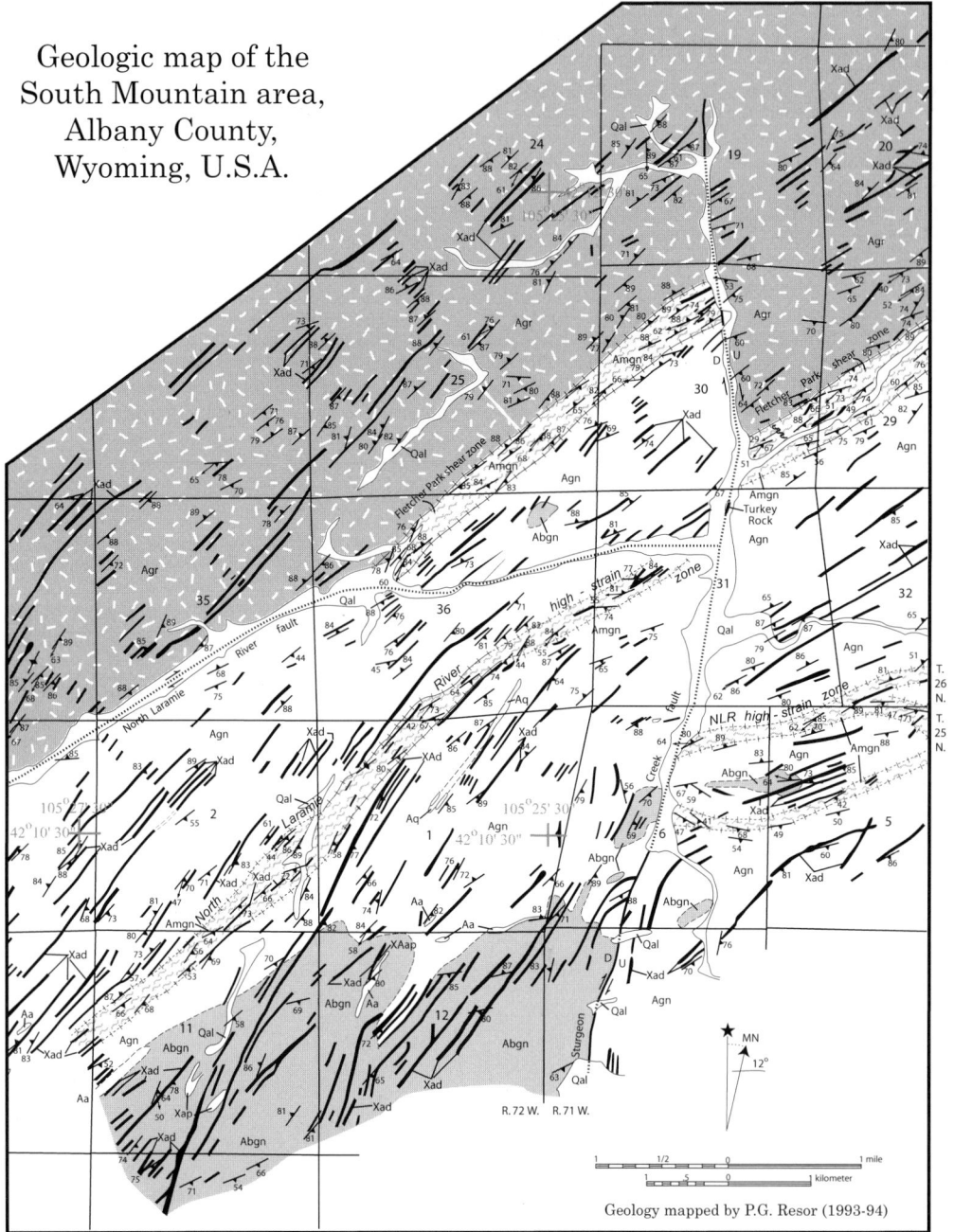

Geology mapped by P.G. Resor (1993-94)

EXPLANATION

SYMBOLS

Lithological boundary, dashed where approximate

Gradational boundary, dashed where approximate

Fault, dotted where concealed; **D** on downthrown side and **U** on upthrown side, arrows indicate relative displacement along fault

Strike and dip of foliation (left), vertical foliation (right)

Trend and plunge of elongation lineation; combined with foliation

Trend and plunge of hinge line of minor fold

Mylonitic foliation associated with the Laramie Peak shear system

ROCK UNITS

Quaternary deposits

Qal — Quaternary alluvium: Sand and gravel deposits along modern stream valleys

Precambrian rocks

Xad — Diabasic amphibolite dyke (Palaeoproterozoic?). Black- to brown-weathering dykes from <1m to 30-m-thick. Dykes are variably metamorphosed and deformed. A few dykes contain plagioclase megacrysts

Xap — Altered peridotite (Palaeoproterozoic?). Green actinolite–chlorite–magnetite rock

Agr — Granite (Archean to Palaeoproterozoic?). Pink to grey, medium- to coarse-grained granite with subordinate aplite, alaskite, and pegmatite. Commonly contains alkali-feldspar megacrysts; locally contains garnet. Unfoliated to weakly foliated with the exception of small-scale ductile shear zones

Agn — Gneiss (Archean). Similar to Agr but generally foliated, sometimes mylonitic. Contains enclaves of Abgn ranging from cm to 100 m scale

Amgn — Mylonitic gneiss (Archean protolith but Palaeoproterozoic crystal-plastic strain). Strongly foliated and lineated pink to grey mylonitic gneiss commonly with feldspar porphyroclasts. Pinstripe layering is distinctive but not ubiquitous

Aq — Quartzite (Archean). Brown, massive quartzite

Aa — Amphibolite (Archean). Black, massive or compositionally layered fine- to medium-grained amphibolite

Abgn — Migmatitic banded gneiss (Archean). Grey and white to pink banded gneiss with leucosome segregations and dykelets. Compositional layering with a biotite foliation; lineation is generally absent. Compositional layering is commonly folded

Fig. 3. Geological map of the South Mountain area, Albany County, Wyoming, USA. NLR, North Laramie River. Base derived from the South Mountain, Wyoming, US Geological Survey 7.5-minute quadrangle. Adapted from Resor (1996).

Fig. 4. (a) Pinstripe foliation in mylonitic granitic gneiss, Fletcher Park shear zone, (b) Near isocline fold preserved in mylonitic granitic gneiss, Fletcher Park shear zone. See the text for discussion of these features.

history. Such evidence includes: (1) shear-sense indicators (especially feldspar porphyroclasts); (2) strong linear fabric, as well as pervasive foliation development (Fig. 4a); (3) relative grain size reduction compared to inferred protolith rock types; (4) highly elongate grains and grain aggregates; and (5) isoclinal folding of compositional layering and mylonitic foliation (Fig. 4b).

Supracrustal rocks

The largest, continuous exposure of supracrustal rocks in the map area occurs in the Brandel–Owen Creeks area (Figs 1 and 2; part of the Brandel Creek greenstone belt of Snyder *et al.* 1995, fig. 2) and is part of the Bluegrass Creek metamorphic suite (Snyder *et al.* 1995, 1998). This supracrustal sequence is dominated by amphibolite. Other rock types that are part of this supracrustal sequence include porphyroblastic pelitic schist, metamorphosed ultramafic rocks, quartzofeldspathic gneiss (probable felsic metavolcanic and/or wacke protolith), banded iron formation, gedrite schist, marble (scarce) and quartzite (scarce) (Snyder *et al.* 1995). These supracrustal rocks must have been deposited prior to *c.* 2.6 Ga as required by the granitic rocks that intrude

them (Snyder *et al.* 1998). This conclusion is supported by U–Pb zircon ages of 2637 ± 10 and 2729 ± 62 Ma from felsic metavolcanic rocks of the metamorphic suite of Bluegrass Creek (Snyder *et al.* 1998).

Although porphyroblastic pelitic schist is not a common component of the supracrustal sequence, these rocks contain mineral assemblages that provide the best $P–T$ estimates from the area. Patel *et al.* (1999) demonstrated the following key conclusions derived from $P–T$ data collected on metapelitic schists of the Bluegrass Creek metamorphic suite (of Snyder *et al.* 1998): (1) the metapelitic rocks exposed in the Owen Creek area reached $P \approx 7.5$ kb (above the GRAIL reaction; i.e. subassemblage: garnet–rutile–aluminosilicate–ilmenite–quartz); (2) the metapelitic rocks are not migmatitic, so that the peak metamorphic temperature must have been *below* the water-saturated melting curve; and (3) textural evidence of decompression (from >7 to <3 kb) and retrograde re-equilibration is widespread in these schists.

Ultramafic rocks

Scattered bodies of ultramafic rocks occur throughout the central Laramie Mountains (e.g. Snyder *et al.* 1998), and their degree of alteration is variable. Resor (1996) described two small ultramafic bodies in the vicinity of Menter Draw (Fig. 3, sections 11 and 12, T. 25N R. 72W). These bodies occur as isolated pods up to 200 m in length and are completely altered to an assemblage of actinolite–chlorite–magnetite. Fine-grained intergrowths of chlorite and actinolite may be pseudomorphs after original igneous olivine. The relative age relationship between these altered peridotite bodies and the diabasic amphibolite dyke swarm could not be determined. However, in the Elk Park area north of the Fletcher Park shear zone (Fig. 2), Snyder *et al.* (1995) mapped a body of altered peridotite that is intruded by a diabasic amphibolite dyke, but the contact of the peridotite truncates several, older diabasic amphibolite dykes. Thus, these map relations suggest a contemporaneity between the peridotite body and diabasic amphibolite dyke swarm.

Diabasic amphibolite dykes

Diabasic amphibolite dykes are widespread throughout the map area (Figs 2 and 3) and constitute the second most abundant rock type beyond the granitic rocks of the area. These mafic dykes are interpreted to be chiefly members of the *c.* 2.01 Ga Kennedy dyke swarm (Graff *et al.* 1982; Cox *et al.* 2000). The diabasic amphibolite dykes form distinctive low, dark-coloured ridges that can be commonly tracked across the landscape. The dykes vary from <1 m to *c.* 30 m thick. Dyke lengths are also quite variable from less than 100 m to more than 2 km. In Figure 3, we have shown mafic dykes that were actually mapped out and not inferred from scattered outcrops. Thus, our distribution of mafic dykes represents a *minimum* of mafic dykes exposed in the map area. We estimate that mafic dykes probably form at least 15% of the bedrock geology in our map area. Most of the mafic dykes strike roughly NE-ward and dip steeply to the south. Two notable exceptions are a thick dyke found south of Menter Draw (Fig. 3; sections 11 and 14, T. 25N R. 72W) and the dyke exposed in Fletcher Park (see Fig. 2 for exact locality), which was the subject of a geochronological study (Resor *et al.* 1996). These dykes strike northerly.

The diabasic amphibolite dykes were originally pyroxene-bearing diabase. This conclusion is demonstrable in the field as well as under the microscope in that relict pyroxene-bearing diabase occurs sporadically in several of the thicker dykes, including the mafic dyke exposed in Fletcher Park (Fig. 2) (Resor *et al.* 1996). With progressive metamorphism and deformation, the mafic dykes transform from (1) igneous-textured, pyroxene-bearing diabase to (2) massive, relict igneous-textured amphibolite to (3) lineated and foliated amphibolite (commonly an L > S tectonite) and finally to (4) medium-grained, sometimes banded, amphibolite. This mineralogical and textural transformation resembles the classic metamorphic transformation of dolerite to hornblende schist (amphibolite) originally described by Teall (1885) for Scourie dykes affected by Laxfordian shear zones in NW Scotland.

Diabasic amphibolite dykes that are penetratively deformed and are now either amphibolite or hornblende schist and range from virtually pure L tectonite to L–S tectonite with gradations between these end members (Fig. 5a–c). Thus, the strain regime appears to vary from plane strain ($k = 1$) to pure shear of the constrictional type ($k \gg 1$; Flinn 1994). An interesting aspect of some of the deformed diabasic amphibolite dykes is that the dyke contact with wallrock has locally acted as a strain guide with the strongest deformation localized along dyke walls. Thus, the deformation in the dykes is clearly solid state rather than igneous flow.

Several diabasic amphibolite dykes in the study area contain granitic wallrock breccia (Fig. 5d). These breccias can serve as relative strain indicators. In relatively undeformed dykes, the breccia consists of roughly equidimensional, subrounded, granitic clasts within a matrix of mafic dyke rock. In more highly deformed dykes, the clasts are stretched into prolate–triaxial ellipsoids with their long axes parallel to the mineral lineation in the surrounding schistose amphibolitic matrix (e.g. see Bauer *et al.* 1996, figs 28–30).

Structural features and rock fabrics

The structural chronology of the Precambrian rocks in our map area is most easily understood if the structural features and fabrics are ordered relative to the development of the LPSS. Thus, we view all deformation that predates the development of the LPSS as 'pre-shear-zone' deformation. This regional deformation is probably Late Archean in age in our map area, although R.L. Bauer and coworkers (Bauer & Zeman 1997; Pratt *et al.* 1999; Curtis & Bauer 1999, 2000; Tomlin & Bauer 2000) recognized an important Palaeoproterozoic deformation along the eastern margin of the central Laramie Mountains that they related to the Trans-Hudson orogeny. This deformational phase is poorly dated. However, the relative chronology as

Fig. 5. Montage of deformational features and characteristics of members of the *c.* 2.01 Ga Kennedy dyke swarm. (**a**) Moderately steep-plunging hornblende lineation in diabasic amphibolite dyke. (**b**) Transition from strong linear fabric in diabasic amphibolite dyke rock to relatively undeformed pyroxene-bearing diabase. (**c**) Narrow diabasic amphibolite dyke distended in foliation of variably deformed granitic gneiss. (**d**) Xenoliths of granitic rock in diabasic amphibolite dyke. See the text for discussion of these features.

deciphered by Bauer and coworkers indicates that this Palaeoproterozoic deformation predated shear-zone deformation in the LPSS that developed within the time interval of the Medicine Bow orogeny as demonstrated U–Pb radiometric dating (Resor *et al.* 1996). Thus, in a regional sense the shear-zone deformation (and Medicine Bow orogeny) is D_3 (chiefly manifested as S_3 and L_3 of the LPSS), the earlier Palaeoproterozoic deformation is D_2 (Tomlin & Bauer 2000) and the Archean deformation is D_1. Recently, the pre-LPSS regional Palaeoproterozoic deformation has been referred to as the 'Black Hills orogeny' by Chamberlain *et al.* (2002) and Allard (2003), and correlated with Palaeoproterozoic deformation in the Black Hills of South Dakota (Dahl *et al.* 1999).

Superposed on the penetrative fabric formed during the shear-zone deformation are several fold phases broadly referred to as 'post-shear-system deformation'. All of these deformational phases occurred between the late Palaeoproterozoic and early Mesoproterozoic (*c.* 1.4 Ga). Subsequently, much younger brittle deformation associated with the late Cretaceous–early Tertiary Laramide orogeny affected parts of our map area (Figs 2 and 3). The Archean–Mesoproterozoic events in the central Laramie Mountains, Wyoming are summarized in Table 1 and discussed in more detail in the following sections.

Pre-shear-zone structural features and rock fabrics

The key to understanding the pre-shear-zone deformational history is a thorough understanding of the structural development of the migmatitic banded gneiss and its relationship to the younger, Palaeoproterozoic (*c.* 2.01 Ga) diabasic amphibolite dyke swarm. Because intrusion of the dyke swarm predates the *c.* 1.76 Ga penetrative deformation characteristic of the LPSS (Resor *et al.* 1996), the structural style of the pre-LPSS deformation can be deduced in outcrops of deformed migmatitic banded gneiss intruded by non-deformed or weakly deformed diabasic amphibolite dykes. Such a situation is exhibited in Figure 6, which shows a moderately dipping, diabasic amphibolite dyke with a chilled margin cutting across both foliation and folds in the migmatitic banded gneiss. Although not apparent in the photograph, but noted on the sketch, a younger shear-zone deformational fabric developed along the lower contacts of the dyke and overprinted both dyke rock and its gneissic wallrocks. This younger, superposed fabric completely transposes the original gneissic fabric within about 0.5 m of the dyke–wallrock contact.

The rock fabrics that predated the intrusion of the amphibolitic mafic dyke swarm and

Table 1. *Summary of Archean–Mesoproterozoic events in the central Laramie Mountains, Wyoming, USA*

1. Development of tonalitic orthogneisses (protolith of the migmatitic banded gneiss) – Middle Archean(?)
2. Intrusion of early mafic dykes (now amphibolite enclaves in migmatitic banded gneiss) – Archean
3. Pre-LPSS deformation (D_1), metamorphism, anatexis and intrusion of leucogranite dykes – Late Archean(?)
4. Tectonic juxtaposition of migmatitic banded gneiss and Archean supracrustal rocks of the Brandel Creek greenstone belt – Late Archean(?)
5. Emplacement of granitic rocks into the Laramie Peak and Palmer Canyon blocks – Late Archean–Palaeoproterozoic (including the Boy Scout Camp Granodiorite (2051 ± 9 Ma, Snyder *et al.* 1995)
6. Intrusion of members of the *c.* 2.01 Ga Kennedy dyke swarm along the rifted margin of the Archean Wyoming province (Cox *et al.* 2000)
7. Black Hills (Trans-Hudson) orogeny (D_2, Palaeoproterozoic (*c.* 1.8 Ga) as used by Chamberlain *et al.* 2002; also see Tomlin & Bauer 2000 and Allard 2003 for relative structural chronology)
8. Deformation and metamorphism related to early phase of the Medicine Bow orogeny (D_3, 1.78–1.76 Ga).
9. LPSS deformation (*c.* 1.76 Ga (Resor *et al.* 1996), late phase of the Medicine Bow orogeny (D_3)) – broadly transpressive in the regional geological setting but local transtensional effects (Chamberlain 1998); e.g. emplacement of the *c.* 1.76 Ga Horse Creek Anorthosite Complex in the southern Laramie Mountains – see Scoates & Chamberlain (1997), high-temperature mylonitization (S_3) and uplift and decompression (Patel *et al.* 1999) of the Palmer Canyon block.
10. Post-LPSS fold events. These deformational events are part of D_4 and/or D_5 regional deformations related to either a late Palaeoproterozoic regional deformation at *c.* 1722 Ma (Allard 2003) or intrusive emplacement of the Mesoproterozoic (*c.* 1.43 Ga) Laramie Anorthosite Complex (Tomlin 2001)

LPSS, Laramie Peak shear system.

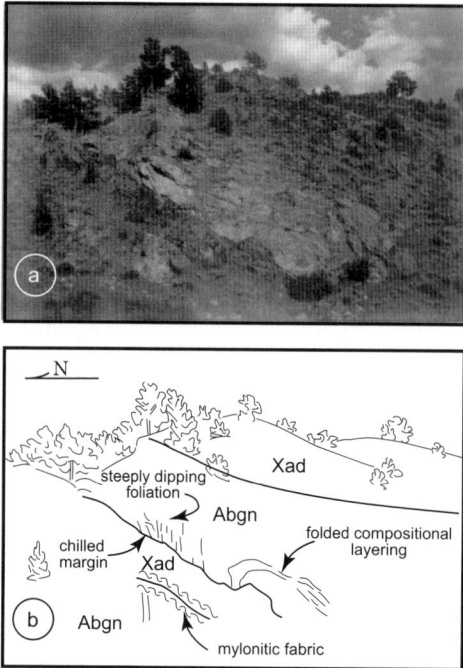

Fig. 6. (**a**) Photograph and (**b**) sketch of outcrop of Archean migmatitic banded gneiss (Abgn) intruded discordantly across folded foliation (banding, S_1) by moderately dipping Palaeoproterozoic diabasic amphibolite dykes (Xad). In this exposure, a shear-zone fabric is developed along the margin of the lower mafic dyke, and is superposed on both the dyke and its migmatitic banded gneiss host rocks. This locality thus indicates that the development of gneissic banding (S_1) and its folding predated the intrusion of the mafic dykes (inferred as members of the c. 2.01 Ga Kennedy dyke swarm). The shear-zone deformation (D_3) is post-dyke swarm, and this deformation has been dated as c. 1.76 Ga by Resor et al. (1996). Sketch by Phyllis A. Ranz from an annotated photograph by A.W. Snoke.

development of the shear-zone deformation are only found in the migmatitic banded gneiss. These pre-LPSS deformation rock fabrics include the development of compositional layering with a parallel biotite foliation and abundant outcrop-scale folding of this layering and the foliation. Compositional layering is defined by alternating biotite-rich and biotite-poor layers. Leucogranite layers may lie subparallel to compositional layering or cross-cut it in intrusive dykelet style. The ubiquitous presence of leucogranite layers of variable orientation gives this rock unit its overall migmatitic appearance. Figure 7a-1 is an equal-area, lower-hemisphere projection that summarizes S-surface data (compositional layering and foliation) from the

migmatitic banded gneiss as exposed in our map area. The data broadly scatter across the stereogram, and there is no simple pattern to the distribution of the data, suggesting a polyphase deformational history. In Figure 7a-2 a subset of the orientation data in Figure 7a-1 is plotted. These data, exclusively from the Owen Creek area, yield a definitive girdle with a fold axis determined as 21°/N48E. These data suggest NW–SE shortening, roughly perpendicular to the shear-zone foliation and presumably a manifestation of the Medicine Bow orogeny.

Shear-zone structural features and rock fabrics

Distribution and Geometry. The shear-zone deformation is heterogeneously distributed across our map area. As previously noted, the LPSS consists of several distinct segments: Garrett–Fletcher Park shear zone, Cottonwood Park shear zone, North Laramie River high-strain zone, as well as several high-strain zones recognized by Allard (2003). The Fletcher Park shear zone (Figs 1–3 and 7b) exhibits the most concentrated shear-strain deformation in the map area and forms a distinct, Palaeoproterozoic deformation front that separates generally non-deformed–weakly deformed granitic rocks (part of the Laramie batholith of Condie 1969, or Laramie Peak block of Patel *et al.* 1999) from the variably deformed igneous and metamorphic rocks of the 'central metamorphic complex' of Condie (1969) or Palmer Canyon block of Patel *et al.* (1999). Scattered evidence of the shear-zone deformation does occur north of the Fletcher Park shear zone (Fig. 7c), and the Cottonwood Park shear zone (Figs 1 and 7d) is the most obvious example of concentrated, penetrative shear strain within the Laramie batholith or Laramie Peak block.

The Fletcher Park shear zone is a steeply dipping, NE-striking, 300–500 m-wide zone of mylonitic gneisses. The zone has been traced approximately 15 km across our map area, but it can be extended about 20 km to the SW to include the Garrett mylonite (shear) zone of Langstaff (1984) and can also be extended approximately 10 km to the NE until it is covered by Tertiary sedimentary rocks. In other words, the Garrett–Fletcher Park shear zone completely transects the Archean rocks of the central Laramie Mountains (Fig. 1). Locally, the Garrett–Fletcher Park shear zone is disrupted by brittle faults that are interpreted as Laramide structural features, related to the uplift of the

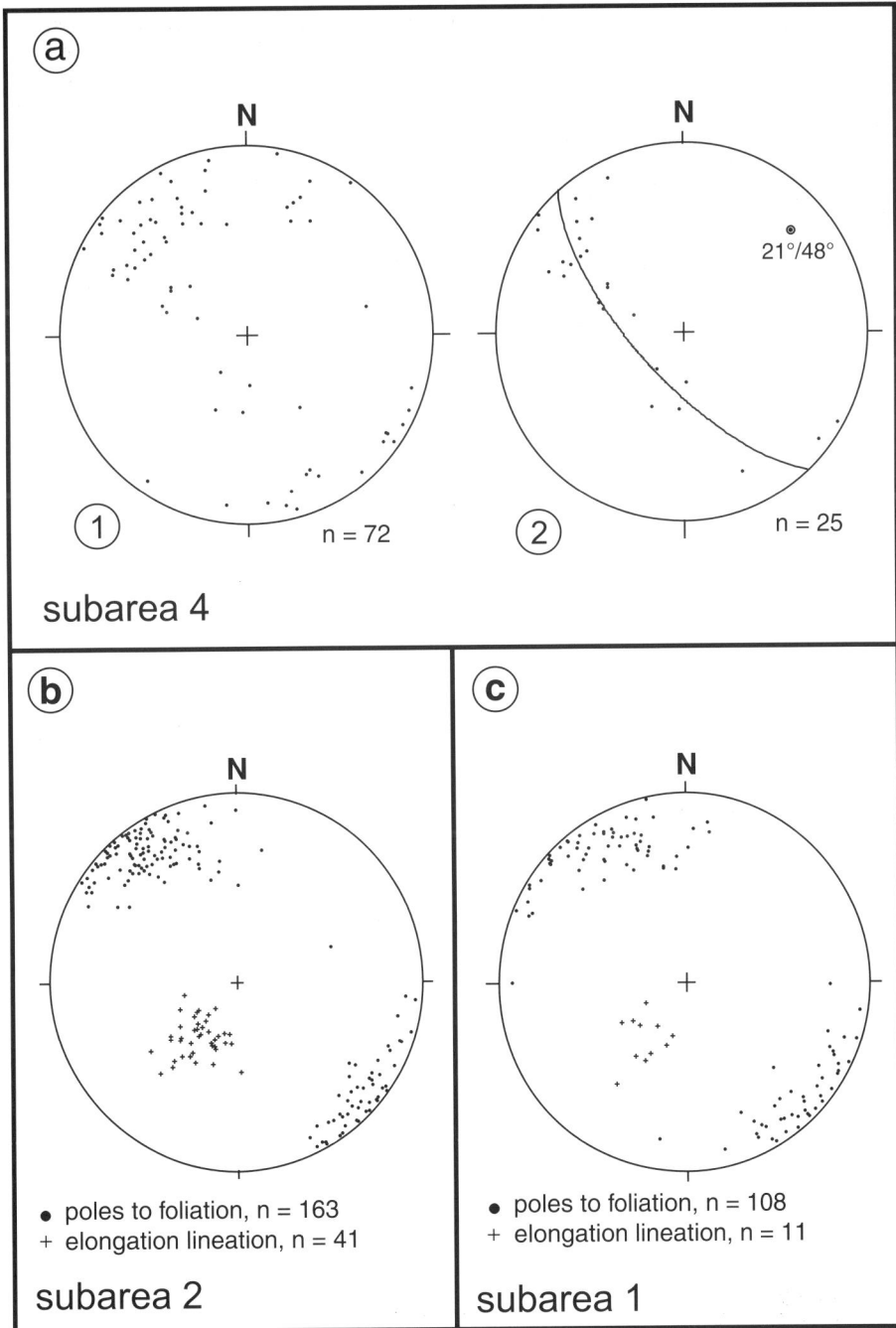

Fig. 7. Montage of equal-area, lower-hemisphere projections of structural data from the Laramie Peak shear system and environs. (**a**) Foliation in Archaean migmatitic banded gneiss (1) and (2). (**b**) Foliation and lineation in Fletcher Park shear zone. (**c**) Foliation and lineation in granitic rocks north of the Fletcher Park shear zone (but exclusive of the Cottonwood Park shear zone, see (d) below). (**d**) Foliation and lineation in the Cottonwood Park shear zone.
(**e**) Foliation and lineation in variably deformed gneissic granite (includes rocks of North Laramie River high strain zone). (**f**) Foliation and lineation in Brandel Creek supracrustal sequence and associated granitic rocks. (**g**) Foliation and lineation in diabasic amphibolite dykes (throughout map area).

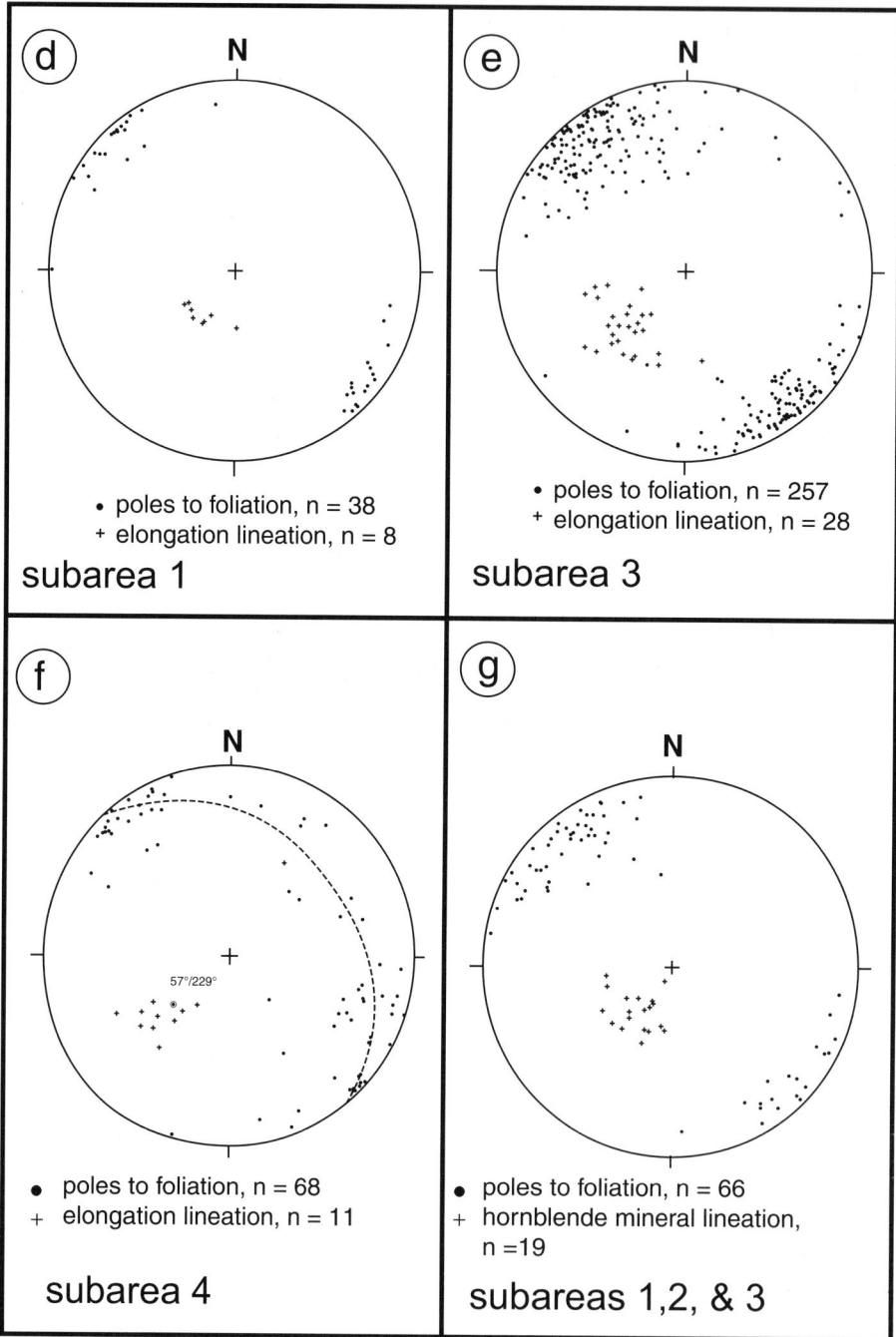

Fig. 7. *Continued.*

range in the late Cretaceous–early Tertiary. Interestingly, all three mapped brittle faults in Figure 2 exhibit a prominent strike-slip component during their movement history.

The Fletcher Park shear zone has gradational boundaries on both its north and south margins. These gradations can be rather sharp, especially along the northern margin, where the transition from weakly deformed granitic rocks (immediately north of the shear zone) to mylonitic gneisses occurs over about 10 m. North of the Fletcher Park shear zone, narrow localized shear zones parallel the Fletcher Park zone, but decrease in size and abundance to the north. The largest shear zone, the Cottonwood Park shear zone (Chamberlain *et al.* 1993), is delineated by an approximately 50 m-wide anastomosing zone of mylonitic granitic gneiss, located about 0.5 km north of the Fletcher Park shear zone (Fig. 1). North of the Cottonwood Park shear zone, localized shear zones are less common and narrower (<5 m wide) and beyond approximately 1.5 km north of the Fletcher Park shear zone the granitic rocks of the Laramie batholith of Condie (1969) are virtually undeformed.

The LPSS deformation is prevalent south of the Fletcher Park shear zone. In this region (the Palmer Canyon block of Patel *et al.* 1999) shear-zone-parallel foliation and lineation are variably developed with localized high-strain zones separated by weakly–moderately deformed blocks (Fig. 7e). Most of these high-strain zones are relatively narrow (<10 m); however, the North Laramie River high-strain zone is a 200–400 m-wide, anastomosing zone of well-foliated to mylonitic gneisses that parallels the Fletcher Park shear zone, approximately 1.5 km to the south. The North Laramie River high-strain zone is in part localized along the contact between granitic rocks of the Laramie River complex of Snyder *et al.* (1995) and supracrustal rocks of the Brandel Creek greenstone belt. This strain localization is apparently controlled by the rheology of the rock types on either side of the contact. The Brandel Creek greenstone belt is dominated by amphibolitic rocks, whereas the granitic rocks are considerably more quartz-rich and thus weaker during amphibolite facies crystal-plastic deformation.

Penetrative fabric elements. The orientation of penetrative fabrics within individual structural domains and rock units is illustrated in a series of stereographic projections (Fig. 7). The map area was subdivided into four structural domains (subareas 1–4) (Fig. 2): (1) map area north of the Fletcher Park shear zone (Fig. 7c); (2) Fletcher Park shear zone (Fig. 7b); (3) variably deformed granitic rocks including the North Laramie River high-strain zone (Fig. 7e); and (4) supracrustal and associated granitic rocks of the Brandel Creek greenstone belt (Fig. 7f). Furthermore, the following units or structural features were also analysed individually: migmatitic banded gneiss (Fig. 7a-1 & a-2), diabasic

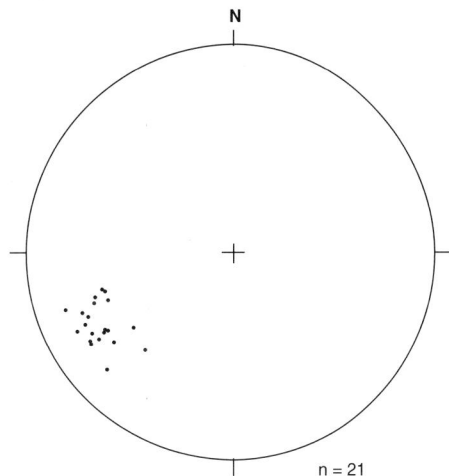

Fig. 8. Late folds in mylonitic shear zone, Murphy Canyon area.

amphibolite dykes (Fig. 7g), Cottonwood Park shear zone (Fig. 7d) and late folds in the Fletcher Park shear zone (Murphy Canyon locality, Fig. 8). Also, a synoptic summary of all linear data from the LPSS is presented in Figure 9.

LPSS fabrics can be defined based on the foliation and lineation in the Fletcher Park shear zone (subarea 2, Fig. 2 and Fig. 7b). The shear-zone fabric consists of a foliation striking $55° \pm 20°$ to the NE, dipping $75° \pm 15°$ to

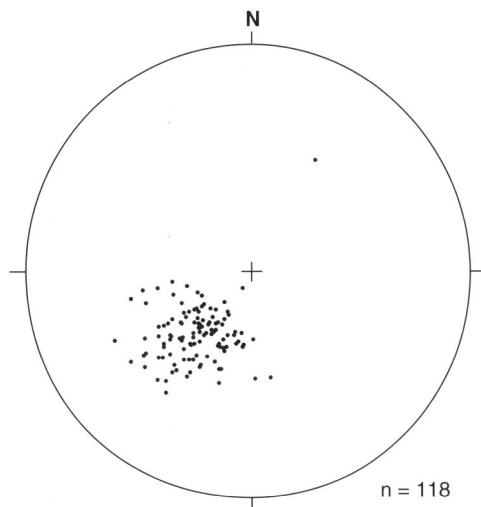

Fig. 9. Synoptic summary of lineation data from the Laramie Peak shear system.

either the SE or NW. This foliation typically contains a lineation trending $220° ± 25°$ and plunging $60° ± 15°$ to the SW. Scatter in the stereogram is interpreted as random variation about the mean but may be due in part to Laramide deformation and/or minor late folding.

Foliations and lineations in granitic rocks north of the Fletcher Park shear zone (subarea 1, Fig. 2 and Fig. 7c) show similar fabric orientations, foliation striking $65° ± 15°$, dipping $70° ± 20°$SE or NW, and a lineation trending $c. 210°$ and plunging to the SW at approximately $55°$. We interpret these fabrics as the northward-decreasing effects of the shear-zone deformation. The Cottonwood Park shear zone (Fig. 7d) also shows the same basic fabric elements, and the orientation of these fabrics fall within the distribution of measurements for the Fletcher Park shear zone.

South of the Fletcher Park shear zone (subareas 3 and 4, Fig. 2) there is evidence for both shear-zone fabrics and presumably Archean fabrics. Gneissic granites (Fig. 7e) primarily show evidence of a shear-zone-parallel foliation (strike $55° ± 30°$, dip $70° ± 20°$) and lineation (trend $230° ± 30°$, plunge $50° ± 20°$); however, there is greater scatter. Migmatitic banded gneiss (Fig. 7a-1) shows a much broader scatter suggestive of large-scale folding. In fact, when data from the vicinity of Owen Creek are isolated they exhibit a well-defined girdle consistent with a gently NE-plunging fold hinge (Fig. 7a-2). A biotite lineation was measured in this vicinity, parallel to the interpreted hinge line. The Brandel Creek supracrustals themselves (Fig. 7f) describe a broad girdle consistent with a SW-ward plunging hinge line parallel to mineral lineations in both supracrustal and granitic rocks, and to several late small-scale folds.

The diabasic amphibolite dykes are particularly illustrative of the predominance of the shear-zone fabric (i.e. Palaeoproterozoic). Figure 7g, an equal-area, lower-hemisphere plot of all data collected from the diabasic amphibolite dykes, shows a foliation and lineation distribution that is once again indistinguishable from the Fletcher Park shear zone (Fig. 7b). In fact, a synoptic plot of all the lineation data collected in the study area (Fig. 9) shows a single cluster of data with the one outlier from near Owen Creek that was previously discussed. These data lead to the interpretation that there has been only one penetrative Palaeoproterozoic deformation in the study area (Fig. 2). This stands in contrast to the work of Bauer and coworkers (Bauer & Zeman 1997; Pratt *et al.* 1999; Curtis & Bauer 1999, 2000; Tomlin & Bauer 2000) to the south

and east where they have found evidence for two penetrative Palaeoproterozoic deformation events.

Microstructures. Microstructures in the rocks of the LPSS provide abundant evidence of deformation under high-temperature conditions and indicate that the mylonitic rocks associated with the shear system are best referred to as 'mylonitic gneisses' (Snoke & Tullis 1998). Feldspar grains commonly exhibit core–mantle structure with large (5–30 mm) grains surrounded by small (0.25 mm) grains, suggesting the importance of subgrain rotation during dynamic recrystallization of these rocks. The importance of grain-boundary migration during deformation of the mylonitic gneisses is suggested by the irregular grain boundaries of the feldspar porphyroclasts. Myrmekitic intergrowths (quartz–sodic plagioclase symplectite) occur in the rims of some of the large alkali-feldspar grains. Simpson & Wintsch (1989) argue that deformation-driven replacement of alkali feldspar by an intergrowth of oligoclase and quartz (myrmekite) is a product of high-temperature (>450 °C) deformation in the presence of an aqueous fluid. Mantled feldspar porphyroclasts (Passchier & Trouw 1996) are common in the mylonitic granitic gneisses, although asymmetric tails (e.g. σ- or δ-type porphyroclasts) are scarce compared to ϕ- or θ-type porphyroclasts. The abundance of symmetrical winged porphyroclasts (i.e. ϕ-type) compared to asymmetric porphyroclasts argues for a general shear (non-coaxial) condition during the high-temperature deformation manifested in the LPSS.

In mylonitic granitic gneisses of the LPSS, quartz commonly occurs in elongated aggregates of irregularly shaped grains with interlocking grain boundaries suggesting widespread grain-boundary migration. However, in weakly deformed granitic rocks of the LPSS, quartz grains are recrystallized but contain subgrains and exhibit undulose extinction. Grain size is highly variable from small ($c. 0.25$ mm) neoblasts with relatively little internal strain to large grains ($c. 2$ mm) that contain abundant subgrains. Grain boundaries of the large quartz grains are highly irregular.

In diabasic amphibolite within high-strain zones of the LPSS, hornblende grains are commonly subidioblastic with a grain-shape orientation; grain boundaries between hornblende grains, as well as other phases, are straight and indicate textural equilibrium. There is no optical evidence of subgrain development within the hornblende grains. Plagioclase becomes increasingly recrystallized with increased deformation in the amphibolitic mafic dykes. In highly strained amphibolitic schists, the original tabular, igneous plagioclase grains have been totally recrystallized to an aggregate of intermixed coarse and much finer grains. These petrographic characteristics indicate recrystallization at high temperatures. The reaction of titanomagnetite to titanite (sphene) (Resor *et al.* 1996) during dynamic recrystallization of the diabasic amphibolite dykes within the LPSS also indicates high-temperature deformation and recrystallization (Frost *et al.* 2000).

Shear-sense indicators throughout the map area yield a consistent south-side-up, sinistral sense of shear (i.e. reverse sense shear). Shear-sense indicators are most prevalent in the Fletcher Park shear zone, but can also be found in the Cottonwood Park shear zone, North Laramie River high-strain zone, and many smaller shear zones distributed throughout the map area. The most common type of shear-sense indicator is sigmoidal feldspar porphyroclasts. Asymmetric feldspar porphyroclasts are best developed in the marginal zones of the Fletcher Park shear zone where deformation-driven recrystallization is moderate. The asymmetry of recrystallized tails on feldspar porphyroclasts, as well as scarce asymmetric quartzose lenses, consistently indicate south-side-up sense of shear. Composite fabrics (S–C–C') are less common, but also indicate an up-lineation shear sense. Local asymmetric folds at a high angle to the lineation are also consistent with south-side-up shearing.

Evidence for a pure shear (flattening) component of the deformation includes the prevalence of shear-zone-parallel foliation even in apparently low strain areas (Fig. 2), the predominance of symmetric feldspar porphyroclasts (ϕ-type porphyroclasts), L > S tectonites and the abundance of lineation-parallel folds. Although lineation-parallel folds are commonly interpreted to form at a high angle to lineation and then rotate into parallelism, such a process would probably form sheath folds (Cobbold & Quinquis 1980; however, see Alsop & Holdsworth 1999 for an alternative interpretation). The lack of sheath folds and abundance of lineation-parallel folds in the LPSS suggest that these folds may have originally formed subparallel to lineation (L_3 in the regional sense). Grujic & Mancktelow (1995) demonstrated that viscosity contrasts between layers (and matrix) are fundamental in the generation of folds subparallel to the principal elongation direction X. However, their analogue experiments also indicated that a flattening component during deformation favoured the development of lineation-parallel folds.

Timing. The broadest constraints that can be placed on the timing of shear-zone deformation come from observations of fabric development in the diabasic amphibolite dykes. Shear-zone deformation clearly post-dates emplacement of the dykes at *c.* 2.01 Ga (Cox *et al.* 2000) and is coincident with high-grade, amphibolite facies metamorphism. Amphibolite facies conditions necessarily predate the uplift and cooling of the Palmer Canyon block to temperatures below approximately 450 °C at *c.* 1745 Ma, as recorded by U–Pb apatite cooling ages (Patel *et al.* 1999). Resor *et al.* (1996) argued that an U–Pb titanite age of *c.* 1763 Ma from a diabasic amphibolite dyke deformed in the Fletcher Park shear zone provides a more direct constraint on the shear-zone deformation based on structural arguments for syntectonic growth at or below the diffusion-based closure temperature. Cox *et al.* (2000) subsequently found a similar age for metamorphic titanite from a deformed dyke within the Palmer Canyon block (their sample DC-6). The earliest evidence for Palaeoproterozoic high-grade metamorphism is an U–Pb age for metamorphic zircon from the southern edge of the Palmer Canyon block of *c.* 1778 Ma (Harper 1997).

Post-shear-system deformation in the Precambrian

Bauer and coworkers (Pratt *et al.* 1999; Curtis & Bauer 1999, 2000; Tomlin & Bauer 2000) and Allard (2003) have recognized several fold phases that post-date the development of the LPSS. These fold phases are best developed south or east of our map area (Figs 2 and 3). These various fold phases are either related to a poorly understood event at *c.* 1722 Ma that Allard (2003) correlated with the Dakotan orogeny of Chamberlain *et al.* (2002) or the emplacement of the Laramie Anorthosite Complex (Tomlin 2001). Pratt *et al.* (1999) recognized a series of upright, NE-trending folds that they referred to as the 'George Creek fold event'. Tomlin (2001) concluded that the 'George Creek fold event' was related to the emplacement of the Mesoproterozoic (*c.* 1.43 Ga) Laramie Anorthosite Complex, and thus is unrelated to the LPSS or Medicine Bow orogeny. Tomlin & Bauer (2000) referred to this deformational event as regional D_4 and concluded that deformational features related to Palaeoproterozoic D_2 and D_3 were reoriented by this younger Proterozoic deformation. Another post-LPSS fold event was called the 'Open Fold Event' by Pratt *et al.* (1999), but the age of this deformational event is uncertain. Pratt et al. (1999) described this event as late oroclinal bending and considered it associated with the Medicine Bow orogeny. Allard (2003), however, argued that the 'Open Fold Event' is younger than c. 1722 Ma based on new geochronometric data and overprinting relationships, and thus post-dates the Medicine Bow orogeny.

Evaluation of some tectonic models

The LPSS is a 10 km-thick zone of heterogeneous general shear (non-coaxial) strain that defines the northern margin of the Palmer Canyon block of the central Laramie Mountains. This uplifted tectonic block, chiefly comprised of rocks of the Archean Wyoming province but extensively intruded by various Palaeoproterozoic rocks ranging in composition from ultramafic to granitic, was regenerated during the Palaeoproterozoic Medicine Bow orogeny (Chamberlain et al. 1993; Patel et al. 1999). The Palaeoproterozoic age (c. 1.76 Ga) of the Fletcher Park shear zone (Resor et al. 1996) suggests that the whole LPSS developed within the time interval of the Medicine Bow orogeny (1.78–1.74 Ga).

Several tectonic models have been devised to explain the uplift of basement rocks in a foreland setting during crustal contraction. Possible end-member models for the uplift of a basement block include: (1) tilted block uplift with a frontal listric fault zone that soles into a deep-crustal décollement; (2) en-masse 'pop-up' uplift bounded by oppositely vergent thrust (reverse) faults; and (3) crustal-scale duplex consisting of a stack of crystalline thrust sheets each bounded by a ductile fault (shear) zone (Fig. 10). To date only models 1 and 2 have been evaluated in regard to the uplift of the Palmer Canyon block (Chamberlain et al. 1993; Patel et al. 1999), and we further evaluate these hypotheses in light of our structural analysis of the LPSS. In addition, we also comment on the implications of the crustal-scale duplex model and suggest that additional structural studies of the Palmer Canyon block are necessary to fully test this model.

Chamberlain et al. (1993) originally suggested a block-uplift model to explain the development of the LPSS in conjunction with an arc–continental margin collision along the Cheyenne belt during the Palaeoproterozoic (see Chamberlain et al. 1993, fig. 5). This model involved the uplift of the Palmer Canyon block along a crustal-scale reverse fault that recorded at least 10 km of differential, vertical displacement across the LPSS. Patel et al. (1999) further refined the block-uplift model by emphasizing

that their P–T data indicated no significant variation across the Palmer Canyon block, and thus they argued against a rotational component for the hanging-wall block (i.e. model 1 above and see Fig. 10a). If the displacement is chiefly along a deep-rooted, frontal thrust fault, substantial rigid-body rotation of the hanging-wall block is required to achieve structural balance and to avoid 'space problems', and is well documented in late Cretaceous–Eocene, Laramide basement-involved uplifts of the Rocky Mountain foreland (e.g. Erslev 1986). To minimize potential rotation of the hanging-wall block, Patel et al. (1999) postulated a complementary back-thrust along the southern margin of the Palmer Canyon block (i.e. en-masse 'pop-up' uplift (model 2 and Fig. 10b) similar to a common thrust-belt structural feature discussed by Butler 1982, fig. 16b). Tangible evidence for the back-thrust interpretation is not available, because the back-thrust is inferred to occur in an area now intruded by the Mesoproterozoic Laramie Anorthosite Complex and Sherman batholith. Patel et al. (1999) also suggested that the uplift of the Palmer Canyon block occurred late in the Medicine Bow orogeny during a transpressive tectonic regime. The en-masse 'pop-up' uplift model is compatible with a crustal-scale shear zone (i.e. LPSS), a prominent metamorphic discontinuity between the hanging-wall and foot-wall blocks, the lack of apparent metamorphic variation across the uplifted high-grade hanging-wall block and the age of shear-zone deformation determined from the Fletcher Park shear zone (Resor et al. 1996). One ad hoc element of the 'pop-up' uplift model is the back-thrust inferred to bound the uplifted block along its southern margin (Patel et al. 1999).

A key observation that supports an en-masse uplift of the Palmer Canyon block is the apparent lack of metamorphic-grade variation across the block suggesting that significant hanging-wall rotation (i.e. a fundamental characteristic of model 1, Fig. 10a) did not occur despite significant displacement based on contrasting metamorphic grade across the frontal (northern) shear zone (i.e. Garrett–Fletcher Park shear zone). Unfortunately, detailed P–T determinations do not exist across the entire Palmer Canyon block. Furthermore, the Palaeoproterozoic deformational and P–T history of the Palmer Canyon block is complicated by the proximity of the Trans-Hudson Orogen as manifested in the eastern Laramie Mountains, as well as Hartville uplift (Sims 1995; Bauer & Zeman 1997; Pratt et al. 1999; Curtis & Bauer 2000; Tomlin & Bauer 2000). There is considerable controversy concerning the age and magnitude

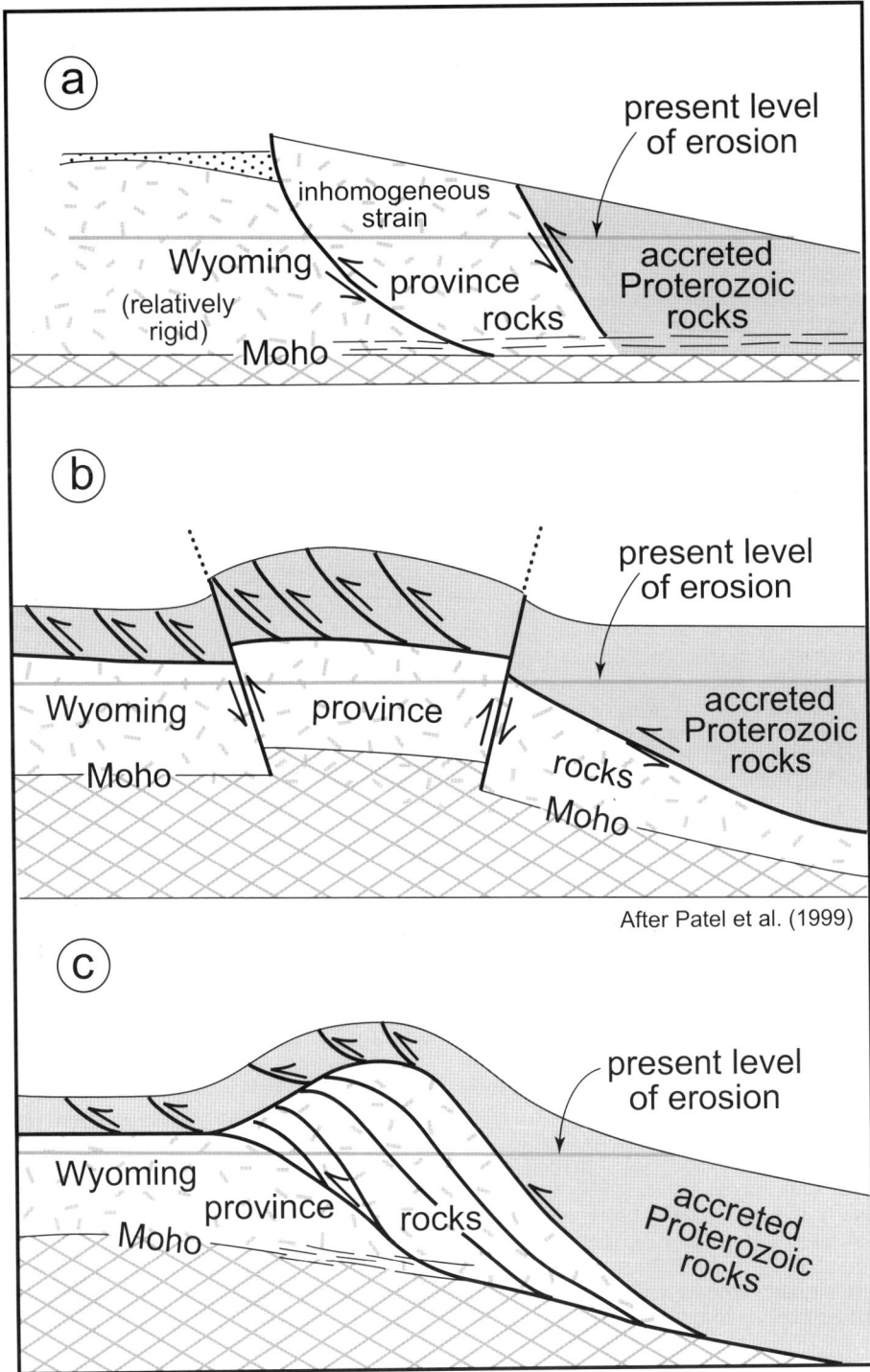

Fig. 10. Some tectonic models for the evolution of the Laramie Peak shear system. (**a**) Tilted basement block bounded by frontal listric thrust fault that soles into a deep-crustal décollement. (**b**) En-masse 'pop-up' uplift bounded by oppositely vergent thrust (reverse) faults. (**c**) Crustal-scale duplex consisting of a stack of crystalline thrust sheets each bounded by a ductile fault (shear) zone.

of the Trans-Hudson orogenic effects in the Laramie Mountains and environs (Edson 1995; Tomlin & Bauer 2000; Hill & Bickford 2001; Chamberlain *et al.* 2002; Allard 2003; Sims & Stein 2003). In the light of only scattered *P–T* determinations and potential overprinting of orogenic events in the Palmer Canyon block, the role of large-scale rotation of the Palmer Canyon block during the development of the LPSS has not been fully evaluated.

Although the en-masse 'pop-up' uplift model is supported by a diverse data set, another structural model that may also explain the geological relationships is a crustal-scale duplex consisting of imbricate slices of basement rocks bounded by anastomosing ductile shear zones rooted into a deep-crustal décollement zone. The heterogeneous nature of the shear strain associated with the LPSS raises the possibility that such heterogeneity could exist throughout the Palmer Canyon block. A ductile thrust stack model could explain the distribution of deformation across the LPSS and the high-grade block to the south; however, it fails to explain the consistent uniform grade of metamorphism to the south (although our previous comments about this observation also apply here). In the crustal-scale duplex model, the Fletcher Park shear zone marks the frontal thrust of a thick-skinned imbricate thrust belt (Fig. 10c). Thus, the Fletcher Park shear zone would be a major basement ramp rooted into a deep-crustal décollement zone. In the Laramie Mountains, the frontal reverse fault does mark a significant metamorphic discontinuity; however, metamorphic grade should also increase across the high-grade block in the hanging wall as additional ductile thrusts increase the cumulative displacement to the south. If the crustal-scale duplex model is applied to the uplift of whole Palmer Canyon, the distribution of the shear-zone deformation could be much more widespread than just the presently delineated LPSS. Although structural analysis is not available throughout the entire Palmer Canyon block, the geological mapping of the late George L. Snyder (e.g. Snyder *et al.* 1995, 1998) and structural studies by Allard (2003) suggest that the shear-zone deformation is restricted to the northern margin of the Palmer Canyon block (i.e. basically the LPSS as defined in this paper).

Given apparent problems with all simple uplift models, we favour a tectonic model for the uplift of the Palmer Canyon block and development of the LPSS by a combination of en-masse block uplift coupled with internal, heterogeneous general shear (non-coaxial) manifested by the development of high-strain zones and complex folding. The Garrett–Fletcher Park shear zone is interpreted as the frontal fault of the uplift (i.e. a major basement ramp) that rooted into a deep-crustal décollement zone that must have underlain the entire Palmer Canyon block during Palaeoproterozoic orogenesis. This décollement zone in turn would have rooted southward (in present geographic co-ordinates) into the evolving collisional zone that must have separated the Wyoming and Colorado provinces (now manifested by the Cheyenne belt, see Duebendorfer & Houston 1987). Internal deformation within the Palmer Canyon block would be manifested by heterogeneous general shear strain and folding. In such a model the ad hoc, S-directed back-thrust suggested by Patel *et al.* (1999) would not be needed to achieve overall uplift of the Palmer Canyon block, whereas widespread (albeit heterogeneous) Palaeoproterozoic general shear and folding is required. The detailed structural studies by Allard (2003) immediately to the south and east of our map area support such a complex, heterogeneous Palaeoproterozoic deformational history.

On the deformation of sialic crust during an arc–continental margin collision

The central Laramie Mountains provide an unusual opportunity to study how sialic crust responded during an arc–continental margin collision. A modern example of such a tectonic phenomena is the convergence of Australian continental lithosphere with the Banda arc (Snyder *et al.* 1996). In regard to the Laramie Mountains example, it is inferred, based on regional geological relationships and geochronometric studies, that the Archean Wyoming province was partially subducted beneath a N-facing (in present geographic co-ordinates) oceanic arc (i.e. rocks of the Colorado province). Because the dip of subduction was directed southward (again in present geographic co-ordinates), there was no synorogenic magmatism associated with the underthrusted continental margin (i.e. Archean Wyoming province). Deformation in the Wyoming province was then purely mechanical in nature and not associated with magmatic softening. The chief manifestation of this situation is heterogeneous Palaeoproterozoic strain distributed throughout the partially subducted Wyoming province (Palmer Canyon block). Locally this strain was intense, forming high-temperature, mylonitic gneisses that involved crystal-plastic deformation mechanisms (e.g. dislocation and diffusion creep). The requisite thermal regime for these chiefly thermally activated deformation mechanisms was facilitated by tectonic thickening related to the emplacement of a Palaeoproterozoic oceanic arc and related rocks structurally above the Wyoming province during the Medicine Bow orogeny (Chamberlain 1998).

A fundamental characteristic of the deformation of the Archean Wyoming province during the Medicine Bow orogeny is strain partitioning and localization (Snoke & Resor 1999). Strain is commonly localized along original compositional (i.e. rheological) boundaries. This strain localization varies from outcropscale, e.g. along the intrusive contact between an amphibolitic mafic dyke and its quartzofeldspathic host rock, to map-scale examples such as the NW margin of the Brandel Creek greenstone belt and the localization of the North Laramie River high-strain zone. We propose that these contacts may have acted as strength defects (Poirier 1980) acting to initialize ductile flow due to geometrical stress concentration and/or chemical and reaction softening (White et al. 1980) associated with localization of aqueous fluid flow. Progressive deformation and amphibolite facies metamorphism of original diabase dykes and the replacement of alkali feldspar by myrmekite demonstrate the importance of aqueous fluids in the deformation process. Developing high-strain zones may have further localized strain through grain-size reduction associated with dynamic recrystallization (Handy 1989) and structural softening (White et al. 1980) associated with the development of compositional layering in pinstripe mylonites and amphibolitic schists.

Strain localization in the Garrett–Fletcher Park shear zone is not clearly influenced by rheological contrasts, although this shear zone may have grown from smaller localized zones in a process similar to the one outlined by Christiansen & Pollard (1997). In this mapscale example, the high strain of the Garrett–Fletcher Park shear zone is the structural front for Palaeoproterozoic regeneration of the Archean Wyoming province. Although strain associated with the LPSS occurs north of the Garrett–Fletcher shear zone, this shear zone is clearly a significant structural, metamorphic and geochronological boundary. This important boundary therefore possesses all the essential characteristics of a true orogenic front. Furthermore, this basic conclusion suggests that the Garrett–Fletcher Park shear zone may be an example of a pro-shear zone (Quinlan et al. 1993; Beaumont & Quinlan 1994; Beaumont et al. 1994) rooted into a deep-crustal detachment that developed during the Palaeoproterozoic oceanic arc–continental margin collision inferred as the tectonic setting of the Medicine Bow orogeny (Chamberlain 1998).

In summary, we propose that strain partitioning and localization rather than broad-scale flowage at lower crustal depth are common processes in the absence of synorogenic magmatism. Deformation of continental crust in such cases is characterized by the development of distributed zones of general shear (non-coaxial), and some of these shear zones may transect the entire crust. A well-documented Phanerozoic example of a crustal-scale fault/shear zone that developed in absence of synorogenic magmatism is the Laramide-age Wind River thrust of the Wyoming Rocky Mountain foreland (Smithson et al. 1979; Sharry et al. 1986). McKenzie et al. (2000) proposed that one manifestation of the absence of lower-crustal flow is the preservation of crustal thickness variations. Interestingly, a recent seismic study across the Cheyenne belt suture (developed during the 1.78–1.74 Ga Medicine Bow orogeny) indicates that locally thickened crust is associated with this ancient suture (Crosswhite & Humphreys 2003). Thus, crustal thickening associated with the Medicine Bow orogeny, and in part manifested by the LPSS, is apparently still manifested in the overall crustal structure (Moho topography) of this part of the Rocky Mountains.

This research was partially funded by US National Science Foundation grants EAR-9205825 and EAR-9706296 awarded to K.R. Chamberlain, A.W. Snoke and B.R. Frost. Numerous individuals have shared with us their observations and thoughts regarding the geology of the Laramie Mountains over the past decade or more; these individuals especially include S.T. Allard, R.L. Bauer, K.R. Chamberlain, B.R. Frost C.D. Frost and the late G.L. Snyder. Reviews and comments on an earlier version of our manuscript by R.L. Bauer, G.I. Alsop, S. Giorgis, W.A. Sullivan, S.T. Allard and an anonymous reviewer greatly helped us prepare a more concise final version of our manuscript. This contribution is dedicated to Plato and Sam, our faithful, but now departed, canine companions during our fieldwork.

References

ALLARD, S.T. 2003. *Geologic evolution of Archean and Paleoproterozoic rocks in the northern Palmer Canyon block, central Laramie Mountains, Albany County, Wyoming*. PhD dissertation, University of Wyoming, Laramie.

ALSOP, G.I. & HOLDSWORTH, R.E. 1999. Vergence and facing patterns in large-scale sheath folds. *Journal of Structural Geology*, **21**, 1335–1349.

AXEN, G.J., SELVERSTONE, J., BYRNE, T. & FLETCHER, J.M. 1998. If the strong crust leads, will the weak crust follow? *GSA Today*, **8**(12), 1–8.

BAUER, R.L. & ZEMAN, T. 1997. Evidence for Paleoproterozoic deformation and metamorphism of the southeastern margin of the Archean Wyoming province during the Cheyenne belt collision and Trans-Hudson orogeny. *Geological Society of America Abstracts with Programs*, **29**(6), A-408.

BAUER, R.L., CHAMBERLAIN, K.R., SNOKE, A.W., FROST, B.R., RESOR, P.G. & GRESHAM, A.D. 1996. Ductile deformation of mid-crustal Archean rocks in the foreland of Early Proterozoic orogenies, central Laramie Mountains, Wyoming. *In*: THOMPSON, R.A., HUDSON, M.R. & PILLMORE, C.L. (eds) *Geologic Excursions in the Rocky Mountains and Beyond: Field Trip Guidebook for the 1996 Annual Meeting of the Geological Society of America*. Colorado Geological Survey, Special Publications, 44.

BEAUMONT, C. & QUINLAN, G. 1994. A geodynamic framework for interpreting crustal-scale seismic-reflectivity patterns in compressional orogens. *Geophysical Journal International*, **116**, 754–783.

BEAUMONT, C., FULLSACK, P. & HAMILTON, J. 1994. Styles of crustal deformation in compressional orogens caused by subduction of the underlying lithosphere. *Tectonophysics*, **232**, 119–132.

BEAUMONT, C., JAMIESON, R.A., NGUYEN, M.H. & LEE, B. 2001. Himalayan tectonics explained by extrusion of a low-viscosity crustal channel coupled to focused surface denudation. *Nature*, **414**, 738–742.

BLOCK, L. & ROYDEN, L. 1990. Core complex geometries and regional scale flow in the lower crust. *Tectonics*, **9**, 557–567.

BOIS, C. & ECORS SCIENTIFIC PARTY. 1990. Major geodynamic processes studied from ECORS deep seismic profiles in France and adjacent areas. *Tectonophysics*, **173**, 397–410.

BOWRING, S.A. & KARLSTROM, K.E. 1990. Growth and stabilization of Proterozoic continental lithosphere in the southwestern United States. *Geology*, **18**, 1203–1206.

BUTLER, R.W.H. 1982. The terminology of structures in thrust belts. *Journal of Structural Geology*, **4**, 239–245.

CHAMBERLAIN, K.R. 1998. Medicine Bow orogeny: Timing of deformation and model of crustal structure produced during continent–arc collision, ca. 1.78 Ga, southeastern Wyoming. *Rocky Mountain Geology*, **33**, 259–277.

CHAMBERLAIN, K.R. & BOWRING, S.A. 2000. Apatite–feldspar U–Pb thermochronometer: a reliable, mid-range (∼450 °C), diffusion-controlled system. *Chemical Geology*, **172**, 173–200.

CHAMBERLAIN, K.R., BAUER, R.L., FROST, B.R. & FROST, C.D. 2002. Dakotan Orogen: continuation of Trans-Hudson Orogen or younger, separate suturing of Wyoming and Superior Cratons? *Geological Association of Canada–Mineralogical Association of Canada, Abstracts*, **27**, 18–19.

CHAMBERLAIN, K.R., PATEL, S.C., FROST, B.R. & SNYDER, G.L. 1993. Thick-skinned deformation of the Archean Wyoming province during Proterozoic arc–continent collision. *Geology*, **21**, 995–998.

CHRISTIANSEN, P.P. & POLLARD, D.D. 1997. Nucleation, growth and structural development of mylonitic shear zones in granitic rock. *Journal of Structural Geology*, **19**, 1159–1172.

COBBOLD, P.R. & QUINQUIS, H. 1980. Development of sheath folds in shear regimes. *Journal of Structural Geology*, **2**, 119–126.

CONDIE, K.C. 1969. Petrology and geochemistry of the Laramie batholith and related metamorphic rocks of Precambrian age, eastern Wyoming. *Geological Society of America Bulletin*, **80**, 57–82.

CONDIE, K.C. 1982. Plate-tectonics model for Proterozoic continental accretion in the southwestern United States. *Geology*, **10**, 37–42.

CONDIE, K.C. & SHADEL, G.A. 1984. An Early Proterozoic volcanic arc succession in southeastern Wyoming. *Canadian Journal of Earth Science*, **21**, 415–427.

COX, D.M., FROST, C.D. & CHAMBERLAIN, K.R. 2000. 2.01-Ga Kennedy dike swarm, southeastern Wyoming: Record of a rifted margin along the southern Wyoming province. *Rocky Mountain Geology*, **35**, 7–30.

CROSSWHITE, J.A. & HUMPHREYS, E.D. 2003. Imaging the mountainless root of the 1.8 Ga Cheyenne belt suture and clues to its tectonic activity. *Geology*, **31**, 669–672.

CURTIS, D.J. & BAUER, R.L. 1999. Variations in nappe-related fabric orientations during Paleoproterozoic ductile reworking of Archean basement, central Laramie Mountains, southeastern Wyoming. *Geological Society of America Abstracts with Programs*, **31**(7), A–177.

CURTIS, D.J. & BAUER, R.L. 2000. Progressive variations in nappe-related fold-hinges formed during the Paleoproterozoic Trans-Hudson orogeny, central Laramie Mountains, southeastern Wyoming. *Geological Society of America Abstracts with Programs*, **32**(5), A–7.

DAHL, P.S., HOLM, D.K., GARDNER, E.T., HUBACHER, F.A. & FOLAND, K.A. 1999. New constraints on the timing of Early Proterozoic tectonism in the Black Hills (South Dakota), with implications for docking of the Wyoming province with Laurentia. *Geological Society of America Bulletin*, **111**, 1335–1349.

DEWEY, J.F. & BIRD, J.M. 1970. Mountain belts and the new global tectonics. *Journal of Geophysical Research*, **75**, 2625–2647.

DUEBENDORFER, E.M. & HOUSTON, R.S. 1987. Proterozoic accretionary tectonics at the southern margin of the Archean Wyoming craton. *Geological Society of America Bulletin*, **98**, 554–568.

EDSON, J.D. 1995. *Early Proterozoic ductile deformation along the southeastern margin of the Wyoming province in the central Laramie Range, Wyoming*. MS thesis, University of Missouri, Columbia.

ERSLEV, E.A. 1986. Basement balancing of Rocky Mountain foreland uplifts. *Geology*, **14**, 259–262.

FLINN, D. 1994. Essay review: kinematic analysis – pure nonsense or simple nonsense? *Geological Journal*, **29**, 281–284.

FROST, B.R., CHAMBERLAIN, K.R. & SCHUMACHER, J.C. 2000. Sphene (titanite): phase relations and role as a geochronometer. *Chemical Geology*, **172**, 131–148.

GANS, P.B. 1987. An open-system, two-layer crustal stretching model for the eastern Great Basin. *Tectonics*, **6**, 1–12.

GRAFF, P.J., SEARS, J.W., HOLDEN, G.S. & HAUSEL, W.D. 1982. Geology of the Elmers Rock Greenstone Belt, Laramie Range, Wyoming. Geological Survey of Wyoming, Report of Investigations, 14.

GRUJIC, D. & MANCKTELOW, N.S. 1995. Folds with axes parallel to the extension direction: an experimental study. *Journal of Structural Geology*, **17**, 279–291.

HANDY, M.R. 1989. Deformation regimes and the rheological evolution of fault zones in the lithosphere: the effects of pressure, temperature, grain-size, and time. *Tectonophysics*, **163**, 119–152.

HARPER, K.M. 1997. *U–Pb age constraints on the timing and duration of Proterozoic and Archean metamorphism along the southern margin of the Archean Wyoming craton*. PhD dissertation, University of Wyoming, Laramie.

HAUSER, E., POTTER, C. ET AL. 1987. Crustal structure of eastern Nevada from COCORP deep seismic reflection data. *Geological Society of America Bulletin*, **99**, 833–844.

HILL, B.M. & BICKFORD, M.E. 2001. Paleoproterozoic rocks of central Colorado: Accreted arcs or extended older crust? *Geology*, **29**, 1015–1018.

HOFFMAN, P.F. 1989. Precambrian geology and tectonic history of North America. *In*: BALLY, A.W. & PALMER, A.R. (eds) *The Geology of North America – An overview*. Geological Society of America, The Geology of North America, **A**, 447–512.

HOUSTON, R.S. 1993. Late Archean and Early Proterozoic geology of southeastern Wyoming. *In*: SNOKE, A.W., STEIDTMANN, J.R. & ROBERTS, S.M. (eds) *Geology of Wyoming*. State Geological Survey of Wyoming Memoir, **5**, 78–116.

HOUSTON, R.S., DUEBENDORFER, E.M., KARLSTROM, K.E. & PREMO, W.R. 1989. A review of the geology and structure of the Cheyenne belt and Proterozoic rocks of southern Wyoming. *In*: GRAMBLING, J.A. & TEWKSBURY, B.J. (eds) *Proterozoic Geology of the Southern Rocky Mountains*. Geological Society of America Special Paper, **235**, 1–12.

HOUSTON, R.S. ERSLEV, E.A., ET AL. 1993. The Wyoming province. *In*: REED, J.C., JR. ET AL. (eds) *Precambrian: Conterminous U.S.* Geological Society of America, The Geology of North America, **C-2**, 121–170.

JACKSON, J. 2002. Strength of the continental lithosphere: Time to abandon the jelly sandwich? *GSA Today*, **12**, 4–10.

JOHNSON, R.C. & HILLS, F.A. 1976. Precambrian geochronology and geology of the Box Elder Canyon area, northern Laramie Range, Wyoming. *Geological Society of America Bulletin*, **87**, 809–817.

KARLSTROM, K.E. & BOWRING, S.A. 1988. Early Proterozoic assembly of tectonostratigraphic terranes in southwestern North America. *Journal of Geology*, **96**, 561–576.

KARLSTROM, K.E. & HOUSTON, R.S. 1984. The Cheyenne belt: Analysis of a Proterozoic suture in southern Wyoming. *Precambrian Research*, **25**, 415–446.

KARLSTROM, K.E. & CD-ROM WORKING GROUP. 2002. Structure and evolution of the lithosphere beneath the Rocky Mountains: Initial results from the CD-ROM experiment: *GSA Today*, **12**(3), 4–10.

KARLSTROM, K.E. & WILLIAMS, M.L. 2006. Nature and evolution of the middle crust: heterogeneity of structure and process due to pluton-enhanced tectonism. *In*: BROWN, M. & RUSHMER, T. (eds) *Evolution and Differentiation of the Continental Crust*. Cambridge University Press, Cambridge, In press.

KLEMPERER, S.L., HAUGE, T.A., HAUSER, E.C., OLIVER, J.E. & POTTER, C.J. 1986. The Moho in the northern Basin and Range province, Nevada, along the COCORP 40°N seismic-reflection transect. *Geological Society of America Bulletin*, **97**, 603–618.

KLEPEIS, K.A., CLARKE, G.L., GEHRELS, G. & VERVOORT, J. 2004. Processes controlling vertical coupling and decoupling between the upper and lower crust of orogens: results from Fiordland, New Zealand. *Journal of Structural Geology*, **26**, 765–791.

KRUSE, S., MCNUTT, M., PHIPPS-MORGAN, J., ROYDEN, L. & WERNICKE, B. 1991. Lithospheric extension near Lake Mead, Nevada: A model for ductile flow in the lower crust. *Journal of Geophysical Research*, **96**, 4435–4456.

LANGSTAFF, G.D. 1984. *Investigation of Archean metavolcanic and metasedimentary rocks of Sellers Mountain, west-central Laramie Mountains, Wyoming*. MS thesis, University of Wyoming, Laramie.

LOVE, J.D. & CHRISTIANSEN, A.C. 1985. *Geologic Map of Wyoming*. scale 1 : 500 000. US Geological Survey, Reston, Virginia.

MACCREADY, T., SNOKE, A.W., WRIGHT, J.E. & HOWARD, K.A. 1997. Mid-crustal flow during Tertiary extension in the Ruby Mountains core complex, Nevada. *Geological Society of America Bulletin*, **109**, 1576–1594.

MCKENZIE, D. & JACKSON, J. 2002. Conditions for flow in the continental crust. *Tectonics*, **21**, 1055, doi:10.1029/2002TC001394.

MCKENZIE, D., NIMMO, F., JACKSON, J.A., GANS, P.B. & MILLER, E.L. 2000. Characteristics and consequences of flow in the lower crust. *Journal of Geophysical Research*, **105**, 11 029–11 046.

MOLNAR, P. 1988. Continetal tectonics in the aftermath of plate tectonics. *Nature*, **335**, 131–137.

NELSON, B.K. & DEPAOLO, D.J. 1985. Rapid pro-
duction of continental crust 1.7 to 1.9 b.y. ago:
Nd isotopic evidence from the basement of the
North American mid-continent. *Geological
Society of America Bulletin*, **96**, 746–754.

NORTHRUP, C.J. 1996. Structural expressions and
tectonic implications of general noncoaxial flow
in the midcrust of a collisional orogen: The northern Scandinavian Caledonides. *Tectonics*, **15**,
490–505.

PASSCHIER, C.W. & TROUW, R.A.J. 1996. *Microtectonics*. Springer, Berlin.

PATEL, S.C., FROST, B.R., CHAMBERLAIN, K.R. &
SNYDER, G.L. 1999. Proterozoic metamorphism
and uplift history of the north-central Laramie
Mountains, Wyoming, USA. *Journal of Metamorphic Geology*, **17**, 243–258.

POIRIER, J.P. 1980. Shear localization and shear
instability in materials in the ductile field.
Journal of Structural Geology, **2**, 135–142.

PRATT, M.L., BAUER, R.L. & TOMLIN, K.P. 1999.
Deep-seated Paleoproterozoic oroclinal bending
in Archean basement rocks along the eastern
margin of the central Laramie Mountains, southeastern Wyoming. *Geological Society of America
Abstracts with Programs*, **31**(7), A–107.

QUINLAN, G., BEAUMONT, C. & HALL, J. 1993. Tectonic model for crustal seismic reflectivity patterns
in compressional orogens. *Geology*, **21**, 663–666.

REED, J.C., Jr, BICKFORD, M.E., PREMO, W.R.,
ALEINIKOFF, J.N. & PALLISTER, J.S. 1987.
Evolution of the Early Proterozoic Colorado
province: Constraints from U–Pb geochronology.
Geology, **15**, 861–865.

RESOR, P.G. 1996. *Nature and timing of deformation
associated with the Proterozoic Laramie Peak
shear zone, Laramie Mountains, Wyoming*. MS
thesis, University of Wyoming, Laramie.

RESOR, P.G., CHAMBERLAIN, K.R., FROST, C.D.,
SNOKE, A.W. & FROST, B.R. 1996. Direct dating
of deformation: U–Pb age of syndeformational
sphene growth in the Proterozoic Laramie Peak
shear zone. *Geology*, **24**, 623–626.

SCOATES, J.S. & CHAMBERLAIN, K.R. 1997. Orogenic
to post-orogenic origin for the 1.76 Ga Horse Creek
Anorthosite Complex, Wyoming USA. *Journal of
Geology*, **105**, 331–343.

SEGERSTROM, K., WEISNER, R.C. & KLEINKOPF, M.D.
1977. *Mineral Resources of the Laramie Peak
Study Area, Albany and Converse Counties,
Wyoming*. US Geological Survey Bulletin,
1397-B.

SHARRY, J., LANGAN, R.T., JOVANOVICH, D.B., JONES,
G.M., HILL, N.R. & GUIDISH, T.M. 1986.
Enhanced imaging of the COCORP seismic line,
Wind River Mountains. *In*: BARAZANGI, M. &
BROWN, L. (eds) *Reflection seismology: A global
perspective*. American Geophysical Union,
WASHINGTON, DC, 223–236.

SIMPSON, C. & WINTSCH, R.P. 1989. Evidence for
deformation-induced K-feldspar replacement by
myrmekite. *Journal of Metamorphic Geology*, **7**,
261–275.

SIMS, P.K. 1995. *Archean and Early Proterozoic
tectonic framework of north-central United States
and adjacent Canada*. US Geological Survey
Bulletin, **1904**, Chapter T.

SIMS, P.K. & STEIN, H.J. 2003. Tectonic evolution of
the Proterozoic Colorado province, Southern
Rocky Mountains: A summary and appraisal.
Rocky Mountain Geology, **38**, 183–204.

SMITHSON, S.B., BREWER, J.A., KAUFMAN, S.,
OLIVER, J.E. & HURICH, C.A. 1979. Structure of
the Laramide Wind River uplift, Wyoming, from
COCORP deep reflection data and from gravity
data. *Journal of Geophysical Research*, **84**,
5955–5972.

SNOKE, A.W. & RESOR, P.G. 1999. Strain partitioning
and other complexities associated with a Proterozoic deformation front, Laramie Mountains,
Wyoming. *Geological Society of America
Abstracts with Programs*, **31**(7), A–259.

SNOKE, A.W. & TULLIS, J. 1998. An overview of fault
rocks. *In*: SNOKE, A.W., TULLIS, J. & TODD, V.R.
(eds) *Fault-related Rocks*. Princeton University
Press, Princeton, NJ, 3–18.

SNYDER, D.B. 2002. Lithospheric growth at margins of
cratons. *Tectonophysics*, **355**, 7–22.

SNYDER, D.B., PRASETYO, H., BLUNDELL, D.J.,
PIGRAM, C.J., BARBER, A.J., RICHARDSON, A. &
TJOKOSAPROETRO, S. 1996. A dual doubly
vergent orogen in the Banda Arc continental–arc
collision zone as observed on deep seismic reflection profiles. *Tectonics*, **15**, 34–53.

SNYDER, G.L., BUDAHN, J.R. ET AL. 1998. *Geologic
Map, Petrochemistry, and Geochronology of the
Precambrian Rocks of the Bull Camp Peak quad-
rangle, Albany County, Wyoming*. US Geological
Survey Miscellaneous Investigations Series Map.
I–2236.

SNYDER, G.L., SIEMS, D.F. ET AL. 1995. *Geologic Map,
Petrochemistry, and Geochronology of the Pre-
cambrian Rocks of the Fletcher Park–Johnson
Mountain area, Albany and Platte counties,
Wyoming*. US Geological Survey Miscellaneous
Investigations Series Map, **I–2233**.

SPEED, R.C. & SLEEP, N.H. 1982. Antler orogeny and
foreland basin: A model. *Geological Society of
America Bulletin*, **93**, 815–828.

SUPPE, J. 1987. The active Taiwan mountain belt. *In*:
SCHAER, J.-P. & RODGERS, J. (eds) *The Anatomy
of Mountain Ranges*. Princeton University Press,
Princeton, NJ, 277–293.

TEALL, J.J.H. 1885. The metamorphosis of dolerite
into hornblende-schist. *Quarterly Journal of the
Geological Society, London*, **41**, 133–145.

TOMLIN, K.P. 2001. *Multiphase Proterozoic deformation along the southeastern margin of the
Archean Wyoming province: A structural analysis
of the northeastern Elmers Rock greenstone belt,
central Laramie Mountains, SE Wyoming*. MS
thesis, University of Missouri, Columbia.

TOMLIN, K.P. & BAUER, R.L. 2000. Multiphase
Proterozoic reworking of the southeastern margin
of the Archean Wyoming province: Evidence
from the northern Elmers Rock greenstone belt,
central Laramie Mountains, Wyoming. *Geological
Society of America Abstracts with Programs*,
32(7), A-454.

TYSON, A.R., MOROZOVA, E.A., KARLSTROM, K.E., CHAMBERLAIN, K.R., SMITHSON, S.B., DUEKER, K.G. & FOSTER, C.T. 2002. Proterozoic Farwell Mountain–Lester Mountain suture zone, northern Colorado: Subduction flip and progressive assembly of arcs. *Geology*, **30**, 943–946.

VAN SCHMUS, W.R., BICKFORD, M.E. *ET AL.* 1993. Transcontinental Proterozoic provinces. *In*: REED, J.C., Jr. & 6 others (eds) *Precambrian: Conterminous U.S.* Geological Society of America, The Geology of North America, **C-2**, 171–334.

WERNICKE, B. 1990. The fluid crustal layer and its implications for continental dynamics. *In*: SALISBURY, M.H. & FOUNTAIN, D.M. (eds) *Exposed Cross-sections of the Continental Crust.* NATO Advanced Studies Institute. Kluwer, Dordrecht, 506–544.

WHITE, S.H., BURROWS, S.E., CARRERAS, J., SHAW, N.D. & HUMPHREYS, F.J. 1980. On mylonites in ductile shear zones. *Journal of Structural Geology*, **2**, 175–187.

Seismic and aseismic weakening effects in transtension: field and microstructural observations on the mechanics and architecture of a large fault zone in SE Tibet

M. A. EDWARDS[1] & L. RATSCHBACHER[2]

[1]*Structural Processes Group, Department of Geological Sciences, University of Vienna, Althanstrasse 14, Vienna A-1090, Austria (e-mail: michael.edwards@univie.ac.at)*
[2]*Institut für Geologie, TU-Bergakademie Freiberg, D-09596 Freiberg, Germany*

Abstract: Fault-zone surveying and microstructural analyses focus on a transtensionally strained section of a >100 km-long, 5–15 km-wide, active, sinistral strike-slip W–E fault – the Damxung–Jiali Shear Zone (DJSZ). Deformation fabric superposition, palaeostress and neotectonic measurements reveal a progressive dominance of the transtension coaxial component. Vertical flattening and broadening of the DJSZ results in a 10–20 km-wide pull-apart depression with reciprocal emergence of flanking highlands that expose abandoned fault-zone domains that are hundreds of metres by tens of km in size. Polylithological fault rocks preserve a suite of frictional and viscous deformation to *c.* 350 °C. Amongst the fault strands, intensity and distribution of grain size reduction from comminution, widespread solution transfer, intragranular plasticity and recrystallization is heterogeneous. A regular pattern of pseudotachylite-coated surfaces short-cutting strong volcanic mylonite domains between weak creeping calcite domains indicates seismic with aseismic strain, and is interpreted as velocity-weakening evidence. Re-brecciation of solution-transfer accommodation-assisted S–C-cataclasite domains document switches from frictional to viscous strain and back again. The co-location of brittle and plastic constituent behaviour for varied fault lithologies plus our interpretations on deformation temperatures suggests major overlap for the frictional and viscous depth ranges. This indicates a very broad and strongly velocity- and fluid-dependent 'brittle–ductile' transition.

A clearer understanding of the behaviour of intracrustal faulting in general, in particular the factors governing frictional displacement characteristics, as well fault-zone growth and overall lifespan, are central to the study of modern tectonics and the behaviour of the crust in failure. The faulting realm where seismogenic (i.e. frictional) displacement or nucleation is primarily governed by pressure-sensitive deformation mechanisms associated with brittle failure, dilatancy and frictional sliding (e.g. Brace & Byerlee 1966), and how this realm mechanically passes in space and time into the viscous (and nominally aseismic) creep realm (Byerlee 1978; Sibson 1983; Scholz 1988; Gratier & Gamond 1990), is a key target in fault zone studies. Much attention has been focused over the past decade or two on understanding the spatial and temporal breadths of, and fluctuations in, the frictional–viscous transition in nature and the degree to which, for example, frictional

strength maxima in the upper crust may be (dramatically) diminished in size and depth by the weakening effects of fluid interaction, grain size and phase transitions, and fabric anisotropy in general, as well as the questions of which are the load-bearing elements or frictionally stable elements in polymineralic fault rocks and in multiple lithology shear zones (e.g. Schmid & Handy 1991; Hacker 1997; Handy *et al.* 1999; Stewart *et al.* 2000; Holdsworth *et al.* 2001). The manifestations of such 'weakening' processes in terms of: (1) changes in strain rate with strength over time; and (2) stick–slip v. stable sliding (e.g. Dieterich 1972; Scholz 1998) remain difficult to transfer from the laboratory to nature; constraints on ideal fault rock weakening, interslip recovery and rate/state-variable characteristics are measured/experimentally determined within necessarily limited boundary conditions of fault rock type, texture, mineralogy, grain size, fluid content and displacement rate/step uniformity

From: BRUHN, D. & BURLINI, L. (eds) 2005. *High-Strain Zones: Structure and Physical Properties.*
Geological Society, London, Special Publications, **245**, 109–141.
0305-8719/05/$15.00 © The Geological Society of London 2005.

(e.g. Logan *et al.* 1979; Rutter *et al.* 1986; Marone *et al.* 1990; Dieterich 1992; Karner *et al.* 1997; Bos & Spiers 2001). Although, the general rate- and state-dependent friction laws can account for (to a first approximation, but see Bos & Spiers 2002*a*) the full range of fault behaviour observed in nature (from aseismic or 'silently creeping' (e.g. Miller *et al.* 2002; Nadeau & McE- villy 2004) to fully coupled fault displacement (e.g. Scholz 1998)), the composition and architec- ture of active fault zones are typically hidden at depth and are inaccessible beyond subsurface imaging and drilling. Meanwhile, exhumed fault-zone studies are hampered by notoriously poor preservation of exposed fault zones due to enhanced preferential weathering. There are sur- prisingly few detailed descriptions of internal fault-zone architecture and these are mainly from highly localized, narrow ($\ll 1$ km), low- asymmetry, parallel-boundary, transpressional fault zones that typically involve one principal host lithology (although cf. Wibberley & Shima- moto 2003), such as the East California Shear Zone or the San Andreas and its relatives (Chester & Logan 1986; Caine *et al.* 1996; Miller 1996) or the slightly broader Carboneras Fault in the Betics (Faulkner *et al.* 2003). In trans- tensional fault systems where zone boundaries are non-parallel and shear-zone asymmetry is high (e.g. triclinic), kinematics of flow have been shown to be significantly more complicated than for simple San Andreas type examples (e.g. Sanderson & Marchini 1984; Jiang & White 1995; Dewey 2002) and it is likely that the mechanics of such fault zones are also somewhat different and may be quite complex (due, for example, to the effects of crustal thinning and shear-zone broadening). To our knowledge, the deformation characteristics, mechanics and overall architecture of multilithology, transten- sional, non-parallel-boundary fault zones hosted in the frictional crust have received almost no attention. We therefore focus on a such a fault zone that is fortuitously well preserved and well exposed; we describe the fault-zone architecture, rock type and meso- and microstructure in an obli- quely diverging section of a $\gg 100$ km left-later- ally displacing 'transcurrent' wrench fault in Tibet, the Damxung–Jiali Shear Zone (DJSZ). In the DJSZ, a range of lithologies, transtension- ally sheared since the Miocene(?) are exhumed (from temperatures of up (down) to greenschist facies) in now-abandoned fault strands, giving an opportune view of the range of fault products down through the frictional–viscous transition of the crust and revealing how their deformation on the field and microstructural scale varies with lithology and geometry over space and time. We describe these fault products and then speculate about the reasons for their development and distri- bution, as well as the implications for seismic or aseismic weakening specific to a transtensional setting.

Geological setting

Regional setting

The Tibet plateau together with the Himalaya (*c.* 3500 × 1500 km) represents the accommo- dation of convergence since India first collided with Asia at *c.* 65–55 Ma (Patriat & Achache 1984; Rowley 1998). This orogenic system (Fig. 1, inset) has incorporated up to 2000 km (Achache *et al.* 1984; Besse *et al.* 1984) of con- tinental crust into the southern margin of Asia as India has continued its northward advance. This deformation, together with prior deformation from Andean/accretionary type orogenesis and terrane addition, has resulted in a near doubling of normal crustal thickness through some mixture of, for example, underthrusting, bulk shortening and widespread fold–thrust defor- mation (e.g. Argand 1924; Dewey & Burke 1973; Powell & McConaghan 1973; Owens & Zandt 1997). This input of continental crustal mass into the collisional system is associated with, and is probably extensively accommodated by (at least in the upper crust), E-directed crustal escape whereby major portions of the Tibet plateau are being displaced out into SE China (e.g. Tapponnier *et al.* 1982; although see also Clark & Royden 2000). All this fresh, young deformation does indeed make Tibet the prover- bial 'natural laboratory' for fault-zone mechanics studies; a series of major transcurrent faults (Fig. 1, inset), some of a magnitude possibly akin to the San Andreas with displacements of several 100 km, have formed over the course of the plateau growth (Tapponnier & Molnar 1977; Leloup *et al.* 1995; van der Woerd *et al.* 1998) and the DJSZ (Fig. 1) is part of such a fault system.

Local setting

The DJSZ is situated in a region of active seismi- city in the Lhasa block (northern part of Fig. 1), a pre-collisional terrane in Tibet (Chang *et al.* 1986; Dewey *et al.* 1988). The rocks around the area of the DJSZ provide the fault protolith and consist of predominantly Mesozoic(?) passive margin and intra- to back-arc basin rocks (Coward *et al.* 1988; Kidd *et al.* 1988; Leeder *et al.* 1988; Pan & Kidd 1992). Regional fold and thrust trains trending ESE–WNW are

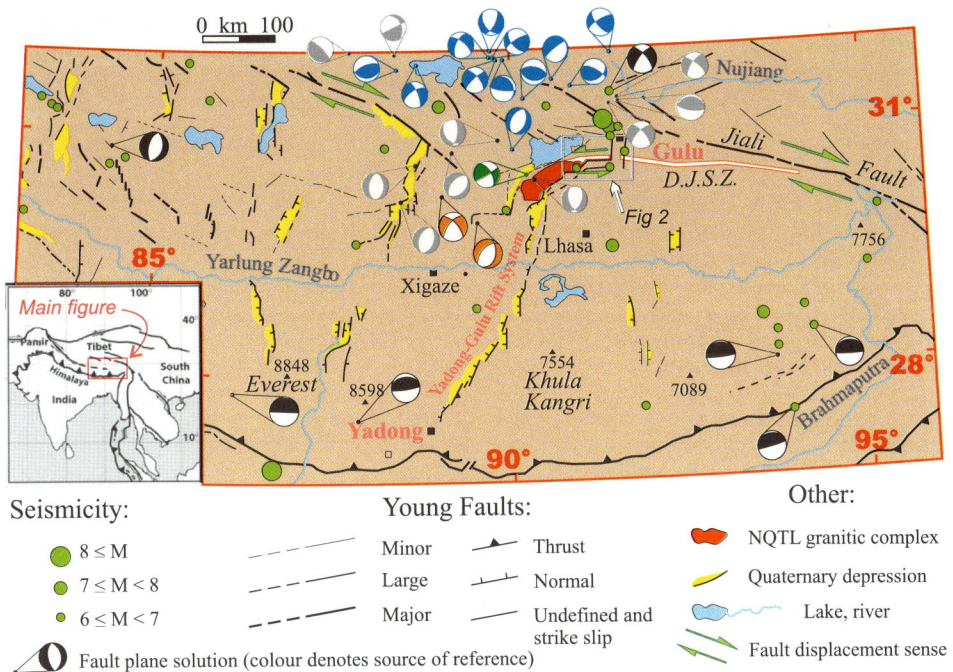

Fig. 1. Neotectonic map of greater region of SE Tibet after Armijo *et al.* (1986) and Edwards *et al.* (2001). ESE-trending en échelon fault suites present all along top of map are parallel to, post-date and have focused on the structural grain of the fault and thrust belt in northern part of map area. Major villages and Himalayan peaks are shown for orientation. The thrust fault present along the base of the map is the main frontal thrust to the Himalayan collision. Earthquake data: green spots, major seismicity from the Institute of Geophysics via Armijo *et al.* (1986); focal mechanism solutions (lower-hemisphere, compressional quadrants filled) are: black, Molnar & Chen (1983); orange, subcrustal events of Armijo *et al.* (1986); blue, INDEPTH III locations computed by Langin *et al.* (2003); grey, Harvard CMT solutions or other sources (Verma & Reddy 1988; Molnar & Lyon-Caen 1989; Randall *et al.* 1995); green, Mechie *et al.* 2004. Processed INDEPTH III data are restricted to 87.5–92.5°E and >31.0°N. Principal trace of the Jiali Fault after Armijo *et al.* (1989) and Tapponnier *et al.* (2001). Trace of DJSZ (Damxung–Jiali Shear Zone) is based on this study, Armijo *et al.* (1989) and interpolation using GeoTopo DEM data.

comprised of thrust nappe repetitions of Jurassic–Cretaceous (volcani-)clastic rocks and are topped by thick Aptian–Albian carbonate platform sheets – the most competent member of the regional deformation (Coward *et al.* 1988; Leeder *et al.* 1988; Edwards *et al.* 2001). The prior deformation in the DJSZ protolith is therefore relatively minor, and only in locally tight fold hinges (folding is otherwise open) is deformation enhanced with some pressure solution in limestone hinges and more intense intersecting cleavage and fine fracturing in mudstones. Lithological packages in the regional rocks are otherwise intact. These are sinistrally transposed and reworked where the DJSZ cuts through the region (Fig. 1).

The DJSZ connects the 100s km-long Jiali fault in the east (Tapponnier *et al.* 1986; Armijo *et al.* 1989) with the Nyainqentanghla (NQTL, a granitoid cored range) in the west.

The DJSZ was hitherto reported (and named) in reconnaissance from two locations by Armijo *et al.* (1989); from our area of study (Fig. 2) and a locality *c.* 100 km to the east. In our area of study, the NQTL range forms the NW shoulder (i.e. topographic flanking boundary) of the NE–SW-trending Nyinzhong Graben (Fig. 2), a portion of the Yadong–Gulu Rift System (Fig. 1, YGRS). The Nyinzhong Graben links NE into the E–W-trending Damxung Corridor and it, in turn, links into the N–S-trending Gulu Rift (the principal northern terminus of the YGRS). The Damxung Corridor is a sinistral, Quaternary(?) material-filled, roughly rhomb-shaped pull-apart basin that is formed by a local releasing bend segment in the active DJSZ. Active (or recent) tectonic features that are related to the DJSZ are present throughout here and on up into the north, in the rift shoulder of the Damxung Corridor (Fig. 2). The rift shoulder

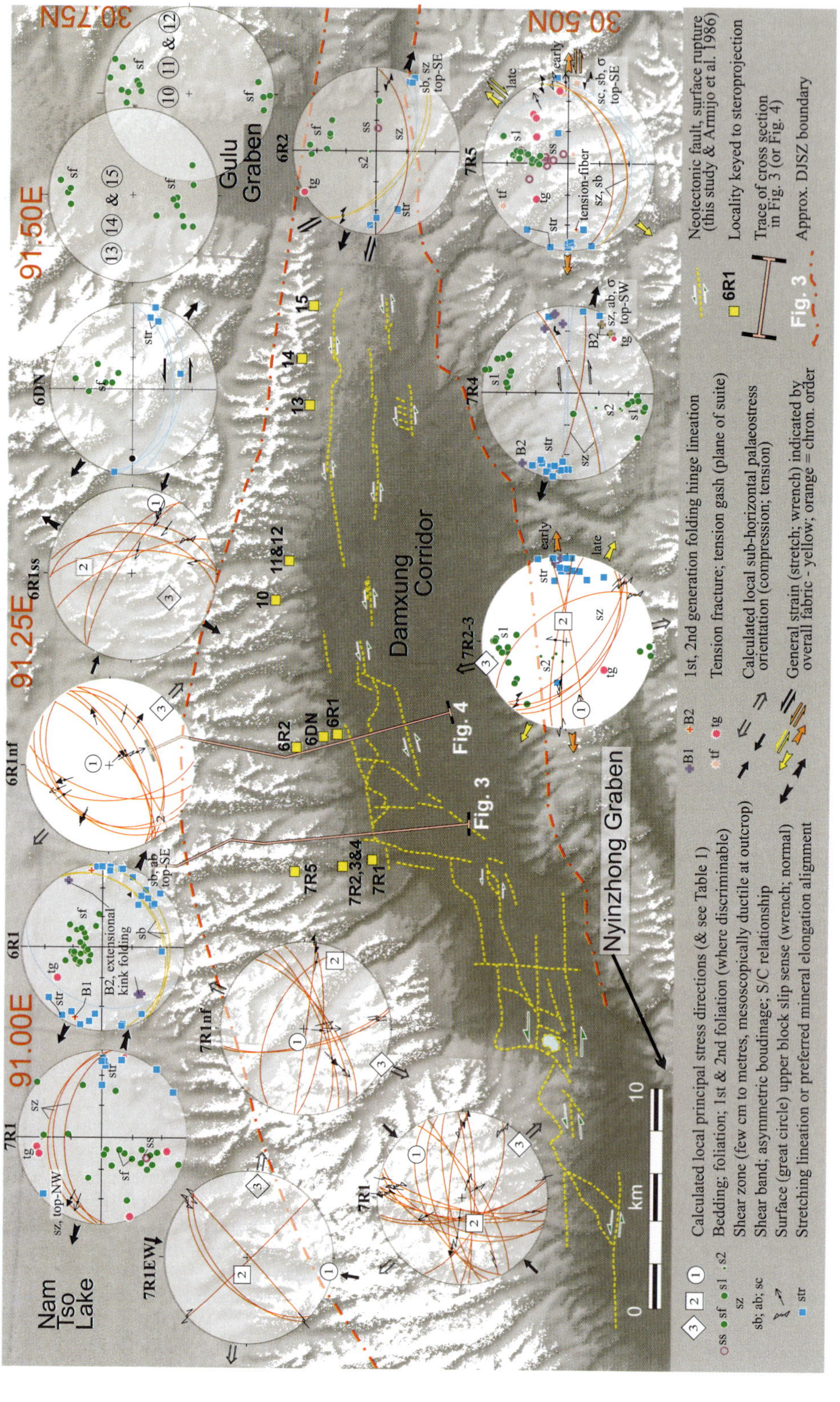

Nam Tso Lake
91.00E
91.25E
91.50E
30.75N
30.50N
Gulu Graben
Damxung Corridor
Nyinzhong Graben
Fig. 3
Fig. 4
7R1
7R1EW
7R1nf
6R1
6R1nf
6R1ss
6DN
6R2
7R2-3
7R4
7R5
7R2,3&4
10
11&12
13
14
15
(13) (14) & (15)
(10) (11) & (12)
B2, extensional kink folding
sz, top-NW
sb, ab top-SE
sb, sz top-SE
sc, sb, σ top-SE
sz, ab, σ top-SW
tension-fiber
early
late
0
10
km
Calculated local principal stress directions (& see Table 1)
Bedding; foliation; 1st & 2nd foliation (where discriminable)
Shear zone (few cm to metres, mesoscopically ductile at outcrop)
Shear band; asymmetric boudinage; S/C relationship
Surface (great circle) upper block slip sense (wrench; normal)
Stretching lineation or preferred mineral elongation alignment
1st, 2nd generation folding hinge lineation
Tension fracture; tension gash (plane of suite)
Calculated local sub-horizontal palaeostress orientation (compression; tension)
General strain (stretch; wrench) indicated by overall fabric - yellow; orange = chron. order
Neotectonic fault, surface rupture (this study & Armijo et al. 1986)
Locality keyed to steroprojection
Trace of cross section in Fig. 3 (or Fig. 4)
Approx. DJSZ boundary

comprises an approximately E–W trending chain of foothills (500–1500 m of relative relief) where now-abandoned, exhumed strands of the 'historic' DJSZ are present throughout. The fault rocks are well exposed by a series of steep drainage valleys that fortuitously run approximately N–S, at a high angle through the shear zone. Between the NE NQTL and the Gulu Rift (Fig. 2), the DJSZ is exposed for c. 80 km, continuing beyond here for more than 150 km to the main Jiali fault (Armijo et al. 1989; Edwards et al. 2000). The effective western limit of the historic DJSZ is not surveyed, but seems to be complexly involved with, and partly obliterated by, the NE margin of the NQTL granitic complex whose protracted intrusion history (D'Andrea et al. 2001; Kapp et al. 2002) is probably synkinematic pull-apart related (Edwards et al. 2000). For the remainder of this paper, when describing our observations, we use the term DJSZ to refer to the study area (Fig. 2).

Overall structure

From west to east, the structural thickness and surface exposure width of the exhumed DJSZ changes significantly. The plan view outcrop is more than 10 km in the region of section A (Fig. 2), and less than 5 km at our traverse numbers 14 and 15. The structural fabric data (mixed stereoplots in Fig. 2) show the main shear-zone overall fabric; moderately to steeply N- or S-dipping, trending approximately E–W. The regional ESE–WNW structural grain of the surrounding folded and thrust country rocks are sinistrally transposed (by 20–30° from ESE–WNW to E–W) and further shortened where they are overprinted by the shear zone. The degree of DJSZ shear must be quite large; thrust repetition packages with 10–15 km of separation outwith the shear zone are attenuated to less than 2 km of separation where repetitions are still recognizable.

Although most exhumed DJSZ rocks seem continuous from west to east, there are numerous occurrences of dismemberment or discontinuity, particularly in the east. This is due to: (1) local excision by internal shears of a few kilometres in length; (2) km-scale boudinage; and (3) possible original sedimentary depocentre boundaries. There is gentle wavelength anastomosing (trending horizontally E–W, as well as vertically) that is frequently associated with partitioning of fault rocks into apparent higher and lower strain domains. We use the term 'domain' in a general sense to refer to a region within the DJSZ of distinct lithology, deformation style and intensity. From approximate surface shape and pattern interpolating between visited outcrops, the DJSZ domains are boudin or lens-like and are usually a few 100 m wide, a few kilometres long, but may be less than 1 km in width and more than 20 km long. It is unclear how continuous at depth the domains are.

The DJSZ portion corresponding to the Damxung Corridor pull-apart basin is 5–15 km wide, bounded by a range of sinistral-oblique faults (see below). Project INDEPTH (International Deep Profiling of Tibet and the Himalayas) seismic reflection profiling was conducted through this area (Alsdorf et al. 1998; Cogan et al. 1998). Forward modelling of inverted ray traces indicates velocities commensurate with an unconsolidated graben fill thickness of <1 km (Cogan et al. 1998) represented in sections A and B (Figs 3 and 4). Below here is less clear; no offset of reflector packages are observed on the INDEPTH shear-zone-normal reflection profiles. We speculate on the deeper structure of the DJSZ in the Discussion.

Fault-zone rock types

Rocks that are deformed and reworked within the exhumed DJSZ are: (1) staurolite schist and other calc-schists; (2) marbles and dolomites; (3) metapelites, metagreywackes and other detrital metaclastics; (4) cataclasites and tectonic breccias; (5) conglomerate; (6) acid metavolcanics; (7) ultramafic rocks; and (8) mineral veins, travertines, pseudotachylites and other

Fig. 2. Shaded contour map of NE YGRS area showing neotectonic data and stereographical representations of structural data for the DJSZ. The topographic interval between 4000 and 6000 is differentiated (dark grey–white). Lats./longs. are in decimal degrees. All structural measurements for a given locality (7R1, 7R13 and 7R14, etc.) are shown together on individual equal-area stereo projections, except for palaeostress data separated according to determined faulting regime (e.g. 6R1nf, etc.). nf, normal faulting; ss, strike-slip. Poles to planes are shown for all planar data except fault data and mesoscopic shear zones. For fault plane data, great circles are fault surfaces, straie are shown as arrows pointing in the direction of hanging-wall displacement. Confidence level of slip sense determination is shown by arrow head style: full, certain; open, reliable; half, poor, absent, very poor. For stress-inversion treatment see Appendix A.

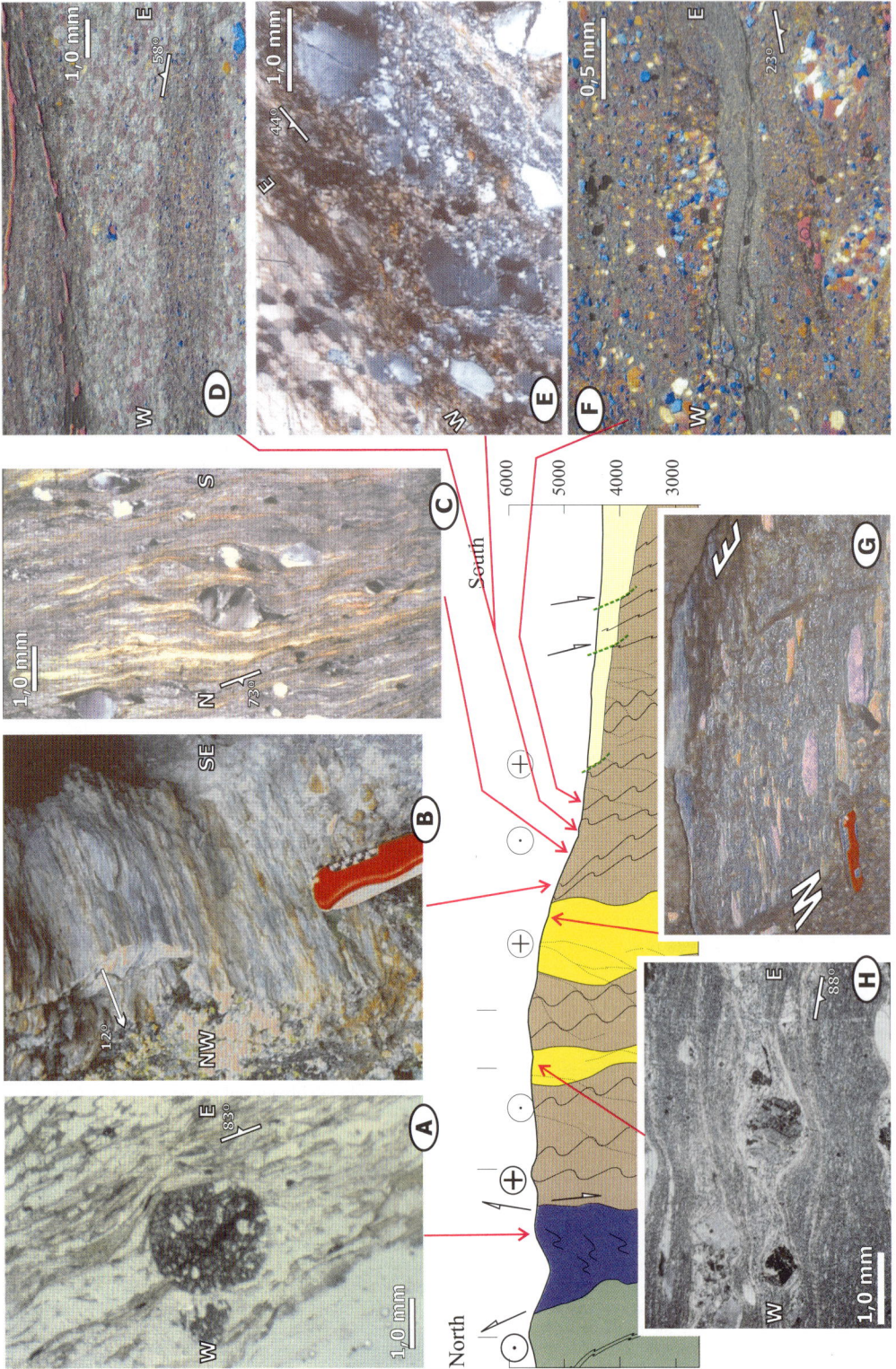

Fig. 3. Cross-section (line of trace noted in Fig. 2): colours are (left to right): pale green, general Cretaceous(?) country rock; deep blue, staurolite schist; brown, undifferentiated DJSZ rocks; rich yellow, stretched pebble conglomerate; pale yellow, unconsolidated fill of Damxung pull-apart graben. Fabric is schematically illustrated: dashed or solid black line represents weak or strong foliation, respectively (the exception is intense schistosity in the schist slice – see image A and Fig. 4 image A). Buckling (schematically shown) of main foliations and minor crenulation is related to general vertical shortening. Open half-arrows show the relative displacement across faults and shear zones. Field and optical photo(micro)graphs (all thin sections were originally cut perpendicular to main foliation, parallel to main the DJSZ mineral lineation): (**A**) Inter-kinematic (late, but predates final deformation) garnet growth in amphibolite–garnet–staurolite schist. Schist locally forms the northern margin of the DJSZ. Mineral lineation, 72° towards 137°; foliation, 83° towards 078°; plane polarized light. No unequivocal sense of shear. (**B**) Looking NE onto the steep main foliation surface typical of a marble domain outcrop in carbonate mylonite north of 7R3. Lineation (12° towards 286° – defined here by alternating clay-rich against clay-poor compositional layering (locally prolate rod forming) due to high (locally constrictional) strain – see also image D)) is subparallel to the fold hinge (12° towards 277° – tens of cm wavelength, few cm amplitude). Folding and local crenulation of steep foliation (26° towards 161°) visible here is second order to 2–5 m scale buckling (axial surface = S_2) present across the whole outcrop (cf. Fig. 6b) that is associated with transtensional general vertical shortening present throughout the DJSZ. Third-order folding on this surface is local and due to solution crenulation intersecting with strong stretching lineation. The steep fracture surface truncating the left side of the prominent (convex to viewer) hinge area in the upper middle part of the image is typical of a brittle suite developed at a high angle to a later stage stretching axis that is displaced dextrally (antithetic to DJSZ: cf. Fig. 2 stereoplot 7R2-3). These carbonate mylonites alternate (domains tens to hundreds of metres in size) with phyllonitic metapelite and mylonite metavolcanics in packages between purple–green conglomerate (yellow fill on the cross-section). The intensity of strain in the domains fluctuates but is not proportional to the distance from the pull-apart basin margin. The Swiss knife is *c.* 3 cm wide. (**C**) Intensely strained phyllonitic metapelite near 7R3. The mineral lineation is 09° towards 078°; foliation, 73° towards 338°; crossed polars. Note the strong crystallographic preferred orientation (CPO), the compositional banding of micas with fine-grained quartz and the relative rigid behaviour of large (0.5–2.0 mm) quartz porphyroclasts with minor recrystallization deformation forming tails. Note the preservation of originally straight crack boundaries (two clasts were possibly originally a single grain), yet the absence of renewed cracking/continued dilatational strain processes (cf. image E). Kinematic indicators (e.g. asymmetry of porphyroclasts tails, C′ surfaces and S–C relationships) in these rocks are >80% sinistral. (**D**) Extensively dynamically recrystallized coarser (100–300 μm) and finer (<100 μm) grain size, pure and impure (increased clay component), respectively, calcite compositional layering within the marble of the carbonate mylonite domain near 7R1. C- and C′-oriented very-fine-grained clay-concentration surfaces (here in purple) with a strong CPO indicate the enhanced role of fluid-assisted solution transfer accommodated deformation and volume loss. Lineation, 14° towards 103°; foliation, 58° towards 025°; crossed polars with gypsum plate. C–C′ and S–C relationships favour a sinistral sense of shear. (**E**) 'dirty' portion of the carbonate domain from near 7R1 (i.e. adjacent to image D) characterized by sideritic alteration, large (0.5–2.5 mm) quartz grains and heterogeneously concentrated clay minerals. Calcite microdomain in the upper-right hand corner is quite pure and typical of large (1–2 × 5–15 mm) oval-shaped (bio?)clasts that are dynamically recrystallizing at tips to form prolate-type rods. Lineation, 13° towards 106°; foliation, 44° towards 029°; crossed polars. Fractured grain offsets, C–C′ relationships and mica 'fish' suggest a sinistral sense of shear. Irregular distribution of brittle-fractured quartz grains (not well depicted here) is possibly the result of local to switches from well to poorly drained conditions (fluid flow cyclicity? – see text). 7R1 is near the margin of Damxung pull-apart basin where enhanced fracturing, solution and fluid products are more common. (**F**) Alternately strongly and weakly foliated cataclasite south of 7R1 showing renewed tectonic brecciation and extensive solution transfer deformation accommodation with substantial material loss (cf. Fig. 4 images D & F, and Fig. 5 images A–D). Quartzite lithic fragments (<0.5 mm) and (derived thereof?) quartz grain-bearing weakly foliated cataclasite lithic fragments are re-brecciated juxtaposing foliated/unfoliated and coarse/fine/heterogeneously distributed grain size microdomains. Thick (tens of μm) dark solution seams are chlorite/clay concentration sites and mark sharp contrasts between microdomains of varying quartz grain proportions and grain size distribution. The gypsum plate reveals general weak/absence of quartz CPO (cf image F). Lineation (weak) 17° towards 297°; foliation, 23° towards 207°; crossed polars, gypsum plate. Trustworthy kinematic indicators are absent but the obliquity of the pressure solution foliation to the overall DJSZ fabric is consistent with the shortening direction expected for sinistral progressive transtension. (**G**) Prolate stretch in purple-green matrix conglomerate cropping out in the riverbed at 7R4. Ignimbrite tuff, quartzite and rare dolomite clasts, as well as a gritty matrix, are visible. A steep fracture surface (running off the upper-left margin of the image from the Swiss knife) intersects several clasts for 3D control of prolate strain (see Fig. 6C inset). The peppering of very bright specks is not a matrix component (camera flash reflection from river). The Swiss knife is 11 cm long. (**H**) Acid metavolcanic mylonite north of 6R2. Fine (50–100 μm)-grained, load-supporting groundmass is deforming via continued viscous flow of quartz grains (+micas), interpreted to be fluid-assisted diffusion creep sensitive to grain size. K-Feldspar porphyroclasts (1–5 mm) are brittlely deforming and passively spinning (*c.* 1 mm grain in the middle of the images is cracking and forming V-pull-apart (Hippertt 1993) and simple shearing along growth twins), showing mainly brittle behaviour with possible phase alteration at the microcrack boundaries, indicating maximum deformation temperatures of *c.* 350 °C. Large grain size distribution may be partly inherited from protolith preventing piezometric estimate of the flow stress (see text). Outcrop distribution of domains of this rock is the key indicator of velocity weakening (see text). Lineation is 04° towards 273°; foliation is 88° towards 197°; crossed polars. Kinematic indicators in these rocks are strikingly sinistral.

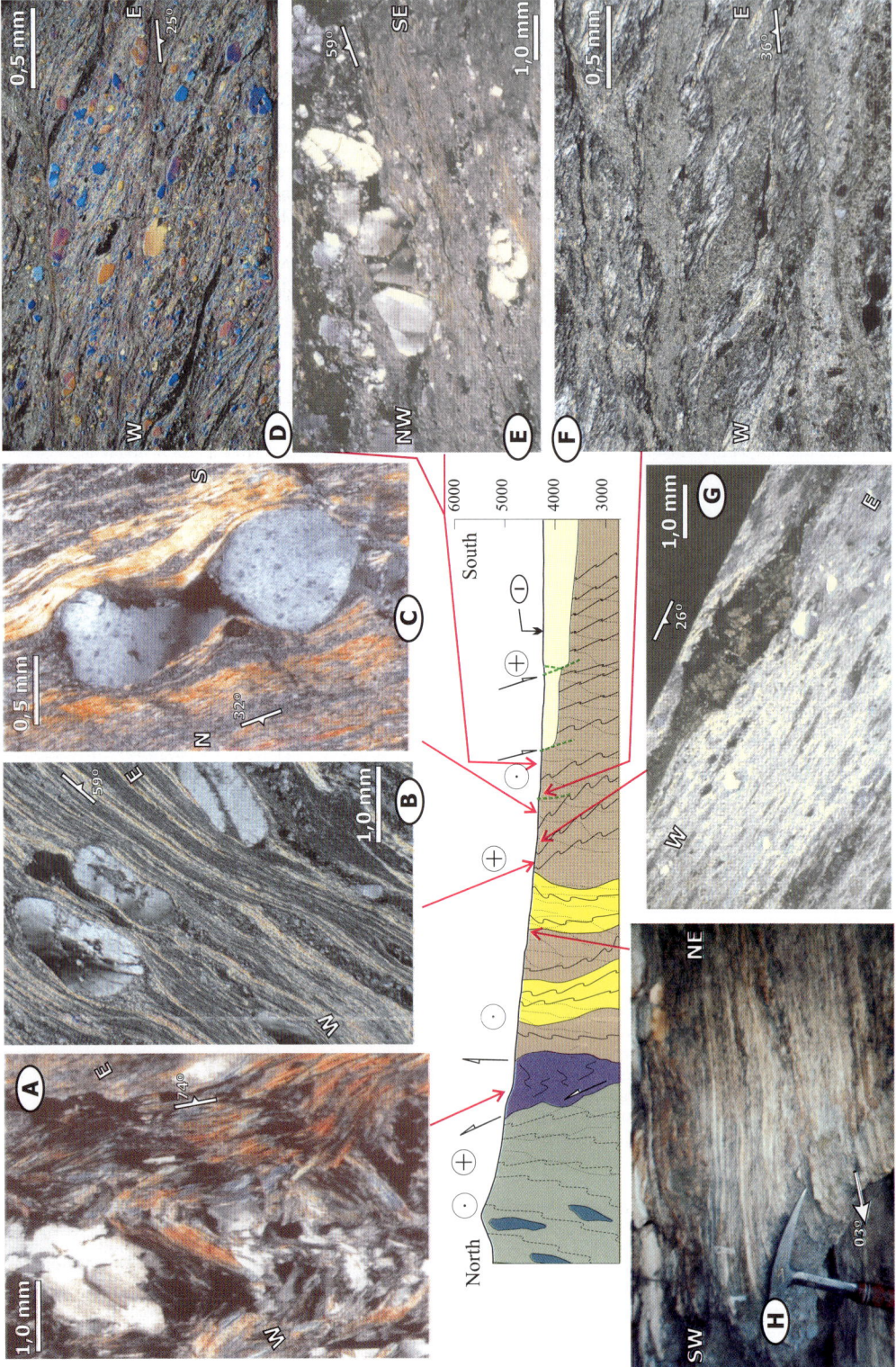

fluid/seismicity products. Examples of the most common of these are displayed in photo(micro)graphs in Figures 3 and 4, and keyed to locations on the cross-sections (A and B) where domains of a given lithology crop out. They are listed here in their order of appearance on Figure 3. On the cross-sections, only the schists and the conglomerate are differentiated (by fill colour) from the default 'DJSZ' rocks, as these are the only rocks that form large, clear layers of sufficiently uniform thickness and continuity to be confidently correlated across several valleys in the field, as well as on remotely sensed imagery.

Staurolite schist and other calc-schists

In many of the DJSZ valleys muscovite- and calc-schist float is present and rare tens of metres by metre-scale lozenges are found in outcrop (often severely altered). In the western part of the DJSZ, in the upper parts of the range north of locations 7R5 and 6R2 on traverses number 7 and 6 (Figs 3 and 4), respectively, is a 1–3 km wide, >10 km long slice of amphibolite–garnet–staurolite-schist, retrogressed from sillimanite grade (image A in Fig. 3). The staurolite-schist is notably higher grade than the other DJSZ rocks and has, in addition, undergone up to

---

Fig. 4. Cross-section (line of trace noted in Fig. 2): colours are as for Fig. 3 except: turquoise pods within the country rock (pale green fill) are altered and dismembered ophiolitic ultramafic bodies. Field and optical photo(micro)graphs (all thin sections were originally cut perpendicular to the main schistosity, parallel to the main mineral lineation). (**A**) 2–3 mm wavelength DJSZ-related (re)folding of the pre-existing deformation fabric in polydeformed staurolite schist (here retrogressed from sillimanite grade) at the margin of DJSZ speculated to be exhumed as a positive flower structure (the thrust arrows on the cross-section indicating the relative upthrow of a deep blue coloured slice). DJSZ has fortuitously not 'reset' micas that preserve Cretaceous cooling (155 Ma Ar/Ar: Copeland 1990) associated with prior deformation event. Coincidence in space of two 150 Ma separated shear products suggests possible crustal fault reactivation (see text). Lineation, 51° towards 031°; foliation, 74° towards 118°; crossed polars. No unequivocal sense of shear. (**B**) Beautifully mylonitized portions of acid volcanics. Feldspar grains have partial and through-going cracks while dynamically recrystallized fine-grained quartz and mica form very continuous, strong CPO, evenly spaced compositional (+grain size variable) banding that is probably undergoing fluid-assisted grain size sensitive flow. Lineation is 08° towards 277°, foliation is 59° towards 204°, crossed polars. Kinematic indicators are overwhelmingly sinistral. (**C**) The mylonitic/phyllonitic metapelite south of 6R2 showing strong CPO and quartz-mica composition astride cracking feldspar grains. Lineation, 17° towards 117°; foliation, 32° towards 226°; crossed polars. Kinematic indicators in these rocks are >80% sinistral. (**D**) Beautiful 'brittle realm' S–C fabric in cataclasite near 6DN. Brittly fractured quartz grains with heterogeneous grain size distribution (tens to 200 μm) and relatively weak CPO (revealed by gypsum plate) sit within stretched-out microdomains of chlorite + muscovite + sericite alternating with very fine-grained clay minerals with pressure-solution-induced clay-concentration surfaces defining S and C orientations. Lineation is 14° towards 105°, foliation is 25° towards 193°, crossed polars, gypsum plate. Kinematic indicators are unequivocally sinistral. (**E**) Phyllonitic metapelite with heterogeneously deformed microdomains in a downthrown fault block near the margin of the DJSZ pull-apart. Mica and fine-grained (10–100 μm) quartz-rich layers show strong CPO, dynamic recrystallization and grain size reduction. Locally present adjacent to the microdomain of heterogeneously distributed clay and siderite are large (0.5–2.0 mm) brittly fractured, angular and non-rotated quartz grains (possibly reflecting fluctuations in fluid pressures/drainage conditions in this area – see the text). Lineation is 08° towards 277°, foliation is 59° towards 204°, crossed polars. Kinematic indicators are sinistral where unequivocal. (**F**) A portion of foliated cataclasite recording dramatic evidence of brittle–viscous and back to brittle behaviour. Sericite concentration together with new growth (+scattered quartz grains) define mica-rich S–C foliation (seen as white and bluish-grey obliquely-oriented lensoids in the upper and middle portions of the photograph) due to deformation of the former cataclasite – similar to image D, above. S–C foliated cataclasites enjoyed subsequent re-brecciation forming new (unfoliated) fine-grained cataclasite microdomains (seen here in green/brown) locally bearing dispersed, fine (50–100 μm) brittly fractured quartz grains. Fresh cataclasite material forms 200–400 μm-wide bands running horizontal (i.e. E–W) across the photograph, as well as infilling portions running diagonally NW–SW in the centre of the photograph (where pre-existing foliated cataclasites have now been pulled apart and separated parallel to the pre-existing S–C fabric, thereby forming lithic fragments of S–C foliated cataclasite in cataclasite!) Lineation, 08° towards 126°; foliation, 36° towards 031°; crossed polars. Kinematic indicators within pre-existing foliated cataclasites are overwhelmingly sinistral but inadequate within new cataclasite portions. (**G**) Millimetre-scale compositional banding in a rather pure portion of dynamically recrystallized carbonate mylonite near 6R2. Note the 1–4 mm-wide clay and siderite band, and the distribution of not significantly deformed quartz grains. Lineation, 18° towards 116°; foliation, 26° towards 203°; crossed polars. Kinematic indicators are weak but sinistral. (**H**) Centimetre-scale composition banding in carbonate mylonite north of the stretched pebble conglomerate. Carbonate and dolomite domains in the western DJSZ are discontinuous along-strike, forming km-long lenses ('megaboudins'). Lineation, 03° towards 062°; foliation, 28° towards 146°. Kinematic indicators are weak but sinistral. Hammer head is 15 cm.

three progressive deformations. Fabric is spectacular but nowhere are unequivocal kinematic
indicators developed. The youngest deformation
phase is DJSZ-related and has involved reworking (100s μm wavelength folding in image A of
Fig. 4) of an earlier shear fabric via viscous deformation mechanisms (kinking and recrystallization of grains). There is an absence of evidence
for pervasive brittle processes. This rock forms
the northern boundary to the DJSZ only locally,
and the DJSZ boundary is otherwise formed by
the above described, relatively undeformed,
regionally widespread sedimentary rocks.

Muscovite fibres that are reworked by the
DJSZ gave Ar/Ar ages of 155 Ma (Copeland
1990) suggesting the pre-DJSZ shear fabric is
latest Jurassic; i.e. the shear-zone rocks cooled
from sillimanite-grade temperatures through to
argon closure temperatures (McDougall &
Harrison 1988) about this time. The spatial
coincidence with, and local reworking by, the
young (Miocene?–Holocene) DJSZ of a zone
of latest Jurassic shear suggests to us that the
DJSZ may be in part a reactivated feature (see
Discussion).

Marbles and dolomites

Extensive mylonitized carbonate rocks are
present throughout the DJSZ. In the central and
eastern parts of the exposed DJSZ, the regional
Aptian–Albian carbonate platform members
are recognizably present as a c. E–W-trending,
suite of strands, tens of metres wide, that are
almost continuous (i.e. unbroken) across the
region and are interlayered with the other DJSZ
rocks. All the DJSZ carbonate rock members
are mylonitically deformed involving dramatic
attenuation in thickness (with respect to the
regional protolith), sinistral shear, stretch
(image B in Fig 3), extensive grain size reduction
through dynamic recrystallization and solution
transfer (image D in Fig. 3; image G in Fig. 4).
Strong compositional banding and prolate
stretch rodding (image B in Fig. 3; image H in
Fig. 4) define an approximately E–W fabric
with subhorizontal lineation. Clay concentration
on solution surfaces (image D in Fig. 3) and
Fe-rich siderite oxidation alteration (image E in
Fig. 3; image G in Fig. 4) are also widespread.

We infer that the preserved deformation mechanisms (solution transfer and recrystallization)
are viscous regimes. The presence of clay concentration on solution surfaces, sideritic alteration and evidence for solution transfer are all
interpreted as indicators of fluid influence in
these rocks.

Metapelites, metagreywackes and other detrital metaclastics

A range of phyllonitic and mylonitic rocks that
are derived from pelites, greywackes and other
detrital-rich clastics of the regional sedimentary
sequences are present throughout the DJSZ
(image C in Fig. 3) as discontinuous domains
of tens–hundreds of metres by a few kilometres. We use the term phyllonite to mean a
phyllosilicate-rich mylonite that has the lustrous
sheen of a phyllite in accordance with the IUGS
subcommission on structural nomenclature.
These mylonitic rocks typically have a strong
approximate E–W-trending foliation with a
well-defined subhorizontal stretching lineation
that is formed by alternating mica- and quartz-
rich compositional banding that ranges from hundreds of micrometres to <1 mm in width. This
texture is due to recrystallization that has apparently caused extensive grain size reduction based
on protolith appearances and presence of larger
grain sizes in less recrystallized microdomains
(e.g. quartz-grain-rich region in image E in
Fig. 3). Frequently, especially in outcrops near
the pull-apart basin margin, isolated groups of
quartz porphyroclasts, ranging from a few
100 μm to a few millimetres in diameter, are
present where internal cracking, grain sliding
and rotation are preserved (i.e. dilatational behaviour). Overall, the micas exhibit a strong crystallographic preferred orientation (CPO) with
crystallographically deformed or recrystallized
micas kinking and bending around porphyroclasts that have remained relatively undeformed
later in the deformation (image C in Fig. 4). Porphyroclast and flow perturbation asymmetry, as
well as shear bands and crack offsets, indicate
predominantly sinistral sense of shear.

We infer that the evenly banded mica and fine-
grained quartz microdomains (the majority of the
metapelites) represent grain-scale viscous behaviour for their recorded history. The extensive
phyllosilicate presence interstitial to the more
quartz-rich microdomains may have enhanced
grain boundary diffusion processes (possibly
with the assistance of fluids) allowing viscous
behaviour to continue at lower temperatures
(e.g. Farver & Yund 1999). The overall less
common   quartz-porphyroclast-rich   microdomains that record dilatation are more frequent
nearer the normal/oblique brittle faulting at
pull-apart basin margins (see below) and we
suggest that this provides evidence for the role
of fluids in the promotion of brittle behaviour.
Dilatation may also be influenced by (in this
case, the absence of) the phyllosilicates; dilatational microstructures in these rocks are more

common in the mica- and clay-mineral-poor areas. This may have kept porosity higher than in mica- and clay-rich areas, thereby promoting higher individual grain–grain contact stresses (as well as fluid pressure fluctuations) and so cracking.

Cataclasites and tectonic breccias

A range of cataclasites, gouge areas and fault-zone brecciated rocks in general are present throughout the DJSZ with somewhat discontinuous and heterogeneous along-fault distribution. We term all these rocks cataclasites and use additional qualifying terms where appropriate. Cataclasite domains range from <1 m up to >100 m in width, and have along-fault lengths that range from stubby lozenges a few tens of metres long to well-defined strands several kilometres long. The cataclasites vary from coarse breccias to intensely fine-grained gouges that range in fabric type from non-foliated to very spectacular polyphase foliation. Most of the cataclasites are sufficiently cohesive or well indurated to form proud-weathered craggy outcrops. Cohesion in the gouge-like cataclasite portions, however, may frequently be secondary (see below).

Image F in Figure 3 shows a cataclasite example incorporating a range of microfabrics that are typical of the DJSZ brecciated fault products. Tabular or lensoidal microdomains define differences in: (1) distribution of grain sizes (tens to hundreds of micrometres); (2) whether matrix- or grain-supported (large grain–grain contacts are visible in much of this example: cf. matrix-supported cataclasite in Fig. 5A); (3) intensity of any foliation (e.g. the strong foliation in a band of fine-grained gouge in the middle of the photograph is caused by wavy clay seam laminations and the abrupt planar termination of bounding microdomains: cf. very weak foliation of isolated solution surfaces and clay-rich seams in Fig. 5A); (4) clasts shape (here clasts are angular–subangular with visibly fractured grain edges); and (5) the presence and integrity of lithic fragments of sandstone or quartzite (i.e. more intact in the lower (southern) part of the photograph, more shattered apart in the centre of the photograph). The scattered quartz grains have no visibly significant CPO (as far as is discernable by gypsum plate – we have not conducted any electron microscopy). The abrupt nature of the contact of shattered quartzite lithic fragment microdomains with the very fine-grained gouge and the clay concentration surfaces within the gouge must certainly represent major material loss by solution (and may

also represent a displacement surface, possibly of slip direction oblique to the plane of view). The very-fine-grained gouge is most probably a locally renewed (i.e. younger) cataclasis but could also be an original fine-grained portion of the protolith.

Many instances of the cataclasites are pervasively and regularly foliated, and in several examples (e.g. image D in Figs 4 and 5B) a magnificant sinistral S–C fabric is present (e.g. la Berthé *et al.* 1979; Lister & Snoke 1984; Hanmer & Passchier 1991). In these rocks, S-surfaces are defined by approximately parallel, solution/clay concentration seams, mica with crystallographic preferred 'fish' orientation (due to bending, kinking, lattice plane slip or some other non-recrystallizational lattice deformation: cf. Bell & Wilson 1981), and by distributed quartz grains with preferred alignment but with weak–absent CPO. The C-surfaces are microshears defined either by transposing or termination/truncation of the S-fabric (in 'classic' S–C pattern) with locally clay-rich solution(?) and/or sliding surfaces. S–C fabrics reported for cataclastic rocks are extremely rare (see Discussion).

In addition to the S–C fabric in the cataclasites, a further unusual microstructural feature that is present in many of the DJSZ cataclastic fault rocks is re-brecciation of pre-existing foliation. A spectacular example of this is shown in image F in Figure 4, where cataclasite material comprising foliation-absent, fine-grained matrix ($\pm$ scattered quartz grains) is infilling portions of the rock between hitherto foliated, mica-rich, S–C fabric microdomains. In some cases, two or three separate deliveries of new cataclasite material portions are recognizable from grain size distribution differences. These earlier-formed, mica-rich microdomains have been torn apart in the original S- and C-orientations, as evidenced by the truncation and separation of micas that are otherwise still in recognizable crystallographic or visual continuity. Thus, there are lithic fragments of S–C foliated cataclasite *in* cataclasite. Figure 5C shows a similar microstructural relationship, but with fresh cataclasite apparently in a C' (Berthé *et al.* 1979; or shear band, Platt & Vissers 1980) orientation. There are numerous other occurrences of re-brecciation of previously foliated cataclasite material, although not always involved with reworking of S–C fabric. Figure 5D shows widespread re-brecciation (as opposed to re-brecciation localizing in microdomains in the above examples) with a variety of cataclasite lithic fragments (of varying foliation intensity, solution seam development, grain size distribution

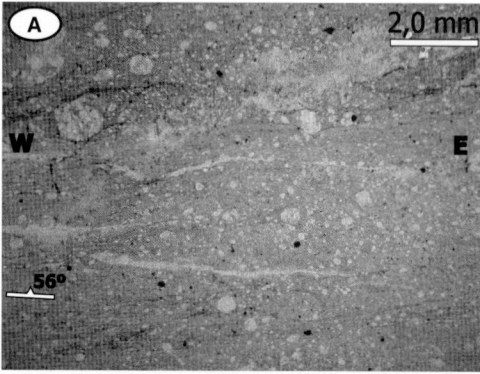

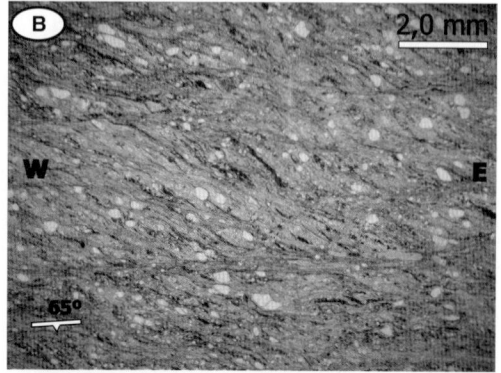

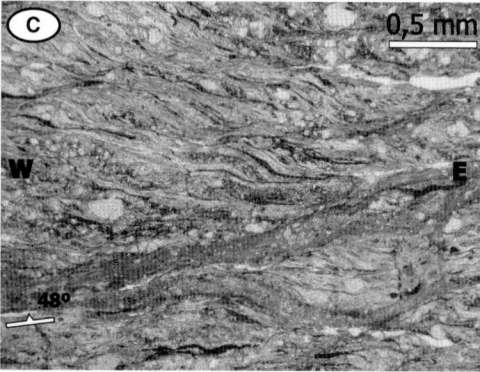

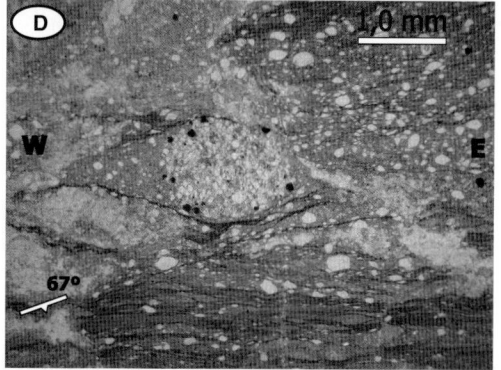

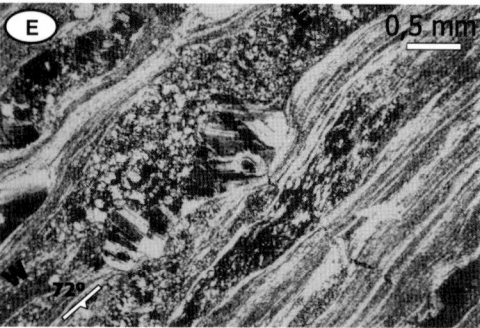

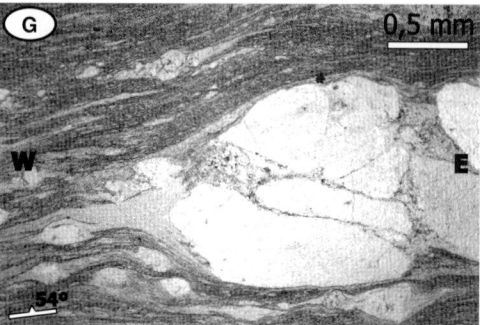

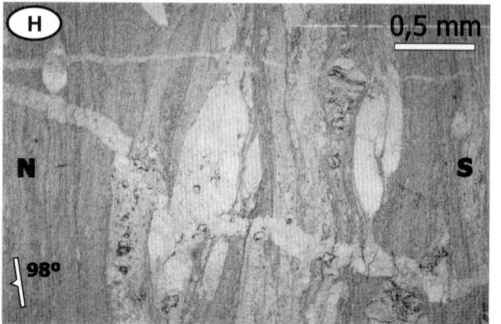

and matrix proportion), as well as lithic fragments of shattering quartzite/sandstone.

Orientation of foliation (where pervasive) is less uniform in orientation in the cataclasites than in other DJSZ rocks, but is overall E–W trending. Lineation is for the most part weak or absent, but subhorizontal where present (e.g. in the S–C fabric cataclasites). In many cases, foliation is affected by later mm–cm wavelength-scale gentle buckling or crenulation. Hinges and intersection lineations related to this phenomenon pitch gently–moderately NW or SE.

The cataclastic rocks show an interesting mixture of dilatational behaviour (i.e. cataclasis) and various creep behaviour mechanisms. To what extent this was, respectively, seismic and aseismic deformation is less clear. We resume this topic in the Discussion.

Conglomerate

Stratigraphic or structural repetition has lead to at least two main strands of a green and purple grit matrix- to grain-supported conglomerate that persists apparently unbroken across the area for >80 km from where it is obliterated by the NQTL intrusive margin in the west to the Gulu rift in the east. Unit thickness varies from several hundreds of metres to >1 km. Pebbles comprise volcanic fragments, quartzite, marble and dolomite (image G in Fig. 3). Foliation in the conglomerate varies in intensity depending on the mica or clay-mineral content, but is overall steep and E–W trending. Original bedding surfaces are also frequently recognizable subparallel to bedding. Lineation is weak, subhorizontal and defined overall by orientation trends of prolately stretched pebbles. In the heart of the DJSZ, the conglomerate shows visible variation from more prolate to more oblate strains. Statistically significant measurements of pebbles in clast-supported portions of the DJSZ conglomerate were possible in three outcrop stations, giving k values of 0.65–2.47 (see Fig. 6A inset). Axial-normally fractured (Fig. 6A) and occasionally sinistrally sheared pebbles are more common in the western DJSZ. Further features that are present within the conglomerate close to the NQTL intrusive margin are trans-clast fractures a few millimetres wide that cross several pebbles and matrix. These

---

Fig. 5. Crossed polars and plane polarized light photomicrograph detail of (A–D) a range of cataclasites from throughout the DJSZ and (E–H) highly mylonitized acid metavolcanics from central and eastern DJSZ. All thin sections were originally cut perpendicular to the main foliation, parallel to the main mineral lineation (when present). Foliation symbol with dip and approximate compass orientations given. (**A**) Large area photomicrograph view of weakly foliated cataclasite near traverse 10. Note the beginnings of clay concentration with pressure solution type cleavage. The lack of strong foliation and the small proportion of visible clasts may indicate that the rock is original gouge and the current cohesion (rock is well indurated) is not primary. No pervasive lineation; foliation, 70° towards 356°. Plane polarized light. (**B**) Further example of cataclasite deformed under semi-brittle conditions to produce a strong S–C fabric from traverse 13. Lineation is 19° towards 252°; foliation is 65° towards 168°. Plane polarized light. Kinematic indicators are unequivocally sinistral. (**C**) Fault rock from traverse 11. Cataclasite subsequently deformed under mica growth and solution transfer creeping conditions to produce a strong S–C fabric thereafter re-brecciated (evidenced by fresh 0.1–0.2 mm band of cataclasite in the C′ orientation (i.e. subhorizontal dipping to left) in the lower part of the photograph – cf. Fig. 4 images D & F). Lineation is 19° towards 272°; foliation is 48° towards 348°. Plane polarized light. Kinematic indicators are unequivocally sinistral. (**D**) Re-brecciated fault rock north of 7R1 indiscriminately juxtaposing sandstone and siltstone lithic fragments of former gouge and cataclasite containing a strong solution-accommodated foliation, and against fabric-less areas. Lithic fragments distribution and individual quartz grain distribution is quite heterogeneous. Lineation (weak), 17° towards 255°; foliation, 67° towards 167°; plane polarized light. (**E & F**) Brittle feldspar behaviour while quartz undergoes extensive climb and glide-led dynamic recrystallization, indicating deformation temperatures of <350 °C (see the text). In (**F**) brittle shearing of feldspar porphyroclast along the growth of twin planes is focus for extensional C′ orientation (not rotation bookshelf faulting!). Both samples from traverse 13: lineation, 39° towards 224°; foliation, 72° towards 166°; crossed polars, both images. Kinematic indicators unequivocally sinistral. (**G**) Large (1 × 2 mm) fractured feldspar porphyroclast with localization of cracking/milling together with dynamically recrystallized quartz at broad grain surface contacts. Otherwise extensive dynamic recrystallization grain size reduction of quartz and mica shows foliation development due to compositional banding tens of μm wide. Porphyroclast trail asymmetry indicates sinistral sense of shear. Lineation, 03° towards 274°; foliation, 54° towards 353°; plane polarized light. Kinematic indicators are overwhelmingly sinistral. (**H**) Quartz-filled vein cross-cutting the mylonitic foliation is due to brittle fracturing that is part of later general vertical shortening associated with the progression of transtension and the Damxung Corridor pull-apart basin growth. Note that the apparent dextral offset of the vein on the left side of image is probably due to a pressure solution seam whose shortening axis would be consistent with subvertical shortening and opening of the Damxung pull-apart. Lineation, 28° towards 269°; foliation, 68° towards 351°; plane polarized light. Kinematic indicators in mylonite are sinistral.

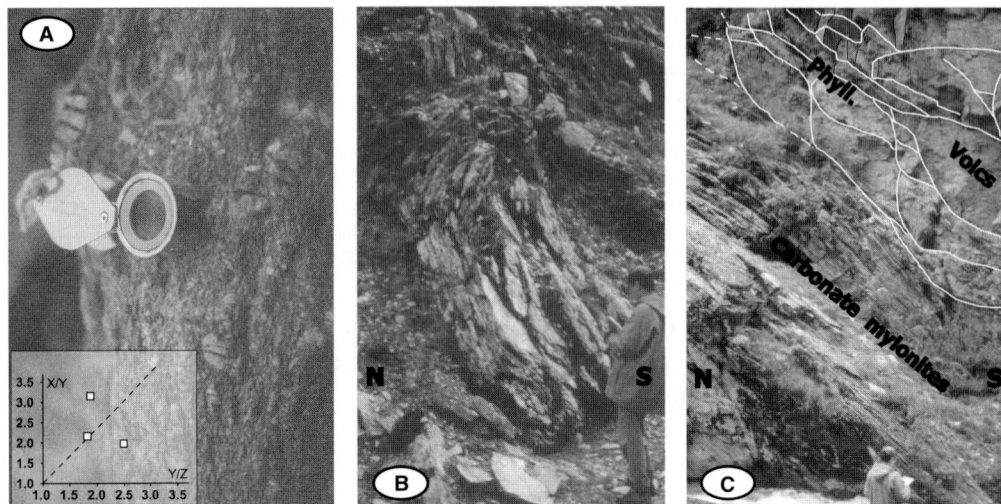

Fig. 6. Outcrop appearance of exhumed DJSZ. (**A**) Axial-normal fracturing in slightly prolately stretched volcanic and quartzite pebbles in the matrix-supported conglomerate in western DJSZ. Axis of stretch is typically inclined at 5–15° (clockwise rotational offset in overhead view) to the km-scale shear zone fabric. Closer to the intrusive margin of the NQTL, similarly fractured pebbles are rotated by sinistral shear. Hand lens is 25 mm. Inset Flinn plot uses data from non-fractured pebbles in the conglomerate cropping out in the cross-section in Fig. 3 (image G). Data are: 7R5a 2.14:1.08:0.43, $\varepsilon_s = 1.13$, $\nu = 0.14$, $k = 0.65$; 7R5b, 2.05:0.95:0.51, $\varepsilon_s = 0.98$, $\nu = -0.11$, $k = 1.34$; 7R4, 2.66:0.84:0.45, $\varepsilon_s = 1.28$, $\nu = -0.29$, $k = 2.47$. (**B**) Looking east to 2–5 m wavelength buckling ($= s_2$) of mylonitic metavolcanics caused by late-stage general vertical shortening and the gradual dominance of the coaxial transtension component. Fold shown here has subsequently failed by tightening and subsequent shearing of the upper hinge (hinge axis, 12° towards 096°). Outcrop is close to the pull-apart basin margin in traverse 11 where fluid flow has probably enhanced the local intensity of brittle fracturing along with generating fluid products; quartz-filled tension gashes, chloritic alteration, tufa and travertine (see the text). Person for scale. (**C**) Looking east to moderately S-dipping, even, planar sheets of carbonate mylonite near 6R1 (labelled, middle left to lower-right corner of image). Structurally above (in minor fault (tens of m² surface) contact) with anastomosing foliation phyllonitic metapelite strands (labelled 'Phyll.') encompassing megaboudin of mylonitic metavolcanics (upper right, labelled 'Volcs'). Below the dolomitic mylonite, section grades downwards into buckled and crenulated limestone mylonites. Person for scale.

are typically quartz or leucocratic phase filled and are frequently rotated by sinistral angular shear. Both fracturing phenomenon are regarded as evidence for the role of fluids in promoting fracture in the DJSZ.

Acid metavolcanic mylonites

Mylonites formed from volcanic protolith of typically acid chemistry are present in all the DJSZ sections visited. They form domains, tens of metres wide and hundred of metres to 1 km long, that define subparallel or broadly anastomosing through-going strands that are somewhat continuous relative to the carbonates or cataclasite domains. The acid metavolcanic mylonites consist of a fine-grained (<100 μm) quartzose and micaceous continuous-interconnected matrix that supports isolated porphyroclasts of quartz and feldspar that range from hundreds of micrometres to >3 mm in size (image H in Fig. 3).

Spectacular foliation and lineation (approximately E–W trending, subhorizontally plunging, respectively) of the matrix is caused by micaceous and quartz-rich compositional separation as well as microdomains of differing quartz grain size (image B in Fig. 4). Coarser (50–150 μm) quartz grains in the matrix have visibly undergone dynamic recrystallization grain size reduction, separating away from larger porphyroclasts and being disposed into the mylonitic foliation (Fig. 5E). The feldspar porphyroclast population largely underwent temperature- and strain-rate-insensitive deformation in the form of cracking as well as spinning (passive rigid-body rotation). Cracking is either: (1) subcritical microcrack growth (e.g. Mitra 1984; Atkinson & Meredith 1987), as evidenced by frequent patchy extinction in larger feldspar clasts; or (2) throughgoing cracking ($\pm$shear offset) frequently localized along anisotropies provided by growth twin planes that have been exploited as surfaces of preferential weakness (Fig. 5F). In many cases these are recognizably oriented at intermediate–low angles to the assumed local σ_1 (the principal stress direction) and thereby have been activated as sliding surfaces. Some feldspar grains preserve original volcanic protolith features such as corrosion on the walls of groundmass-filled embayments in feldspar clasts; a possible indication of some original chemical disequilibrium with the surrounding melt. All the metavolcanic mylonites preserve a wide range of clear kinematic indicators such as σ- and δ-type porphyroclast systems (relative directions of recrystallized grains in trails away from porphyroclast pressure shadows), discrete shear fractures, mica shape and orientation asymmetry ('mica fish') that are sinistral where unequivocal.

We infer that within the matrix, grain-size-sensitive viscous flow operated via some mixture of diffusion creep–accommodated granular flow as is typical for a fine-grained quartz + phyllosilicate polymineralic aggregate (e.g. Brodie & Rutter 1987; Handy 1989). In view of the ample evidence for fluid presence in the other DJSZ fault rocks and mica in this rock, we speculate that diffusion processes in the volcanic mylonites were partly Coble (grain circumference) mechanism, and were strongly fluid and phyllosilicate enhanced (e.g. Farver & Yund 1999). In contrast to the plastically deforming quartz, we infer that the brittle deformation in the feldspar grains represents pressure-sensitive behaviour; the concomitant brittle feldspar–plastic quartz suggests maximum deformation temperatures of c. 350 °C (Simpson 1985; Tullis & Yund 1985; Hirth & Tullis 1994). Taken together with the dynamic recrystallization deformation conditions for the quartz grains, we conclude overall deformation in these rocks proceeded at greenschist facies temperatures.

Although the ongoing deformation has almost certainly furthered grain size reduction in the acid metavolcanic mylonites, it is not clear to what extent the present (recrystallized) matrix framework grain size can be related to the differential flow stress magnitude during recrystallization (e.g. Ord & Christie 1984); from undeformed protolith examples of the acid volcanics we have seen, some of the large grain size distribution between the feldspars and the matrix is partly inherited from the original porphyritic texture.

Ultramafic rocks

Rare, spatially limited occurrences are present of (a few metres in width, a few tens of metres in length) deformed and highly weathered ultramafic rocks, including metagabbro and metaharzburgite, with minor–severe alteration in the form of serpentine and talc in fracture networks and cm-scale shear zones. Although shear zones in these rocks occasionally preserve sense-of-shear evidence in talc and chlorite slickenfibres, orientations are inconsistent or chaotic. These are regarded as co-deformed relics of a regional widespread highly dismembered ophiolite (the Dongqiao: Girardeau *et al.* 1984; Chang *et al.* 1986). They behaved in a very weak manner, but are proportionally minor and we regard them as insignificant in the overall weakening story of the DJSZ.

Fluid and seismicity/fracture-fill products

The other rock types present in the DJSZ are: (1) mineral precipitates; (2) travertine-type deposits; and (3) pseudotachylites.

Throughout the DJSZ, mineral precipitates are present in veins, tensions gashes and so on. These include calcite, quartz (e.g. Fig. 5D) and other leucocratic material that are all present as thicker (cm-scale) accumulations, while chlorite, talc (rare, and restricted to the ultramafic rocks) and biotite (rare, and restricted to the NQTL margin) are millimetre–centimetre in thickness. There is not really a systematic trend to the distribution of the thicker phases, but quartz is much more abundant than calcite, even in the carbonates that may indicate that fluid sources were non-local (see Discussion). The orientation and kinematic relevance of these features is reported in the subsection on 'Brittle trends in the exhumed DJSZ'.

A range of tufa and travertine deposits are frequently present within the southern portions of many of the N–S valleys that incise through the Damxung Range, in each case cropping out in the lowermost parts adjacent to the normal and oblique faults that represent significant downthrow (south side) and constitute the boundary of the Damxung Corridor pull-apart basin. Characteristics vary from pale grey to white, massive to bedded, powdery and weak (malleable) to porous, bedded and strong (brittle). Mostly, the deposits (even the bedded variety) form rather chaotic mounds several metres high and tens of metres in length – almost always with a W–E long axis, that is approximately parallel to the strike of the related fault(s). Sometimes instances of tavertine are restricted to, and built up within, fissures associated with these faults; they have not 'spilled out' onto the erosion surface. Tufa and travertine deposits are more common on the downthrown footwall. We infer that all the tufa and travertine deposits are related to elevated thermal fluids, possibly exploiting the most efficient pressure gradient along fracture paths that acted as channelways (e.g. Hancock *et al.* 1999).

Pseudotachylites are present throughout the exposed DJSZ. Two main types occur. Type 1 are present within the volcanic and more quartz-rich metapelite mylonites (and occasionally within the foliated cataclasites) as recurring pseudotachylite-rich sections with structural thicknesses of metres to tens of metres that house numerous foliation-parallel, 0.5–2.5 mm-thick pseudotachylite layers with millimetre–centimetre spacing between repetitions. In the frequent cases where there is late millimetre–centimetre-scale gentle wavelength buckling or other crenulation that is pervasive in the host rock across the outcrop overall (as described above), the type 1 pseudotachylite layers are co-buckled providing a key timing constraint (they cannot be younger than the buckling). The type 1 pseudotachylites are black, partly glassy and inherently fine grained, with up to 100 μm diameter rounded fragments of quartz, feldspar and other pieces of apparent host rock. Type 2 pseudotachylites post-date type 1 and are found almost exclusively in the volcanics and metapelite, forming 0.1–1 cm (rarely thicker) dark, shiny coatings on planar surfaces. They are observed to be both parallel with, and (unlike type 1) to cross-cut, the original fabric surfaces and later crenulation surfaces of the various host rock mylonites.

In two cases, occurrences >5 mm show sub-millimetre-scale banding that we speculate are due to the successive accumulation of frictional melt (see below). Both pseudotachylite types may consist of a range of matrix microtextures that typically show heterogeneous devitrification with spherulites on the micron scale. Some pseudotachylite examples contain up to 100 μm diameter angular–subrounded feldspar and quartz grain fragments, presumably comminuted from the original wallrock. Pseudotachylite coated surfaces are at least tens of metres in vertical dimension and may have lengths that are much larger (outcrop sizes limit constraint); nowhere were they observed to cross-cut the slickenfibre/slickenside bearing brittle fault plane suites (for palaeostress). Obvious slickenlines are rarely preserved in the pseudotachylites themselves. Injection into adjoining cracks or other neighbouring fractures was not observed, possibly suggesting a limitation of available frictional melt volumes (if there was true melting).

Although a frictional melt origin for shear-zone pseudotachylites has been demonstrated in both field and experimental examples (e.g.

Sibson 1975; Maddock 1983), there remains the question of whether all such pseudotachylites represent frictional melting or sometimes fine cataclasis instead (e.g. Spray 1995). The partial preservation of glassy areas in the devitrified matrix of the type 2 pseudotachylite is good evidence for a melting origin. The co-location of type 1 with type 2 pseudotachylites in the metapelites and volcanics encourages us that they were generated by the same process. Moreover, recent work by Lin (1999*a*) suggests that very fine cataclasite, in addition to true melt pseudotachylites, may also represent some measure of co-seismic strain and the key differences lie in the strain rate. We therefore regard both our pseudotachylites as: (1) accelerated frictional slip engendered; and (2) providing a view into the distribution and seismogenic processes occurring in the Earth's upper crust. We resume this topic in the Discussion.

Fault-zone fabrics and fracture patterns

Figure 2 stereoplots show structural data from our field measurements of the DJSZ that illustrate the main elements of the DJSZ fabric, including mesoscopic (i.e. outcrop scale) shear zones and faults. The trends represented by these data are described in the following sections.

Foliation and lineation trends in the exhumed DJSZ

As noted above, the principal foliation is carried mainly by the mylonites and intermittently by the conglomerates and cataclasites. The data plots in Figure 2 (cf. Fig. 6B & c) show how foliation dips moderately to steeply N (to NE) or S (to SE). In the western part of the section, moderately dipping foliation is more common (e.g. shallow SSW foliation dips indicated by green dot clusters in 7R5 and 6R1; broad range of NNE foliation dips indicated by long cluster of green dots in 7R1). The principal foliation is commonly accompanied by a strong mineral stretching and elongation lineation that is typically subhorizontal, trending E–W to NW–SE (blue squares in all data plots). Much of the fluctuation in dip of foliation is due to an over-printing component of regional buckling, corresponding to an overall vertical shortening/flattening that is associated with approximately E–W-trending axes that syn- to post-date the main foliation- and lineation-forming part of the finite deformation. This (tens of metres scale) folding is more strongly expressed in the carbonates and cataclasites (e.g. 7R1 is predominantly cataclasites and carbonates). This buckling is, in places, accompanied by second (tens of centimetres to a few metres) and third (mm–cm) order folding (e.g. Fig. 6B and image B in Fig. 3, respectively), again the local expression of which fluctuates. In cases (e.g. Fig. 6B) hinges are tightened to become fully fractured (closer to margins of pull-apart basin); this is likely to be part of a greater role of macroscopically 'brittle' (i.e. fracturing prevailing over folding) deformation due to continued exhumation. There is also an overall, tens to hundreds of metres' wavelength, E–W-trending gentle anastomosing (e.g. fluctuation from NNW–SSE to NNE–SSW in steeply dipping foliation as indicated by clusters of green dots in 7R2–7R3 to 7R4, respectively). A lower order of this is also frequently seen at the metre is tens of metres scale. In many cases, anastomosing mylonite strands encompass lozenges, a few metres thick, and pods of lesser-deformed metapelite, as well as carbonate and cataclasite (e.g. Fig. 6C).

Fold hinge lineations are frequently subparallel to the mineral stretching lineation. In a few key areas, there is a consistent development of a small angle between the main stretching lineation and the overprinting fold axes; fold axes are always rotationally offset clockwise with respect to the stretching lineation (viewed from overhead). Additionally present, on individual foliation surfaces, are visible progressive changes of stretching lineations (e.g. alignment of discrete, separated millimetre-long muscovite blades subparallel to continuous alignments of quartz and biotite grains). Thirdly, there are local S–C fabrics (with sinistral or top-to-the-east offset sense) and crenulation/solution surface–folding intersection lineations that post-date earlier portions of the folding. Finally, in a few outcrops, fold axes are boudinaged with a sinistral displacement sense. These phenomena all provide a strong indication of progressive transtension that led to an increasing dominance of the coaxial component of the deformation (due to the cumulative effects of coaxial flow on pure shear strain in contrast to simple shear: e.g. Sanderson & Marchini 1984; Fossen et al. 1994; Dewey 2002).

Brittle trends in the exhumed DJSZ

We have measured (plots in Fig. 2) and characterized a range of brittle deformation phenomena in the exhumed DJSZ. The palaeostress analyses (the largest data set) are discriminated and presented on separate plots from the other structural phenomena. Full procedural details for palaeostress analyses are given in Appendix A. Calculated values are shown in Table 1. These data correspond to planar fractures with constrainable displacement sense. The overall trend is consistent with sinistral transtension, although there are a few fault sets that are not; station 6R1ss, for example. In this anomalous case, the proximity of station 6R1 to the pull-apart margin (note that local foliation is tilted more to the horizontal than elsewhere: cf. cluster of green foliation poles in 6R1 to the left of 6R1ss in Fig. 2) may have co-tilted the fault plane set. Many of the steep faults show orientations that are beautifully consistent with the Neotectonic faults mapped in the pull-apart basin (Fig. 2, and see below).

Present in a few locations are Ramsay & Graham (1970)-type shear zones, planar features a few centimetres wide, of effectively continuous deformation corresponding to centimetre to <10 cm displacement (typically in the foliated cataclasites or more phyllonitic metapelites). Also, hydrothermally precipitated biotite is present, restricted to near the NQTL intrusive margin. Approximately 1–4 cm, c. E–W-trending, NW- and SE-dipping fracture surfaces in conjugate pair orientation are filled with thick, kinked accumulations of biotite. Both these features, together with conjugate sets of tension fibre suites, tension gashes and other planar fracture sets (omitted from palaeostress calculations due to the absence of recognizable unique displacement sense), were used for an overall assessment of kinematic trends as shown in the stereoplots (Fig. 2). They typically indicate a general NNW–SSE extension and are, in most cases, consistent with sinistral oblique-divergence kinematics and transtension. Anomalous data, however, are found in the western part of the study area. Here, offset on early, approximately E–W oriented, moderately S-dipping mesoscopic shear zones (locally chlorite-filled) is sinistral while later features (e.g. tension gashes, tension fibre sets), as well as reactivation of these shear zones, corresponds to c. NE–SW dextral movement (e.g. 7R5). We regard the dextrally displaced features as antithetic to the overall DJSZ and, although this younger, NE-directed, dextral displacement is not in the ideal R' (i.e. R₂ or antithetic Riedel) orientation with respect to the DJSZ boundaries, we assume that this is due to reorientation. This is predicted as part of rotation with (or in this case against) the DJSZ vorticity with continued transtension (e.g. Dewey 2002).

Microscopic brittle features include submillimetre-scale, quartz-filled fractures that are present in the mylonitic volcanics and metapelites

Table 1. *Location of stations and parameters of the deviatoric stress tensor, DJSZ stations*

Site	Lithology	Latitude	Longitude	Method	No. of data	σ_1 (°)	σ_2 (°)	σ_3 (°)	F	R
6R1nf	Metavolcanics (?K–T)	30°35.662'	91°11.468'	NDA	11 11	005 74	225 13	133 10	14	0.5
6R1ss	Metavolcanics (?K–T)	30°35.662'	91°11.468'	P-B-T	6 6	105 16	002 37	214 49		
7R1	Metavolcanics (?K–T)	30°34.908'	91°06.896'	NDA	17 17	043 27	243 63	137 08	21	0.8
7R1EW	Metavolcanics (?K–T)	30°34.908'	91°06.896'	NDA	4 4	189 02	286 73	098 17	8	0.5
7R1nf	Metavolcanics (?K–T)	30°34.908'	91°06.896'	NDA	9 9	295 76	114 14	204 00	23	0.7
7R2-3	Metavolcanics (?K–T)	30°35.35'	91°06.9'	GRIT	9 8	254 20	081 70	345 02	27	0.3

For methods used to calculate stress tensors, see Appendix. P–B–T, pressure-tension method; NDA, numeric dynamic analysis technique; GRIT, random grid search approach; n.c., principal subhorizontal stress orientation obtained from visual inspection. In the 'No. of data' column, the first number is number of measurements, and the second number is number of measurements used for calculation. For $\sigma_1 - \sigma_3$, azimuth (first number) and plunge (second number) of the principal stress axes are given. The stress ratio R is $(\sigma_2 - \sigma_3)(\sigma_1 - \sigma_3)^{-1}$ (where 1 is uniaxial confined extension and 0 is uniaxial confined compression). The fluctuation F gives the average angle between the measured slip and the orientation of the calculated theoretical shear stress. K, Cretaceous; T, Tertiary.

(e.g. Fig. 5D). They overprint the main mylonitic fabric and are thought to represent 'healed', synmylonitization microveins. They are subsequently crenulated/buckled by later strain related to vertical shortening, and are fully cross-cut by planar brittle faults, such as those used for the palaeostress calculations. They are orientated at smaller angles to the shear zone (c. ESE–WNE) consistent with the overall horizontal extension direction. The relative temporal point of generation of these microveins in the overall transtension is well constrained by the cross-cutting relationships. These give important evidence for fracturing during otherwise nonfrictional flow in the mylonites. We interpret this phenomenon as fracture in the presence of pressurized fluids during higher grade deformation of the DJSZ. The vein geometry of these we take as evidence for fluid pressures being nearly as high as the magnitude of minimum principal stress. The phenomenon of brittle features that were formed synchronously with (i.e. at the same crustal conditions as) the 'ductile' structures is discussed below.

Neotectonics in the active DJSZ

Although in the exposed DJSZ there are faults that are certainly young, only those faults that deform the Quaternary fill in the Damxung Graben pull-apart are unequivocally neotectonic (Fig. 2). The geometries of faults in the pull-apart basin, as well as their overall pattern, varies through the Damxung Graben.

In the central and eastern graben faults are typically orientated ENE–WSW and are in the approximate R_1-shear or Y-shear (i.e. Cloos 1955; Tchalenko 1970; Logan et al. 1979) orientations; we cannot be specific as the absolute orientation of the DJSZ boundary is ill-defined here. Graben-bounding faults at the edge of the exposed DJSZ are displacing obliquely-normal; top-to-south plus sinistral. This is identifiable from a comparison of: (1) slickenlines (scour and abrasion trails) where present on steep fault surfaces (that are kept exposed by seasonal high water in the rivers in the N–S-trending valleys); with (2) relay ramp and off-stepping fault scarps tens of metres long by 2–5 m high, in addition to minor offsets (tens to <200 m long) of active and abandoned streams along fault surface traces that are exposed along the tops of alluvial and gravel fans adjoining the range front. We did not survey fault scarps in detail, but were able to identify an overall younging of fault surfaces towards the front of the Quaternary fans based on: (1) the hierarchy of freshness and preservation of faults and debris and/or inverse hierarchy of travertine-type deposits in faults and deforming wedge material in the range front; and (2) the relative ages recognizable from active and abandoned stream offsets. Control on the degree of obliquity (i.e. net kinematic contribution normal displacement components) of the ENE–WSW suite of faults is impeded towards the central portions of the basin because stream channels are rarer and alluvial fans become less distinct (being gradually replaced by general graded fill). Off-steps across rhomb-shaped marshy elongated oval depressions (tens to hundreds of metres in length, <50 m wide) indicate that DJSZ-parallel wrench displacement components are still sinistral and significant. Within these, we observed younger(?) ground breaks connecting into the depressions where, in a few cases, clear shutter ridges were present suggesting recent seismic slip, consistent with the wider pattern of seismicity in this area (Fig. 1, cf. Langin et al. 2003). The ENE–WSW suite remains the dominant trend; there are only a few faults or well-developed ground breaks formed at a high angle (i.e. R' or antithetic Riedel trends). These are more common in the western part of the Damxung Graben and are to be expected for pull-apart system in typical wrench tectonics (e.g. Wilcox et al. 1973).

In the western part of the graben, the sinistral ENE–WSW suite swings to become E–W. Numerous antithetic dextral faults at a high angle (i.e. R' – approximately N–S) are present and are well developed, often with a significantly greater normal displacement component than wrench component (as far as was discernible in the faults we examined). There is good consistency with the brittle faulting data for the exhumed DJSZ (Fig. 2). Exposed alluvial scarps are also much larger than in the eastern Damxung Graben (several tens of metres in height), and alluvial fans are also steeper and thicker. Faulting-derived bedrock scarps above fans are marked by triangular facets frequently >100 m high. Sag depressions are much shorter than in the east (but possibly significantly deeper, cf. Sims et al. 1999) due to the intersection of the E–W faults with the well-developed, antithetic Riedel N–S faults. Sag depression pot holes and ponds are frequently located on isolated 'shelves' or platforms that have been subsequently abandoned by progressive relocation of fault scarps out into the basin. Armijo et al. (1986) present further data that document the neotectonic sinistral pull-apart development. We have included their main mapped structures in our Figure 2.

The observed significant variation in Neotectonic faulting style from west to east (i.e. increase in faulting at high angles to the DJSZ, and apparently greater offsets) is evident from fault groupings on the map (Fig. 2). We interpret this to be related to the change in geometry and orientation of the DJSZ boundaries from west to east. Non-uniformity in fault-zone boundaries has a profound influence on strain styles in transtension (see the section on Kinematic Interpretations).

Overall distribution of deformation

Distribution of deformation within the fault zone is complex and localization is somewhat cryptic. At the fault-zone margins, we observed no uniform gradient in deformation. The southern margin is obscured by the Damxung Corridor pull-apart alluvia and fill; the only conspicuous gradient in intensity of deformation is the heightened brittle fracturing closer to the margin. For the northern margin of the DJSZ, nowhere was an actual outcrop observed where the fault rock–country rock contact was present. The examined fault rock to country rock outcrops were separated by at least tens of metres. The country rock is either limestone–siltstone of the regional DJSZ protolith or, in one case, a post-collisional(?) unconformably overlying a limestone breccia 'red bed'. Although these rocks show fracturing and fault-slip styles consistent with young sinistral transtension and pull-apart opening (Edwards et al. 2001), no pervasive strain is present that can be associated with a significant contribution to displacement of the DJSZ. Within the exposed DJSZ itself, the corresponding rocks at the would-be margin show very variable strain. Two cases are the polydeformed staurolite schists whose latest strain contribution is via re-working in the DJSZ (see above). In other instances, deformation conditions range from moderately mylonitized metapelite to heterogeneously deformed, fine-grained, cohesive tectonic breccia (with clay concentration foliation). These outcrops reveal that, at least for the north, the shear-zone margin is not characteristic of the classic Ramsay & Graham (1970) steady-state deformation shear-zone with a uniform distribution and change in deformation intensity.

Localization of strain within the DJSZ also is uncharacteristic when compared to other large, seismogenic fault zones (e.g. Brock & Engelder 1977; Chester & Logan 1986). There is no concentration of deformation into a single zone or strand suite away from which deformation intensity steadily decays and there is no zone containing *the* ultracataclasite fault core, foliated gouge, etc. (e.g. Chester et al. 1993). Zones of clear localization are recognizable in many locations across the DJSZ, but they are very minor: portions of the cataclasites have zones of inwards-increasing strain marked by gradients in S–C fabric intensity, density of (auto-)lithic fragments and overall clay content. The carbonates have zones where rodding and foliation gradually becomes finer (<1 to <1 mm) and apparent discontinuous deformation surfaces are formed in the centre of these zones. These are, however, never large and do not continue over significant distances (≪100 m). Clay concentration on these surfaces suggests their possible role in fault drainage/fluid conduit systems. The (type 1) millimetre–centimetre interfoliated pseudotachylite layers in the metapelites and volcanics are, of course, also discontinuous deformation surfaces, but, again, the localization represented by an individual pseudotachylite layer is probably only minor (<100 m?). Similarly, in the pull-apart basin, the active buried portion of the DJSZ also does not appear to have responded to ongoing and protracted shear by central localization and short-cutting of splays or connecting up discrete en echelon elements to form an predominant through-going fault; the patterns and distribution of surface fractures and faults, and other neotectonic features, are not suggestive of fault join up and intense localization (Fig. 2).

The most striking localization phenomenon in the DJSZ is present in numerous examples of otherwise anastomosing high strain strands (see above). The strands short-cut across less deformed domains of (typically) metapelite, as well as carbonates and (less frequently) metapelite and foliated cataclasite domains. In such cases, these are frequently accompanied or bridged by tens to hundreds of metres of fracture surfaces coated with the thick (type 2) pseudotachylites described above. This is a key observation in our seismic weakening hypothesis (see the Discussion).

Mechanical interpretations

The DJSZ is a large polylithological fault zone involving extensive deformation, fabric transposition and grain size reduction. This has arisen through: (1) comminution-generated general cataclasis; and (2) grain scale plasticity-generated mylonitization, and their attendant phase change/reaction softening/general rheological transformation, representing what must be a dramatically weakened crustal zone with regards to both shear or frictional strength, as well as flow strength (e.g. White & Knipe 1978; Rutter & Brodie 1988; Hobbs et al. 1990; Wintsch et al. 1995). We attempt to interpret the complex

temporal and spatial (trans-fault and depth distribution) interplay of these, respectively, (normal) stress- and strain-rate sensitive processes in our study of seismic v. aseismic deformation and weakening mechanisms.

Fluid activity

In characterizing weakening mechanisms we first examine fluids. Understanding the role and distribution of fluids in the DJSZ is a key part of the rheological behaviour interpretation. There is much evidence for fluid activity associated with the various deformation mechanisms and alteration rock types at a range of crustal depths and conditions throughout the DJSZ history.

The cataclasites, carbonates and metapelites often show major solution-led material loss as well as alteration and growth (e.g. siderite, clay concentration on solution surfaces/seams), all certainly due to fluids. Interestingly, in those areas (on the micro-mesoscopic scale) from which large volumes of material are being dissolved and removed there is a striking absence of companion precipitation 'sinks'; nearby areas that are removed from the stress concentration and sited, for example, in strain shadows, microdomains and so on. Moreover, where 'sinks' are present further away (on the outcrop scale) in neighbouring domains as mineral precipitate-filled hydraulic fractures, quartz is much more abundant than calcite even in the carbonate-rich domains, suggesting an abundant, but non-local, silica-rich source (cf. Beach 1977). Both of these fluid precipitation phenomena may indicate that fluid flow conditions throughout the DJSZ were transiently and/or heterogeneously quite well drained or 'open system' in the main deformation fabric forming period. Taken together with the vein relationships that suggest fluid pressures were close to magnitudes of minimum principal stress, it may be that high pore fluid pressures kept certain dilatant sites open at certain times (e.g. Johnson & McEvilly 1995). The later fluid products, such as the travertines, are consistent with this; such deposits are restricted to a discontinuous, DJSZ-parallel strip along the pull-apart bounding faults, suggesting that fluid flow focuses (or preferentially exploits) the largest pressure gradient along the highest permeability fracture conduit paths that acted as aggregate channelways (e.g. Jamtveit & Yardley 1997), and suggesting a recent history of high volume fluid throughput for the DJSZ. This preferential focusing or 'compartmentalization', of fluids into the steep, E–W-trending, DJSZ-parallel directions is further evidenced by the striking N–S heterogeneity (described above) but W–E continuity in fluid-related deformation amongst the E–W-trending fault strand domains. This is consistent with the strong fabric anisotropy of the (in particular, phyllosilicate-rich) mylonites and (foliated and unfoliated) cataclasites; the trans-fault fabric changes delineate constituent fault strands comprising: (1) domains of poorly sorted, (once) porous fault rocks and discrete through-going fractures of relatively high permeability that are separated by; (2) phyllonites and strongly-foliated cataclasites with strong alignment of phyllosilicate minerals into fault-parallel orientations; or (3) intact domains of carbonate or volcanic mylonites. Such anisotropy in fault rock fabric distribution is well known from other phyllosilicate-rich fault zones (e.g. Caine *et al.* 1996) to cause fault-zone permeability that is strongly favoured in the fault-parallel direction, due to complex, fluid conduit/barrier systems. An interpretation of overall DJSZ permeability pattern and magnitude/maintenance of high fluid pressures is, however, precluded by the certainly large permeability variations associated with the contrasting protolith lithologies and complex fault geometry (transtension-related? – see below), an issue addressed in other large, multistrand fault zones (e.g. Faulkner *et al.* 2003; Wibberley & Shimamoto 2003). Moreover, it is not required that massive fluid flux occurs (occurred) at depth in the DJSZ. Fluid storage in fault zones is tricky to assess (e.g. Faulkner & Rutter 2001), but studies from other high-grade shear zones that were hitherto thought to be examples of substantial fluid throughput have been shown to instead be major fluid storage loci (e.g. Le Hébel *et al.* 2002).

Present-day evidence (shown in Fig. 7 – see below) for the non-local (silica-rich?) source(s) for the fluids include: (1) a major highly-conductive S-dipping layer 5–10 km beneath the exposed DJSZ interpreted by Li *et al.* (2003) as fluid (graphite is permissible but implausible); (2) seismic reflection profiling 'bright spots' (impedance contrasts of large negative polarity) at 10–15 km depth beneath Damxung Graben (Brown *et al.* 1996) that must represent at least some fluid (Gaillard *et al.* 2004); and (3) thermally elevated fluids at Yanbaijan (cf. Francheteau *et al.* 1984), a commercial hydrothermal plant 50 km to the south along the Nyinzhong Graben (Fig. 2) that is typical of fields associated with high fluid pressure and permeability fluctuation (e.g. Boullier *et al.* 2001). This scenario of locally fluctuating fluid flow, hydrostatic pressure and varied permeability is consistent with repeated DJSZ seismicity that is evidenced from the neotectonic, hydraulic and frictional (e.g. pseudotachylites, discussed below) features, and is expected in large fault zones (e.g. Sibson *et al.* 1975).

Depth distribution of deformation mechanisms and weakening phenomena

Figure 7 shows our interpretations and assumptions of the depth distribution of key mechanical parameters and fault-zone characteristics. We use this to present our interpretations of the deformation mechanisms and weakening behaviour of the main fault rocks at different depths. We allow a 30 °C km^{-1} geothermal gradient; slightly steeper that typical continental crust but conservative in view of the higher heat flow (Francheteau *et al.* 1984) at the commercial plant 50 km away and the proximity to the multiphase intrusion body (the NQTL).

The deformation mechanisms recorded by the main DJSZ rock types provide an exciting view of the breadth of the frictional–viscous transition. The re-brecciated cataclasites with *in situ* preservation of deformation steps recording brittle–plastic (S–C) and back again are a spectacular example of *temporal* frictional–viscous transition (and not depths oscillations). To our knowledge, they are a hitherto undescribed phenomenon. Even (non-re-brecciated) S–C fabric in cataclasites or gouge descriptions emerge very rarely in the literature; recent work from Japan (Takagi & Kobayashi 1996; Lin 1999*b*) details S–C fabric in cataclasites in granitic protolith rocks (albeit with much denser rock fragments and textural resemblance to a protomylonite). Our S–C cataclasites are a key observation of viscous rheology distributed simple shear at low temperature (due to solution recrystallization in the clays ± crystallographic deformation in the micas). Deformation temperatures in the S–C cataclasite may be 150–200 °C based on the intracrystallographic slip deformation in muscovite and biotite (e.g. Lin 1999*b*) but otherwise CPO is absent in quartz. In other cataclasites (typically where strain is less homogeneous) foliation (of varied intensity) is only associated with new clay growth and extensive material transfer (often loss) through solution, and intracrystallographic slip in micas is absent suggesting even lower temperatures. This interpretation is consistent with the work of Takagi & Kobayashi (1996) who invoked plastic behaviour of clay minerals formed under wet condition at a near-surface crustal levels. Numerous previous authors have also noted the role of fluids in production (Wintsch *et al.*

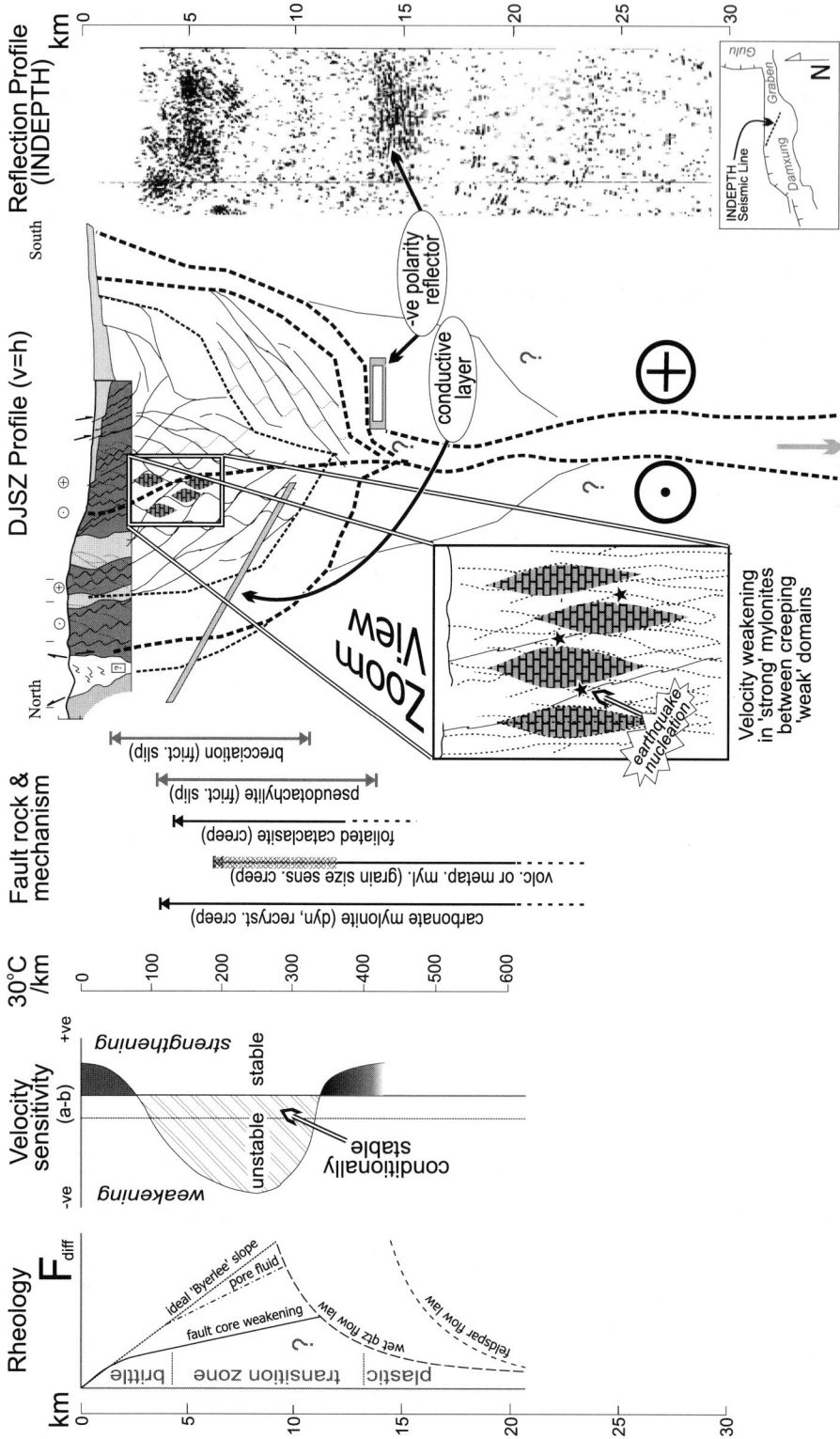

1995) and deformation (Rutter 1983; Stewart *et al.* 2000) of phyllosilicate at lower temperatures in support of a shallow onset of viscous behaviour. Most importantly, quantitative constraints for this have recently been derived by Bos & Spiers (2002*b*) who determined an upper boundary of pressure-solution-controlled flow in fine-grained fault rock that they calculated for, amongst others, transcurrent faulting. Using their Byerlee intercept values, modified with our geothermal gradient, we interpret the foliated cataclasites to have formed (i.e. creep occurred) at ≥ 5 km depth (from where we take our upper limit of the frictional viscous transition in the rheology profile). A lower limit for creep in the cataclasites is moot; creep will prevail far into the deeper crust, even beyond the onset metamorphism.

For the cataclasite frictional behaviour, although more speculative, we take ≤ 10 km as a lower limit (roughly coinciding with 300 °C). This value is slightly greater than (as we assume a faster strain rate than) the Bos & Spiers (2002*b*) peak strength drop-off for a 10^{-10} strain rate (which is slow for the velocity step-up type dilatational behaviour assumed for cataclasite formation). Below this, plasticity is a likely constituent for this rock type as a whole, at any realistic strain rate. The upper limit is effectively the surface (although the

'pull-apart' stress field may limit true brecciation to ≥ 2–3) km.

For the quartzo-micaceous mylonite (i.e. the volcanics plus the metapelites), we speculatively position the upper limit of creep in the mylonites at 200 °C, *c.* 5 km depth. We base this on the very fine grain size and likelihood of fluid- and phyllosilicate-enhanced grain-boundary diffusion (cf. Mitra 1984; Farver & Yund 1999), which can maintain creep behaviour in the absence of true intragranular plasticity. For the lower limit, although in the quartzo-micaceous mylonite, the observed constituent plastic behaviour occurred at ≤ 350 °C (i.e. in the rocks presently at the surface) based on concomitant brittle feldspar and plastic quartz behaviour (described above), creep in the mylonites will clearly prevail far into the deeper crust, as intimated by main black arrow; the hatched box shows only the permissible depths of formation only for those rocks exposed today (*c.* 12 km) and is not an end-member constraint in our weakening model. It is possible that the phyllonitic mylonite presence may become denser until 'maxing out' around the base of the 'seismogenic zone' (i.e. the base of the unconditionally stable field and pseudotachylite field – see below). This has been recently reported for the Kii Peninsula in Japan (Wibberley *et al.* 2003).

Fig. 7. Diagrammatic profiles illustrating mechanics and architecture of the DJSZ. All profiles to the same depth scale are shown on the left and right (depth is relative to the surface of Tibet *c.* 4300 m-a-s-l (metres above sea level) at the lowest point of pull-apart). We encourage the reader to use a ruler/straight-edge to assist reading-off and comparing the depth between the profiles. Rheological profile: schematic strength/depth profile of typical aggregate (i.e. polymineralic) quartzo-feldspathic rock, illustrating changeover from (and modifying influences on) 'brittle' (i.e. purely stress-sensitive) to 'plastic' (i.e. intra-crystalline plasticity governed) rheology. Closely dashed outer line has ideal 'Byerlee' frictional coefficient slope. Solid line lower part is the depth-dependent change in pore fluid factor of Streit (1997). Dotted and dashed line is the dramatic fault core weakening from fluid phase alteration and solution transfer (e.g. White *et al.* 1980; Holdsworth 2004). Quartz and feldspar flow law curves are schematic. The question mark denotes uncertainties of rheology (i.e. semi-brittle realm) in the brittle to viscous strength transition (i.e. from onset of quartz plasticity to onset of feldspar plasticity). 30° km^{-1} profile: real geothermal gradient may be steeper (see the text). Velocity sensitivity profile: shows stability/instability velocity dependence according to the rate/state-variable friction parameter $a - b$ (i.e. slowness law, diagram after Scholz 1988; Scholz 1998). Rock type and mechanism profile: arrowed lines show the depth area across which the various DJSZ fault rocks will undergo either viscous (black) or frictional (grey) sliding. Note the extremely shallow upper limit for viscous (i.e. aseismic) slip for general mylonites and foliated cataclasites (see text). Hatched pattern overlain on the mylonites arrow represents the observed limit of deformation temperature (i.e. <350 °C $\equiv$ brittle feldspar) for exposed DJSZ mylonites, the main black arrow indicates the entire possible range. DJSZ profile: rough interpretation of the geometry of the main fault splays of the DJSZ. The key constraint is a strong reflector with negative reflection polarity (i.e. 'bright spot') interpreted to be fluid filled shallowing out of fault splays. The question marked splays signify uncertainty of whether the DJSZ must dramatically localize at *c.* 15 km depth. The zoom view is a cartoon of the velocity weakening and earthquake nucleation hypothesis in 'strong' quartzo-feldspathic mylonite domains between creeping weak domains. Stars indicate a range of possible nucleation points (only one activates at a given time), corresponding to the multiple fault strand and spaced multiple localization architecture of the DJSZ (see the text). Seismic profile (INDEPTH): portion C of Tib11 deep seismic reflection profile from Project INDEPTH. See Alsdorf (1998) for processing details. The location of the line in the Damxung Graben is shown in the inset at the foot of profile.

The deformation temperatures in the carbonate mylonites is difficult to asses from basic microscopy but they are known to be weak (at suitable strain rates recrystallization in experiment is observed at room temperature), deforming via steady creep for a given differential flow stress, with possible augmentation by mica and clay minerals within the impure carbonate (e.g. Rutter 1999; Brodie & Rutter 2000; Herwegh & Jenni 2001). Wherever they are present alongside the brittlely deformed cataclasites and type 1 pseudotachylites, the carbonates are plastically deformed indicating that they exhibit viscous rheology high into the shallow crust. We nominally put a 100 °C upper limit on their flow.

Because the interpretation of the pseudotachylite depths of formation is more conjectural, we first examine rheology from the point of view of frictional stability. The rheological profile in Figure 7 (somewhat schematic) has been constructed for the combined DJSZ rocks and includes lithostatic load dependence and flow law curves. The solid line curve is drawn after Holdsworth (2004) and Bos & Spiers (2002b), is much weakened from the (dashed line) classic 'Byerlee' (1978) or a fluid-modified version thereof (from Streit 1997), and is consistent with the weakening evidence we observe in the load-bearing region of the crust. The quartz and feldspar flow law curves are highly schematic. The very broad depth interval containing the transition between constituent brittle and constituent plastic behaviour (in both cases, for the combined fault rock package at natural strain rates) is indicated by the boundaries on the locations at depth for fault rock type and mechanisms interpreted above, and is consistent with recent studies (cf. Holdsworth 2004). For a mature, polylithological fault like the DJSZ, however, a single (aggregate) strength is somewhat limited in parameterizing deformation and weakness (cf. Scholz 1998), and the concept of velocity dependence of frictional stability is better suited.

The Figure 7 velocity sensitivity profile (after Scholz 1988) embodies the general rate-variable/state-variable constitutive law for friction in materials. This incorporates the relationship (following a change in displacement/step-up in velocity) between (a) direct velocity (creep) effects on friction and (b) subsequent decay of friction thereafter; the Dietrich–Ruina law (or 'Slowness' law), the rate and state variable (cf. Scholz 1998). Central to the stability–instability concept is the time-with-displacement history on friction (e.g. compaction–adhesion–healing with dilatation–gaining of porosity: Bos & Spiers 2002a) and, when $(a - b) < 0$

(i.e. velocity weakening), how this relates to effective normal stress $(\sigma_{eff} = \sigma_n - P_{fluid})$. Because $(a - b)$ does incorporate shear strength, the potentially unstable (velocity-weakening) realm is situated mainly in the 'brittle' crust and the lower limit corresponds roughly to the onset of effective plastic constituent behaviour in quartz. Following Scholz (2002) we have drawn the lower limit as shown in Figure 7, modified for our geothermal gradient and lower frictional viscous transition in the rheological profile (e.g. enhanced fracturing in the otherwise creeping phyllonitic metapelites). Earthquakes will only nucleate in the unstable field; those propagating into the unstable field will persist, those into the stable field will terminate. The depth interval of the potential velocity-weakening realm corresponds, at least in part, to pseudotachylite formation depths. We therefore use this field as our outline for the pseudotachylite formation depth interval but draw the lower limit deeper, to c. 350 °C (deeper than the onset of intracrystalline plasticity in quartz), following Scholz (2002) who noted that pseudotachylites frequently develop below the lower limit of the potential velocity-weakening realm. This interval corresponds to the seismogenic zone for the DJSZ, and is typical for continental wrench faulting; the Parkfield section of the San Andreas, for example, has a depth interval of seismicity of 4–12 km (Marone & Scholz 1988).

Although earthquake propagation termination in the stable realm below the theoretical maximum possible depths for velocity weakening is to be expected, due to negative stress drops here, seismogenic slip velocities must presumably be realized in this realm due to the distance required for deceleration of large/rapid slip earthquakes propagating from above. The concept of earthquakes propagating from above has been put forward by Scholz (2002) to explain the co-location of pseudotachylites in otherwise viscous deformation governed (i.e. confining pressure insensitive) mylonites. Hitherto, Hobbs et al. (1986) suggested this to be due to ductile instability (an idea we take up in the Discussion).

Kinematic interpretations

Main foliation and lineation kinematic trends

We assume that in the strongly lineated mylonites, L > S, there is some dominance of constrictional strain (e.g. Hancock et al. 1984). This is not unexpected in progressive transtension (Sanderson & Marchini 1984 – note these authors only referred to transpression but with negative notation for divergent-oblique settings). The LS fabric is typically inclined at $5-25°$ (clockwise rotational offset in the overhead view) to the oblique to the DJSZ shear-zone boundary (albeit imprecisely constrained – red dashed line in Fig. 2) at an angle consistent with the overall kinematics for sinistral transtension (e.g. Tikoff & Fossen 1999). For the most part, the otherwise steep foliation planes of the LS fabric have not rotated significantly into horizontal orientations as might be expected from the overall crustal thinning in bulk transtension. This may indicate that the intermediate incremental stretch (λ_2) and/or the intermediate instantaneous stretching axis (ISA$_2$) for this, the mylonite-forming part of the finite strain, remained vertical during this time and/or the kinematic vorticity number (ω_k) was high ($0.8 < \omega_k < 1$) when compared to some 'obliquity parameter' (by this we mean the key angle that may be, depending on the methodology, either (1) the size of the divergence angle, α, associated with the oblique flow apophysis (e.g. Fossen & Tikoff 1998) or (2) the boundary velocity vector, **v** (e.g. Lin *et al.* 1998) of the instantaneous flow field (Ramberg 1975; Means *et al.* 1980)). In the visibly lower grade foliated cataclasites that were formed later in the deformation history (and presumably higher in the crust), there is only a weak or even absent L-fabric and, moreover, there is a greater expression of local vertical shortening-related buckling. This suggests to us that, over time, progressive transtension caused a gradual reorientation into near vertical of λ_3 (the minimum incremental stretch) or ISA$_3$ or both. Finally, the late-stage brittle normal faulting pattern that is pervasive through the exposed DJSZ, as well as in the Quaternary-filled Damxung Graben, is quite typical for rhomb-shaped basin pull-apart in oblique-divergence kinematics (e.g. Basile & Brun 1999). Conjugate fracture sets indicate NNW–SSE extension, while the palaeostress calculations (for the brittle features in Fig. 2 and Table 1) show sometimes σ_2 (the intermediate principal stress direction) vertical and sometimes σ_1 (the maximum principal stress direction). σ_3 (the minimum) fluctuates from WNW to NWN. Assuming some rough correspondence between the stress ellipse and the incremental strain, the palaeostress data provide further consistency with our interpretation of the overall transtension history. We therefore assume that by the final (present) stages λ_3 is vertical, the 'obliquity parameter' is $>20°$ (rock compressibility and material loss notwithstanding) and λ_1 is probably $\gg20°$. This progressive

kinematic pattern is predicted due to the cumulatively effective coaxial component of the deformation (on pure shear, not simple shear) at more advanced stages in the shear-zone displacement history (e.g. Fossen & Tikoff 1998).

The above is something of a general idea as to how our total data set is accounted for in transtension deformation, and we recognize that in complex shear (triclinic, dilating, etc.) linear fabrics can frequently occur at highly oblique or otherwise non-intuitive directions (e.g. Jiang & Williams 1998). An estimate of the true finite strain or a comparison of our observed features with those predicted by the various models for mono- and triclinic three-dimensional shear is, however, outside the scope of this paper. We emphasize that, in addition to the non-uniform shear-zone boundaries, the polylithological nature of the DJSZ makes the fault zone very heterogeneous. In addition, the apparently massive volume loss from these rocks to outwith the shear zone further obscures the real value of ω_k and the overall angle of principal finite or instantaneous stretch; Passchier (1991) has shown the profound role of volume change on ω_k values and stretching rates.

Sinistral shear sense

The historical and active DJSZ is clearly a sinistral feature, as evidenced by the regional transposition, the prevailing kinematic indicators in all fault rocks, the palaeostress data and the neotectonic features. Nearby resolved focal mechanisms (Mechie *et al.* 2004) (Fig. 1) are consistent with the overall sinistral shear sense in NE NQTL. There are, however, rare anomalous examples of DJSZ-parallel, dextral displacement within the fault zone (these are separate from the expected antithetic Riedel neotectonic features and SW–NE slip axis dextral shear zones that are detailed above). We attribute this to either: (1) differential stretching rates and amounts between domains (or microdomains) causing transient reversals in the local flow (e.g. Means 1989, stretching faults; Simpson & De Paor 1997); or (2) unrecognized overturning via tight third-order folding that is part of the subvertical shortening. We emphasize that these local, DJSZ-parallel, dextral shear examples have no consistent association with a particular lithology or metamorphic grade, or any other characteristic of the DJSZ (cf. interlayered domains of sinistral and dextral shear evidencing synconvergence extension in the W. Himalayas; Argles & Edwards 2002) and they do not represent any key event in the overall deformation history.

Discussion

Velocity-dependent, fluid-dependent?

In the DJSZ the co-location of densely spaced, recurring pseudotachylites (both types) within mylonites of quartz plastic constituent rheology, and alongside plastically deformed carbonates, and cataclasites, occurs throughout the exposed rocks, thereby highlighting that overlap of the seismogenic depth interval with the viscous creep depth interval is common. This co-location was sufficiently deep to result in the type 1 pseudotachylites being co-buckled with the host fault rocks and subsequently cross-cut by the type 2 pseudotachylites (and various other brittle features, see above) as they gradually rose in the exhumation process. Pseudotachylites found alongside viscous-creep mylonites are to be expected near the base of the seismogenic zone (above). What is crucial to our observations, however, is that pseudotachylites are found alongside coevally viscously deforming rocks for a long way up into the crust, effectively the entire potential velocity-weakening realm (the 'velocity sensitivity' realm of Scholz (1988) but with the new caveat that unlimited viscous creep can occur throughout this realm. We suggest here that fluid is effectively the other governing factor that will influence whether weakening (i.e. localization of displacement) is seismic or aseismic.

Consistent with this large frictional viscous overlap that we propose is the widespread spatial co-existence of dilatational- and creep-generated structures observed in the cataclasites (typically within a few metres when not co-occurring on a microscale as in the spectacular in situ re-brecciation). We anticipate repeated switching between the respective frictional and viscous deformation mechanisms that are overlapping at the transition depth interval. A likely cause is sudden change in σ_{eff} due to sudden fluctuations in P_{fluid}, associated with the fluid flow detailed above. Whereas fluid presence at low P_{fluid} is expected to facilitate creep in all the fault rock types (as detailed above), a sudden increase in P_{fluid} will dramatically favour dilatational processes, providing a transient excursion into the frictional realm through transient lowering of the effective normal stress.

The multiple strand geometry of the DJSZ provides the framework for repeated switching of slip localization during overall fault-zone displacement. Although the DJSZ is migrating to the south, the re-brecciated cataclasites show that on single fault strands there has been frictional displacement, followed by creep, possible locking-up and then further displacement, providing evidence of repeated 'in-strand' switching of slip localization (and so militating against the alternative suggestion that each strand is active by itself for a single, discrete period after which it becomes permanently abandoned as displacement migrates to the south in a regular progression). This raises the possibility that for an overall setting of displacement and lithostatic load for the fault zone, individual strands alone may slip while others may creep or lock up (and therefore create/renew adhesion and frictional strength from the last dilatational event). Therefore a slipping strand will govern the effective strength of the displacing fault zone (and there must be changes in the finite difference between shear stress amongst the strands). A system of competition or reciprocation to accommodate far-field displacement gradients will probably occur, with slip migrating to a neighbour depending on the change in state (fluid pressure, chemistry, cohesion, etc.) in a given strand. This was intuitively recognized by Hobbs et al. (1986) as their 'ductile instabilities'. This problematically circumvents part of the rate-friction and state friction laws and was noted (on the much smaller scale of ring-shear healing experiments) by Bos & Spiers (2002a).

We suggest that the aforementioned 'compartmentalization' of fluids probably also plays a crucial role in the repeated switching of slip localization. If there is very limited 'cross-talk' between compartments due to permeability anisotropies then local P_{fluid} can be dramatically heterogeneous, as may be the local σ_{eff}.

Syn-creep seismogenesis model

Figure 7 is our speculation regarding earthquake nucleation in the DJSZ, as follows: the frequent occurrence of type 1 and 2 pseudotachylite-coated fracture surfaces throughout the DJSZ is always located within the mylonitic metavolcanics and metapelites. Outcrop constraints indicate to us that the pseudotachylite surfaces shortcut and join together domains that have a fine grain size that is volumetrically very abundant relative to any porphyroclasts within; overall, these rocks are defined by a framework of interconnected weak phase layers containing structurally isolated relative stronger porphyroclasts (e.g. Handy et al. 1999). Pseudotachylite-coated fracture surfaces are seen nowhere in the carbonate mylonites, and only very rare occurrences are present in the tectonic breccias/secondary cohesion (adhesive wear) cataclasites.

The 'zoom view' in Figure 7 shows a cartoon depth section through the DJSZ in which the carbonate megaboudins (stippled grey – middle depth interval) and cataclasite (white patches devoid of pattern – upper depth interval and at surface) are separated by various anastomosing through-going strands of quartzo-feldspathic rock (the DJSZ mylonitic metavolcanics and metapelites). The carbonates and cataclasites are steadily creeping while stress strength builds up in the areas marked by the stars that then become subject to velocity weakening and are the focal points for earthquake nucleation (probably triggered by a fluid kick-in and drop in σ_{eff}). Several stars and possible nucleation points are shown but only one can activate at a given time. This is part of the multiple fault strands and spaced multiple localization architecture of DJSZ, as opposed to a single plane of localization.

Typical lithospheric scale fault?

It is not clear how atypical the width and broadening character of the DJSZ is. The lack of localization in a single principal strand is different to other seismogenic, identifiable lithospheric-scale wrench faults such as the Punchbowl Fault (Chester & Logan 1986). Studies of lithospheric-scale, seismogenic fault zones with demonstrated fault growth broadening are rather rare. One example, the transcurrent Carboneras Fault in southern Spain (a diffuse plate boundary of the Betic Orogen), has >40 km displacement and cuts the entire crust but nevertheless has an average width of <3 km (Faulkner et al. 2003). Meanwhile, the DJSZ width (active + exhumed) in places approaches 20 km. Taken together with the identified length (>180 km) and displacement (probably ≫30 km: Edwards et al. 2001), basic fault scaling trends (e.g. Cowie & Scholz 1992) allow that the DJSZ is also lithospheric-scale. We are, however, unsure if this can be expected. The multiple DJSZ (re-)location observed at the surface is inferred to be part of the kinematics of transtension, whereby the active surface strands migrate southwards. Consequently, the exposed DJSZ would form the northern branches of a SSE-dipping splay in a governing crustal shear structure (Fig. 7). Accommodating this, we incorporate the positions of the highly conductive (fluid?) layer and the area's 'bright spot' in our interpretation of the subsurface fault-zone architecture. We have assumed that each of the two strong geophysical observations of fluids coincides with a principal fault-zone strand/splay of corresponding geometry. These are thereupon suggested to flatten and sole into a principal fault strand whose position

corresponds to the base of the above proposed seismogenic zone and the lower termination of the velocity instability depth interval (Scholz 1988), where we have inferred the onset of true plastic constituent behaviour. The significant narrowing and localization represented by our speculation of the fault-zone architecture at >15 km depth is neither constrained nor strongly argued for. Competing hypotheses for what lies below this general depth include: (1) DJSZ termination into free Poiseuille crustal flow (i.e. Clark & Royden 2000); (2) DJSZ continuation to ≫65 km total depth, led by a mantle lithosphere shear structure (i.e. Tapponnier et al. 2001); or (3) an abandoned jelly sandwich (i.e. Jackson 2002) – meaning DJSZ termination at the Moho. In view of the magnitude and regional extent of the DJSZ, we might favour hypothesis (2).

Reactivation?

Mature fault zones are thought to be persistent weaknesses in crustal dynamics, and fault reactivation seems to be a characteristic feature of the continental lithosphere that is probably central to many cases of fault localization on a range of scales (e.g. White et al. 1986; Holdsworth et al. 1997). In particular, areas of continental crust such as terranes and microcontinents are quite susceptible to exploitation of former strain accommodation zones in that, because their buoyant quartzo-feldspathic bulk lithology can only be subducted with difficulty, they are repeatedly shunted around and reworked between the stages of large continent accretion and collision (e.g. Dewey et al. 1988; Hoffman 1988). Our area of study, the northern Lhasa Block, has something of a polydeformational history (since its breakaway from Gondwana it docked onto the Eurasian southern margin prior to, and was subsequently 'sandwiched' along with, the India–Asia collision: (Chang et al. 1986; Coward et al. 1988). The coincidence, locally, of the northern DJSZ boundary with an exposure of a tectonic slice preserving latest Jurassic shear suggests to us that there may be reactivation of an earlier formed inherently weak shear zone. We speculate that some type of pre-existing weakness has played a role in the localization of the DJSZ as follows.

In the ongoing collisional kinematics, intense crustal deformation is accompanied by the development and gradual northwards growth of the Tibet plateau, as well as eastward crustal escape, both of which are accommodated by numerous major strike slip faults (Tapponnier et al. 2001). In analogue and numerical models of Indian plate indentation and Tibet plateau escape, the presence of steep weak zones in the pre-collisional medium results in (from an early stage) their dramatic and continued exploitation by major wrench faulting in the collisional process (Cobbold & Davy 1988; Tapponnier *et al.* 2001). A deep-seated, crustal strike-slip shear zone within the Lhasa Block that was formed (for some reason) during pre-collisional tectonic activity provided the weak point for initiation of a tear (once the India indentation was significantly underway) that propagated to the east to accommodate part of the escape kinematics as the present DJSZ.

Conclusions

The Damxung–Jiali Shear Zone is an active crustal scale, >150 km-long, 5–15 km-wide, sinistral transtensional fault that exposes fault a range of protolith domains. Mesoscopic and microscopic characteristics of the fault-zone rocks indicate protracted weakening and deformation from depths associated with upper greenschist facies to low-temperature conditions of near-surface exhumation with co-spatial and coeval switches in brittle and plastic (and back again) constituent strain throughout these depths. The widespread role of fluids and the intermittent periods of acceleration to seismic slip with switching between multiple strands is evidenced throughout the fault rocks. There is apparently a large depth range overlap between frictional and viscous mechanisms in all fault rocks and significant spatial and temporal co-existence of aseismic and seismic DJSZ creep. Weakening mechanisms in the Damxung–Jiali Shear Zone are regarded as fluid- and velocity-sensitive.

This research was supported in part by the German Research Council (DFG grant Ra442/12-1). M.A. Edwards is supported through a EU-Marie Curie Research Fellowship (contract no. HPMF-CT2002-01703). We thank Bob Holdsworth, Paul Kapp and Michelle Markley for very constructive reviews. All of the members of both Project INDEPTH and the Structural Processes Group, Vienna are acknowledged for free and open discussion, feedback and sharing of data and ideas, and helping with logistics, sampling, measuring. We are particularly grateful to W. Kidd for unpublished mapping data, J. Mechie, W. Langin and E. Sandvol for original seismology location/magnitude data, and B. Grasemann for help with data analyses and exemplary generosity in sharing of ideas and interpretations. Editors D. Bruhn & L. Burlini are thanked for their encouragement. E. Rutter and M. Handy prompted M.A. Edwards to look at DJSZ frictional behaviour. We dedicate this work to the memory of our departed friend K.D. Nelson, head of Project INDEPTH, whose character and leadership were an inspiration to us all.

Appendix A

Methods of fault-slip analysis and definition of stress-tensor groups

Fault-slip data were collected from individual outcrops or groups of small outcrops termed 'stations' – an area up to quarry size with uniform lithology. Sense of slip along the faults was deduced from kinematic indicators, for example offset markers, fibrous minerals grown behind fault steps, Riedel shears, and tension gashes. Because errors in slip sense determination may have severe effects on principal stress axes calculations, a confidence level from 1 to 4 was assigned to each slip sense datum. These levels are recorded in the style of the arrowheads expressing the slip direction of the hanging-wall block in the fault slip data diagrams, thus allowing judgment of the quality of the database. Surface morphology of the slickenline features (e.g. fibre- or stylolite-coated or polished) and fault size, classified qualitatively based on an estimate of the displacement and the lateral extent of the fault, were recorded. The aim was to discriminate first-order faults and to enable a comparison of faults measured in outcrops with those inferred from mapping. Indications of multiple slip were recorded, and the relative chronology was used for separation of heterogeneous raw data fault sets into subsets. Overprinting relationships such as consistent fault superposition, overgrowths of differently oriented fibres, or fibres with changing growth direction guided the assignment of the subsets to relative age groups. The raw data usually contain several fault slip sets with incompatible slip sense but with consistent grouping and consistent overprinting relationships, which were used as the geological constraint for separation into subsets. Note, however, that the subsets may still contain incompatible data. The latter are included in the stereoplots but excluded from the calculations of the stress axes. Faulting was always evaluated in relation to folding and any sign of post-faulting folding was recorded (e.g. rotated conjugate fault sets with one principal stress direction normal to tilted bedding). Rotations around bedding and foliation have been applied to some faults sets, but minor rotations (<20°) were not performed, as we are more interested in the kinematics/dynamics of regional faulting than in exact stress-field orientations. We used the computer program package of Sperner *et al.* (1993) and

Ratschbacher & Sperner (1994) for fault-slip analysis to calculate the orientation of principal stress axes and the reduced stress tensors (e.g. Angelier 1984). Out of this package, we obtained stress axes by the 'pressure-tension (P–B–T) axes' method and calculated stress tensors by the 'numerical dynamic analysis'. In addition to stress orientation, the computation of the reduced stress tensor determines the ratio R, which expresses the relationship between the magnitudes of the principal stresses. Extreme values of R correspond to stress ellipsoids with $r_2 = r_3$ ($R = 0$) or $r_1 = r_2$ ($R = 1$). The quality and the quantity of field data determined the selection of the method used for calculation. The P–B–T axes method was used with scarce data and where insufficient time was available in the field for careful analysis of fault and striae characteristics.

References

ACHACHE, J., COURTILLOT, V. & ZHOU YAO, X. 1984. Paleogeographic and tectonic evolution of southern Tibet since Middle Cretaceous time; new paleomagnetic data and synthesis. *Journal of Geophysical Research*, **B89**, 10 311–10 339.

ALSDORF, D.E., BROWN, L., NELSON, K.D., MAKOVSKY, Y., KLEMPERER, S. & ZHAO, W. 1998. Crustal deformation of the Lhasa Terrane, Tibet Plateau from Project INDEPTH deep seismic reflection profiles. *Tectonics*, **17**, 501–519.

ANGELIER, J. 1984. Tectonic analysis of fault slip data sets. *Journal of Geophysical Research*, **B89**, 5835–5848.

ARGAND, E. 1924. Le tectonique de L'Asie. *Comptes Rendus, Congrès Internationale Geologique*, **XIII** (1922), 171–372.

ARGLES, T.W. & EDWARDS, M.A. 2002. First evidence for high-grade, Himalayan-age synconvergent extension recognised within the western syntaxis, Nanga Parbat, Pakistan. *Journal of Structural Geology*, **24**, 1327–1344.

ARMIJO, R., TAPPONNIER, P., MERCIER, J. & HAN, T. 1986. Quarternary extension in Southern Tibet: Field observations and tectonic implications. *Journal of Geophysical Research*, **B91**, 13 803–13 872.

ARMIJO, R., TAPPONNIER, P. & HAN, T. 1989. Late Cenozoic right-lateral strike-slip faulting in Southern Tibet. *Journal of Geophysical Research*, **B94**, 2787–2838.

ATKINSON, B.K. & MEREDITH, P.G. (eds). 1987. *The Theory of Subcritical Crack Growth with Applications to Minerals and Rocks*. Academic Press, London.

BASILE, C. & BRUN, J.P. 1999. Transtensional faulting patterns ranging from pull-apart basins to transform continental margins: an experimental investigation. *Journal of Structural Geology*, **21**, 23–37.

BEACH, A. 1977. Vein arrays, hydraulic fractures and pressure solution structures in a deformed flysch sequence, S.W. England. *Tectonophysics*, **40**, 201–225.

BELL, I.A. & WILSON, C.J.L. 1981. Deformation of biotite and muscovite: TEM microstructure and deformation model. *Tectonophysics*, **78**, 201–228.

BERTHÉ, D., CHOUKROUNE, P. & JEGOUZO, P. 1979. Orthogneiss, mylonite and non coaxial deformation of granites: the example of the South Armorican shear zone. *Journal of Structural Geology*, **1**, 31–43.

BESSE, J., COURTILLOT, V., POZZI, J.P., WESTPHAL, M. & ZHOU, Y.X. 1984. Palaeomagnetic estimates of crustal shortening in the Himalayan thrusts and Zangbo Suture. *Nature*, **311**, 621–626.

BOS, B. & SPIERS, C.J. 2001. Experimental investigation into the microstructural and mechanical evolution of phyllosilicate-bearing fault rock under conditions favouring pressure solution. *Journal of Structural Geology*, **21**, 1187–1202.

BOS, B. & SPIERS, C.J. 2002a. Fluid-assisted healing processes in gouge-bearing faults: insights from experiments on a rock analogue system. *Pure and Applied Geophysics*, **159**, 2537–2566.

BOS, B. & SPIERS, C.J. 2002b. Frictional-viscous flow of phyllosilicate-bearing fault rock: Microphysical model and implications for crustal strength profiles. *Journal of Geophysical Research*, **107**, 101 029–101 038.

BOULLIER, A.-M., OHTANI, T., FUJIMOTO, K., ITO, H. & DUBOIS, M. 2001. Fluid inclusions in pseudotachylytes from the Nojima fault, Japan. *Journal of Geophysical Research*, **106**, 21 965–21 979.

BRACE, W.F. & BYERLEE, J.D. 1966. Stick-slip as a mechanism for earthquakes. *Science*, **153**, 990–992.

BROCK, W.G. & ENGELDER, T. 1977. Deformation associated with the movement of the Muddy Mountain Overthrust in the Buffington Window, southeastern Nevada. *Geological Society of America Bulletin*, **88**, 1667–1677.

BRODIE, K.H. & RUTTER, E.H. 1987. The role of transiently fine-grained reaction products in syntectonic metamorphism; natural and experimental examples. *Canadian Journal of Earth Sciences*, **24**, 556–564.

BRODIE, K.H. & RUTTER, E.H. 2000. Deformation mechanisms and rheology: why marble is weaker than quartzite. *Journal of the Geological Society, London*, **157**, 1093–1096.

BROWN, L.D., ZHAO, W. ET AL. 1996. Bright spots, structure, and magmatism in southern Tibet from INDEPTH seismic reflection profiling. *Science*, **274**, 1688–1690.

BYERLEE, J.D. 1978. Friction of rocks. *Pure and Applied Geophysics*, **116**, 615–626.

CAINE, J.S., EVANS, J.P. & FORSTER, C.B. 1996. Fault zone architecture and permeability structures. *Geology*, **24**, 1025–1028.

CHANG, C., CHEN, N. ET AL. 1986. Preliminary conclusions of the Royal Society and Academia Sinica 1985 geotraverse of Tibet. *Nature*, **323**, 501–507.

CHESTER, F.M. & LOGAN, J.M. 1986. Implications for mechanical properties of brittle faults from observations of the Punchbowl fault zone, California. *Pure and Applied Geophysics*, **124**, 79–106.

CHESTER, F.M., EVANS, J.P. & BIEGEL, R.L. 1993. Internal structure and weakening mechanisms of the San Andreas Fault. *Journal of Geophysical Research*, **B98**, 771–786.

CLARK, M.K. & ROYDEN, L.H. 2000. Topographic ooze: Building the eastern margin of Tibet by lower crustal flow. *Geology*, **28**, 703–706.

CLOOS, E. 1955. Experimental analysis of fracture patterns. *Geological Society of America Bulletin*, **66**, 241–256.

COBBOLD, P.R. & DAVY, P. 1988. Indentation tectonics in nature and experiment. 2. Central Asia. *Bulletin of the Geological Institutions of Uppsala, New Series*, **14**, 143–162.

COGAN, M.J., NELSON, K.D., KIDD, W.S.F., WU, C. & INDEPTH TEAM. 1998. Shallow structures of the Yadong–Gulu Rift, southern Tibet, from refraction analysis of Project INDEPTH common midpoint data. *Tectonics*, **17**, 46–61.

COPELAND, P. 1990. *Cenozoic Tectonic History of the Southern Tibetan Plateau and Eastern Himalaya; Evidence From $^{40}Ar/^{39}Ar$ Dating.* State University of New York, Albany, NY.

COWARD, M.P., KIDD, W.S.F. ET AL. 1988. The structure of the 1985 Tibet Geotraverse, Lhasa to Golmud. *Philosophical Transactions of the Royal Society of London*, **A327**, 307–336.

COWIE, P.A. & SCHOLZ, C.H. 1992. Physical explanation for the displacement–length relationship of faults using a post-yield fracture mechanics model. *Journal of Structural Geology*, **14**, 1133–1148.

D'ANDREA, J., HARRISON, T.M., GROVE, M. & LIN, D. 2001. The thermal evolution of the Nyainqentanglha Shan; evidence for a long-lived, lower crustal magmatic system in southern Tibet. *Journal of Asian Earth Sciences*, **19**, 12–13.

DEWEY, J.F. 2002. Transtension in arcs and orogens. *International Geology Review*, **44**, 402–439.

DEWEY, J.F. & BURKE, K.C. 1973. Tibetan, Variscan, and Precambrian basement reactivation: products of continental collision. *Journal of Geology*, **81**, 683–692.

DEWEY, J.F., SHACKLETON, R.M., CHENFA, C. & YIYIN, S. 1988. The tectonic evolution of the Tibetan plateau. *Philosophical Transactions of the Royal Society of London*, **A-327**, 379–413.

DIETERICH, J.H. 1972. Time dependent friction in rocks. *Journal of Geophysical Research*, **77**, 3690–3697.

DIETERICH, J.H. 1992. Earthquake nucleation on faults with rate- and state-dependent strength. *Tectonophysics*, **211**, 115–134.

EDWARDS, M.A., RATSCHBACHER, L. & INDEPTH TEAM. 2000. Role of pre-existing weaknesses in the pre-collisional 'Andean Margin' (S. Lhasa Block) of the Tibet Plateau. *American Geophysical Union, EOS Transactions*, **81**, F186.

EDWARDS, M.A., RATSCHBACHER, L.R. ET AL. 2001. Shortening within the northern Lhasa Block at ca. 90 degrees E; assessing strain contributions from the India–Asia collision. *Geological Society of America, Abstracts with Programs*, **33**, 328.

FARVER, J.R. & YUND, R.A. 1999. Oxygen bulk diffusion measurements and TEM characterisation of a natural ultramylonite: implications for fluid transport in mica-bearing rocks. *Journal of Metamorphic Geology*, **17**, 669–683.

FAULKNER, D.R. & RUTTER, E.H. 2001. Can the maintenance of overpressured fluids in large strike–slip fault zones explain their apparent weakness? *Geology*, **29**, 503–506.

FAULKNER, D.R., LEWIS, A.C. & RUTTER, E.H. 2003. On the internal structure and mechanics of large strike-slip fault zones; field observations of the Carboneras Fault in southeastern Spain. *Tectonophysics*, **367**, 235–251.

FOSSEN, H. & TIKOFF, B. 1998. Extended models of transpression and transtension, and application to tectonic settings. *In*: JONES, R.R. & HOLDSWORTH, R.E. (eds) *Continental Transpressional and Transtensional Tectonics*. Geological Society, London, Special Publications, **135**, 15–33.

FOSSEN, H., TIKOFF, B. & TEYSSIER, C. 1994. Strain modeling of transpressional and transtensional deformation. *Norsk Geologisk Tidsskrift*, **74**, 134–145.

FRANCHETEAU, J., JAUPART, C., SHEN, X., KANG, W.L.D., BAI, J., WEI, H. & DENG, H. 1984. High heat flow in southern Tibet. *Nature*, **307**, 32–36.

GAILLARD, F., SCAILLET, B. & PICHAVANT, M. 2004. Evidence for present day leucogranite pluton growth in Tibet. *Geology*, **32**, 801–804.

GIRARDEAU, J., MARCOUX, J., ALLEGRE, C.J. ET AL. 1984. Tectonic environment and geodynamic significance of the Neo-Cimmerian Donqiao Ophiolite, Bangong–Nujiang suture zone, Tibet. *Nature*, **307**, 27–31.

GRATIER, J.P. & GAMOND, J.F. 1990. Transition between seismic and aseismic deformation in the upper crust. *In*: KNIPE, R.J. & RUTTER, E.H. (eds) *Deformation, Mechanisms, Rheology and Tectonics*. Geological Society, London, Special Publications, **54**, 461–474.

HACKER, B.R. 1997. Diagenesis and the fault-valve seismicity of crustal faults. *Journal of Geophysical Research*, **102**, 24 459–24 467.

HANCOCK, P.L., CHALMERS, R.M.L., ALTUNEL, E. & CAKIR, Z. 1999. Travitonics: using travertines in active fault studies. *Journal of Structural Geology*, **21**, 903–916.

HANCOCK, P.L., KLAPER, E.M., MANCKTELOW, N.S. & RAMSAY, J.G. 1984. Planar and linear fabrics of deformed rocks. *Journal of Structural Geology*, **6**, 1–215.

HANDY, M.R. 1989. Deformation regimes and the rheological evolution of fault zones in the lithosphere; the effects of pressure, temperature, grain-size and time. *Tectonophysics*, **163**, 119–152.

HANDY, M.R., WISSING, S.B. & STREIT, J.E. 1999. Frictional–viscous flow in mylonite with varied bimineralic composition and its effect on lithospheric strength. *Tectonophysics*, **303**, 175–191.

HANMER, S. & PASSCHIER, C.W. 1991. *Shear Sense Indicators: A Review.* Geological Survey of Canada Paper, **90**.

HERWEGH, M. & JENNI, A. 2001. Granular flow in polymineralic rocks bearing sheet silicates: new evidence from natural examples. *Tectonophysics*, **332**, 309–320.

HIPPERTT, J.F.M. 1993. 'V'-pull-apart microstructures: a new shear-sense indicator. *Journal of Structural Geology*, **15**, 1393–1403.

HIRTH, G. & TULLIS, J. 1994. The brittle-plastic transition in experimentally deformed quartz aggregates. *Journal of Geophysical Research*, **99**, 11 731–11 747.

HOBBS, B.E., ORD, A. & TEYSSIER, C. 1986. Earthquakes in the ductile regime? *Pure and Applied Geophysics*, **124**, 309–336.

HOBBS, B.E., MÜHLHAUS, H.B. & ORD, A. 1990. Instability, softening and localisation of deformation. *In*: KNIPE, R.J. & RUTTER, E.H. (eds) *Deformation Mechanisms, Rheology and Tectonics.* Geological Society, London, Special Publications, **54**, 73–78.

HOFFMAN, P.F. 1988. United states of America, the birth of a craton; early Proterozoic assembly and growth of Laurentia. *Annual Review of Earth and Planetary Sciences*, **16**, 543–603.

HOLDSWORTH, R.E. 2004. Weak faults – rotten cores. *Science*, **303**, 181–182.

HOLDSWORTH, R.E., BUTLER, C.A. & ROBERTS, A.M. 1997. The recognition of reactivation during continental deformation. *Journal of the Geological Society, London*, **154**, 73–78.

HOLDSWORTH, R.E., STEWART, M., IMBER, J. & STRACHAN, R.A. 2001. The structure and rheological evolution of reactivated continental fault zones: a review and case study. *In*: MILLER, J.A., HOLDSWORTH, R.E., BUICK, I.S., & HAND, M. (eds) *Continental Reactivation and Reworking.* Geological Society, London, Special Publications, **184**, 115–137

JACKSON, J. 2002. Strength of the continental lithosphere: time to abandon the jelly sandwich? *Geological Society of America Today*, **12**, 4–10.

JAMTVEIT, B. & YARDLEY, B.W.D. 1997. Fluid flow and transport in rocks: an overview. *In*: JAMTVEIT, B. & YARDLEY, B.W.D. (eds) *Fluid Flow and Transport in Rocks.* Chapman & Hall, London, 1–14.

JIANG, D. & WHITE, J.C. 1995. Kinematics of rock flow and the interpretation of geological structures with particular reference to shear zones. *Journal of Structural Geology*, **17**, 1249–1266.

JIANG, D. & WILLIAMS, P.F. 1998. High strian zones: a unified model. *Journal of Structural Geology*, **20**, 1105–1120.

JOHNSON, P.A. & McEVILLY, T.V. 1995. Parkfield seismicity: fluid driven? *Journal of Geophysical Research*, **100**, 12 937–12 950.

KAPP, J., HARRISON, M., GROVE, M., KAPP, P., DING, L. & LOVERA, O. 2002. Structural constraints on the evolution of the Nyainqentanglha massif, southeastern Tibet. *American Geophysical Union, EOS Transactions*, **83**, T51B–1144.

KARNER, S.L., MARONE, C. & EVANS, B. 1997. Laboratory study of fault healing and lithification in simulated fault gouge under hydrothermal conditions. *Tectonophysics*, **277**, 41–55.

KIDD, W.S.F., PAN, Y. *ET AL.* 1988. Geological mapping of the 1985 Chinese–British Tibetan (Xizang–Qinghai) Plateau geotraverse route. *Philosophical Transactions of the Royal Society of London*, **A327**, 287–305.

LANGIN, W.R., BROWN, L.D. & SANDVOL, E.A. 2003. Seismicity of central Tibet from Project INDEPTH III seismic recordings. *Bulletin of the Seismological Society of America*, **93**, 2146–2159.

LE HÉBEL, F., GAPAIS, D., FOURCADE, S. & CAPDEVILA, R. 2002. Fluid-assisted large strains in a crustal-scale decollement (Hercynian Belt of South Brittany, France). *In*: DE MEER, S., DRURY, M.R., DE BRESSER, J.H.P. & PENNOCK, G.M. (eds) *Deformation Mechanisms, Rheology and Tectonics; Current Status and Future Perspectives.* Geological Society, London, Special Publications, **200**, 85–101.

LEEDER, M.R., SMITH, A.B., YIN, J., CHANG, C., SHACKLETON, R.M. & DEWEY, J.F. 1988. Sedimentology, palaeoecology and palaeoenvironmental evolution of the 1985 Lhasa to Golmud geotraverse. *Philosophical Transactions of the Royal Society of London*, **A327**, 107–143.

LELOUP, P.H., LACASSIN, R. *ET AL.* 1995. The Ailao Shan–Red River shear zone (Yunnan, China), Tertiary transform boundary of Indochina. *Tectonophysics*, **251**, 3–10.

LI, S., UNSWORTH, M.J., BOOKER, J.R., WEI, W., TAN, H. & JONES, A.G. 2003. Partial melt or aqueous fluid in the mid-crust of southern Tibet? Constraints from INDEPTH magnetotelluric data. *Geophysical Journal International*, **153**, 289–304.

LIN, A. 1999a. Roundness of clasts in pseudotachylytes and cataclastic rocks as an indicator of frictional melting. *Journal of Structural Geology*, **21**, 473–478.

LIN, A. 1999b. S–C cataclasite in granitic rock. *Tectonophysics*, **304**, 257–273.

LIN, S., JIANG, D. & WILLIAMS, P.F. 1998. Transpression (or transtension) zones of triclinic symmetry: natural example and theoretical modelling. *In*: JONES, R.R. & HOLDSWORTH, R.E. (eds) *Continental Transpressional and Transtensional Tectonics.* Geological Society, London, Special Publications, **135**, 41–57.

LISTER, G.S. & SNOKE, A.W. 1984. S–C mylonites. *Journal of Structural Geology*, **6**, 617–638.

LOGAN, J.M., FRIEDMAN, M., HIGGS, N.G., DENGO, C.A. & SHIMAMOTO, T. 1979. *Experimental Studies of Simulated Gouge and their Application to Studies of Natural Fault Zones.* US Geological Survey Open-file Report, **79/239**, 305–343.

MADDOCK, R.H. 1983. Melt origin of fault-generated pseudotachylytes demonstrated by textures (Outer Hebrides Thrust Zone, Scotland). *Geology*, **11**, 105–108.

MARONE, C. & SCHOLZ, C.H. 1988. The depth of seismic faulting and the upper transition from stable to unstable slip regimes. *Geophysical Research Letters*, **15**, 621–624.

MARONE, C., RALEIGH, C.B. & SCHOLZ, C.H. 1990. Frictional behavior and constitutive modeling of simulated fault gouge. *Journal of Geophysical Research*, **B95**, 7007–7025.

McDOUGALL, I. & HARRISON, T.M. 1988. *Geochronology and Thermochronology by $^{40}Ar/^{39}Ar$ Method.* Oxford University Press, Oxford.

MEANS, W.D. 1989. Stretching faults. *Geology*, **17**, 893–896.

MEANS, W.D., HOBBS, B.E., LISTER, G.S. & WILLIAMS, P.F. 1980. Vorticity and non-coaxiality in progressive deformation. *Journal of Structural Geology*, **2**, 371–378.

MECHIE, J., SOBOLEV, S.V. *ET AL.* 2004. Precise temperature estimation in the Tibetan crust form seismic detection of the alpha-beta quartz transition. *Geology*, **32**, 601–604, doi:10.1130/G20367.1.

MILLER, M.G. 1996. Ductility in fault gouge from a normal fault system, Death Valley, California. A mechanism for fault-zone strengthening and relevance to paleoseismicity. *Geology*, **24**, 603–606.

MILLER, M.M., MELBOURNE, T., JOHNSON, D.J. & SUMNER, W.Q. 2002. Periodic slow earthquakes from the Cascadia subduction zone. *Science*, **295**, 2423–2425.

MITRA, G. 1984. Brittle to ductile transition due to large strains along the White Rock Thrust, Wind River Mountains, Wyoming. *Journal of Structural Geology*, **6**, 51–61.

MOLNAR, P. & CHEN, W.P. 1983. Focal depths and fault plane solutions of earthquakes under the Tibetan plateau. *Journal of Geophysical Research*, **B88**, 1180–1196.

MOLNAR, P. & LYON-CAEN, H. 1989. Fault plane solutions of earthquakes and active tectonics of the Tibetan Plateau and its margins. *Geophysical Journal International*, **99**, 123–153.

NADEAU, R.M. & McEVILLY, T.V. 2004. Periodic pulsing of characteristic microearthquakes on the San Andreas fault. *Science*, **303**, 220–222.

ORD, A. & CHRISTIE, J.M. 1984. Flow stresses from microstructures in mylonitic quartzites of the Moine thrust zone, Assynt area, Scotland. *Journal of Structural Geology*, **6**, 639–654.

OWENS, T.J. & ZANDT, G. 1997. Implications of crustal property variations for models of Tibetan Plateau evolution. *Nature*, **387**, 37–43.

PAN, Y. & KIDD, W.S.F. 1992. Nyainqentanglha shear zone; a late Miocene extensional detachment in the southern Tibetan Plateau. *Geology*, **20**, 775–778.

PASSCHIER, C.W. 1991. The classification of dilatant flow types. *Journal of Structural Geology*, **13**, 101–104.

PATRIAT, P. & ACHACHE, J. 1984. India–Eurasia collision chronology has implications for crustal shortening and driving mechanism of plates. *Nature*, **311**, 615–621.

PLATT, J.P. & VISSERS, R.L.M. 1980. Extensional structures in anisotropic rocks. *Journal of Structural Geology*, **2**, 397–410.

POWELL, C. & McCONAGHAN, P.J. 1973. Plate Tectonics and the Himalayas. *Earth and Planetary Science Letters*, **20**, 1–12.

RAMBERG, H. 1975. Particle paths, displacement and progressive strain applicable to rocks. *Tectonophysics*, **28**, 1–37.

RAMSAY, J.G. & GRAHAM, R.H. 1970. Strain variation in shear belts. *Canadian Journal of Earth Sciences*, **7**, 786–813.

RANDALL, G.E., AMMON, C.J. & OWENS, T.J. 1995. Moment tensor estimation using regional seismograms from a Tibetan Plateau portable network deployment. *Geophysical Research Letters*, **22**, 1665–1668.

RATSCHBACHER, L. & SPERNER, B. 1994. Stress und Strain in Tübingen oder ein Computerlabor zur Erarbeitung von Stress- und Strain. *Göttinger Arbeiten zur Geologie und Paläontologie*, **Sb1**, 27–29.

ROWLEY, D.B. 1998. Minimum age of initiation of collision between India and Asia north of Everest based on the subsidence history of the Zhepure Mountain section. *Journal of Geology*, **106**, 229–235.

RUTTER, E.H. 1983. Pressure solution in nature, theory and experiment. *Journal of the Geological Society, London*, **140**, 725–740.

RUTTER, E.H. 1999. On the relationship between the formation of shear zones and the form of the flow law for rocks undergoing dynamic recrystallization. *Tectonophysics*, **303**, 147–158.

RUTTER, E.H. & BRODIE, K.H. 1988. The role of tectonic grain size reduction in the rheological stratification of the lithosphere. *Geologische Rundschau*, **77**, 295–308.

RUTTER, E.H., MADDOCK, R.H., HALL, S.H. & WHITE, S.H. 1986. Comparative microstructures of natural and experimentally produced clay-bearing fault gouges. *Pure and Applied Geophysics*, **124**, 3–30.

SANDERSON, D. & MARCHINI, R.D. 1984. Transpression. *Journal of Structural Geology*, **6**, 449–458.

SCHMID, S.M. & HANDY, M.R. 1991. Towards a genetic classification of fault rocks: geological usage and tectonophysical implications. *In*: MULLER, D.W., McKENZIE, J.A., & WEISSERT, H. (eds) *Controversies in Modern Geology, Evolution of Geological Theories in Sedimentology, Earth History and Tectonics*. Academic Press London, 339–361.

SCHOLZ, C.H. 1988. The brittle-plastic transition and the depth of seismic faulting. *Geologische Rundschau*, **77**, 319–328.

SCHOLZ, C.H. 1998. Earthquakes and friction laws. *Nature*, **391**, 37–42.

SCHOLZ, C.H. 2002. *The Mechanics of Earthquakes and Faulting*. Cambridge University Press, New York.

SIBSON, R.H. 1975. Generation of pseudotachylyte by ancient seismic faulting. *Royal Astronomical Society Geophysical Journal*, **43**, 775–794.

SIBSON, R.H. 1983. Continental fault structure and the shallow earthquake source. *Journal of the Geological Society, London*, **140**, 741–767.

SIBSON, R.H., MOORE, R.M. & RANKIN, A.H. 1975. Seismic pumping – a hydrothermal transport mechanism. *Journal of the Geological Society, London*, **131**, 653–659.

SIMPSON, C. 1985. Deformation of granitic rocks across the brittle–ductile transition. *Journal of Structural Geology*, **7**, 503–511.

SIMPSON, C. & DE PAOR, D.G. 1997. Practical analysis of general shear zones using the porphyroclast hyperbolic distribution method: An example from the Scandinavian Caledonides. *In*: SENGUPTA, S. (ed.) *Evolution of Geological Structures in Micro- to Macro-scales*. Chapman & Hall, London, 169–184.

SIMS, D., FERRILL, D.A. & STAMATAKOS, J.A. 1999. Role of a ductile decollement in the development of pull-apart basins: Experimental results and natural examples. *Journal of Structural Geology*, **21**, 533–554.

SPERNER, B., RATSCHBACHER, L. & OTT, R. 1993. Fault-striae analysis: a Turbo Pascal program package for graphical presentation and reduced stress tensor calculation. *Computers and Geosciences*, **19**, 1361–1388.

SPRAY, J.G. 1995. Pseudotachylyte controversy: Fact or friction? *Geology*, **23**, 1119–1122.

STEWART, M., HOLDSWORTH, R.E. & STRACHAN, R.A. 2000. Deformation processes and weakening mechanisms within the frictional–viscous transition zone of major crustal-scale faults: insights from the Great Glen Fault Zone, Scotland. *Journal of Structural Geology*, **22**, 543–560.

STREIT, J.E. 1997. Low frictional strength of upper crustal faults; a model. *Journal of Geophysical Research*, **B102**, 24 619–24 626.

TAKAGI, H. & KOBAYASHI, K. 1996. Composite planar fabrics of fault gouges and mylonites; comparative petrofabrics. *Journal of the Geological Society of Japan*, **102**, 170–179.

TAPPONNIER, P. & MOLNAR, P. 1977. Active faulting and tectonics in China. *Journal of Geophysical Research*, **82**, 2905–2930.

TAPPONNIER, R., PELTZER, G., LE DAIN, A.Y., ARMIJO, R. & COBBOLD, P. 1982. Propagating extrusion tectonics in Asia; new insights from simple experiments with plasticine. *Geology*, **10**, 611–616.

TAPPONNIER, P., PELTZER, G. & ARMIJO, R. 1986. On the mechanics of the collision between India and Asia. *In*: COWARD, M.P. & RIES, A.C. (eds) *Collisional Tectonics*. Geological Society, London, Special Publications, **19**, 115–157.

TAPPONNIER, P., XU, Z., ROGER, F., MEYER, B., ARNAUD, N., WITTLINGER, G. & YANG, J. 2001. Oblique stepwise rise and growth of the Tibet Plateau. *Science*, **294**, 1671–1677.

TCHALENKO, J.S. 1970. Similarities between shear zones of different magnitudes. *Geological Society of America Bulletin*, **81**, 1625–1640.

TIKOFF, B. & FOSSEN, H. 1999. Three-dimensional reference deformations and strain facies. *Journal of Structural Geology*, **21**, 1497–1512.

TULLIS, J. & YUND, R.A. 1985. Dynamic recrystallization of feldspar: A mechanism for ductile shear zone formation. *Geology*, **13**, 238–241.

VAN DER WOERD, J., RYERSON, F.J. ET AL. 1998. Holocene left-slip rate determined by cosmogenic surface dating on the Xidatan segment of the Kunlun Fault (Qinghai, China). *Geology*, **26**, 695–698.

VERMA, R.K. & REDDY, Y.S.K. 1988. Seismicity, focal mechanisms and their correlation with the geological/tectonic history of the Tibetan Plateau. *Tectonophysics*, **156**, 107–131.

WHITE, S., BURROWS, S.E., CARRERAS, J., SHAW, N.D. & HUMPHREYS, F.J. 1980. On mylonites in ductile shear zones. *Journal of Structural Geology*, **2**, 175–187.

WHITE, S.H. & KNIPE, R.J. 1978. Transformation and reaction-enhanced ductility in rocks. *Journal of the Geological Society, London*, **125**, 513–516.

WHITE, S.H., BRETAN, P.G. & RUTTER, E.H. 1986. Fault zone reactivation: kinematics and mechanisms. *Philosophical Transactions of the Royal Society, London*, **A-317**, 81–97.

WIBBERLEY, C.A.J. & SHIMAMOTO, T. 2003. Internal structure and permeability of major strike-slip fault zones: the Median Tectonic Line in Mie Prefecture, Southwest Japan. *Journal of Structural Geology*, **25**, 59–78.

WIBBERLEY, C.A.J., HIROSE, T. & SHIMAMOTO, T. 2003. Permeability structure of large fault zones and consequences for fluid-controlled fault dynamics. http://www.casp.cam.ac.uk/news/ wibberley-abs1.htm (web publication).

WILCOX, R.E., HARDING, T.P. & SEELY, D.R. 1973. Basic wrench tectonics. *American Association of Petroleum Geologists Bulletin*, **57**, 74–96.

WINTSCH, R., CHRISTOFFERSON, R. & KRONENBERG, A.K. 1995. Fluid-rock reaction weakening of fault zones. *Journal of Geophysical Research*, **100**, 13 021–13 032.

Magnetic fabric and microstructure of a mylonite: example from the Bitterroot shear zone, western Montana

D. SIDMAN[1], E. C. FERRÉ[2], C. TEYSSIER[1] & M. JACKSON[3]

[1]*Department of Geology and Geophysics, University of Minnesota, MN 55455, USA (e-mail: sidm0001@umn.edu)*

[2]*Department of Geology, Southern Illinois University, Carbondale, IL 62901, USA*

[3]*Institute for Rock Magnetism, University of Minnesota, MN 55455, USA*

Abstract: The Bitterroot shear zone, SW Montana, is a mylonitic detachment that developed by strain localization during the Palaeocene–Eocene orogenic collapse of this part of the North American Cordillera. Anisotropy of magnetic susceptibility (AMS) data from two transects across the shear zone and into the granitic footwall demonstrate the continuity between the low to high-temperature solid-state fabric in the shear zone and the magmatic fabric developed in the footwall granite. This fabric gradually and smoothly rotates from E-dipping in the shear zone to W-dipping in the footwall granites, forming an arch over 10 km wide. Furthermore, the mineral fabric of both paramagnetic and ferrimagnetic minerals is consistent with the AMS fabric, displaying the same arching, which is interpreted to have developed by a rolling-hinge process in the footwall granites during activation of the Bitterroot shear zone. The AMS method thus stands out as a robust indicator of fabric over a wide range of deformation conditions.

The anisotropy of magnetic susceptibility (AMS) is a versatile and fast method for analysing both quantitatively and qualitatively the magmatic and solid-state fabric of granitic rocks (Hrouda 1982; Rochette *et al.* 1992; Borradaile & Henry 1997; Bouchez 1997). Much work has been carried out to understand the complex relationships between the several magnetic carriers commonly present in granitic rocks and the resulting magnetic fabric (e.g. Borradaile 1991; Rochette *et al.* 1992; Housen *et al.* 1995; Grégoire *et al.* 1995; Archanjo *et al.* 1995). These studies reveal that: (1) the magnetic contribution of ferrimagnetic minerals, such as magnetite, can result from both shape anisotropy and distribution anisotropy, the latter of which can be constructive or destructive, depending on the spatial relationships of one or more grains; (2) because para- and ferrimagnetic minerals, such as biotite and magnetite, respectively, crystallize before diamagnetic minerals, such as quartz and feldspar, the magnetic fabric commonly reveals the overall magmatic fabric in igneous rocks; and (3) the variation of magnetic fabric with strain magnitude is poorly understood; however, AMS may track, at least qualitatively, the approximate shape of the finite-strain ellipsoid, although these comparisons are certainly rather complex (e.g. Lüneberg *et al.* 1999). Quantifying finite strain is an important step in determining the deformation history of a region; however, to use AMS as a method for determining finite strain, it is first necessary to understand the precise relationship between magnetic fabric and mineral fabric.

The Bitterroot shear zone (BSZ) was chosen for this study because: (1) its kinematic history and fabric are relatively well defined; (2) a wide range of deformation fabrics (from low-temperature solid state to magmatic) are preserved in a relatively short distance, so the relationship between deformation and AMS can be easily studied; and (3) both paramagnetic and ferromagnetic (*sensu lato*) minerals are present, which makes it possible to compare microstructural data with AMS. In this paper, we present the results of this AMS analysis and its implications for the mineral fabric in the BSZ.

Geological history

The BSZ, western Montana, forms the westernmost edge of the Bitterroot metamorphic core complex, a N–S-trending, asymmetric massif approximately 50 km wide and 100 km long (Foster *et al.* 2001) (Fig. 1). Its asymmetry is due to a larger amount of unroofing in the east than in the west. The core complex is composed

From: BRUHN, D. & BURLINI, L. (eds) 2005. *High-Strain Zones: Structure and Physical Properties.*
Geological Society, London, Special Publications, **245**, 143–163.
0305-8719/05/$15.00 © The Geological Society of London 2005.

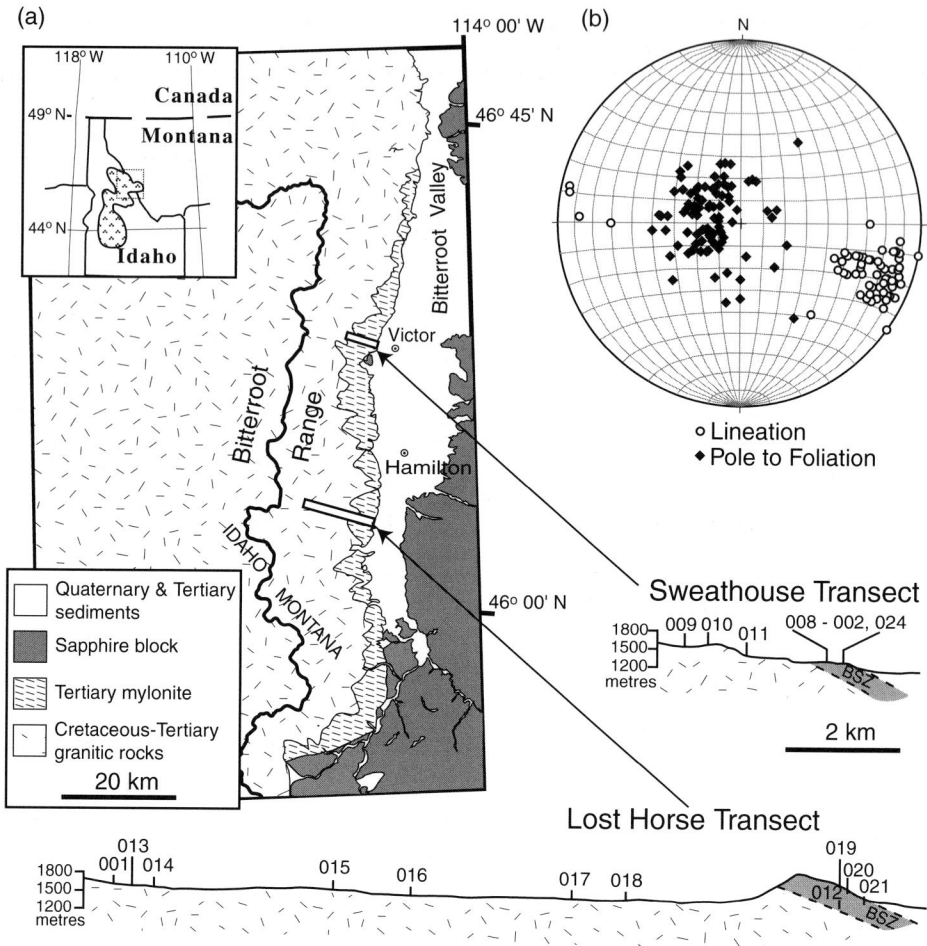

Fig. 1. (**a**) Geological setting of the Bitterroot shear zone (modified from Foster *et al.* 2001), with cross-sections (no vertical exaggeration) showing 22 AMS sample locations. (**b**) Lower-hemisphere, equal-area stereonet with field foliation and lineation measurements.

primarily of the Idaho–Bitterroot Batholith, which began forming during Mesozoic–Palaeogene continental thickening, and metamorphosed Proterozoic Belt Supergroup sediments (Chase 1973; Hyndman 1980; Kerrich & Hyndman 1986; Hyndman & Meyers 1988; Foster *et al.* 2001). Between 85 and 70 Ma, intermediate–mafic plutons intruded (Armstrong *et al.* 1977; Shuster & Bickford 1985; Foster *et al.* 2001). Widespread partial melting resulted in the intrusion of granitic plutons between 65 and 53 Ma (Foster & Fanning 1997; Foster *et al.* 2001) that formed a large fraction of the middle crust. Orogenic collapse, thinning and extension caused the exhumation of the core complex. Differential exhumation from west to east resulted from the eastward motion of the Sapphire block that occurred between 52 and 45 Ma along the BSZ (Foster & Fanning 1997; Foster *et al.* 2001).

A wide range of deformation fabrics is preserved across the shear zone. In the deepest rocks – the footwall of the BSZ – a magmatic fabric is preserved in the Palaeogene granitic rocks. Magmatic foliation and lineation are faint in the field and require petrographic analysis to be identified. Moving east across the shear zone, high-temperature and then low-temperature, solid-state deformation become apparent. Mylonitic deformation appears in zones with increasing frequency as one nears the brittle eastern edge (top) of the BSZ. This mylonite is approximately 500 m thick, and displays dominantly E-dipping foliation and a lineation trending approximately 110° (Fig. 1).

Towards the top of the BSZ, fabric is characterized by a low-temperature, cataclastic overprint of the mylonitic fabric (Hyndman & Meyers 1988; Foster 2000). The BSZ ends abruptly at a brittle fault that separates the Bitterroot block from the Sapphire block, defining the Bitterroot Valley (Foster & Raza 2002).

Sampling and analytical methods

Sampling

Core samples were collected from 22 stations using a portable drill and a stainless steel drill bit. Most outcrops in the granite and the high-temperature mylonite are fresh; magnetic minerals in most core samples were therefore unaltered below a few millimetres from the surface. In some cases, large oriented blocks were collected and subsequently drilled in the laboratory. The 22 stations, approximately evenly spaced, are located in two ENE–WNW traverses across the BSZ, one located in Sweathouse Creek canyon and the other in Lost Horse Creek canyon (Fig. 1). In the Sweathouse Creek traverse, samples came predominantly from a large quarry that facilitated sample collection. Where possible, especially in the shear zone, field lineation and foliation were also measured. In the laboratory, the 22 drill cores were sliced into 199 samples, 25 mm in diameter and 22 mm in length, for AMS analysis.

Anisotropy of magnetic susceptibility

The magnetic susceptibility, K, is defined as the ratio between the induced magnetization of the specimen and the inducing magnetic field. This magnetization disappears when the field is relaxed. The anisotropy of magnetic susceptibility (AMS) is described by a symmetric second-rank tensor with three principal axes ($K_1 \geq K_2 \geq K_3$). K_1 is the magnetic lineation, and K_3 is normal to the magnetic foliation (Tarling & Hrouda 1993). The bulk magnetic susceptibility, K, is the arithmetic mean of the principal susceptibilities [$K_m = 1/3\ (K_1 + K_2 + K_3)$]. The total anisotropy can be expressed by P_j and the ellipsoid shape by the parameter T (Jelínek 1978).

$$P_j = \exp\left(\sqrt{2\left[(\eta_1 - \eta)^2 + (\eta_2 - \eta)^2 + (\eta_3 - \eta)^2 \right]} \right)$$

$$\eta = \frac{\eta_1 + \eta_2 + \eta_3}{3}$$

$$T = [2(\eta_2 - \eta_3)/(\eta_1 - \eta_3)] - 1 \quad \eta_i = \ln K_i.$$

Ferrimagnetic minerals, such as magnetite, display mostly a magnetostatic anisotropy that is a direct reflection of their shape (Tarling & Hrouda 1993). Paramagnetic minerals, such as biotite, and antiferromagnetic minerals, such as haematite, display a magnetocrystalline anisotropy that is related to the crystallographic lattice orientation. In euhedral crystals the grain shape follows the lattice orientation. In practice, however, it is whole-rock samples that are measured for AMS, not individual mineral grains. Therefore, it is not only the AMS of individual grains, but also their preferred orientation that governs the AMS of the rock. Further, ferromagnetic grains interact with other grains, both constructively and destructively, which affects the bulk AMS of a particular sample (e.g. Hargraves et al. 1991; Siegesmund et al. 1995). However, Cañón-Tapia (2001) suggested that the role of these interactions is minimal in magmatic rocks. Because of the complex interactions among AMS, grain shape and crystallography, AMS data are useful for understanding rock fabric only *after* the magnetic mineralogy is well characterized.

AMS measurements were conducted on a Kappabridge Magnetic Susceptibility Bridge (Geofyzika Model KLY2, frequency 920 Hz, sensitivity 4×10^{-8} SI, AC field 300 A m^{-1} RMS, temperature approximately 300 K) at the Institute for Rock Magnetism (University of Minnesota). The magnetic mineralogy was characterized by using: (1) a Vibrating Sample Magnetometer (VSM) (Princeton Measurements, sensitivity 5×10^{-9} A m^2, temperature c. 292 K); (2) a Magnetic Properties Measurement System (MPMS) (Quantum Designs Magnetic Properties Measurement System cryogenic susceptometer, temperature range 20–300 K, sensitivity 1×10^{-11} A m^2, magnetic field up to 2.5 tesla (T)); and (3) a JEOL JXA-8900R electron probe microanalyser (15 kV accelerating voltage, 20 nA probe current, 1 μm probe diameter, with the following crystals: LIF (Fe Kα, Mn Kα, Cr Kα, Ca Kα), TAP (Al Kα, Mg Kα, Si Kα) and PET (Ti Kα)).

Particle analysis

From the two transects of the Bitterroot shear zone (Fig. 1), a small subset of drill core samples was selected for particle analysis on the basis of: (1) spatial distribution – relatively even spacing across the transects; (2) type of dominant magnetic carrier – both para- and ferrimagnetic samples – that was known from the VSM measurements; and (3) conditions of deformation – from magmatic state to solid state, so that the effect of this transition on the magnetic fabric could be studied. The purpose of the particle analysis was to compare mineral fabric with the magnetic fabric already known from AMS measurements, and therefore to evaluate the effectiveness of AMS in measuring rock fabric developed in a wide range of temperature conditions. To accomplish this, thin sections were made from 11 core samples, oriented as nearly as possible into the K_1-K_3 plane. From field observations of the BSZ, it is clear that the geometry and kinematics of this shear zone are consistent with plane strain (well-developed foliation and lineation). Furthermore, as will be shown by the shape of the AMS ellipsoid, this shear zone probably developed under nearly plane-strain conditions. Therefore, comparing three-dimensional (3D) AMS fabric to 2D shape fabric in the K_1-K_3 principal plane may be justified (Passchier & Trouw 1996).

Thin sections of the ferrimagnetic samples (eight of the 11) were analysed with an electron microprobe to obtain backscattered electron images of all magnetite grains (Fig. 2). From these images, using the NIH Image software, the aspect ratios, orientations and relative sizes of the particles were precisely measured. Assuming that relative cross-sectional area is indicative of grain volume, very small grains in the K_1-K_3 plane are likely to contribute less, by volume, to the overall magnetic fabric or to be corners of larger grains, not necessarily representative of the overall mineral fabric. Therefore, by giving greater weight to the larger particles in the thin section, a more accurate particle fabric was obtained. For particle analysis of biotite,

orientations were measured using an optical microscope. While this type of particle analysis does not allow for weighting the relative contribution of individual particles based on particle size, the biotite particles were both greater in number and more uniform in size in a given thin section than were the magnetite particles. Therefore, this method was adequate for comparing biotite mineral fabric with magnetic fabric. Fabric ellipses aspect ratios were then created from the particle orientation data according to the method described by Benn & Allard (1988), using the normalized orientation tensor (Harvey & Laxton 1980), where x and y are the direction cosines of the major axes of mineral grains:

$$T = \frac{1}{N} \left(\begin{array}{cc} \sum x^2 & \sum xy \\ \sum yx & \sum y^2 \end{array} \right).$$

Orientation of fabric ellipses was calculated by averaging the orientations of particles.

Results

Magnetic mineralogy

To determine the dominant magnetic carrier, the hysteresis properties of a representative subset (103 of 199) of the drill core samples were measured in high field on the VSM. Most specimens displayed a hysteresis indicating the presence of ferromagnetic phases. The high field slope represents the sum of the diamagnetic and paramagnetic susceptibilities. In granitic protoliths the diamagnetic contribution to the magnetic susceptibility has been estimated

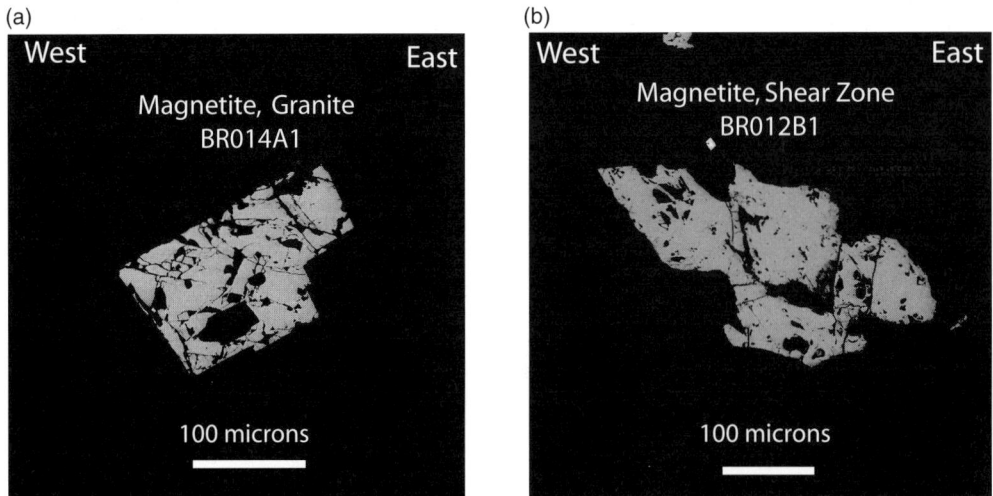

(a)

West   East

Magnetite, Granite
BR014A1

100 microns

(b)

West   East

Magnetite, Shear Zone
BR012B1

100 microns

Fig. 2. Backscattered electron images of two magnetites from samples (**a**) BR014, core A1 and (**b**) BR012, core B1, from the Lost Horse transect (Fig. 1).

around -14×10^{-6} SI (Rochette *et al.* 1992). Once the diamagnetic contribution is subtracted from the high field slope the difference is the paramagnetic susceptibility only. Most samples were dominantly ferromagnetic, but a few were dominated by paramagnetic minerals. This difference in magnetic mineralogy is attributed to compositional variation within the Bear Creek pluton in the Lost Horse profile and different small volume intrusions in the Sweathouse Creek profile (Foster pers. comm.).

Petrographic investigations indicate the presence of biotite (main paramagnetic mineral) and magnetite (main ferromagnetic mineral). To further identify and characterize the magnetite, its grain size, chemical composition and Verwey transition – the low-temperature crystallographic transition from cubic to orthorhombic (Verwey & Haayman 1941) – were measured. The Verwey transition was determined during low-temperature experiments on five specimens using the MPMS instrument. The change in remanence at 120 K, characteristic of the Verwey transition in magnetite, was systematically observed in these specimens. The hysteresis parameters were plotted in a Day *et al.* (1977) graph (Fig. 3) and show that most of the magnetite grains are multidomain. Chemical composition of ferrimagnetic minerals was determined from electron microprobe analyses, using the scheme of Schumacher (1991) to separate the ferric iron from ferrous iron in the microprobe weight per cent data. Ferrimagnetic samples outside the shear zone were found to contain pure magnetite and titanohematite (most of which contains exsolution bands of ilmenite), while those within the shear zone contain only magnetite. Susceptibility of titanohematite is several orders of magnitude lower than that of magnetite, so magnetite dominates the AMS fabric for all ferrimagnetic samples.

Anisotropy of magnetic susceptibility (AMS)

The AMS data from both transects (Fig. 4, Table 1) show a well-defined magnetic fabric, even in the more weakly deformed granite (sites BR001, BR013–BR018, Lost Horse transect, Fig. 1). K_1 (magnetic lineation) generally trends at about $110°$, and the $K_1–K_2$ plane (magnetic foliation) strikes approximately N–S, which is generally consistent with field measurements of mineral lineation and foliation (Fig. 1). The most notable feature from *both* AMS transects is the gradual rotation of magnetic fabric from W-dipping to E-dipping, and the rotation of K_1 from W-plunging to E-plunging, although the trend remains a fairly consistent $110°$. This arching of the magnetic fabric is more fully pronounced in the longer Lost Horse Creek transect (Fig. 5). In both cases, however, magnetic foliation and lineation rotate smoothly and continuously from the magmatically deformed granites through the intermediate- to low-temperature to brittlely deformed rocks of the shear zone.

The AMS parameters, K_m, P_j and T_j have been plotted in the diagrams of Figure 6. The degree of anisotropy, P_j, increases with the magnetic susceptibility, K_m, and this increase becomes particularly visible around 10^{-2} SI and above (Fig. 6a). The distribution of K_m appears to be bimodal, with a gap in K_m around 5000×10^{-6} SI. The maximum paramagnetic susceptibility in similar granites (Rochette *et al.* 1992) has been estimated to be 200×10^{-6} SI. Thus, the bimodal distribution cannot be accounted for by the separation between paramagnetic and ferromagnetic types. Instead it may be controlled by compositional differences between specimens. The shape of the magnetic ellipsoid (T) is more oblate than prolate for the majority of samples (Fig. 6a & b); however, magnetic fabrics in the shear zone tend to be more oblate than in the granite (Fig. 6b).

Microstructures and particle analysis

Eleven core samples that had already been measured for AMS were cut into thin sections for particle analysis (Fig. 7). These samples, which span across both transects, display microstructures indicative of various deformation conditions from magmatic to low-temperature and even brittle deformation. In the west,

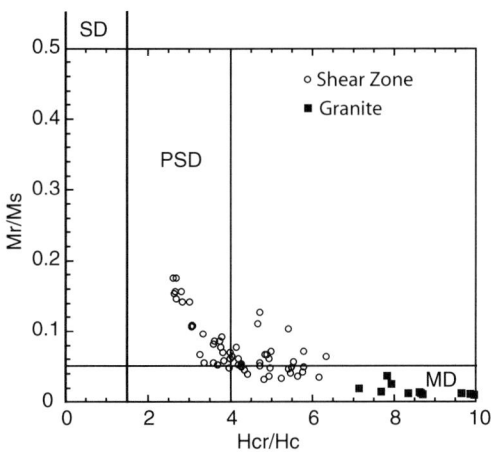

Fig. 3. Hysteresis properties of 77 ferromagnetic drill core specimens (10 cm³) (Day *et al.* 1977). Magnetite carrying the AMS fabric is in the multidomain range. SD, single domain; PSD, pseudo-single domain; MD, multidomain.

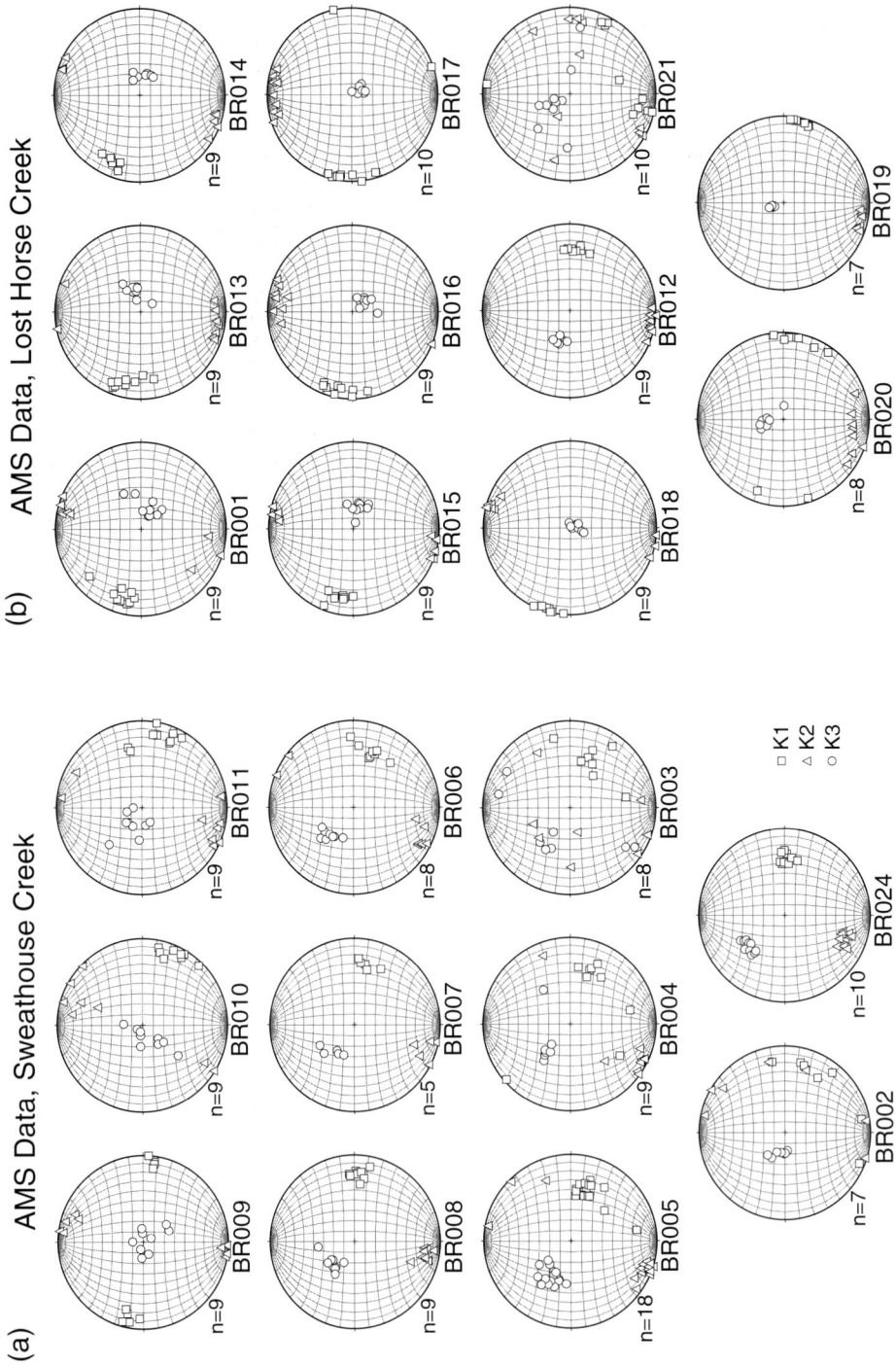

Fig. 4. Results of AMS analyses for (**a**) the Sweathouse transect and (**b**) the Lost Horse transect (Fig. 1), presented from west (top left) to east (bottom right) on equal-area, lower-hemisphere stereonets. Magnetic lineation trends consistently WNW–ESE, and foliation progressively rotates from W-dipping to E-dipping.

Table 1. *Low-field magnetic data for the Bitterroot shear zone*

Site	K_{LF} (10^{-6} SI)	LF-normed principal susceptibilities			LF-anisotropy factors			LF-AMS principal directions					
		K_1	K_2	K_3	P_j	L	T	K_1 (°)		K_2 (°)		K_3 (°)	
								dec	inc	dec	inc	dec	inc
BR001A1	12 330	1.254	0.969	0.778	1.617	1.294	−0.080	286	16	17	3	116	74
BR001A2	7663	1.241	0.971	0.788	1.578	1.278	−0.080	289	16	198	3	99	73
BR001A3	10 540	1.214	0.962	0.824	1.478	1.261	−0.199	281	14	12	6	123	75
BR001B1	2337	1.085	1.056	0.859	1.292	1.028	0.764	318	20	218	26	81	56
BR001B2	2426	1.223	0.934	0.843	1.470	1.309	−0.448	293	20	23	1	115	70
BR001B3	1308	1.150	1.004	0.846	1.361	1.146	0.110	288	29	186	22	64	53
BR001C1	9525	1.232	0.939	0.829	1.501	1.311	−0.368	276	21	12	16	136	63
BR001C2	12 520	1.184	0.994	0.823	1.440	1.191	0.039	279	16	13	13	139	69
BR001C3	10 390	1.177	1.003	0.820	1.439	1.174	0.116	278	27	11	6	112	62
BR002A1	88	1.029	1.024	0.947	1.097	1.005	0.875	202	7	109	25	307	64
BR002A2	1980	1.098	1.045	0.858	1.299	1.050	0.603	129	12	36	12	263	73
BR002A3	468	1.066	1.014	0.921	1.160	1.051	0.319	121	26	30	1	298	64
BR002A4	300	1.094	1.010	0.897	1.221	1.083	0.196	170	10	77	17	290	70
BR002B1	674	1.089	1.042	0.869	1.270	1.045	0.608	103	18	12	5	267	71
BR002B2	1173	1.144	1.033	0.824	1.400	1.108	0.376	106	20	197	3	294	70
BR002B3	446	1.107	1.046	0.847	1.326	1.058	0.578	80	18	172	5	278	71
BR003A1	509	1.110	0.994	0.896	1.240	1.117	−0.035	125	15	253	66	30	18
BR003A2	1305	1.052	0.999	0.949	1.108	1.053	−0.004	126	52	269	32	11	18
BR003A3	1195	1.067	1.030	0.903	1.192	1.036	0.575	104	43	199	5	295	47
BR003A4	1425	1.050	1.036	0.914	1.165	1.014	0.794	169	34	59	27	300	44
BR003B1	512	1.101	1.006	0.892	1.235	1.094	0.142	112	30	336	52	215	22
BR003B2	587	1.197	0.984	0.819	1.464	1.216	−0.030	118	43	214	6	310	47
BR003B3	798	1.068	1.028	0.904	1.190	1.039	0.544	110	38	316	49	211	13
BR003B4	943	1.074	1.036	0.890	1.220	1.036	0.621	77	19	174	18	303	63
BR004A1	294	1.042	1.012	0.945	1.105	1.029	0.406	111	33	206	8	309	55
BR004A2	355	1.053	1.003	0.944	1.115	1.049	0.121	319	3	227	41	52	49
BR004A3	314	1.052	1.004	0.944	1.115	1.048	0.144	98	34	199	15	309	52
BR004B1	374	1.068	1.015	0.918	1.167	1.052	0.330	124	35	216	3	311	55
BR004B2	330	1.092	0.989	0.919	1.188	1.103	−0.143	111	31	205	6	305	58
BR004B3	424	1.048	1.001	0.952	1.101	1.047	0.043	167	29	68	16	313	56
BR004C1	267	1.057	1.001	0.943	1.121	1.056	0.047	212	33	215	6	314	56
BR004C2	446	1.052	1.013	0.935	1.128	1.039	0.347	110	42	207	8	306	47
BR004C3	442	1.133	1.007	0.860	1.318	1.125	0.145	115	25	208	8	314	64
BR005A1	195	1.070	1.015	0.915	1.172	1.054	0.322	96	46	201	14	304	40
BR005A2	177	1.212	0.982	0.806	1.505	1.234	−0.031	138	46	45	3	313	44
BR005A3	3004	1.686	0.922	0.392	4.580	1.829	0.172	110	32	210	16	323	53
BR005A4	363	1.235	0.915	0.849	1.488	1.349	−0.600	122	27	216	7	320	62
BR005B1	264	1.137	1.019	0.844	1.352	1.117	0.259	112	43	203	2	295	47
BR005B2	434	1.128	1.013	0.859	1.317	1.113	0.215	110	44	201	1	292	46
BR005B3	782	1.086	1.014	0.900	1.211	1.071	0.269	104	45	201	7	298	44
BR005C1	140	1.175	1.051	0.774	1.542	1.118	0.466	99	34	195	9	298	54
BR005C2	110	1.148	1.055	0.797	1.466	1.088	0.537	108	33	200	4	297	57
BR005C3	105	1.138	1.054	0.809	1.432	1.080	0.551	104	32	198	7	300	57
BR005D1	818	1.078	0.996	0.926	1.164	1.083	−0.049	104	40	11	4	276	49
BR005D2	722	1.099	1.005	0.897	1.225	1.094	0.117	107	38	197	0	288	52
BR005D3	1421	1.330	0.989	0.681	1.967	1.344	0.116	108	27	204	11	313	60
BR005D4	1271	1.071	1.015	0.914	1.176	1.055	0.325	106	39	197	0	287	51
BR005E1	2403	1.099	1.025	0.875	1.263	1.072	0.388	96	39	198	14	304	47
BR005E2	1845	1.085	1.042	0.873	1.260	1.041	0.633	101	45	196	5	290	44
BR005E3	162	1.057	1.034	0.910	1.176	1.022	0.709	170	23	68	25	296	55
BR005E4	395	1.168	1.028	0.804	1.462	1.136	0.316	96	41	195	10	296	48
BR006A1	2686	1.198	1.050	0.752	1.620	1.142	0.431	108	39	210	15	317	47
BR006A2	2126	1.309	0.978	0.713	1.846	1.338	0.042	113	30	22	2	288	60

(continued)

Table 1. *Continued*

Site	K_{LF} (10^{-6} SI)	LF-normed principal susceptibilities			LF-anisotropy factors			LF-AMS principal directions					
		K_1	K_2	K_3	P_j	L	T	K_1 (°)		K_2 (°)		K_3 (°)	
								dec	inc	dec	inc	dec	inc
BR006A3	2060	1.090	1.064	0.846	1.324	1.024	0.812	130	33	38	3	304	57
BR006A4	1664	1.091	1.023	0.886	1.237	1.067	0.376	95	29	195	16	311	56
BR006B1	1707	1.161	1.011	0.829	1.404	1.148	0.180	112	34	206	6	305	56
BR006B2	2363	1.292	0.962	0.746	1.740	1.343	−0.073	88	20	190	28	328	54
BR006B3	1372	1.096	1.025	0.879	1.254	1.069	0.394	109	38	208	11	311	50
BR006B4	3270	1.152	1.010	0.838	1.378	1.140	0.176	109	32	209	15	320	54
BR007A1	203	1.212	0.981	0.807	1.504	1.235	−0.039	100	30	192	4	289	60
BR007A2	169	1.166	0.983	0.851	1.372	1.186	−0.081	94	24	198	28	330	51
BR007B1	630	1.116	1.044	0.840	1.347	1.068	0.535	116	29	208	3	304	61
BR007B2	78	1.174	1.034	0.792	1.497	1.135	0.358	103	35	205	17	317	50
BR007B3	105	1.122	0.985	0.893	1.258	1.140	−0.144	96	28	192	11	300	59
BR008A1	1197	1.172	1.007	0.821	1.431	1.165	0.144	96	33	194	12	301	54
BR008A2	1008	1.072	1.011	0.916	1.172	1.060	0.254	102	13	200	32	352	55
BR008A3	872	1.123	0.977	0.901	1.250	1.149	−0.262	93	23	193	21	322	58
BR008B1	1553	1.222	1.014	0.764	1.608	1.206	0.203	99	27	195	12	308	60
BR008B2	891	1.069	1.042	0.889	1.221	1.026	0.716	91	19	188	19	320	63
BR008B3	1028	1.108	0.994	0.898	1.234	1.116	−0.039	96	20	193	18	322	62
BR008C1	1862	1.213	1.024	0.763	1.603	1.184	0.272	93	21	186	8	297	68
BR008C2	1216	1.143	1.047	0.810	1.431	1.091	0.495	86	23	185	20	311	59
BR008C3	926	1.126	0.995	0.879	1.282	1.131	0.006	87	25	186	17	306	59
BR009A1	27	1.045	0.999	0.956	1.093	1.046	−0.024	280	8	190	2	86	81
BR009A2	17	1.056	1.011	0.933	1.134	1.045	0.292	98	7	188	1	288	83
BR009A3	42	1.028	1.004	0.967	1.064	1.024	0.227	94	0	184	10	2	80
BR009B1	13	1.082	1.023	0.896	1.214	1.057	0.409	98	10	7	6	244	78
BR009B2	50	1.053	1.005	0.942	1.118	1.047	0.173	98	15	8	1	275	75
BR009B3	24	1.040	1.017	0.943	1.108	1.022	0.549	285	6	15	6	151	82
BR009C1	23	1.060	1.002	0.939	1.129	1.058	0.075	283	17	20	20	156	63
BR009C2	37	1.059	0.998	0.943	1.123	1.061	−0.015	285	22	22	15	143	62
BR009C3	45	1.063	1.000	0.937	1.135	1.063	0.041	273	13	182	2	84	77
BR010A1	54	1.040	0.990	0.970	1.074	1.051	−0.432	134	5	44	3	285	84
BR010A2	55	1.041	1.005	0.954	1.093	1.036	0.202	113	7	20	24	219	65
BR010A3	60	1.048	1.007	0.945	1.110	1.040	0.236	120	10	21	43	220	45
BR010B1	62	1.030	0.994	0.975	1.057	1.036	−0.293	119	7	211	17	6	72
BR010B2	59	1.031	0.993	0.976	1.058	1.038	−0.350	121	5	212	2	319	84
BR010B3	66	1.038	0.989	0.973	1.070	1.050	−0.504	120	10	30	3	282	80
BR010C1	57	1.034	1.011	0.955	1.085	1.024	0.415	102	9	9	17	218	71
BR010C2	61	1.041	1.007	0.951	1.096	1.034	0.266	104	14	9	18	230	67
BR010C3	53	1.041	1.006	0.953	1.093	1.034	0.233	108	20	16	4	275	70
BR011A1	1822	1.070	1.036	0.894	1.211	1.033	0.642	114	19	204	1	296	71
BR011A2	1148	1.145	1.004	0.851	1.347	1.140	0.117	102	16	195	10	315	71
BR011A3	10 770	1.137	1.033	0.830	1.381	1.101	0.389	114	18	207	7	318	70
BR011B1	5627	1.318	1.027	0.655	2.044	1.283	0.286	99	16	7	6	258	73
BR011B2	11 720	1.516	0.882	0.602	2.571	1.718	−0.172	77	30	172	7	274	59
BR011B3	4158	1.327	1.081	0.592	2.338	1.227	0.493	79	34	191	29	312	42
BR011C1	2516	1.252	0.979	0.770	1.630	1.279	−0.013	120	8	28	13	240	75
BR011C2	3741	1.192	1.005	0.803	1.488	1.187	0.135	113	8	204	11	346	76
BR011C3	3337	1.100	1.017	0.883	1.250	1.082	0.283	100	1	191	8	2	82
BR012A1	7188	1.165	1.004	0.831	1.404	1.160	0.121	89	29	183	6	284	60
BR012A2	7652	1.139	1.014	0.848	1.347	1.123	0.215	85	30	178	5	276	60
BR012A3	7769	1.165	1.011	0.824	1.417	1.152	0.183	93	29	183	0	274	61
BR012B1	5915	1.137	1.030	0.833	1.376	1.104	0.366	102	30	194	2	288	60
BR012B2	5791	1.173	1.032	0.795	1.489	1.137	0.340	96	31	190	5	288	59
BR012B3	4242	1.172	1.014	0.813	1.447	1.156	0.207	93	32	187	6	287	57
BR012C1	7405	1.191	1.008	0.801	1.491	1.182	0.159	97	31	191	6	291	58

(*continued*)

Table 1. *Continued*

Site	K_{LF} (10^{-6} SI)	LF-normed principal susceptibilities			LF-anisotropy factors			LF-AMS principal directions					
		K_1	K_2	K_3	P_j	L	T	K_1 (°)		K_2 (°)		K_3 (°)	
								dec	inc	dec	inc	dec	inc
BR012C2	9344	1.196	1.019	0.785	1.533	1.174	0.239	109	31	202	5	300	59
BR012C3	10 200	1.152	1.048	0.800	1.463	1.100	0.481	100	25	192	5	292	65
BR013A1	8676	1.292	0.983	0.724	1.792	1.314	0.055	289	11	21	7	143	76
BR013A2	14 040	1.331	0.959	0.709	1.887	1.388	−0.041	288	17	195	9	78	70
BR013A3	10 450	1.267	0.969	0.764	1.663	1.307	−0.060	293	14	200	12	70	71
BR013B1	10 370	1.286	0.955	0.759	1.701	1.346	−0.128	291	9	200	8	71	78
BR013B2	10 580	1.337	0.953	0.710	1.894	1.403	−0.070	282	16	189	9	72	71
BR013B3	9079	1.336	0.959	0.706	1.904	1.394	−0.040	282	22	187	11	73	65
BR013C1	13 170	1.296	0.970	0.734	1.772	1.336	−0.020	259	23	349	1	83	67
BR013C2	10 330	1.269	0.998	0.734	1.736	1.272	0.121	275	21	179	13	58	65
BR013C3	8674	1.294	0.977	0.729	1.781	1.324	0.021	268	28	171	14	58	58
BR014A1	6508	1.235	0.998	0.767	1.730	1.238	0.102	291	19	201	3	104	71
BR014A2	9830	1.290	0.961	0.749	1.630	1.343	−0.086	287	19	194	11	75	68
BR014A3	6462	1.257	0.970	0.773	1.754	1.296	−0.067	288	20	20	6	126	69
BR014B1	7467	1.305	0.947	0.748	1.875	1.379	−0.156	295	21	27	4	127	69
BR014B2	11 450	1.327	0.962	0.712	1.834	1.380	−0.034	304	21	212	6	108	68
BR014B3	9470	1.309	0.974	0.717	1.717	1.345	0.017	293	20	202	2	106	70
BR014C1	7089	1.270	0.987	0.743	1.650	1.287	0.060	286	11	194	9	67	75
BR014C2	8624	1.267	0.962	0.771	1.708	1.318	−0.111	292	17	199	8	85	72
BR014C3	9605	1.287	0.956	0.757	1.907	1.346	−0.119	287	21	18	5	120	68
BR015A1	10 910	1.336	0.960	0.704	1.777	1.391	−0.033	278	19	187	3	87	71
BR015A2	9548	1.319	0.933	0.749	1.546	1.414	−0.223	278	23	185	8	77	66
BR015A3	11 520	1.250	0.931	0.819	1.847	1.343	−0.394	285	21	15	2	110	69
BR015B1	9726	1.348	0.911	0.741	1.726	1.480	−0.309	291	6	200	0	107	84
BR015B2	8701	1.327	0.874	0.799	1.620	1.518	−0.645	285	18	195	2	100	72
BR015B3	8686	1.255	0.969	0.776	1.657	1.294	−0.075	292	24	201	4	102	65
BR015C1	7912	1.284	0.935	0.782	1.580	1.374	−0.280	279	24	10	3	106	66
BR015C2	9367	1.239	0.976	0.786	1.482	1.270	−0.049	270	23	6	13	122	63
BR015C3	10 440	1.225	0.940	0.835	1.479	1.303	−0.383	278	21	12	10	125	66
BR016A1	5896	1.181	1.017	0.802	1.388	1.161	0.226	292	11	23	4	134	78
BR016A2	7584	1.167	0.992	0.841	1.524	1.176	0.008	292	14	202	1	109	76
BR016A3	7064	1.234	0.952	0.814	1.785	1.296	−0.247	289	10	20	9	151	77
BR016B1	7092	1.310	0.952	0.738	1.534	1.376	−0.111	281	15	13	8	130	73
BR016B2	6780	1.238	0.951	0.812	1.418	1.302	−0.250	278	12	10	9	136	74
BR016B3	8046	1.191	0.968	0.842	1.879	1.230	−0.198	269	12	3	18	147	68
BR016C1	9022	1.338	0.945	0.717	1.337	1.416	−0.114	270	4	1	7	154	82
BR016C2	6782	1.167	0.950	0.883	1.576	1.228	−0.471	282	4	13	23	183	66
BR016C3	5279	1.260	0.932	0.809	1.551	1.352	−0.359	259	8	351	16	142	72
BR017A1	7211	1.207	1.011	0.782	1.520	1.194	0.183	521	3	341	2	98	87
BR017A2	5565	1.225	0.966	0.809	1.568	1.269	−0.146	78	0	348	6	168	84
BR017A3	7311	1.230	0.984	0.786	1.588	1.250	0.003	253	7	345	12	132	76
BR017B1	8158	1.242	0.975	0.784	1.357	1.273	−0.051	283	4	13	5	153	84
BR017B2	5253	1.166	0.973	0.861	1.599	1.198	−0.192	278	8	9	6	136	79
BR017B3	7217	1.244	0.976	0.780	1.668	1.274	−0.038	269	8	0	8	133	79
BR017B4	9795	1.266	0.974	0.761	1.524	1.300	−0.031	263	0	353	11	173	79
BR017C1	5767	1.220	0.978	0.802	1.467	1.248	−0.056	287	6	18	7	156	80
BR017C2	6183	1.179	1.015	0.807	1.654	1.162	0.209	280	5	11	9	160	79
BR017C3	7406	1.266	0.967	0.768	1.659	1.309	−0.079	278	5	9	11	164	77
BR018A1	10 000	1.259	0.980	0.761	1.674	1.285	0.005	282	3	192	0	95	87
BR018A2	9156	1.268	0.972	0.760	1.766	1.304	−0.036	275	2	185	1	72	88
BR018A3	12 620	1.301	0.959	0.740	1.612	1.356	−0.079	285	5	16	2	147	83
BR018B1	8604	1.234	0.998	0.768	1.575	1.237	0.102	295	1	25	13	199	77
BR018B2	8517	1.251	0.950	0.799	1.536	1.317	−0.229	284	0	14	11	192	79

(continued)

Table 1. *Continued*

Site	K_{LF} $(10^{-6}$ SI)	LF-normed principal susceptibilities			LF-anisotropy factors			LF-AMS principal directions					
		K_1	K_2	K_3	P_j	L	T	K_1 (°)		K_2 (°)		K_3 (°)	
								dec	inc	dec	inc	dec	inc
BR018B3	10 850	1.208	1.003	0.789	1.516	1.205	0.125	290	5	200	3	77	84
BR018C1	6881	1.209	0.992	0.799	1.532	1.219	0.043	286	5	17	6	153	82
BR018C2	10 300	1.205	1.006	0.789	1.591	1.197	0.150	284	0	14	1	180	89
BR018C3	8282	1.239	0.981	0.780	1.130	1.263	−0.011	284	1	14	7	182	83
BR019A1	155	1.051	1.015	0.934	1.132	1.035	0.414	95	4	185	11	342	78
BR019B1	162	1.051	1.017	0.932	1.122	1.033	0.462	97	4	188	10	348	79
BR019C1	172	1.049	1.013	0.938	1.136	1.036	0.372	109	6	200	7	338	80
BR019C2	158	1.055	1.014	0.931	1.122	1.041	0.354	108	8	199	11	342	76
BR019D1	177	1.049	1.013	0.938	1.129	1.036	0.375	105	5	196	8	340	81
BR019D2	165	1.049	1.017	0.933	1.121	1.032	0.470	99	4	190	8	345	81
BR019E1	173	1.048	1.014	0.938	1.107	1.034	0.396	97	4	187	8	341	81
BR020A1	31	1.038	1.019	0.943	1.124	1.019	0.600	114	10	206	12	345	74
BR020A2	34	1.043	1.022	0.935	1.109	1.021	0.630	290	13	199	3	94	77
BR020B1	39	1.037	1.022	0.942	1.134	1.015	0.690	252	6	160	13	7	75
BR020B2	34	1.046	1.024	0.930	1.127	1.021	0.646	94	6	187	20	348	69
BR020B3	31	1.041	1.026	0.933	1.139	1.014	0.747	101	5	193	19	358	71
BR020C1	26	1.054	1.017	0.929	1.107	1.037	0.424	84	3	175	21	347	68
BR020C2	25	1.038	1.019	0.943	1.172	1.018	0.623	91	9	184	20	337	68
BR020C3	24	1.076	1.005	0.919	1.090	1.071	0.137	124	7	215	10	360	77
BR021A1	49	1.032	1.017	0.952	1.105	1.015	0.640	117	12	209	8	332	75
BR021A2	79	1.051	0.999	0.951	1.085	1.052	−0.011	7	5	104	51	273	38
BR021A3	73	1.036	1.008	0.956	1.101	1.027	0.331	183	16	89	14	318	69
BR021B1	56	1.039	1.014	0.947	1.124	1.025	0.462	118	8	209	3	319	81
BR021B2	98	1.057	1.003	0.940	1.234	1.053	0.109	192	3	283	23	95	67
BR021C1	67	1.104	1.000	0.896	1.091	1.104	0.055	192	8	300	66	99	23
BR021C2	66	1.032	1.017	0.951	1.199	1.015	0.629	191	20	96	13	335	66
BR021C3	108	1.083	1.012	0.905	1.091	1.070	0.249	165	40	60	17	312	45
BR021D1	48	1.035	1.013	0.952	1.106	1.021	0.492	114	10	206	11	342	75
BR021D2	69	1.034	1.023	0.943	1.210	1.011	0.765	198	24	100	17	339	60
BR024A1	1072	1.082	1.020	0.898	1.245	1.061	0.368	104	35	208	19	320	48
BR024B1	2394	1.101	1.013	0.886	1.286	1.087	0.228	92	30	203	31	328	44
BR024B2	2465	1.098	1.039	0.864	1.227	1.057	0.541	97	34	210	30	331	41
BR024C1	2352	1.093	1.014	0.893	1.258	1.078	0.254	90	26	198	31	328	47
BR024C2	2704	1.105	1.015	0.880	1.256	1.089	0.253	87	28	195	30	323	46
BR024D1	1952	1.100	1.020	0.880	1.243	1.079	0.321	95	40	207	24	320	40
BR024E1	2382	1.088	1.030	0.882	1.265	1.056	0.483	87	40	198	24	310	41
BR024E2	3674	1.111	1.058	0.891	1.228	1.050	0.556	85	38	195	23	309	43
BR024F1	2394	1.080	1.032	0.888	1.214	1.046	0.537	91	37	197	20	310	46
BR024G1	3083	1.076	1.030	0.894	1.204	1.045	0.527	89	38	201	25	316	42

microstructures are dominantly magmatic, as indicated by triple junctions of grain boundaries as well as graphic and myrmekitic textures in feldspars, with some high-temperature, solid-state overprint, as indicated by quartz subgrains and weak undulatory extinction in the K-feldspar. Temperature corresponding to magmatic deformation is probably in excess of 600–650 °C (Wyllie 1983).

Closer to the shear zone, microstructures display increasing evidence of solid-state deformation. This includes plastic deformation and partial recrystallization in feldspar grains, and more pronounced and continuous shear bands composed of recrystallized quartz grains. Biotite defines an increasingly prominent foliation, and small amounts of later-formed chlorite are parallel to the biotite foliation. An S–C fabric is also apparent in many samples from the shear zone. In the east, within the mylonitic zone, low-temperature, solid-state deformation is evident from more significant brittle deformation of K-feldspar clasts, surrounded by more elongated quartz grains and quartz ribbons and a more fine-grained matrix. In the absence of geothermometric analysis, the temperature range for

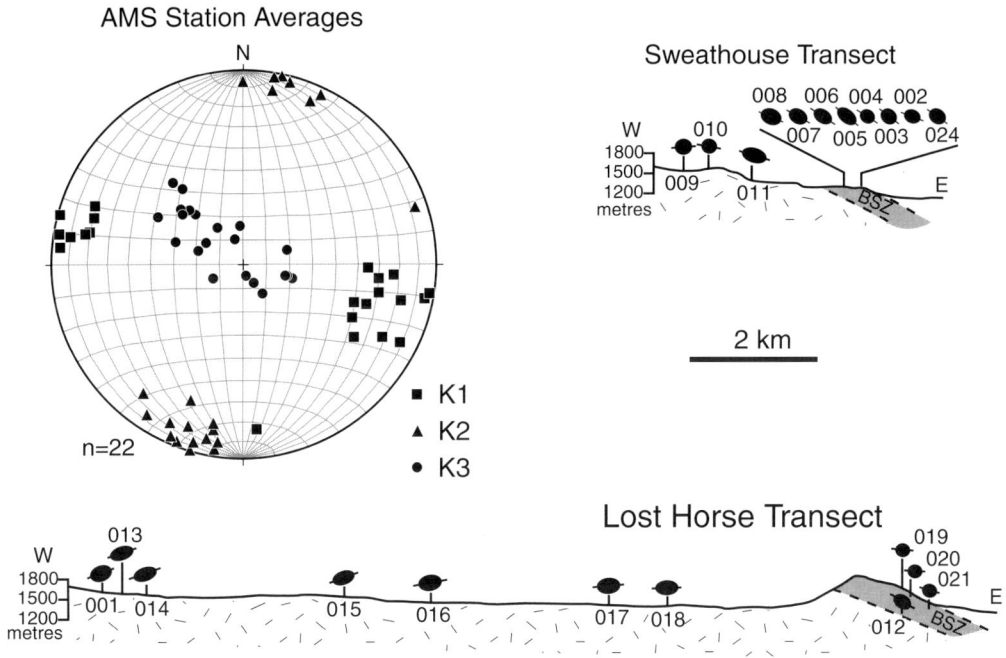

Fig. 5. Averaged AMS data for all 22 stations on equal-area, lower-hemisphere stereonet (complete data in Fig. 4 and Table 1). Geological cross-sections with no vertical exaggeration along Sweathouse Creek and Lost Horse Creek, with black ellipses showing the K_1–K_3 aspect ratio and plunge of magnetic lineation.

ductile, solid-state deformation in the BSZ is not well defined. In this study, low-temperature, solid-state deformation is defined by the onset of plastic deformation in quartz ($c.$ 300 °C), and high-temperature, solid-state deformation is defined by the onset of plastic deformation in feldspar ($c.$ 450 °C). (See Snoke & Tullis 1998 for a review and discussion.)

Samples from within the easternmost, brittle part of the shear zone show similar evidence of low-temperature, solid-state deformation, although the K-feldspar grains are more rounded and the elongated quartz grains in the quartz ribbons diminish in grain size. Sample BR024A1 (Sweathouse transect) is structurally the shallowest and displays brittle deformation, such as large, brittly deformed K-feldspar clasts separated by ultramylonite bands, and elongated quartz grains that have not entirely recrystallized. The shear zone shows a very well defined S–C fabric with top-to-the-east shear-sense indicators.

Particle analysis of both magnetite and biotite grains shows a good correlation between mineral fabric and K_1 (Figs 8–10). Chlorite grains that are apparent in some samples comprise a very small percentage of the total micas (<10%), and are counted with the biotite grains, as they

are parallel. In all thin sections used for particle analysis, every magnetite grain was imaged, although in some cases this is a small number of grains (seven–nine). However, the correlation between mineral fabric and K_1 in those samples is also reasonably close. Samples with the largest number of grains show the best-defined fabric and greatest correlation with AMS fabric (Fig. 8). Particle analysis of biotite grains show a well-defined fabric and a fairly close ($\pm 10°$) correlation with the AMS fabric (Fig. 9). Most samples are dominantly ferrimagnetic (as indicated by VSM measurements, Fig. 11a & b), and thus the biotite mineral fabric does not directly control the AMS fabric; nonetheless, there is a reasonably strong correlation between biotite mineral fabric and AMS fabric in most samples. The gradual variation of fabric from west to east seen in the AMS data is prominent in the particle analysis data (Figs 9 and 10), and the rotating plunge of K_1, from about 10°W in sample BR001A1 to about 35°E in the shear zone, is also seen in the mineral fabric. The strength of the biotite fabric is much more consistent than the magnetite fabric, as shown in the fabric ellipses. Slight variations in the biotite fabric (Fig. 9) may be due to the presence

(a)

(b)

(c)

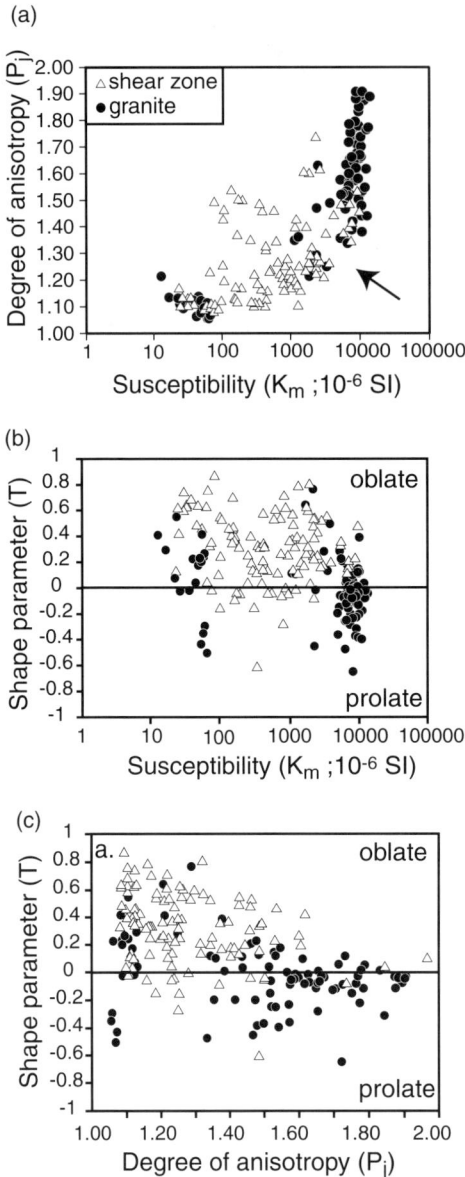

Fig. 6. Relationship between: (**a**) magnetic anisotropy, P_j, and susceptibility, K_m; (**b**) the shape parameter, T, and K_m; and (**c**) T and P_j for all samples ($n = 199$). Shear-zone samples tend to be more oblate than the granite samples, and most of the granite samples have a higher susceptibility. Bimodal distribution of K_m is defined by the gap in (a), marked with an arrow.

of S–C fabrics. In such cases, the biotite fabric ellipse is likely to be weaker (Fig. 10). In general, however, the overall correlation between magnetic and mineral fabrics is strong.

Discussion

Previous studies of deformation in the BSZ concentrated on the well-foliated and lineated shear zone, and demonstrated a consistent fabric along the 100 km-trend of the zone (e.g. Hyndman 1980; Foster *et al.* 2001). The geometric relationship between structures in the BSZ and the Idaho–Bitterroot batholith, however, has not been clearly established due somewhat to the lack of macroscopically visible fabric in the batholith. Our study has attempted to understand this relationship by focusing on both AMS and particle fabric across the metamorphic and strain gradient, and several important results have been established. First, and most significantly, there is a strong correlation between the shape fabric of magnetite and the AMS bulk measurement (Fig. 10). In samples with biotite-dominated (paramagnetic) AMS fabric (e.g. BR019C2 and BR021D1) there is also an excellent agreement between the SPO of biotite and AMS. In addition, the AMS foliation and lineation faithfully track the macroscopic foliation and lineation from field measurements within the shear zone (stereonets, Figs 1b and 5). In the granite, where field measurements of rock fabric are rather difficult to make, the AMS foliation is also well developed. It varies smoothly and continuously from E-dipping just below the shear zone to W-dipping at the western edge of the transects, forming an anticlinal arch. Furthermore, the AMS lineation in the granite remains very close to the N110° trend measured in the shear zone. Therefore, the fabric in the shear zone and the granite appear to be genetically related.

Particle fabric v. AMS

Hysteresis data (Fig. 3) indicate that the AMS is carried dominantly by the magnetite, except in a few localities – both within the shear zone and in the granite – where the magnetic carrier is biotite (Fig. 11a & b). These variations may be controlled by the composition of the granitic protolith. Microstructural observations indicate that magnetite is part of the primary magmatic assemblage of the batholith; furthermore, most magnetite grains are elongate and define a shape fabric that is subparallel to the silicate shape fabric. In the weakly deformed granites, magnetite shape preferred orientation (SPO) results from rigid marker rotation of magnetite grains in the magmatic state, as well as crystal-plastic deformation at high temperature in the quartzo-feldspathic matrix. In the shear zone, magnetite SPO is controlled by low-temperature crystal-plastic deformation. Although cataclastic deformation has clearly affected the silicate

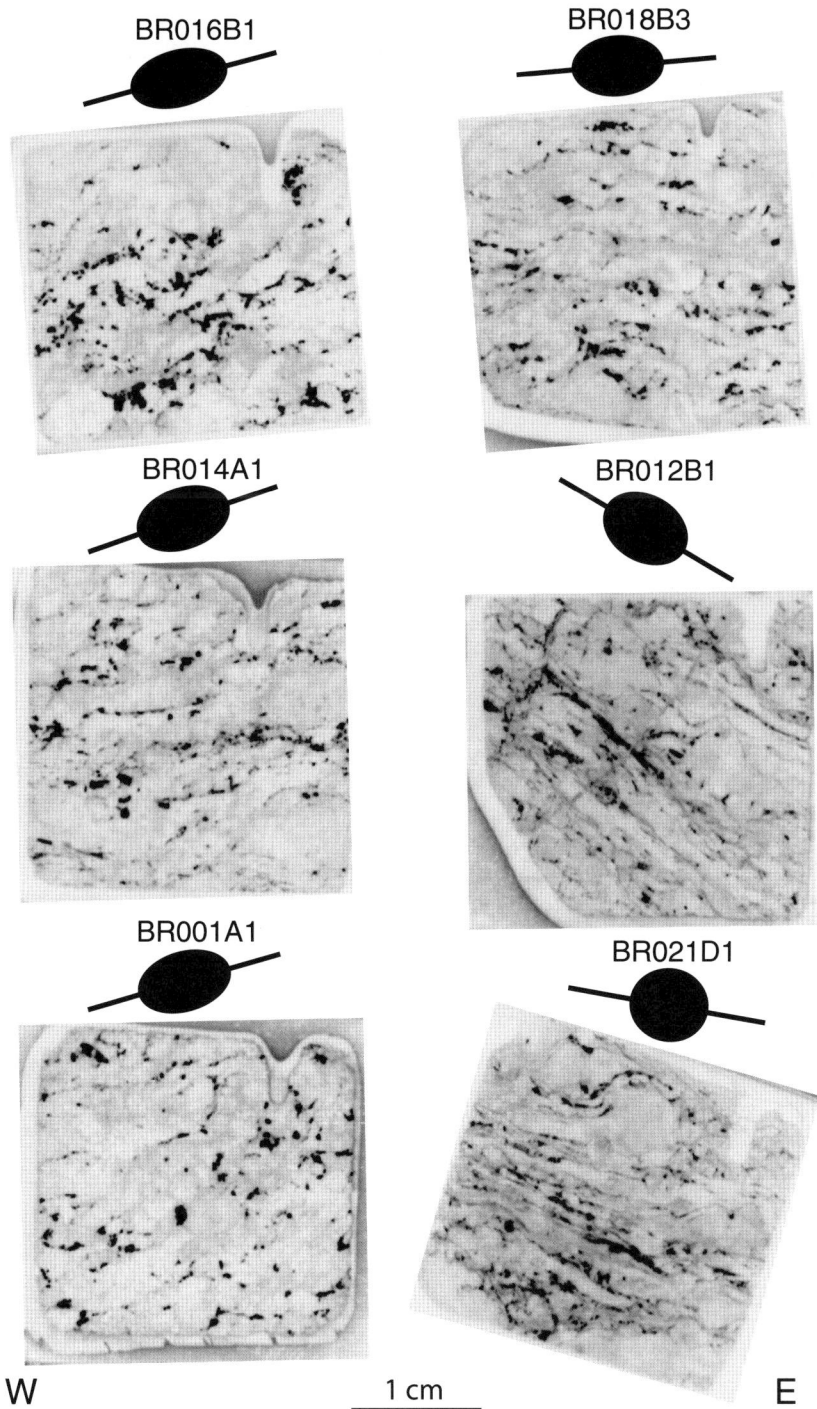

Fig. 7. Comparison of (**a**) rock fabric, as shown in thin sections used for particle analysis, and (**b**) magnetic fabric, from the Lost Horse transect. Thin sections oriented as in the field. Samples shown from west to east, clockwise from lower left. Black ellipses are K_1–K_3 ellipses, constructed from AMS measurements; black arrow is the magnetic lineation (K_1).

BR008B1   BR012B1

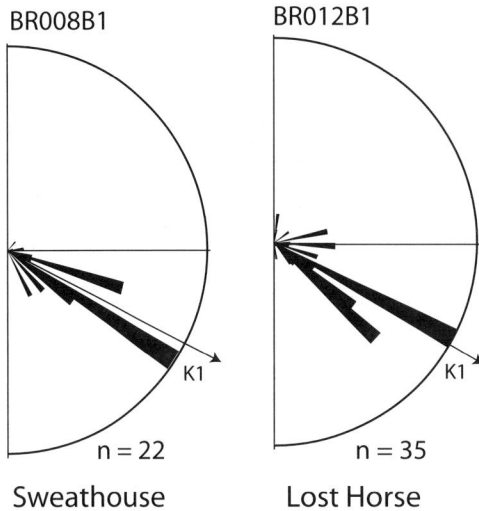

K1   K1

n = 22   n = 35

Sweathouse   Lost Horse

Fig. 8. Rose diagrams oriented in a nearly vertical plane looking north, approximately parallel to the $K_1–K_3$ plane, showing the weighted distribution of magnetite grains in two shear-zone samples. Rose petals in 5° increments.

minerals, magnetite grains seem to have retained their euhedral shape, and thus do not show evidence of cataclasis. However, there is evidence of magnetite grain size variation across the shear zone (Fig. 3). Smaller magnetite grains in the shear zone may result from strain-induced grain size reduction that affects all other matrix phases, such as quartz and feldspar. However, another possibility to explain the range in magnetite grain size may be different original grain sizes resulting from a different protolith.

Although magnetite dominates the AMS fabric in most samples, there is still a strong correlation between biotite fabric and overall AMS fabric. Microstructural observations reveal the presence of an S–C fabric that is strongest in the shear zone and gradually weakens from east to west in the granite. Despite this S–C fabric, K_1 (magnetic lineation) in paramagnetically dominant samples is parallel to the macroscopic lineation (Figs 1b and 10), rather than the intersection lineation of S and C. Aranguren *et al.* (1996) showed K_1 to be an average of S and C in the direction of macroscopic lineation, in contrast to the work of Housen *et al.* (1995) that showed K_1 to be parallel to the S–C intersection lineation.

AMS aspect ratio (P_j) v. bulk

susceptibility (K_m)

The degree of anisotropy, P_j, shows a positive correlation with the low-field magnetic

susceptibility, K_m (Fig. 6a). A similar relationship has often been observed in other case studies (e.g. Archanjo *et al.* 1995; Bouchez 1997) regardless of protolith composition or state of deformation. A number of explanations for this phenomenon have been proposed. Distribution anisotropy (Hargraves *et al.* 1991) may explain an increase in degree of anisotropy due to a strain-related decrease in the average distance between magnetite grains. However, this phenomenon requires that the distance between grains be comparable to grain diameter (Grégoire *et al.* 1995, 1998). Theoretical and experimental studies by Cañón-Tapia (2001) have demonstrated that this process does not contribute significantly to the degree of anisotropy, particularly in felsic rocks. Furthermore, our own observations show that the number of magnetite grains per thin section is relatively small (about 10–20), which precludes the occurrence of distribution anisotropy in the BSZ. The boudinage of competent magnetite grains during low-temperature deformation in the shear zone might also lead to distribution anisotropy in highly strained rocks with a small number of grains. However, in the BSZ, magnetite occurs as widely spaced, non-boudinaged single grains that are unlikely to generate a distribution anisotropy.

In general, K_m is greater in mafic rocks than in felsic rocks due to the higher iron content; therefore, if mafic rocks exhibit a large anisotropy this would lead to a positive correlation between P_j and K_m. Such a relationship between the composition of the protolith and finite strain has been documented in other case studies (e.g. de Saint Blanquat & Tikoff 1997). In the BSZ, we cannot completely rule out the possibility of a lithological control on magnetic anisotropy (P_j), but the range of protolith composition is limited and is unlikely to produce the observed $P_j–K_m$ relationship.

Yet another explanation for the $P_j–K_m$ relationship lies in the demagnetization factor of magnetite, which results in the increase of magnetic anisotropy as the grain size increases, regardless of the grain aspect ratio (Merrill 1977). Preliminary modelling suggests that this effect would account only for an insignificant increase in P_j. Inversely, the variation of magnetite grain size should result in a decrease in magnetic susceptibility as shown, for example, by Nagata (1961). Again, this effect would account only for a minute change in susceptibility for a given degree of anisotropy.

Finally, an alternative hypothesis for the positive correlation between P_j and K_m may be related to the growth characteristics of magnetite. As magnetite forms in a magmatic system, it

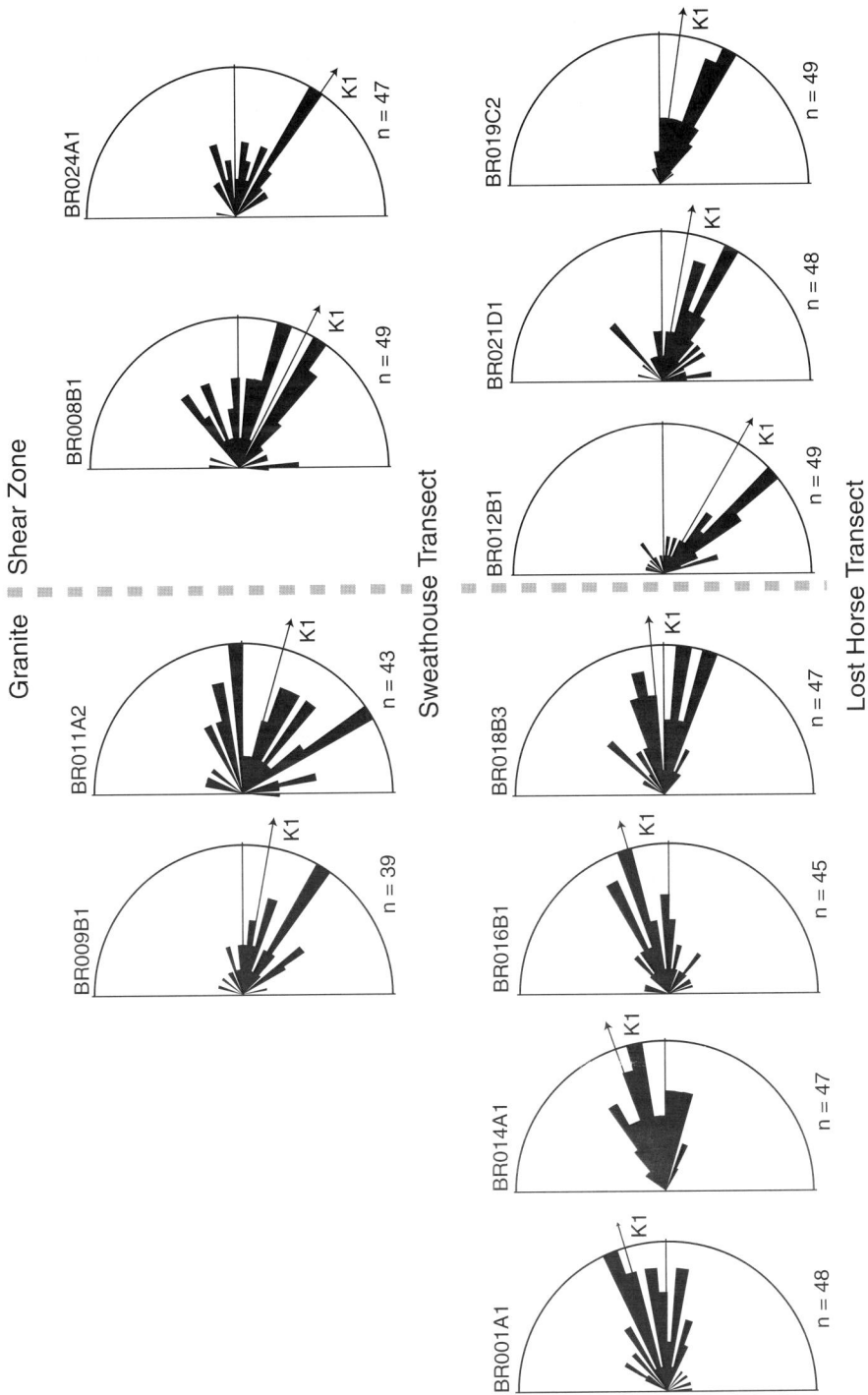

Fig. 9. Rose diagrams oriented in a nearly vertical plane looking north, approximately parallel to the K_1–K_3 plane, showing the distribution of biotite grains in samples from both transects. Rose petals in 5° increments. The Sweathouse transect is considerably shorter than the Lost Horse transect (Fig. 1). In samples BR009B1, BR021D1 and BR019C2 magnetic fabric is dominated by biotite.

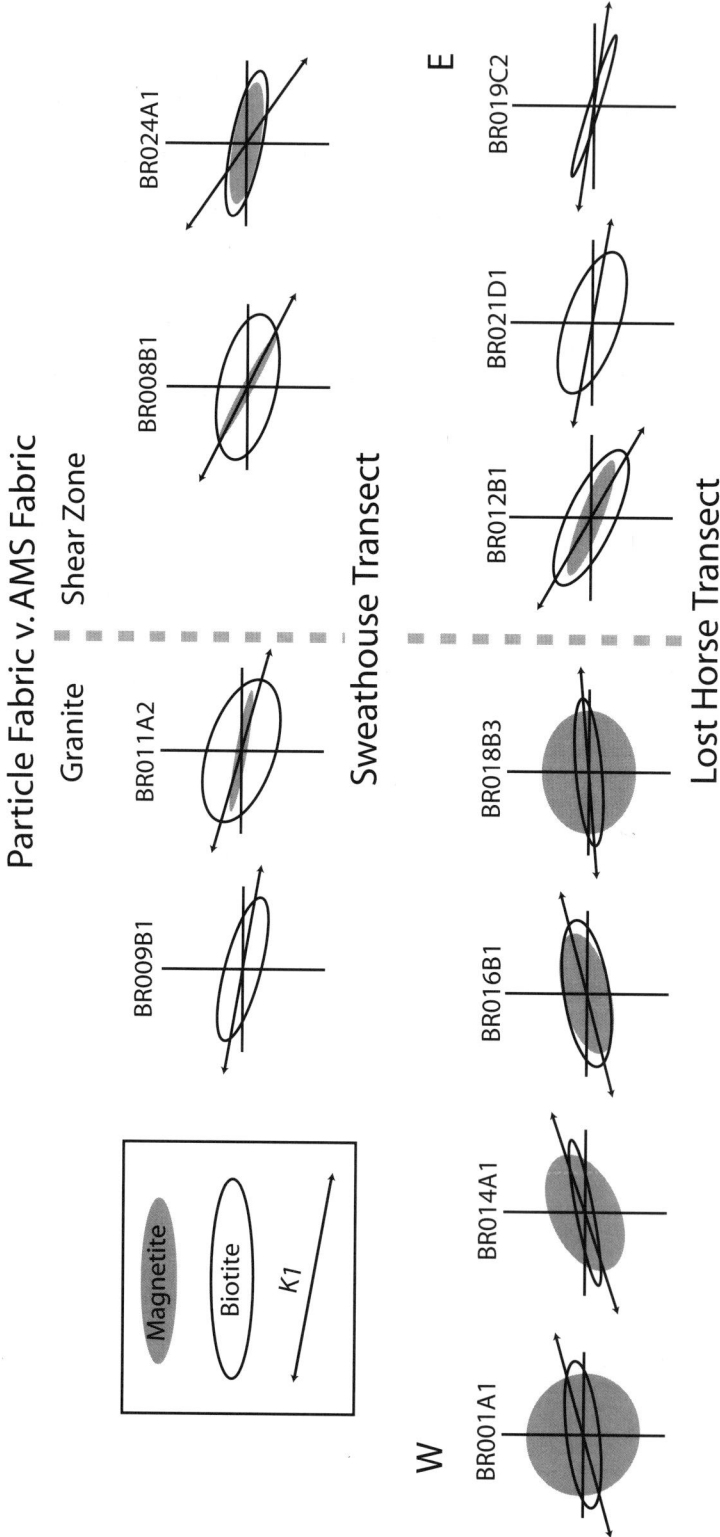

Fig. 10. Comparison of AMS fabric and mineral fabric ellipses, calculated using Benn & Allard (1988). All figures oriented as in the field. Both AMS and particle fabric – particularly in the Sweathouse transect – arch from west to east.

grows at different rates in different crystallographic directions, as other mineral phases (e.g. Lofgren & Gooley 1977; Kirkpatrick 1983). For example, biologically grown magnetite grows preferentially along the [100] axis (Moskowitz et al. 1993). Therefore, as magnetite grains become larger, the magnetic anisotropy increases. Our observations from the BSZ show a P_j–K_m positive correlation in both the granite and the shear zone. Therefore, if the growth characteristics control the P_j–K_m relationship, they would have to be the same in the granite and the shear zone, or, if they developed only in the magmatic state, they would have to be preserved in the shear zone.

Spatial variation of P_j and T

The degree of magnetic anisotropy (P_j) in both paramagnetic and ferromagnetic granites has been interpreted as a strain proxy (e.g. Hrouda 1982; Rochette et al. 1992; Tarling & Hrouda 1993; Borradaile & Henry 1997). Indeed, the AMS in such rocks results from either preferential orientation of rigid markers (Jeffery 1922) or shape-preferred orientation of ferromagnetic minerals (e.g. Archanjo et al. 1995). In the BSZ, P_j varies from 1.05 to 2.00, which is in agreement with the AMS being dominated by a ferromagnetic phase (magnetite) (Fig. 6, Table 1). A few samples from the easternmost part of the Lost Horse and the westernmost part

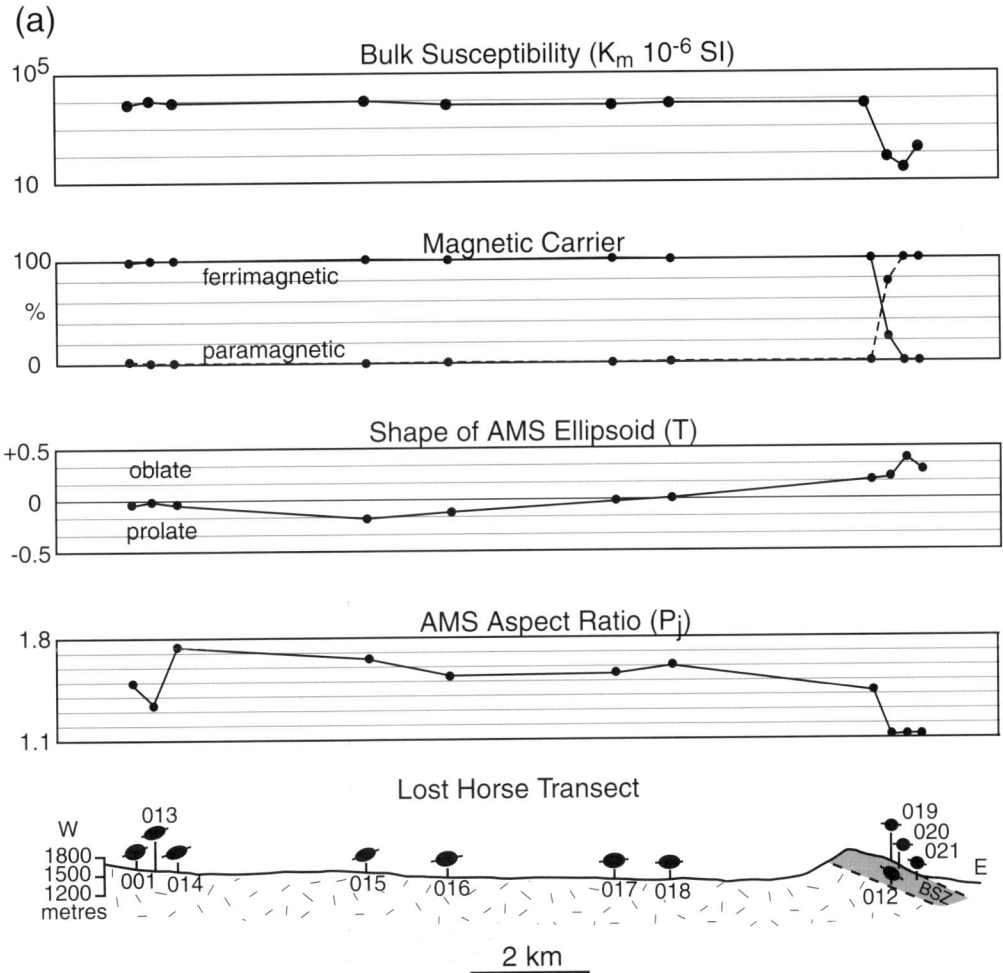

Fig. 11. Spatial distribution of magnetic results in (**a**) the Lost Horse transect and (**b**) the Sweathouse transect. The horizontal axis is distance. Bulk susceptibility is measured in 10^{-6} SI units, and P_j and T are ratios.

(b)

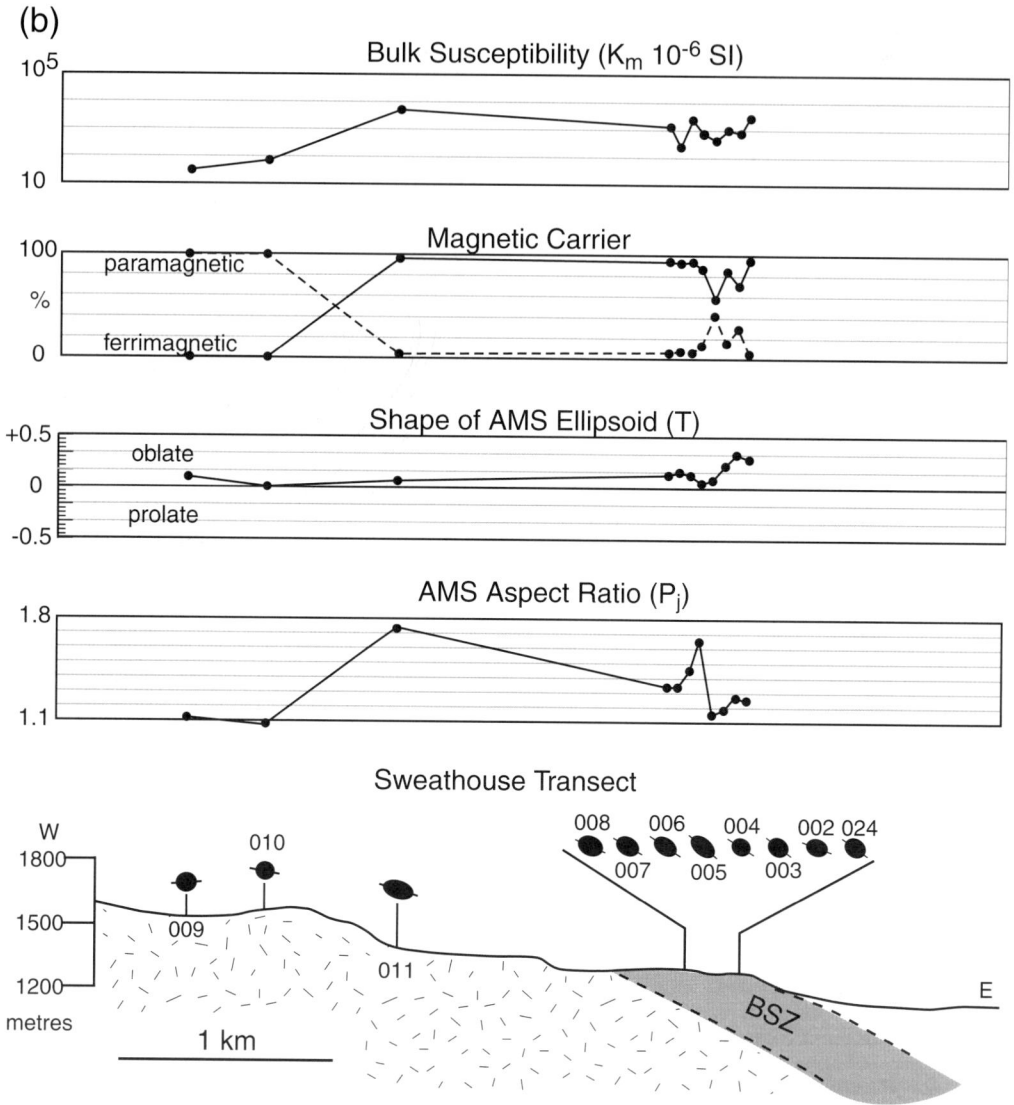

Fig. 11. *Continued.*

of the Sweathouse transects reflect mostly para-magnetic parameters (BR009, BR010, BR019, BR020 and BR021).

Surprisingly, P_j decreases toward the shear zone where deformation is typically mylonitic, whereas increased strain would be expected to lead to higher P_j. A possible explanation for this apparent anomaly might be that a different protolith may host smaller magnetite grains that intrinsically carry a lower P_j. An alternative explanation might be that the deformation mech-anisms that operate in the mylonitic part of the shear zone would not result in an increase in P_j

with increasing strain. For example, strain parti-tioning between a rigid object (e.g. magnetite) and its quartzo-feldspathic matrix may leave the object's shape unaffected by increased shear. A legitimate mechanism that would lead to this behaviour is grain-boundary sliding in which grains rotate with minimum stretch. Mechanical decoupling between rigid marker and host may also prevent a further record of deformation at higher strains.

Similar to P_j, the shape parameter (T) has been previously interpreted as a strain indicator (e.g. Tarling & Hrouda 1993). Shape parameter (T)

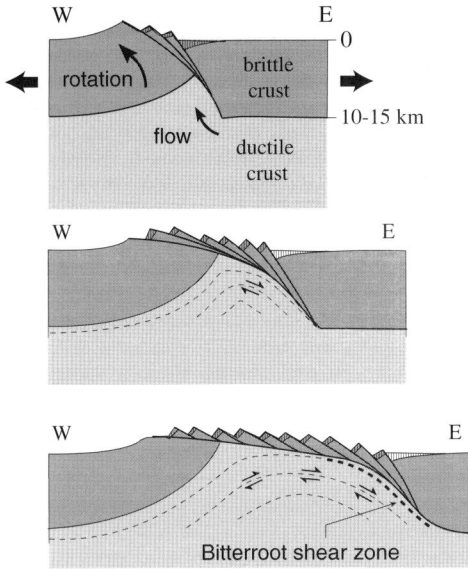

Fig. 12. Rolling-hinge model for exhumation of the Bitterroot metamorphic core complex (modified from Brun & Van Den Driessche 1994).

is dominantly neutral (T ≈ 0) in both low- and high-susceptibility, weakly deformed granites. In contrast, T is clearly oblate in the shear zone (Figs 6b & c and 11a & b). Therefore, a possible explanation would be that the shear zone is characterized by a component of flattening strain, while the less deformed granite flows under plane-strain conditions. Another explanation relates to the low P_j encountered in the magnetite-bearing mylonites, supported by microstructural observations indicating the magnetite grains in such mylonites are relatively equant. Assuming that magnetite does not contribute significantly to the magnetic anisotropy, then the contribution of biotite becomes relatively more important. In such a case, the overall AMS will be governed by the biotite-intrinsic magnetic parameters ($P_j \approx 1.35$, $T \approx 0.95$; Tarling & Hrouda 1993). Therefore, with increasing deformation T shifts toward more oblate fabrics, and P_j decreases.

Tectonic interpretation

The most remarkable result from both AMS and particle analysis is the well-defined, consistent fabric that arches from the undeformed footwall granites into the low-temperature shear-zone mylonites. This arched fabric is clearly associated with the development of the extensional BSZ, and can be explained by the progressive rotation of fabric formed during rolling-hinge exhumation (Fig. 12) (Buck 1988, 1991; Brun & Van Den Driessche 1994; Axen *et al.* 1995). In the absence of map coverage of structural and/or AMS data beyond the two studied traverses, it is not possible to test whether the arching represents a N–S-trending cylindrical structure or a series of domes, as is commonly documented in metamorphic core complexes (Teyssier & Whitney 2002; Teyssier *et al.* 2005).

Based on argon and fission-track thermochronological data, Foster & Raza (2002) determined that the footwall of the BSZ cooled progressively from west to east, which is consistent with exhumation in the rolling hinge of an E-directed extensional system. As the upper crust is fractured by normal faults, the ductile lower crust flows upward into the gap beneath the upper crust (Fig. 12) (Brun & Van Den Driessche 1994). The rolling-hinge model is also well supported by AMS data from this study, which show fabric continuity between the granite and the shear zone. This fabric continuity is expected from the rolling-hinge model as the ductile crust flows upward beneath the detachment system, resulting in progressive exhumation. Therefore, we propose that the model typically used to explain the development of metamorphic core complexes is also appropriate for the flow and exhumation of a magma-dominated core complex.

Conclusion

A combined AMS and particle analysis study across two transects of the BSZ suggests several important results.

- In the BSZ, the AMS foliation and lineation faithfully track the macroscopic foliation and lineation measured in the field. The AMS foliation is well developed in the granite beneath the BSZ where it forms a 10 km-wide anticlinal arch, showing a smooth and continuous variation in orientation from E-dipping in the east to W-dipping in the west. The AMS lineation in the granite remains very close to the N110° trend measured in the shear zone. This structure is consistent with the rolling-hinge model of ductile flow underneath a detachment zone.
- There is a good correlation between the shape fabric of magnetite determined from particle analysis and the bulk AMS. Where AMS is carried by biotite, there is also an excellent agreement between the SPO of biotite and AMS.

- The degree of magnetic anisotropy (P_j) is most pronounced in samples in which the magnetic carrier is magnetite and in the region where magnetite was part of a magmatic or high-temperature fabric. P_j decreases in the upper part of the shear zone, where deformation took place under low-temperature conditions, and magnetite grains display a nearly spherical shape. The shape parameter (T) indicates that the magnetic fabric is more oblate in the mylonite than in the granite; this is consistent with the relatively small contribution of nearly equant magnetite grains to P_j in the mylonite and a commensurate increase in the role of biotite with a strong SPO.

The authors gratefully acknowledge the contributions of the University of Minnesota Graduate School Fellowship, University of Minnesota Department of Geology and Geophysics Summer Research Funds (2000), and the generous facilities of the Institute for Rock Magnetism. This work was also partially supported through NSF-EAR9814669 and NSF-EAR0106953. For his assistance in the field and the laboratory, we thank B. Siwiec, who was supported through the University of Minnesota Summer Internship Program. The constructive comments of D. Foster and C. Lüneburg significantly improved the manuscript.

References

ARANGUREN, A., CUEVAS, J. & TUBÍA, J.M. 1996. Composite magnetic fabrics from *S–C* mylonites. *Journal of Structural Geology*, **18**, 863–869.

ARCHANJO, C.J., LAUNEAU, P. & BOUCHEZ, J.L. 1995. Magnetite fabric vs. magnetite and biotite shape fabrics of the magnetite-bearing granite pluton of Gameleiras (Northeast Brazil). *Physics of the Earth and Planetary Interiors*, **89**, 63–75.

ARMSTRONG, R.L., TAUBENECK, W.H. & HALES, P.O. 1977. Rb–Sr and K–Ar geochronometry of Mesozoic granitic rocks and their Sr isotopic composition, Oregon, Washington, and Idaho. *Geological Society of America Bulletin*, **88**, 397–411.

AXEN, G.J., BARTLEY, J.M. & SELVERSTONE, J. 1995. Structural expression of a rolling hinge in the footwall of the Brenner Line normal fault, eastern Alps. *Tectonics*, **14**, 1380–1392.

BENN, K. & ALLARD, B. 1988. Preferred mineral orientations related to magmatic flow in ophiolite layered gabbros. *Journal of Petrology*, **30**, 925–946.

BORRADAILE, G.J. 1991. Correlation of strain with anisotropy of magnetic susceptibility. *Pure and Applied Geophysics*, **135**, 15–29.

BORRADAILE, G.J. & HENRY, B. 1997. Tectonic applications of magnetic susceptibility and its anisotropy. *Earth Science Reviews*, **42**, 49–93.

BOUCHEZ, J.L. 1997. Granite is never isotropic: an introduction to AMS studies of granitic rocks. *In*: BOUCHEZ, J.L., HUTTON, D.H.W. & STEPHENS, W.E. (eds) *Granite: From Segregation of Melt to Emplacement Fabrics*. Kluwer, Dordrecht, 95–112.

BRUN, J.-P. & VAN DEN DRIESSCHE, J. 1994. Extensional gneiss domes and detachment fault systems: structure and kinematics. *Bulletin of the Geological Society of France*, **165**, 519–530.

BUCK, W.R. 1988. Flexural rotation of normal faults. *Tectonics*, **7**, 959–973.

BUCK, W.R. 1991. Modes of continental lithospheric extension. *Journal of Geophysical Research*, **96**, 20 161–20 178.

CAÑÓN-TAPIA, E. 2001. Factors affecting the relative importance of shape and distribution anisotropy in rocks: theory and experiments. *Tectonophysics*, **340**, 117–131.

CHASE, R.B. 1973. *Petrology of the Northeastern Border Zone of the Idaho Batholith, Bitterroot Range, Montana*. Montana Bureau of Mines and Geology Memoir, **43**, 1–28.

DAY, R., FULLER, M. & SCHMIDT, V.A. 1977. Hysteresis properties of titanomagnetites: grain-size and compositional dependence. *Physics of the Earth and Planetary Interiors*, **13**, 260–267.

DE SAINT BLANQUAT, M. & TIKOFF, B. 1997. Development of magmatic to solid-state fabrics during syntectonic emplacement of the Mono Creek granite, Sierra Nevada Batholith. *In*: BOUCHEZ, J.L., HUTTON, D.H.W. & STEPHENS, W.E. (eds) *Granite: From Segregation of Melt to Emplacement Fabrics*. Kluwer, Dordrecht, 231–252.

FOSTER, D.A. 2000. Tectonic evolution of the Eocene Bitterroot metamorphic core complex, Montana and Idaho. *In*: ROBERTS, S. & WINSTON, D. (eds) *Geologic Field Trips, Western Montana and Adjacent Areas*. Rocky Mountain Section of the Geological Society of America, University of Montana, 1–29.

FOSTER, D.A. & FANNING, C.M. 1997. Geochronology of the northern Idaho batholith and the Bitterroot metamorphic core complex: magmatism preceding and contemporaneous with extension. *Geological Society of America Bulletin*, **109**, 379–394.

FOSTER, D.A. & RAZA, A. 2002. Low-temperature thermochronological record of exhumation of the Bitterroot metamorphic core complex, northern Cordilleran Orogen. *Tectonophysics*, **349**, 23–36.

FOSTER, D.A., SCHAFER, C., FANNING, C.M. & HYNDMAN, D.W. 2001. Relationships between crustal partial melting, plutonism, orogeny, and exhumation: Idaho–Bitterroot batholith. *Tectonophysics*, **342**, 313–350.

GRÉGOIRE, V., DE SAINT BLANQUAT, M., NÉDÉLEC, A. & BOUCHEZ, J.L. 1995. Shape anisotropy versus magnetic interactions of magnetite grains: experiments and application to AMS in granitic rocks. *Geophysical Research Letters*, **22**, 2765–2768.

GRÉGOIRE, V., NEDELEC, A., LAUNEAU, P., DARROZES, J. & GAILLOT, P. 1998. Magnetite grain shape fabric and distribution anisotropy vs rock magnetic fabric: a three-dimensional case study. *Journal of Structural Geology*, **20**, 937–944.

HARGRAVES, R.B., JOHNSON, D.C. & CHAN, C.Y. 1991. Distribution anisotropy: the cause of AMS in igneous rocks? *Geophysical Research Letters*, **18**, 2193–2196.

HARVEY, P.K. & LAXTON, R.R. 1980. The estimation of finite strain from the orientation distribution of passively deformed linear markers: eigen value relationships. *Tectonophysics*, **77**, 1–34.

HOUSEN, B.A., VAN DER PLUIJM, B.A. & ESSENE, E.J. 1995. Plastic behavior of magnetite and high strains obtained from magnetic fabrics in the Parry Sound shear zone, Ontario Grenville Province. *Journal of Structural Geology*, **17**, 265–278.

HROUDA, F. 1982. Magnetic anisotropy of rocks and its application in geology and geophysics. *Geophysical Surveys*, **5**, 37–82.

HYNDMAN, D.W. 1980. Bitterroot dome-Sapphire tectonic block, an example of a plutonic core–gneiss–dome complex with its detached suprastructure. *In*: CRITTENDEN, M.D., CONEY, P.J. & DAVIS, G.H. (eds) *Cordilleran Metamorphic Core Complexes*. Geological Society of America Memoir, **153**, 427–443.

HYNDMAN, D.W. & MEYERS, S.A. 1988. The transition from amphibolite-facies mylonite to chloritic breccia and role of the mylonite in formation of Eocene epizonal plutons, Bitterroot dome, Montana. *Geologische Rundschau*, **77**, 211–226.

JEFFERY, J.B. 1922. The motion of ellipsoidal particles immersed in viscous fluid. *Proceedings of the Royal Society of London*, **A102**, 161–179.

JELÍNEK, V. 1978. Statistical processing of anisotropy of magnetic susceptibility measured on groups of specimens. *Studia Geophyzika et Geodetika*, **22**, 50–62.

KERRICH, R. & HYNDMAN, D. 1986. Thermal and fluid regimes in the Bitterroot lobe-Sapphire block detachment zone, Montana: evidence from $^{18}O/^{16}O$ and geologic relations. *Geological Society of America Bulletin*, **97**, 147–155.

KIRKPATRICK, R.J. 1983. Theory of nucleation in silicate melts. *American Mineralogist*, **68**, 66–77.

LOFGREN, G.E. & GOOLEY, R. 1977. Simultaneous crystallization of feldspar intergrowths from the melt. *American Mineralogist*, **62**, 217–228.

LÜNEBERG, C.M., LAMPERT, S.A., HERMANN, D.L., HIRT, A.M., CASEY, M. & LOWRIE, W. 1999. Magnetic anisotropy, rock fabrics and finite strain in deformed sediments of SW Sardinia (Italy). *Tectonophysics*, **307**, 51–74.

MERRILL, R.T. 1977. Origin of thermoremanent magnetization. *Advances in Earth and Planetary Sciences*, **1**, 53–60.

MOSKOWITZ, B.M., FRANKEL, R.B. & BAZYLINSKI, D.A. 1993. Rock magnetic criteria for the detection of biogenic magnetite. *Earth and Planetary Science Letters*, **120**, 283–300.

NAGATA, T. 1961. *Rock Magnetism*. Maruzen, Tokyo.

PASSCHIER, C.W. & TROUW, R.A.J. 1996. *Microtectonics*. Springer, New York.

ROCHETTE, P., JACKSON, M. & AUBOURG, C. 1992. Rock magnetism and the interpretation of anisotropy of magnetic susceptibility. *Reviews of Geophysics*, **30**, 209–226.

SCHUMACHER, J.C. 1991. Empirical ferric iron corrections: necessity, assumptions, and effects on selected geothermobarometers. *Mineralogical Magazine*, **55**, 3–18.

SHUSTER, R.D. & BICKFORD, M.E. 1985. Chemical and isotopic evidence for the petrogenesis of the northeastern Idaho batholith. *Journal of Geology*, **93**, 727–742.

SIEGESMUND, S., ULLEMEYER, K. & DAHMS, M. 1995. Control of magnetic rock fabrics by mica preferred orientation: a quantitative approach. *Journal of Structural Geology*, **17**, 1601–1613.

SNOKE, A.W. & TULLIS, J. 1998. An overview of fault rocks. *In*: SNOKE, A.W., TULLIS, J. & TODD, V.R. (eds) *Fault-related Rocks: A Photographic Atlas*. Princeton University Press, Princeton, NJ, 3–18.

TARLING, D.H. & HROUDA, F. 1993. *The Magnetic Anisotropy of Rocks*. Chapman & Hall, London.

TEYSSIER, C. & WHITNEY, D.L. 2002. Gneiss domes and orogeny. *Geology*, **30**, 1139–1142.

TEYSSIER, C., FERRÉ, E.C., WHITNEY, D.L., NORLANDER, B., VANDERHAEGHE, O. & PARKINSON, D. 2005. Flow of partially molten crust and origin of detachments during collapse of the Cordilleran Orogen. *In*: BRUHN, D. & BURLINI, L. (eds) *High-Strain Zones: Structure and Physical Properties*. Geological Society, London, Special Publications, **245**, 39–64.

VERWEY, E.J.W. & HAAYMAN, P.W. 1941. Electronic conductivity and transition point in magnetite. *Physics*, **8**, 979–982.

WYLLIE, P.J. 1983. Experimental studies on biotite- and muscovite-granites and some crustal magmatic sources. *In*: ATHERTON, M.P. & GRIBBLE, C.D. (eds) *Migmatites, Melting, and Metamorphism*. Shiva Publishing, Nantwich, 12–26.

Electrical conductivity images of active and fossil fault zones

O. RITTER, A. HOFFMANN-ROTHE, P. A. BEDROSIAN,
U. WECKMANN & V. HAAK

*GeoForschungsZentrum Potsdam, Telegrafenberg, D-14473 Potsdam,
Germany (e-mail: oritter@gfz-potsdam.de)*

Abstract: We compare recent magnetotelluric investigations of four large fault systems: (i) the actively deforming, ocean–continent interplate San Andreas Fault (SAF); (ii) the actively deforming, continent–continent interplate Dead Sea Transform (DST); (iii) the currently inactive, trench-linked intraplate West Fault (WF) in northern Chile; and (iv) the Waterberg Fault/Omaruru Lineament (WF/OL) in Namibia, a fossilized intraplate shear zone formed during early Proterozoic continental collision. These fault zones show both similarities and marked differences in their electrical subsurface structure. The central segment of the SAF is characterized by a zone of high conductivity extending to a depth of several kilometres and attributed to fluids within a highly fractured damage zone. The WF exhibits a less pronounced but similar fault-zone conductor (FZC) that can be explained by meteoric waters entering the fault zone. The DST appears different as it shows a distinct lack of a FZC and seems to act primarily as an impermeable barrier to cross-fault fluid transport. Differences in the electrical structure of these faults within the upper crust may be linked to the degree of deformation localization within the fault zone. At the DST, with no observable fault-zone conductor, strain may have been localized for a considerable time span along a narrow, metre-scale damage zone with a sustained strength difference between the shear plane and the surrounding host rock. In the case of the SAF, a positive correlation of conductance and fault activity is observed, with more active fault segments associated with wider, deeper and more conductive fault-zone anomalies. Fault-zone conductors, however, do not uniquely identify specific architectural or hydrological units of a fault. A more comprehensive whole-fault picture for the brittle crust can be developed in combination with seismicity and structural information. Giving a window into lower-crustal shear zones, the fossil WF/OL in Namibia is imaged as a subvertical, 14 km-deep, 10 km-wide zone of high and anisotropic conductivity. The present level of exhumation suggests that the WF/OL penetrated the entire crust as a relatively narrow shear zone. Contrary to the fluid-driven conductivity anomalies of active faults, the anomaly here is attributed to graphitic enrichment along former shear planes. Once created, graphite is stable over very long time spans and thus fault/shear zones may remain conductive long after activity ceases.

The San Andreas Fault in California, the North Anatolian Fault in Turkey, the Altyn Tagh Fault in China and the Dead Sea Transform in Jordan are prominent examples of high-strain zones and expressions of dynamic processes in the Earth's lithosphere. Such large-scale faults can be traced for hundreds to thousands of kilometres on the Earth's surface. There is growing evidence that some of these faults penetrate the mantle lithosphere, leaving in their wake a region of lithospheric-scale weakness (Henstock *et al.* 1997; Herquel *et al.* 1999; Rümpker *et al.* 2003) while complex zones of localized brittle deformation characterize shearing in the upper (seismogenic) crust. Information about the structure (and dynamics) of these zones derive primarily from surface investigations of active and exhumed fault zones, as well as from geophysical images of the subsurface.

A considerable amount of research has focused on the internal structure of crustal fault zones and the mechanics of faulting (Chester *et al.* 1993; Holdsworth *et al.* 2001; Ben-Zion & Sammis 2003; Faulkner *et al.* 2003). It is well established that the structural and compositional evolution of fault zones is intimately linked to fluids; they may also control the nucleation, propagation, arrest and recurrence of brittle failure (Hickman *et al.* 1995). Fluids can induce fracturing by modifying the strength and constitution of the host rocks. Subsequent deformation creates further pathways for fluids to enter the fault zone, creating a highly permeable conduit for fluid flow (leading to a feedback mechanism).

From: BRUHN, D. & BURLINI, L. (eds) 2005. *High-Strain Zones: Structure and Physical Properties.*
Geological Society, London, Special Publications, **245**, 165–186.
0305-8719/05/$15.00 © The Geological Society of London 2005.

A lithological contrast resulting from the fault's displacement or the formation of gouge due to shearing processes may further create a barrier impeding cross-fault fluid flow (Caine *et al.* 1996).

Surface-based geophysical methods provide invaluable information concerning the nature and distribution of fluids at seismogenic depths (Eberhart-Phillips *et al.* 1995). The most intensively studied example of any fault is the San Andreas Fault, separating the oceanic Pacific plate from the continental North American plate. Unsworth *et al.* (1997) demonstrated that its internal structure can be imaged with magnetotelluric (MT) measurements. Several short profiles across the San Andreas Fault image a highly conductive structure down to a depth of several kilometres, attributed to the circulation of saline fluids within the damage zone of the fault. In this situation, bulk conductivity is dominated by ion transport within the pore space. Fluid transport within the rock opens pores and cracks, increasing the mobility of solutes such as salts, calcite or quartz, thereby increasing the bulk conductivity. Precipitation from hydrothermal fluids, on the other hand, may cement open fractures within the rock and, in turn, lower bulk conductivity.

Exhumed fossil shear zones, in contrast to upper crustal faults, commonly expose structures that originated below the depth of predominantly brittle deformation (although they may have experienced brittle deformation during reactivation). These shear zones can be similarly conductive, but, in these cases, bulk conductivity may be dominated by electron transport, for example in an interconnected graphite network. Data from the KTB deep drill hole and its accompanying experiments revealed a quantitative relation between the shearing process along faults, the formation of graphite and high electrical conductivity (ELEKTB Group 1997). The critical observation is that the shearing process itself can lead to the interconnection of conductive material (Jödicke *et al.* 2004).

Images of electrical conductivity can help decipher the internal architecture of crustal-scale fault zones, and the magnetotelluric method is one of the few tools capable of imaging from the Earth's surface through to the mantle. We start with a short description of MT and its resolution characteristics. The main part of this paper presents MT studies of four crustal-scale fault zones in the geodynamic settings shown in Figure 1, and is not intended to be an exhaustive review. Actively deforming

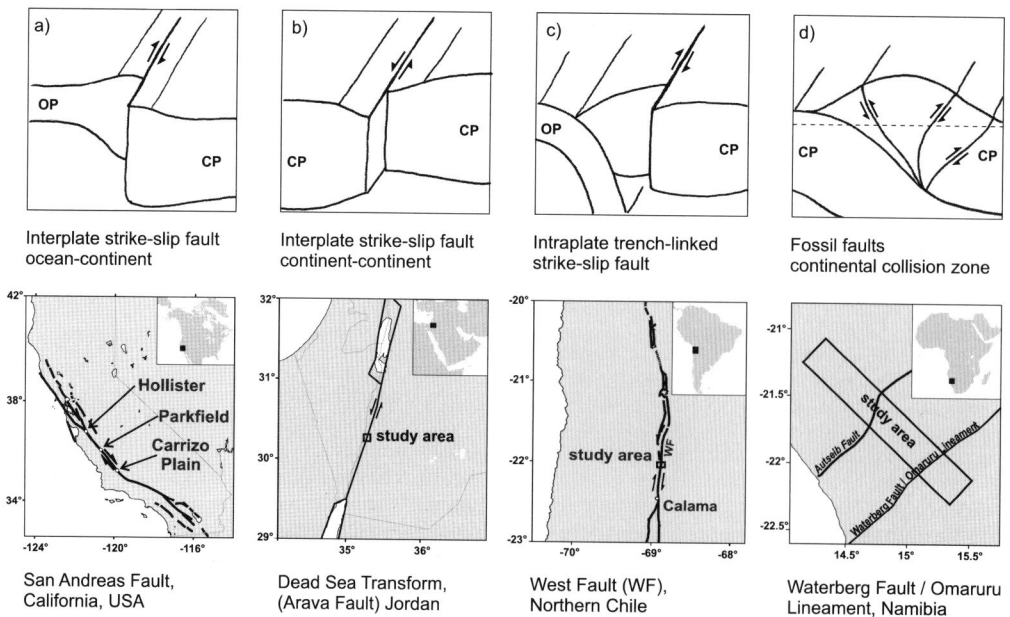

Fig. 1. Geodynamic settings of case studies discussed in this paper. (**a**),(**b**),(**c**) The brittle upper crustal part of these active margins is accessible for investigation. Ductile deformation, operating at greater depth is not easily imaged. (**d**) Fossil shear zones exhumed at the surface (dashed line indicating present surface level) serve as a proxy for lower crustal shearing processes at currently active faults. OP, oceanic plate; CP continental plate.

faults are represented by the ocean–continent interplate San Andreas Fault (SAF) in California and the continent–continent interplate Dead Sea Transform (DST) in Jordan. These fault zones exhibit markedly different conductivity structure. The currently inactive West Fault (WF) in northern Chile is a trench-linked intraplate strike-slip fault associated with the Andean subducting margin. Exposure of the fault's damage zone illuminates how structural damage is linked to electrical conductivity. A fossil representative of a lower crustal shear zone is the Waterberg Fault/Omaruru Lineament (WF/OL) in Namibia, part of a broad Palaeozoic suture. Having experienced several episodes of reactivation, this mechanically weak zone, formerly within the lower crust, exhibits strong electrical anisotropy. Following the presentation of these case studies, both upper and lower crustal structures are discussed, with attention to the hydrological and structural implications these conductivity images provide.

Some background on the magnetotelluric method

The electrical conductivity of rocks. The electrical resistivity (ρ) and its inverse, the electrical conductivity (σ), characterize charge transport within materials. They are intrinsic material properties, independent of sample size. Rocks and rock-forming minerals vary in their electrical properties, with conductivities ranging from 10^6 to 10^{-14} S m^{-1} (Guéguen & Palciauskas 1994). A summary of important crustal constituents and their conductivities is given in Figure 2.

In most sedimentary rocks, currents are propagated primarily by ions within pore fluids. Bulk conductivity is thus strongly dependent on the volume of included fluid and its conductivity, in addition to the size and arrangement of pores. Fluid conductivity itself is temperature-dependent, and further varies with the concentration, mobility and charge of the ions. The relationship between fluid conductivity (σ_{fluid}) and bulk conductivity (σ_{rock}) can be described by the empirical (Archie's Law 1942), given in its simplified form:

$$\frac{\sigma_{fluid}}{\sigma_{rock}} = \phi^{-m} \qquad (1)$$

ϕ is the porosity of the rock raised to the cementation or compaction exponent (m), which typically varies with pore geometry from 1 for spheres to 2 for cracks.

Rocks containing a high concentration of minerals such as sulphides, magnetite and graphite

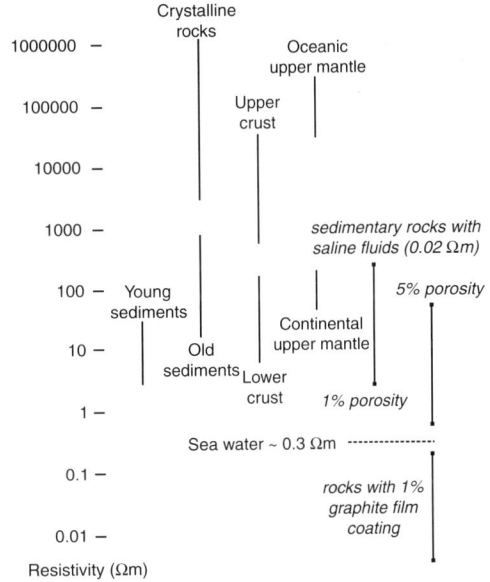

Fig. 2. The range of electrical resistivity in Earth materials. Resistivities of fluid-bearing rocks are largely independent of rock type and calculated based on a fluid resistivity of 0.02 Ωm. Upper and lower resistivity bounds correspond to fluids contained in spherical pores and cracks, respectively. In the case of graphite films, bulk resistivity scales with the grain size of the host rock (after Haak & Hutton 1986).

may exhibit high conductivity if these electronically conducting minerals are interconnected. Even small concentrations of graphite can significantly increase conductivity due to its tendency to form thin continuous films along grain boundaries (Guéguen & Palciauskas 1994). Graphite in metamorphic rocks typically results from the conversion of organic matter through metamorphism or precipitation from natural, carbon-bearing fluids (Luque *et al.* 1998; Wannamaker 2000). Conductivity will increase if shearing processes smear isolated graphite grains into continuous films or press them into interconnected crack systems (Jödicke *et al.* 2004). In addition, graphite coatings lower shear friction and, hence, add to the mobility of faults. Once created, graphite is stable over very long time spans allowing shear zones to remain conductive long after activity ceases. A good example is the Münchberg Gneiss complex in southern Germany, where graphitic remnants along horizontal shear planes probably cause the observed high conductivity anomalies (Ritter *et al.* 1999). Quantifying the effect of graphite or other conductive minerals on bulk conductivity requires mixing laws (Doyen 1988) or network and percolation theories.

The magnetotelluric impedance tensor. The magnetotelluric method is based on the induction of electromagnetic fields in the Earth. Its ultimate aim is to determine the electrical conductivity structure of the subsurface. Observations of orthogonal components of time-varying horizontal electric $\mathbf{E}[mV\,km^{-1}]$ and magnetic fields $\mathbf{B}$ [nT] at the surface of the Earth are related via the components of the impedance tensor $\mathbf{Z}$:

$$\begin{pmatrix} E_x \\ E_y \end{pmatrix} = \begin{pmatrix} Z_{xx} & Z_{xy} \\ Z_{yx} & Z_{yy} \end{pmatrix} \begin{pmatrix} B_x \\ B_y \end{pmatrix}. \qquad (2)$$

The impedance tensor $\mathbf{Z}$ is a frequency-dependent, complex quantity. At a particular frequency (ω), it can be converted into an amplitude (apparent resistivity $\rho_a(\omega)$) and phase ($\phi(\omega)$) relation. In magnetotellurics, frequency ω can be viewed as a rough (non-linear) proxy for depth, as lower frequencies penetrate deeper into the subsurface (skin depth effect). Only in the special case of a homogeneous conductivity structure is the apparent resistivity frequency independent and equal to the Earth's resistivity.

If lateral conductivity variations exist within the subsurface on length scales equal to or greater than the penetration depth of the induced horizontal magnetic fields, a vertical magnetic field component is generated. The geomagnetic response functions $T_x(\omega)$, $T_y(\omega)$ are defined as:

$$B_z = T_x B_x + T_y B_y. \qquad (3)$$

The form of the impedance tensor and the existence of vertical magnetic fields are directly linked to the electric and magnetic field distributions at the Earth's surface, and in turn the (electrical) complexity of the subsurface. Over a one-dimensional (1D) Earth (conductivity varying only with depth) vertical magnetic fields and the diagonal entries of the impedance tensor vanish and the off-diagonal elements carry the same impedance information Z (Fig. 3a & b). The measured impedance tensors over a 2D Earth (conductivity varying with depth and in one lateral direction) have vanishing diagonal elements only if the electromagnetic fields are either recorded or mathematically rotated into a co-ordinate system aligned with geoelectric strike (Fig. 3c & d). The impedance tensor is fully occupied over a 3D conductivity distribution (Fig. 3g) except at particular locations where, due to symmetry, the field distributions and therefore the impedance tensor may appear 1D or 2D. In the presence of electrical anisotropy a fully occupied impedance tensor is observed even if the regional structure is 2D. Only in the rare case of anisotropy striking parallel to the regional strike will the impedance tensor reduce to its 2D form (Fig. 3f).

Tensor rotation and induction arrows. Traditionally, Swift's (1967) method was applied to derive a geoelectric strike angle from the measured impedance tensor by searching for a rotation angle that maximizes the off-diagonal components (see 2D aligned co-ordinate system in Fig. 3c). Swift's parameter can be misleading, however, as the measured electric field may be

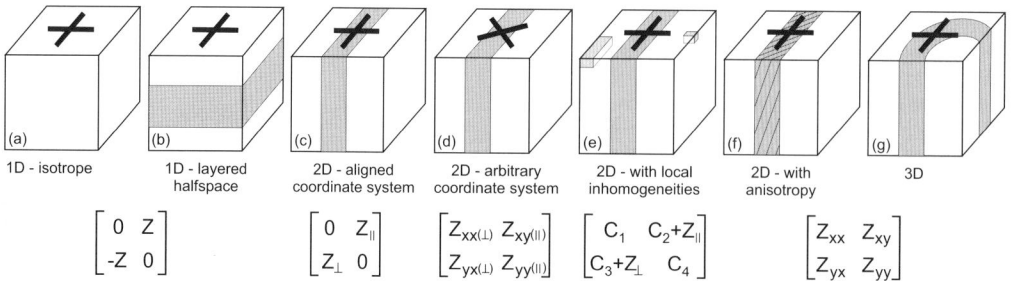

Fig. 3. Symmetries in Earth conductivity structure reflected in the MT impedance tensor ($\mathbf{Z}$). (**a**),(**b**) An homogeneous or 1D Earth structure gives rise to $\mathbf{Z}$ with off-diagonal elements of opposite sign. (**c**) Measurements made over 2D Earth structure giving rise to $\mathbf{Z}$ with unequal off-diagonal elements. The measurements are aligned with earth structure. (**d**) General case of (c) with a fully occupied tensor, i.e. all elements of $\mathbf{Z}$ are non-zero. Case (d) can be mathematically rotated into case (c). (**e**) Two-dimensional Earth structure with localized 3D distortion. Frequency-independent real constants C_1–C_4 represent effects of 3D galvanic distortion and can to some degree be determined through tensor decomposition. (**f**),(**g**) Two-dimensional anisotropic and 3D Earth structure generate fully occupied $\mathbf{Z}$ except in special cases.

severely distorted in the presence of near-surface heterogeneities (compare Fig. 3e). Under these circumstances, tensor decomposition methods can help recover an (undistorted) regional 2D impedance tensor from the measured impedance tensor (Smith 1995; McNeice & Jones 2001, and reference therein). These methods assume that local 3D bodies (distorters) accumulate charges at their edges giving rise to a frequency-independent response intertwined with the (desired) regional 2D impedance tensor. Tensor decomposition thus attempts to separate a real distortion matrix from the measured impedance tensor that contains, among other distortion parameters, information on the strike direction. A consistent frequency-independent strike direction must exist if the 2D assumptions are fulfilled.

An independent determination of strike directions is obtained from geomagnetic response functions, relating horizontal and vertical magnetic fields (see equation 3). The induction vector is a graphical representation of the vertical magnetic field transfer function in which T_x and T_y are viewed as components of a vector and plotted in map view. Induction vectors vanish in the absence of lateral resistivity gradients and thus provide an important means of identifying conductivity contrasts that may be expected across a fault. In the 2D case, real and imaginary induction arrows are parallel or anti-parallel to each other, and parallel to geoelectric strike (see Fig. 4). In this paper, induction vectors are plotted in the Wiese (1962) convention, in which the real parts point away from regions of high conductivity.

Two-dimensional modelling and inversion. In order to determine the 2D conductivity distribution of the subsurface either forward modelling or inversion is applied to apparent resistivity and phase curves. It is at this point that an accurate assessment of the dimensionality of MT data is important. In the case of 2D isotropic Earth structure, the MT equations reduce to fitting two elements of the impedance tensor. While real MT data are seldom purely 2D, methodologies have been established for the analysis of weakly 3D data sets using 2D inversion methods (e.g. Ledo *et al.* 2002). The alternative, a 3D inversion, is in most cases not feasible due to both computational requirements and the areal data coverage required to constrain such an inversion.

All 2D resistivity models, except for the anisotropic models, presented in this paper derive from inversion of MT data using the iterative RLM2DI algorithm (Rodi & Mackie 2001), part of the WinGLink software package (www. geosystem.net/software.htm). RLM2DI finds regularized solutions (Tikhonov Regularization) to the 2D MT inverse problem using the method of non-linear conjugate gradients. Prior to inversion, a finite-difference mesh is created, where row thickness increases with depth (dependent on the frequency range of the data) and column width scales with site spacing.

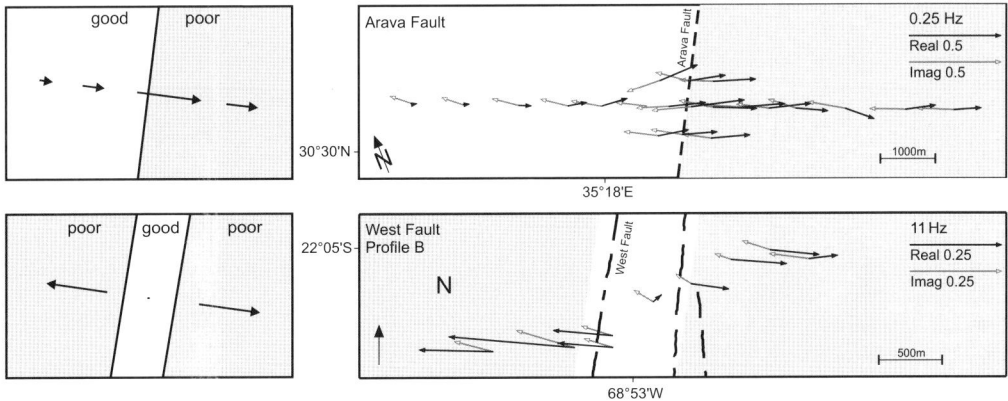

Fig. 4. Expected and observed behaviour of induction vectors (IV) for two different fault structures. In Wiese convention real (black) arrows point away from good conductors; imaginary (grey) arrows are parallel or antiparallel to the real arrows, given 2D Earth structure. Across a conductivity contrast (top panel) real IVs point away from the more conductive side and are largest over the less conductive side. The DST, locally the Arava Fault, is representative of this type of fault structure. A conductive fault zone sandwiched between more resistive units is characterized by IVs pointing away from the fault on both sides and small IVs over the fault zone conductor. The West Fault is representative of this type of fault structure.

A homogeneous resistivity model is commonly employed as a starting model. Alternatively, geological or geophysical constraints may be introduced into the starting resistivity model *a priori*. The inversion algorithm attempts to minimize an objective function expressed as the sum of the normalized data misfit and the smoothness of the model. The trade-off between data misfit and model smoothness is controlled by the regularization parameter τ, determined via several inversion runs. The best value of τ gives rise to monotonic decreases in both data misfit and model smoothness as the inversion proceeds.

The smoothness (Laplacian) regularization described above is but one way to constrain the inherently non-unique MT problem. Regularization can be chosen, for example, to minimize the flatness (gradient) of the model or its deviation from an *a priori* model. Each regularization has its strength and weakness, which must be taken into account when assessing the fit of the inversion model. For the studies presented in this paper, the smoothness regularization produces minimum structure models, in which only structures required by the data are presented. A disadvantage of this approach is that sharp boundaries are sometimes not well recovered during inversion, resulting in a smearing of structures at depth. The assessment of any MT inversion model hinges not only on the root mean square (rms) data misfit, but also on a careful examination of the existence, geometry and resistivity of individual features of the model.

Forward modelling is often used to assess the sensitivity of measured data to modification of part of an inversion model. It is also commonly employed to determine the resolution depth, that is the depth below which changes in conductivity are invisible to measured data. Constrained inversion, in which part of a model is modified and held fixed during subsequent inversion, is further used to constrain the geometry and strength of anomalous structure.

In the case of a 1D Earth, the conductance, $s[S]$, of a layer of thickness, $h[m]$, is the most robust parameter determined by magnetotellurics, and is used to quantitatively compare conductivity anomalies (e.g. Li *et al.* 2003):

$$s = \sigma h. \qquad (4)$$

In the case of 2D and 3D Earth structure, conductance can still be calculated, but will vary with location (Bedrosian *et al.* 2004). For the purpose of this paper, we wish to compare the conductivity anomalies associated with various fault zones. We thus introduce the lateral conductance, a quantity we define as the product of the average width and conductivity of an imaged conductor. In a 1D case, lateral conductance is meaningless, while in a 3D case, it is direction-dependent. It is important to note that while lateral conductance can be used for comparison of fault zones, it is not subject to a conductivity/thickness trade-off. The studies presented herein, for example, can distinguish between a 500 m wide, 1 S m^{-1} zone and a 1 km wide, 0.5 S m^{-1} zone.

Two-dimensional inversion is limited to fitting only two elements of the measured impedance tensor (see Fig. 3c). For data sets exhibiting significant diagonal impedance elements, the 2D anisotropic forward modelling code of Pek & Verner (1997) can be used to model a 3D conductivity tensor if an anisotropic subsurface is a reasonable assumption. By defining a set of three conductivity values and their associated rotation angles, the surface response of any physically possible electrical anisotropy can be calculated. Defining conductivities in different directions primarily reflects the intrinsic or microscopic anisotropy of rocks. However, Eisel & Haak (1999) point out that MT results over a microscopically anisotropic subsurface are indistinguishable from those over structural or macroscopic anisotropy in the form of conductive lamellae. Intrinsic electrical anisotropy may arise from a crystallographic preferred orientation (CPO) in the subsurface, while structural anisotropy is commonly of tectonic or volcanic origin.

Case studies

The San Andreas Fault in California, USA

Formed around 30 Ma ago as the Pacific plate impinged upon the North American plate, the San Andreas Fault (SAF) continues to lengthen as the Mendocino and Rivera triple junctions migrate apart. The SAF exhibits a right-lateral offset of approximately 300 km and currently accommodates *c.* 35 mm year^{-1} of strain across its multiple strands (Irwin 1990; Bennett *et al.* 1999; Argus & Gordon 2001). There is marked seismic variability along the SAF, with some segments characterized by infrequent, large-magnitude earthquakes while others exhibit abundant microseismicity and aseismic creep (Allen 1968).

Throughout central California, the SAF juxtaposes granitic rocks of the Salinian block against the Franciscan complex, a Mesozoic subduction complex. MT profiles (c. 20 km in length) acquired near Hollister (in 1999) and Parkfield (in 1997) are considered here (see Fig. 1a for locations). The SAF near Hollister creeps, while at Parkfield it is in transition between a creeping segment to the north and a locked segment to the south. MT data were recorded in the frequency range 100–0.001 Hz at 80 (50) sites at Hollister (Parkfield). Calculation of geoelectric strike is consistently within 5° of geological strike at all but the lowest frequencies (Unsworth et al. 1999; Bedrosian et al. 2004).

Two-dimensional resistivity models, derived via inversion of MT and vertical magnetic field data, are shown in Figure 5. At both locations, a strong resistivity contrast is imaged between the Salinian granites SW of the fault and a highly conductive zone to the NE. This boundary is further delineated by seismicity at a depth of between 2 and 7 km. At Hollister, the prominent zone of high conductivity is loosely bound between the San Andreas and Calaveras faults, extending to mid-crustal depths beneath the SAF (Bedrosian et al. 2004). At Parkfield, anomalous conductivity is confined to a zone centred on the SAF and extending from the surface to a depth of 2–5 km, based on constrained inversion studies by Unsworth & Bedrosian (2004). To the NE, a diffuse zone of high conductivity is imaged.

The prominent fault zone conductors (FZC) imaged at Hollister and Parkfield coincide with zones of reduced seismic velocity and enhanced V_p/V_s (where V_p/V_s is the ratio of compressional wave velocity to shear-wave velocity). (Thurber et al. 1997, 2003; Catchings et al. 2002). Furthermore, the depths and geometries of the low-velocity zones are similar to those of the imaged FZCs.

The most plausible explanation for the anomalously high conductivity at both locations is saline fluids within a fractured region surrounding the fault (Unsworth et al. 1997; Bedrosian et al. 2004). However, Park et al. (2003) alternatively propose that the sediments of the Parkfield syncline may explain the observed FZC at Parkfield; however, an additional MT profile 6 km to the south imaged both a FZC and the Parkfield syncline (Unsworth et al. 2000), suggesting the syncline cannot solely explain the high conductivity of the fault zone. The case for fluids is supported by high salinities measured in wells close to the fault. Using Archie's law (equation 1), porosities of 9–30

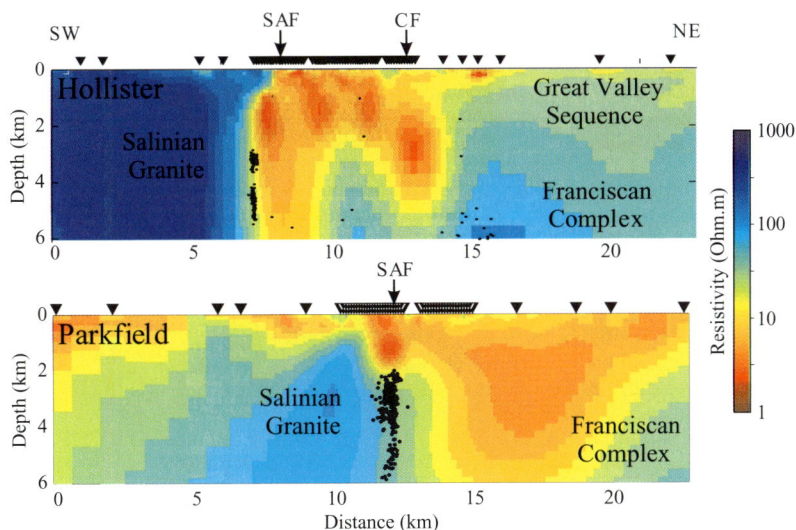

Fig. 5. Hollister and Parkfield resistivity models after Bedrosian et al. (2002) and Unsworth & Bedrosian (2004), respectively. Black dots indicate earthquake hypocentres within 1 km of each profile. Inverted triangles indicate site locations, while arrows denote surface traces of the San Andreas (SAF) and Calaveras (CF) faults. Hypocentres are from the Northern California Seismic Network, UC Berkeley High Resolution Seismic Network and the Parkfield Area Seismic Observatory.

and 15–35% have been estimated for Parkfield and Hollister, respectively (Unsworth *et al.* 1997; Bedrosian *et al.* 2002).

The Dead Sea Transform in the Arava Valley, Jordan

The Dead Sea Transform (DST) is a left-lateral transform fault separating the continental African and Arabian plates (Garfunkel & Avraham 1996). It extends over 1000 km from the Red Sea in the south to the Taurus collision zone in the north (Fig. 1b). The fault has been active, more or less continuously, since its origin in the Miocene with a total left-lateral motion of 105 km (Freund *et al.* 1970) and a present-day slip rate of 4 ± 2 mm year^{-1} (Klinger *et al.* 2000a, Wdowinski *et al.* 2004). MT studies were part of the multidisciplinary DESERT (DEad SEa Rift Transect) project, a

300 km-long profile traversing Israel, Jordan and the Palestinian territories.

Figure 6 shows images obtained from (a) magnetotelluric modelling and (b) seismic tomography across the Arava Fault (AF), a local expression of the DST (Ritter *et al.* 2003a). The total length of the near-vertical seismic reflection line was 100 km; MT data were recorded along the innermost 10 km of this profile, centred on the AF. A dense site spacing of 100 m in the centre of the MT profile was supplemented by more widely spaced sites near the profile ends. Strike angle analysis, using the method of Smith (1995) reveals a geoelectric strike of N17°E, in close agreement with the surface strike of the AF (N15°E). This strike angle is further confirmed by induction vectors, as shown in the upper panel of Figure 4.

For the shallow crust, the inversion model reveals a highly conductive layer from the surface to a depth of approximately 100 m on

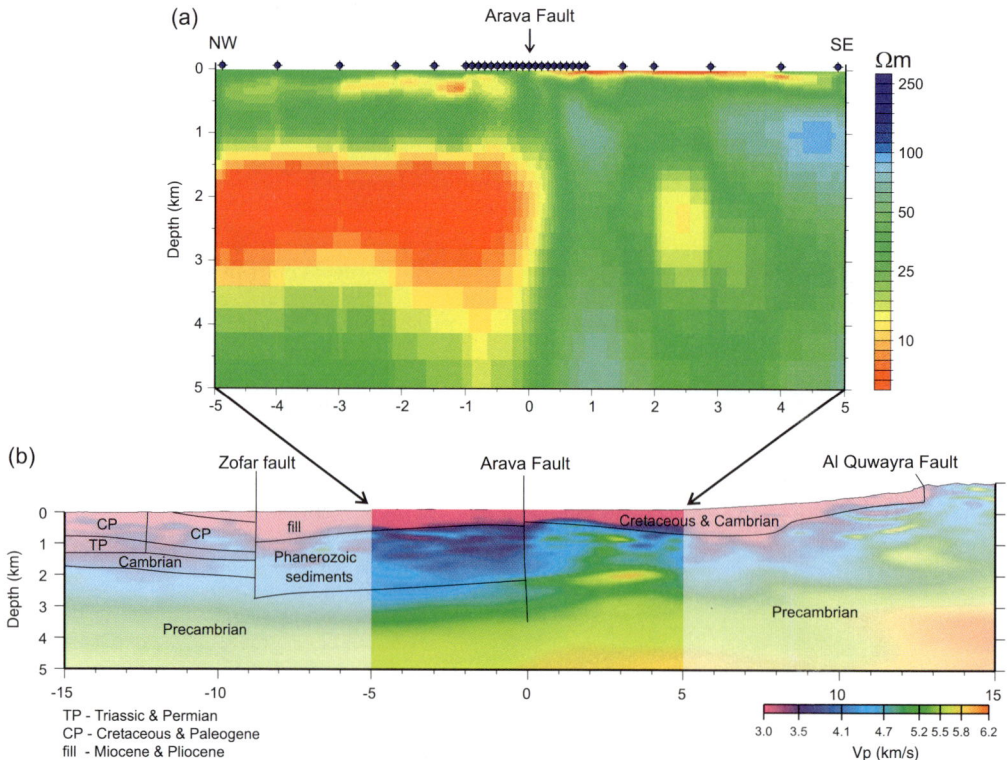

Fig. 6. Coincident magnetotelluric and seismic tomography sections crossing the Arava Fault (AF) modified after Ritter *et al.* (2003a). (**a**) Resistivity model surrounding the AF. Diamonds depict MT site locations. (**b**) Seismic tomography section with overlain structural interpretation taken from Ritter *et al.* (2003a). Only the upper 5 km of the crust are shown, as resolution degrades at larger depths due to limited profile length (MT) and diminished ray coverage (seismics). Both model sections indicate significant lateral changes of physical properties across the AF.

the eastern side of the profile. However, the most prominent feature on the MT image is a conductive half-layer confined to west of the fault and beginning at a depth of approximately 1.5 km. The surface trace of the AF correlates with a sharp vertical conductivity boundary at the eastern edge of this feature. The high conductivity may be due to brines in porous sedimentary rocks; saline waters have been found in a lower Cretaceous aquifer at a depth of approximately 1000 m in a drill hole c. 14 km west of the AF (Ritter et al. 2003a). The seismic image reveals a strong increase in P-wave velocities (to values exceeding $5 \, \mathrm{km \, s^{-1}}$) east of the AF, where the MT model indicates higher resistivities. The seismic velocities are consistent with crystalline basement rocks; however, the observed resistivities (50–250 Ωm) are unusually low for unfractured crystalline rocks. Both the seismic and MT observations may be explained by fractured crystalline rocks with interconnected fluid-bearing veins.

The West Fault in northern Chile. The Precordilleran Fault System (PFS), classified as a trench-linked strike-slip fault, trends subparallel to the northern Chilean margin for more than 2000 km (Fig. 1c). It was initiated in an Eocene magmatic arc in response to oblique convergence of the Nazca and South American plates. The West Fault (WF) forms the main branch of the northern PFS. It records approximately 40 km of left-lateral displacement, although the sense of slip has seen several reversals throughout the fault's evolution. Belmonte (2002) reports little seismicity in recent times.

MT profiles (2.5 and 4 km long) were recorded where the fault cuts an inclined plain of Quaternary alluvial deposits (Hoffmann-Rothe 2002; Hoffmann-Rothe et al. 2004). The profiles are separated along-strike by 3.5 km, and have site spacings between 100 and 300 m. Data were recorded in the frequency range 1000–0.001 Hz. Dimensionality and distortion analysis reveal a geoelectric strike direction of N10°E for the frequency range 1000–0.1 Hz, consistent with the strike of the surface trace. This strike direction is supported by the induction vectors shown in the lower panel of Figure 4. At greater skin depths (frequencies <0.1 Hz) the conductivity structure becomes more complex, as different strike directions begin to effect the data; modelling and inversion are thus limited to frequencies above 0.1 Hz.

Two-dimensional inversion models for both profiles are shown in Figure 7. The most prominent and robustly imaged structure is a subvertical zone of high conductivity located east of the surface trace. Hypothesis testing using forward modelling and inversion confine the depth of the FZC to about 1500 m (Hoffman-Rothe 2002). Independent evidence from fracture density mapping reveals an approximately 1000 m-wide damage zone (Hoffmann-Rothe 2002). Comparison of structural and MT results (inset in Fig. 7) shows a good spatial correlation of the FZC with a c. 400 m wide zone of intensely comminuted material that has undergone pronounced fluid alteration (Janssen et al. 2002, 2004). Along profile A (see Fig. 7) the FZC is about 300 m wide and bounded to the west by a resistive block. This resistive block corresponds to Jurassic limestones and Triassic andesites that outcrop at locations west of the fault and are not present along the southern profile (B). Subsidiary faults and fractures preserved in the alteration zone of profile A are steeply inclined c. 70° to the east (Hoffman-Rothe et al. 2004); a similar dip angle is required for the FZC of profile A based on constrained inversion studies. Together this suggests that fluids within the fault/fracture mesh give rise to the high conductivity. This scenario is further supported by the presence of a spring that surfaces along the trace of the fault. Groundwater salinity measured in adjacent boreholes and mines, in conjunction with the observed conductivity of the FZC, results in porosity estimates of 20% at the surface, reducing to 1% at a depth of 200 m (Archie's law).

The Waterberg Fault/Omaruru Lineament in Namibia. The Namibian Damara Belt is part of a system of convergent orogens formed during the amalgamation of South Gondwana. It is the remnant of continental collision between the Kongo and Kalahari cratons in the early Cambrian (Miller 1983). During this orogeny, older extensional faults were reactivated as thrusts that now separate zones of different stratigraphic and structural evolution, metamorphic grade, distribution of plutonic rocks and geochronology (Daly 1986, 1989). Subsequently, the Damara Belt was eroded down to form a new basement on which Permian marine sediments were deposited. The major tectono-stratigraphic lineaments, including the Autseib Fault (AuF) and the Waterberg Fault/Omaruru Lineament (WF/OL), can be traced for hundreds of kilometres, and subsequently continued to control deformation of the Namibian crust in the Cretaceous (Holzförster et al. 1999; Raab et al. 2002) (Fig. 1d).

This fossil fault zone, although quite dissimilar to the previous examples of active regimes, illustrates a high-strain zone that is also marked

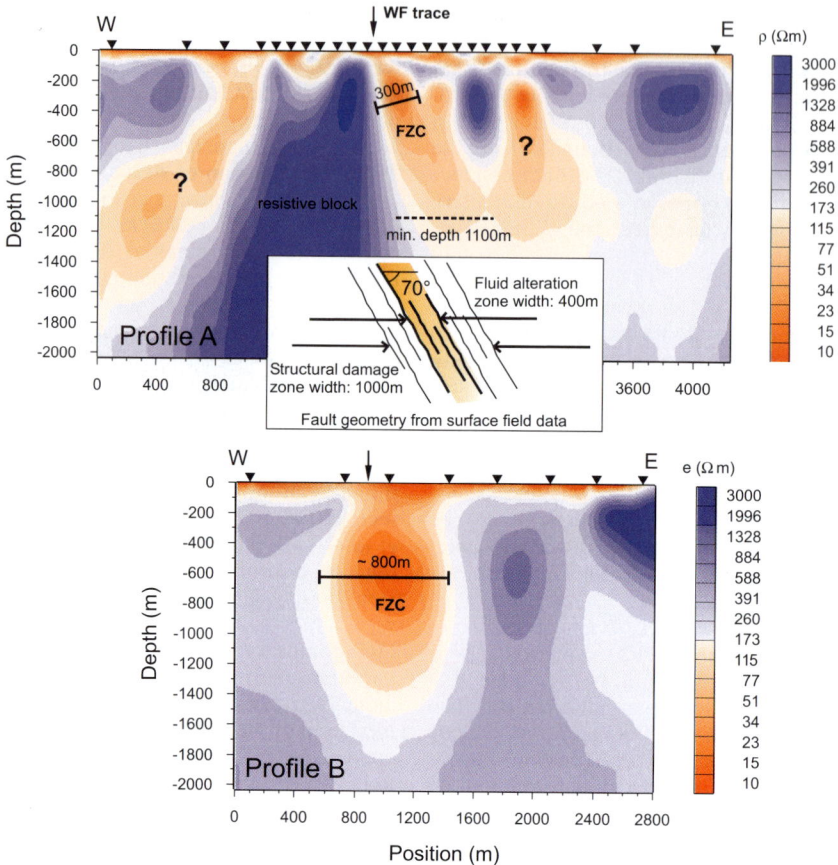

Fig. 7. Resistivity models of the West Fault modified after Hoffmann-Rothe (2002). Sites are depicted by inverted triangles. The most prominent feature on both profiles is a zone of high conductivity adjacent to the fault (labelled FZC). The inset depicts results from structural mapping along profile A, extrapolated to greater depth. The correlation of dip angles of the FZC and the fault/fracture mesh suggests that high conductivity arises from fluids within the fault/fracture mesh. Question marks indicate poorly constrained model features. Profile B is located 3.5 km further south of profile A.

by high conductivity. MT data were collected on both regional (200 km) and local (20 km) scales, with the majority of the 110 sites focused on the local profiles, where site spacings ranged from 500 to 2000 m.

The regional-scale MT model, shown in Figure 8a, provides insight into the deep structure of the Damara Belt (Ritter *et al.* 2003*b*). A generally resistive upper crust is imaged that at depths of 15 ± 2 km transitions to a conductive lower crust confined to the central part of the profile. The resistive upper crust is furthermore pierced by two subvertical conductors, whose locations coincide with the surface expression of the AuF and WF/OL, suggesting that they represent the continuation of these shear zones at depth. Both structures appear to terminate at

the lower crustal conductor, the top of which is interpreted as a lower crustal detachment.

The local MT study consisted of two dense profiles crossing the WF/OL with additional off-profile sites to provide areal coverage of the lineament zone (Weckmann *et al.* 2003*a*). The acquired data show fully occupied 3D impedance tensors that originate from a more complicated Earth structure than the 2D lineament structure observed at the surface. Based on a new imaging method (Weckmann *et al.* 2003*b*) that transforms the entire impedance tensor into resistivity ellipses, 3D effects can be attributed to a 10 km-wide zone of anisotropic conductivity in the shallow crust. Figure 8(b) illustrates a simplified 2D anisotropic model of the WF/OL and its close surroundings that reproduces the general

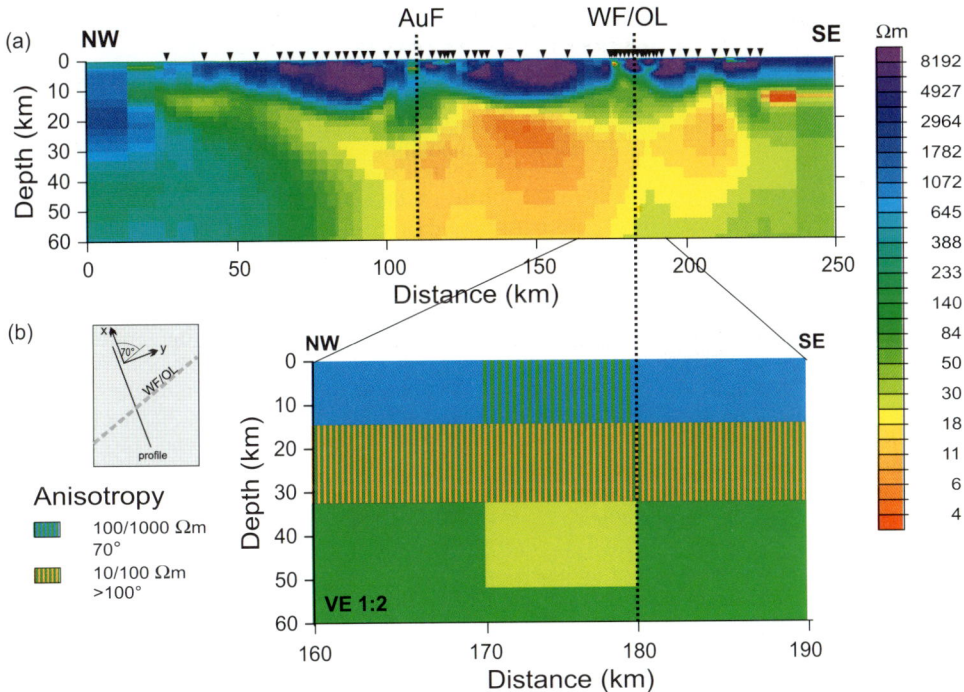

Fig. 8. Resistivity models of the Damara belt, Namibia modified after Ritter *et al.* (2003*b*) and Weckmann *et al.* (2003*a*) using the same colour scale. (**a**) Two-dimensional regional model of the lithosphere. Main features of the resistivity model include the general high conductivity of the mid-crust in the central part of the profile and the narrow, subvertical conductivity anomalies attributed to basement shear zones. AuF, Autseib Fault; WF/OL, Waterberg Fault/Omaruru Lineament. (**b**) Two-dimensional anisotropic model of the WF/OL: in the upper layer, a 10 km-wide and 14 km-deep anisotropic block (anisotropy striking subparallel to the strike of the WF/OL) is sandwiched between two resistive blocks. Beneath is a second anisotropic layer with a poorly defined strike. A 20 Ωm block embedded within the bottom half-space is required to model the low-frequency induction vectors.

characteristics of the data (Weckmann *et al.* 2003*a*). Despite its simple block structure, the model can resolve the internal structure and minimal depth extent of the fault, as the unusual shape of the observed MT responses considerably reduces the number of equivalent models. The data can only be fit if the anisotropic zone reaches to a depth of at least 14 km. The ratio of orthogonal to fault-parallel conductivity is 1:10, with a strike direction of N70°E, subparallel to the WF/OL. An anisotropic lower crustal layer is also necessary, with an anisotropy direction differing by at least 30° from that of the upper crustal layer. The tectonic significance of this layer is unknown. A 20 Ωm block is further embedded within the bottom half space and is required to model the induction vectors at long periods.

The regional and the local models of the WF/OL look quite dissimilar; however, this is primarily due to the disparate resolving power of the applied techniques. Isotropic 2D inversion tries to fit two components (see Fig. 3c) of the impedance tensor at all sites over the entire frequency range, but cannot explain the observed diagonal components of the impedance tensor. Two-dimensional anisotropic forward modelling, on the other hand, can model the entire 3D impedance tensor, but is limited to finding generalized models that reproduce the basic characteristics of the data. The lateral and depth extents of the anisotropic block coincide very well with the width of the subvertical conductor(s) in the regional model. In addition, the depth of the lower crustal conductor in the regional model agrees with the depth to the lower anisotropic layer in the local model. The resistivities of both models are also comparable; in the regional model they represent an azimuthal average of the anisotropic resistivity of the local model.

Fluids cannot be ruled out as a cause of the observed crustal conductivity; however, there is no supporting field evidence for hydrothermal alteration along the WF/OL. In contrast, graphite-bearing marbles are widespread in the area and suggest that shearing could create an interconnected graphite network on kilometre length scales (Liu *et al.* 2002).

Discussion

The shallow picture

Images of fault zone conductors (FZC) are characteristic of a number of faults throughout the world. Examples include the San Andreas Fault at Parkfield (Unsworth *et al.* 2000) and Hollister (Bedrosian *et al.* 2002), the West Fault in Chile (Hoffmann-Rothe 2002) and the Yamasaki fault in Japan (Electromagnetic Research Group for the Active Fault 1982). A FZC is absent, however, along the active Arava Fault in Jordan (Ritter *et al.* 2003*a*) and is only weakly expressed along the San Andreas Fault at Carrizo Plain (Unsworth *et al.* 1999). What are the major controls for the extent of enhanced conductivity within and surrounding the damage zone of a fault? In order to address this question we must consider the complex interplay between electrical conductivity, lithology, fluid supply, permeability and fault geometry as related to the faults described in this paper (see Table 1).

The San Andreas and West faults are characterized by FZC with widths of hundreds of meters and depths ranging from 1.5 to over 8 km (Table 1). The spatial extents of these FZCs are in rough accordance with other studies. At the West Fault, fracture density mapping shows strain distributed over a broad zone, with a region of intense shearing whose width is on the order of the FZC width (Hoffmann-Rothe 2002). Along the SAF, seismic tomography studies (Thurber *et al.* 2003, 1997; Catchings *et al.* 2002) image zones of reduced V_p and enhanced V_p/V_s coincident with the imaged FZCs and of similar scale and geometry. In addition, fault zone trapped wave studies estimate a damage zone of approximately 150 m

width at a location about 15 km SE of the Parkfield MT study (Li *et al.* 2004).

The fault zones under consideration can also be examined in light of their lateral conductances (see the earlier subsection on 'Two-dimensional modelling and inversion'). Bedrosian *et al.* (2002) note a FZC with an approximately 600 S lateral conductance at Hollister. A more modest conductance of about 250 S exists at Parkfield, whereas at Carrizo Plain, a mere 20 S is observed (Unsworth *et al.* 1999). As discussed in the section 'The San Andreas Fault in California, USA', the high lateral conductances at Hollister and Parkfield are best explained by saline fluids within and surrounding the damage zone of the fault. In contrast, the lateral conductance at Carrizo Plain can be adequately explained by a narrow zone of deformation with only small quantities of fluids present (Unsworth *et al.* 1999). Meteoric waters, transported towards the fault zone by topography, sufficiently explain the observed lateral conductance of 5 S at the West Fault.

The lack of an imaged FZC at the AF only shows that a SAF-type FZC is not required to fit the data. Constrained inversion can be used to provide limits on the maximum conductance that can be added to the model in Figure 6. Within 200 m of the fault trace and from the surface to a depth of 1500 m, the average resistivity is of the order of 20 Ωm, giving rise to a 'background' conductance of 20 S for this 400 m-wide zone. How much additional conductance can be added to this model and still fit the data? Two-dimensional inversions based on this model, but with the *a priori* inclusion of a vertical FZC, show that at most 20 S can be added between the surface and a depth of 400 m, 40 S between depths of 400 and 1000 m, and 80 S between 1000 and 1500 m. Anything greater is incompatible with the measured data, meaning that a FZC of any significance could only be 'hidden' beneath the fault at depths below 1500 m. At these depths, however, a FZC

Table 1. *Comparison of fault-zone conductivity structure (modified after Hoffmann-Rothe 2002). Findings from Carrizo Plain (Unsworth* et al. *1999) are listed in addition to studies discussed in this paper*

	Arava Fault	West Fault	San Andreas Fault		
			Hollister	Parkfield	Carrizo Plain
FZC imaged?	No	Yes	Yes	Yes	Yes
Recent activity	Active	Currently inactive	Creeping	Small earthquakes, creeping	Locked since 1875
Width of FZC	Not resolved	300–400 m	750 m	750 m	<300 m
Depth of FZC	Not resolved	*c.*1.5 km	*c.*8 km	2–5 km	*c.*3 km
Lateral conductance	Not resolved	*c.*5 S	*c.*600 S	*c.*250 S	*c.*20 S

would effectively merge with the prominent conductive half-layer and could not be resolved.

Seismicity provides another means by which to compare the faults under consideration. On a regional scale, the distribution of seismicity along the SAF correlates well with the lateral conductance and spatial extent of the FZC (see Table 1). Hollister exhibits abundant microseismicity, pronounced aseismic creep, a distinct absence of earthquakes with magnitude greater than 5 and the area of highest lateral conductance (of six published MT profiles along ths SAF). Carrizo Plain, in contrast, with its modest lateral conductance, has no creep and suffers large, damaging earthquakes such as the 1857 Fort Tejon quake ($\mathbf{M} = 7.8$; Ellsworth 1990). Parkfield lies in transition between the active, creeping segment to the north and the locked segment to the south, and, consequently, the imaged FZC is intermediate between Hollister and Carrizo Plain with respect to lateral conductance and spatial extent. The West Fault, believed to be currently inactive, is characterized by a low lateral conductance. The few seismic events that have been located at depths between 5 and 18 km cannot clearly be attributed to the Precordilleran Fault System (Belmonte 2002).

The DST within the Arava Valley is nearly devoid of recent seismicity (Ambraseys & Jackson 1998). This conclusion must be tempered, however, by the fact that regional seismic networks have only been in operation since the 1980s, and even then have a detection threshold of magnitude 2.5 (Klinger et al. 2000b). Nevertheless, several magnitude 7 events can be confidently attributed to this segment of the fault within the last millennia and suggest a recurrence interval of about 250 years (Klinger et al. 2000b). Based on this incomplete catalogue, we infer that seismic behaviour along the DST most closely resembles the SAF at Carrizo Plain. The fact that a FZC of modest conductance is imaged at Carrizo Plain while none is seen along the DST may be related to the lower strain rates (2–6 v. 33–39 mm year^{-1}) and cumulative slip (105 v. c. 300 km) attributed to the DST within the Arava Valley (Irwin 1990; Bennett et al. 1999; Argus & Gordon 2001).

Structural and hydrogeological implications. At the SAF, DST and WF, fluids are proposed to explain the upper crustal zones of high conductivity within or surrounding the fault zones. In conceptual fault zone models, large-scale faults can be broken down into three structural, mechanical and hydrological units (see Fig. 9): (i) the protolith, or undeformed country rock; (ii) the damage zone, a broad (up to hundreds of metres) highly permeable zone with an

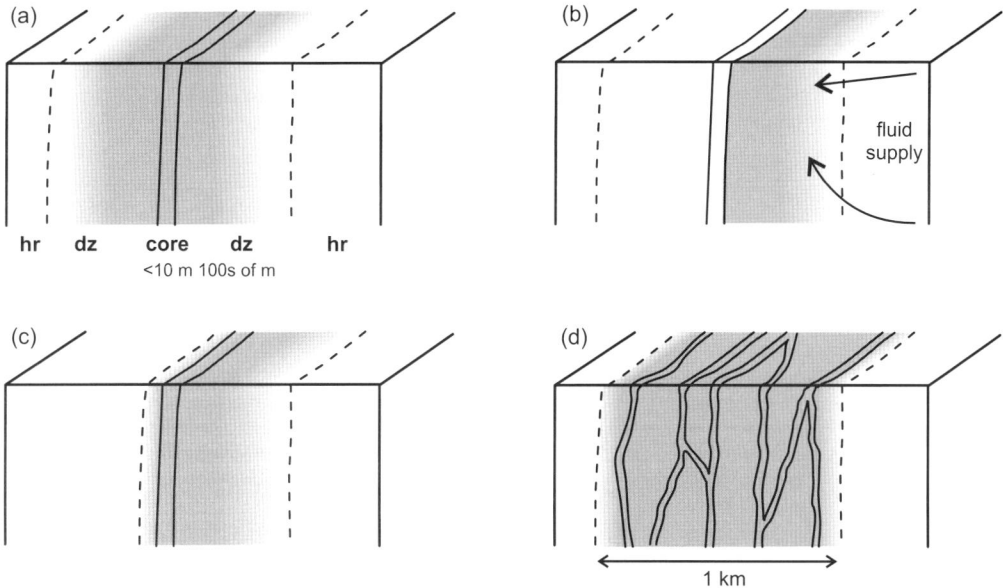

Fig. 9. Interrelation of the conceptual model of large-scale strike-slip fault zones with electrical conductivity images. The FZCs are shaded in grey, arrows indicate fluid flow. hr, undeformed host rock (protolith); dz, damage zone. Dashed lines indicate the gradual transition from damage zone to host rock. See the text for discussion.

increased concentration of fractures, faults and veins; and (iii) the fault core, a narrow (centimetres–metres), often impermeable zone of highly deformed rock where the majority of slip is accommodated (Chester & Logan 1986; Caine *et al.* 1996; Evans *et al.* 1997). The fluid transport properties of faults are primarily controlled by fault geometry and permeability contrasts. Fluid transport in faults can be classified as localized conduit, distributed conduit, localized barrier or a combined conduit barrier (Caine *et al.* 1996).

In what way are the observed conductivity structures related to the above-mentioned hydrogeological classes or architectural units of fault zones? The FZCs imaged on three parallel profiles at Parkfield (Unsworth *et al.* 2000) suggest that, at least locally, they form a continuous along-strike feature and a possible conduit for fluid flows (Gudmundsson 2000). Studies of fault gouge from a depth of 400 m support this possibility, concluding that the SAF near Hollister is permeable to fluids at shallow depths (O'Neil 1984). The MT results from Hollister, where two profiles spaced 20 km apart image similar FZCs, further suggest that the FZC, although varying in strength and extent, may be a continuous feature throughout much of central California.

The sharp resistivity contrast along the western edge of the FZCs (Fig. 5) additionally suggests that the SAF acts as a barrier to cross-fault fluid flow transport (Bedrosian *et al.* 2002). A strong lithological contrast clearly exists across the fault and there is little geochemical evidence of fluid mixing. Irwin & Barnes (1975) noticed an abundance of chloride-rich springs on the eastern (Franciscan) side of the SAF in contrast to the western (Salinian) side. If these saline waters had infiltrated the granites SW of the fault, one would expect significantly lower resistivities than the 200–1000 Ωm observed west of the fault. Whether the fluid barrier is due to the lithology contrast, a mineralized fault seal or a combination of the two cannot be resolved from the conductivity models.

The West Fault paints a similar picture to that of the SAF. The FZC appears continuous along-strike and is sharply terminated along its western edge. A zone of alteration acts as a sink for surficially derived meteoric water. The continuity of the FZC, as seen on the two profiles, suggests a potential along-strike fluid conduit. Springs emanating along the fault trace indicate that cross-fault fluid flow transport is hampered.

The conductivity model along the Arava Fault (Fig. 6) reveals two highly conductive zones cut abruptly by the AF, including the prominent

conductor at a depth of between 1.5 and 3 km west of the fault. These truncated conductors are also imaged on parallel profiles, and can be traced for more than 10 km along-strike (not shown). A lithological change (across the fault) from Phanerozoic sequences (limestones–sandstones) to Precambrian metamorphic basement may be the cause of the deeper conductivity contrast; however, the near-surface conductors, on opposite sides of the fault, are in similar lithology (alluvial fan deposits). This suggests that an impermeable fault-seal may be arresting cross-fault fluid flow transport at shallow depths. In addition, the interconnected fluid-bearing veins, posited to exist within the Precambrian basement (see the section on 'The Dead Sea Transform in the Arava Valley, Jordan'), do not appear linked to the deep conductor west of the fault. Thus, a fault-seal may be restricting fluid transport at greater depths as well.

The distinct lack of a FZC associated with the Arava Fault may imply that a broad damage zone does not exist. According to fault-scaling laws, however, it is reasonable to expect a damage zone up to 1 km wide for the Arava Fault, based on its displacement of approximately 105 km (Scholz 1987; Scholz *et al.* 1993). What could give rise to this apparent discrepancy?

- The alluvial fan deposits at the surface record only a short time span of faulting and hence little displacement has accumulated. In this case, we would expect a FZC at greater depths (below 1 km), where basement rocks are cut. Constrained inversions indicate that a FZC at that depth cannot be ruled out, although we can exclude the existence of a shallow FZC (above 1.5 km).

- Deformation is partitioned among parallel, yet undetected, fault strands. Seismic sections suggest the existence of parallel faults, but neither seismic profiles nor surface observations support a substantial strike-slip component.

- There is no FZC or it is too narrow to be resolved with MT, i.e. it has a width of less than 50 m. This conclusion is supported by the fault zone trapped wave study of Haberland *et al.* (2003) that argues for a low-velocity fault zone of 3–10 m in width (with 40–50% velocity reduction relative to the country rock). Furthermore, the field record of deformation along the fault trace, where accessible, does not exhibit a broad or intensively fractured damage zone.

The last suggestion is our preferred interpretation. It implies that strain is extremely localized along shear planes within a very narrow damage zone. Processes of fault hardening and weakening within fault zones control whether they grow in width or remain localized throughout time (Mitra & Ismat 2001). It is therefore conceivable that the AF has not experienced episodes of complete strength recovery, whereas the WF and the SAF have gone through repeated cycles of healing, strength reloading and subsequent failure, causing the formation of broad zones of structural deformation. Alternatively, where a prominent strength contrast exists across the fault, structural deformation predominantly effects the weaker block and is distributed over a wider range (for example, the SAF at Hollister). Hence, the lack of a FZC at the AF may indicate that there exists little strength contrast between the eastern and western side of the fault. In summary, the spatial extent of the FZC, where present, appears to reflect the size of the zone of intense structural deformation, and as such may be indicative of the degree of strain localization within the fault.

The relationship between seismicity and the imaged FZCs offers some insight into the dynamics of faulting. Along the San Andreas Fault, seismicity is confined to the edge (base) of the imaged FZCs at Hollister (Parkfield) (Fig. 5). This spatial separation between seismicity and the FZC suggests that fluids may inhibit seismicity in the upper crust. Unsworth et al. (2000) proposed that FZCs composed of fluid-saturated fault breccia are too weak to accumulate the shear stresses necessary to undergo brittle failure. This hypothesis, however, requires that fluids within the fault zone are hydraulically conductive (at hydrostatic pressure), as overpressured fluids tend to induce rather than inhibit seismicity. Studies by Townend & Zoback (2000) find this to be the case, and infer that interconnected faults and fractures help to maintain hydrostatic pressure.

Based on the discussion until now, is it possible to relate the imaged conductivity anomalies to fault zone architecture? Magnetotellurics provides one piece of the puzzle, imaging some or all of a fault's damage zone while remaining blind to the narrow fault core. Of the faults studied, the imaged FZCs are best explained by fluid-filled fractures, and thus the MT images also provide indirect information on fluid supply and fault zone permeability. Seismicity, on the other hand, can pinpoint the core of a fault, but provides no information about the damage zone of a fault. It is the spatial relationship between the active shear plane (defined by

seismicity and/or the fault trace) and the imaged FZC that can be used to shed light on the extent, symmetry and hydrogeology of a fault. Figure 9, based on this idea, presents a classification of fault zone architecture within the framework of this study.

- 'Classical' symmetric fault zone (Fig. 9a). Given a symmetric deformation zone, seismicity and the surface trace would align with the centre of the FZC. The FZC thus reflects the entire permeable damage zone. Fracture density mapping would further reveal a symmetric damage profile peaked over the fault core. This scenario is similar to what is seen at Parkfield, although structural confirmation is unavailable due to a lack of outcrop.
- Symmetric damage zone with an impermeable fault seal (Fig. 9b). Seismicity and the surface trace will in this case be aligned with one edge of the FZC, rather than the centre. As in the previous case, however, structural mapping would reveal a symmetric fracture distribution. Owing to an impermeable fault core, only the side of the damage zone penetrated by fluids is imaged with MT. The existence and location of the FZC can thus be directly linked to local fluid supply. This type of image is relevant to a number of the locations discussed in this paper. Along the West Fault, the surface trace locates at the edge of the imaged FZC, a symmetric fracture distribution is found and fluid supply is clearly one-sided (controlled by topography). A similar conductivity image is found at the Arava Fault, where an impermeable seal is also believed to be in place. In contrast, however, the associated layer of high conductivity is not related to a damage zone. The San Andreas Fault at Hollister is, similar to the West Fault, characterized by a fault core aligned with the edge of the FZC. There is further evidence that fluids are supplied from the east side of the fault plane, where the FZC is imaged. Unlike at the WF, however, the absence of structural mapping precludes us from attributing these observations to an impermeable fault seal.
- Asymmetric fault zone (Fig. 9c). An asymmetric fault zone, in which damage is concentrated largely on one side of the fault, is plausible where a pronounced strength contrast exists across the fault. Although not common in the literature, an asymmetric damage zone has been suggested for the Punchbowl Fault in southern California

(Schulz & Evans 2000). Concerning our observations, the spatial relation between seismicity and the FZC would be similar to the previous case shown in Figure 9b. The FZC would correlate with the mechanically weaker, and probably more permeable, lithology. Structural mapping would provide a clear means of differentiating between this and the previous case. The San Andreas Fault at Hollister may fall into this class of fault zone. There is a known lithology contrast between the Salinian granite and the highly deformed Franciscan Complex, within which the FZC is imaged. In the absence of further information, however, it is not possible to attribute the observations at Hollister to this or the previous case.

• Anastomosing fault zone (Fig. 9d). Faulkner *et al.* (2003) expand on the conceptual model of large strike-slip fault zones, drawing attention to fault zones in phyllosilicate-rich material that exhibit broad damage zones composed of anastomosing strands of fault gouge bands (Fig. 9d). The MT image in this case is difficult to predict, and will depend on both the relative permeability within the fault zone and the spacing and lateral extent of gouge bands. It is,

however, possible that such a fault zone could exhibit strong electrical anisotropy, similar to what we observe at the WF/OL in Namibia.

The deeper picture

The discussion has until now focused on the upper crustal structure of active fault zones, such as the SAF and DST, and the currently inactive WF. In order to understand better the dynamics of faulting and its relation to lithospheric driving forces, a whole-fault picture is needed. Figure 10 presents an adaptation of Sibson's (1983) view on the downward continuation of major fault zones, together with a range of models inferred from published regional studies. An open question centres on whether shear zones remain localized in the ductile regime and retain a constant width (Fig. 10a) or broaden with increasing depth (Fig. 10c), consistent with the increased role of dislocation creep at elevated temperatures. Molnar *et al.* (1999) suggest that continuous deformation may occur within the ductile regime, implying a vertical decoupling around the brittle–ductile transition (Fig. 10d).

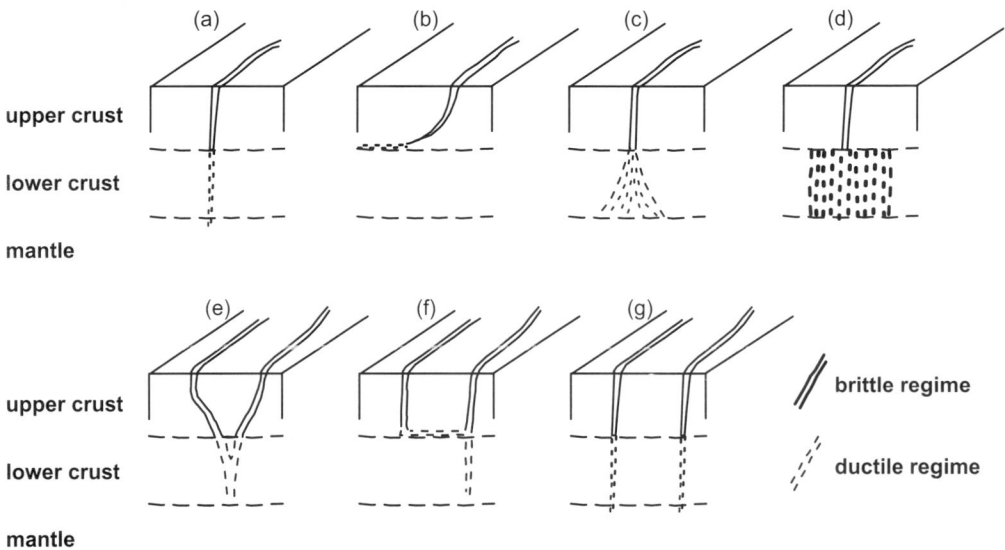

Fig. 10. A range of end-member models for ductile shearing in the lower crust and upper mantle: (**a**) Localized deformation cutting the lithosphere (e.g. Norris & Cooper 2003), (**b**) listric fault soling into lower crustal detachment, (**c**) broadening deformation governed by lower crustal rheology and temperature (e.g. Hanmer 1988), (**d**) continuous instead of localized deformation of lower crust and mantle lithosphere (e.g. Molnar *et al.* 1999). Models discussed for the SAF system of parallel upper crustal fault strands: (**e**) distributed upper crustal strain with a common ductile root (Parsons & Hart, 1999), (**f**) upper crustal faults connected via décollement (Brocher *et al.* 1994) and (**g**) singular through-going faults (Henstock *et al.* 1997).

Seismology provides the only other means to image *in situ* lower crustal shear zones; however, these images are not always in strict accordance with one another. A case in point is the linkage of subparallel strands of the SAF at depth. Studies within and north of the San Francisco Bay area argue for faults merging in the lower crust (Fig. 10e), connected by a mid-crustal detachment (Fig. 10f), and remaining parallel and separate throughout the entire crust (Fig. 10g) (Parsons & Hart 1999; Brocher *et al.* 1994; Henstock *et al.* 1997).

To assess the structure of lower crustal shear zones with MT we face the difficulty of decreasing resolution with increasing depth, as well as the masking effect of upper crustal conductivity. Electromagnetic studies across the Alpine Fault in New Zealand Southern Alps (Ingham & Brown 1998; Wannamaker *et al.* 2002) show quite clearly that observed conductivity anomalies coincide with fault zones in the upper crust. However, the deeper structures of the Alpine Fault are clouded, and the lower crust as imaged appears generally conductive but unstructured. A way to overcome this resolution obstacle is to study exhumed fossil shear zones as analogues for the ductile roots of modern fault zones. This is, of course, a well-known strategy among structural geologists, the investigation the Ivrea–Verbano zone in the Italian southern Alps (Rutter *et al.* 1995) serving as a representative example.

The WF/OL, one of the major fault zones within the Damara Orogen, originated in a collisional setting and has been reactivated several times. Metamorphic facies indicate that the area surrounding the WF/OL was once at mid-crustal depths. The conductivity model in Figure 8b indicates that the WF/OL is not a single, narrow fault but rather a 10 km-wide anisotropic fault zone with high conductivity parallel to strike. In general, MT cannot distinguish between intrinsic and structural anisotropy (see the subsection on 'Two-dimensional modelling and inversion'); however, graphite-bearing marbles in the vicinity suggest that graphite films or lamellae may exist along shear planes within the lineament. Causes for high conductivity in lower-crustal shear zones are discussed in some detail in Wannamaker *et al.* (2002) for the Alpine Fault – a modern analogue of the WF/OL. It is the stability of graphite over long time spans combined with its high conductivity (when connected by shearing) that makes graphite an ideal contrast medium for studying the structure of fault zones at depth (Boerner *et al.* 1996; Wannamaker 2000).

The depth of the modelled anisotropic zone in Figure 8b must be at least 14 km. Considering that the upper crust has already been eroded,

this suggests the fault zone penetrated throughout the entire crust. The conductivity structure of the WF/OL thus supports a model with strain localized within a zone of moderate width in the lower crust and falls somewhere between the models depicted by Figure 10c & d. The width of 10 km of this zone is in the range of observed and imaged mylonitic shear zones (Henstock *et al.* 1997; Herquel *et al.* 1999; Norris & Cooper 2003 and reference therein), and suggests localized strain compared to the discontinuous deformation model in Figure 10d. The anisotropy in electrical conductivity implies that within this zone strain remains concentrated on subvertical anastomosing shear zones of relatively narrow width (semi-continuous deformation of the middle and lower crust).

Sutures and their investigations can thus provide insights into the former ductile lower crust (e.g. Banks *et al.* 1996; Eisel & Haak 1999; Eisel *et al.* 2001; Santos *et al.* 1999, 2002; Almeida *et al.* 2001; Tauber *et al.* 2003). However, to unravel the lower crustal structure of former times requires high-resolution images of the present upper crust, and only with a sufficiently high areal site coverage can we constrain the presence and nature of electrical anisotropy. Often MT studies in basement complexes, such as the Canadian shield (Jones 1999; Jones *et al.* 2001, 2003), focus on the regional conductivity distribution. While these experiments resolve deep lithospheric anomalies, the structure of fossil shear zones in the upper crust remains poorly resolved.

Conclusions

Electrical images from the San Andreas Fault in California, the West Fault in Chile and the Arava Fault in Jordan show both similarities and differences in their subsurface structure of the upper crust.

Typical of many of the young (as compared to fossil) faults investigated in this paper are subvertical regions of high conductivity, associated with the faults' damage zones. Fluids distributed within the fracture network of the damage zone generally explain the observed high conductivity. At least for the SAF, the MT images reveal a correlation between FZC magnitude (lateral conductance) and seismicity. The creeping segment of the SAF at Hollister, associated with the highest fault zone conductance, is characterized by abundant microseismicity and a lack of strong earthquakes. In contrast, the locked segment near Carrizo Plain, where a low-conductance fault zone is imaged, suffers large and damaging earthquakes. Within this framework, the shallow and weak FZCs of the currently inactive WF and the locked SAF at Carrizo Plain may be imaging a closing fracture network across the entire fault zone, i.e. cementation and sealing processes are in operation.

The image of the AF in Jordan appears to be an exception, as it shows a distinct lack of an electrically conducting deformation zone at its centre. The lack of a FZC in the brittle crust is coupled with the fault acting as a barrier to cross-fault fluid transport due to an impermeable fault seal, a lithological contrast across the fault or some combination of the two. At the SAF and WF, the widths of the FZCs are of the order of a few hundred metres, in agreement with independent estimates of damage-zone width. According to Mitra & Ismat (2001), fault zones can grow in width due to repeated cycles of healing and reloading with successive rupturing. In contrast, the narrow, metre-scale damage zone at the AF is the result of a faulting mechanism where strain is extremely localized, apparently over long periods.

Care must be taken when interpreting electrical images of fault zones as conductivity structures are not invariably linked to specific architectural units within the fault. A reasonably comprehensive whole-fault picture is best obtained by combining MT images (fluids in fracture nets) with information from seismology (fault core or active slip plane) and structural geology (fracture density). A FZC may image either the entire damage zone including the fault's core (e.g. the SAF at Parkfield) or some part of it (e.g. the WF in Chile or the SAF at Hollister). The narrow core of a fault is beyond the resolution power of MT. However, taking into account the sealing potential of fault gouge, the position of a fault core can be inferred for at least some of the examined faults.

Fossil shear zones, often exhumed from midcrustal depths, can provide insight into former lower crustal shear zones. In the absence of large volumes of fluids, as suggested for the other fault systems, graphite films, interconnected through shearing, can cause the observed high electrical conductivity. To resolve structural details of fossil shear zones, however, requires high-resolution images of today's upper crust. The fossil Waterberg Fault/Omaruru Lineament in Namibia is imaged as a 10 km-wide and 14 km-deep zone of anisotropic conductivity. The present level of exhumation suggests that the WF/OL penetrated the entire crust as a relatively narrow shear zone.

The examples presented in this paper illustrate the utility of the magnetotelluric method in imaging fault zone structure within the brittle crust. High-resolution images are obtained if data are gathered with a dense site spacing on the order of 100 m, ideally with detailed areal coverage of the target region. The conductivity structures associated with fault zones in the brittle crust reflect a complex interplay between lithology, fluid supply, permeability and fault geometry. This combination can best be untangled for an individual fault zone using additional information from structural geology and seismology.

We thank D. Faulkner, P. Wannamaker and an anonymous reviewer for their critical and constructive comments, which have helped to clarify this manuscript. The existence of the Geophysical Instrument Pool Potsdam and Electromagnetic Consortium for Studies of the Continents made possible the high-resolution studies discussed herein. One of the authors (P. A. Bedrosian) was funded by the Alexander von Humboldt Foundation.

References

ALLEN, C. 1968. The tectonic environments of seismically active and inactive areas along the San Andreas fault system. *In*: DICKINSON, W. & GRANTZ, A. (eds) *Proceedings of Conference on Geologic Problems of the San Andreas Fault System.* 70–82, Stanford University Publications, Standord.

ALMEIDA, E., POUS, J. *ET AL.* 2001. Electromagnetic imaging of transpressional tectonics in SW Iberia. *Geophysical Research Letters*, **28**, 439–442.

AMBRASEYS, N.N. & JACKSON, J.A. 1998. Faulting associated with historical and recent earthquakes in the Eastern Meaditerranean region. *Geophysical Journal International*, **133**, 390–406.

ARCHIE, G.E., 1942. The electrical resistivity log as an aid in determining some reservoir characteristics. *Transactions of the American Institute of Mining, Metallurgical and Petroleum Engeneers*, **146**, 54–62.

ARGUS, D. & GORDON, R. 2001. Present day motion across the Coast Ranges and San Andreas fault system in central California. *Geological Society of America Bulletin*, **113**, 1580–1592.

BANKS, R.J., LIVELYBROOKS, D., JONES, P., & LONGSTAFF, R. 1996. Causes of high crustal conductivity beneath the Iapetus Suture Zone in Great Britain. *Geophysical Journal International*, **124**, 433–455.

BEDROSIAN, P.A., UNSWORTH, M.J. & EGBERT, G. 2002. Magnetotelluric imaging of the creeping segment of the San Andreas Fault near Hollister. *Geophysical Research Letters*, **29**, 10.1029/2001GL014119.

BEDROSIAN, P., UNSWORTH, M., EGBERT, G. & THURBER, C. 2004. Geophysical images of the creeping segment of the San Andreas Fault: Implications for the role of crustal fluids in the earthquake process, *Tectonophysics*, **385**, 137–158.

BELMONTE, A. 2002. *Crustal seismicity, structure and rheology of the upper plate between the Pre-Cordillere and the volcanic arc in northern Chile (22°S–24°S)*. Ph.D. thesis, Free University, Berlin. http://www.diss.fu-berlin.de/2002/202/

BEN-ZION, Y. & SAMMIS, C.G. 2003. Characterization of fault zones. *Pure and Applied Geophysics*, **160**, 677–715.

BENNETT, R., DAVIS, J. & WERNICKE, B. 1999. Present-day pattern of Cordilleran deformation in the western United States. *Geology*, **27**, 371–374.

BOERNER, D., KURTZ, R. & CRAVEN, J., 1996. Electrical conductivity and paleo-proterozoic foredeeps. *Journal of Geophysical Research*, **101**, 13 775–13 791.

BROCHER, T.M., MCCARTHY, J. *ET AL.* 1994. Seismic evidence for a lower-crustal detachment beneath San Francisco Bay, California. *Science*, **265**, 1436–1439.

CAINE, J.S., EVANS, J.P. & FORSTER, C.P. 1996. Fault zone architecture and permeability structure. *Geology*, **24**, 1125–1128.

CATCHINGS, R., RYMER, M., GOLDMAN, M., HOLE, J., HUGGINS, R. & LIPPUS, C. 2002. High-resolution seismic velocities and shallow structure of the San Andreas fault zone at Middle Mountain, Parkfield, California. *Bulletin of the Seismological Society of America*, **92**, 2493–2503.

CHESTER, F.M. & LOGAN, J.M. 1986. Implications for mechanical properties of brittle faults from observations of the Punchbowl fault zone, California. *Pure and Applied Geophysics*, **124**, 79–106.

CHESTER, F.M., EVANS, J.P. & BIEGEL, R.L. 1993. Internal structure and weakening mechanisms of the San Andreas Fault. *Journal of Geophysical Research*, **98**, 771–786.

DALY, M.C. 1986. Crustal shear zones and thrust belts: their geometry and continuity in Central Africa. *Philosophical Transactions of the Royal Society of London*, **317(A)**, 111–128.

DALY, M. 1989. Rift basin evolution in africa: the influence of reactivated steep basement shear zones. *In*: COOPER, G. (ed.) *Inversion Tectonics*, Geological Society, London, Special Publications, **44**, 309–334.

DOYEN, P. 1988. Permeability, conductivity, and pore geometry of sandstone. *Journal of Geophysical Research*, **93**, 7729–7740.

EBERHART-PHILLIPS, D., STANLEY, W.D., RODRIGUEZ, B.D., LUTTER, & W.J. 1995. Surface seismic and electrical methods to detect fluids related to faulting. *Journal of Geophysical Research*, **100**, 12 919–12 936.

EISEL, M. & HAAK, V. 1999. Macro-anisotropy of the electrical conductivity of the crust: a magnetotelluric study of the German Continental Deep Drilling Site (KTB). *Geophysical Journal International*, **136**, 109–122.

EISEL, M., HAAK, V., PEK, J. & CERV, V. 2001. A magnetotelluric profile across the KTB surrounding: 2D and 3D modelling results. *Journal of Geophysical Research*, **106**, 16 061–16 079.

ELECTROMAGNETIC RESEARCH GROUP FOR THE ACTIVE FAULT. 1982. Low electrical resistivity along an active fault. *Journal of Geomagnetism and Geoelectricity*, **34**, 103–127.

ELEKTB GROUP. 1997. KTB and the electrical conductivity of the crust. *Journal of Geophysical Research*, **102**, 18 289–18 305.

ELLSWORTH, W.L. 1990. Earthquake history, 1769–1989. *In*: WALLACE, R. (ed.) *The San Andreas Fault System, California*. US Geological Survey, Professional Paper, **1515**, 153–188.

EVANS, J.P., FORSTER, C.B. & GODDARD, J.V. 1997. Permeability of fault-related rocks and implications for hydraulic structure of fault zones. *Journal of Structural Geology*, **19**, 1393–1404.

FAULKNER, D.R., LEWIS, A.C. & RUTTER, E.H. 2003. On the internal structure and mechanism of large strike-slip fault zones: field observations of the Caboneras fault in southeasten Spain. *Tectonophysics*, **367**, 235–251.

FREUND, R., GARFUNKEL, Z., ZAK, I., GOLDBERG, M., WEISSBROD, T. & DERIN, B. 1970. The shear along the Dead Sea rift. *Philosophical Transactions of the Royal Society of London*, **267**, 117–126.

GARFUNKEL, Z. & AVRAHAM, Z.-B. 1996. The structure of the Dead Sea basin. *Tectonophysics*, **266**, 155–176.

GUDMUNDSSON, A. 2000. Active fault zones and groundwater flow. *Geophysical Research Letters*, **27**, 2993–2996.

GUÉGUEN, Y. & PALCIAUSKAS, V. 1994. *Introduction to the Physics of Rocks*. Princeton University Press, Chichester, West Sussex.

HAAK, V. & HUTTON, V.R.S. 1986. Electrical resistivity in continental lower crust. *In*: DAWSON, J.B. (ed.) *The Nature of the Lower Continental Crust*. Geological Society, London, Special Publications, **24**, 35–49.

HABERLAND, C., AGNON, A. *ET AL.* 2003. Modeling of seismic guided waves at the Dead Sea Transform. *Journal of Geophysical Research*, 10.1029/2002JB002309.

HANMER, S. 1988. Great Slave Lake shear zone, Canadian Shield: reconstructed vertical profile of a crustal-scale fault zone. *Tectonophysics*, **149**, 245–264.

HENSTOCK, T., LEVANDER, A. & HOLE, J. 1997. Deformation in the lower crust of the San Andreas Fault system in northern California. *Science*, **278**, 650–653.

HERQUEL, G., TAPPONNIER, P., WITTLINGER, G., MEI, J. & DANIAN, S. 1999. Teleseismic shear wave splitting and lithospheric anisotropy beneath and across the Altyn Tagh fault. *Geophysical Research Letters*, **26**, 3225–3228.

HICKMAN, S.H., SIBSON, R. & BRUHN, R. 1995. Mechanical involvment of fluids in faulting. *Journal of Geophysical Research*, **100**, 12 831–12 840.

HOFFMANN-ROTHE, A. 2002. *Combined structural and magnetotelluric investigation across the West Fault zone in the Andes of northern Chile*. Ph.D. thesis, University of Potsdam. http://www.gfz-potsdam.de/bib/pub/str0212/0212.htm.

HOFFMANN-ROTHE, A., RITTER, O. & JANSSEN, C. 2004. Correlation of electrical conductivity and structural damage at a major strike-slip fault in northern Chile. *Journal of Geophysical Research*, 10.1029/2004JB003030.

HOLDSWORTH, R.E., HAND, M., MILLER, J.A. & BUICK, I.S. 2001. Continental reactivation and reworking: an introduction. *In*: MILLER, J.A., HOLDSWORTH, R.E., BUICK, I.S. & HAND, M. (eds) *Continental Reactivation and Reworking.* The Geological Society, London, Special Publications, **184** 1–12.

HOLZFÖRSTER, F., STOLLHOFEN, H. & STANISTREET, I.G. 1999. Lithostratigraphy and depositional environments in the Waterberg–Erongo area, central Namibia, and correlation with the main Karoo Basin, South Africa. *Journal of African Earth Sciences*, **29**, 105–123.

INGHAM, M. & BROWN, C. 1998. A magnetotelluric study of the Alpine Fault, New Zealand. *Geophysical Journal International*, **135**, 542–552.

IRWIN, W. 1990. Geology and plate-tectonic development, *In*: WALLACE, R. *The San Andreas Fault System, California.* US Geological Survey Professional Paper, **1515**, 61–80.

IRWIN, W. & BARNES, I. 1975. Effect of geologic structure and metamorphic fluids on seismic behavior of the San Andreas fault system in central and northern California. *Geology*, **3**, 713–716.

JANSSEN, C., HOFFMANN-ROTHE, A., TAUBER, S. & WILKE, H. 2002. Internal structure of the Precordilleran fault system (Chile) – insights from structural and geophysical observations. *Journal of Structural Geology*, **24**, 123–143.

JANSSEN, C., LÜDERS, V. & HOFFMANN-ROTHE, A. 2004. Contrasting styles of fluid–rock interaction within the West Fissure Zone in northern Chile. *In*: ALSOP, G.I., HOLDSWORTH, R.E., MCCAFFREY, K.J.W. & HAND, M. (eds) *Flow Processes in Faults and Shear Zones.* Geological Society, London, Special Publications, **224**, 141–160.

JÖDICKE, H., KRUHL, J.H., BALLHAUS, C., GIESE, P. & UNTIEDT, J. 2004. Syngenetic, thin graphite-rich horizons in lower crustal rocks from the Serre San Bruno, Calabria (Italy), and implications for the nature of high-conducting deep crustal layers. *Physics of the Earth and Planetary Interiors*, **141**, 37–58.

JONES, A.G. 1999. Imaging the continental upper mantle using electromagnetic methods. *Lithos*, **48**, 57–80.

JONES, A.G., FERGUSON, I.J., CHAVE, A., EVANS, R. & MCNEICE, G. 2001. Electrical lithosphere of the Slave craton. *Geology*, **29**, 423–426.

JONES, A.G., LEZAETA, P., FERGUSON, I.J., CHAVE, A.D., EVANS, R.L., GARCIA, X. & SPRATT, J. 2003. The electrical structure of the Slave craton. *Lithos*, **71**, 505–527.

KLINGER, Y., AVOUAC, J.P., KARAKI, N.A., DORBATH, L., BOURLES, D. & REYSS, J.L. 2000a. Slip rate on the Dead Sea transform fault in northern Araba valley (Jordan). *Geophysical Journal International*, **142**, 755–768.

KLINGER, Y., AVOUAC, J.P., KARAKI, N.A., DORBATH, L. & TISNERAT, N. 2000b. Seismic behaviour of the Dead Sea fault along Araba

valley, Jordan. *Geophysical Journal International*, **142**, 769–782.

LEDO, J., QUERALT, P., MARTI, A. & JONES, A. 2002. Two-dimensional interpretation of three-dimensional magnetotelluric data; an example of limitations and resolution. *Geophysical Journal International*, **150**, 127–139.

LI, S., UNSWORTH, M.J., BOOKER, J., WEI, W., TAN, H. & JONES, A. 2003. Partial melt or aqueous fluid in the mid-crust of southern Tibet? Constraints from INDEPTH magnetotelluric data. *Geophysical Journal International*, **153**, 289–304.

LI, Y.G., VIDALE, J. & COCHRAN, E. 2004. Low-velocity damaged structure of the San Andreas fault at Parkfield from fault-zone trapped waves. *Geophysical Research Letters*, **31**, 10.1029/2003GL019044.

LIU, J., WALTER, J. & WEBER, K. 2002. Fluid-enhanced low-temperature plasticity of calcite marble: Microstructures and mechanisms. *Geology*, **30**, 787–790.

LUQUE, F.J., PASTERIS, J.D., WOPENKA, B., RODAS, M. & BARRENECHA, J.F. 1998. Natural fluid-deposited graphite: mineralogical characteristics and mechanisms of formation. *American Journal of Science*, **298**, 471–498.

MCNEICE, G.W. & JONES, A.G. 2001. Multisite, multifrequency tensor decomposition of magnetotelluric data. *Geophysics*, **66**, 158–173.

MILLER, R. 1983. *The Pan-African Damara Orogen of the South West Africa/Namibia.* Geological Society of South Africa, Special Publications, **11**, 431–515.

MITRA, G. & ISMAT, Z. 2001. Microfracturing associated with reactivated fault zones and shear zones: what can it tell us about deformation history? *In*: HOLDSWORTH, R.E., STRACHAN, R.A., MAGLOUGHLIN, J.F. & KNIPE, R.J. (eds) *The Nature and Tectonic Significance of Fault Zone Weakening.* Geological Society, London, Special Publications, **186**, 113–140.

MOLNAR, P., ANDERSON, H.J. *ET AL.* 1999. Continuous deformation versus faulting through the continental lithosphere of New Zealand. *Science*, **286**, 516–519.

NORRIS, R.J. & COOPER, A.F. 2003. Very high strains recorded in mylonites along the Alpine Fault, New Zealand: implications for the deep structure of plate boundary faults. *Journal of Structural Geology*, **25**, 2141–2157.

O'NEIL, J. 1984. Water-rock interactions in fault gouge. *Pure and Applied Geophysics*, **122**, 440–446.

PARK, S.K., THOMPSON, S.C., RYBIN, A., BATALEV, V. & BIELINSKI, R. 2003. Structural constraints in neotectonic studies of thrust faults from the magnetotelluric method. *Tectonics*, **22**, 10.1029/2001TC001318.

PARSONS, T. & HART, P.E. 1999. Dipping San Andreas and Hayward faults revealed beneath San Francisco Bay, California. *Geology*, **27**, 839–842.

PEK, J. & VERNER, T. 1997. Finite-difference modelling of magnetotelluric fields in two-dimensional anisotropic media. *Geophysical Journal International*, **132**, 535–548.

RAAB, M.J., BROWN, R.W., GALLAGHER, K., CARTER, A. & WEBER, K. 2002. Late Cretaceous reactivation of major crustal shear zones in northern Namibia: constraints from apatite fission track analysis. *Tectonophysics*, **349**, 75–92.

RITTER, O., HAAK, V., RATH, V., STEIN, E. & STILLER, M. 1999. Very high electrical conductivity beneath the Münchberg Gneiss area in southern Germany: Implications for horizontal transport along shear planes. *Geophysical Journal International*, **139**, 161–170.

RITTER, O., RYBERG, T., WECKMANN, U., HOFFMANN-ROTHE, A., ABUELADAS, A., GARFUNKEL, Z. & DESERT RESEARCH GROUP. 2003a Geophysical images of the Dead Sea Transform in Jordan reveal an impermeable barrier for the fluid flow. *Geophysical Research Letters*, **30**, 1741, doi:10.1029/2003GL017541.

RITTER, O., WECKMANN, U., VIETOR, T. & HAAK, V. 2003b. A magnetotelluric study of the Damara Belt in Namibia 1. regional scale conductivity anomalies. *Physics of the Earth and Planetary Interiors*, **138**, 71–90, doi:10.1016/S0031–9201(03)00078–5.

RODI, W. & MACKIE, R.L. 2001. Nonlinear conjugate gradients algorithm for 2D magnetotelluric inversion. *Geophysics*, **66**, 174–187.

RÜMPKER, G., RYBERG, T., BOCK, G. & DESERT SEISMOLOGY GROUP. 2003. Boundary-layer mantle flow under the Dead Sea transform fault inferred from seismic anisotropy. *Nature*, **425**, 497–501.

RUTTER, E., BRODIE, K. & EVANS, P. 1995. Structural geometry, lower crustal magmatic underplating and lithospheric stretching in the Ivrea–Verbano zone, northern Italy. *Journal of Structural Geology*, **15**, 647–662.

SANTOS, F., POUS, J., ALMEIDA, E., QUERALT, P., MARCUELLO, A., MATIAS, H. & VICTOR, L. 1999. Magnetotelluric survey of the electrical conductivity of the crust across the Ossa Morena zone and South Portugese zone suture. *Tectonophysics*, **313**, 449–462.

SANTOS, F.A.M., MATEUS, A., ALMEIDA, E.P., POUS, J. & MENDES-VICTOR, L.A. 2002. Are some of the deep crustal conductive features found in SW Iberia caused by graphite? *Earth and Planetary Science Letters*, **201**, 353–367.

SCHOLZ, C.H. 1987. Wear and gouge formation in brittle faulting. *Geology*, **15**, 493–495.

SCHOLZ, C.H., DAWERS, N.H., YU, J.-Z., ANDERS, M.H. & COWIE, P.A. 1993. Fault growth and fault scaling laws: preliminary results. *Journal of Geophysical Research*, **98**, 21 951–21 961.

SCHULZ, S.E. & EVANS, J.P. 2000. Mesoscopic structure of the Punchbowl fault, southern California and the geologic and geophysical structure of active strike-slip faults, *Journal of Structural Geology*. **22**, 913–930.

SIBSON, R.H. 1983. Continental fault structure and the shallow earthquake source. *Journal of the Geological Society, London*, **140**, 741–767.

SMITH, J. 1995. Understanding telluric distortion matrices. *Geophysical Journal International*, **122**, 219–226.

SWIFT, C. 1967. *A magnetotelluric investigation of an electrical conductivity anomaly in the southwestern United States*. PhD thesis, M.I.T., Cambridge, MA.

TAUBER, S., BANKS, R., RITTER, O. & WECKMANN, U. 2003. A high-resolution magnetotelluric survey of the Iapetus Suture Zone in southwest Scotland. *Geophysical Journal International*, **153**, 548–568.

THURBER, C., ROECKER, S., ELLSWORTH, W. CHEN, Y., LUTTER, W. & SESSIONS, R. 1997. Two-dimensional seismic image of the San Andreas fault in the northern Gabilan Range, central California: evidence for fluids in the fault zone. *Geophysical Research Letters*, **24**, 1591–1594.

THURBER, C.H., ROECKER, S., ROBERTS, K., GOLD, M., POWELL, L. & RITTGER, K. 2003. Earthquake locations and three-dimensional fault zone structure along the creeping section of the San Andreas fault near Parkfield, CA: preparing for SAFOD. *Geophysical Research Letters*, **30**, 10.1029/2002GL016004.

TOWNEND, J. ZOBACK, M.D. 2000. How faulting keeps the crust strong. *Geology*, **28**, 399–402.

UNSWORTH, M. BEDROSIAN, P.A. 2004. Electrical resistivity structure at the SAFOD site from magnetotelluric exploration. *Geophysical Research Letters*, **31**, 10.1029/2003GL019405.

UNSWORTH, M.J., BEDROSIAN, P., EISEL, M., EGBERT, G.D. & SIRIPUNVARAPORN, W. 2000. Along strike variations in the electrical structure of the San Andreas fault at Parkfield, California. *Geophysical Research Letters*, **27**, 3021–3024.

UNSWORTH, M.J., EGBERT, G.D. & BOOKER, J.R. 1999. High resolution electromagnetic imaging of the San Andreas fault in central California. *Journal of Geophysical Research*, **104**, 1131–1150.

UNSWORTH, M.J., MALIN, P.E., EGBERT, G.D. & BOOKER, J.R. 1997. Internal structure of the San Andreas fault at Parkfield, California. *Geology*, **25**, 359–362.

WANNAMAKER, P.E. 2000. Comment on The petrologic case for a dry lower crust by Bruce W.D. Yardley and John W. Valley. *Journal of Geophysical Research*, **105**, 6057–6064; 10.1029/1999JB900324.

WANNAMAKER, P.E., JIRACEK, G.R., TODT, J.A., CALDWELL, T.G., GONZALES, V.M., MCKNIGHT, J.D. & PORTER, A.D. 2002. Fluid generation and pathways beneath an active compressional orogen, the New Zealand Alps, inferred from magnetotelluric data. *Journal of Geophysical Research*, **107**, 10.1029/2001JB000186.

WDOWINSKI, S., BOCK, Y. *ET AL.* 2004. GPS measurements of current crustalmovements along the Dead Sea Fault. *Journal of Geophysical Research*, **109**, 10.1029/2003JB002640.

WECKMANN, U., RITTER, O. & HAAK, V. 2003a. A Magnetotelluric study of the Damara Belt in Namibia 2. internal structure of the Waterberg Fault/Omaruru Lineament. *Physics of the Earth *and Planetary Interiors*, **138**, 91–112, doi:10.1016/S0031–9201(03)00079–7.

WECKMANN, U., RITTER, O. & HAAK, V. 2003*b*. Images of the magnetotelluric apparent resistivity tensor. *Geophysical Journal International*, **155**, 456–468.

WIESE, H. 1962. Geomagnetische Tiefentellurik Teil II: die Streichrichtung der Untergrundstrukturen des elektrischen Widerstandes, erschlossen aus geomagnetischen Variationen. *Geofisica pura e applicata*, **52**, 83–103.

High-temperature and pressure seismic properties of a lower crustal prograde shear zone from the Kohistan Arc, Pakistan

L. BURLINI[1], L. ARBARET[1,2], G. ZEILINGER[3] & J.-P. BURG[1]

[1]Institute of Geology, ETH-Z, Sonneggstrasse, 5 CH-8092 Zurich, Switzerland
(e-mail: Burlini@erdw.ethz.ch; Jean-pierre.Burg@erdw.ethz.ch)
[2]Present address: Department Sciences de la Terre, Rue de Chartres,
45067 Orléans cédex 2, France (e-mail: Laurent.Arbaret@univ-orleans.fr)
[3]Present address: Institute of Geological Sciences, University of Bern, Baltzerstrasse 1–3,
CH-3012 Bern, Switzerland (e-mail: zeilinger@geo.unibe.ch)

Abstract: Anastomosing shear zones exposed in the lower crust of the Kohistan palaeo-island arc in Pakistan have parageneses indicating increasing pressure during deformation. Therefore, they represent a rare example of strain localization during crustal thickening. We investigated the seismic properties of a sheared gabbro with constant bulk chemistry across one of these shear zones. The compressional wave velocity was measured at confining pressures of up to 0.5 GPa and temperatures of up to 600 °C. The density, the average V_p (compressional P-wave velocity) and the acoustic impedance at room temperature increase from the undeformed protolith through the strain gradient to the intensely sheared mylonite. The seismic anisotropy is largest for the strain gradient.

The V_p dependence of the velocities on temperature is much higher in the mylonite than in the protolith. The acoustic impedance contrast between protolith and mylonite is high enough to generate seismic reflections at low temperatures, but not at high temperatures. This suggests that the seismic reflectivity may also depend on the temperature. Consequently, ductile shear zones can be detected in reflection profiles of crusts with low geothermal gradients, and may be transparent in crusts with high geothermal gradients.

One important question for the geological interpretation of seismic reflection profiles is the geological nature of reflectors, knowing that seismic reflectivity is produced by acoustic impedance contrasts (e.g. Sheriff & Geldart 1995). Hypotheses formulated in the literature refer in most cases to lithological changes or to the presence of partial melts or pressurized fluids (e.g. Rudnick & Fountain 1995). Particular attention was paid to ductile shear zones (e.g. Fountain *et al.* 1994). Christensen & Szymanski (1988) investigated the Brevard shear zone and, comparing their laboratory results with reflection seismic profiling, concluded that shear zones can enhance reflectivity because they bring different lithologies in contact and also because of their intense foliation and lineation. On the latter point, it has also been argued that the seismic anisotropy variation can even produce reflections from the same lithology (e.g. Rey *et al.* 1994; Kazanehdari *et al.* 1998). In all documented cases, seismic anisotropy increased with the amount of strain induced by the coeval development and strengthening of the lattice preferred orientation (LPO).

Most of previous work investigated seismic properties at high pressures and room temperature, with the exception of Kern & Wenk (1990) who investigated a mylonite formed under retrogressive greenschist conditions. These authors found that the pressure and temperature derivatives of the velocities are similar from the protolith to the mylonite, and are almost identical in the three orthogonal directions investigated. This implies that the seismic reflectivity between the protolith and the upper crustal mylonite does not significantly change with pressure and temperature.

In order to verify this assertion for lower crustal mylonites, we investigated the pressure and temperature dependence of the seismic property across a shear zone that developed in

From: BRUHN, D. & BURLINI, L. (eds) 2005. *High-Strain Zones: Structure and Physical Properties.*
Geological Society, London, Special Publications, **245**, 187–202.
0305-8719/05/$15.00   © The Geological Society of London 2005.

'prograde' metamorphic conditions in the lower crustal section of the Kohistan Arc (Pakistan). We selected a shear zone across which the bulk rock composition is constant but the mylonite paragenesis reflects pressure conditions higher than those of the gabbroic protolith.

In a collision zone, crustal thickening can be associated at depth with strain localization into anastomosing thrust shear zones. These prograde shear zones are often obliterated by later thermal and tectonic events during exhumation. In the Kohistan Arc, anastomosing shear zones form a pervasive pattern in the so-called Sarangar gabbro (Treloar *et al.* 1990; Arbaret *et al.* 2000; Burg *et al.* 2005) (Fig. 1). Parageneses indicating increasing pressure during shear deformation suggest crustal thickening (Zeilinger 2002; Arbaret & Burg 2003). The gabbro equilibrated at about 700 °C and 0.7 Gpa, and underwent shearing at slightly higher pressures (about 1 GPa, 700 °C, Fig. 2). Ductile shearing began during the magmatic stage (the U/Pb age on zircon from the gabbro is 98 ± 0.4; Schaltegger *et al.* 2002), and propagated until at least 83 Ma ago (Ar–Ar age on amphiboles; Treloar *et al.* 1989). Miller & Christensen (1994) began the seismic characterization of samples collected from the same area, but focused on the relationship between bulk chemistry, compressional and shear wave velocities, and density up to 1 GPa confining pressure but only at room temperature. They did not document the shear zones and their seismic anisotropy.

In this work, we measured density at room conditions and seismic properties up to 0.5 GPa

and 600 °C on the protolith gabbro, on the gabbro mylonite as well as on the sheared gabbro in the strain gradient between the protolith and the mylonite. The protolith gabbro and the mylonite are the end members for strain and dynamic recrystallization. We report the major element bulk chemistry to document the constant bulk composition from the protolith to the mylonite. We also describe the microstructure and the mineral chemistry for the complete characterization of the samples used for the seismic investigation. These data were used for the determination of the pressure and temperature of equilibration of both the protolith and the mylonite, and indicate that later retrogression did not occur in these samples.

Sample description

The studied shear zone was collected *c.* 2 km east of Patan along the jeep-road Patan–Chor Nala close to Shalkanabad at GPS point E 73°01′37″ N 35°07′01″ (Fig. 1). This gently NE-dipping shear zone is *c.* 50 cm thick, and contains stretched plagioclase veins and numerous plagioclase porphyroclasts. Sense of shear is top-to-the-SW. Both walls display continuous strain gradients from the 'Sarangar' Gabbro protolith to the central mylonite. The hanging-wall gradient was sampled in one single piece (sample 90/02/01) for this work, and the protolith, the gradient zone and the mylonite were cored in the laboratory. From each position (circles in Fig. 3) three mutually orthogonal cores (15 mm in diameter and 30 mm in length)

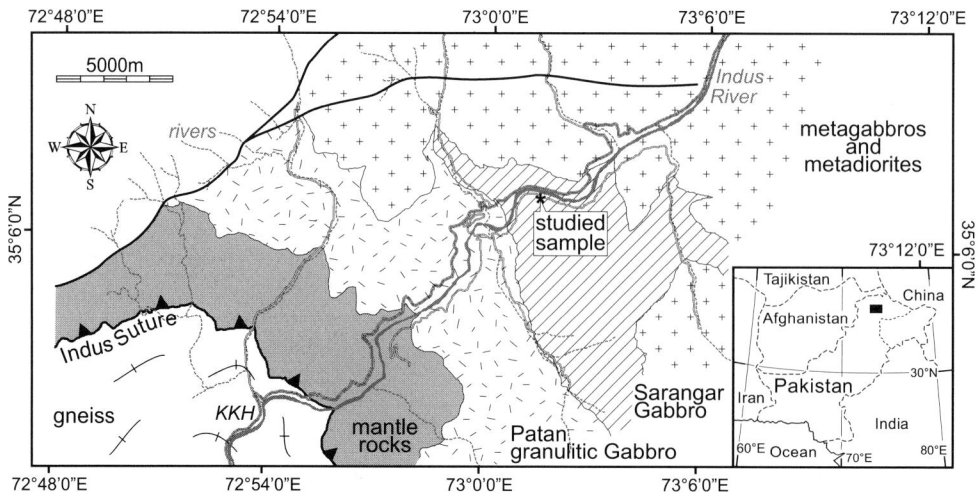

Fig. 1. Location map of the studied shear-zone sample 90/02/01.

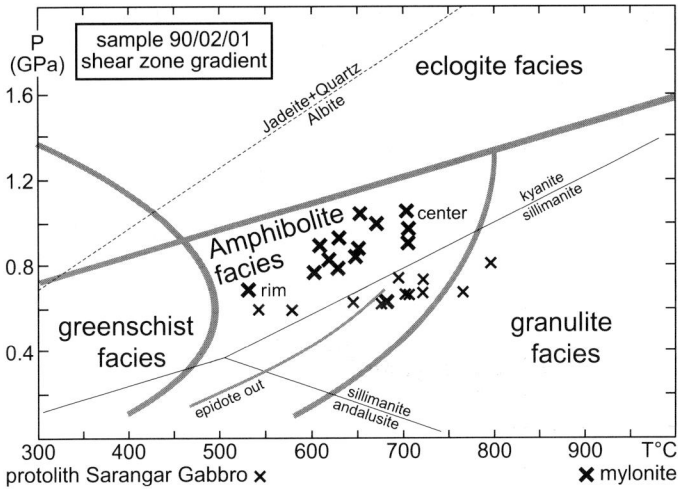

Fig. 2. Thermobarometry from garnet–plagioclase–clinopyroxene–quartz assemblages in the studied mylonite and protolith sample portions (from Zeilinger 2002). The mylonite recorded pressures c. 0.25 GPa higher than the host Sarangar gabbro. Centre and rim refer to positions of analyses in mineral grains.

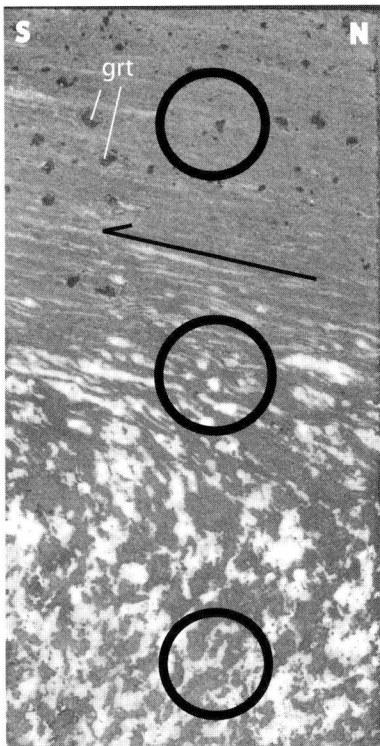

Fig. 3. Sample 90/02/01. Black circles: drilled cores. Note abundant garnet (grt) in the mylonite. Long edge of the picture is about 10 cm long.

were drilled in order to determine the directional dependence of the seismic wave velocities. The foliation and the lineation of the mylonite make the reference frame, so that X cores are parallel to the mineral/stretching lineation, Z cores are normal to the mylonitic foliation and the Y cores are normal to the lineation and parallel to the foliation. As the gradient foliation is curved, the Z core in the gradient is actually oblique to the foliation.

Previous petrological descriptions of the Sarangar gabbro agree that lenses surrounded by anastomosing shear zones preserved pyroxene-bearing metagabbro (Treloar *et al.* 1990) called pyroxene granulite by Yoshino *et al.* (1998). The dominant mineral composition includes amphibole, pyroxene, plagioclase, rutile, ilmenite, epidote and quartz. Grain size, mineral assemblages and orientations of minerals change rapidly along sections a few centimetres in length from undeformed gabbro to mylonite.

It is worth noting that the bulk composition does not change from the protolith to the mylonite (Table 1), which suggests that mylonitization occurred without volume loss and that deformed rocks behaved as a chemically closed system. Arbaret & Burg (2003) reached the same conclusion on a neighbouring shear zones.

The undeformed gabbro

The medium-grained protolith gabbro is rather homogeneous and has a weak magmatic fabric (Fig. 4a & b). The magmatic paragenesis is

Table 1. *Major-element chemical composition of the Sarangar gabbro*

	Protolith (wt%)	Mylonite (wt%)
SiO_2	52.22	53.45
TiO_2	0.72	0.78
Al_2O_3	17.63	17.44
Fe_2O_3	9.62	9.28
MnO	0.17	0.15
MgO	5.28	4.62
CaO	9.69	9.45
Na_2O	2.52	2.87
K_2O	0.40	0.30
P_2O_5	0.12	0.11
Loss on ignition	0.90	0.92
Total	99.29	99.39

comprised of clinopyroxene, orthopyroxene, plagioclase and amphibole, with accessory oxides, ilmenite, rutile, epidote, zircon and quartz. This magmatic assemblage indicates crystallization under granulite facies conditions (Yamamoto 1993). The hypidioblastic orthopyroxene is progressively replaced by clinopyroxene. The sporadic occurrence of poikiloblastic garnet may be due to the reaction: orthopyroxene + plagioclase = garnet + quartz + clinopyroxene (Yoshino *et al.* 1998).

The granulite facies assemblage is overprinted by the amphibolite facies assemblage: hornblende + plagioclase + quartz. Interstitial brown–green hornblende, associated with quartz, occurs as rims between pyroxene and plagioclase (Fig. 4b). Yoshino *et al.* (1998) concluded that hornblende and quartz are probably produced by the reaction: pyroxene + plagioclase + H_2O = hornblende + quartz (Fig. 4c). Plagioclase shows an increase of symplectitic structures with pyroxene closer to the mylonite (Fig. 4c & d).

In the protolith no shape preferred orientation (SPO) is visible macroscopically and microscopically, no LPO of plagioclase and pyroxene were detected using the compensator under cross-polars. On the other hand, the newly grown amphibole in the gradient zone shows a crystallographic preferred orientation (CPO) when observed with the compensator (qualitative observation). Moreover, a strong SPO of grains and grain cluster (glomeroblastic structure) can be recognized (Fig. 4c & d).

The mylonite

No pyroxene was found in the mylonite. The amphibolite facies assemblage is: hornblende +

plagioclase + garnet ± rutile ± quartz. Garnet is abundant, showing poikiloblastic growth (Fig. 4e & f). Some garnets have helicitic rutile inclusions, suggesting synkinematic growth. The amphibole grains underline the fabric and are cut by small quartz veins. Rounded grains of epidote eventually contain quartz inclusions. Ilmenite is sparsely preserved as inclusions in brownish hornblende, pointing to the retrograde reaction: ilmenite + plagioclase + quartz = rutile + garnet. Plagioclase and quartz aggregates constitute macroscopic porphyroclasts (*c.* 2 mm). Chlorite and muscovite overgrowing hornblende are late, greenschist facies phases.

Both plagioclase and amphibole show a SPO and, using the compensator under cross-polars, there is an evident LPO. Moreover the (110) cleavage planes of amphibole almost systematically lies parallel to the lineation in the X–Z section (Fig. 4e) suggesting again a strong CPO.

Mineral chemistry

Plagioclase composition in the protolith Sarangar gabbro is An_{55-70} (Table 2, Fig. 5). Only few grains show a lower anorthite component, around An_{25}. In the mylonite, plagioclase has two different compositions, one with An_{25-45} and one with An_{70-80}. Composition ranges from the protolith and the mylonite are not overlapping, pointing to a compositional change related to shearing and/or later alteration in the shear zone.

Garnet occurs only sporadically in the protolith gabbro (Table 2, Fig. 6). An analysis with $Alm_{48-51}Prp_{30}Grs_{18-21}$ (Yoshino *et al.* 1998; sample P5) is similar to that of the garnets in the underlying 'Patan' granulitic gabbro (Fig. 1). In the mylonite, garnet compositions ($Alm_{55-69}Prp_{15-23}Grs_{11-25}Adr_{01-04}$) are poorer in pyrope content than the protolith garnet. Profiles across garnets are commonly flat. Clinopyroxenes ($En_{32-38}Fs_{19-26}Wo_{38-45}$) and orthopyroxenes ($En_{48-60}Fs_{35-52}Wo_{01-15}$) coexist in the protolith (Table 2, Fig. 7).

Amphiboles of the protolith gabbro (Table 2, Fig. 8) plot in the fields of tschermakitic hornblende, tschermakite, ferro-tschermakite and ferroan pargasite; a subsidiary group plots in the field of actinolitic hornblende (Fig. 8), whose composition is typical for epidote-amphibole and greenschist facies conditions. Amphibole of the mylonite plots into the tschermakite and ferroan–pargasite fields, with few analyses within the magnesio-hornblende and actinolitic hornblende fields (Fig. 8). Pargasite and tschermakite are the main phases of the mylonite matrix. Hornblende compositions from the protolith and the mylonite overlap.

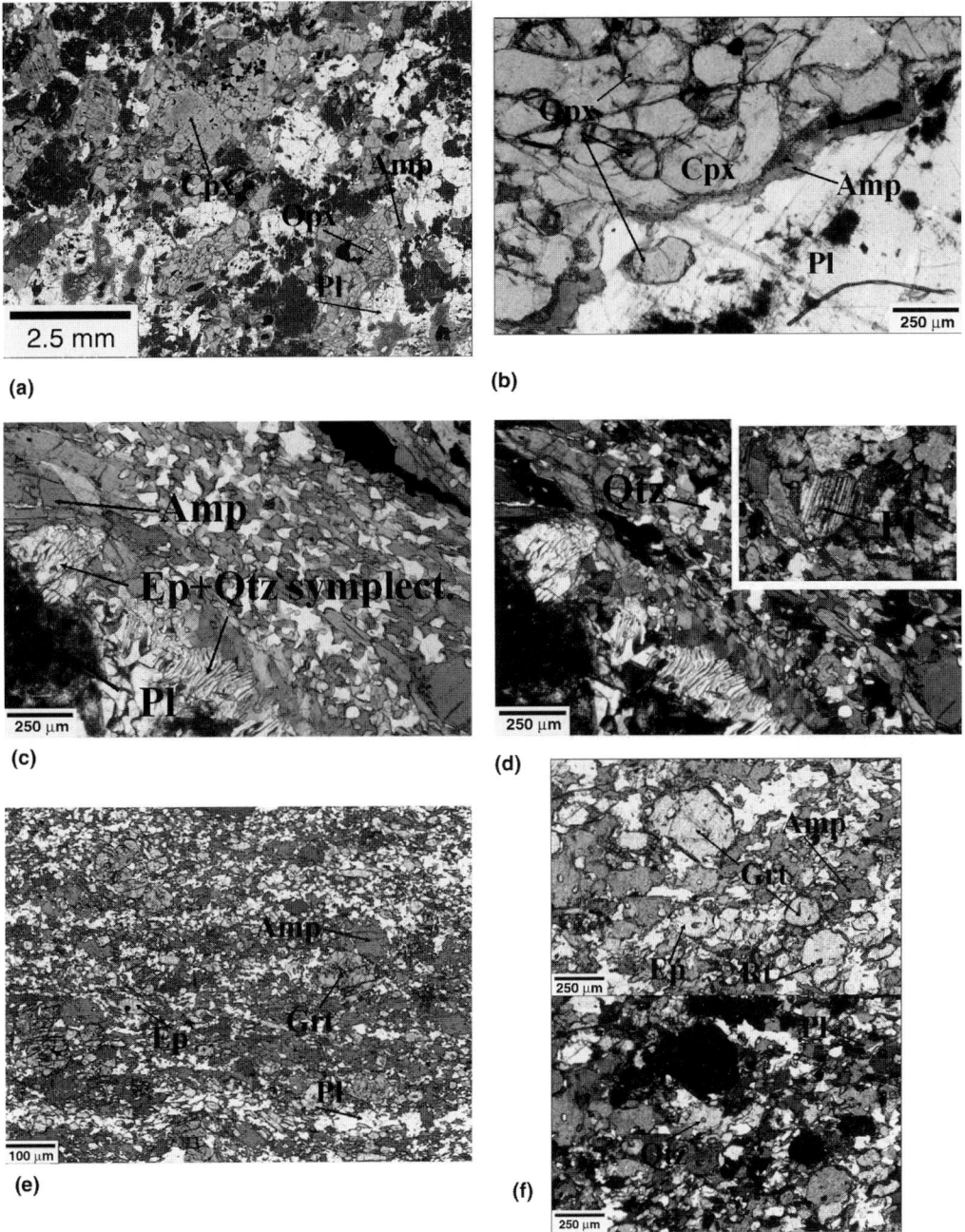

Fig. 4. Sarangar gabbro, from protolith to mylonite. Microphotographs (a)–(c), (e) & (f) top are with parallel polars; (d) and (f) bottom with crossed polars on the same area as (c) and (f) bottom. (a) Overview of the protolith. The magmatic structure is still well preserved. (b) Small scale of the protolith. Note the thin rim of amphibole between the plagioclase and clinopyroxene. (c) Gradient zone. Small-scale microphotograph with parallel polars and (d) cross-polars. Note the symplectites of epidote and quartz. (e) Overview of the mylonite. The long edge of the figure is parallel to the lineation and the short edge is normal to the foliation. (f) Small scale of the mylonite. Mineral abbreviations according to Kretz (1983).

Table 2. *Representative (averaged, n is the number of analyses) chemical composition of plagioclase, clinopyroxene (CPX) and orthopyroxene (OPX), garnet, amphibole of the undeformed (undef) and mylonitic Sarangar gabbro*

	Plagioclase		CPX	OPX	Garnet	Amphibole	
	undef	mylonitic	undef	undef	mylonitic	undef	mylonitic
SiO_2	53.5	59.12	49.79	48.87	37.32	39.24	40.49
TiO_2	0.01	0.05	0.35	0.08	0.07	0.54	0.66
Cr_2O_3	0.01	0.01	0.09	0.02	0.02	0.02	0.03
Al_2O_3	28.32	25.14	3.79	2.61	21.3	16.76	16.14
Fe_2O_3	0.07	0.09	0	0.79	1.43	5.7	4.07
FeO	0.05	0.15	11.97	27.61	25.13	12.44	11.68
MnO	0.01	0.03	0.33	0.74	1.91	0.26	0.12
MgO	0.01	0.01	10.81	16.07	4.17	7.18	9.02
NiO	0.02	0.02	0.02	0.02	0.01	0.01	0.02
CaO	11.62	7.09	20.85	0.8	7.73	10.74	11.45
Na_2O	4.87	6.39	0.66	0.06	0.04	1.45	1.82
K_2O	0.17	0.08	0.01	0.02	0.01	1.29	0.63
Cl						0	0
H_2O						1.96	1.99
Total	98.65	98.18	98.67	97.69	99.14	97.58	98.1
Mean of n	35	59	18	23	94	27	100

Biotite and muscovite are rare and occur in the mylonite, sparsely in its vicinity. Rare chlorite, filling fractures in pyroxene or replacing amphibole, was formed during local retrogression. Epidote of the protolith shows vermicular growth and rutile inclusions indicative for fast crystallization at solidus conditions. In the mylonite, epidote concentration is higher in garnet-free domains, suggesting that the Fe availability was decisive for the growth of both epidote and garnet.

Experimental technique

The velocity of compressional elastic waves was measured using the pulse transmission technique (Birch 1960) at room temperature and at confining pressures up to 0.3 GPa, using a Heard-type

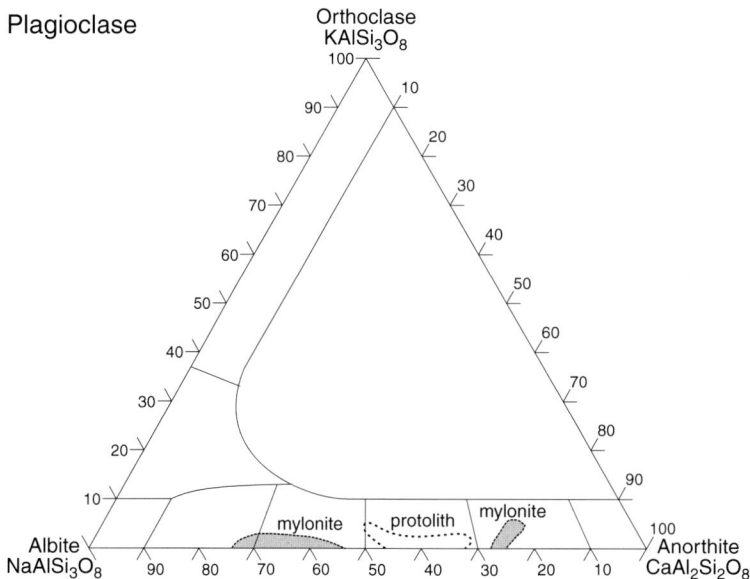

Fig. 5. Mineral chemistry of plagioclase.

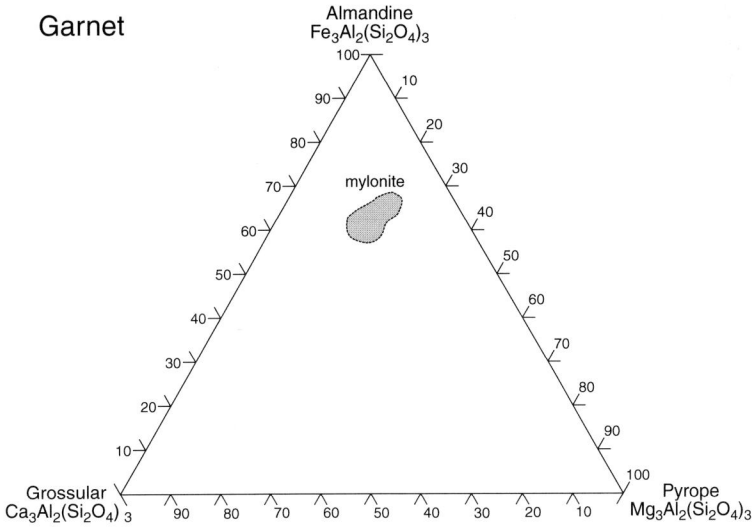

Fig. 6. Mineral chemistry of garnet.

pressure vessel (experimental technique described in Burlini & Kunze 2000). The set-up is sketched in Figure 9. Here we describe the technique used for the measurements at temperatures up to 600 °C and pressures up to 0.5 GPa using an internally heated gas medium apparatus (Paterson Rig).

The high-temperature measurements were conducted in the Paterson apparatus using the sample assembly reported in Figure 10. As the piezoelectric transducers cannot undergo high temperatures, a buffer rod of zirconia-ceramic was introduced between the sample and the transducer at each side. The column causes a

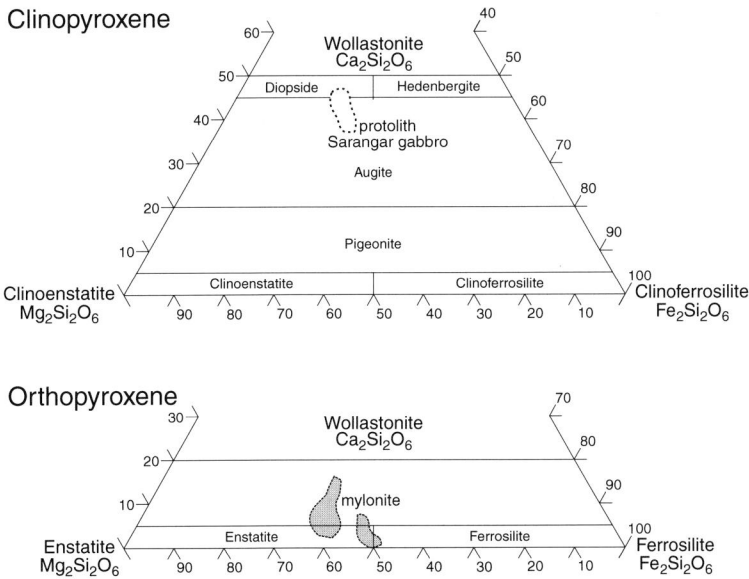

Fig. 7. Mineral chemistry of pyroxenes. Nomenclature after Morimoto (1988).

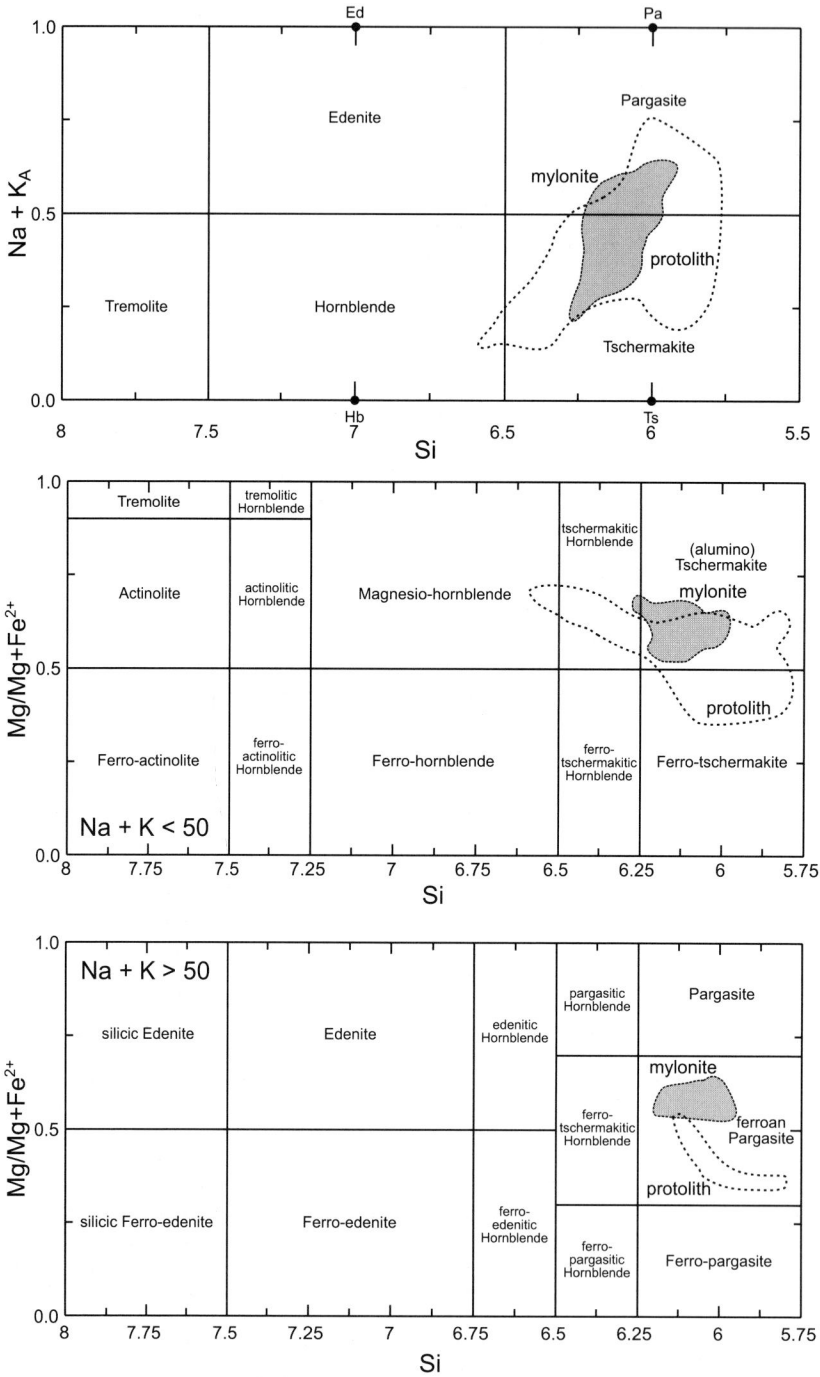

Fig. 8. Mineral chemistry of amphibole: the uppermost diagram shows the ranges of the measured amphiboles expressed as numbers of $Na + K_A$ and Si atoms per formula unit, whereas the two lower diagrams use Si v. $Mg/Mg + Fe^{2+}$ in the case where $Na + K$ is minor (centre) or higher (bottom) than 50. Nomenclature after Leake (1978).

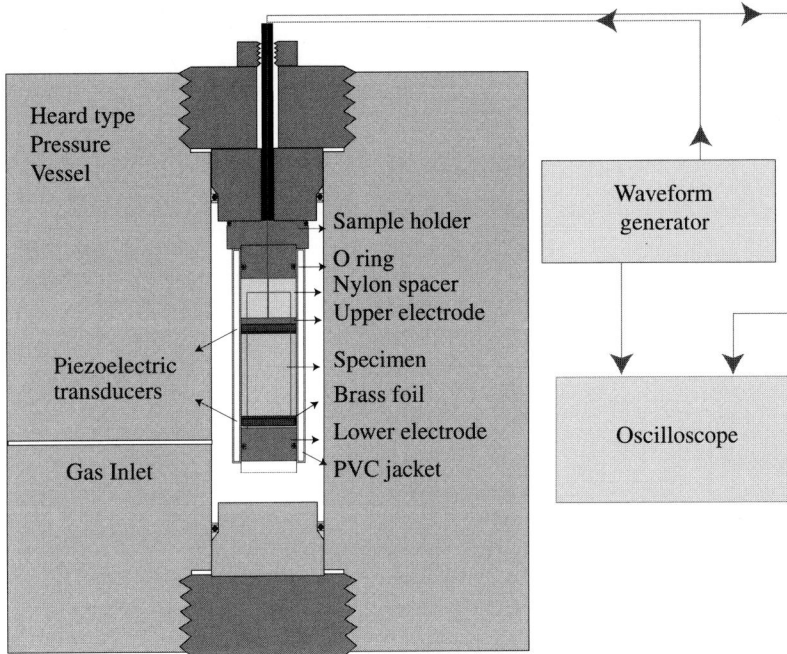

Fig. 9. Experimental set-up for the room temperature and high-pressure measurements.

time delay that was calibrated using a synthetic sapphire crystal cut with the (0001) axis parallel to the cylinder axis as standard. The time delay is about 24.5–35.5 μs depending primarily on temperature and on pressure.

Measurements were performed at two frequencies: 1.2 and 0.05 MHz. At 1.2 MHz the most difficult task was to determine the time of flight, as the beginning of the seismic signal is characterized by low-amplitude oscillations. Therefore, absolute measurements were difficult to make. On the other hand, it is possible to focus on the change in time position of a peak with pressure and temperature, with the aim to obtain the pressure and temperature derivatives. In practice, during a test, the first strong peak was selected and followed for the entire experiment. The absolute value of the velocities at room temperature and pressures up to 0.3 GPa was determined using the Heard apparatus. The measurements at 0.05 MHz are affected by large uncertainties (about ± 0.1 μs) in the time of flight. We chose the first rise at all pressures and temperatures (absolute measurements). The results are reported in Figures 11 (measurements at room temperature) and 12 (V_p as a function of temperature). The size of the symbols is proportional to the uncertainties in the time frame.

Temperature was measured using two K-type thermocouples (within + 1 K) placed directly on the top and bottom sides of the sample. The difference between the two thermocouples at the highest temperature was <5 K. The confining pressure was measured with a manganin coil pressure transducer with a precision of about 1 MPa at 0.5 GPa.

The elastic wave velocities were measured with increasing confining pressure by approximately 0.05 GPa increments until around 0.5 GPa, and with decreasing confining pressure by 0.05 GPa decrements. After the pressure cycle, both pressure and temperature were raised in steps of 0.1 GPa and 100 °C until the maximum pressure and temperature was reached, then cooled down to room conditions, again in steps of about 100 °C, and finally the pressure was released in steps of about 0.05 GPa. Using this specific pressure–temperature ($P-T$) path avoided thermal cracking of the samples, which were recovered intact after the tests.

Results

Measurements at room temperature

The confining pressure–velocity curves of the protolith, gradient and mylonite at room

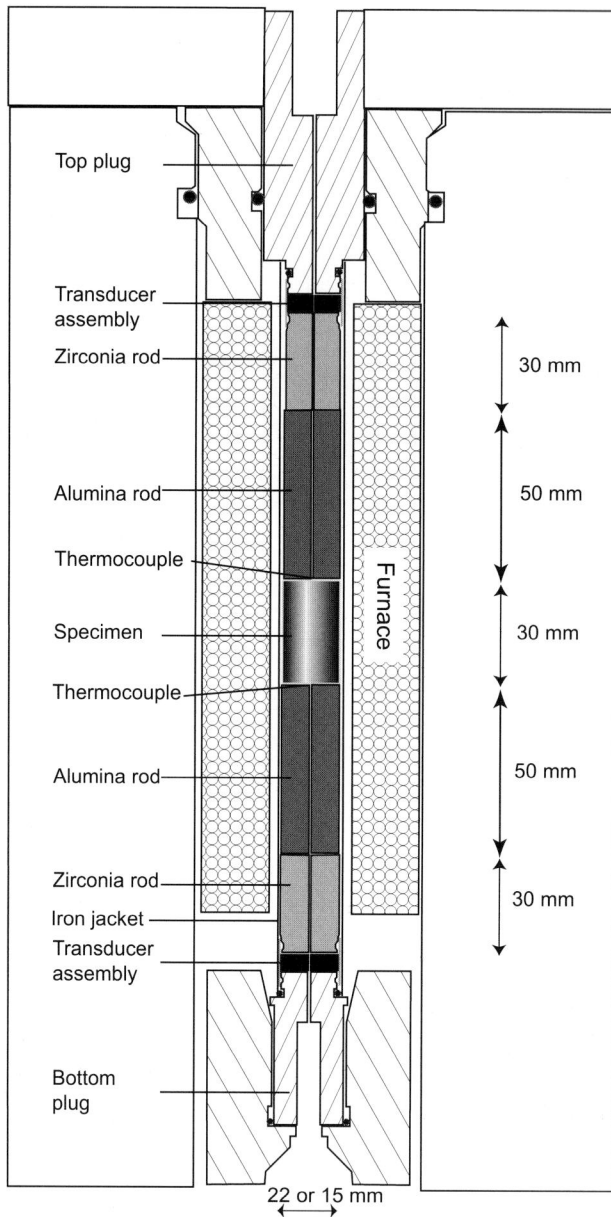

Top plug

Transducer
assembly

Zirconia rod

30 mm

Alumina rod

50 mm

Thermocouple

Furnace

Specimen

30 mm

Thermocouple

Alumina rod

50 mm

Zirconia rod

Iron jacket

30 mm

Transducer
assembly

Bottom
plug

22 or 15 mm

Fig. 10. Experimental set-up for the high-temperature and high-pressure measurements.

temperature are reported in Figures 11 and 13. The non-linear part of the curves is attributed to crack and pore closure, whilst the linear part (at high pressure) reflects the intrinsic seismic properties of the rock (Birch 1961) and corresponds, therefore, to the maximum velocity in the selected direction of the rock type under consideration.

The velocity measured during pressurization is always lower than that measured during depressurization because some of the cracks or pore spaces do not immediately reopen during depressurization (hysteresis). According to Burke (1987), measurements made during pressurization are not reproducible, whilst those during depressurization are reproducible within

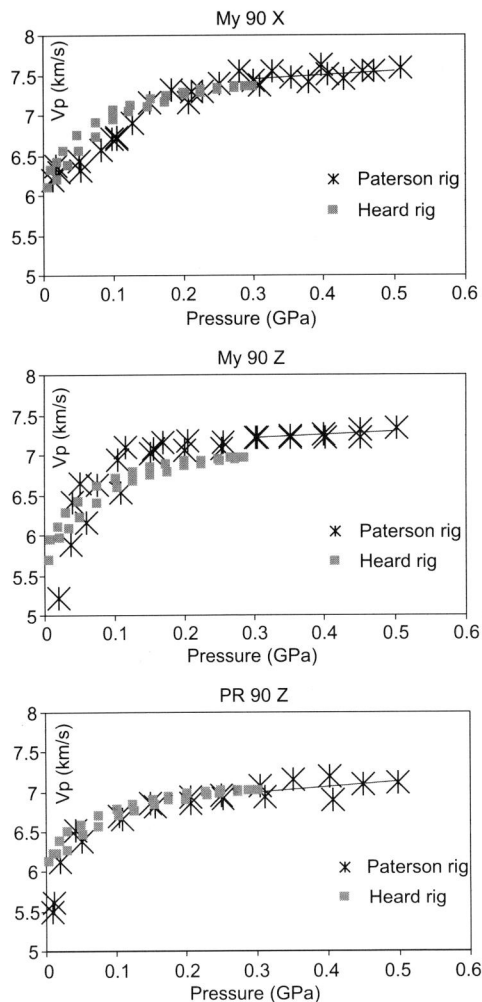

Fig. 11. Experimentally measured V_p velocities in function of confining pressure for the protolith (PR90) and the mylonite (M90) at room temperature. X, parallel to lineation; Z, orthogonal to foliation. Measurements on Heard and Paterson rigs are both displayed for comparison.

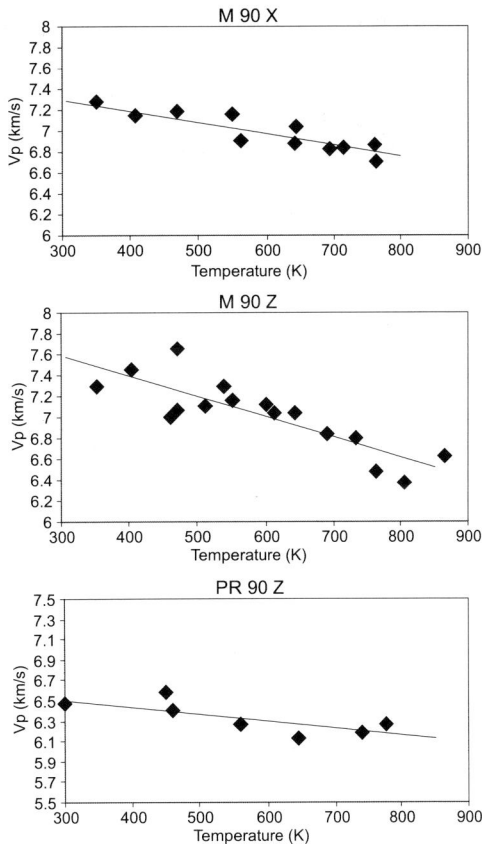

Fig. 12. Variation of V_p velocities with temperature. Data recalculated to 0.5 GPa. Same samples and definition as for Figure 9. Note that the slope in the Z direction of the mylonite is stronger than in the other cases.

measurement error limits. Because sample preparation may have induced minor cracking, the most reliable experimental data, for the present purpose, are those obtained during decompression.

At 0.3 GPa confining pressure (Table 3) the protolith had a V_p of almost 7.0 km s^{-1} in the three measured directions, with the maximum V_p for the Z core. This sample did not display any significant anisotropy (Fig. 13). Cracks had closed at pressures of >0.25 GPa, where the V_p–confining pressure curve became linear.

The V_p in the gradient zone and in the mylonite at 0.3 GPa confining pressure was highest for the X cores and lowest for Z cores (Fig. 13). The gradient zone displays a triaxial distribution of the V_p: the velocity in the Y-core direction is almost equal to the mean between the X and Z direction velocities. The V_p in the mylonite is axially symmetric (transverse isotropy), with V_pX similar to $V_pY > V_pZ$. Cracks in the protolith and mylonite closed at around 0.25 GPa confining pressure. It is worth noting that the average V_p increases from the protolith through the gradient zone to reach its maximum in the mylonite.

The directional dependence of the wave velocity is direct evidence of seismic anisotropy. The intrinsic seismic anisotropy is defined as $A = 100\% \times (V_{p\,max} - V_{p\,min})/V_{p\,mean}$. The maximum and minimum velocities were taken from the three measured directions, and may

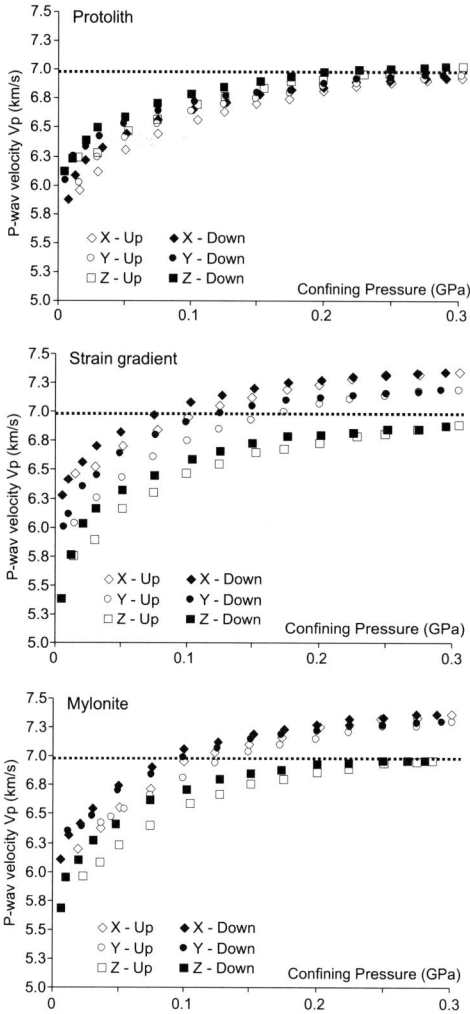

Fig. 13. Experimentally determined seismic properties of the protolith, gradient zone and mylonite at confining pressures up to 0.3 GPa. X direction is parallel to lineation, Y is normal to lineation and parallel to foliation and Z is normal to foliation, Open symbols represent measurements during pressurization, closed symbols during depressurization. The dashed line corresponds to the V_p of the protolith at high confining pressures. Note that the protolith is isotropic (V_p is equal in the three directions), and shows an orthorhombic anisotropy in the gradient zone and a transverse isotropy in the mylonite.

Table 3. V_p at 0.3 GPa pressure and room temperature, density, anisotropy and acoustic impedance of the various samples in the three directions of measurements. Velocities are in $km\ s^{-1}$, anisotropy is in % and acoustic impedance is in $kg\ s^{-1}\ m^{-2}$

	Average	X	Y	Z
Protolith				
V_p	6.96	6.93	6.95	7.02
Density	3.00	3.01	2.99	2.99
Anisotropy	1.28	–	–	–
Acoustic impedance	21.25	20.85	20.74	21.01
Gradient				
V_p	7.13	7.34	7.18	6.88
Density	3.04	3.05	3.04	3.05
Anisotropy	6.39	–	–	–
Acoustic impedance	21.71	22.38	21.79	20.97
Mylonite				
V_p	7.20	7.36	7.29	6.96
Density	3.08	3.09	3.08	3.07
Anisotropy	5.63	–	–	–
Acoustic impedance	22.19	22.72	22.47	21.37

the mylonite), which is almost isotropic from a seismic point of view, may be responsible for this reduction. The LPO of the plagioclase and amphibole grains probably caused this anisotropy.

Measurements at high temperature

The dependence of V_p on the temperature is reported in Figure 12. Velocities were recalculated at room pressure using the linear regression of the data measured between 0.3 and 0.5 GPa on the same sample during the pressure cycle. It is known that velocities decrease with temperature (e.g. Ji *et al.* 2002), but generally decrease similarly in the different directions. In the mylonite, however, the velocity decrease with temperature is different if measured parallel or perpendicular to foliation. In the X direction the derivative of V_p with temperature is 1.1×10^{-3} $(km\ s^{-1} K^{-1})$, whilst in the Z direction the same derivative is almost twice as large (Table 4). Moreover, the temperature derivatives of the protolith are twice as small as those of the mylonite. Finally, the decrease in velocities with temperature was, in the three cases, linear in the range investigated, and velocities measured with decreasing temperature were similar to those measured with increasing temperature, within the experimental resolution, suggesting that no mineral reactions or phase transition occurred during the experiments.

not represent the absolute extreme velocities in any direction. The seismic anisotropy is less than 2% in the protolith, is maximal in the strain gradient (6.4%) and reduces slightly in the mylonite (5.6%). Garnet (10% in volume in

Table 4. V_p *pressure and temperature derivatives. The pressure derivative (in km s^{-1} GPa^{-1}) were calculated in the range 0.25–0.5 GPa. The temperature derivatives were calculated from room conditions to up to 600 °C*

	My90 X	My90 Z	PR90 Z
dV_p/T	-1.07×10^{-3}	-1.93×10^{-3}	-6.75×10^{-4}
dV_p/P	5.12×10^{-1}	4.46×10^{-1}	6.89×10^{-1}

Discussion

The study of the seismic properties as a function of temperature and pressure of a shear zone from the Sarangar gabbro in northern Pakistan leads to the following findings.

Seismic anisotropy

The measurements made under different pressure and temperature conditions, and in different directions, show a strong increase of seismic anisotropy from the protolith to the strain gradient. This anisotropy is likely to be produced by the build-up of a LPO due to both the oriented growth of amphibole and the crystal–plastic deformation of plagioclase during shearing, as in other upper amphibolite–granulite facies tectonites. For example, in the clinopyroxene-bearing amphibolites and gabbros from the Ivrea zone, the plagioclase LPO has the (010)* axis concentrated normal to the foliation (the maximum V_p direction of a single crystal), whilst the (001)* axis (close the minimum V_p direction of a single crystal) tends to girdle

parallel to the foliation (Barruol & Kern 1996; Barruol & Mainprice 1993). The amphibole, on the other hand, has the [001] axis (the fast V_p direction of a single crystal) parallel to the mineral lineation and the (100)* axis (the slow direction of V_p in a single crystal) normal to foliation (Siegesmund *et al.* 1991; Barruol & Kern 1996). Therefore, plagioclase and amphibole give rise to destructive interferences, and the shape of the bulk anisotropy will depend on the dominant mineral phase and its fabric. Barruol & Mainprice (1993) reported that the intensity of the amphibole LPO is four times larger than that of the plagioclase in a tectonite from the Ivrea zone.

Moreover, Arbaret & Burg (2003) reported the amphibole LPO measured on a mylonite from the same set of shear zones as the sample reported here. This amphibole LPO has a point maxima of the a^* axis normal to the foliation plane and a girdle of the c-axes parallel to the foliation plane, with the maximum parallel to the mineral lineation. From this fabric, we have calculated the V_p distribution (Fig. 14) assuming a random distribution of plagioclase, garnet and

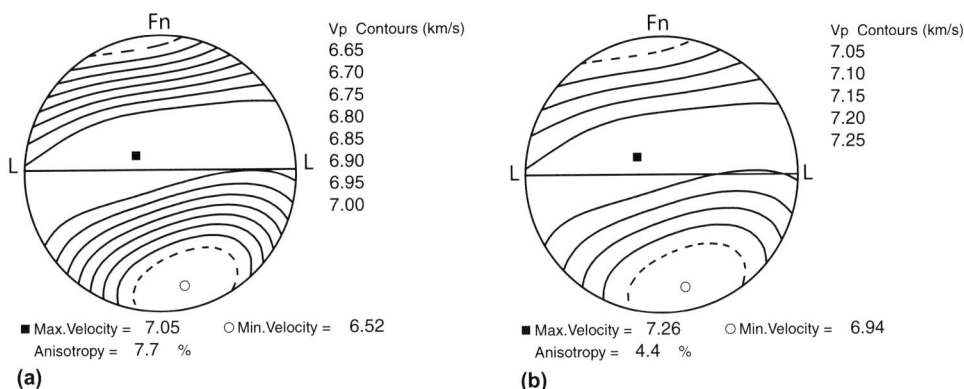

Vp Contours (km/s)
6.65
6.70
6.75
6.80
6.85
6.90
6.95
7.00

■ Max.Velocity = 7.05   ○ Min.Velocity = 6.52
Anisotropy = 7.7 %

(a)

Vp Contours (km/s)
7.05
7.10
7.15
7.20
7.25

■ Max.Velocity = 7.26   ○ Min.Velocity = 6.94
Anisotropy = 4.4 %

(b)

Fig. 14. Lower-hemisphere stereoprojection of the distribution of V_p calculated using the programs of Mainprice (1990), using the Voigt averaging scheme. Fn indicate normal to foliation and L the lineation. Contours are in km s^{-1}. (**a**) Velocities calculated using the LPO of the amphibole from mylonite reported in Albaret & Burg (2003). (**b**) Velocities calculated using the estimated modal composition of the rock (60% amphibole, 30% plagioclase and 10% garnet), the PLO of the amphibole and the isotropic stiffness tensors calculated for aggregates of plagioclase (An$_{56}$) and garnet.

epidote. The calculated anisotropy is similar to the anisotropy of the mylonite measured here, and the V_p distribution is consistent with the measurements. These results imply that the seismic anisotropy and V_p distribution are controlled by the LPO of the amphibole, and that the plagioclase and the epidote reduce the seismic anisotropy.

Reflectivity in low geothermal gradient crustal sections

At room temperatures, the average V_p increases progressively and slightly from the protolith to the mylonite (Fig. 15). The density, hence the acoustic impedance, also increases from the protolith to the mylonite. On the other hand, the anisotropy jumps from the protolith to the strain gradient, and remains nearly constant across the mylonite. The implications are that the mylonite can be observed on a seismic reflection profile provided it is thick enough (e.g. a quarter of the wavelength of the seismic waves; Sheriff & Geldart 1995), i.e. 50–100 m for standard surveys. A thinner, single mylonite cannot be detected, but if it is bounded by strain gradients, then its total thickness is increased. According to Burg *et al.* (2005), nearly all zones in the Kohistan cross-section fulfil such conditions. These shear zones are generally gently dipping. Therefore, reflection profiling of the present-day setting would investigate a situation close to

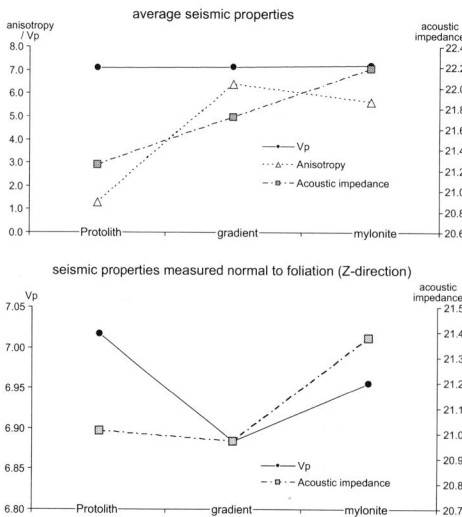

Fig. 16. Seismic properties at 600 °C and 0.5 GPa. V_p in km s^{-1}, anisotropy in % and acoustic impedance in kg s^{-1}m^{-2}.

the anisotropic case depicted in Figure 15b. In this case, the acoustic impedance measured for V_p propagating normal to foliation is less important than in the 'isotropic' case, if we consider the mean velocities, as in Figure 15a. This means that the seismic reflectivity due solely to seismic anisotropy would hardly be detectable.

Reflectivity in high geothermal gradient crustal sections

The results for the high-temperature case (Fig. 16) may be applicable, for instance, to shear zones placed in thermal peak conditions during collision. At 0.5 GPa and 600 °C (i.e. between 15 and 20 km in depth), the temperature-dependent decrease in velocities from protolith to mylonite reduces the reflection coefficient from 0.04 to 0.01 for the 'isotropic case' and remains about 0.02 for the anisotropic case, even if the seismic anisotropy increases from the protolith to the mylonite (about 11%). This implies that the mylonite, no matter how thick it is, will be seismically transparent. Therefore this type of shear zones cannot contribute to the reflectivity of the lower crust during peak metamorphic conditions.

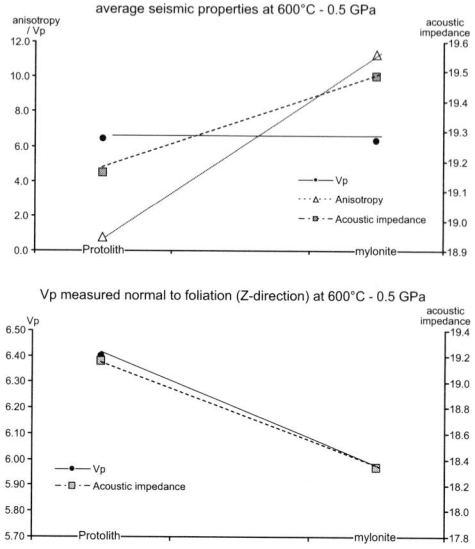

Fig. 15. Measured seismic properties from the protolith to the mylonite at room temperature. V_p is in km s^{-1}, anisotropy in % and acoustic impedance in kg s^{-1} m^{-2}.

Conclusions

Density and V_p of a mylonitic gabbro and its protolith were measured both at high pressures and high temperatures. The main results are summarized below.

- At low temperatures density, average V_p and acoustic impedance increased from the protolith through the strain gradient zone to the mylonite, whilst seismic anisotropy is at a maximum in the gradient zone.
- The seismic anisotropy is primarily controlled by the LPO of the amphibole.
- V_p varies with temperature in different ways for the protolith and the mylonite. In particular, the decrease in velocity with temperature is more pronounced for the mylonite than for the protolith. Moreover, the temperature derivatives also depend on the propagation direction in the mylonite, which results in an increase of the seismic anisotropy from 6% at room temperature to 11% at high pressure and temperature.
- Seismic anisotropy alone on these rocks cannot give rise to detectable reflection in seismic profiling except in the case of a very low thermal gradient.

Field work was supported by the Swiss NF grant 20-49372.96. Laboratory equipment and analyses were supported by the R'equip NF grant 2160-053289.98 and the ETH grant 02150/41-2704.5. F. Pirovino is thanked for the thin section preparation and U. Gerber for help with microphotographs. K. Kunze provided the LPO data and D. Mainprice the calculation of seismic properties along with fruitful discussions. Two anonymous reviewers and the guest editor (D. Bruhn) helped to significantly improve the manuscript.

References

ARBARET, L. & BURG, J.-P. 2003. Complex flow in lowest crustal, anastomosing mylonites: Strain gradients in a Kohistan gabbro, northern Pakistan. *Journal of Geophysical Research*, **108**, art. no. 2467; doi: 10.1029/2002JB002295.

ARBARET, L., BURG, J.-P., ZEILINGER, G., CHAUDHRY, N., HUSSAIN, S. & DAWOOD, H. 2000. Pre-collisional anastomosing shear zones in the Kohistan arc, NW Pakistan. *In*: KHAN, M.A., TRELOAR, P.J., SEARLE, M.P. & JAN, M.Q. (eds) *Tectonics of the Nanga Parbat Syntaxis and the Western Himalaya*. Geological Society, London, Special Publications, **170**, 295–311.

BARRUOL, G. & MAINPRICE, D. 1993. 3-D seismic velocities calculated from lattice-preferred orientation and reflectivity of a lower crustal section – examples of the Val Sesia section (Ivrea zone, northern Italy). *Geophysical Journal International*, **115**, 1169–1188.

BARRUOL, G. & KERN, H. 1996. seismic anisotropy and shear-wave splitting in lower-crustal and upper-mantle rocks from the Ivrea zone –

experimental and calculated data. *Physics of the Earth and Planetary Interiors*, **95**, 175–194.

BIRCH, F. 1960. The velocity of compressional waves in rocks to 10 kilobars, Part 1. *Journal of Geophysical Research*, **65**, 1083–1102.

BIRCH, F. 1961. Velocity of compressional waves in rocks to 10 kilobars, Part 2. *Journal of Geophysical Research*, **66**, 2199–2224.

BURG, J.-P., ARBARET, L., CHAUDHRY, N.M., DAWOOD, H., HUSSAIN, S. & ZEILINGER, G. 2005. Shear strain localization from the upper mantle to the middle crust of the Kohistan Arc (Pakistan). *In*: BRUHN, D. & BURLINI, L. (eds) *High-Strain Zones: Structure and Physical Properties*. Geological Society, London, Special Publications, **245**, 25–38.

BURKE, M.M. 1987. *Compressional wave velocities in rocks from the Ivrea–Verbano and Strona–Ceneri zones, Southern Alps, northern Italy: Implications for models of crustal structure*. MS thesis, University of Wyoming, Laramie.

BURLINI, L. & KUNZE, K. 2000. Fabric and seismic properties of a calcite mylonite. *Physics and Chemistry of the Earth*, **25**, 133–139.

CHRISTENSEN, N.I. & SZYMANSKI, D.L. 1988. Origin of reflections from the Brevard fault zone. *Journal of Geophysical Research – Solid Earth*, **93**, 1087–1102.

FOUNTAIN, D.M., HURICH, C.A. & SMITHSON, S.B. 1984. Seismic reflectivity of mylonite zones in the crust. *Geology*, **12**, 195–198.

FOUNTAIN, D.M., BOUNDY, T.M., AUSTRHEIM, H. & REY, P. 1994. Eclogite-facies shear zones – deep-crustal reflectors. *Tectonophysics*, **232**, 411–424.

JI, S., WANG, Q. & XIA, B. 2002. *Handbook of Seismic Properties of Minerals, Rocks and Ores*. Polytechnic International Press, Montréal.

KAZANEHDARI, J., RUTTER, E.H., CASEY, M. & BURLINI, L. 1998. The role of crystallographic fabric in the generation of seismic anisotropy and reflectivity of high strain zones in calcite rocks. *Journal of Structural Geology*, **20**, 293–299.

KERN, H., & WENK, H.R. 1990. Fabric-related velocity anisotropy and shear-wave splitting in rocks from the Santa Rosa mylonite zone, California. *Journal of Geophysical Research – Solid Earth*, **95**, 11 213–11 223.

KRETZ, R. 1983. Symbols for rock-forming minerals. *American Mineralogist*, **68**, 277–279.

LEAKE, B.E. 1978. Nomenclature of amphiboles. *Mineralogical Magazine*, **42**, 533–563.

MAINPRICE, D. 1990. An efficient FORTRAN program to calculate seismic anisotropy from the lattice preferred orientation of minerals. *Computers and Geosciences*, **16**, 385–393.

MILLER, D.J. & CHRISTENSEN, N.I. 1994. Seismic signature and geochemistry of an island-arc – a multidisciplinary study of the Kohistan accreted terrane, northern Pakistan. *Journal of Geophysical Research*, **99**, 11 623–11 642.

MORIMOTO, N. 1988. Nomenclature of pyroxenes. *Mineralogical Magazine*, **52**, 535–550.

REY, P.F., FOUNTAIN, D.M. & CLEMENT, W.P. 1994. P-wave velocity across a noncoaxial ductile shear zone and its associated strain gradient – Consequences for upper crustal reflectivity. *Journal of Geophysical Research – Solid Earth*, **99**, 4533–4548.

RUDNICK, R.L. & FOUNTAIN, D.M. 1995. Nature and composition of the continental-crust – a lower crustal perspective. *Reviews in Geophysics*, **33**, 267–309.

SCHALTEGGER, U., ZEILINGER, G., FRANK, M. & BURG, J.-P. 2002. Multiple mantle sources during island arc magmatism: U-Pb and Hf isotopic evidence from the Kohistan arc complex, Pakistan. *Terra Nova*, **14**, 461–468.

SHERIFF, R.E. & GELDART, L.P. 1995. *Exploration Seismology*, 2nd edn. Cambridge University Press, Cambridge.

SIEGESMUND, S., VOLLBRECHT, A. & NOVER, G. 1991. Anisotropy of compressional wave velocities, complex electrical-resistivity and magnetic-susceptibility of mylonites from the deeper crust and their relation to the rock fabric. *Earth and Planetary Science Letters*, **105**, 247–259.

TRELOAR, P.J., REX, D.C. *ET AL.* 1989. K–Ar and Ar–Ar geochronology of the Himalayan collision in NW Pakistan – Constraints on the timing of suturing, deformation, metamorphism and uplift. *Tectonics*, **8**, 881–909.

TRELOAR, P.J., REX, D.C. *ET AL.* 1990. The evolution of the Kamila Shear Zone, Kohistan, Pakistan. *In*: SALLISBURY, M.H. & FOUNTAIN, D.M. (eds) *Exposed Cross-sections of the Continental Crust*. Kluwer Academic Press, Amsterdam, 175–214.

YAMAMOTO, H. 1993. Contrasting metamorphic P–T–time paths of the Kohistan granulites and tectonites of the western Himalayas. *Journal of the Geological Society, London*, **150**, 843–856.

YOSHINO, T., YAMAMOTO, H., OKUDAIRA, T. & TORIUMI, M. 1998. Crustal thickening of the lower crust of the Kohistan arc (N. Pakistan) deduced from Al zoning in clinopyroxene and plagioclase. *Journal of Metamorphic Geology*, **16**, 729–748.

ZEILINGER, G. 2002. *Structural and geochronological study of the lowest Kohistan complex, Indus Kohistan region in Pakistan, NW Himalaya*. PhD thesis, ETH Zurich.

Damage and recovery of calcite rocks deformed in the cataclastic regime

A. SCHUBNEL[1,2], J. FORTIN[2], L. BURLINI[3] & Y. GUÉGUEN[2]

[1]*Lassonde Institute, University of Toronto, 170 College Street, Toronto, Ontario, Canada, M5S 3E3 (e-mail: alexandre.schubnel@utoronto.ca)*
[2]*Laboratoire de Géologie, Ecole Normale Supérieure, 24 rue Lhomond, 75005 Paris, France*
[3]*Experimental Rock Deformation Laboratory, ETH, Sonneggstrasse 5, 8092 Zürich, Switzerland*

Abstract: Compressional and shear wave velocities have been measured during the experimental deformation of Carrara marble and Solnhofen limestone in the cataclastic regime, both in dry and wet conditions at room temperature. Measurements were performed under hydrostatic conditions (up to 260 MPa confining pressure and 10 MPa pore pressure) during triaxial loading (at the constant strain rate of 10^{-5} s^{-1}) as well as during differential stress relaxation. During a full cycle, our results show that the seismic velocities first increase as effective mean stress increases. However, when the stress onset of cataclastic deformation was reached, elastic velocities showed rapid decrease due to stress-induced damage in the rock. During stress relaxation tests we observed an increase of elastic velocities with time, which suggested a fast 'recovery' of the microstructure. A substantial and rapid drop in the velocities occurred when reloading, suggesting that the previous 'recovery' was only transient. Subsequent relaxation tests showed other marked increases in velocities. These experimental results suggest that during the deformation of low-porosity calcite-rich rocks, dilatant (crack opening and frictional sliding) and compaction micromechanism (pore closure) compete. Evolutions of elastic properties (mainly sensitive to crack density) and macroscopic volumetric strain (more sensitive to porosity) are therefore not systematically correlated and depend on the strain rate, the solid stress conditions and the pore pressure.

Interest in the brittle–ductile transition has increased considerably in recent years, in large part due to the fact that the maximum depth of seismicity corresponds to a transition in the crust and in the upper mantle from seismogenic brittle failure to aseismic cataclastic flow, i.e. from localized to homogeneous deformation. The mechanics of the transition depends both on some extrinsic variable (state of solid stress, pore pressure, temperature, fluid chemistry and strain rate) and intrinsic parameters (crack and dislocation density, modal composition of the rock or porosity, for example). The deformation mechanisms operative during the transition occur on scales ranging from microscopic to macroscopic, and have a profound influence on the spatio-temporal evolution of stress and deformation during the earthquake cycle, as well as in the coupling of crustal deformation and fluid transport.

Limestones and marbles, even the ones with very low porosity like Carrara marble, have been studied widely in the past as they can undergo a brittle–plastic transition at room temperature for confining pressures easily attainable in the laboratory (Robertson 1955; Paterson 1958; Heard 1960; Rutter 1974). Such a behaviour is probably due to the fact that calcite requires relatively low shear stresses to initiate twinning and dislocation glide, even at room temperature (Turner *et al.* 1954; Griggs *et al.* 1960). In consequence, many previous studies have already documented the mechanical behaviour of both Carrara marble (Fredrich *et al.* 1989, 1990) and Solnhofen limestone (Robertson 1955; Heard 1960; Byerlee 1968; Edmond & Paterson 1972; Rutter 1972; Fisher & Paterson 1992; Renner & Rummel 1996; Baud *et al.* 2000). These studies have demonstrated that porosity change and failure mode are intimately

From: BRUHN, D. & BURLINI, L. (eds) 2005. *High-Strain Zones: Structure and Physical Properties.* Geological Society, London, Special Publications, **245**, 203–221.
0305-8719/05/$15.00 © The Geological Society of London 2005.

related. In the cataclastic flow regime, Baud *et al.*
(2000) showed that the pore space may either
dilate or compact in response to an applied stress.

In the present study, compressional and shear
wave velocities have been measured during
both hydrostatic and triaxial experiments performed on Carrara marble and Solnhofen limestone at room temperature. Our new set of data
show that during cataclastic deformation,
elastic wave velocities show large variations.
Damage (a decrease in the effective elastic properties) and apparent recovery of those properties
are transient phenomena that depend mainly on
the the stress history, the pore pressure and the
strain rate. Elastic properties, macroscopic
strain and post-deformation microstructural
analysis were put together and may have implications for the understanding of fault gouge
behaviour and, thus, direct consequences on the

understanding of fault zones and the earthquake
cycle.

Experimental set-up

Here we describe for the first time the triaxial cell
installed at the Laboratoire de Géologie of Ecole
Normale Supérieure (ENS) (Paris, France). This
is a prototype that was designed and constructed
by the company Geodesign, based in Roubaix,
France.

Description of the vessel

The Geodesign triaxial cell can reach a confining
pressure of 300 MPa (Fig. 1) and the confining
medium is oil. The confining pressure is servo-
controlled with an accuracy of 0.1 MPa. Axial
load is achieved through an auto-compensated

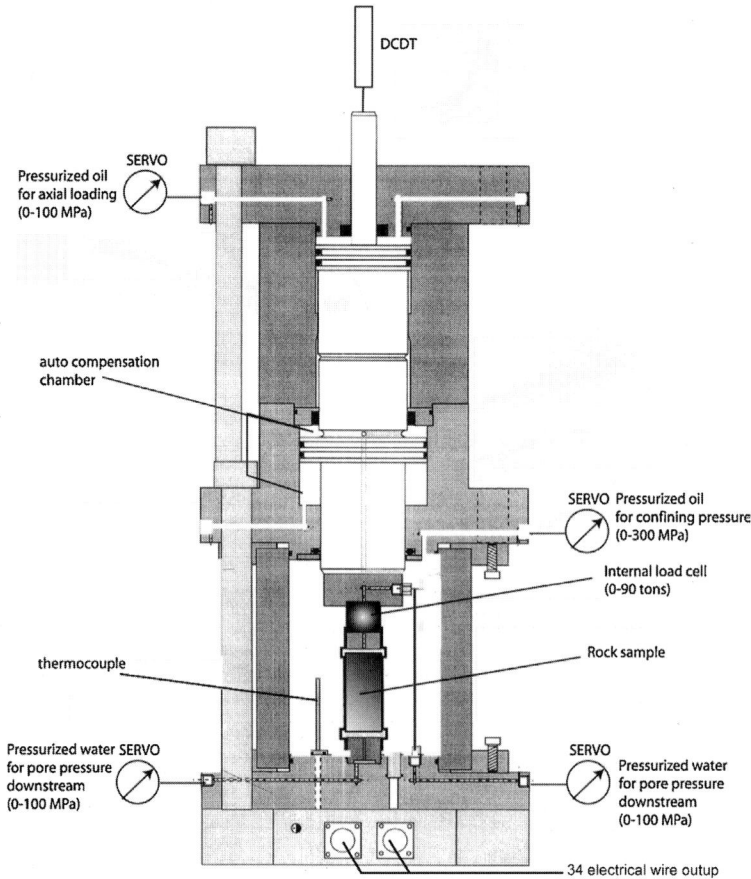

Fig. 1. Schematic diagram of the triaxial high-pressure cell installed at the Laboratoire de Géologie of Ecole Normale
Supérieure of Paris (France).

(i.e. that does not move when confining pressure varies) hydraulic piston that can be servo-controlled in either strain or stress. Maximum load is 90 tons, which corresponds to 717 MPa for a 40 mm-diameter sample. An internal load cell, manufactured by AMC automation, enables measurement of the applied load directly on top of the sample with an accuracy of approximately 1 MPa. Minimum strain rate is 10^{-6} s^{-1} and is monitored externally through two LVDTs (Linear Variable Differential Transformers) placed on top of the piston, outside the vessel. Minimum axial loading rate is 0.01 MPa s^{-1} and is servo-controlled using two pressure transducers.

Both confining and axial pressure systems are driven by the same hydraulic ram (0–35 MPa) along with two intensifiers: 35–300 MPa for the confining pressure, 35–100 MPa for the axial pressure. Pore pressure is driven by two precision volumetric pumps manufactured by Maxitechnologies. Pore fluid is introduced into the sample through hardened steel end pieces placed on the top and bottom of the rock sample. Maximum pore pressure in the system is 100 MPa. Both pumps can be controlled either in pressure (0.01 MPa precision), in flow (minimum flow is 0.1 h^{-1}) or in volume (precision is approximately 0.005 cm^3).

The main advantage of the ENS triaxial apparatus is its 34 electric feedthroughs that can allow simultaneous measurements of seismic velocities in several directions, as well as other properties across the sample (for example local strains). Finally, a thermocouple enables the temperature in the oil chamber to be monitored.

Sample set-up and preparation

Four samples (two of Carrara marble and two of Solnhofen limestone) were cored perpendicular to the bedding. All samples measured 80 mm in length and 40 mm in diameter. The end surfaces

were rectified and polished to ensure homogeneous loading and minimum friction during testing. The porosity of each sample was measured using a double saturation technique. Four parallel flat surfaces were machined along the cylinder of the sample at 90° to one another. Compressional and shear wave velocities were measured in each direction under dry atmospheric pressure conditions. Average initial porosity and velocity values of Carrara marble and Solnhofen limestone used in this study are summarized in Table 1.

During an experiment, axial and radial strains were measured directly on the sample's surface using strain gauge pairs (TML FLA-20, Tokyosokki). Each of which was mounted in a 1/4 Wheatstone bridge; strain measurement accuracy was approximately 10^{-6}. Volumetric strain was calculated using strain gauges according to the formula $\varepsilon_v = \varepsilon_a + 2 \times \varepsilon_r$, where ε_a is the axial strain measurement from longitudinal strain gauge, ε_r is the radial strain measurement from horizontal strain gage and ε_v the calculated volumetric strain.

P, SV and SH elastic wave velocities were measured perpendicular to compression axis, along diameters of the sample, using couples of source–receiver lead-zirconate piezoceramic transducers (PZTs) (Fig. 2). PZTs (PI255, PI ceramics, 1 MHz eigenfrequency) were glued directly onto the flat surfaces of each sample and positioned with approximately 0.5 mm accuracy, whilst the distance between opposite (paired) PZTs from which the velocities were calculated was measured within 0.01 mm. Compressional PZTs were 5 mm diameter discs, 1 mm thick; Shear PZTs were plates (5 × 5 × 1 mm). Pulse was generated by a Sofranel source generator (up to 370 V at 1 MHz frequency).

For each velocity measurement, more than 200 waveforms were stacked on a digital oscilloscope, in order to increase the signal/noise

Table 1. *Composition, initial porosity, grain size and elastic wave velocities at atmospheric pressure under dry conditions of Carrara marble and Solnhofen limestone used in this study*

	Carrara marble	Solnhofen limestone
Composition	>99% calcite	>99% calcite
Initial porosity (%)	0.5	4.5
Grain size (μm)	c. 150	c. 5
Mean initial, V_p (km s^{-1})	5.92	5.64
Mean initial, V_s (km s^{-1})	3.23	3.05
Incompressibility, K (dynamic, GPa)	57.5	50.4
Poisson ration, ν (dynamic)	0.29	0.29
Initial P wave anisotropy (%)	<1	c. 2

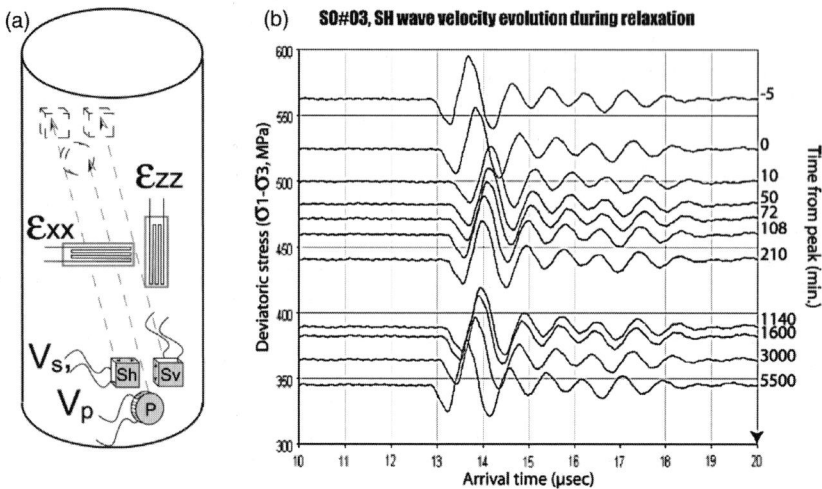

Fig. 2. (a) Schematic view of prepared sample with attached strain gauges and PZT sensors. Elastic wave velocities were measured along diameters. All samples measured 80 mm in length, 40 mm in diameter. An example of obtained waveforms for SH waves is shown in (**b**).

ratio. In such conditions, the absolute velocity error bar was of the order of a few per cent, but relative error in between two consecutive measurements was lowered to 0.5% using a double-picking technique. An example of obtained waveform recordings is shown in Figure 2b. Inside the vessel, the sample was covered with a Neoprene jacket that insulated it from the confining oil.

Experimental procedure

Two experiments (So#02 and Ca#01) were carried out under dry conditions at a confining pressure of to 200 MPa on Solnhofen limestone and Carrara marble, respectively. Two additional experiments (So#03 and Ca#02) were carried out in the saturated regime in which rock samples were first subjected to 15 MPa confining pressure and 10 MPa pore pressure for a minimum of 2 days in order to attain full saturation. Confining pressure was then raised up to 260 MPa, and the sample was left overnight for pore pressure equilibration before the triaxial loading cycle.

All triaxial cycles performed in this study were strain-rate controlled at a constant strain rate of 10^{-5} s^{-1}. During each experiment several relaxation tests were also performed in order to investigate possible recovery of the sample. During the stress relaxation tests, the piston position was locked and the differential stress decreased due to deformation of the sample. These relaxation tests lasted from 1 h to several days in the case of saturated samples Ca#02 and So#03.

Experimental results

Figure 3 illustrates stress–strain curves for the four tests. Plastic deformation was reached for smaller differential stress in Carrara marble than in Solnhofen limestone. Hardening

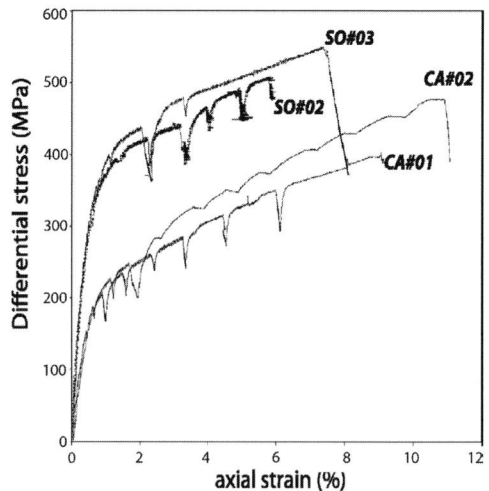

Fig. 3. Differential stress v. axial strain in the four tests presented in this paper. Ca#01 and Ca#02 were experiments performed on Carrara marble in dry ($P_c = 200$ MPa) and wet ($P_c = 260$ MPa, $P_p = 10$ MPa) conditions, respectively. So#02 (dry, $P_c = = 200$ MPa) and So#03 (wet, $P_c = 260$ MPa, $P_p = 10$ MPa) were experiments performed on Solnhofen limestone.

coefficient increases with confining pressure and figure shows that several relaxation tests were performed during the testing of each sample.

During the four experiments, strains, stresses and elastic wave velocities were continuously measured. This section presents a non-exhaustive compilation of our results in terms of coupled evolution of elastic wave velocities and volumetric strain v. effective mean stress, differential stress and time (Elastic wave velocities were measured perpendicular to the main compressive axis, and S wave velocities refer to the average between S_h and S_v wave measurements).

Volumetric strain and elastic wave velocities v. effective mean stress

We will refer to onset of dilatancy C' and onset of compaction C^* (as defined by Wong et al. 1997) as the stress onset (in differential stress for a given effective mean stress) at which macroscopic inelastic dilatancy and compaction were first observed. The onset of cataclastic dilatancy $C^{*\prime}$ is defined as the stress value at which cataclastic deformation turned from macroscopically compactive to dilatant.

Figure 4 shows the coupled evolution of volumetric strain and elastic wave velocities as a

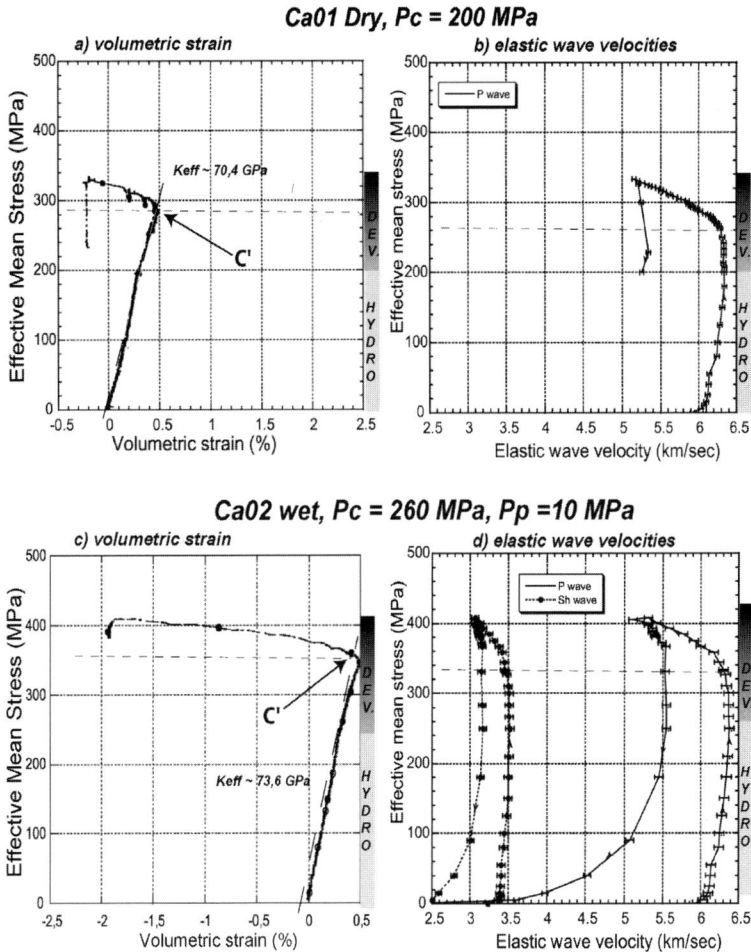

Fig. 4. Coupled evolution of volumetric strain and elastic wave velocities as a function of effective mean stress $[(\sigma_1 + 2\sigma_3)/3 - P_p]$ for experiments performed on Carrara marble. The elastic wave velocities and volumetric strain of the experiment Ca#01 (dry, $P_c = 200$ MPa) are reported in (**a**) and (**c**), respectively, and those of experiment Ca#02 (wet, $P_c = 260$ MPa, $P_p = 10$ MPa) are reported in (**b**) and (**d**) respectively. On the right-hand side of the diagram are reported the conditions at which the measurements were taken. Hydro, hydrostatic; dev, under differential stress. C' indicates the onset of cataclastic dilatancy.

function of effective mean stress for experiments performed on Carrara marble. The effective mean stress is $\mathbf{P} = [(\sigma_1 + 2\sigma_3)/3 - P_p]$, where σ_1 stands for the axial stress, σ_3 is the confining pressure and P_p the pore pressure; $\mathbf{P}$ is the first stress invariant and allows the representation of the hydrostatic and the differential stresses together on the same plot. The bar on the right-hand side of Figure 4 represents which part of the plot corresponds to hydrostatic stress and which to differential.

Figure 4a & b illustrate, respectively, the volumetric strain and V_p measured during the experiment Ca#01 (dry, $P_c = 200$ MPa). Up to 275 MPa effective mean stress, the rock mechanical response was elastic and the volumetric strain increased linearly. Static effective compressibility of Carrara marble in the dry regime was equal to 70.4 GPa (Fig. 4a). When onset of cataclastic dilatancy C' was reached (at 275 MPa), the rock started to dilate. Inelastic dilation was also accompanied by non-negligible strain hardening. Figure 4b shows that during the linear elastic phase, P wave velocity exhibits a slight increase due to crack closure. At 200 MPa, P wave reached a plateau value equal to 6.3 km s^{-1} which corresponds to the Voigt–Reuss–Hill average reported for P wave velocity in calcite non-porous aggregates (see for example Birch 1961). Macroscopic dilatancy was associated with a large linear decrease in P wave velocity that actually started a little before C'. One per cent of macroscopic dilation resulted in a decrease of more than 20% in P wave velocities. Therefore, elastic wave velocity measurements appear more sensitive to dilatancy than macroscopic volumetric strain. Very interestingly, V_p increased when the differential stress was removed.

Figure 4c & d show the volumetric strain and V_p and V_s, respectively, v. effective mean stress in experiment Ca#02 (wet, $P_c = 260$ MPa, $P_p = 10$ MPa). Comparison of Figure 4a & c shows that the effect of water on the static effective elastic incompressibility of Carrara marble is small. Effective compressibility increased to 73.6 GPa (+5%). This is compatible with an initially small and mainly crack-related porosity of Carrara marble (c. 0.5%). The onset of dilatancy C' was reached at 340 MPa effective mean stress. Strain hardening of the sample was less marked than during the dry experiment, whereas macroscopic dilation was much larger. Cataclastic dilatancy was also associated with linear decrease of both P and S wave velocities (Fig. 4d). Total decrease of P wave velocity reached 20% after more than 2% dilation. Comparing this experiment with the previous dry

experiment shows that P wave velocity undergoes a very similar reduction, although the sample dilated more and the strain-hardening coefficient was smaller. This could simply be due to a saturation effect.

The sample was left for relaxation for more than 3 days. During those three days, differential stress slowly decreased, and both P and S wave velocities showed significant increases. After removing the confining pressure, the elastic wave velocities decreased drastically as newly formed cracks re-opened. P wave velocity reached a final value of only 3.3 km s^{-1} at 5 MPa effective confining pressure. Once taken out of the pressure vessel, the sample showed less than 10% seismic anisotropy and no macroscopic localization band.

Figure 5 shows the coupled evolution of volumetric strain and elastic wave velocities as a function of effective mean stress for experiments performed on Solnhofen limestone. In Figure 5a & b, respectively, volumetric strain and P wave velocities are represented for experiment So#02 (dry, $P_c = 200$ MPa). The rock mechanical response was first elastic and the static effective compressibility of dry Solnhofen limestone was equal to 50.3 GPa (Fig. 5a), which is comparable to what has been previously observed by Baud *et al.* (2000). At 290 MPa effective mean stress, the onset of compaction C^* was reached and the rock started to compact inelastically. The onset of cataclastic dilatancy $C^{*'}$ was attained at 335 MPa. Whether the sample was compacting or dilating, cataclastic deformation was associated with significant strain hardening. Figure 5b shows that in the elastic regime, P wave velocity increased linearly and reached a maximum value of 5.75 km s^{-1}. Compaction was not accompanied by an increase in velocity, but conversely, by a decrease. P wave velocity increased instantaneously when the differential stress was unloaded, whereas unloading of the hydrostatic stress was accompanied by a large decrease in P wave velocity due to crack opening. When retrieved, the sample showed no strong elastic wave anisotropy (less than 5%) as well as no macroscopic localization band.

Figures 5c & d show, respectively, the evolution of volumetric strain and of P and S velocities as a function of effective mean stress in experiment So#03 (wet, $P_c = 260$ MPa, $P_p = 10$ MPa). Comparing Figure 5a & c, it is possible to observe that the presence of water increased significantly the static effective elastic compressibility of Solnhofen limestone (62.5 GPa, +12%). Water had a smaller increasing effect on P wave velocities, as can be seen when comparing Figure 5b & d, which is

So02 dry, Pc = 200 MPa

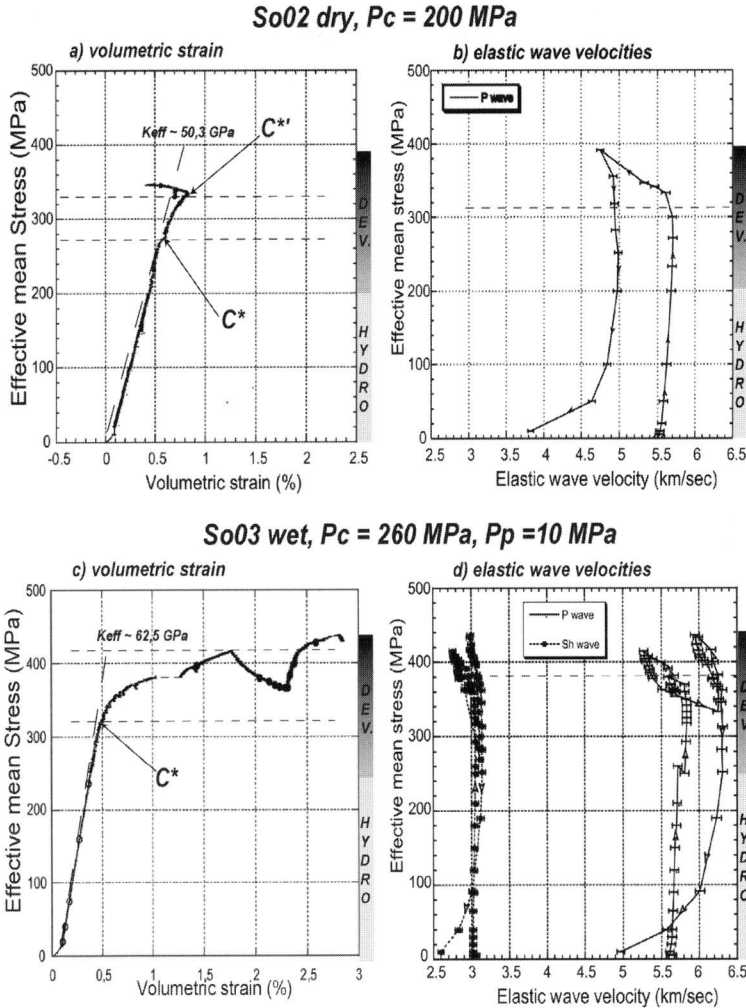

a) volumetric strain

Keff ~ 50,3 GPa

C*'

C*

b) elastic wave velocities

P wave

Fig. 5. Coupled evolution of volumetric strain and elastic wave velocities as a function of effective mean stress $[(\sigma_1 + 2\sigma_3)/3 - P_p]$ for experiments performed on Solnhofen limestone. The elastic wave velocities and volumetric strain of the experiment So#02 (dry, P_c = 200 MPa) are reported in (**a**) and (**c**), respectively, and those of experiment So#03 (wet, P_c = 260 MPa, P_p = 10 MPa) are reported in (**b**) and (**d**), respectively. On the right-hand side of the diagram are reported the conditions at which the measurements were taken. Hydro, hydrostatic; dev, under differential stress. C^* stands for the onset of cataclastic compaction, while $C^{*'}$ indicates the onset of cataclastic dilatancy.

compatible with the fact that, contrary to Carrara marble, the initial porosity in Solnhofen is quite large (c. 4.4%) and mainly equant.

Figure 5c shows that the onset of compaction C^* was reached at 310 MPa effective mean stress. Strain hardening was more important than during the dry experiment. However, no macroscopic cataclastic dilatancy was ever observed. Figure 5d shows that, at first, compaction was not associated with any variation in elastic wave velocities. However, P and S wave

velocities both started to decrease at 360 MPa effective mean stress. The decrease was smaller than during the dry experiment. The sample was then left to relax for 3 days. As differential stress slowly decreased, the sample compacted. P wave velocities increased to finally reach the value of 6.3 km s^{-1}, i.e. the reported Voigt–Reuss–Hill average for P wave velocity in non-porous calcite aggregates. Such behaviour was also observed on S waves. The sample was then reloaded. Velocities started to decrease again at 370 MPa. Inelastic compaction started when the maximum effective mean stress previously reached during the first cycle was retrieved, i.e. at 410 MPa. The sample was left again for relaxation and after 3 days velocities recovered completely to their maximum value. These results will be discussed more extensively in the next section.

When hydrostatic stress was unloaded, P and S velocities started to decrease as cracks were opening again. Once removed form the pressure vessel, the sample showed no strong elastic wave anisotropy (less than 5%) or macroscopic localization band.

Differential stress relaxation and elastic recovery

During relaxation, the piston position was locked. Differential stress $Q = (\sigma_1 - \sigma_3)$ decreased due to the sample deformation, while pore pressure and confining pressure were kept constant. Samples were left relaxing from 1 h in the case of dry experiments to several days in the case of saturated ones. Because of the machine design, the piston could not be perfectly maintained in position and therefore, some exponentially decreasing small amount of axial strain continued while the relaxation tests were performed (as can be seen in Fig. 7c) – in all the following plots, axial, radial and volumetric strains were measured using strain gauges. As a consequence and to be perfectly right, deformation taking place in the sample was not only composed of true relaxation, but also of a small component of creep. However, in the rest of this paper, we will refer to those tests as relaxation phases anyway. Seismic implications of these results will be discussed in the next section.

Results obtained while Solnhofen limestone and Carrara marble were relaxing in the dry regime (So#02 and Ca#01) are shown in Figure 6. In both experiments, relaxation took place after approximately 5% axial strain. Figure 6a shows the evolution of P wave velocity (plain symbols) and differential stress (empty symbols) v. time. On the figure, the differential stress scale on the right is inverted for easier reading and the timescale is linear. Solnhofen sample So#02 was left for relaxation for 1 h (bottom of Fig. 6a). We can observe that: (1) P wave velocity reached a minimum value when differential stress was maximum; and (2) P wave velocity increased as differential stress was decreasing. Total P wave velocity increase during relaxation was greater than 3%. Differential stress and P wave velocity curves are very

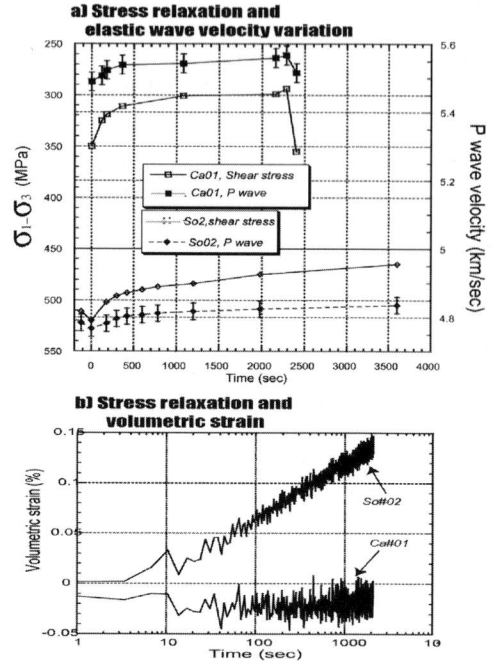

Fig. 6. Solnhofen limestone and Carrara marble relaxation experiments in the dry regime (So#02 and Ca#01). In both cases, relaxation took place after approximately 10% axial strain. Figure 6a shows the evolution of both P wave velocity (plain symbols) and differential stress (empty symbols) v. time. In (**a**) the differential stress scale on the right-hand side is inverted for easier reading and the timescale is linear. (**b**) Shows the evolution of volumetric strain (measured using strain gauges) in both samples during the same tests. Time is in logarithmic scale.

well anti-correlated, so that the stress dependence of P wave velocity is clearly demonstrated.

In the same way, Carrara marble sample Ca#01 was left for relaxation for 45 min (top of Fig. 6a) after approximately 5% axial strain. Again, P wave elastic velocity increased while differential stress decreased, and total P wave velocity increase this time was no larger than 2%. Stress dependence of P wave velocity is clearly observed. More surprising was that when differential stress increased again to the peak value prior to relaxation, P wave velocities immediately resumed to their prior minimum values. It can therefore be concluded that during cataclastic deformation of both Solnhofen limestone and Carrara marble in dry conditions, P wave velocity exhibits an 'elastic like' behaviour, i.e. reversible. Figure 6b shows the evolution of volumetric strain in both samples during the same relaxation tests. Time is in logarithmic scale. During relaxation, Solnhofen sample So#02 compacted, while Carrara marble sample Ca#01 showed no volumetric change. Hence, elastic wave properties are shown to be much more sensitive to small stress variations than the macroscopic strain.

Results obtained during two different relaxation tests performed on wet Carrara marble sample Ca#02 are shown in Figure 7. The first and second relaxation events took place after approximately 2.5 and 10% axial strain, respectively. Figure 7a shows the evolution of both P wave (empty symbols to be read on the left-hand scale) and S wave (plain symbols to be read on the right-hand scale) v. time. We can see that during both tests P and S wave velocities increased monotonically with time in a logarithmic scale. Increase in velocities was greater during the second relaxation phase, and reached approximately 6% and 3% for P and S wave,

respectively. This was certainly due to larger prior axial strain (c. 10%). Consequently, damage in the rock had been more important before the second test started.

Evolution of differential stress v. time is shown in Figure 7b. During both relaxation phases, differential stress decreased monotonically in the semi-log space. Again, the slope was steeper during the second relaxation, i.e. when the rock had experienced more damage. Figure 7c shows the evolution of axial strain and volumetric strain v. time during both tests. It is important to remember that at such high effective confining stress, Carrara marble porosity is close to zero and mainly crack-related. This explains why, although the sample shortened significantly (approximately 0.4% in 3 days), volumetric strain remained small. The evolution of volumetric strain can be schematically described as follow: a first phase where the rock continued

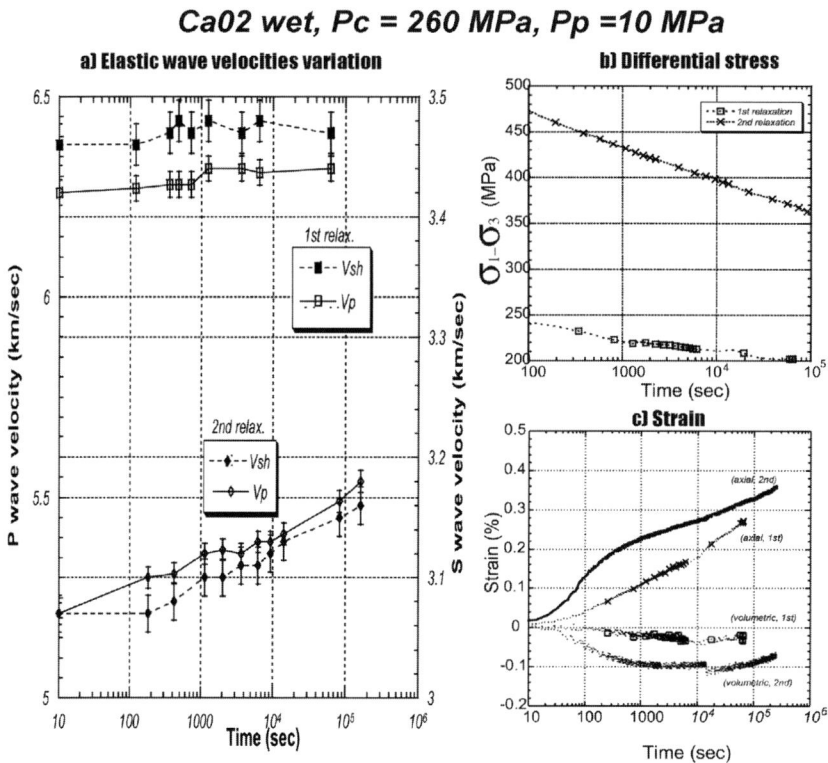

Fig. 7. Carrara marble sample Ca#02 relaxing in the wet regime. First and second relaxation phases took place after approximately 2.5 and 10% axial strain, respectively. (a) shows the evolution of both P wave (empty symbols to be read on the left-hand scale) and S wave (plain symbols to be read on the right-hand scale) elastic velocities v. time. Evolution of differential stress is shown in (b) while (c) shows the evolution of both axial strain and volumetric strain (measured using strain gauges). Time scales are logarithmic.

to dilate, an equilibrium phase where compaction and dilatancy balanced, finally followed by a compaction phase.

Results obtained during two relaxation tests performed on wet Solnhofen limestone sample So#03 after approximately 2.5 and 10% axial strain, respectively, are shown in Figure 8. Figure 8a shows the evolution of P wave (empty symbols to be read on the left-hand scale) and S wave (plain symbols to be read on the right-hand scale) v. time in logarithmic scale. During the first relaxation, increase in velocities was more or less linear with time. Elastic wave velocity increase was also more important during that first test than during the second relaxation, and variation in velocities reached more than 25% (from approximately 5 to 6.3 km s^{-1} in the case of P waves). During the second relaxation, behaviour resembled much of what has been seen previously on Carrara marble: P and S wave velocities increase monotonously and

velocity variations reached approximately 6 and 3% for P and S_h waves, respectively. Figure 8c shows that compaction was more important (almost 0.3%) during the first relaxation than during the second relaxation (0.1%). However, decrease in differential stress was comparable (*c.* 150 MPa) and monotonous in both cases, as seen in Figure 8b.

Comparing this figure with Figure 5 indicates that compaction in Solnhofen limestone can be associated both with an increase or decrease in elastic wave velocities. At low strain rates (typically during a relaxation test) compaction is slow and wave velocities are increasing. At fast strain rates (typically during triaxial loadings) compaction is much faster but wave velocities decrease due to increasing damage in the rock. This observation highlights that dilatancy and compaction are transient during cataclastic deformation, probably because their associated mechanisms are in competition. As a consequence, and

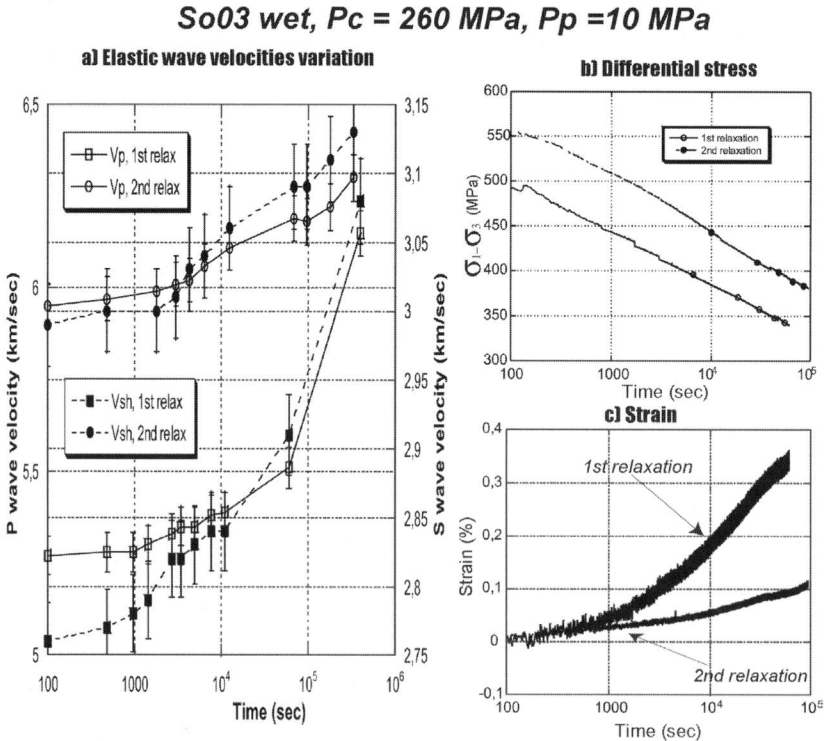

Fig. 8. Solnhofen sample So#03 when relaxing in the wet regime. First and second relaxation phases took place after approximately 2.5 and 10% axial strain, respectively. (**a**) shows the evolution of both P wave (empty symbols to be read on the left-hand scale) and S wave (plain symbols to be read on the right-hand scale) elastic velocities v. time in a logarithmic scale. Evolution of differential stress v. time in logarithmic scale is shown in (**b**), while (**c**) shows the evolution of both axial strain and volumetric strain (measured using strain gauges).

because elastic properties are much more sensitive to cracks than to pores, macroscopic strain and elastic wave velocities are not systematically correlated. Such observation is clear throughout Figures 5–7, and seems to be quite common in porous rocks as it has also been reported recently in high porosity sandstones by Fortin *et al.* (2004). When the fully compacted state is reached, P and S wave velocities should attain a maximum value of approximately 6.3 and 3.2 km s^{-1}, respectively, which are the reported Voigt–Reuss–Hill averages for calcite non-porous aggregates. This stage was apparently reached for sample So#03 and from then, its behaviour resembles that of Carrara marble. This observation is clear when comparing data from the second relaxation tests in Figures 7 and 8.

Microstructural analysis

Figure 9 presents micrographs obtained in the scanning electron microscopy (SEM) of the intact Solnhofen sample (Fig. 9a & b), a sample deformed dry So#02 (Fig. 9c & d) and a sample deformed wet So#03 (Fig. 9e & f). The compressive axis is vertical on all the pictures. The initial porosity of the intact rock appears to be quite heterogeneous, with both equant and elliptical pores. Grain boundaries are serrated and hardly visible, even at high magnification, as in Figure 9b. When deformed in the dry regime (Fig. 9c & d), the initial porosity disappeared and was replaced by a large density of small intergranular cracks located at grain boundaries. In Figure 9d, grain boundaries appear to be opened. Cracks are short (a few tens of micrometres) with large apertures and aspect ratios (between 10^{-1} and 10^{-2}). Furthermore, cracking is well distributed and non-localized, and shows no evidence of anisotropy in its distribution and orientation pattern. Such observation is in agreement with post-deformation elastic wave velocity measurements. In the sample deformed in the wet regime (Fig. 9e & f), initial porosity decreased even more drastically and almost no spherical pores can be seen. Crack density is much lower than in the dry case. Cracks are still located at grain contacts but their aperture appear much smaller (comparing with Fig. 9d). One might also argue that some evidence of dissolution can be seen at several grain contacts. Such features, however, might have been pre-existent in the rock. Figure 10 presents two optical micrographs of Solnhofen samples So#02 and So#03 obtained in cross-polarized light. Irrespective of whether deformation occurred in the dry or wet regime,

calcite crystals exhibit extensive twinning that does not pre-exist in the intact rock samples.

Figure 11 presents two optical micrographs of Carrara marble sample Ca#02. The sample was deformed in water-saturated conditions, and on both pictures the axis of compression is vertical. The microstructures of the deformation are quite heterogeneous and takes place both on the intergranular and intragranular scales. As observed in Solnhofen limestone, many cracks are located at grain boundaries and are generally never longer than one or two times the grain size. Voids initiating at the triple junction between grains show evidence of frictional sliding along grain boundary. On the intragranular scale, mechanical twins are present in every single grain at a very high density on both pictures. Bent and kinked twins are also present, as can be seen in Figure 11a, and as was already reported by Fredrich *et al.* (1989). Twin geometries are quite complex, with two and sometimes three sets of active twins in a single grain (see Fig. 11b). In general and depending, of course, on every crystal orientation, twins seem to be oriented more or less vertically. Geometries strongly suggestive of interaction between brittle and plastic deformation mechanisms can be particularly well observed in Figure 11a as the bending of twins often nucleates an intragranular crack. Tullis & Yund (1992) have reported similar observations on feldspar aggregates undergoing cataclastic deformation at a temperature of 300 °C. Again, microstructural observations suggest a complex interplay between plastic deformation and microcracking.

Features observed under the optical microscope and the SEM indicate that deformation mechanisms active during the experiments included: twinning, microcracking and frictional sliding along fractured grain boundaries and, possibly, pressure solution (when saturated). Macroscopic strain and elastic properties evolution depend on the interplay of each of these mechanisms. As previously reported in Carrara marble by Fredrich *et al.* (1989), twinning activity and dislocation glide in calcite grains are likely to be largely responsible for crack nucleation at grain boundaries. Indeed, because of different crystallographic orientations between neighbouring crystals, stress concentrations due to differential strain may facilitate crack nucleation and trigger frictional sliding along fractured grain boundaries. Because of stress concentration effects, granular flow would tend to concentrate around large pores and interplay of several mechanisms (twinning, crack opening and frictional sliding) would therefore be needed to induce macroscopic compaction.

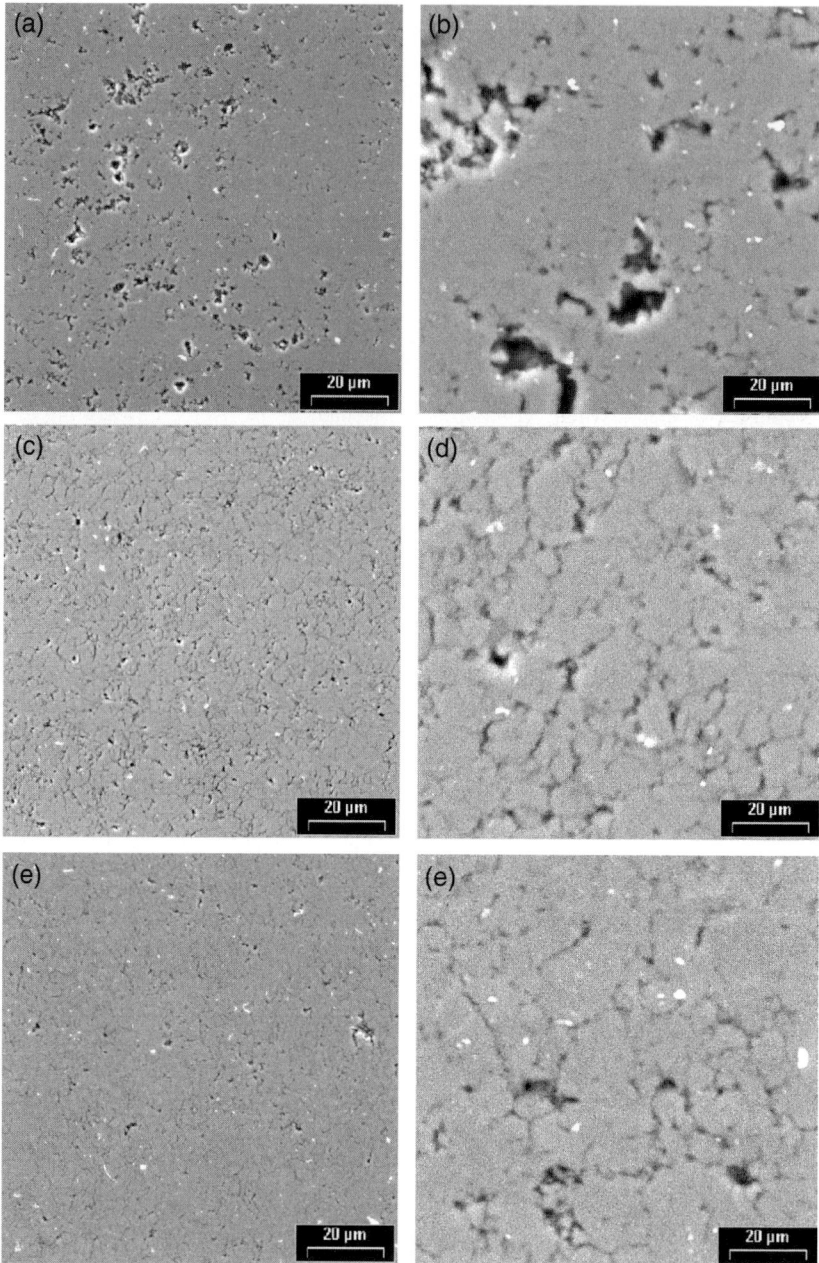

Fig. 9. Series of secondary electron SEM images of (**a**) and (**b**) intact, (**c**) and (**d**) deformed dry (So#02), and (**e**) and (**f**) deformed wet (So#03) Solnhofen samples. The compressive axis is vertical on the pictures. Scale is 20 μm in (a), (c) and (e), and 5 μm in (b), (d) and (f).

Mechanical analysis

In this section we analyse our results in the light of previous studies performed on Solnhofen limestone and Carrara marble.

Damage and dilatancy in carbonate rocks

Onset of crack propagation and frictional sliding. Figure 12 compares the onsets of dilatancy for both Solnhofen limestone and

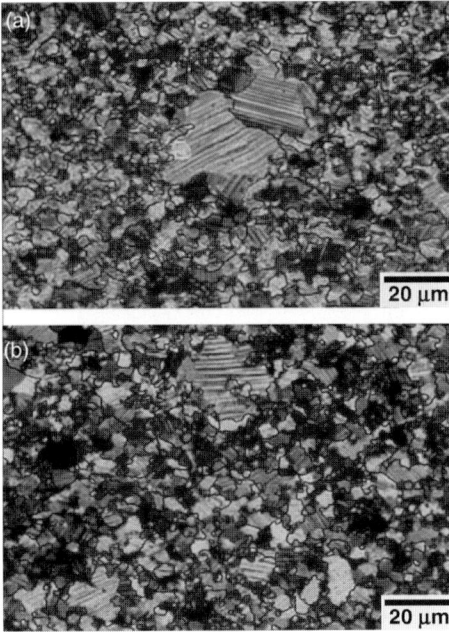

Fig. 10. Optical micrographs obtained in cross-polarized light of Solnhofen samples (**a**) So#02 and (**b**) So#03 deformed in the dry and wet regimes, respectively. Compression direction is vertical in all photographs. Note the twins oriented at a high angle to the compression direction in the largest grains. These twins developed during the experiments, as the undeformed material does not contain this type of twin.

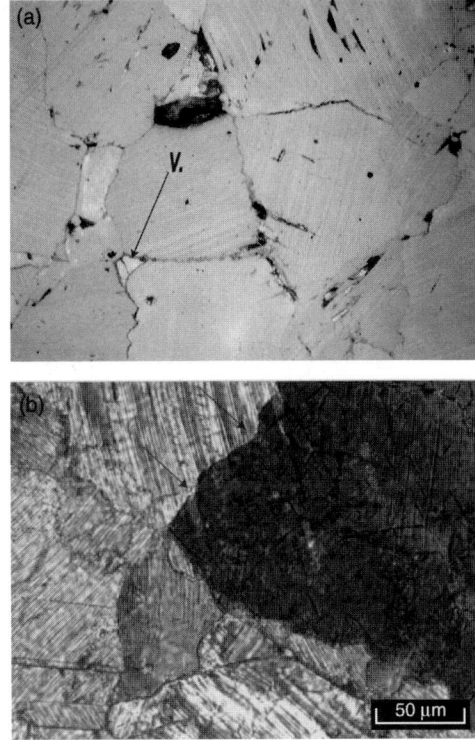

Fig. 11. Optical micrographs in (**a**) reflected light and (**b**) cross-polarized light of deformed Carrara marble sample Ca#02. In both photomicrographs, the axis of compression is vertical. In (a) V points to a void at a grain triple junction. In (b) the arrows point towards evidence of slip in the geometry of a grain boundary (top right-hand corner)

Carrara marble determined in this study with those reported by various authors in the literature (see Table 2). In Figure 12, it appears clear that maximum compressive stress σ_1 at onset of dilatancy evolves linearly with confining pressure. The onset of dilatancy is also much smaller in Carrara marble than in Solnhofen limestone.

Using the well-established wing crack model (Ashby & Sammis 1990; Fredrich *et al.* 1990; Baud *et al.* 2000) has proved to be of particular help in understanding the onset of crack propagation in rocks under compression. In axisymmetric compression, and for a randomly oriented distribution of cracks, the stress conditions for a tensile wing crack to propagate in mode I from a sliding crack tip are (see, for example, Lehner & Kachanov 1996):

$$\sigma_1 = \frac{\sqrt{1+\mu^2}+\mu}{\sqrt{1+\mu^2}-\mu}\sigma_3 + \frac{\sqrt{3}}{\sqrt{1+\mu^2}-\mu}\frac{K_{Ic}}{\sqrt{\pi l}}$$

$$(1)$$

where σ_1 and σ_3 stand for the axial stress and confining pressure, respectively. μ is the internal friction coefficient of the material, K_{Ic} the critical intensity factor (or fracture toughness) for a crack to propagate in mode I and l is the initial length of the pre-existing crack or default. In such a simple model, but widely used in geomechanics, the slope of the dilatancy envelope depends solely on μ. Assuming that μ is constant, the dilatant envelope is a straight line (in the space (σ_1, σ_3)) whose origin depends on the surface energy of the crack and on its length. According to Figure 12, best fit of equation (1) gives a value for the internal friction coefficient equal to 0.54 and 0.56 for Solnhofen limestone and Carrara marble, respectively. Such results are comparable with former values obtained by Ashby & Sammis (1990) ($\mu = 0.55$) and Baud *et al.* (2000) ($\mu = 0.53$) for Solnhofen limestone, and very close to the

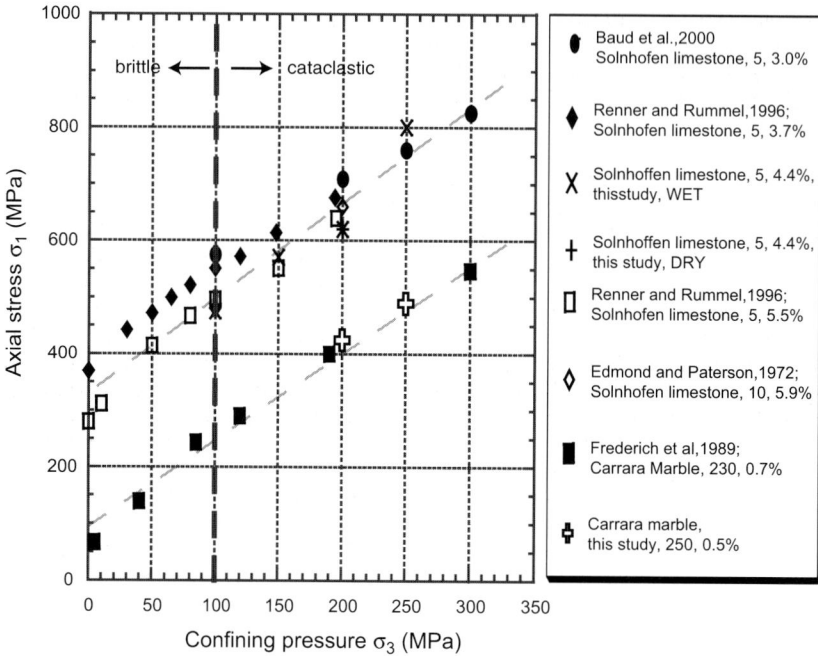

Fig. 12. Comparison in $\sigma_1 - \sigma_3$ space of the onsets of dilatancy for Solnhofen limestone and Carrara marble, as reported in the literature and this study.

value found by Fredrich *et al.* (1990) ($\mu = 0.5$) in Carrara marble. The best fit for $K_{Ic}/\sqrt{\pi l}$ was found to be 124 MPa for Solnhofen limestone and 27 MPa for Carrara marble. Since both rocks are composed of 99% calcite, K_{Ic} should be equal in both rocks. In our case, assuming that $K_{Ic} = 0.2$ MPa m$^{1/2}$, as reported by Atkinson & Avdis (1980), we find an initial crack length of $2l \approx 3$ μm and $2l \approx 70$ μm for Solnhofen limestone and Carrara marble, respectively. The initial crack length ratio is approximately 20 when the grain size ratio between Solnhofen limestone and Carrara marble is about 30. These results are compatible with our microstructural analysis. In the case of Solnhofen limestone, initial porosity also seems

to be a key issue and dilatancy occurs at lower stress with increasing porosity. Such an observation has already been reported by Baud *et al.* (2000) in terms of peak stress and could be explained by the fact that initial damage in the rock is likely to scale with initial porosity.

Cataclastic deformation and crack nucleation. Calcite is well known to require relatively low shear stresses to initiate twinning and slip, even at room temperature: Turner *et al.* (1954) and Griggs *et al.* (1960) obtained values of the order of 145–216 MPa for *f*-slip and *r*-slip dislocation glide, respectively. In addition, *e*-twinning in calcite crystal has been reported to be triggered for even smaller stresses.

Table 2. *Compilation of stress values at which the onset of dilatancy C′ and threshold of macroscopic dilatancy C*′ *were observed in Carrara marble and Solnhofen limestone, respectively, during our experiments*

	Confining pressure P_c (MPa)	Pore pressure P_p (MPa)	Effective mean stress **P** (MPa)	Differential stress **Q** (MPa)
C'	Carrara		marble	
Ca#01	200	dry	275	225
Ca#02	260	10	340	270
$C^{*\prime}$	Solnhofen		limestone	
So#2	200	dry	335	405

As shown by our microstructural analysis, during cataclastic deformation, cracks seem to nucleate preferably at grain contacts and boundaries. Baud et al. (2000) suggested that microcracking could be due to dislocation pile-ups. In such a model, if the dislocations pile-up at an obstacle (for example at a grain boundary), tensile stress concentrations nucleate a crack (Stroh 1957; Wong 1990).

On the other hand, microstructural observations reported by Fredrich et al. (1990) and ourselves in this paper showed that interplay between twinning and crack nucleation are intense during deformation. At least one source of cracking might be twin intersections, as seen in Figure 11, that could depend inversely on grain size, as is suggested when comparing twin density between Figures 10 and 11.

However, two paradoxes can be pointed out. First, we observed throughout our experiments the decoupling of macroscopic strain and elastic properties. For example, during our experiment Ca#02 (Fig. 4c & d), dilatancy was observed to be very large (2%) with small effects on elastic properties. Second, and as noted originally by Baud et al. (2000), the data suggest a slight dependence of $C^{*\prime}$ on pressure, whereas dislocation pile-up models or twin–crack interaction models predict it to be pressure independent (for initiation shear stress). Figure 12 seems to suggest that if cracks can be nucleated by dislocation pile-up or by twinning, damage and dilatancy is nevertheless due to frictional sliding on these newly formed crack surfaces.

Compaction mechanics and recovery in Solnhofen limestone

Effect of initial porosity and water. In their study of compaction and failure modes in Solnhofen limestone, Baud et al. (2000) performed experiments in the dry regime only, on samples with an average porosity of 3% and an average grain size of 5 µm. As previously mentioned, our experiments were performed on a block whose initial porosity was 4.4%, both in dry and wet conditions. Four additional experiments were performed on samples coming from the same block (Schubnel 2003), in the wet regime, at the State University of New York in Stony Brook. Table 3 compiles the values of effective mean stress **P** and differential stress **Q** at which the onset of compaction C^* (as defined by Wong et al. 1997) were observed in that block.

A comparison with data previously obtained by Baud et al. (2000) is plotted in Figure 13. It shows that water enhances compaction greatly. At a given confining pressure, decrease in differential stress at the onset of compaction can be as large as 150 MPa. This has already been shown by Rutter (1986) and could be explained in three different ways: (1) the presence of water increases single crystal plasticity; (2) pressure solution takes place and accounts for the early observed compaction; and (3) water has a lubricating effect.

Variation in porosity also seems to have a small reduction effect on the onset of compaction: C^* for experiment So#03 falls a little lower than the compactive envelope previously obtained by Baud et al. (2000). The latter observation is confirmed by a recent general study performed on the compaction of several carbonate rocks (Vajdova et al. 2003).

Baud et al. (2000) and Vajdova et al. (2004) modelled the behaviour of limestones using Curran & Carroll's (1979) model of porous void compaction. In this model, compaction occurs when a pressure-independent plastic yield criterion, σ_Y (or von Mises criterion), is reached. Carroll (1980) predicted that initial yield in hydrostatic compression occurs at the macroscopic pressure:

$$P^* = \frac{2}{3}\sigma_Y\left[1 - \frac{2\mu\phi}{2\mu + \sigma_Y(1 - \phi)}\right] \quad (2)$$

Table 3. *Compilation of stress values at which the onset of compaction C^* was observed in Solnhofen limestone. In our block of Solnhofen limestone, initial porosity was 4.4%. The superscripts * indicates experiments that have been performed in the Rock Mechanics Laboratory of SUNY at Stony Brook*

Solnhofen limestone	Confining pressure P_c (MPa)	Pore pressure P_p (MPa)	Effective mean stress **P** (MPa)	Differential stress **Q** (MPa)
C^*				
So#12*	160	10	225	225
So#2	200	dry	290	270
So#07*	210	10	270	210
So#3	260	10	310	180
So#11*	390	10	410	90

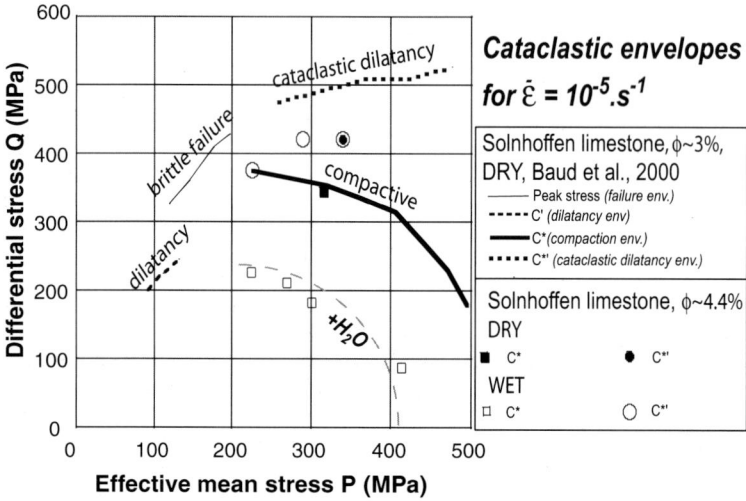

Fig. 13. Comparison of our results to those previously obtained by Baud *et al.* (2000). **P** is the effective mean stress and **Q** the differential stress.

where μ is the shear modulus and ϕ the initial porosity. Baud *et al.* (2000) concluded that, in order for Curran & Carroll's (1979) model to provide a reasonable fit of Solnhofen limestone's compactive envelope, σ_Y needed to be larger (*c.* 975 MPa) than experimentally determined values for single-crystal plasticity in calcite. When water is present, we found the value of σ_Y to decrease to approximately 600 MPa, which, in any case, is still much larger than experimentally determined values for single-crystal plasticity in calcite. It is interesting to add that investigating the compaction of dry Tavel limestone (9% porosity, 5 μm average grain size), Vajdova *et al.* (2004) showed that Curran & Carroll's (1979) model overestimated the yield criterion in the same way. In light of this paradox, one can suggest that because the von Mises criterion requires five independent slip systems (Paterson 1969) for homogeneous plastic flow to occur in the proximity of a pore surface, the plastic yield criterion σ_Y was likely to be larger than the one observed for single slip systems. However, and as shown on Figure 10, twinning of calcite was observed. Paterson (1969, 1978) showed that when e-twinning in calcite is activated, the combination of twinning and only one slip mode is equivalent to five independent slip systems. The problem remains therefore unanswered but two hypothesis can be drawn: (1) plastic deformation mechanisms like twinning and dislocation glide are grain size sensitive; and (2) operative mechanisms also involve

pressure-dependant frictional processes. The observation of evidence for frictional sliding along grain boundaries in our microstructural analysis and the observed effect of water on compactive yield stresses may point towards the latter solution.

Recovery of elastic properties. Cyclic loading in the brittle regime is characterized generally by a large hysteresis between loading and unloading. However, there is an almost elastic region when changing the state of stress, with very little hysteresis and almost constant P and S wave velocities. Our experimental observations of P and S wave evolutions in low-porosity calcite rocks deforming in the cataclastic regime showed that such a 'deadband', typical of hysteresis cycles where the velocities remained constant during loading and unloading, does not exist. On the contrary, the apparent elastic properties of the rock showed a high stress dependency and varied immediately when the stress state was changed.

The simplest option to explain apparent recovery of elastic properties is to assume that it is partly due to back-sliding on cracks. If so, the onset of back-sliding on a crack is simply given by:

$$\Delta\tau = \mu\Delta\sigma_n \qquad (3)$$

where $\Delta\tau$ and $\Delta\sigma_n$ correspond to the variations in shear and normal stresses on the crack surface after the loading changed direction. In the case of an axisymmetric unloading, Kachanov

(1982*a*, *b*) showed that the macroscopic envelope of the 'dead band' for a family of randomly oriented cracks is a hyperbola in the (σ_1, σ_3) space. At effective confining pressures higher than 200 MPa, and assuming a friction coefficient of approximately 0.5 for calcite, frictional back-sliding on crack surfaces cannot be initiated before axisymmetric extension state is reached and, therefore, back-sliding cannot account for any observations of recovery.

However, if we assume that the friction coefficient is a state variable (as in the framework of Rate and State Friction; see, for example, Marone 1998), recovery can be due to contact strengthening. Under contact loading, particle surface may deform inelastically (for example due to dissolution of crack asperities or visco-elastic interpenetration of crack surfaces) and cohesion between contacting surfaces may increase. In such a way, strengthening derives from growth of contact area (Dietrich & Kilgore 1994) via chemically assisted mechanisms (Frye & Marone 2002). In the laboratory, contact healing has been widely observed and was shown to depend on the logarithm of contact duration (e.g. Beeler & Tullis 1997), which has also been observed in our experiments. However, in our relaxation experiments, inferring internal friction coefficient cannot be straightforward and further analysis would be needed to quantify the exact variations of friction coefficient with time.

Conclusions and implications for fault zones

In the cataclastic regime, our experimental results shows that during the deformation of low-porosity calcite-rich rocks, dilatant (crack opening and sliding, void creation) and compactive (pressure solution, pore closure) micromechanisms are not exclusive and may interact. In the case of an initially porous rock like Solnhofen limestone, the evolution of apparent elastic properties (mainly sensitive to crack density) and macroscopic volumetric strain (more sensitive to equant porosity) is not systematically correlated and depends on the strain rate, the solid stress conditions and the pore pressure. Our data suggest that twinning activity and dislocation pile-up in calcite polycrystals participate in the nucleation of cracks at grain boundaries. Macroscopic compaction occurs when frictional sliding is triggered on those boundary cracks.

During our experiments, damage and recovery of P–S elastic wave velocities were observed to be transient phenomena that are associated with distinct elastic wave velocity variations. In our experiments, these variations could be large and fast, especially in the presence of fluid (>10% in a couple of days, at room temperature). Crack nucleation and propagation was shown to be responsible for increasing damage, while recovery of elastic properties is thought to be due to re-strengthening of grain contacts.

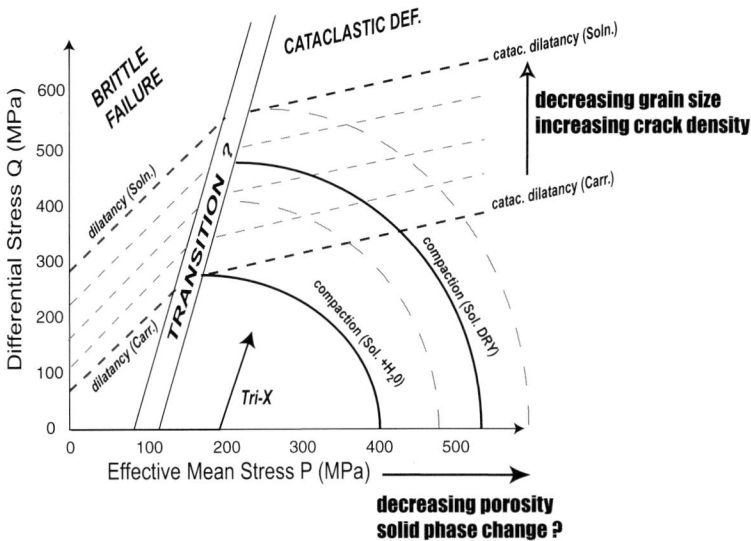

Fig. 14. Schematical deformation map in the principal stress space for calcite rocks according to our experimental observations. In the principal stress space, we recall that the triaxial loading path has a slope equal to 3.

Recent *in situ* velocity monitoring field studies (Crampin *et al.* 2002) where temporal changes of both P and S velocities were observed before earthquakes and volcanic eruptions could be explained by similar mechanisms.

As fault gouges are highly complex materials, experimentalists need to find good analogues for simple and relevant laboratory experiments. Figure 14 summarizes schematically the deformation map in the principal stress space for calcic rocks according to our experimental observations. In this space, we recall that the triaxial loading path has a slope equal to 3. In the cataclastic domain, we see that the dilatant and compactive envelopes are overlapping. The observed macroscopic strain was shown to be a function of both strain rate and pore pressure, and the dominant mechanism is selected depending on its kinetics. On one hand, increasing mean stress seems to favour porosity decrease (and possible solid-phase changes like the calcite–aragonite phase transition). On the other hand, increasing differential stress is generally accompanied by grain size reduction and an increase in the crack density.

One of the main parameters controlling the earthquake cycle is known to be not only the differential stress but also the pore pressure (Rice 1992). Owing to the new fracture network and increase in fault gouge permeability, poro-elasticity predicts a viscous release of the pore pressure during the co- and post-seismic phases. Such variations in pore pressure suggest that the stress path a gouge material experiences during the seismic cycle is likely to be much more horizontal than regular triaxial paths in the **PQ** plot. This last observation highlights the limitations of the 'regular triaxial' stress path when trying to understand fault gouge behaviour or the brittle–ductile transition. Recent analysis of the seismic cycle by Miller (2002) and Shapiro *et al.* (2003) are pointing towards that way. As a consequence, we believe it is necessary to perform experiments along different stress paths (for example fluid pressurization) in order to elucidate the natural failure processes and their interplay. Further experimental investigations of the cataclastic regime using acoustic emission localization techniques, could, potentially, provide insights.

A. Schubnel would like to thank warmly Professor R.P. Young for supporting this research under an NSERC grant and for his many instructive comments. This work was initiated when L. Burlini was a visiting scientist in the Laboratoire de Géologie of ENS Paris. The authors would also like to thank Professor B. Evans and an anonymous reviewer, whose comments helped improve this paper greatly. Also to be thanked are V. Vajdova, W. Zhu and T.-F. Wong for very useful and instructive discussions when A. Schubnel performed additional experiments in the Rock Physics Laboratory at SUNY Stony Brook under an NSF/CNRS co-operation program. This work also benefited from discussions with many scientists. Among them, we would like to thank particularly B. Thompson, K. Mair, J. Hazzard and Pierre Bésuelle. G. Marolleau and T. Descamps were indispensable for their technical help on the rig.

References

ASHBY, M.F. & SAMMIS, C.G. 1990. The damage mechanics of brittle solids in compression. *Pure and Applied Geophysics*, **133**, 489–521.

ATKINSON, B.K. & AVDIS, V. 1980. Fracture mechanics parameters of some rock-deforming minerals determined using an indentation technique. *International Journal of Rock Mechanics and Mining Science, Geomechanical Abstract*, **17**, 383–386.

BAUD, P., SCHUBNEL, A. & WONG, T.-F. 2000. Dilatancy, compaction and failure mode in Solnhofen limestone. *Journal of Geophysical Research*, **105**, 19 289–19 320.

BEELER, N.M. & TULLIS, T.E. 1997. The roles of time and displacement in velocity-dependent volumetric strain of fault zones. *Journal of Geophyscial Research*, **102**, 22 595–22 609.

BIRCH, F. 1961. The velocity of compressional waves in rocks up to 10 kbars, part 2. *Journal of Geophysical Research*, **66**, 2199–2224.

BYERLEE, J.D. 1968. Brittle–ductile transition in rocks. *Journal of Geophysical Research*, **73**, 4741–4750.

CRAMPIN, S., VOLTI, T., CHASTIN, S., GUDMUNDSSON, A. & STEFANSSON, R. 2002. Indication of high pore-fluid pressure in a seismically-active fault zone. *Geophysical Journal International*, **151**, F1–F5.

CURRAN, J.H. & CARROLL, M.M. 1979. Shear stress enhancement of void compaction. *Journal of Geophysical Research*, **84**, 1105–1112.

DIETERICH, J.H. & KILGORE, B.D. 1994. Direct observation of frictional contacts – new insights for state dependent properties. *Pure and Applied Geophysics*, **143**, 283–302.

EDMOND, J.M. & PATERSON, M.S. 1972. Volume changes during the deformation of rocks at high pressure. *International Journal of Rock Mechanics and Mining Science*, **9**, 161–182.

FORTIN, J., SCHUBNEL, A. & GUÉGUEN, Y. 2005. Elastic wave velocities and permeability evolution during compaction of Bleuswiller sandstone. *International Journal of Rock Mechanics and Mining Science*, in press.

FREDRICH, J.T., EVANS, B. & WONG, T.-F. 1989. Micromechanics of the brittle to plastic transition in Carrara marble. *Journal of Geophysical Research*, **94**, 4129–4145.

FREDRICH, J.T., EVANS, B. & WONG, T.-F. 1990. Effect of grain size on brittle and semi-brittle strength: implications for micromechanical modelling of failure in compression. *Journal of Geophysical Research*, **95**, 10 907–10 920.

FISHER, G.J. & PATERSON, M.S. 1992. Measurement of permeability and storage capacity in rocks during deformation at high temperature and pressure. *In*: EVANS, B. & WONG, T.-F. (eds) *Fault Mechanics and Transport Properties of Rocks*. Academic Press, San Diego, CA, 213–252.

FRYE, K.M. & MARONE, C. 2002. Effect of humidity on granular friction at room temperature. *Journal of Geophysical Research*, **107**, art. no. 2309.

GRIGGS, D.T., TURNER, F.J. & HEARD, H.C. 1960. Deformation of rocks at 500° to 800°. *In*: GRIGGS, D.T. & HANDIN, J. (eds) *Rock Deformation*. Geological Society of America Memoir, **79**, 39–104.

HEARD, H.C. 1960. Transition from brittle fracture to ductile flow in Solnhofen limestone as a function of temperature, confining pressure and interstitial fluid pressure. *In*: GRIGGS, D.T. & HANDIN, J. (eds) *Rock Deformation*. Geological Society of America Memoir, **79**, 193–226.

KACHANOV, M. 1982*a*. A microcrack model of rock inelasticity, part I: Frictional sliding on microcracks. *Mechanics of Materials*, **1**, 19–27.

KACHANOV, M. 1982*b*. A microcrack model of rock inelasticity, part II: Propagation of microcracks. *Mechanics of Materials*, **1**, 29–41.

LEHNER, F. & KACHANOV, M. 1996. On modelling 'wing-cracks' forming under compression. *International Journal of Fracture*, **77**, R69.

MARONE, C. 1998. Laboratory-derived friction laws and their application to seismic faulting. *Annual Review of Earth and Planetory Science*, **26**, 643–696.

MILLER, S.A. 2002. Properties of large ruptures and the dynamical influence of fluids on earthquakes and faulting. *Journal of Geophysical Research*, **107**, 2182.

PATERSON, M.S. 1958. Experimental deformation and faulting in Wombeyan marble. *Geological Society Bulletin*, **69**, 465–476.

PATERSON, M.S. 1969. The ductility of rocks. *In*: ARGON, A.S. (ed.) *Physics and Crystal Strength*. MIT Press, Cambridge, MA.

PATERSON, M.S. 1978. *Experimental Rock Deformation: The Brittle Field*. Springer, New York.

RENNER, J. & RUMMEL, F. 1996. The effect of experimental and microstructural parameters on the transition from brittle failure to cataclastic flow of carbonate rocks. *Tectonophysics*, **258**, 151–169.

RICE, J.R. 1992. Fault stress states, pore pressure distributions and the weakness of the San Andreas fault. *In*: EVANS, B. & WONG, T.-F. (eds) *Fault Mechanics and Transport Properties of Rocks*. Academic Press, London, 475–503.

ROBERTSON, E.C. 1955. Experimental study of the strength of rocks. *Geological Society of America Bulletin*, **66**, 1275–1314.

RUTTER, E.H. 1972. The effects of strain-rate chages on the strength and ductility of Solnhofen limestone at low temperatures and confining pressures. *International Journal of Rock Mechanics and Mining Science*, **9**, 183–189.

RUTTER, E.H. 1974. The influence of temperature, strain rate and interstitial water in the deformation of calcite rocks. *Tectonophysics*, **22**, 331–334.

RUTTER, E.H. 1986. On the nomenclature of failure transitions in rocks. *Tectonophysics*, **122**, 381–387.

SCHUBNEL, A. 2003. *Mécanismes de la dilatance et de la compaction dans les roches de la croûte*. Thèse de doctorat de l'Institut de Physique du Globe de Paris.

SHAPIRO, S.A., PATZIG, R., ROTHERT, E. & RINDSCHWENTNER, J. 2003. Triggering of seismicity by pore-pressure perturbations: Permeability-related signatures of the phenomenon. *Pure and Applied Geophysics*, **160**, 1051–1066.

STROH, A.N. 1957. A theory of the fracture of metals. *Advances in Physics*, **6**, 418–465.

TULLIS, J. & YUND, R.A. 1992. The brittle–ductile transition in feldspar aggregates: and experimental study. *In*: EVANS, B. & WONG, T.-F. *Fault Mechanics and Transport Properties of Rocks*. Academic Press, San Diego, CA, 89–117.

TURNER, F.J., GRIGGS, D.T. & HEARD, H.C. 1954. Experimental deformation of calcite crystals. *Geological Society America Bulletin*, **65**, 883–934.

VAJDOVA, V., BAUD, P. & WONG, T.-F. 2004. Compaction, dilatancy and failure in porous carbonate rocks. *Journal of Geophysical Research*, **109**, art. B05204.

WONG, T.-F. 1990. Mechanical compaction and the brittle-ductile transition in porous sandstones. *In*: KNIPE, R.J. & RUTTER, R.H. (eds) *Deformation Mechanisms, Rheology and Tectonics*. Geological Society, London, Special Publications, **54**, 111–112.

WONG, T.-F., DAVID, C. & ZHU, W. 1997. The transition from brittle faulting to cataclastic flow in porous sandstones: Mechanical deformation. *Journal of Geophysical Research*, **102**, 3009–3025.

Effects of bedding and foliation on mechanical anisotropy, damage evolution and failure mode

PATRICK BAUD[1], LAURENT LOUIS[2,4], CHRISTIAN DAVID[2],
GEOFFREY C. RAWLING[3] & TENG-FONG WONG[4]

[1]*Institut de Physique du Globe (CNRS/ULP), 5 rue René Descartes,
67084 Strasbourg, France (e-mail: pbaud@eost.u-strasbg.fr)*
[2]*Université de Cergy Pontoise, UMR CNRS 7072, Cergy-Pontoise, France*
[3]*New Mexico Bureau of Geology and Mineral Resources,
Socorro, New Mexico, USA*
[4]*Department of Geosciences, State University of New York at Stony Brook,
Stony Brook, NY 11790-2100, USA*

Abstract: In this study we review recent advances in our understanding of anisotropy in rocks, focusing on dilatant and compactant failure in sandstones and in a foliated metamorphic rock. In sandstones, the anisotropy can be associated with bedding, as in the Rothbach sandstone, or it can also be due to shape anisotropy of the grains and/or the pores, as in the Bentheim sandstone. Two scenarios are proposed for the development of anisotropy in these two end members. In a metamorphic rock with strong foliation like the Four-mile gneiss, it has been commonly observed that the brittle strength is minimum at a foliation angle of about 30°–45°. A damage mechanics model is proposed that underscores the dominant role of biotite foliation in the development of microcracking. In contrast it is often observed in sandstones with strong bedding that the strength is minimum in the direction parallel to bedding. New results for the Rothbach sandstone showed that compared to parallel-to-bedding samples: (i) in the brittle faulting regime the perpendicular-to-bedding samples have both a higher strength and dilatancy stress; and (ii) in the cataclastic flow regime the compactive yield envelope for the perpendicular-to-bedding samples expands significantly towards higher stress values.

Significant anisotropy in mechanical behaviour and failure strength may arise from planar rock fabrics such as bedding in sedimentary rocks, cleavage in slates and preferred orientation, and/or arrangement of minerals and cracks in crystalline igneous and metamorphic rocks. Elastic anisotropy of a rock can be related to its fabric, a seismic manifestation of which is shear-wave splitting (e.g. Barruol & Mainprice 1993). Textural anisotropy can also result in pronounced anisotropy of tensile (e.g. Nova & Zaninetti 1990) and compressive (e.g. Donath 1964; Borg & Handin 1966; Vernik *et al.* 1992*a*; Shea & Kronenberg 1993) strength that may be associated with different failure modes and deformation mechanisms, depending on how stress is applied relative to the anisotropy planes. At the borehole scale mechanical anisotropy and anisotropic rock strength can significantly influence the morphology and interpretation of wellbore breakout as well as inference of *in situ* stress (Vernik *et al.* 1992*b*).

The orientation dependence of brittle strength in an anisotropic rock such as slate, gneiss, phyllite, schist, amphibolite and shale has been intensively studied. Traditionally, the strength anisotropy is analysed by introducing a 'plane of weakness' into the empirical Coulomb criterion (Jaeger & Cook 1979) or modified Griffith criterion (Walsh & Brace 1964). These models are limited in that they are strictly only criteria for the activation of slip on a critically oriented 'plane of weakness' or crack and thus cannot predict the ultimate stress for compressive failure of a brittle rock, which is attained after the crack has nucleated, propagated and coalesced to form a macroscopic fault (Paterson 1978). In this study we review recent advances in our understanding of the anisotropy of dilatant and compactant failure. How are the damage development and brittle faulting controlled by the interplay of textural anisotropy and dilatancy? While damage mechanics models (e.g. Horii & Nemat-Nasser 1986; Kemeny &

From: BRUHN, D. & BURLINI, L. (eds) 2005. *High-Strain Zones: Structure and Physical Properties.*
Geological Society, London, Special Publications, **245**, 223–249.
0305-8719/05/$15.00 © The Geological Society of London 2005.

Cook 1987; Ashby & Sammis 1990) have successfully captured the progressive development
of dilatant failure in an isotropic rock, can they
be modified to describe anisotropic failure?

While bedding represents one type of planar
anisotropy, it should be noted that anisotropy
may also derive from the preferred alignment
of inequant voids in a sedimentary rock. What
are the geometric attributes associated with
these two types of anisotropy and how are they
manifested in the physical properties? To illustrate this point, we first contrast the petrophysical
properties of the Rothbach and Bentheim sandstones, which show relatively strong and weak
bedding anisotropies, respectively. Measurements of ultrasonic velocity, electrical conductivity, permeability and magnetic susceptibility
were conducted on samples cored in three
orthogonal directions. In parallel microstructural
observations and X-ray computed tomography
(CT) measurements were performed, and a
methodology is developed whereby the relative
contributions of bedding and pore space anisotropy can be inferred. These petrophysical data
provide the context for the subsequent discussion
of mechanical anisotropy, damage evolution and
failure mode.

We next consider the relatively compact Four-
mile gneiss. The brittle strength of this foliated
rock had been investigated by Gottschalk *et al.*
(1990), and with constraints from new data on
the onset of dilatancy and micromechanics we
extended the damage mechanics model of
Ashby & Sammis (1990) to analyse the crack
nucleation around a pre-existing weak phase
and the influence of the biotite foliation on
subsequent damage accumulation.

Finally, we consider related anisotropy issues
in sedimentary rocks, in which bedding is pervasive as a primary structure resulting from deposition. To what extent is the phenomenology of
brittle failure in an anisotropic sedimentary
rock similar to that in a foliated metamorphic
rock? In such a porous rock the pore space may
dilate or compact in response to an applied deviatoric stress field, and the brittle–ductile transition is sensitively dependent on the interplay
of dilatancy and compaction. How does
bedding influence the development of compactive yield, and are the effects comparable to
those on the development of dilatant failure and
brittle faulting? To what extent can bedding
anisotropy influence the development of strain
localization and failure modes associated with
the brittle–ductile transition? We chose the
Adamswiller, Rothbach and Bentheim sandstones that had been investigated by David
et al. (1994), Wong *et al.* (1997) and Louis

et al. (2003). Pertinent published data and new
results from this study are reviewed to address
some of these questions.

In this paper we will adopt the convention that
compressive stresses and compactive strains (i.e.
shortening and porosity decrease) are positive,
and denote the maximum and minimum (com-
pressive) principal stresses by σ_1 and σ_3, respectively. The pore pressure will be denoted by
P_p, and the difference between the confining
pressure ($P_c = \sigma_2 = \sigma_3$) and pore pressure will
be referred to as the 'effective pressure', P_{eff}.
The effective mean stress $(\sigma_1 + 2\sigma_3)/3 - P_p$
will be denoted by P and the differential stress
$\sigma_1 - \sigma_3$ by Q.

Bedding, foliation and anisotropy of physical properties

Traditionally, the bedding and foliation anisotropies are inferred from observations on hand
specimens or characterized by microstructural
observations. However, recent advances have
been made in the use of petrophysical measurements (such as elastic moduli, magnetic
susceptibility, electrical conductivity and permeability) for the characterization of anisotropic
behaviour of rocks in relation to both matrix and
pore space. The matrix (or solid phase) of a rock
can be anisotropic because of layering or preferred mineral orientation, associated, for
example, with magmatic flow in igneous rocks,
water current during deposition of sedimentary
rocks, mineral growth or pressure solution in
response to an anisotropic stress field. The pore
space distribution can be anisotropic because of
the sedimentation processes controlled by
gravity that often result in transversely isotropic
rock formations, depositional processes driven
by water currents and the presence of preferentially oriented cracks within or between the minerals. In the latter case, the cracks appear mainly
following non-isotropic stress conditions or in
the course of loading–unloading episodes.

The best way to evaluate the anisotropy of any
physical property in the laboratory is to work on
rock samples with a spherical shape (Vickers &
Thill 1969; Hrouda *et al.* 1993), thus avoiding
uncertainties due to rock heterogeneity between
multiple samples and presenting always the same
shape to the measuring apparatus (geometry
of contact, constant volume or distance of investigation). However, owing to the difficulty of
machining spheres, it is generally easier to
work on cylindrical cores. In the latter case, the
optimal conditions only occur in the plane
perpendicular to the core axis (i.e. across

diameters): this, however, makes the dimension of the problem fall to 2. Despite this limitation, the ease of obtaining such a shape and its relevance with regards to several kinds of measurements (magnetic susceptibility, electrical conductivity, acoustic velocities, permeability) led us to work on cylinders. We drilled in each block three samples in orthogonal directions: one perpendicular to the bedding and two within the bedding plane in random orientations. As shown in Figure 1, samples are oriented with respect to the block and to the bedding: the X and Y samples have their core axis within the bedding plane, whereas the Z sample is perpendicular to it. The size of the drilled cylinders is approximately 22.5 mm long by 25 mm diameter, which corresponds to the standard size for magnetic susceptibility studies in the laboratory at Cergy-Pontoise. We present in this paper our results on the Bentheim and the Rothbach sandstones.

Anisotropy of the elastic properties of dry and saturated rock

We measured the time of flight for P waves travelling across eight different diameters with an angular offset of 22.5° between each measurement. This was done first on dry samples, then on the same samples saturated with water, in both cases at room conditions. Taking into account the error on the travel time readings on the oscilloscope and on sample length, the standard error for the measurements on dry samples is ± 0.03 km s^{-1}, and ± 0.02 km s^{-1} for the measurements on water-saturated samples. For the three orthogonal samples, only 21 measurements out of the total 24 are independent because each sample has two common geographical directions with the others (Fig. 1).

We show in Figure 2a the velocity data as a function of azimuth for three orthogonal samples of dry Bentheim sandstone. The modal composition, porosity and grain size for the Bentheim sandstone are compiled in Table 1. The petrographic analysis does not indicate appreciable bedding anisotropy or planar fabric in this yellowish sandstone (Van Bareen et al. 1990; Klein et al. 2001), and therefore the appreciable anisotropy of approximately 10% in P wave velocity (Fig. 2a) was somewhat unexpected. Notice that the low anisotropy for the Z sample tells us that the Bentheim sandstone can be considered as a transversely isotropic rock formation. The measurements conducted on water-saturated Bentheim sandstone samples showed appreciable reduction of velocity anisotropy (Fig. 2b). Integrating these two sets of data, Louis et al. (2003) attributed the velocity anisotropy to geometric attributes of the pore space that were described by a set of oblate

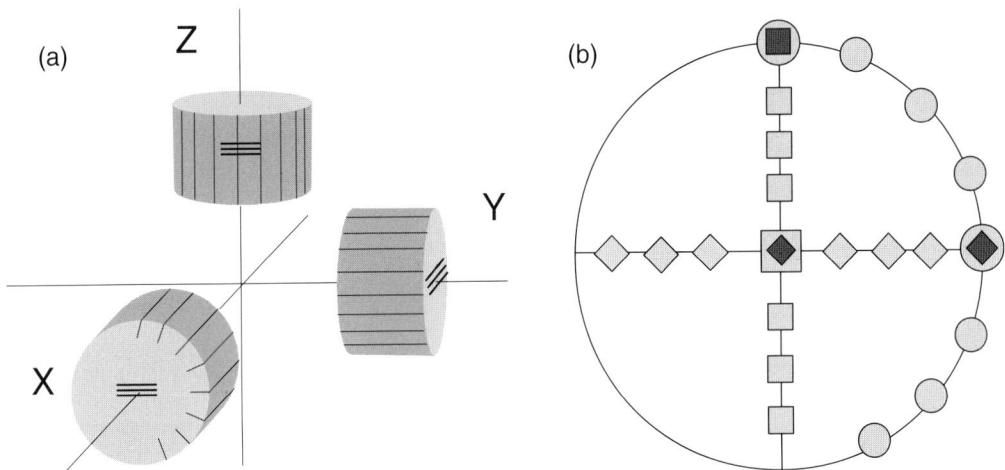

Fig. 1. (a) Orientation of the three sampled elements. The velocity measurements are made across several diameters with an angular spacing of 22.5° (the black lines on the cylindrical surfaces correspond to the contact lines with the ultrasonic transducer). The bedding plane (white lines) corresponds to the XY plane. (b) Stereographic plot (equal-area, lower-hemispheric projection) of the 21 measured positions. Eight directions of measurement are available in each plane (XY, XZ, YZ). The overlapping symbols show the directions that are common for each subset of two samples.

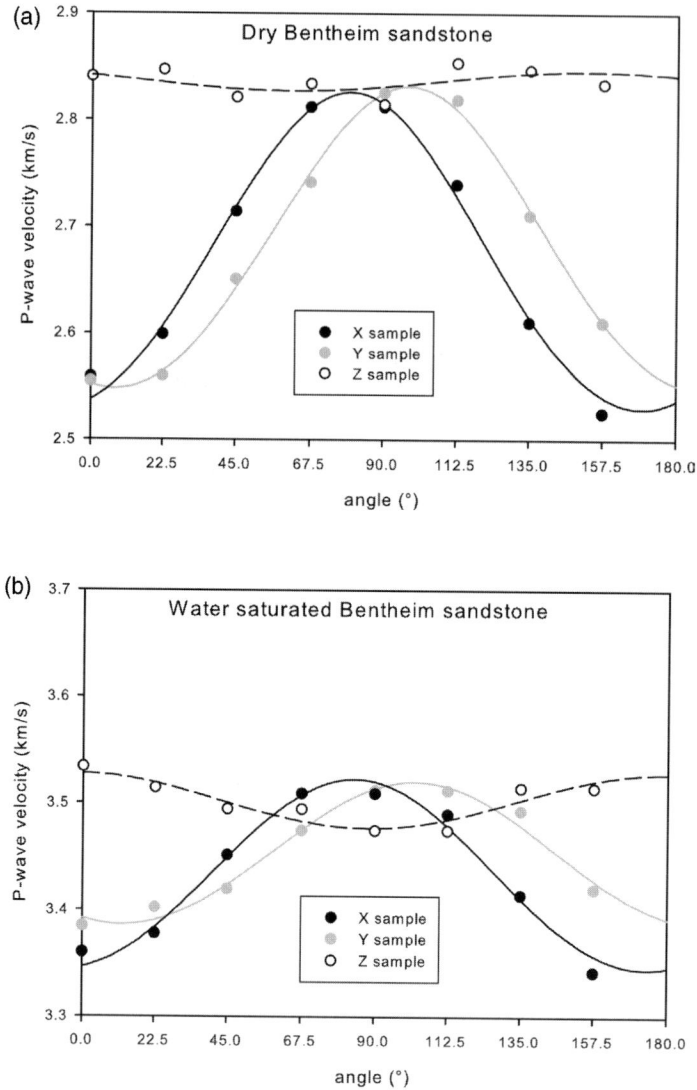

Fig. 2. (a) P wave velocity data in dry Bentheim sandstone samples *X*, *Y* and *Z*, and corresponding fitting curves assuming that the velocity data can be described by a second-order tensor. (b) same as (a) for water-saturated Bentheim sandstone.

ellipsoids embedded in an elastic matrix. This model, as well as pertinent electrical conductivity and permeability data, are discussed in the next sections.

We show in Figure 3 the velocity data as a function of azimuth for orthogonal samples of dry and saturated samples of the Rothbach sandstone. The modal composition, porosity and grain size of this reddish Triassic sandstone from the Vosges Mountains, eastern France are compiled in Table 1. The Rothbach Formation

was probably deposited in channel conditions, as cross-laminations of about 1 cm in height are observed. Owing to these laminations the Rothbach sandstone appears heterogeneous and anisotropic, with a bedding clearly visible in the block. Given the bedding anisotropy of Rothbach sandstone, the velocity anisotropy measured in the dry samples (Fig. 3a) is somewhat lower than expected. Nevertheless, pronounced anisotropy was observed in the saturated samples for which the acoustic signals were of better

Table 1. *Petrophysical description of the anisotropic rocks*

Rock	Porosity (%)	Equivalent grain radius (mm)	Modal analysis	Reference
Bentheim sandstone	24.50 ± 0.18	0.1–0.3	quartz 95%, kaolinite 3%, feldspar 2%	Louis *et al.* (2003)
Rothbach sandstone	21.70 ± 0.83	0.23	quartz 68%, feldspar 16%, oxides and micas 3%, clays (mostly illite) *c.* 12%	Louis *et al.* (2003)
Four-mile gneiss	0.5–0.9	0.025–0.8 (quartz) 0.005–0.55 (microcline)	quartz 29.0%, plagioclase 46.1%, microcline 14.8%, biotite 9.0%, muscovite *c.* 1.0%	Gottschalk *et al.* (1990); Rawling *et al.* (2002)
Adamswiller sandstone	22.15–23.60	0.05	quartz 44%, K feldspar 38%, mica 12%, chlorite 6%	Millien (1993); Gatelier *et al.* (2002)
Navajo sandstone	23.7–24.2	0.04–0.08	quartz 65–71%, poly. quartz 5–8%, feldspar 1%	Dunn *et al.* (1973)
Kayenta sandstone	15.9–16.7	0.05–0.14	quartz 71–77%, poly. quartz 5–7%, feldspar 1–2%, calcite 1%	Dunn *et al.* (1973)
Cutler sandstone	9.2–15.4	0.03–0.06	quartz 55–61%, calcite 22–28%, poly. quartz 3–5%, feldspar 1–3%, micas 2%	Dunn *et al.* (1973)
Navajo sandstone	4.6–5.6	0.05–0.09	quartz 66–72%, calcite 19–24%, poly. quartz 2–4%, feldspar 1–3%	Dunn *et al.* (1973)

quality (Fig. 3b). Notice that the velocity data are not fully compatible with transverse isotropy. In fact, when looking at the data in detail, a triaxial fabric is found, with the maximum velocity axis oblique with respect to the bedding pole. The increase in velocity anisotropy when saturating the rock samples with water is not supported by a possible pore space anisotropy as in Bentheim sandstone. It was shown that the velocity data for the Rothbach sandstone can be interpreted by a granular model that accounts for anisotropy of cemented contacts associated with bedding (Louis *et al.* 2003). This model, as well as pertinent magnetic measurements, are discussed in the next sections.

Anisotropy of magnetic susceptibility, permeability and electrical conductivity

In addition to acoustic velocity measurements, we also investigated the anisotropy of magnetic susceptibility (AMS), permeability and electrical conductivity in Bentheim and Rothbach sandstones.

Magnetic susceptibility studies are versatile in that the full tensor can be determined on one single cylindrical sample; they provide a measurement of induced magnetization in all minerals, combining the contribution of diamagnetic, paramagnetic and ferromagnetic phases. The resultant magnetic tensor at the scale of the sample is thus the sum of individual intrinsic magnetic tensors corresponding to each mineral in the rock. The Bentheim sandstone presents an almost isotropic susceptibility with negative values. This can be explained by the dominant presence of quartz, known to be diamagnetic and the grains of which are considered to be isotropic (see Table 1). Conversely, the AMS tensor obtained for Rothbach sandstone is quite well defined, with a mean susceptibility that is positive paramagnetic. Both clays and oxides (Table 1) can carry such a susceptibility. The

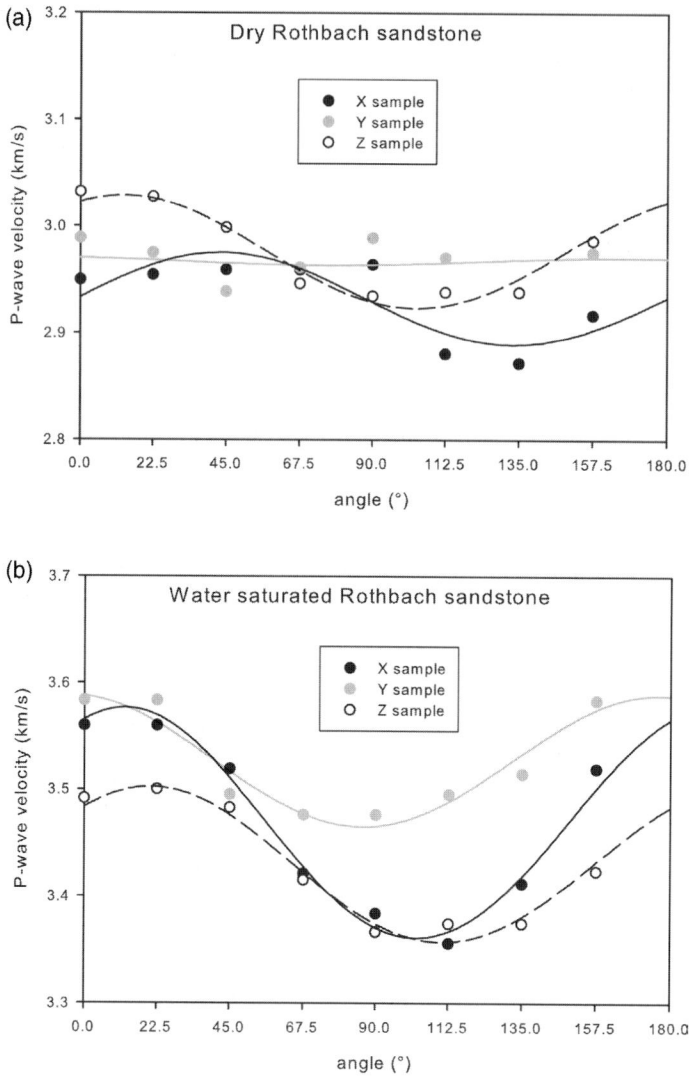

Fig. 3. (**a**) P wave velocity data in dry Rothbach sandstone samples *X*, *Y* and *Z*, and corresponding fitting curves assuming that the velocity data can be described by a second-order tensor. (**b**) same as (a) for water-saturated Rothbach sandstone.

three principal susceptibility values are given in Table 2, and we found that the maximum and intermediate axes are almost parallel to the bedding plane. This corresponds to a typical sedimentary fabric that is developed under sedimentary compaction (Hrouda 1982). It is worth noting that slight obliquity corroborates the presence of cross-bedding in the Rothbach sandstone.

The permeability was measured on large, water-saturated samples (4 cm in diameter, 8 cm in length) at very low effective pressure

(<2 MPa), using the steady-state flow method (Metz 2002). The data show (Table 2) that both sandstones are anisotropic but the magnitude of the variations observed is different. Indeed, we found that for the Rothbach sandstone the bedding effect results in a dramatic permeability decrease (from about 140×10^{-15} down to 27×10^{-15} m^2) when comparing fluid flow in a direction parallel to bedding to fluid flow in the direction perpendicular to bedding.

This can be easily understood if one considers the heterogeneity in the rock microstructure for

Table 2. *Anisotropy of magnetic susceptibility, permeability and electrical conductivity. For AMS the data correspond to the principal axes of the full tensor averaged on several samples (~bp' means that the axis is almost in the bedding plane). For the permeability and formation factor, the given values correspond to measurements for the X, Y and Z samples*

	Magnetic susceptibility (10^{-6})	Sample	Permeability (10^{-15} m^2)	Formation factor
Bentheim sandstone	Min -1.28	X	991	11.5 ± 0.2
	Int -1.20	Y	1214	
	Max -1.04	Z	813	13.4 ± 0.2
Rothbach sandstone	Min 26.6	X	149	
	Int 27.58 (approximately bp)	Y	132	22.6 ± 0.5
	Max 27.92 (approximately bp)	Z	27	

that sandstone. We carried out a CT-scan study using a medical scanner to investigate the density distribution in a Rothbach sandstone sample. Figure 4a clearly shows the presence of several narrow dense bands (dark) alternating with less dense zones (bright), which reveals that the laminations visible at the rock surface correspond to areas where the grain packing has a much higher density, probably associated with a reduction in porosity. This is confirmed in Figure 4b, which shows a porosity map obtained from a thin section analysis on a surface of about 60 mm². We built this map by moving, step by step, a 100×100 pixel size window on the image, characterizing each time the average porosity, ϕ, in the window. As in Figure 4a, we can see the succession of high-porosity zones (white areas, $\phi > 20\%$) followed by low-porosity bands (dark grey areas, $\phi < 10\%$). When fluid flows in the direction perpendicular to these bands it has to go through these compact zones, which results in a low permeability. However, when fluid flows in a direction parallel to the bands it can bypass these regions and most of the flow will be confined in the high-porosity regions, hence an overall larger permeability. This is consistent with our permeability measurements in Rothbach sandstones.

Another important feature of the pore geometry can induce the permeability variations, namely the roughness of the solid–fluid interface. Figure 5a represents a SEM micrograph that shows that clays are present almost everywhere as a coating material on the grain surfaces. This results in an enhanced surface roughness.

(a)

(b)

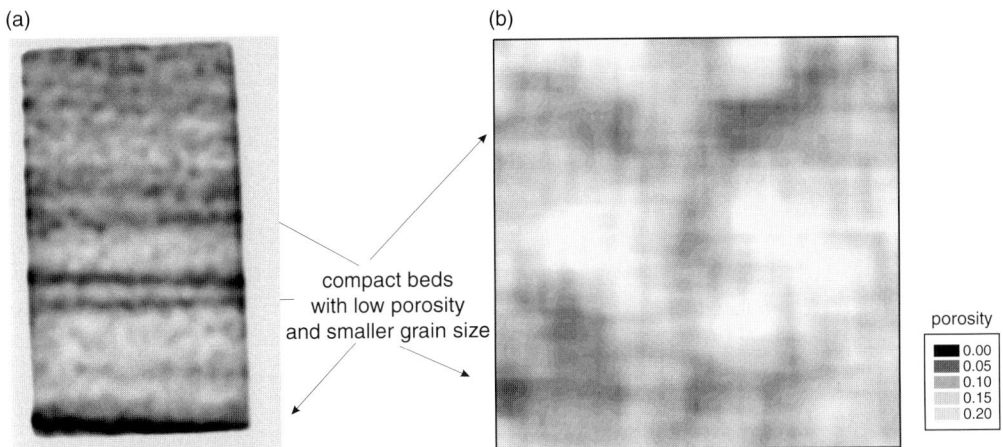

compact beds
with low porosity
and smaller grain size

porosity
0.00
0.05
0.10
0.15
0.20

Fig. 4. (**a**) X-ray slice obtained on a Rothbach sandstone sample using a medical scanner. One can clearly recognize the presence of small horizontal layers of higher density (dark regions) inside the rock sample. The size of the sample is about 4 cm in height. (**b**) Map of porosity obtained on a thin section in Rothbach sandstone. The area covered by the analysis is a square with 7.8 mm length.

Notice, however, that the chemical analysis that we carried out under the SEM tells us that clays are also present as cementing material at the grain contacts: this will have strong consequences on the mechanical behaviour and elastic properties of the Rothbach sandstone, as will be discussed later. Another way to image surface roughness is by using the confocal laser scanning microscope (Menéndez *et al.* 2001), a technique that allows for the three-dimensional (3D) reconstruction of the rock pore space at the scale of a few grains. This was performed from sets of images obtained at various depths on an epoxy-injected sample, and one of the reconstructions is shown in Figure 5b. In this figure the pores appear as the opaque phase, whereas the grains are transparent. Owing to the submicron resolution, one can see clearly the complex topography of the pore–grain interface associated with the presence of coating clays. It is, however, more difficult to detect clays as cementing material from such images. The consequence is that the large surface roughness will lead to a reduction in permeability and, depending on the spatial distribution of clays in the sandstone, this may also lead to anisotropic transport properties (as well as anisotropic elastic properties and rock strength, as will be discussed later). For the Bentheim sandstone, the higher permeability in the bedding plane is consistent with the pore anisotropy, as discussed in the next sections, because the pore connectivity is enhanced for oblate ellipsoids aligned along the bedding plane.

The formation factor data in Table 2 were derived from our measurements of electrical conductivity on samples saturated successively with brines with increasing salinity (Louis *et al.* 2003). As for the permeability we do not have enough information to define the full tensor. For the Bentheim sandstone, the data are consistent with transverse isotropy: the formation factor is significantly larger in the direction perpendicular to the bedding (z direction), which shows that electrical transport is enhanced within the bedding plane. This is in agreement with the pore anisotropy model, which will be discussed in the next section. Surprisingly, we found no significant anisotropy for the electrical properties of the Rothbach sandstone. Although one can imagine that the clays can have a specific influence on electrical conductivity, we have no definitive explanation for that result which is in disagreement with our observations on the permeability anisotropy.

Anisotropic inclusion and cemented granular models

In many respects, the Bentheim and Rothbach sandstones represent two end members. To model the elastic anisotropy of the Bentheim sandstone Louis *et al.* (2003) considered the simple case of elastic anisotropy that arises from non-spherical pores embedded within an isotropic solid phase. The velocity data in Figure 2 indicate that the Bentheim sandstone is more compliant in the Z direction and stiffer in the (X, Y) plane, which suggests an inclusion model made up of oblate ellipsoids, with their short axes parallel to the Z direction in order to

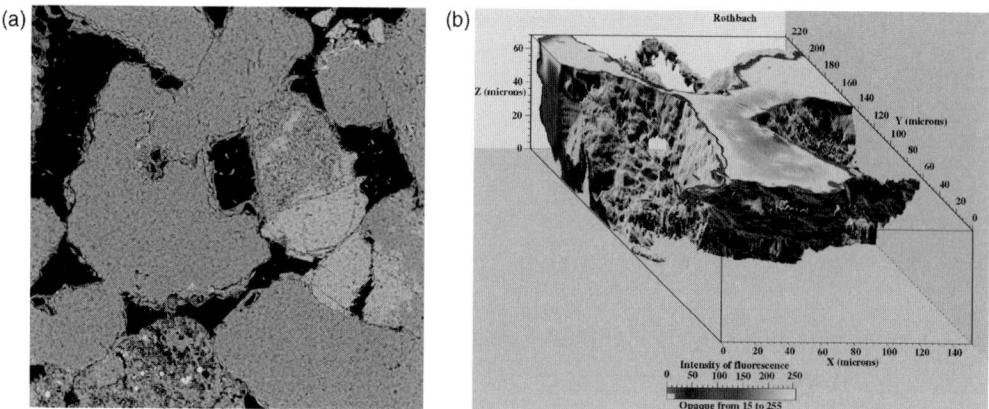

Fig. 5. (**a**) SEM image in one of the compact layers in Rothbach sandstone. Notice the clay coating on free surfaces, but also at grain contacts. (**b**) Three-dimensional reconstruction of the pore space in a Rothbach sandstone sample confocal scanning laser microscopy images. Notice the roughness at the grain surface associated with the presence of a coating of clay.

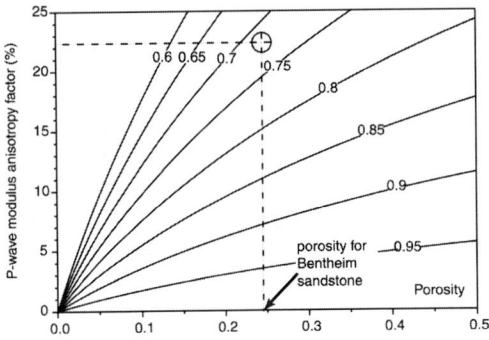

Fig. 6. Variations of the P-wave modulus anisotropy factor as a function of porosity predicted by the Kachanov (1993) model. The plain curves correspond to different pore aspect ratios. The model predicts a pore aspect ratio between 0.7 and 0.75 for the Bentheim sandstone.

produce an overall anisotropic media. Kachanov's (1993) model was used to arrive at expressions for the Young's modulus and the Poisson ratio for a homogeneous solid containing empty cavities of ellipsoidal shape. Without going into the details (see Louis *et al.* 2003), the model predicts the evolution of the P wave modulus, defined as $M = \rho V_p^2$ (where V_p is the speed of compressional waves and ρ is the density), as a function of the rock porosity and pore aspect ratio. The results are shown in Figure 6 where the elastic modulus anisotropy is plotted as a function of porosity for different values of the pore aspect ratio. Taking into account the measured porosity for the Bentheim sandstone, the model predicts that the pore aspect ratio should range somewhere between 0.7 and 0.75, a moderate value that seems to be realistic for a sandstone in that

porosity range. However, the velocity data for the Rothbach sandstone are not supported by pore space anisotropy. For that sandstone, a model was proposed (Louis *et al.* 2003) in which the anisotropy in elastic properties is associated with the anisotropic distribution of cement at grain contacts, the pores being isotropic in the model. The starting point of the model was the observation that at the microscopic scale, the Rothbach sandstone looks very heterogeneous and anisotropic, with the presence of small dense bands in which clays are concentrated. If we assume that the clays play a major role as cementing material between adjacent grains, the consequence is that statistically there are more cemented bonds in the bedding plane than in directions oblique to that plane. This was conceptualized by adapting the Dvorkin & Nur (1996) model that allows for the calculation of the effective elastic moduli of granular frameworks made of spheres with radius R cemented by a material with known elastic properties. Our input was to consider that the length of the cemented bonds is not constant, but varies in direction with a maximum value a_{max} in the bedding plane and a minimum value a_{min} in the direction perpendicular to the bedding plane. The input parameters of the model (Fig. 7a) include the mean grain radius R, and the bulk and shear moduli of the grains (K_{grain}, G_{grain}) and of the cementing material (K_{cement}, G_{cement}). All these parameters have been derived from the rock composition and microstructural studies (Wong *et al.* 1997). To account for the absolute value of the P wave velocity and its variation with azimuth, the length of the cemented contacts has to range between 0.55 and 0.65 times the mean grain size, which gives a contact length ratio a_{min}/a_{max} equal to 0.85

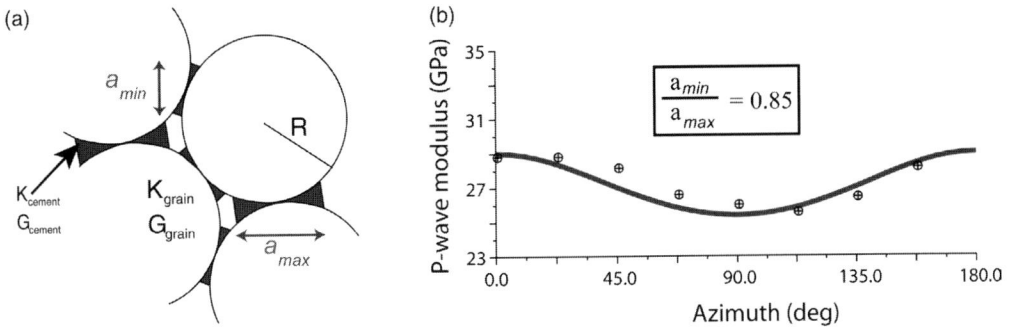

Fig. 7. (a) Sketch representing the anisotropy in cementation length in the Rothbach sandstone. (b) Variation of the P wave modulus for the saturated Rothbach sandstone v. orientation. The solid line shows the prediction of the modified cemented grains scheme of Dvorkin & Nur (1996). The anisotropy is obtained by applying a variable cement distribution. The data are consistent with a cement radius ratio of $\chi = 0.85$.

(Fig. 7b). Therefore, we can see that a mild anisotropy in cemented contact length is enough to
explain our observation on the P wave velocity
anisotropy in Rothbach sandstone. As the
maximum susceptibility axis is oriented in
the bedding plane, the cement distribution in
the model is also in agreement with the magnetic
susceptibility data. Although the model used
does not provide any information on the rock
strength, one would expect that the cement
distribution also leads to a higher strength in
the direction perpendicular to bedding.

Anisotropy of dilatancy and brittle strength

We have underscored that anisotropy of physical
properties can arise from bedding and foliation in
the matrix as well as pore space anisotropy.
When such an anisotropic rock is deformed,
additional complexity arises from stress-
induced anisotropy due to damage evolution
and strain localization. While many empirical
studies have been conducted on the influence of
foliation and bedding on the brittle strength, not
much is known on the micromechanics and
damage evolution during dilatant failure in an
anisotropic rock. Even less is known about mechanisms associated with compaction and strain
localization in anisotropic porous rocks. We
will first summarize recent research on the development of dilatant failure in a compact foliated
rock, and then review what is known about the
influence of bedding on the brittle failure of
porous sedimentary rocks.

Influence of foliation on development of dilatancy and brittle faulting in the Four-mile gneiss

In metamorphic rocks foliation is pervasive as a
secondary structure that is associated with surfaces defined by discontinuities, shape preferred
orientation of inequant minerals (which often
coincide with the lattice preferred orientation)
or laminar mineral aggregates (Hobbs *et al.*
1976). For several decades numerous studies
have been conducted on the brittle strength of
such rocks as a function of the angle β that the
specimen axis makes with the foliation. The
strength anisotropy, often manifested by a
minimum in the peak stress at $\beta = 30°–45°$
and maxima at $\beta = 0°$ and $90°$ (Fig. 8a:
Donath's 1972 data on phyllite; Fig. 8b:
Niandou *et al.* 1997 data for shale), has been
observed in relatively compact rocks such as
slate, phyllite, and schist (Donath 1972; Nasseri

et al. 1997), and gneiss and amphibolite
(Vernik *et al.* 1992*a*), as well as a porous rock
such as shale (McLamore & Gray 1967;
Niandou *et al.* 1997).

A deeper understanding of the mechanics of
brittle failure requires more detailed investigation of the inelastic behaviour and microstructural evolution. Rawling *et al.* (2002) conducted
such a study on the Four-mile gneiss, and we will
review their data that are pertinent to the present
discussion. The predominant phyllosilicate is
biotite that occurs as isolated grains, and
defines a strong foliation and a lineation within
the foliation (Gottschalk *et al.* 1990). The
samples were cored in five orientations within a
plane perpendicular to the macroscopic foliation
and containing the lineation. The vacuum-dried
samples were triaxially compressed at confining
pressures ranging from 50 to 300 MPa. A petrophysical description of the rocks is given in
Table 1.

The samples all failed by the formation of a
single through-going fault. After attaining a
peak stress, each of the samples underwent an
unstable stress drop. Figure 9a & b show the critical stress C' for the onset of dilatancy and the
peak stress as functions of P_C and β. While the

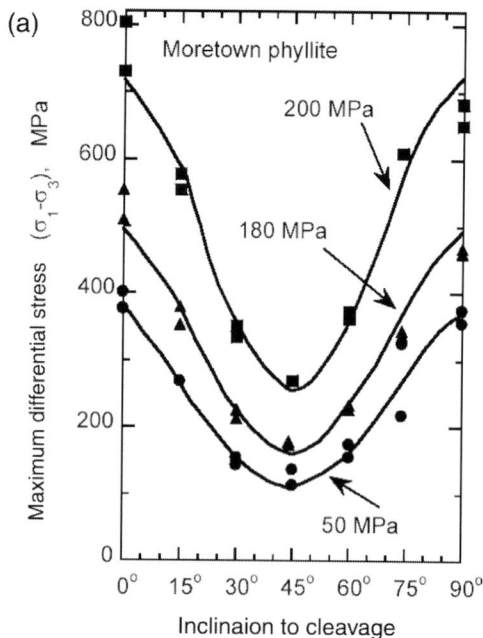

Fig. 8. (**a**) Peak stress as a function of bedding angle
β for selected data on phyllite from Donath (1972).
(**b**) Peak stress as a function of bedding angle β for
selected data on shale from Niandou *et al.* (1997).

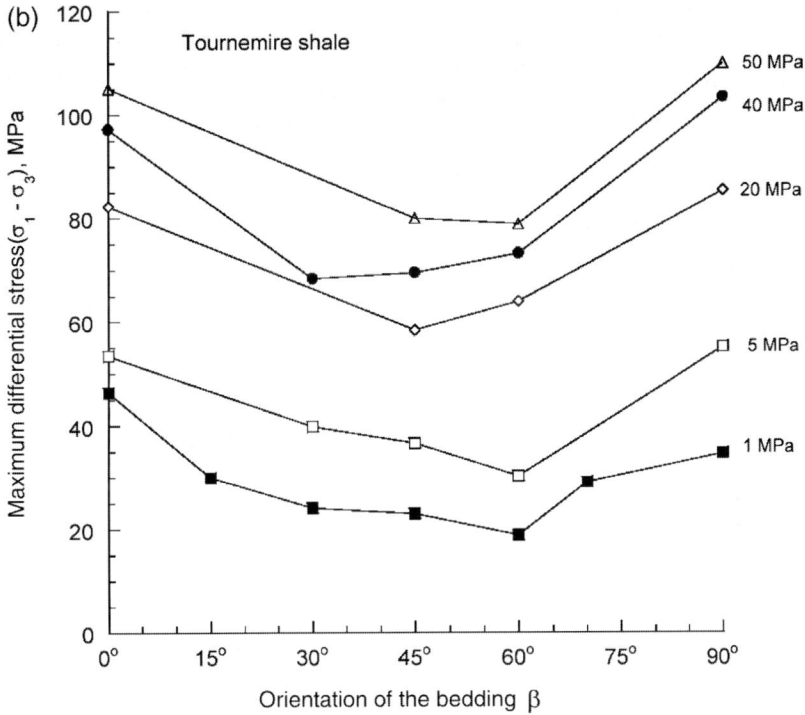

Fig. 8. *Continued.*

anisotropy in peak stress or C' was relatively small at $P_C = 50$ MPa, it is considerable at $P_C = 300$ MPa. The peak stresses follow the trend with a minimum at $\beta = 30°–45°$, and maxima at $\beta = 0°$ and $90°$, typical of texturally anisotropic rocks (Fig. 8).

A key finding here is that the onset of dilatancy and peak stresses follow qualitatively similar trends in anisotropy, with concomitant variation of C' and brittle strength as functions of foliation orientation and confining pressure. The foliation exerts significant control over the onset of dilatancy as well as brittle fracture. Even though such a positive correlation between dilatancy and strength anisotropies is implicitly assumed in the interpretation of wellbore breakout and inference of *in situ* stress (Vernik *et al.* 1992*b*), there had been a paucity of data that provide the mechanical basis for this plausible assumption.

Anisotropic nucleation of wing cracks in a foliated rock

Scanning electron microscope (SEM) observations were conducted on the deformed

samples to explore the damage evolution during dilatant failure of this foliated rock. The most significant observation is that microcracks nucleated around biotite grains oriented favourably for frictional slip on cleavage (Fig. 10a). Shear deformation in the biotite would cause a stress concentration at the end of the grain, which was relieved by the formation of extensile 'wing cracks'. Such a scenario would be most favoured at $\beta = 30°$ and $45°$, resulting in an earlier onset of dilatancy at these orientations. While the foliation provides a plane of weakness for nucleation of these extensile cracks, the ultimate failure develops from the coalescence of a multiplicity of such wing cracks (Fig. 10b) in a scenario analogous to that documented for the relatively isotropic Westerly granite (Tapponier & Brace 1976; Wong 1982) and San Marcos gabbro (Wong & Biegel 1985).

In many aspects, this scenario is captured by the 'sliding wing crack' model (Horii & Nemat-Nasser 1986; Ashby & Hallam 1986; Kemeny & Cook 1987). If the resolved shear stress on an inclined crack, which may be identified as a cleavage crack in a biotite grain, exceeds the frictional strength, slip occurs and tensile stress concentration develops at the tips

(a)   onset of dilatancy (*C'*)

(b)   peak stress

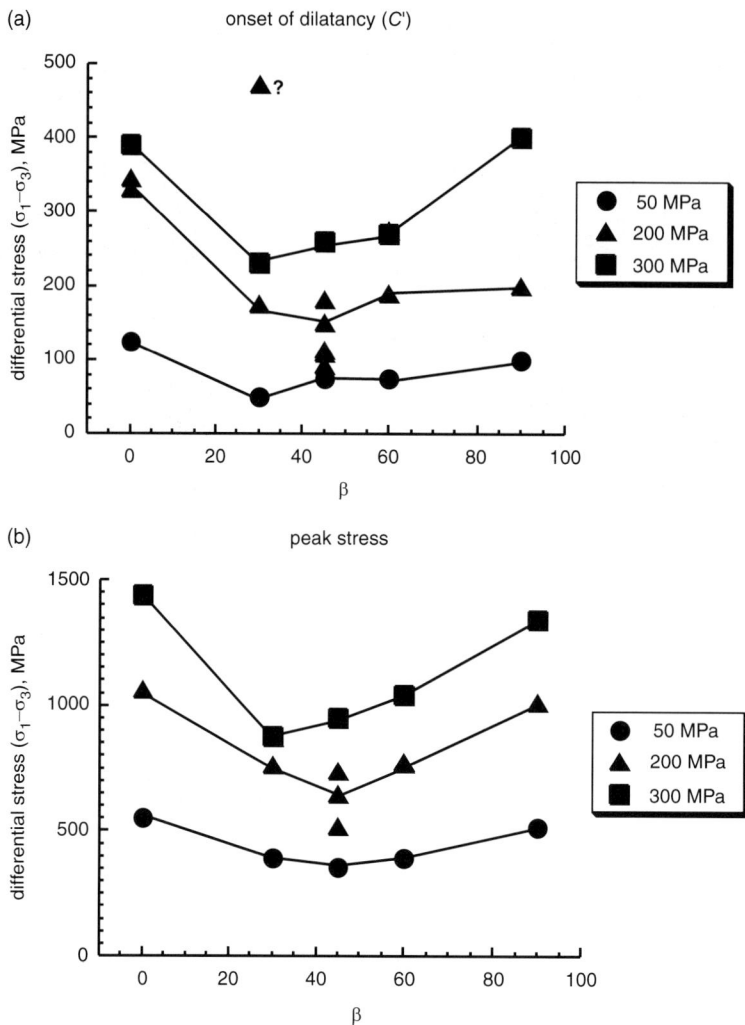

Fig. 9. (**a**) Stress for the onset of dilatancy and (**b**) peak stress as functions of foliation angle β and confining pressure in Four-mile gneiss.

of the inclined sliding crack (Fig. 11). Extensile wing cracks nucleate and propagate along a direction subparallel to σ_1. To analyse this first stage of damage development, consider a crack of length $2a$ inclined at an arbitrary angle γ to σ_1 (Fig. 11). When frictional slip occurs on this inclined crack, the stress concentrations at its tips may induce 'wing cracks' to nucleate at an angle of about 70° to the sliding crack. As summarized by Rawling *et al.* (2002), conventionally the sliding cracks are assumed to be randomly oriented, and such an 'isotropic nucleation model' has its limitations. Because biotite has a

relatively low frictional coefficient (Horn & Deere 1962), frictional slip will be activated first in biotite if favourably oriented cleavages are available. However, in the biotite grains the potential sliding surfaces will not be randomly oriented, as assumed in the 'isotropic nucleation' model. A majority will have orientations close to $\gamma = \beta$, the macroscopic foliation angle, corresponding to a 'plane of weakness' (Jaeger & Cook 1979; Walsh & Brace 1964). Accordingly Rawling *et al.* (2002) derived an 'anisotropic nucleation' condition (with γ fixed and equal to β) which predicts that the wing cracks will first

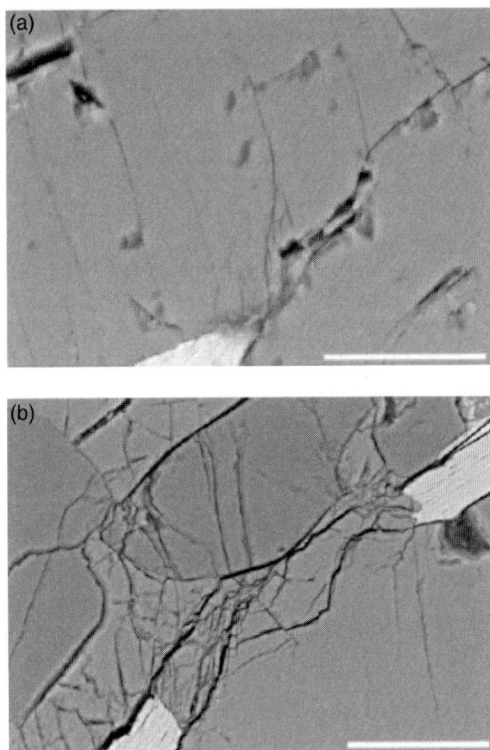

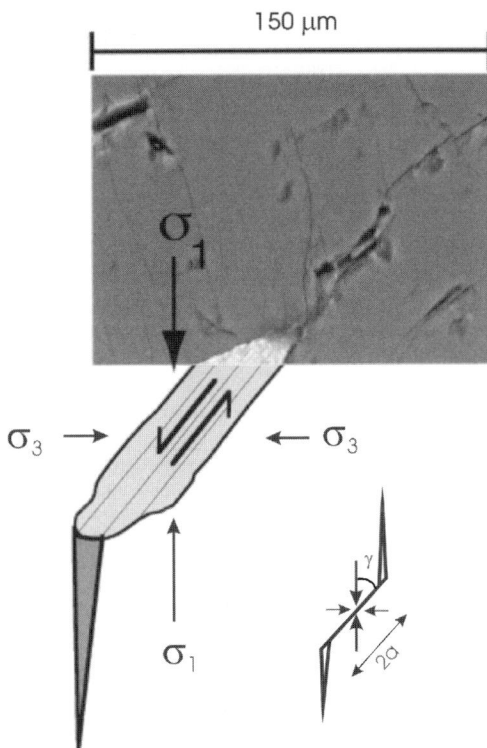

Fig. 10. (a) SEM image of a sample deformed to just beyond the onset of dilatancy C' showing nucleation of wing cracks from the tip of a biotite grain (lighter area in the lower part of the micrograph). Direction of σ_1 was vertical, and scale bar is 50 μm. (b) SEM image of a sample deformed near the peak stress showing crack coalescence and brecciation between adjacent biotite grains. Arrays of subvertical cracks are also present. Direction of σ_1 was vertical, and the scale bar is 100 μm.

Fig. 11. Schematic diagram of a wing crack nucleated from a sliding crack. Geometric parameters of the model are also defined.

nucleate when the principal stresses are related by:

$$\sigma_1 = m\sigma_3 + c \tag{1}$$

with

$$m = \frac{\sin 2\beta + \mu(1 + \cos 2\beta)}{\sin 2\beta - \mu(1 - \cos 2\beta)} \tag{2a}$$

and

$$c = \frac{\sqrt{3}}{\sin 2\beta - \mu(1 - \cos 2\beta)} \frac{K_{1C}}{\sqrt{\pi a}}. \tag{2b}$$

If we identify the onset of dilatancy with the nucleation of wing cracks, then C' should

correspond to the critical stress state given by equation (1). As the stresses follow linear trends, the slope and intercept can be determined by linear regression. From the slope m, we can use equation (2a) to infer the coefficient μ for frictional sliding on the inclined crack surface. As shown in Figure 12 the μ values so inferred for the anisotropic model are somewhat lower than those calculated from the isotropic model, with smaller scatter. Values inferred for the relatively isotropic Westerly granite and San Marcos gabbro are also shown in Figure 12. For all angles, the inferred values for Four-mile gneiss (with 9% biotite) are lower than that of Westerly granite (with 5% biotite). For the intermediate angles, our inferred μ values are comparable to that of San Marcos gabbro (with 12% biotite) and within the range (0.26–0.31) determined for frictional sliding on cleavage surfaces of biotite (Horn & Deere 1962). These data indicate that mica content influences the brittle failure process.

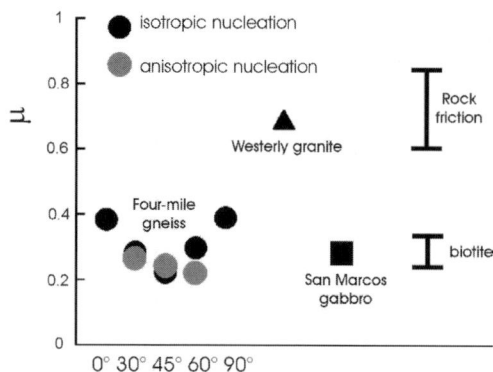

Fig. 12. Friction coefficient as a function of foliation angle β from dilatancy onset data on the basis of the wing crack model using isotropic and anisotropic nucleation conditions. Friction coefficients for Westerly granite (from data of Brace *et al.* 1966) and San Marcos gabbro (from data of Hadley 1973) were calculated using the isotropic nucleation model. For comparison, experimental values for rock friction (Byerlee 1978) and biotite friction (Horn & Deere 1962) are shown.

Damage mechanics of dilatant failure in foliated rocks

The initial propagation of a wing crack is stable, but the mutual interaction of the stress fields of multiple wing cracks may lead to instability, which corresponds to the onset of shear localization and macroscopic fracture. For this second stage, Ashby & Sammis' (1990) two-dimensional damage mechanics model was adopted to analyse the influence of mica content and foliation orientation on brittle strength. The key damage parameter in this model is the crack density $D = \pi(\lambda + a\cos\gamma)^2 N_A$, where λ is the length of the wing crack and N_A is the number of sliding cracks per unit area initially present. Before wing cracks nucleate, the length $\lambda = 0$ and therefore the initial damage is given by $D_o = \pi(a\cos\gamma)^2 N_A$. If one specifies the material parameters D_o, $K_{IC}/\sqrt{\pi a}$ and μ, then the evolution of the principal stress σ_1 as a function of damage D at a fixed confining stress σ_3 can be calculated using the Ashby & Sammis (1990) model. In the brittle regime, the damage accumulation is manifested first by strain hardening and then by strain softening. The critical stress state at which instability occurs is identified as the peak value at the transition from hardening to softening for each curve. Repeating the calculation for different values of fixed σ_3 allows one to map out the brittle failure envelope in the principal stress space. To a first approximation this failure envelope for the wing crack model

(Baud *et al.* 2000*a*, *b*) can be described by a linear relation

$$\sigma_1 = A(\mu, D_o)\sigma_3 + B(\mu, D_o) K_{IC}/\sqrt{\pi a}. \quad (3)$$

If triaxial compression data for the onset of dilatancy and peak stress follow the linear trends described by equations (1) and (3), then the slopes and intercepts of the two sets of stress data provide four constraints for inferring the three parameters D_o, $K_{IC}/\sqrt{\pi a}$ and μ.

Rawling *et al.* (2002) used the onset of dilatancy data and equation (1) to constrain μ, and the peak stress data and equation (3) to constrain D_o and $K_{IC}/\sqrt{\pi a}$. Their inferred values of $K_{IC}/\sqrt{\pi a}$ fall in a relatively narrow range (of 80–95 MPa) and do not show any systematic trends with foliation (Fig. 13). If we make the plausible assumption (e.g. Fredrich *et al.* 1990) that the sliding crack length $2a$ can be approximated by the average grain size, then the fracture toughness K_{IC} is inferred to range from 4.5 to 5.3 MPa m$^{1/2}$, comparable to the high end of experimental values for silicate rocks (Atkinson & Meredith 1987).

While the values of D_o (Fig. 13) are comparable to those obtained by Ashby & Sammis (1990) for granite, eclogite, dunite and gabbro, there is a trend for the initial damage to be higher in the intermediate range of foliation angles, indicating a negative correlation with the peak stress. This implies that the reduction of brittle strength in the intermediate range of foliation angle arises from an enhancement of initial damage, as well as reduction of friction coefficient along the mica cleavages. How does foliation influence the damage state? Rawling *et al.* (2002) interpreted the initial damage to be specifically from two contributions: (1) a set of pre-existing microcracks with random orientation; and (2) a set of cleavage cracks in mica grains preferentially oriented along the foliation angle. The minimum D_o values (of 0.08 and 0.09 at $\beta = 0°$ and 90°, respectively) correspond to the first set, with negligible contribution from mica cleavages. The enhanced D_o values for intermediate β angles arise from the additional contributions from the favourably oriented mica cleavages.

As the damage mechanics model predicts that the brittle strength decreases with increasing initial damage D_o, and if, indeed, the initial damage D_o includes an important contribution from cleavage cracks in mica grains, then the experimental observation that the strength of a foliated rock decreases with increasing mica content can be explained by a positive correlation between the damage and mica content.

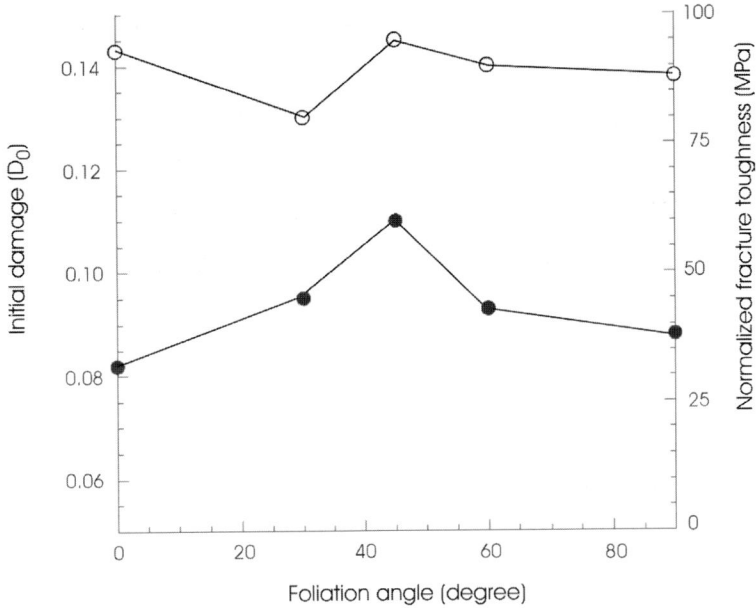

Fig. 13. Inferred value of initial damage (solid circles) and normalized fracture toughness $K_{IC}/\sqrt{\pi a}$ (open circles) as functions of foliation angle of the Four-mile gneiss.

To quantify this correlation Rawling *et al.* (2002) analysed experimental data for the strengths of eight gneisses as a function of mica content f_m (for $\beta = 45°$ and $P_C = 200$ MPa) from Shea & Kronenberg (1993) and our study (Fig. 9). For each gneiss sample, the damage mechanics model (with $\mu = 0.24$ and $K_{IC}/\sqrt{\pi a} = 95$ MPa, values appropriate for Four-mile gneiss with $\beta = 45°$) was used to infer the D_o value that corresponds to the experimentally determined strength for $\sigma_3 = 200$ MPa. Figure 14 shows an approximately linear correlation between the initial damage D_o so inferred and mica content f_m, except for two samples that had anomalously low strengths. Shea & Kronenberg (1993) reported that one of the samples had 3% calcite, which suggests that it may have been altered, perhaps along macroscopic fractures. A linear regression (excluding these two points) gives $D_o = D_o^m f_m + D_o^c$ with $D_o^m = 1.04$ and $D_o^c = 0.06$. In light of the microstructural observations that pre-existing damage is usually associated with cleavage planes in mica acting as sliding cracks, the correlation between D_o and f_m is reasonable. In fact, if one assumes that the damage state in each mica grain in all of the gneisses is about the same, then D_o^m can be interpreted as an 'intrinsic' damage parameter that is relatively constant in all of the grains, whereas D_o^c is damage unrelated to mica content,

representing a population of randomly oriented cracks in the other minerals or grain-boundary cracks. It is of interest to note that this estimate of D_o^c is comparable to the minimum values of D_o (0.08 and 0.09) independently inferred for $\beta = 0°$ and $90°$ in the Four-mile gneiss (Fig. 13), which we have attributed to the same set of pre-existing microcracks with random orientation. Rawling *et al.* (2002) showed that it requires just one cleavage crack in every two mica grains to explain the inferred value of $D_o^m = 1.04$.

Influence of bedding on brittle strength of porous sandstones

The Four-mile gneiss corresponds to the end member of a relatively compact rock, with minimal contribution to mechanical anisotropy from the pre-existing porosity. It has an interconnected porosity of 0.5–0.9% and crack porosity of 0.2–0.3%, which is expected to be closed under elevated pressures (Rawling *et al.* 2002). The damage mechanics model underscores the dominant role of biotite foliation (in the solid matrix) in the nucleation, propagation and coalescence of microcracking during dilatant failure.

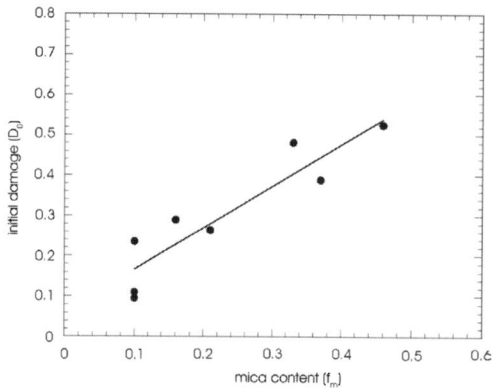

Fig. 14. Inferred value of initial damage as a function of mica content. Two data points have been excluded. The line from linear regression is also shown.

In contrast, if a significant portion of the pore space remains open then damage evolution may be fundamentally different from that of the relatively compact gneiss. Indeed, data for porous sandstones show that the anisotropy in brittle strength and dilatancy follow trends quite different from those in Figures 8 or 9, and three distinct features can be noted. First, the brittle strength for samples cored perpendicular to bedding is systematically higher than those cored parallel to bedding. Our data for Rothbach sandstone at effective pressures up to 20 MPa are shown in Figure 15. As discussed earlier the anisotropy

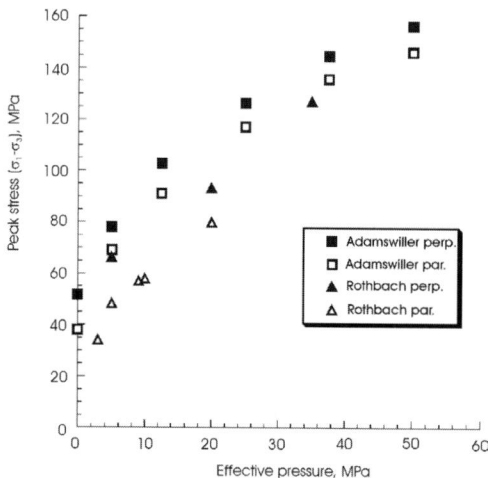

Fig. 15. Peak stress for Rothbach sandstone (this study) and Adamswiller sandstone (Millien 1993). The open symbols are for samples cored parallel to bedding, and the close symbols are for samples cored perpendicular to bedding.

of physical properties in this rock is primarily due to bedding. Millien's (1993) data for Adamswiller sandstone at confining pressures up to 50 MPa are also plotted. Like the Rothbach sandstone this sandstone is from the Vosges region, with bedding anisotropy due to the directional arrangement of the mica minerals. Its P wave velocity parallel to bedding is 15% higher than that perpendicular (Gatelier et al. 2002), and a petrophysical description is given in Table 1. For these two Vosges sandstones, appreciable anisotropy in the brittle strength was observed in the pressure ranges investigated. At effective pressures up to 5 MPa the reduction of the strength for samples parallel to bedding was up to 30%, but seems to stabilize around 7% for the higher pressures. That the anisotropy effect of bedding decreases with increasing effective pressure is in contrast to the strength anisotropy in a foliated rock that tends to increase with increasing pressure (Figs 8a and 9).

Second, the brittle strength of a porous sandstone as a function of bedding angle seems not to follow the general trend (with a minimum at intermediate angles $\beta = 30°-45°$) illustrated in Figure 8. Because a pronounced minimum of brittle strength is commonly observed at intermediate bedding angles in shale (Fig. 8b), there is sometimes the misconception that such a trend universally applies to all porous sedimentary rocks. It was probably Dunn et al. (1973) who first pointed out that the strength anisotropy in porous sandstone seems not to follow this trend. Petrophysical descriptions of their samples are compiled in Table 1, and selected data are plotted in Figure 16a. As their sandstone samples were quite heterogeneous and had significant porosity variation, Dunn et al. (1973) were not able to draw a definitive conclusion. In contrast, as the porosity variation (22.15–23.60%) was very small in Millien's (1993) samples, she was able to observe a consistent trend in strength anisotropy at each confining pressure investigated, with a gradual decrease from a maximum for σ_1 normal to bedding to a minimum for σ_1 parallel to bedding (Fig. 16b). Similar data were recently presented by Gatelier et al. (2002), who also show that the gradual change in brittle strength correlates with concomitant change in the dilatancy stress (Fig. 16b). Unlike the foliated rocks or shale, in the porous sandstone neither the onset of dilatancy nor the peak stress dip to a minimum at intermediate bedding angles.

Third, even though the dilatancy and peak stresses depend on the bedding angle, the failure mode and fault angle θ (between the macroscopic shear fracture and σ_1) seem to be

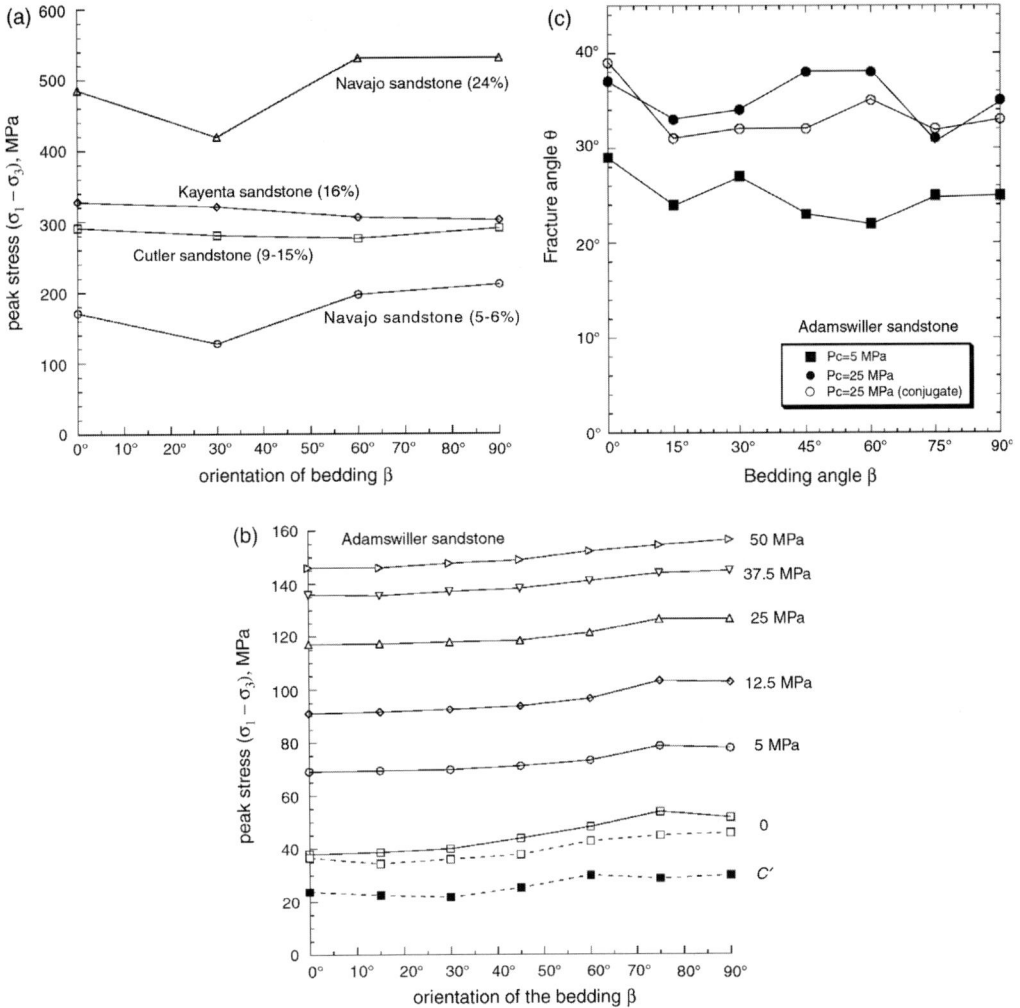

Fig. 16. (**a**) Peak stress as a function of bedding angle β for selected data of Dunn *et al.* (1973). (**b**) Peak stress and onset of dilatancy C' as a function of bedding angle β in Adamswiller sandstone. Data of Millien (1993) are connected by the solid curves, and data of Gatelier *et al.* (2002) are connected by the dashed curves. (**c**) Fault angle θ as a function of bedding angle β for selected data of Millien (1993) on Adamswiller sandstone.

the same for samples at different orientations. For Rothbach sandstone we did not discern any systematic difference in fault angles for samples parallel and perpendicular to bedding. Millien (1993) measured the fault angles of her Adamswiller sandstone samples, and the data for θ at confining pressures of 5 and 25 MPa are shown in Figure 16c. A single shear band was observed in each of the samples deformed at 5 MPa with an average fault angle of 25°, whereas conjugate shear bands were observed at 25 MPa with an average angle of 35°.

At each pressure the fault angle θ did not show any systematic variation with bedding angle β.

Given the notable differences discussed above, one would consider it unlikely that a micromechanical model developed for foliated rocks can apply to a porous sandstone. Because there is a paucity of mechanical and microstructural data for the latter, it may be premature at this point to develop a detailed model in the absence of better mechanical constraints. It is also necessary to have a better understanding of the micromechanics of failure in anisotropic sandstones, as well as other porous sedimentary rocks. Recent microstructural studies (e.g. Menéndez *et al.* 1996; El Bied *et al.* 2002; Mair *et al.* 2002; Bésuelle *et al.* 2003) have focused on sandstone samples cored perpendicular to the bedding. The microstructural observations of deformed samples show that dilatancy and brittle faulting arise from the interplay of breakage of lithified grain contacts, relative grain movement, as well as Hertzian fractures emanating from grain contacts (Fig. 17). A realistic model should capture these key micromechanical processes. Because the failure mode and fault angle do not systematically change with bedding angle (Fig. 16c), these processes probably operate in a qualitatively similar manner for different orientations, and the rather subtle changes in critical stresses (Fig. 16b) may arise from the anisotropy in

grain contact geometry and strength discussed earlier (Fig. 7a).

Bedding anisotropy and the brittle–ductile transition in porous sandstones

At elevated pressures a porous sandstone undergoes compactant failure that may be localized or delocalized. Here we present new mechanical data and microstructural observations on Rothbach sandstone samples that had been cored parallel and perpendicular to bedding. The experimental methodology was identical to that of Wong *et al.* (1997) and Bésuelle *et al.* (2003), and the new data will be compared with these previous studies to provide insights into the effect of bedding anisotropy on

$P_{eff} = 20$ MPa

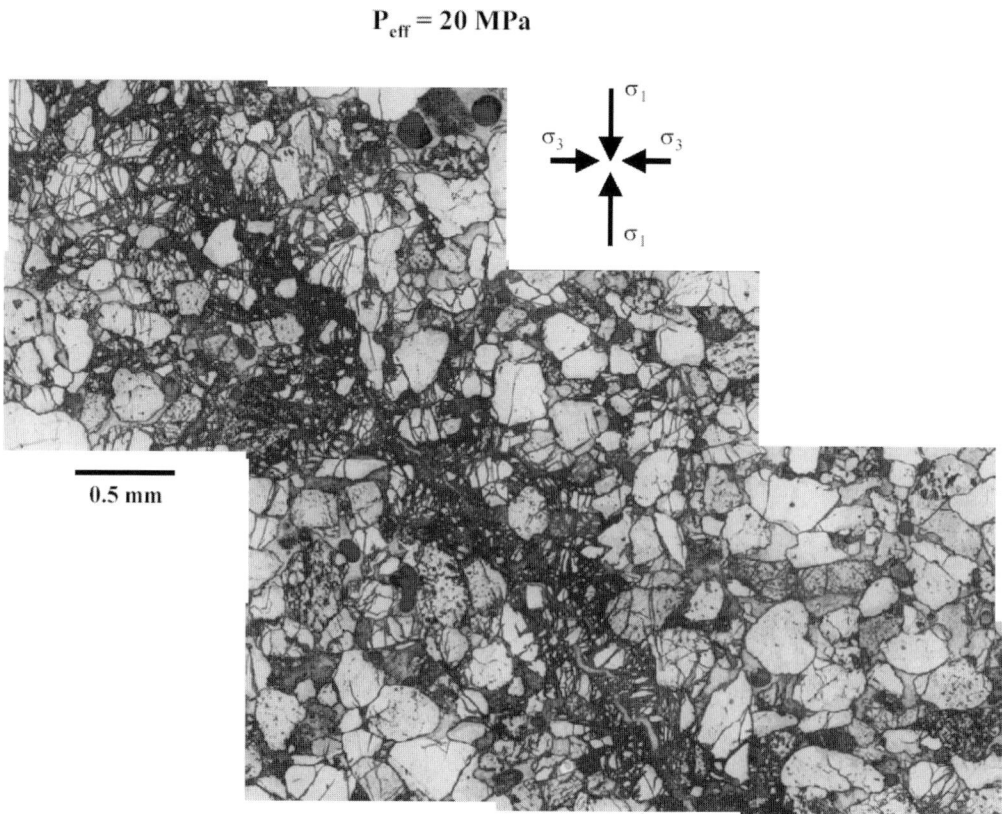

Fig. 17. Mosaic of optical (reflection) micrographs showing part of the shear band that developed in a sample of Rothbach sandstone cored perpendicular to bedding and deformed in the brittle faulting regime.

shear-enhanced compaction, failure mode and strain localization.

Shear-enhanced compaction and anisotropic yield caps

Figure 18a & b present the differential stress as a function of axial strain for Rothbach sandstone samples cored parallel and perpendicular to bedding, respectively. The samples saturated with distilled water were deformed at a confining pressure of 140 MPa and a pore pressure of 10 MPa. While an overall trend of strain hardening was shown in either sample, the stress level attained in the perpendicular-to-bedding sample was appreciably higher.

To underscore the compactive failure behaviour, we plot in Figure 19a & b the effective mean stress as a function of porosity change in these two experiments. In a conventional triaxial compression experiment the non-hydrostatic and hydrostatic loadings are coupled, and if the axial stress increases by an increment $\Delta\sigma_1$, while the confining and pore pressures are maintained constant, then the effective mean stress P and differential stress Q would increase by the amounts $\Delta\sigma_1/3$ and $\Delta\sigma_1$, respectively. If porosity change is solely controlled by the hydrostatic stresses and independent of the differential stress (which is valid in a poroelastic material), then the

triaxial data (solid curves) in Figure 19 should coincide with the hydrostat (dashed curves). Any deviation from the hydrostat would imply that the porosity change in a triaxial compression experiment depends on not only the effective mean stress, but also the deviatoric stresses. It is noted in Figure 19a & b that the triaxial curve coincided with the hydrostat up to the critical stress state C^*, beyond which there was an accelerated decrease in porosity in comparison to the hydrostat. At stress levels beyond C^* the deviatoric stress field provided significant contribution to the compactive strain, and this phenomenon of 'shear-enhanced compaction' (Wong et al. 1997) is attributed to the inception of grain crushing and pore collapse in the sandstone (Menéndez et al. 1996).

Table 3 compiles data of the compactive yield stress C^* for the onset of shear-enhanced compaction in Rothbach sandstone for the two orthogonal orientations, including data of Wong et al. (1997) and Bésuelle et al. (2003). Our new C^* data for parallel-to-bedding samples are in good agreement with the preliminary data of Wong et al. (1997). There is considerable scatter among the data for perpendicular-to-bedding samples: while our new data seem to be in better agreement with those of Wong et al. (1997), the C^* data of Bésuelle et al. (2003) tend to be somewhat lower. This apparent discrepancy may arise because the yield stress for $\beta = 90°$ is

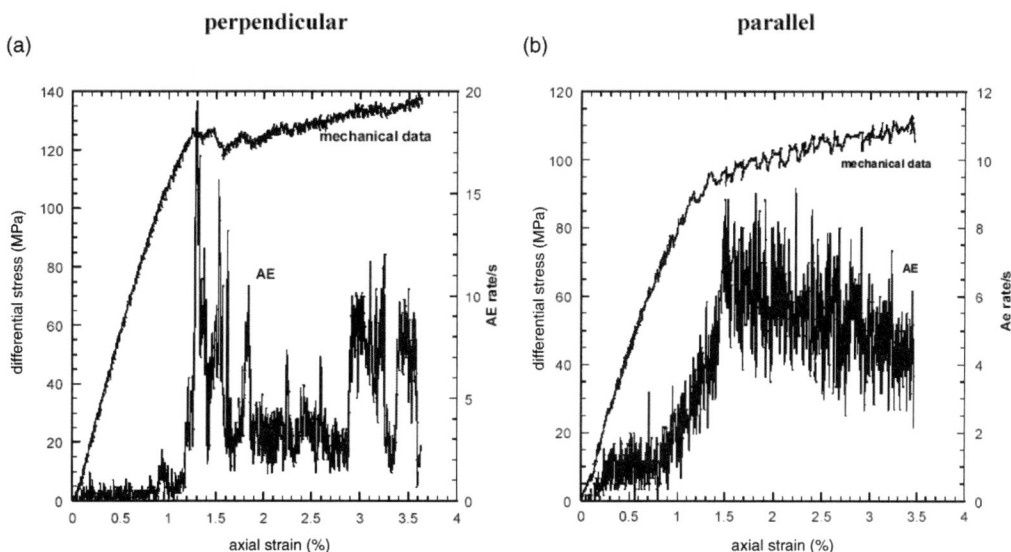

Fig. 18. Principal stress difference and rate of acoustic emissions per s v. axial strain for triaxially deformed samples of Rothbach sandstone at 130 MPa of effective pressure, for samples cored (**a**) perpendicular to bedding and (**b**) parallel to bedding. Principal stress σ_1 was along the axial direction.

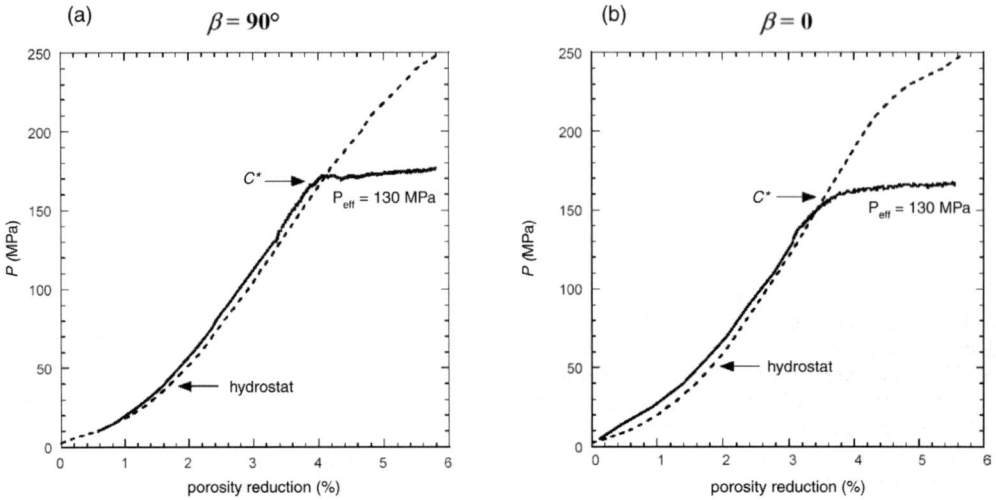

Fig. 19. Effective mean stress as a function of porosity reduction of Rothbach sandstone deformed at 130 MPa of effective pressure for samples cored (**a**) perpendicular to bedding and (**b**) parallel to bedding. For reference, the hydrostats are shown as dashed curves.

sensitive to slight deviation in orientation and, as shown by the CT images in Figure 4 here and figures 3 and 5 of Bésuelle *et al.* (2003), it is quite difficult to obtain cores that are exactly perpendicular to bedding.

Notwithstanding the data scatter, there is an overall trend for the compactive yield stress to be higher for perpendicular-to-bedding samples, as illustrated in Figure 20. Wong *et al.* (1997)

have shown that the yield stresses for the onset of shear-enhanced compaction in porous sandstones can be described by an elliptical cap

$$\frac{(P/P^* - \xi)^2}{(1 - \xi)^2} + \frac{(Q/P^*)^2}{\delta^2} = 1 \qquad (4)$$

with P^* denoting the critical effective pressure for inelastic yield under hydrostatic loading. Our data can be fitted such two elliptical caps,

Table 3. *Critical stress state C^* at the onset of shear-enhanced compaction in Rothbach sandstone samples in two orthogonal orientations*

Effective pressure (MPa)	Differential stress $Q = \sigma_1 - \sigma_3$ (MPa)	Effective mean stress $P = (\sigma_1 + 2\sigma_3)/3 - P_p$ (MPa)
Perpendicular to bedding		
35	127	77
55	128	98
90	122	130
110	136	155
130	125.5	172
150	126	192
165	90	225
Parallel to bedding		
50	90	80
55	96	88
90	99	124
115	101	149
130	88	161
140	84	168
200	30	210

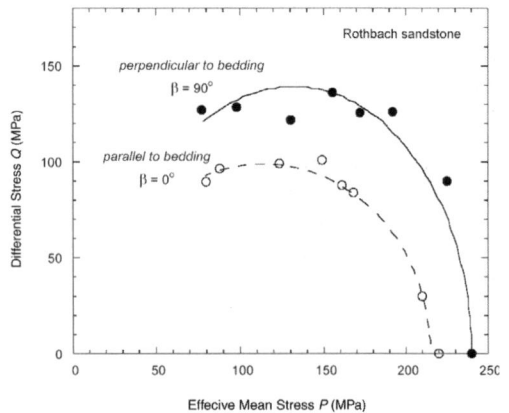

Fig. 20. Elliptical compactive yield envelopes (equation 4) that fit the data for the onset of shear-enhanced compaction for Rothbach sandstone oriented perpendicular (solid symbols) and parallel (open symbols) to bedding.

with $\xi = 0.53$, $\delta = 0.46$, and $P^* = 215$ MPa parallel to bedding and $\xi = 0.55$, $\delta = 0.58$, and $P^* = 240$ MPa perpendicular to bedding. The anisotropy in compactive yield is manifested by differences in three parameters δ, ξ and P^*.

Our sandstone data demonstrate that the effect of bedding on compactive yield is similar to that on dilatancy and brittle strength, in that the critical stresses for the onset of shear-enhanced compaction and dilatancy, as well as the peak stresses for brittle faulting, are consistently higher for perpendicular-to-bedding samples than corresponding stresses for parallel-to-bedding samples. These features should be accounted for in analysing the deformation and failure in porous sandstone formations. An intriguing question that is unresolved is whether the orientation dependence of the compactive yield stress C^* follows the same trend as that for the brittle strength, with a gradual decrease from a maximum for $\beta = 90°$ to a minimum for $\beta = 0°$ (Fig. 16b). It is desirable to conduct a future study to clarify this important question.

Acoustic emission activity and compaction localization

Optical microscopy observations were conducted to characterize the spatial distribution of cracking and damage localization in the failed samples. In selected samples the spatial distribution of damage was quantified following the methodology of Wu et al. (2000) and Bésuelle et al. (2003). The area centrally located in a petrographic thin section was divided into 10×29 subregions, each of which has an area of 1.63×1.22 mm^2. For each subregion the images were acquired in reflected light at a magnification of $\times 100$, and we counted the number of crack intersections with a test array of five parallel lines (spaced at 0.33 or 0.24 mm apart) in two orthogonal directions parallel and perpendicular to σ_1, respectively. If we denote the linear intercept density (number of crack intersections per unit length) for the array oriented parallel to σ_1 by $P_L^{\parallel}$, and that for the perpendicular array by $P_L^{\perp}$, then 290 pairs of stereological parameters were determined that map out the spatial evolution of damage- and stress-induced anisotropy.

The spatial distribution of damage in a triaxially compressed sample cored perpendicular to bedding is expected to be axisymmetric. From geometric probability it can be shown (Underwood 1970) that the crack surface area per unit volume (S_V) is given by

$$S_V = \frac{\pi}{2} P_L^{\perp} + \left(2 - \frac{\pi}{2}\right) P_L^{\parallel} \tag{5a}$$

and the anisotropy of crack distribution is characterized by the parameter

$$\Omega_{23} = \frac{P_L^{\perp} - P_L^{\parallel}}{P_L^{\perp} + (4/\pi - 1)P_L^{\parallel}}. \tag{5b}$$

that represents the ratio between the surface area of cracks parallel to σ_1 and the total crack surface area.

For a triaxially compressed sample cored parallel to bedding, to the extent that the damage can be approximated as a system of surfaces with a preferred planar orientation, then the crack surface area per unit volume is given by (Underwood 1970)

$$S_V = P_L^{\perp} + P_L^{\parallel} \tag{6a}$$

and the anisotropy of crack distribution is characterized by

$$\Omega_{23} = \frac{P_L^{\perp} - P_L^{\parallel}}{P_L^{\perp} + P_L^{\parallel}}. \tag{6b}$$

Figures 21 and 22 contrast the spatial distribution of damage (in terms of the specific crack surface area S_V) in the two oriented samples deformed at effective pressure of 130 MPa (Figs 18 and 19). Baud et al. (2004) recently established a number of connections between failure mode and acoustic emission (AE) activity. The AE rates measured in these experiments are also shown in Figure 18a & b. The stereological data in Figure 21 were previously presented by Bésuelle et al. (2003), who emphasized the development of several elongate bands with intense damage that are subparallel to bedding and cut through the sample. Such a localized structure that is subperpendicular to σ_1 represents a 'compaction band' (Olsson 1999; Issen & Rudnicki 2000). Baud et al. (2004) recently distinguished between compaction bands that extend laterally over only a few (say ≤ 3) grains and ones that extend laterally over many grains. The former category has been documented in the Bentheim sandstone (Wong et al. 2001) and is referred to as a 'discrete' compaction band, and the latter category (Fig. 21) is referred to as a 'diffuse' compaction band. The development of these diffuse compaction bands was associated with several distinct surges in AE activity (Fig. 18a). In Rothbach sandstone the bedding

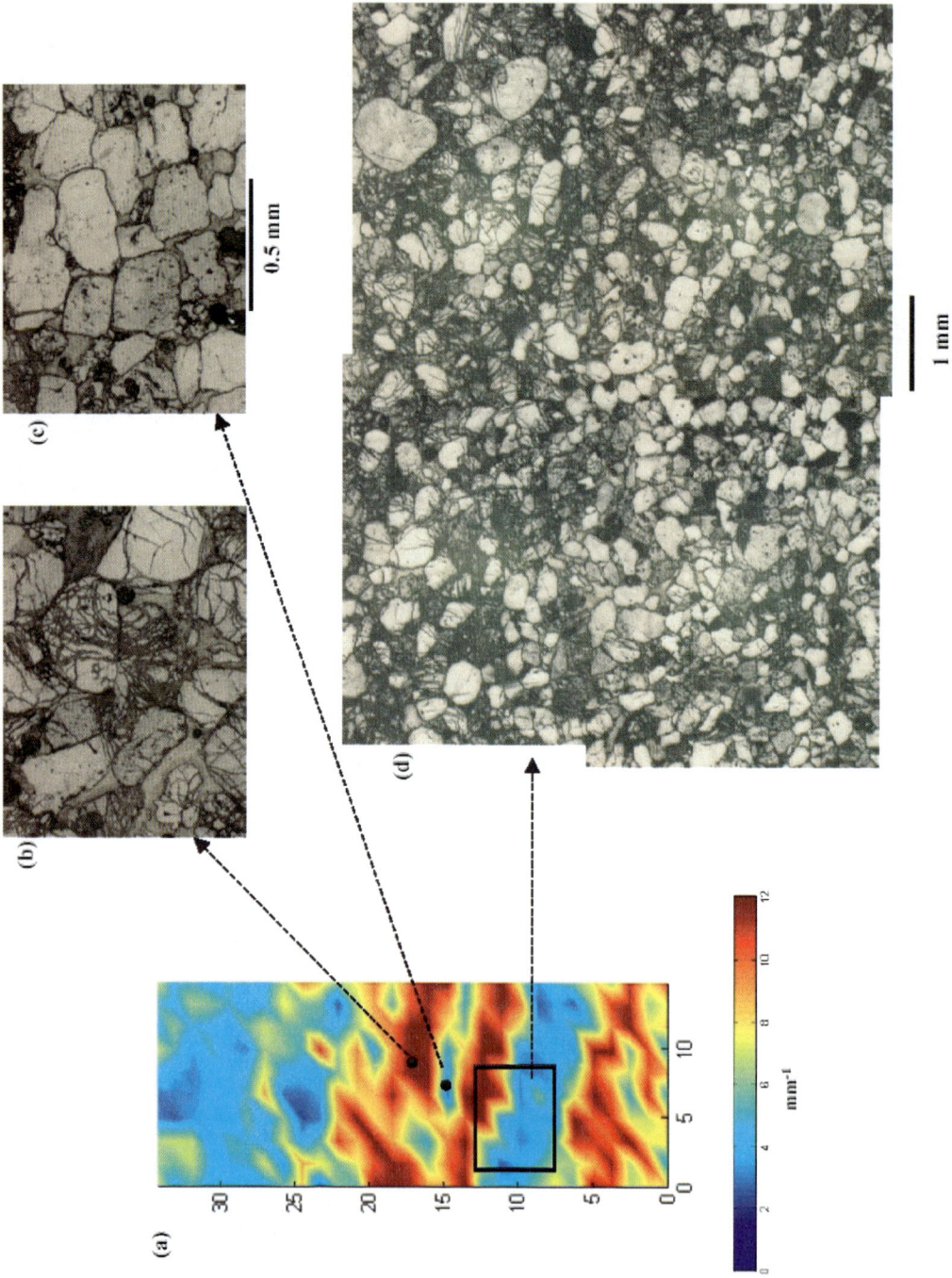

Fig. 21. (**a**) Spatial distribution of specific crack surface area in a sample of Rothbach sandstone cored perpendicular to bedding and deformed at $P_{eff} = 130$ MPa. (**b**) Detail of grain crushing in a diffuse compaction band. (**c**) Relatively undamaged area between two compaction bands. (**d**) Mosaic of micrographs showing the transition between a compaction band and an undamaged zone. Principal stress σ_1 was along the axial direction.

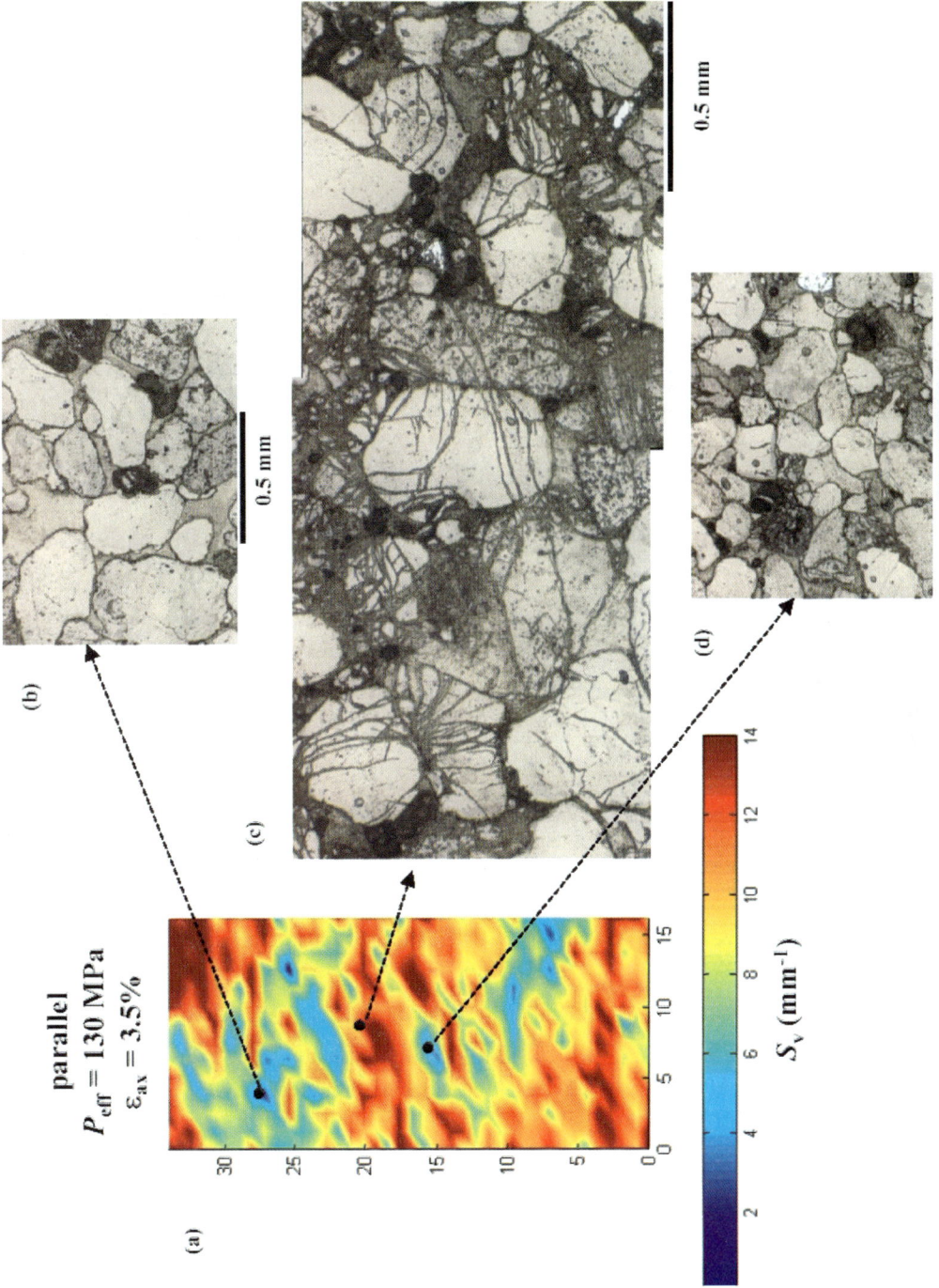

Fig. 22. (a) Spatial distribution of specific crack surface area in a sample of Rothbach sandstone cored parallel to bedding and deformed at $P_{eff} = 130$ MPa. (b) Relatively undamaged area in a more porous layer of the sample. (c) Extensive grain crushing in a cluster of intense damage. (d) Relatively undamaged area in a compact layer of the sample. Principal stress σ_1 was along the axial direction.

anisotropy is manifested by alternating layers of relatively compact and porous materials (Fig. 4), and it is likely that damage first initiated near an interface and then propagated laterally to form a relatively diffuse compaction band between bedding planes.

In the parallel-to-bedding sample, relatively short segments of elongate damage subperpendicular to σ_1 were observed. The overall damage was distributed more homogeneously in that, although elongate clusters had developed, they did not propagate all the way across the sample, possibly because the bedding planes inhibited their continuous growth (Fig. 22). The AE activity did not show any discrete surges (Fig. 18b), indicating that the diffuse mosaic of compaction localization developed from progressive accumulation of damage in the form of intense grain crushing. In contrast, little damage was observed outside the localized structures (Fig. 22). Overall, the specific crack surface area, S_V, in the parallel-to-bedding sample seems to be higher and the anisotropy factor, Ω_{23}, lower than those in their counterparts in the perpendicular-to-bedding sample. However, as several assumptions on damage anisotropy were made in deriving the stereological relations given in equations (5) and (6), we are inclined not to push such a quantitative comparison further in the absence of more comprehensive data.

At lower effective pressures strain localization developed by conjugate shear bands at high angles. In this study we observed conjugate shear bands at high angles for both orientations in samples deformed at effective pressures between 55 and 90 MPa. The compaction localization mode and damage distribution are qualitatively similar for both orientations, indicating that the bedding orientation seems not to have a major influence on the failure mode. As this failure mode for the perpendicular-to-bedding samples was described by Bésuelle *et al.* (in their figs 9 and 10), we will not repeat the details here. The stress–strain curves and AE activity of these samples are shown in Figure 23. For each of the perpendicular-to-bedding samples, the initial surge of AE rate seems to be more pronounced than that for a parallel-to-bedding sample.

Summary and conclusions

Bedding in sedimentary rock and foliation in metamorphic rock can readily be resolved by detailed field and microstructural observations. Both are expected to result in appreciable anisotropy in various petrophysical properties. However, it should be emphasized that in a rock with almost negligible apparent bedding or foliation, other mechanisms exist that may induce significant anisotropy. Synthesizing experimental data on the variations of ultrasonic P wave velocities, magnetic susceptibility, permeability and electrical conductivity in dry and/or saturated samples of Bentheim and Rothbach sandstones, we contrast the interplay of bedding and pore shape and underscore two different scenarios for the development of anisotropy in a porous sandstone. Notwithstanding the absence of appreciable bedding in the Bentheim sandstone, enhancement of electrical conductivity and permeability were observed in the parallel-to-bedding samples. The inference is that the significant anisotropy arises from the inequant shape of the porosity. In contrast, insignificant anisotropy in electrical conductivity but large anisotropy in permeability were observed in the Rothbach sandstone. In addition, the elastic anisotropy was surprisingly smaller, although the bedding is quite evident on visual inspection of the samples. Strong anisotropy in magnetic susceptibility indicates that it is controlled by the matrix, with clay and oxide grains preferentially aligned along the bedding planes. Two different microstructural models were employed to quantify these two anisotropy scenarios.

In a metamorphic rock with strong foliation it is generally observed that significant anisotropy in the brittle strength is manifested by a minimum in the peak stress at foliation angles $\beta = 30°–45°$ and maxima at $\beta = 0°$ and $90°$ with respect to σ_1. Recent measurements on the Four-mile gneiss demonstrate that the critical stress C' at the onset of dilatancy follows a similar anisotropy trend. Microstructural observations reveal the dominant role of biotite foliation in the nucleation, propagation and coalescence of microcracking during dilatant failure. The wing crack model originally developed for isotropic material can be modified to account for the anisotropic behaviour. The anisotropic nucleation and damage mechanics model captures the micromechanics of brittle failure and explains the effect of mica content on the brittle strength of foliated rocks.

Because a pronounced minimum of brittle strength is commonly observed at intermediate bedding angles in shale (Fig. 8b), there is sometimes the misconception that such a trend universally applies to all porous sedimentary rocks. Our synthesis of data for two porous sandstones shows that neither the onset of dilatancy nor the peak stress dip to a minimum at intermediate bedding angles. Instead, a maximum and minimum in the brittle strength (and C') were

perpendicular **parallel**

(a)

(c)

(b)

(d)

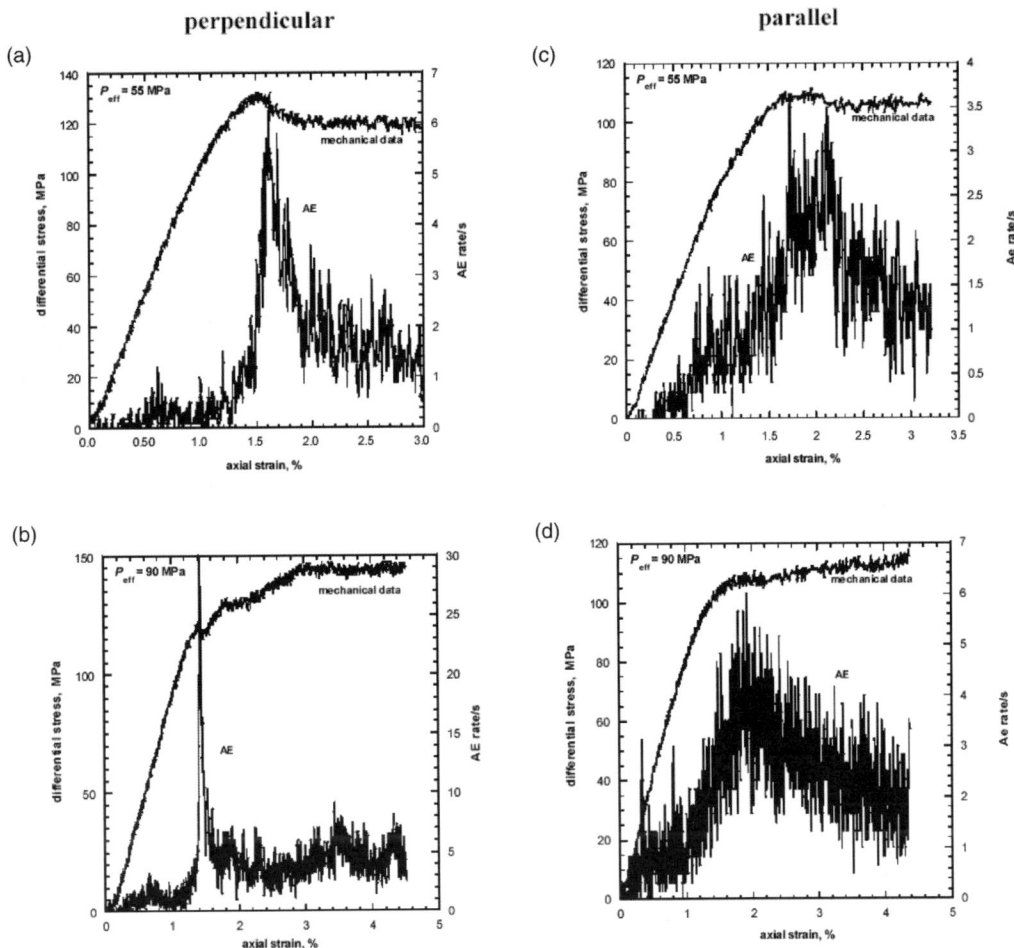

Fig. 23. Principal stress difference and rate of acoustic emissions per s v. axial strain for samples of Rothbach sandstone cored perpendicular to bedding and triaxially deformed at (**a**) 55 MPa and (**c**) 90 MPa of effective pressure, and for samples of Rothbach sandstone cored parallel to bedding and triaxially deformed at (**b**) 55 MPa and (**d**) 90 MPa of effective pressure. Principal stress σ_1 was along the axial direction.

observed in the perpendicular- and parallel-to-bedding samples, respectively. In the intermediate angles the brittle strength increases progressively from the minimum at $\beta = 0°$ to the maximum at $90°$, with concomitant change in C'. Even though these critical stresses are sensitive to the bedding angle, the failure mode and fault angle seem to be the same for samples at different orientations.

New data on the compactive yield stress C^* (associated with the onset of shear-enhanced compaction) of the Rothbach sandstone show that it is consistently higher for perpendicular-to-bedding samples than for parallel-to-bedding samples. Microstructural observations indicate

that overall the modes of failure and compaction localization are similar for the two orientations, with subtle differences that cannot be elucidated without more detailed investigation. An intriguing question that is also unresolved is whether the orientation dependence of C^* follows the same trend as that for the brittle strength, with a gradual decrease from a maximum for $\beta = 90°$ to a minimum for $\beta = 0°$.

We thank P. Bésuelle for drawing our attention to the comprehensive data on Adamswiller sandstone presented in A. Millien's thesis. We have also benefited from discussions with him, Y. Guéguen, B. Menéndez and V. Vajdova. Many thanks to both reviewers for their

248   P. BAUD *ET AL.*

contructive comments. The permeability data were obtained by V. Metz in her first-year graduate study at the University of Cergy-Pontoise. Within the frame of a collaborative research with the University of Oviedo (French–Spanish program PICASSO), we conducted the X-ray analysis of the rock samples at the hospital in Mieres thanks to A. Rodríguez Rey, V. Ruiz de Argandoña and C. Celorio, and the confocal microscopy study with the help of A. Martínez Nistal. The research at Stony Brook was partially supported by the Office of Basic Energy Sciences, Department of Energy under grant DE-FG02-99ER14996 and National Science Foundation under grant INT9815570. The research at Cergy-Pontoise was partially funded by a research grant from Gaz de France.

References

ASHBY, M.F. & HALLAM, S.D. 1986. The failure of brittle solids containing small cracks under compressive stress states. *Acta Metallurgica*, **34**, 497–510.

ASHBY, M.F. & SAMMIS, C.G. 1990. The damage mechanics of brittle solids in compression. *Pure and Applied Geophysics*, **133**, 489–521.

ATKINSON, B.K. & MEREDITH, P.G. 1987. Experimental fracture mechanics data for rocks and minerals. *In*: ATKINSON, B.K. (ed.) *Fracture Mechanics of Rock*. Academic Press, San Diego, CA, 477–525.

BARRUOL, G. & MAINPRICE, D. 1993. A quantitative evaluation of the contribution of crustal rocks to the shear-wave splitting of teleseismic SKS waves. *Physics of the Earth and Planetary Interior*, **78**, 281–300.

BAUD, P., SCHUBNEL, A. & WONG, T.-F. 2000a. Dilatancy, compaction and failure mode in Solnhofen limestone. *Journal of Geophysical Research*, **195**, 19 289–19 303.

BAUD, P., ZHU, W. & WONG, T.-F. 2000b. Failure mode and weakening effect of water on sandstone. *Journal of Geophysical Research*, **105**, 16 371–16 390.

BAUD, P., KLEIN, E. & WONG, T.-F. 2004. Compaction localization in porous sandstones: spatial evolution of damage and acoustic emission activity. *Journal of Structural Geology*, **26**, 603–624.

BÉSUELLE, P., BAUD, P. & WONG, T.-F. 2003. Failure mode and spatial distribution of damage in Rothbach sandstone in the brittle-ductile transition. *Pure and Applied Geophysics*, **160**, 851–868.

BORG, I. & HANDIN, J. 1966. Experimental deformation of crystalline rocks. *Tectonophysics*, **3**, 249–368.

BRACE, W.F., PAULDING, B. & SCHOLZ, C.H. 1966. Dilatancy in the fracture of crystalline rocks. *Journal of Geophysical Research*, **71**, 3939–3954.

BYERLEE, J.D. 1978. Friction of rocks. *Pure and Applied Geophysics*, **116**, 615–626.

DAVID, C., WONG, T.-F., ZHU, W. & ZHANG, J. 1994. Laboratory measurement of compaction-induced permeability change in porous rock: implications for the generation and maintenance of pore pressure excess in the crust. *Pure and Applied Geophysics*, **143**, 425–456.

DONATH, F.A. 1964. Strength variation and deformational behaviour in anisotropic rock. *In*: JUDD, W.R. (ed.) *State of Stress in the Earth's Crust*. Elsevier, New York, 281–297.

DONATH, F.A. 1972. Effects of cohesion and granularity on deformational behaviour of anisotropic rock. *In*: DOE, B.R. & SMITH, D.K. (eds) *Studies in Mineralogy and Precambrian Geology*. Geological Society of America Memoir, **135**, 95–128.

DUNN, D.E., LA FOUNTAIN, L.J. & JACKSON, R.E. 1973. Porosity dependence and mechanism of brittle fracture in sandstones. *Journal of Geophysical Research*, **78**, 2403–2417.

DVORKIN, J. & NUR, A. 1996. Elasticity of high porosity sandstones: Theory for two North Sea data sets. *Geophysics*, **61**, 1363–1370.

EL BIED, A., SULEM, J. & MARTINEAU, F. 2002. Microstructure of shear zones in Fontainebleau sandstone. *International Journal of Rock Mechanics and Mining Sciences*, **39**, 917–932.

FREDRICH, J.T., EVANS, B. & WONG, T.-F. 1990. Effect of grain size on brittle and semi-brittle strength: Implications for micromechanical modelling of failure in compression. *Journal of Geophysical Research*, **95**, 10 729–11 358.

GATELIER, N., PELLET, F. & LORET, B. 2002. Mechanical damage of an anisotropic porous rock in cyclic triaxial tests. *International Journal of Rock Mechanics and Mining Sciences*, **39**, 335–354.

GOTTSCHALK, R.R., KRONENBERG, A.K., RUSSELL, J.E. & HANDIN, J. 1990. Mechanical anisotropy of gneiss: Failure criterion and textural sources of directional behavior. *Journal of Geophysical Research*, **95**, 21 613–21 634.

HADLEY, K. 1973. Laboratory investigation of dilatency and motion on fault surfaces at low confining pressures. *In*: KOVACH, R.L. & NUR, A. (eds) *Proceedings of the Conference on Tectonic Problems of the San Andreas Fault System*. Stanford University, Stanford, 427–435.

HOBBS, B.E., MEANS, W.D. & WILLIAMS, P.F. 1976. *An Outline of Structural Geology*. Wiley, New York.

HORII, H. & NEMAT-NASSER, S. 1986. Brittle failure in compression: splitting, faulting and brittle-ductile transition. *Philosophical Transactions of the Royal Society of London*, **319**, 337–374.

HORN, H.M. & DEERE, D.U. 1962. Frictional characteristics of minerals. *Geotechnique*, **12**, 319–335.

HROUDA, F. 1982. Magnetic anisotropy of rocks and its application in geology and geophysics. *Geophysical Surveys*, **5**, 37–82.

HROUDA, F., ZDENEK, P. & WOHLGEMUTH, J. 1993. Development of magnetic and elastic anisotropies in slates during progressive deformation. *Physics of Earth and Planetary Interior*, **77**, 251–265.

ISSEN, K.A. & RUDNICKI, J.W. 2000. Conditions for compaction bands in porous rock. *Journal of Geophysical Research*, **105**, 21 529–21 536.

JAEGER, J.C. & COOK, N.G.W. 1979. *Fundamentals of Rock Mechanics*, 3rd edn. Chapman and Hall, London.

KACHANOV, M. 1993. Elastic solids with many cracks and related problems. *Advances in Applied Geophysics*, **30**, 259–445.

KEMENY, J.M. & COOK, N.G.W. 1987. Crack models for the failure of rocks in compression. *Proceedings of the International Conference on Constitutive Laws for Engineering Materials*, **2**, 879–887.

KLEIN, E., BAUD, P., REUSCHLÉ, T. & WONG, T.-F. 2001. Mechanical behaviour and failure mode of Bentheim sandstone under triaxial compression. *Physics and Chemistry of the Earth (A)*, **26**, 21–25.

LOUIS, L., DAVID, C. & ROBION, P. 2003. Comparison of the anisotropic behaviour of undeformed sandstones under dry and saturated conditions. *Tectonophysics*, **370**, 193–212.

MAIR, K., ELPHICK, S.C. & MAIN, I.G. 2002. Influence of confining pressure on the mechanical and structural evolution of laboratory deformation band. *Geophysical Research Letters*, **29**, 10.1029/2001GL013964.

MCLAMORE, R. & GRAY, K.E. 1967. The mechanical behaviour of anisotropic sedimentary rocks. *Journal of Engineering for Industry (Transactions of the American Society of Mechanical Engineers, Series B)*, **89**, 62–73.

MENÉNDEZ, B., DAVID, C. & MARTÍNEZ NISTAL A. 2001. Confocal scanning laser microscopy applied to the study of void networks in cracked granite samples and in cemented sandstones. *Computer & Geosciences*, **27**, 1101–1109.

MENÉNDEZ, B., ZHU, W. & WONG, T.-F. 1996. Micromechanics of brittle faulting and cataclastic flow in Berea sandstone. *Journal of Structural Geology*, **18**, 1–16.

METZ, V. 2002. *Perméabilité et déformation de roches granulaires*. Rapport de DEA–Master degree, Université de Cergy-Pontoise.

MILLIEN, A. 1993. *Comportement anisotrope du grès des Vosges: élasto-plasticité, localisation, rupture*. Thèse de Doctorat en Sciences, Université Joseph Fourier, Grenoble.

NASSERI, M.H., RAO, K.S. & RAMAMURTHY, T. 1997. Failure mechanism in schistose rocks. *International Journal of Rock Mechanics and Mining Sciences*, **34**, 460.

NIANDOU, H., SHAO, J.F., HENRY, J.P. & FOURMAINTRAUX, D. 1997. Laboratory investigation of the mechanical behaviour of Tournemire shale. *International Journal of Rock Mechanics and Mining Sciences*, **34**, 3–16.

NOVA, R. & ZANINETTI, A. 1990. An investigation into the tensile behavior of a schistose rock. *International Journal of Rock Mechanics and Mining Sciences*, **27**, 231–242.

OLSSON, W.A. 1999. Theoretical and experimental investigation of compaction bands in porous rock. *Journal of Geophysical Research*, **104**, 7219–7228.

PATERSON, M.S. 1978. *Experimental Rock Deformation – The Brittle Field*. Spinger, New York.

RAWLING, G.C., BAUD, P. & WONG, T.-F. 2002. Dilatancy, brittle strength and anisotropy of foliated rocks: Experimental deformation and micromechanical modeling. *Journal of Geophysical Research*, **107**, 2234, doi:10.1029/2001JB000472.

SHEA, W.T. & KRONENBERG, A.K. 1993. Strength and anisotropy of foliated rocks with varied mica contents. *Journal of Structural Geology*, **15**, 1097–1121.

TAPPONIER, P. & BRACE, W.F. 1976. Development of stress-induced microcracks in Westerly granite. *International Journal of Rock Mechanics and Mining Sciences*, **13**, 103–112.

UNDERWOOD, E.E. 1970. *Quantitative Stereology*. Addison Wesley, Reading, MA.

VAN BAREEN, J.P., VOS, M.W. & HELLER, H.J.K. 1990. *Selection of outcrop samples*. Internal report, Delft University of Technology.

VERNIK, L., LOCKNER, D. & ZOBACK, M.D. 1992a. Anisotropic strength of some typical metamorphic rocks from the KTB pilot hole, Germany. *Science of Drilling*, **3**, 153–160.

VERNIK, L., ZOBACK, M.D. & BRUDY, M. 1992b. Methodology and application of the wellbore breakout analysis in estimating the maximum horizontal stress magnitude in the KTB pilot hole. *Science of Drilling*, **3**, 161–169.

VICKERS, B.L. & THILL, R.E. 1969. A new technique for preparing rock spheres. *Journal of Scientific Instrumentation*, **2**, 901–902.

WALSH, J.B. & BRACE, W.F. 1964. A fracture criterion for brittle anisotropic rock. *Journal of Geophysical Research*, **69**, 3449–3456.

WONG, T.-F. 1982. Micromechanics of faulting in Westerly granite. *International Journal of Rock Mechanics and Mining Sciences*, **19**, 49–64.

WONG, T.-F. & BIEGEL, R. 1985. Effects of pressure on the micromechanics of faulting in San Marcos gabbro. *Journal of Structural Geology*, **7**, 737–749.

WONG, T.-F., DAVID, C. & ZHU, W. 1997. The transition from brittle faulting to cataclastic flow in porous sandstones: Mechanical deformation. *Journal of Geophysical Research*, **102**, 3009–3025.

WONG, T.-F., BAUD, P. & KLEIN, E. 2001. Localized failure modes in a compactant porous rock. *Geophysical Research Letters*, **28**, 2521–2524.

WU, X.Y., BAUD, P. & WONG, T.-F. 2000. Micromechanics of compressive failure and spatial evolution of anisotropic damage in Darley Dale sandstone. *International Journal of Rock Mechanics and Mining Sciences*, **37**, 143–160.

Experimental deformation of flint in axial compression

DAVID MAINPRICE[1] & MERVYN PATERSON

Research School of Earth Sciences, The Australian National University,
PO Box 4, Canberra, ACT 2601, Australia
[1]*Present address: Laboratoire de Tectonophysique,*
Université Montpellier II, Place E. Bataillon,
34950 Montpellier cédex 04, France

Abstract: The experimental deformation of flint, a water-rich (1–2 wt%) fine-grained (1 μm) microcrystalline quartz, has been studied using a gas-confining medium apparatus at 300 MPa confining pressure in the temperature range 500–1000 °C. In constant strain-rate axial compression tests a mechanically unstable behaviour with peaked and undulating stress–strain curves, especially at the faster strain rates, is manifest. The rheology of these tests can be approximately described by the traditional power-law equation with a stress exponent (n) of between 3 and 5, and an apparent activation energy (Q) of 108 kJ mol^{-1}. However, the strain-independent ('steady-state') equation is only a partial description of the data. The mechanical properties were studied at lower stresses using stress relaxation testing, this method has the advantage of recording specimen response over a very small strain interval (c. 1%), i.e. nearly constant structure. These tests revealed that in a high stress regime above 100 MPa the rheology was only weakly dependent on strain and very similar to the constant strain-rate behaviour. Below 100 MPa the rheology can be described by a power law with $Q = 64$ kJ mol^{-1} and $n = 1$. The low-stress behaviour is extremely sensitive to specimen strain, the strain rate decreasing with increasing strain for a given stress.

The optical microstructure reveals that the mechanical instabilities are related to localized displacement zones, sometimes en echelon, consisting of relatively large grains (10 μm). The important reduction of the integral infrared absorption of deformed specimens, a 98% reduction of the initial water content, suggests that a pore fluid was developed by specimen dehydration in the sealed (undrained) assembly. The presence of pores or bubbles decorating grain boundaries and microfractures normal to the piston–specimen interface both testify to the presence of the pore fluid during deformation. The pore fluid pressure may have approached the confining pressure resulting in a near-zero effective pressure. Transmission electron microscopy revealed the presence of many Brazil micro-twins parallel to grain boundaries. The twins are thought to be growth defects indicating rapid grain growth during deformation. Dislocations in the basal plane were observed with a $1/3 \langle a \rangle$-type Burgers vector. Dislocations were also observed in the grain boundaries, possibly caused by grain-boundary sliding in the low stress regime.

The complete lattice preferred orientation (LPO) of homogeneously deformed specimens has been determined by X-ray texture goniometry. The inverse pole figures have a strong concentration of compression axes parallel, and a weaker concentration normal, to the c-axis. The local c-axis pole figures in heterogeneously deformed specimens were characterized by the photometric technique. The c-axis pole figures in the bulk are identical to that determined by X-rays in homogeneously deformed specimens. In the displacement zones the fabrics are asymmetric with respect to the zone boundaries. The asymmetry is not consistent with the shear sense imposed by the deformation geometry and the usually accepted dislocation slip models. Fabric development by a grain growth mechanism in which the mechanically nucleated micro-twins (Brazil law) favour growth of grains with low-energy twin grain-boundary orientation relationships with their neighbours is proposed. The mechanism is consistent with the microstructure (twins parallel to grain boundaries, low density of dislocations on glide planes) and the c-axis pole figures. In natural deformations under similar conditions (e.g. subsolidus granites in the β-quartz field) we suggest caution in using fabric asymmetry as it may not be a dependable indicator of shear sense, as grain growth and dislocation slip will give rise to different interpretations.

From: BRUHN, D. & BURLINI, L. (eds) 2005. *High-Strain Zones: Structure and Physical Properties.* Geological Society, London, Special Publications, **245**, 251–276.
0305-8719/05/$15.00 © The Geological Society of London 2005.

The high-temperature mechanical properties of quartz polycrystals are poorly understood because of our lack of understanding of the hydrolytic weakening process. It has been shown that the solubility of water in quartz, a key factor for our understanding of the weakening processes, is not strongly dependent on hydrostatic pressure in truly hydrostatic conditions, both experimentally (Kronenberg *et al.* 1986) and theoretically (Paterson 1986). Here we report a series of rheological experiments on flint (a water-rich microcrystalline quartz) conducted at 300 MPa confining pressure, and at temperatures of between 500 and 1000 °C in gas deformation apparatus equipped with a fully internal load cell (Paterson 1970).

The fine grain size (1 μm) and hydroxyl content ($1000 \, H/10^6$ Si under the experimental conditions) make flint a good analogue for fine-grained quartz mylonites. The majority of previous experimental studies have concentrated on the development of crystal preferred orientation (CPO) (Wenk *et al.* 1967; Green *et al.* 1970; Kern 1977) and microstructure (Masuda & Fujimura 1981), with the exception of the recent work by Schmocker *et al.* (2003). Rheological studies of other fine-grained rocks have revealed changes of flow regime at low stresses (Schmid *et al.* 1977; Karato *et al.* 1986) and the importance of such changes in quartz mylonites has been the subject of much speculation (e.g. White 1976). The only published rheological study of fine-grained quartz polycrystals in axial compression at various confining pressures has revealed non-equilibrium effects (Kronenberg & Tullis 1984) due to water diffusing into initially dry quartz, perhaps assisted by fracture under non-hydrostatic conditions (Kronenberg *et al.* 1986; FitzGerald *et al.* 1991). Our experiments have been conducted on an initially supersaturated quartz polycrystal that, as we will demonstrate, has dehydrated on reaching the experimental conditions. Little is known about the kinetics of diffusion of the defect responsible for hydrolytic weakening, although some progress has been made in defining the equilibrium conditions (e.g. Cordier & Doukhan 1989) and structure of the defect (e.g. Heggie & Jones 1987; Cordier *et al.* 1994). It is now clear that the presence of water in quartz as high-pressure inclusions greatly facilitates the nucleation and multiplication of dislocations (e.g. McLaren *et al.* 1989) and it is has been established that dislocation climb becomes an important mechanism when quartz is supersaturated with water (e.g. Cordier & Doukhan 1989). In our experiments we started with a supersaturated material, which eliminated

most excess water by grain-boundary sweeping (grain growth). Perhaps the sweeping process has allowed our samples to approach more closely a chemical equilibrium than would be possible on a laboratory timescale by starting with an undersaturated ('dry') material and relying on diffusion to regulate the equilibrium concentration (cf. Kronenberg & Tullis 1984; Post & Tullis 1998).

Our study has revealed that a flow regime exists at geological stress levels (above a critical weakening temperature) characterized by the following:

- the constitutive equation of the form $\dot{\varepsilon} \propto \exp(B\sigma)$ or $\dot{\varepsilon} \propto \sigma^n$, where $n \simeq 5$ describes the data;
- an apparent activation energy of $Q_1 = 108 \, \mathrm{KJ \, mol^{-1}}$;
- the stress–strain curves show pronounced stress maxima and/or oscillatory behaviour at the faster strain rates, giving way to an almost constant flow stress at slower strain rates;
- the strain to the yield point increases with decreasing strain rate (see Chopra & Paterson 1981 for the opposite observation in Dunite);
- extensive grain growth observable in petrographic thin section;
- observation of 'growth accidents' in the form of microtwins, dislocation loops and straight dislocations normal to grain boundaries;
- high heterogeneous dislocation densities adjacent to some grain boundaries;
- strong point maximum CPO of [0001] axes parallel to the compression direction and a weaker girdle normal to it.

Further at the lower stresses (6–100 MPa) investigated by stress relaxation testing another flow regime was characterized by:

- the constitutive equation of the form $\dot{\varepsilon} \propto \sigma^n$, where $n \simeq 1$;
- an apparent activation energy of $Q_2 = 64 \, \mathrm{KJ \, mol^{-1}} \simeq 2/3 Q_1$.

In both regimes the grains remained equiaxed and showed no tendency for flattening.

Additional dehydratation experiments were conducted to assess the importance of hydrolytic weakening in deformation experiments conducted in undrained conditions. Many observations indicated that the specimens have dehydrated significantly on reaching test conditions.

Experimental details

All deformation experiments were conducted in an apparatus described by Paterson (1970). It uses argon gas as the confining medium and is fitted with an internal furnace and load cell. The temperature gradient over the length of the specimen was minimized to within $\pm 5\,^{\circ}\mathrm{C}$ and measured by a thermocouple 2 mm from the upper face of the specimen. The temperature gradient was verified by periodic calibrations in which the temperature profile was measured over the length of a hollow dummy aluminium oxide specimen. All specimens were cores 7 mm in diameter and 14 mm in length, drilled from a single nodule of flint in the same azimuth. The nodule was collected and kindly supplied by Dr J. Starkey from the Cretaceous Chalk of Northern France. Specimens were oven-dried at $110\,^{\circ}\mathrm{C}$ prior to sealing in copper jackets with a wall thickness of 0.25 mm. Except where specified, solid aluminium oxide spacers were used so that the specimens were isolated from the argon confining gas and air at room pressure (Fig. 1).

The differential stresses were determined using a cross-sectional area calculated on the assumption that the specimen deforms with a constant volume as a right circular cylinder. The usual corrections for apparatus distortion and the load borne by the copper jacket were made in deriving the stress–strain curves. Tests were conducted at constant piston displacement rate, corresponding to constant strain rate when a constant stress is achieved. Additional stress relaxation tests were performed by turning off the motor at the end of constant strain-rate experiments and locking the loading piston. The strain rate at a given time was calculated from the load v. time trace by

$$\dot{\varepsilon} = \frac{\mathrm{d}\varepsilon}{\mathrm{d}t} = -\frac{1}{L_o}\frac{\mathrm{d}L}{\mathrm{d}t}\left(\frac{L}{A_o E_s} + \frac{1}{K_a}\right)$$

where $\dot{\varepsilon}$ is strain rate, $\mathrm{d}L/\mathrm{d}t$ the rate of change of load, L_o is the specimen length at the start of relaxation, A_o is the sectional area of the specimen at the start of the relaxation, E_s is the Young's modulus of the specimen under the test conditions and K_a is the stiffness of the apparatus.

For optical studies of deformed specimens, polished thin sections were prepared after impregnation under vacuum with epoxy resin from a cut containing the core axis. Similar sections were also prepared for transmission electron microscopy (TEM) studies; areas of interest were selected with the aid of an optical microscope, thinned by ion bombardment and examined in a JEOL 200B microscope operating at 200 kV. By suitable calibration the orientation of the compression axis in the electron micrographs was always known.

Polished thin sections, typically 0.1 mm thick, were used for cryogenic (78 K) infrared study of

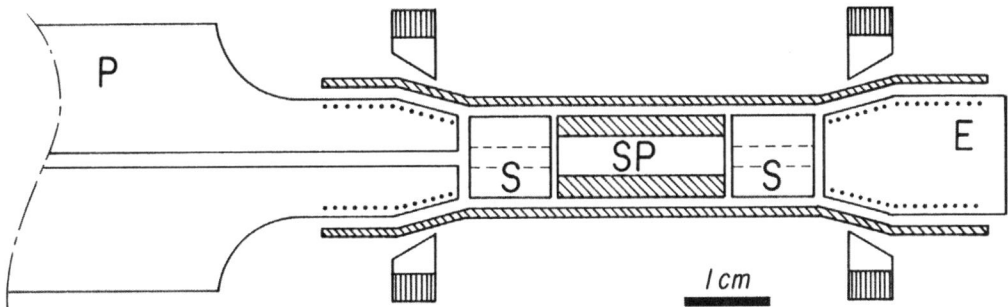

········ Polished surface
P   Piston
S   Spacer
SP   Specimen
E   End piece

▢ Al_2O_3
▨ Cu
▥ TZM

Fig. 1. Sample assembly.

the hydroxyl content of as received and deformed samples. A Perkin-Elmer 180 spectrophotometer fitted with a cold-finger cryostat was used. The presence of araldite from impregnation in deformed specimens caused interference in the $2000-4000$ cm^{-1} region. Complete removal of the araldite from deformed specimens using a mixture of CH_3OH and CH_2Cl_2 in the proportions 1:5.3 resulted in specimens too weak for further manipulation. Hence, the spectra of a specimen with the araldite partly removed will be reported here. In the as-received specimen the background was level; however, in the deformed specimens a sloping background appeared. The sloping of the background became increasingly severe as the araldite was progressively removed, suggesting that scattering from pores previously filled with araldite was chiefly responsible. The background was partly restored by saturating the specimen with an infrared transparent liquid hexachlorobutadiene under vacuum and then placing two infrared transparent CaF_2 plates on each face of the sample to reduce liquid loss during the recording of the spectrum under vacuum in the cryostat. The background was removed by fitting a straight line to region $3800-4000$ cm^{-1} and extrapolating the base line to 2600 cm^{-1}. It should be emphasized that these spectra can only be regarded as semi-quantitative; nevertheless, they provide valuable information.

Infrared spectroscopy

The application of infrared spectroscopy to the study of quartz single crystals has enabled the concentration of the OH species to be determined (e.g. Blacic 1975; Paterson 1982). The OH species can be subdivided on the basis of their cryogenic infrared absorption (Kekulawala et al. 1978) into three distinct classes: structurally bound, 'gel' and molecular OH species. The 'gel' species is characterized by a broad absorption, which is insensitive to the temperature of measurements (normally room temperature and liquid nitrogen, 78 K) and is isotropic in the synthetic single crystal. The 'gel' species are thought to be responsible for the hydrolytic weakening effect in natural amethyst and synthetic single crystals (Kekulawala et al. 1978) at moderate confining pressures. Near infrared spectrophotometry has revealed that the 'gel' species is due to the presence of finely dispersed water molecules (e.g. Aines et al. 1984). The infrared measurements allow us to compare the hydroxyl concentrations with other quartz polycrystals (e.g. Mainprice & Paterson 1984) and

previously published data for single crystals (see Paterson 1982 for a review).

Spectra were recorded at room and liquid nitrogen temperature (Fig. 2) but were found to be identical, consistent with a 'gel'-dominated spectra. For as-received samples the infrared absorption coefficient was 145 cm^{-1} in the $3200-3400$ cm^{-1} wave number region, with another prominent band at 3620 cm^{-1}. The spectrum is rather 'flat-topped', possibly indicating that the IR detector has saturated (A. Kronenberg pers. comm.), in which case the infrared absorption coefficient of 145 cm^{-1} is a minimum estimate. The absorption coefficient at 3000 cm^{-1} is about a quarter of its value in the $3200-3400$ cm^{-1} region. The impregnation of the deformed specimens with Araldite epoxy resin results in interference by strong aliphatic C–H stretching modes near 2900 cm^{-1} and broad absorption in the $3300-3500$ cm^{-1} region. The infrared spectrum of specimen 3259 (lowest flow stress recorded in constant strain-rate mode, 10^{-5} s^{-1}, 1000 °C) shows a drastically reduced absorption coefficient in the region $3200-3400$ cm^{-1} of about 14 cm^{-1}, which would be even less if the remaining Araldite absorption is subtracted (c. 3 or 4 cm^{-1}). Schmocker et al. (2003) have reported similar values for specimens heat-treated at 1300 K (1027 °C) and confining pressure of 350 MPa.

While specimen 3259 is mechanically the weakest tested (Table 1, Fig. 3d), its water content (1500 ± 500 H/10^6 Si) after deformation is less than 2% of the starting material value ($90\,000 \pm 30\,000$ H/10^6 Si); hence, a large quantity of water was driven out of the specimen on going to test conditions and possibly during the test. The specimens were

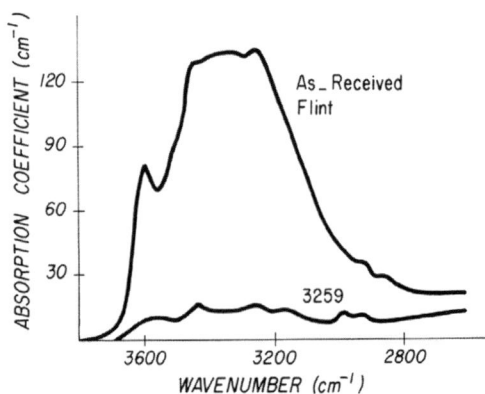

Fig. 2. Infrared spectra recorded at 78 K (liquid nitrogen).

Table 1. *Flint deformation experiments*

Run no.	Temperature (°C)	Flow stress (MPa)	Strain rate (s^{-1})	Final strain (%)	Comments
3195	900	134	1.73×10^{-5}	13.1	2 h cook at 900 °C, 300 MP at 900 °C, 300 MPa No fault
3230	1000	108	1.19×10^{-5}	21.7	No fault
3237	900	275	1.49×10^{-4}	18.1	Stress drop. Strong fossil fabric. Strain rate at 9.3%. Fault at 33° to σ_1
3240	900	212	1.64×10^{-4}	30.6	Small yield drop. Fault at 33° to σ_1
3241	900	266	1.71×10^{-3}	32.0	Stress drop. Strain rate at 20.2% strain quoted. Fault at 27° to σ_1
3242	900	228	1.39×10^{-4}	28.9	Small stress drop (4 MPa) Fault at 24°–31° to σ_1
3244/1	800	202	1.38×10^{-4}	37.9	1 at 5% strain, first part of strain, first part of strain rate stepping test. Fault at 32° to σ_1
3245/1	800	157	1.81×10^{-5}		1 at 9.8%
3245/2	800	252	1.76×10^{-4}	30.6	2 at 15.4%
3245/3	800	285	1.68×10^{-4}		3 at 26.8%. Fault at 28° to σ_1
3246/1	700	176	1.63×10^{-5}	27.3	1 at 6.8%
3246/2	700	209	1.68×10^{-4}		2 at 13.5% Fault at 28° to σ_1
3247	900	195	1.69×10^{-4}	18.1	Small stress drop. Fault at 30° to σ_1
3248	1000	140	1.68×10^{-4}	21.0	Undulating stress. No fault
3249	1000	215	1.54×10^{-4}	18.0	Small stress drop. Fault at 26° to σ_1
3250/1	1000	224	1.60×10^{-3}	25.9	1 at 9.7%
3250/2	1000	147	1.54×10^{-4}		2 at 20.7% No fault
3252	1000	147	1.59×10^{-4}	25.8	En échelon zones at 28° to σ_1
3253	1000	153	1.56×10^{-4}	21.2	En échelon zones at 26° to σ_1
3254	900	184	1.56×10^{-4}	19.9	Fault at 32° to σ_1
3255	800	225	1.68×10^{-4}	24.5	Fault at 26° to σ_1
3256	700	234	1.44×10^{-4}	9.5	Peak stress at 2.8% strain Fault at 23° to σ_1
3257	600	468	1.64×10^{-4}	7.4	Fault at 31° to σ_1
3258	600	282	1.56×10^{-5}	9.6	Peak stress at 4.7% strain Fault at 58° to σ_1
3259	1000	70	1.75×10^{-5}	25.5	No fault
3260	800	166	1.72×10^{-5}	18.9	No fault
3261	700	172	1.77×10^{-5}	23.4	Fault at 20° to σ_1
3262	700	184	1.67×10^{-5}	23.9	Fault at 26° to σ_1
3263	800	122	1.66×10^{-5}	25.6	No fault
3264	900	121	1.54×10^{-5}	27.0	En échelon zones at 25° to σ_1
3265	900	310	1.49×10^{-5}	16.2	Heat treated for 70 h at 900 °C, room pressure
3266	900	161	1.64×10^{-5}	24.1	Heat treated for 20 h at 900 °C, room pressure
3305	500	717	1.65×10^{-5}		

deformed in the presence of a pore fluid with a pressure, which may have been close to the confining pressure (300 MPa). At pore pressures approaching 300 MPa it is likely that the sample assembly (Fig. 1) vented some of this pore fluid.

Rheological results

Variations in stress–strain behaviour were observed over the entire range of conditions studied. For example at 1000 °C (Fig. 3) and the fastest strain rate (10^{-3} s^{-1}) all specimens showed a pronounced stress drop at 10^{-4} s^{-1} followed by undulating stress–strain behaviour, whereas at 10^{-5} s^{-1} a nearly constant flow stress was observed. The strain required to reach the stress maximum progressively increases as the strain rate is decreased. Below 700 °C the maximum strength rapidly increases with decreasing temperature (Fig. 4) for a strain rate of 10^{-5} s^{-1} and similar trends are followed

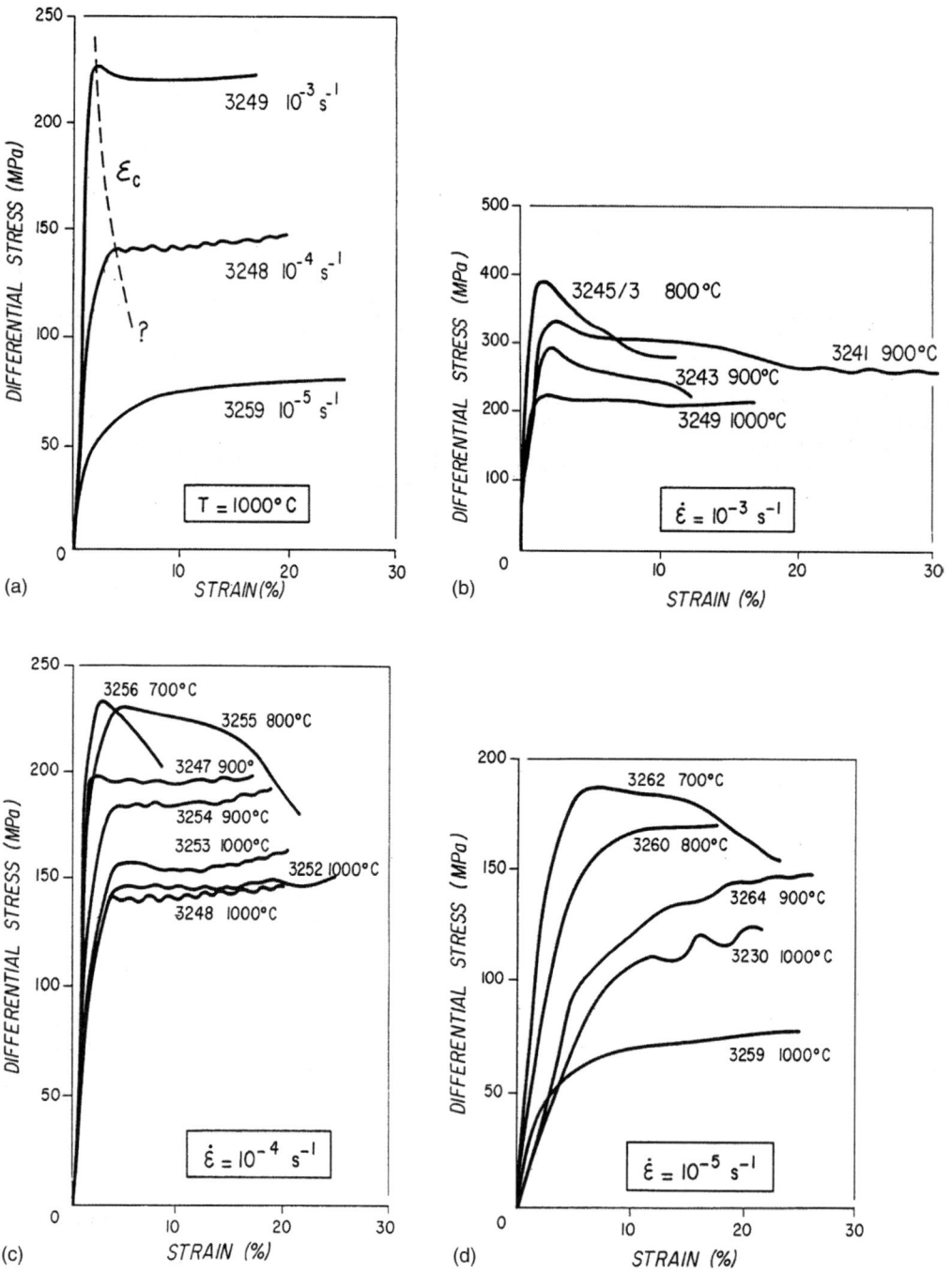

Fig. 3. Stress–strain curves: (**a**) 1000 °C. (**b**) Strain rate of 10^{-3} s^{-1}. (**c**) Strain rate of 10^{-4} s^{-1}. (**d**) Strain rate of 10^{-5} s^{-1}.

at faster strain rates. Cores with the greatest degree of homogeneity in colour were chosen for deformation experiments in the hope that this might reflect a homogeneity in impurity content and microstructure. However one specimen (3237) showed a strength outside the normal specimen variability of this material (Fig. 5). Subsequent examination of a thin

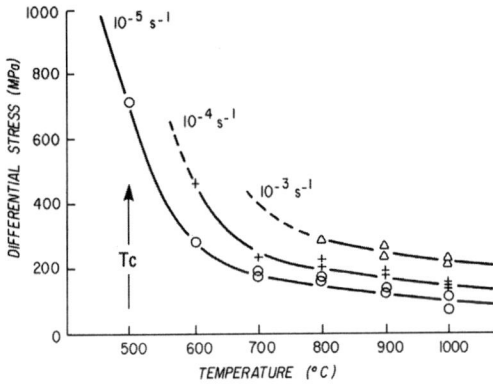

Fig. 4. Maximum differential stress in the constant strain-rate tests as a function of temperature for various strain rates. Note the rapid increase in stress at 500 °C and a strain rate of 10^{-5} s^{-1}, which is characteristic of the critical weakening temperature T_c.

section of this specimen revealed that a large part of this core had a fossil replacement texture, which possessed a very strong c-axis lattice orientation.

Strain-rate-stepping tests have been conducted to study the nature of the yield phenomena. Stepping tests where the strain rate has been progressively increased (e.g. 3245, Fig. 6a) and decreased (e.g. 3250, Fig. 6b) show that yield points depend on the imposed strain rate rather than the deformation history.

Reproducible data from the constant strain rate and strain-rate-stepping tests have been used to constrain a least-squares fit to the exponential and power-law rheological descriptions (Table 2). In this analysis no data below 700 °C have been included, a point that will be discussed

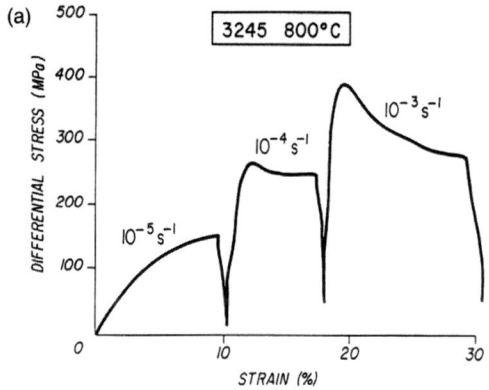

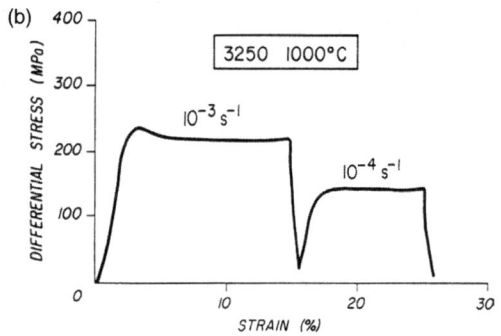

Fig. 6. History dependence of stress–strain behaviour. (**a**) Increasing strain rate steps at 800 °C. (**b**) Decreasing strain rate steps at 1000 °C.

later. The stress level for a given experiment was taken to be the peak stress if no constant value was achieved. The least-squares fit to

$$\dot{\varepsilon} = K \exp(-Q/RT) \exp(B\sigma)$$

is shown in Figure 7 where the exponential law gives a curved line in a plot with axes of log (stress) v. log (strain rate). The data are closely grouped for a given strain rate, temperature and stress with the exception of experiments at 1000 °C and 10^{-5} s^{-1} (runs 3230 and 3259, cf. Fig. 3d), where the difference has been exaggerated by the log stress scale (cf. Fig. 4). The values of the constants K, B and Q are given in Table 2 with errors stated at two standard deviations (2SD) (95% confidence level). The least-squares fit to the power-law description

$$\dot{\varepsilon} = A \exp(-Q/RT)\sigma^n$$

is shown in Figure 7, where the power law gives a straight line in a plot with axes of log (stress) v. log (strain rate). Both the exponential and power laws provide a good description of the

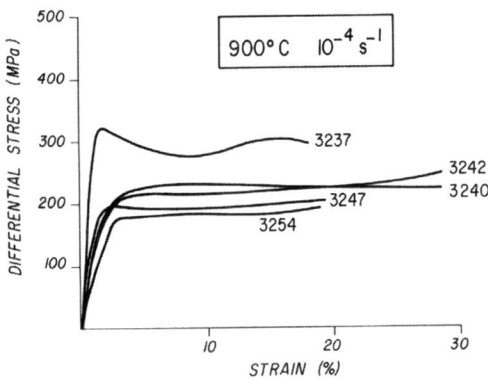

Fig. 5. Reproducibility of the stress–strain behaviour at 900 °C, 10^{-4} s^{-1}. Note the high flow stress for sample 3237.

Table 2. *Exponential and power-law constants*

Data set	(A) Exponential-law fits				(B) Power-law fits			
	Figure no.	Log K (s^{-1})	Q (kJ mol^{-1})	B (MPa^{-1})	Figure no.	Log A (s^{-1} MPa^{-n})	Q (kJ mol^{-1})	n
Constant strain rate	7	-1.61 ± 0.32	108.48 ± 15.24	0.0145 ± 0.0014	7	-10.80 ± 0.45	112.04 ± 21.85	5.35 ± 0.75
High stress regime stress relaxation	10	-1.55 ± 0.17	105.15 ± 7.49	0.0123 ± 0.0012	10	-6.21 ± 0.17	108.22 ± 7.86	3.03 ± 0.30
Low stress regime stress relaxation	10	-3.42 ± 0.45	67.54 ± 12.78	0.0147 ± 0.0030	10	-4.87 ± 0.26	64.21 ± 7.31	1.24 ± 0.14
Constant strain rate plus high-stress regime stress relaxation	7 and 10	-1.56 ± 0.38	109.73 ± 11.92	0.0144 ± 0.0010	7 and 10	-8.94 ± 0.64	117.09 ± 20.40	4.60 ± 0.56

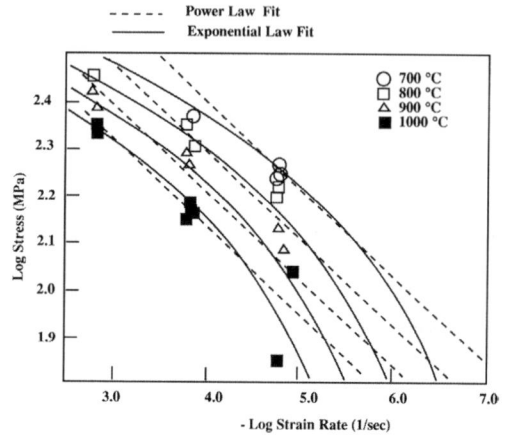

Fig. 7. Exponential and power-law empirical flow law descriptions of the rheological behaviour.

data in the experimental range. These treatments serve only for comparison with other published data, the descriptions are entirely empirical with no physical basis. As can be seen from Figure 3, the form of the stress–strain behaviour changes over the range of experimental conditions of strain, strain rate and temperature. Hence, the details of the micromechanics also changes precluding a 'steady-state' description that may be more appropriate for geological extrapolation (e.g. Paterson & Luan 1990).

Previous experience using our mechanically sealed sample assemblies has shown that water is not completely retained by the assembly (Mainprice & Paterson 1984). To assess whether our experiments were being significantly affected by water loss from the dehydration of the hydroxyl-rich starting material, we performed additional constant strain-rate experiments on previously heat-treated samples. These samples were heated at 900 °C and then deformation at 900 °C, 10^{-5} s^{-1}. Sample 3195 was heated for 2 h (approximately the length of a typical deformation run at 10^{-5} s^{-1}) in the deformation apparatus at the test confining pressure (300 MPa) with vented spacers to allow the maximum fluid loss (Fig. 1). Subsequent deformation showed no difference outside the usual specimen variability with a stress–strain curve that compares favourably with that of untreated sample 3264 (Fig. 8). Further heat treatments were carried out for 20 (run 3266) and 70 (run 3265) h at room pressure in flowing argon. Subsequent deformation tests on these samples revealed progressive strengthening, an increase in work hardening and decrease in the strain required to reach the

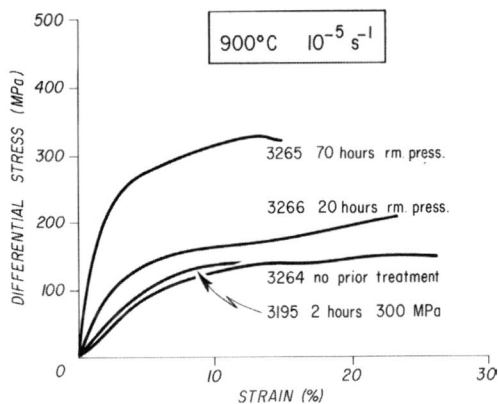

Fig. 8. Stress–strain behaviour at 900 °C and 10^{-5} s^{-1} after various heat treatments.

stress maximum with increasing heating time. The strengthening rate $(d\sigma/dt)$ produced at room pressure is 1.25 MPa h^{-1}, which is about an order of magnitude less than found by Kekulawala *et al.* (1978) of 10.25 MPa h^{-1} for synthetic quartz single crystals at 900 °C.

Stress relaxation tests were conducted in order to explore the rheology of flint at lower stresses than those studied by constant strain-rate testing and to study the rheological behaviour of a single specimen at different strains or stages of microstructural development. The tests were carried out at either approximately 10% (Fig. 9a, also Fig. 6a) or 15% strain (Fig. 9b, also Fig. 6b) achieved during strain-rate stepping or a (single) constant strain-rate deformation of individual samples (Fig. 9c & d).

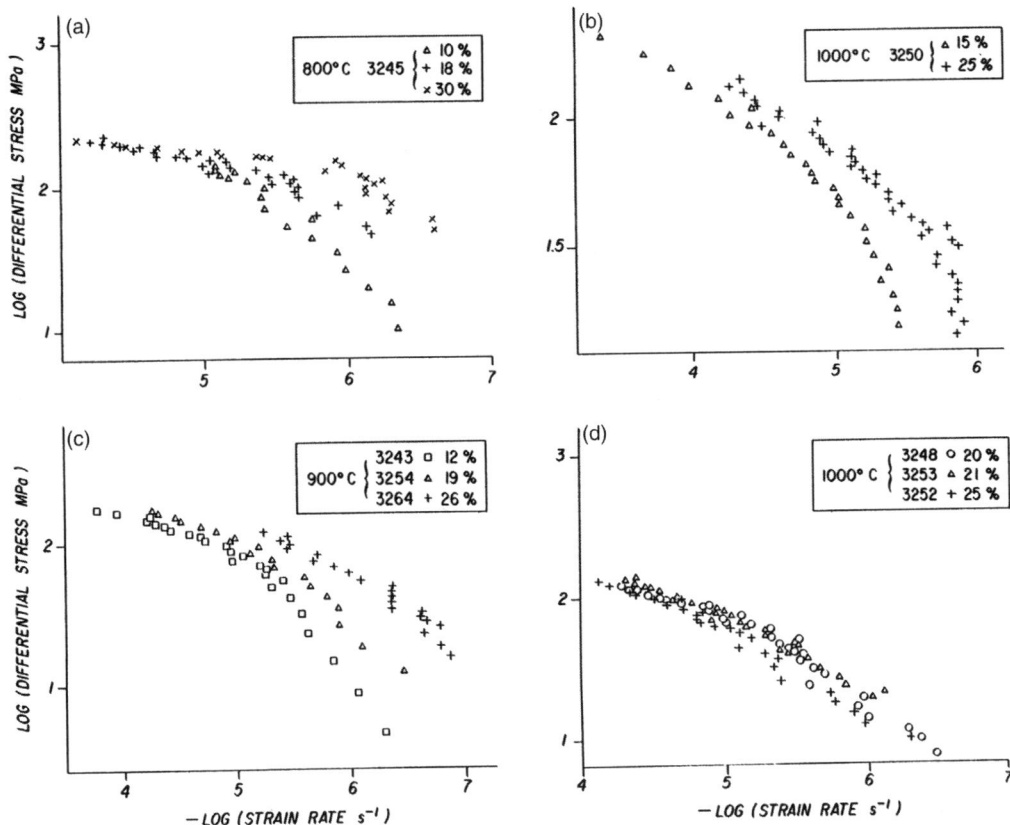

Fig. 9. Stress-relaxation response after deformation at a constant strain rate to various strains. (**a**) One specimen at various strains 800 °C. (**b**) One specimen at two strains, 1000 °C. (**c**) Three different specimens at various strains. (**d**) Three different specimens after a similar strain (20–25%).

Stress relaxations carried out on single samples show (Fig. 9a & b) that specimens become stronger for a given strain rate as the specimen strain increases, this effect is much more pronounced at stress levels below 100 MPa than above it. At higher stresses the effect is barely discernable (Fig. 9a), particularly at the lower temperatures (800 and 900 °C). Similar trends are evident even for relatively small increments of strain (10%; Fig. 9b). For tests conducted on several individual samples (Fig. 9c) the trend is the same as long as the strain increments between samples is greater than 5% to overcome the problems associated with specimen variability (Fig. 9d).

The knee in the log stress v. log strain-rate plots of relaxation test data indicates a change in flow regime below approximately 100 MPa. At 800 °C the transition occurs at 100 MPa with the high-stress regime being insensitive to strain rate with a stress exponent (n) assuming a power-law description of between 5 and 10 (Fig. 9a). At 1000 °C the transition has moved to a lower stress of 60 MPa with $n = 3$ in the high-stress regime. In the low-stress regime $n = 1$ in all cases.

To further quantify the two regimes data has been taken from five relaxation tests at similar strains of 23 (3262), 21 (3255), 25 (3263), 19 (3254) and 20 (3248) covering the temperature range 700–1000 °C (Fig. 10). The data have been fitted to two power-law descriptions, which provide an adequate description in each regime (Table 2). The flow-regime regressions were performed independently, hence assuming *a priori* the mechanisms in each regime are independent, i.e. the mechanism giving the fastest strain rate for a given stress provides the total strain-rate contribution.

Microstructure

The microstructure of as-received is heterogeneous on the grain scale with an average grain size of between 1 and 2 μm. Occasionally grains up to 10 μm are observed. Inclusions with radiating fibrous structure 50 μm in diameter are present. The grain size varies from

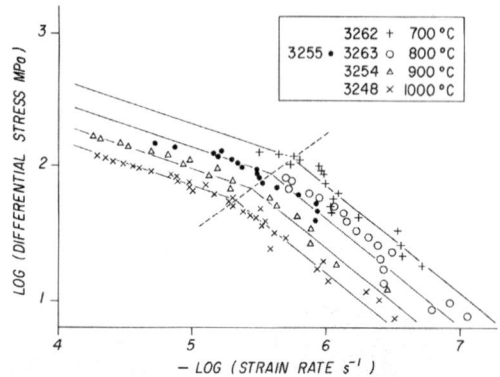

Fig. 10. Stress-relaxation behaviour as a function of temperature for five different specimens after a similar strain in constant strain-rate mode (19–25%, see text). Note the high stress regime and low stress regime for which the solid back lines are the power-law least-squares fits (see Table 2 for constants).

place to place in the nodule in a patchy manner, never being larger than 5 μm.

The heterogeneity of deformed specimens is a pronounced feature at all conditions except those with low-flow stresses (i.e. high-temperature runs 900 and 1000 °C, and the slowest strain rate 10^{-5} s^{-1}). There is abundant evidence for coarsening of the grain size from the very fine initial grain size (1 μm). There is also an indication of grain growth in all heat-treated specimens, although less pronounced in those heat-treated at room pressure. Preliminary measurements of the recrystallized grain size as a function of applied stress have also been made.

At the lower temperatures and faster strain rates, grain coarsening is associated with faults, which have clearly defined offsets within the specimens (Fig. 11a). The larger grain size areas appear lighter in plane polarized light (PPL) due to the lower density of grain boundaries that are decorated by a high density of scattering centres (probably fluid inclusions or pores at grain-boundary triple points) in the fine-grained matrix. Conjugate zones of displacement with a dihedral angle of about 56° are frequently not as well developed as the main fault.

Fig. 11. Optical microstructure. (**a**) Specimen 3258 (600 °C, 10^{-4} s^{-1}, 234 MPa) in plane polarized light (PPL). Note the displacement of fossil fragment and light regions (large grain size) around the faults. (**b**) Specimen 3242 (900 °C, 10^{-4} s^{-1}, 228 MPa). Specimen–spacer interface at the bottom of the figure. The change of angle of en echelon displacement zones (light grey) is shown from the bottom to middle of the specimen. (**c**) Specimen 3254 (900 °C, 10^{-4} s^{-1}, 184 MPa) in PPL showing detail of the en echelon displacement zones. Irregular boundaries are marked by a grain growth front and the median fracture (black line) in some zones. (**d**) Specimen 3195 (900 °C, 10^{-5} s^{-1}, 134 MPa), centre of specimen with numerous square grain outlines.

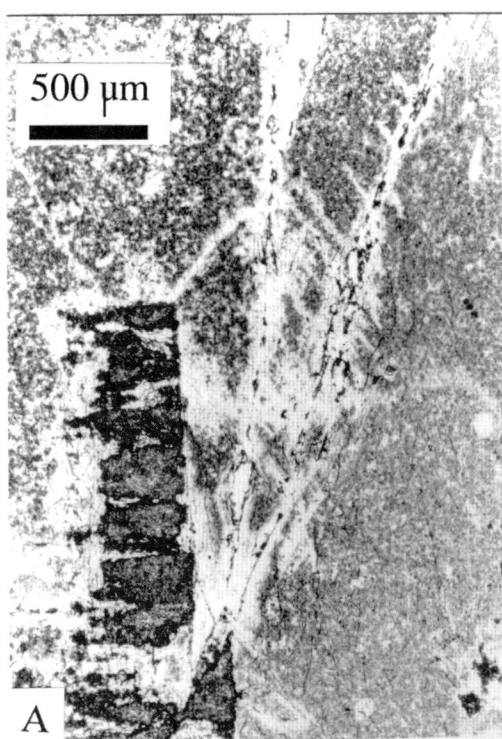

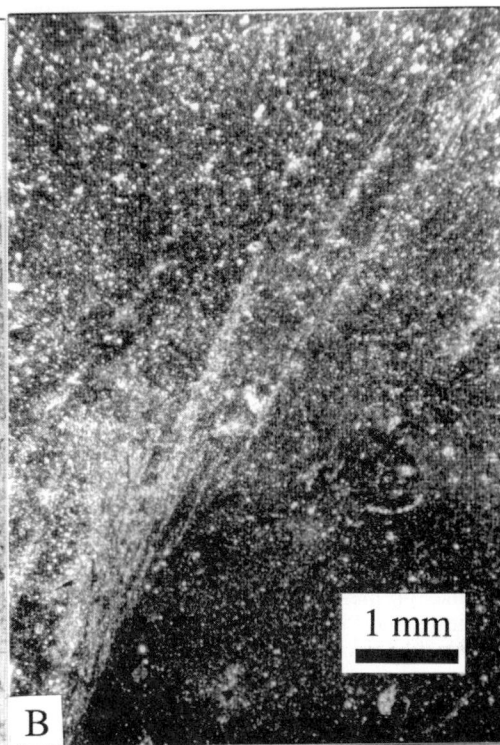

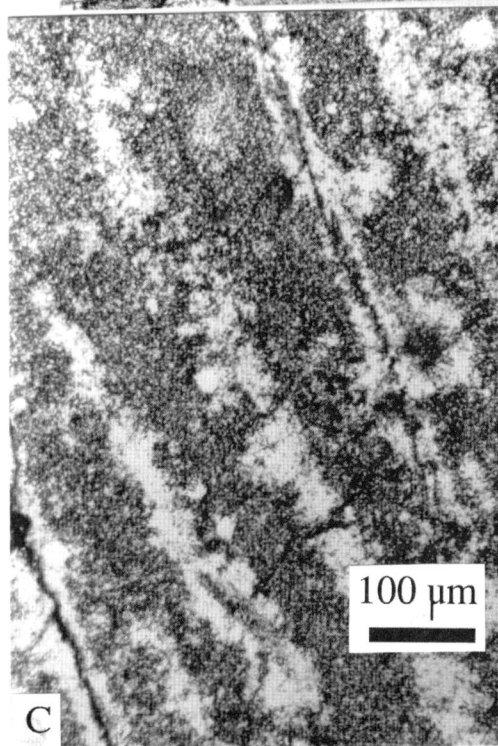

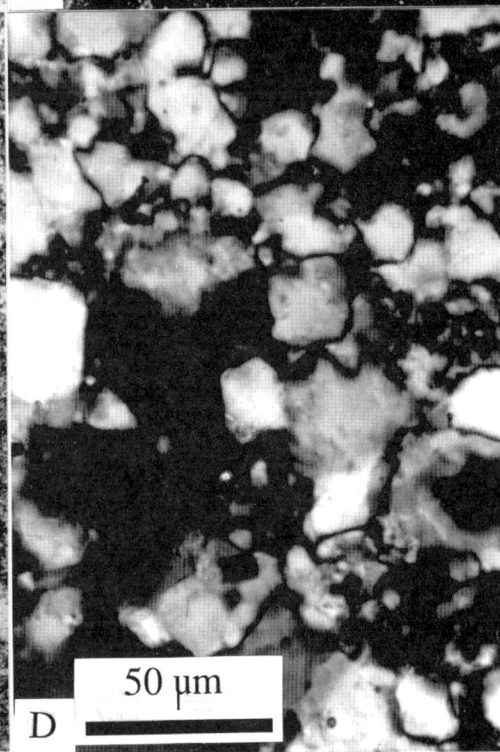

Typically such faults make an angle between 22° and 25° to the compression direction. Significant grain growth is restricted to regions within 50–100 μm of the displacement zones.

At higher temperatures no through-going faults with distinct offsets are observed. However, arrays of en echelon pinnate grain growth zones are present. These zones are common in samples that have undergone undulating stress–strain behaviour. These zones are best developed at the ends of the specimen and become increasingly difficult to observe in the coarse-grained sample centre (Fig. 11b). The zones gradually change orientation from 20° to the compression axis near the spacer contact to 30° near the sample centre (Fig. 11b). Over distances of 2 mm, the zones are linear, although in detail the relationships between individual zones are complex, particularly near the centre of the system where many zones appear to overlap. Details of the zones can be seen in Figure 11c from sample 3254 in which fracture-like or pressure-solution seam-like traces can be seen in PPL at the centre of some zones. Examination in cross-nicols reveals that such zones are composed of grains, approximately 10 μm in diameter, enclosed in a matrix of grains with diameters of 5 μm or less.

Specimens that deformed at low-flow stresses have microstructures that are much more homogeneous, although smaller grain sizes are observed near the specimen–spacer interface. Large porphyoblastic grains about 30 μm in diameter, which often have a square outline, are seen in these specimens. Specimen 3195, heat treated under confining pressure, has a more complete development of the straight grain boundaries and a larger grain size (Fig. 11d) than other samples deformed under similar conditions. In contrast, samples heat-treated at room pressure under dehydrating conditions showed almost no grain growth.

Grain size measurements have been made using the linear intercept method (Exner 1972). Thin sections are often several grain diameters thick, at 20–30 μm. Attempts to produce

Table 3. *Grain size measurements**

Run no.	Stress (MPa)	Grain size (μm)
3230	108	30.5 ± 3.2
3264	121	25.8 ± 2.9
3254	184	20.0 ± 2.2
3249	215	13.8 ± 1.4
3195	134	33.6 ± 3.5
3259	71	31.2 ± 3.6

*Errors quoted at 95% confidence level.

ultra-thin (1–10 μm) sections were unsuccessful (cf. Schmid *et al.* 1977). Therefore measurements have only been made at the centre of the samples with a relatively homogeneous microstructure and were made by focusing on the top surface in reflection. The results (Table 3) show that all measured samples deformed in the as-received condition have a grain size that varies with applied stress, decreasing with increasing stress, except possibly for 3259. Sample 3195, deformed after 2 h of heat treatment at 900 °C, 300 MPa hydrostatic pressure appears to have a grain size that is anomalous for the applied stress, perhaps due to grain growth during hydrostatic annealing.

In TEM the grain size of as-received flint appeared to be somewhat smaller than measured by optical microscopy, at 0.5–0.25 μm, presumably reflecting the presence of grains with parallel *c*-axes. The extremely rapid electron beam damage is 5–10 s at magnifications greater than ×30 000 precluded any detailed study of the as-received material. Although deformed specimens were less radiation sensitive, damage became evident within 20–40 s at magnifications between ×20 000 and ×40 000.

The zones of larger grain size regions of deformed specimens had interlocking square or polyhedral grains (Fig. 12a). Occasional pores were present at grain triple points, which were much more common in the regions of finer grain size. Within the grains the dislocation density was heterogeneous, sometimes being locally high near grain boundaries (Fig. 12d),

Fig. 12. TEM microstructure. (**a**) Specimen 3241. Grains A and B have a square outline with grain boundaries parallel to micro-twins (T in grain A) with rhombohedral planes. Dislocations intersect grain boundaries at right angles. The diffracting vectors $\mathbf{g_a} = (1\bar{0}11)$ and $\mathbf{g_b} = (1\bar{1}0\bar{1})$. (**b**) Specimen 3241. A series of micro-twin boundaries enclose the top third of the grain with form similar to the grain. Dislocations are normal to the grain boundary, and form networks and small loops near the grain centre. The diffracting vector $\mathbf{g} = (1\bar{1}0\bar{1})$. (**c**) Specimen 3259. Curved grain boundary with dislocations and bubbles lying in the boundary plane. On the concave side of the boundary, dislocations intersect the boundary at right angles. Black arrows mark the maximum principal stress direction. (**d**) Specimen 3252. High local dislocation density on the convex side of a grain boundary. Note the straight-line dislocation segments forming 'V' shapes at (d) and bubbles at the boundary triple point (T). Black arrows mark maximum principal stress direction. The diffraction vector $\mathbf{g} = (\bar{1}101)$. Scale bar = 1 μm in all figures.

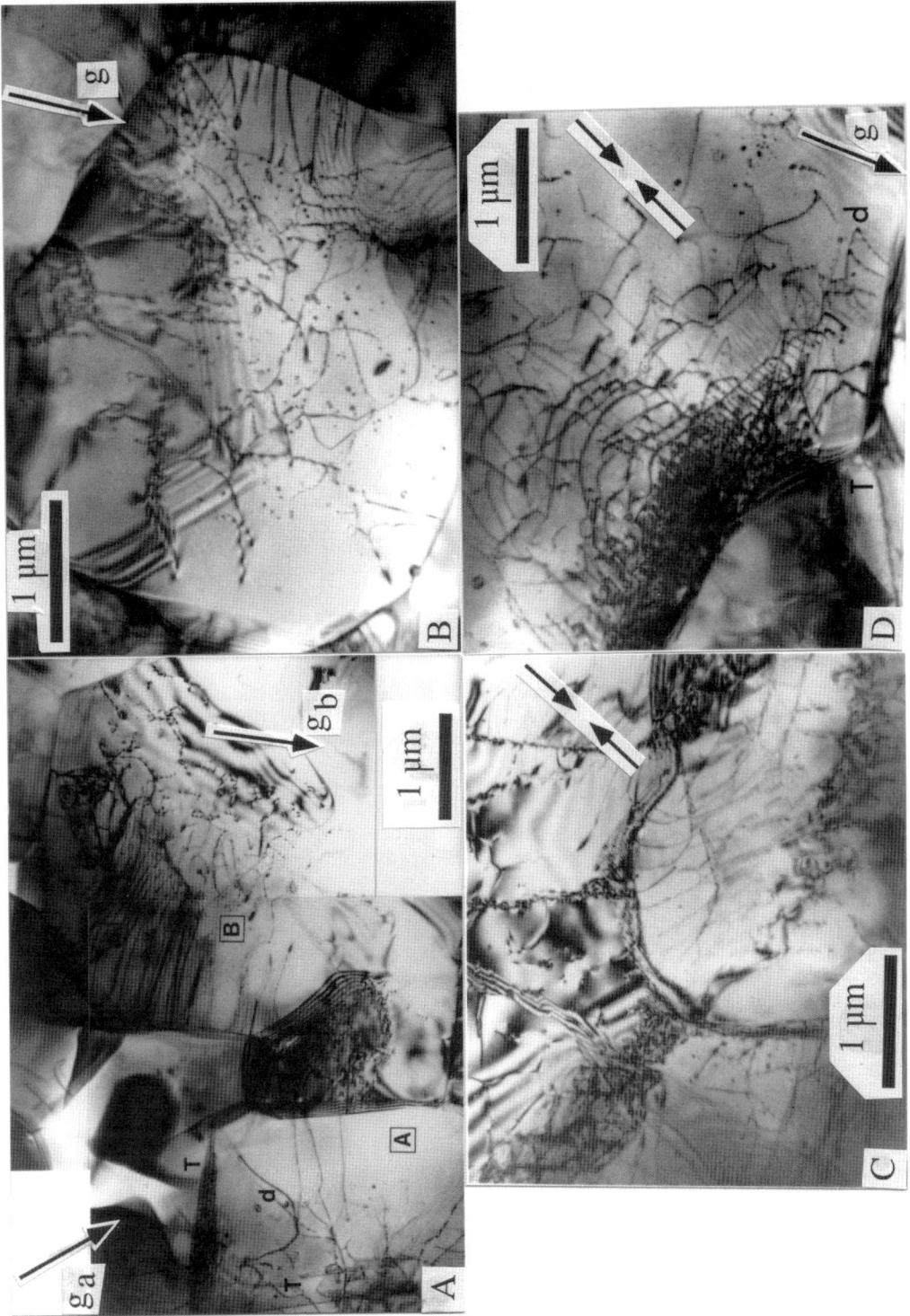

although not always (Fig. 12c). Micro-twins are abundant, characterized by their major rhombohedral **r** ($10\bar{1}1$) composition plane, which was determined by tilting the twin boundary into a vertical orientation. No orientation change was observed across the composition plane. The twin boundary is characterized by no contrast for $\mathbf{g} = (\bar{1}011)$ reflections (Fig. 12a) and asymmetric contrast fringes for $\mathbf{g} = (10\bar{1}1)$ reflections (Fig. 12b), and symmetric fringes for $\mathbf{g} = (11\bar{2}0)$. All the TEM observations are compatible with the Brazil twin law (McLaren & Phakey 1966). The twin composition planes were commonly parallel to grain boundaries (Fig. 12a & b), indicating that grain boundaries were also rhombohedral planes. A low density of dislocations was seen forming small loops and straight dislocations connecting networks in the rhombohedral plane with the grain boundary. These straight dislocations intersect grain boundaries at right angles (Fig. 12b & c). Occasionally lines of voids were observed in twin composition planes in the out-of-contrast imaging conditions $\mathbf{g} = (1\bar{1}01)$ (Fig. 13a). When higher densities were observed the dislocations tended to be straight with line orientations near *a*-axis directions (Fig. 13c). The Burgers vectors of the two types of straight dislocation have been identified using the dislocation image simulation technique (Head *et al.* 1973). The region studied was in a displacement zone of larger grain size. The first type was a dislocation from a region of relatively high density, lying in the (0001) glide plane with a line direction [$5\bar{1}40$] (i.e. 15° from the *a*-axis). The second type was a dislocation in a low-density network in the ($\bar{1}011$) plane with a [$1\bar{1}02$] line direction. Both dislocation types had a $1/3 \langle \mathbf{a} \rangle$-type Burgers vector. In the specimen deformed at the lowest flow stress (3259; 70 MPa) tilt subgrain walls are well developed (Fig. 13b). Occasionally grain boundaries with many dislocations on the interface (Fig. 13c & d) are observed.

Observations in samples heat-treated at room pressure revealed that they had similar microstructure to the untreated specimens deformed at comparable stresses. Dislocations tended to be straight and parallel to the *a*-axes. Pores

were seen at grain-boundary triple points as seen in other samples, but no signs of extensive void formation.

Crystal preferred orientation

The CPO was studied by X-ray texture goniometry and flat-stage photometric optical microscopy. The more homogeneously deformed specimens could be studied by the X-ray technique, whereas the heterogeneously deformed specimens lent themselves to study by the photometric technique.

The complete CPO of axially symmetric specimens was studied by the X-ray texture goniometry technique developed by Baker *et al.* (1969). Profiles of the reflecting planes ($10\bar{1}1$), ($11\bar{2}0$), ($20\bar{2}0$), ($20\bar{2}1$), ($11\bar{2}2$), ($21\bar{2}1$), ($11\bar{2}3$) and (0003) were measured for as-received and deformed specimens from runs 3252 and 3264. All profiles except the weak (0003) reflection, which is within 0.5° of the strong ($11\bar{2}2$), were used in the computer data reduction using the analysis devised by Baker *et al.* (1969) and the weighting factors given by Baker & Wenk (1972) to derive the inverse pole figure (IPF) for each specimen.

The IPF calculated from profiles of as-received flint had a maximum deviation from a uniform distribution of only 0.26, indicating that a uniform distribution of grain orientations was present. The IPFs from specimens 3252 and 3264 had two maxima of *c*-axes parallel to the compression axis, and a subsidiary maximum at right angles to it (Fig. 14a). A minor concentration of negative rhombohedral forms (*z*) is the only asymmetry about the $\mathbf{c}$–$\mathbf{a}_3$ line indicating a 6/mmm symmetry, which reflects the deformation in the β-quartz (point group 622) stability field. The maximum concentration of *c*-axes parallel to the compression axis is 2.6 and 2.9 times in 3252 and 3264, respectively.

c-axis pole figures have been derived from small areas (2 mm²) of the heterogeneously deformed specimens using the photometric technique (Price 1973, 1980). Particular attention has been paid to the displacement zones of larger

Fig. 13. TEM microstructure. (**a**) Specimen 3243. Micro-twin out of contrast for $\mathbf{g} = (1\bar{1}01)$, note the linear contrast features in the twin-boundary plane. Evidence for dislocation multiplication at bubbles is apparent (a, b, c, e). (**b**) Specimen 3259. Well-developed subgrain boundary with a misorientation of 2°. At the intersection of the subgrain and grain boundaries there is an amorphous phase containing bubbles or voids. (**c**) Specimen 3243. Grain with rhombohedral boundaries, and grain-boundary dislocations and bubbles. Straight dislocations (A, B) have $1/3 < \mathbf{a}>$ Burgers vectors in the basal plane. $\mathbf{g} = (1\bar{1}01)$. (**d**) Specimen 3243 illustrating grain-boundary dislocations. Breaks in pattern (marked by black arrows) probably indicate boundary ledges. $\mathbf{g} = (01\bar{1}1)$ in grain b. Scale bar = 1 μm in (a)–(c). Scale bar = 0.5 μm in (d).

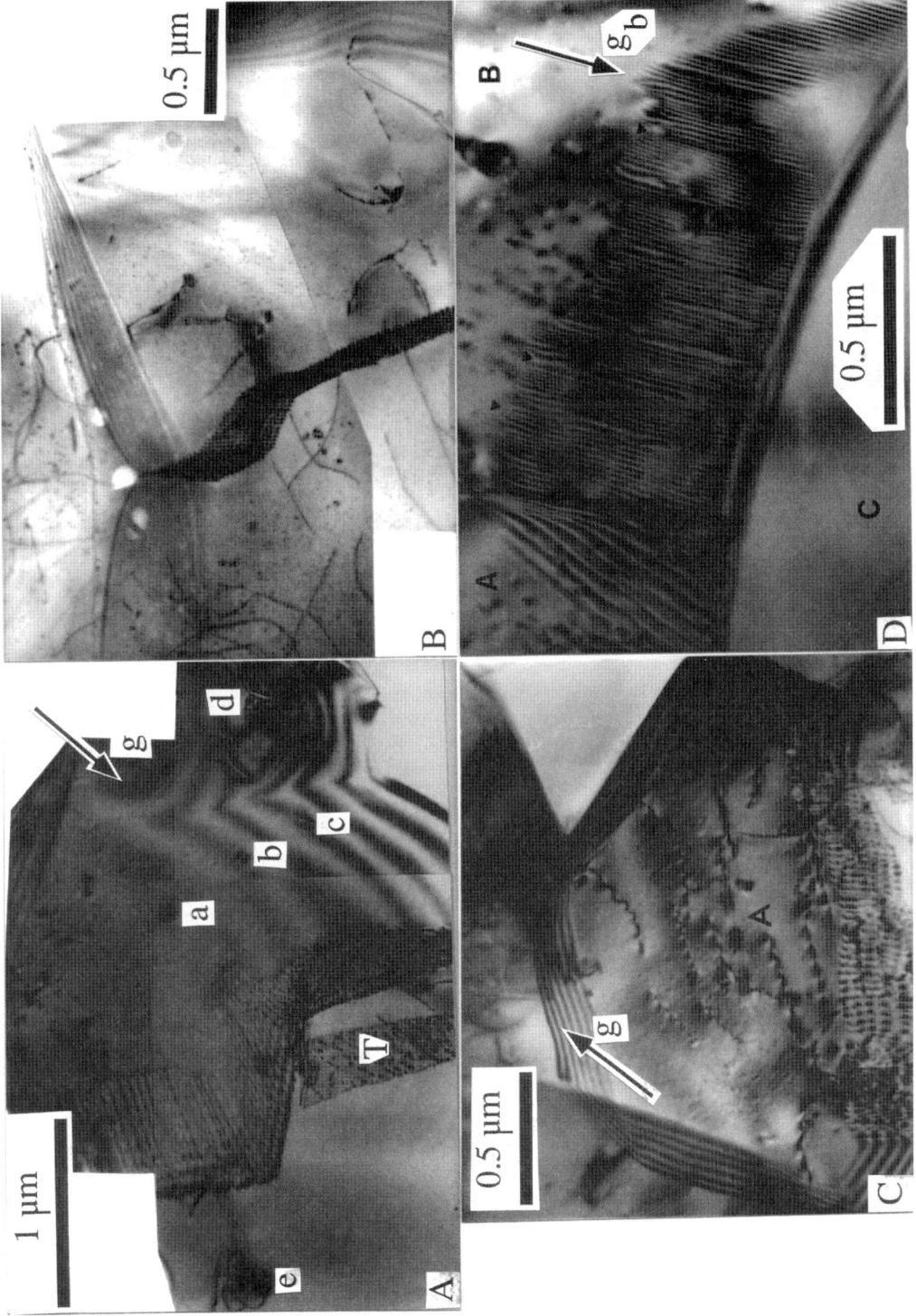

(a)

(b)

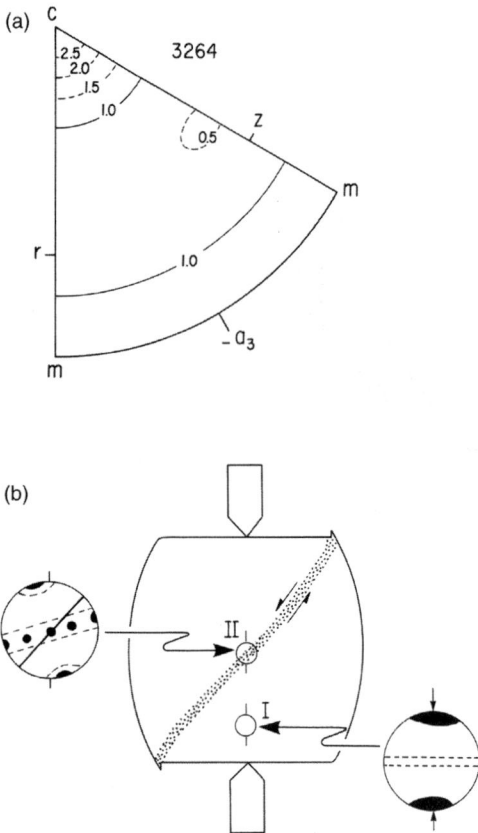

Fig. 14. (**a**) Inverse pole figure for specimen 3264 (900 °C, 10^{-5} s^{-1}, 121 MPa). (**b**) Generalized summary of photometric c-axes pole figures.

respect to the compression direction (Fig. 15, FV3 and FV4). In the bulk the trace of the c-axis fans away from the specimen (axial median line). In detail, along the zone from bottom left (FV3), middle (FV2) to top right (FV4), the maximum intensity in the girdle changes from bottom left, almost uniformly to top right. The high-density region of the pole figures appears to be related to the orientation of the maximum local compressive stress, which in turn is related to the local boundary conditions. At the top and bottom of the sample, friction between the specimen and the Al_2O_3 spacer causes a fanning of the c-axis traces and at the tips of the recrystallized zones (FV3 and FV4), the boundary conditions imposed by the specimen–copper jacket interface results in an asymmetric girdle density.

Discussion

Mechanical behaviour

Constrant strain-rate tests. Tests at the faster strain rates and lower temperatures are characterized by stress maxima or undulating stress–strain curves. Specimens deformed in these conditions have well-developed displacement zones comprised of grains with a relatively large grain size. The zones often have an echelon geometry that changes angle with respect to the principal (macroscopic) axial compressive stress direction as one moves towards the sample centre. This stress–strain behaviour appears to be linked to the development of displacement zones. The sample assembly is mechanically sealed (Fig. 1) from the hydrostatic and room pressure for all runs except 3195, which was vented. Any fluid released during the dehydration of the specimen during the rise of temperature and pressure to test conditions, or at test conditions, is trapped in the sample assembly during the test. The axial fractures near the end pieces and pores or bubbles in the samples indicate the presence of this fluid during the deformation. Hence, the tests were probably conducted at non-zero pore fluid pressure (P_p), which may have been near the confining pressure (P_c) value (300 MPa); thus, at nearly zero effective pressure ($P_c - P_p$). In such conditions it is not surprising to observe that displacement zones have geometry more consistent with a frictional sliding origin than a truly ductile shear. Note, for example, that the displacement zones are not at 45° to the compressive stress as might be expected for plastic shear or shear localization in porous material at high effective pressure. The appearance of a stress

grain size and their c-axis fabric compared to that of the bulk of the specimen. A schematic representation of the results (Fig. 14b) for homogeneously deformed specimens shows that the bulk has a c-axis fabric identical to that shown by the homogeneously deformed specimens (Fig. 14a), and that the fabric in the zones of larger grain size has the same symmetry but is rotated by about 25° with respect to the bulk fabric. The c-axis fabric in the displacement zones shows a difference in concentration, with the point maxima having a lower concentration than the girdle.

Detailed study of the c-axis fabric of specimen 3242 using the photometric technique reveals considerable fabric variation. The fabric in the bulk (Fig. 15, FV1) is identical to that measured by X-rays in the homogeneously deformed samples (3252 and 3264). However, in the displacement zones, the fabric is rotated with

SPECIMEN 3242

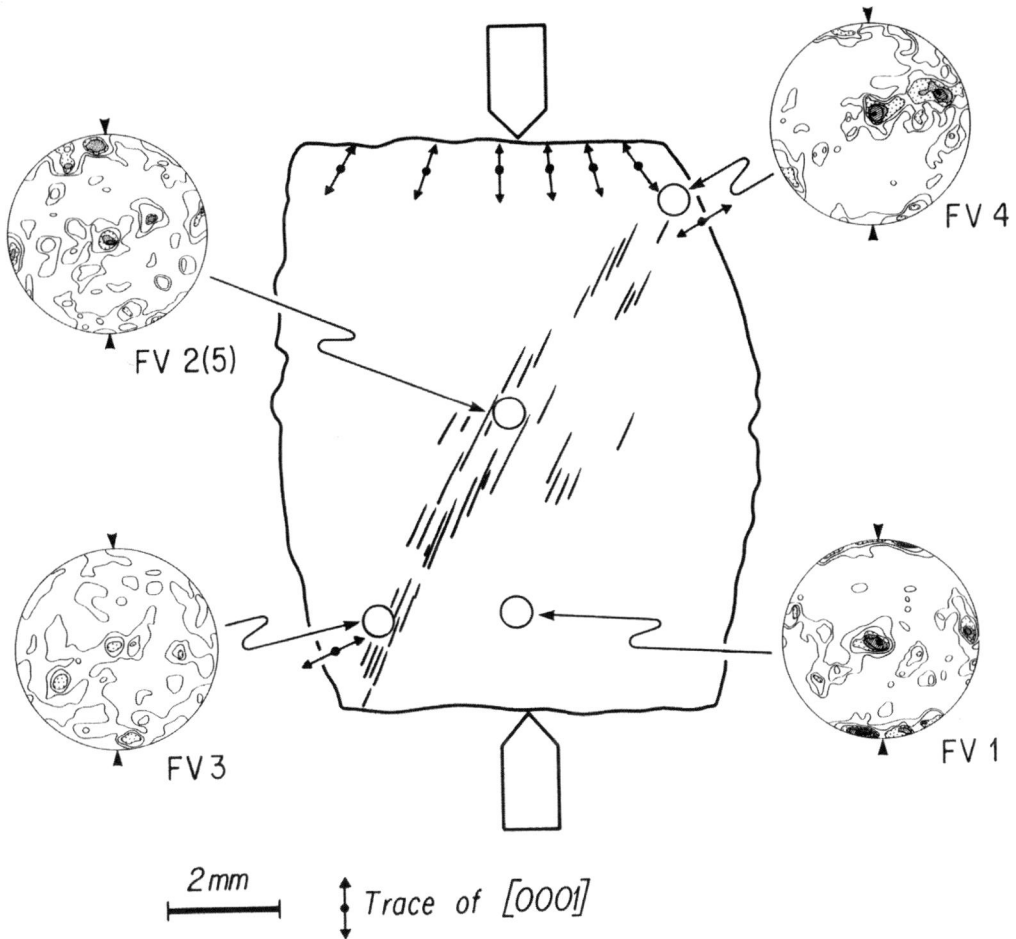

FV 4

FV 2(5)

FV 3

FV 1

2mm

Trace of [0001]

Fig. 15. Detailed photometric investigation of specimen 3242 (900 °C, 10^{-4} s^{-1}, 228 MPa). Circles represent area over which the pole figure measurement was integrated. Note fanning of the trace of the c-axes at the top spacer and the complete change of orientation in the displacement zone.

maximum at fast strain rates is consistent with a frictional sliding origin. The origin of the undulatory stress–strain behaviour is due to the nucleation and propagation of the en echelon zones. Deformation twinning could also give such undulatory stress–strain behaviour, but the gradual nature of the undulations and the low-volume fraction of micro-twins observed appear to rule out this interpretation.

Despite the frictional sliding characteristics of the runs at faster strain rates and lower temperature, certain features at higher temperature are similar to those observed in hot-worked metals.

For example, as strain rate is increased at 1000 °C the value of the strain at peak stress is reduced (Fig. 3a), indicating that the critical strain (ε_c) for the onset of the softening process (localized grain growth or dynamic recrystallization) has been reached at progressively lower strains (cf. Sellars 1978). Similarly, only one stress maximum is observed at the fastest strain rate, whereas the undulatory stress–strain behaviour is observed at slower strain rates (cf. Luton & Sellars 1969). The general similarity in stress–strain behaviour reflects the efficiency of the softening mechanism in both cases.

In our case softening occurs with the development of localized deformation zones with significant grain growth, whereas in metals softening is linked to dynamic recrystallization of the whole sample over a relatively small strain interval.

Stress relaxation tests. The rheology of flint has been explored at low stresses (down to 6 MPa) by using stress relaxation testing. At similar stress levels stress relaxation and constant strain-rate testing has produced similar data (Table 2). Over the lower stress range it reveals a change in the power-law stress exponent (n) from 3.0 ± 0.3 at moderate stresses (60–180 MPa) to 1.2 ± 0.1 at low stresses (6–100 MPa) (Table 2, Fig.10). The change in stress exponent is characteristic of a change in deformation mechanism from dislocation creep ($n \simeq 3$) to diffusive creep ($n \simeq 1$) at low stresses in fine-grained aggregates (e.g. Schmid *et al.* 1977). The deformation mechanisms can be further characterized by noting that the apparent activation energy for the low stress regime ($64.2 \pm 7.3 \text{ kJ mol}^{-1}$) is 60% of the value for the high stress regime ($108.2 \pm 7.9 \text{ kJ mol}^{-1}$), which is typical of the ratio of values reported between diffusive and dislocation-creep mechanisms (Ashby 1972).

The strain-dependent nature of the stress relaxation data (Fig. 9) indicates that the low stress mechanism is much more sensitive to the amount of specimen strain. The diffusive and superplastic mechanism models (e.g. Ashby & Verrall 1973; Gifkins 1976, 1977) predict strain rates that are inversely proportional to the grain size to some power. Hence, an increase in grain size with strain would produce the reduction in strain rate for a given stress observed in the data. However, the situation may not be as simple as this. Many workers have established that the recrystallized grain size produced during hot-working is an inverse function of stress (e.g. Sah *et al.* 1974) (see Table 3), so that the stress level prior to relaxation may determine the grain size at least initially and hence the low-stress rheological behaviour. Further complications may result from the increasing volume fraction of grains that have reached the equilibrium size during progressive deformation. Alternatively, the samples may be progressively dehydrating with time and this may be effecting the low stress rheology either by reducing the intragranular plasticity or reducing grain-sliding boundary by creating obstacles on the boundaries, such as bubbles or pores. We now discuss these aspects in turn.

If dehydration effects the mechanical behaviour on the timescale of our experiments (2–5 h) we should see the effect in the experiment on the sample heat-treated in a vented assembly at 300 MPa confining pressure prior to the constant strain-rate test (3195, Fig. 8). However, there is no significant difference between this test and experiment 3264 with no prior heat treatment. In stress relaxation down to 50 MPa, no difference is seen between untreated (3264) and 20 h heat-treated (3266) specimens in the low stress regime. We conclude that dehydration is not a significant factor at low stress for samples of similar strains prior to relaxation. If the strain-dependent nature of the low stress relaxation data were due to the development of a stress-dependent grain size (d) in the constant strain-rate test prior to the relaxation test, then the low stress diffusive mechanism strain rate ($\dot{\varepsilon} \propto 1/d^m$) would be enhanced by a high stress (resulting in a small grain size) in the preceding constant strain-rate test. The data shown in Figure 9a have been obtained after samples have been subjected to constant strain-rate tests to successively higher stress levels (Fig. 6a). The data indicate that the strain rate is decreased with increasing stress prior to the relaxation. It is concluded that magnitude of the grain size is not the controlling factor (assuming the grain size is a function of stress, see Table 3). The other hypothesis is that the volume fraction of recrystallized grains increases with increasing strain. This would lead to a progressively, larger 'effective' grain size with progressive deformation of the aggregate due to its small initial grain size (less than 1 μm). The relative reduction of the low stress diffusive mechanism strain rate with increased strain (Fig. 9) is compatible with the suggestion that the volume fraction of recrystallized grains increases with strain. The effective grain size due to a change in the volume fraction of recrystallized grains can be evaluated using the expression

$$d_e^m = d_o^m[V(d_r/d_o)^m + (l - V)]$$

where d_e is the effective grain size of the aggregate, d_o is the original grain size, d_r is the recrystallized grain size, V is the volume fraction of recrystallized grains and m is the grain size exponent. The expression has been taken from previous work by Burton (1977), who assumed that the strain-rate contribution of the original and recrystallized grains was equal. The grain size changes from 1 to 10–30 μm during grain growth (Fig. 11). Hence, for an order of magnitude calculation we can assume that $(d_r/d_o) = 10$, which results in an order of magnitude decrease in strain rate after a volume fraction increase of 10% with $m = 2$. The effect of the volume fraction of recrystallized grains can be further evaluated by taking the constitutive equations for superplastic behaviour given by Ashby & Verrall (1973) and Gifkins (1976, 1977), and inserting the effective grain size given by d_e. The constitutive equations for behaviour in the superplastic regime given by Ashby & Verrall (1973) are

$$\dot{\varepsilon}_{DC} = \frac{A_{DC}BD_LG}{KT}\left(\frac{\sigma}{G}\right)^3$$

$$\dot{\varepsilon}_{DA} = \frac{A_{DA}BD_{eff}}{KTd_e}(\sigma - \sigma_o)$$

$$D_{eff} = D_L\left[1 + \frac{3.3w}{d_e}\left(\frac{D_{GB}}{D_L}\right)\right]$$

where $\dot{\varepsilon}_{DC}$ is the dislocation creep rate, A_{DC} and A_{DA} are constants, B is the Burgers vector, D_L is the lattice diffusivity, K is Boltzman's constant, T is the absolute temperature, σ the differential stress, G the shear modulus, $\dot{\varepsilon}_{DA}$ the diffusion-accommodated creep rate, D_{eff} the effective diffusivity, d_e the effective grain size, σ_o is the threshold stress, w the grain boundary width and D_{GB} is the grain-boundary diffusivity. The equations by Gifkins (1976, 1977) are given in Etheridge & Wilkie (1979) as

$$\dot{\varepsilon}_{III} = \frac{A_{III}BD_L\sigma^3}{G^2KT}\left[1 + \frac{0.015}{d_e^{1.32}}\right]$$

$$\dot{\varepsilon}_{IIb} = \frac{A_{IIb}B^3D_{GB}\sigma^2}{Gd_e^2KT}$$

$$\dot{\varepsilon}_{IIa} = \frac{A_{IIa}B^3D_{GB}w\sigma}{d_e^3KT}$$

where $\dot{\varepsilon}_{III}$ is the creep rate given by grain-boundary sliding (GBS) accommodated by dislocation glide and climb throughout the grains, $\dot{\varepsilon}_{IIb}$ is the creep rate given by GBS accommodated by dislocation glide and climb in the grain margins, and $\dot{\varepsilon}_{IIa}$ is the creep rate given by GBS accommodated by grain-boundary diffusion. The observed strain rate is given by the sum of the dislocation and diffusion-accommodated strain rates by assuming two mechanisms are independent. The values of the various parameters were taken to be those given by Etheridge & Wilkie (1979) except that the constants (A_{DC}, A_{DA}, A_{III}, etc.) were adjusted to 'fit' the relaxation data (Figs 10 and 16) at 20% strain $\sigma_o = 3$ MPa, $V = 10\%$, $d_o = 1$ µm and $d_r = 10$ µm. The same values were then used to compare the effect of varying volume fraction with strain (Fig. 17). At 19% strain the 'fit' is good as expected, but at higher and lower

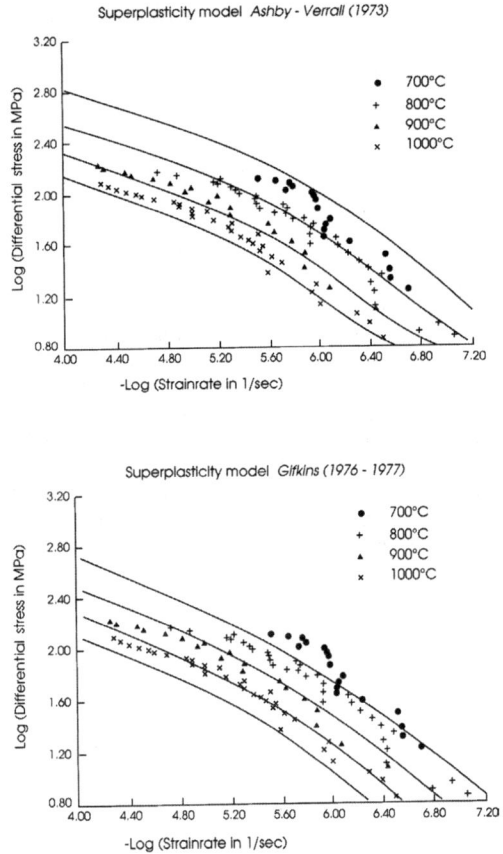

Fig. 16. The superplastic models of (a) Ashby & Verrall (1973) and (b) Gifkins (1976, 1977) fitted to the stress-relaxation data at 19–25% strain (Fig. 10). Solid lines are model values at 700, 800, 900 and 1000 °C assuming 10% volume fraction of larger grain size (10 µm) and an initial grain size of 1 µm.

volume fractions only the Ashby & Verrall (1973) model predicts values consistent with the data (Fig. 17). Consequently, it is concluded that the increasing volume fraction of recrystallized grains or grain growth with strain is responsible for the strain-dependent stress-relaxation behaviour in the low stress regime.

Microstructural evidence for the deformation of limestone by a mechanism requiring a grain-boundary sliding component has been supplied by grain-boundary off-sets produced during the deformation of split cylinders (Schmid et al. 1977). No such experiments have been conducted in the present study. Electron microscopy shows that grain-boundary dislocations are present (Fig. 13c & d). It has been suggested that the movement of grain-boundary

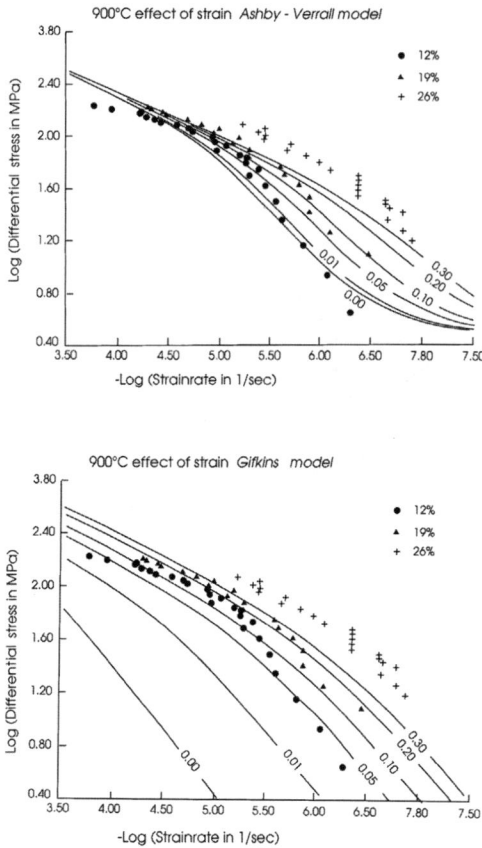

Fig. 17. The effect of specimen strain on the superplastic models of (**a**) Ashby & Verrall (1973) and (**b**) Gifkins (1976, 1977) at 900 °C compared with the stress-relaxation data from runs 3243, 3254 and 3264 at strains of 12, 19 and 26%, respectively. Solid lines are the model values at 0, 0.01, 0.05, 0.10, 0.20 and 0.30 volume fractions of larger grain size (10 μm) and an initial grain size of 1 μm.

dislocations accommodates interfacial sliding (Pond *et al.* 1978). Interfacial sliding can occur on growth defect (or annealing) twins (Wirmark *et al.* 1981), which is consistent with present observations. The periodic arrangement of dislocations in the boundary observed in Figure 13d may be due to boundary movement (Pond *et al.* 1978), to ledges caused by a deviation from an exact lattice coincidence or twin relationship across the boundary (Brandon 1966), or to boundary migration (Gleiter *et al.* 1980; Ray & Smith 1980). Voids or bubbles are common on many boundaries (e.g. Figs 12a and 13b), their presence may result in a threshold stress if they act as obstacles to grain-boundary dislocation motion (Dunlop & Nilsson 1980),

which we have modelled above by setting $\sigma_o = 3$ MPa.

The measurements of grain size indicate that the grain size varies as a function of the differential stress (Table 3, Fig. 18). All the measurements from Table 3 have been least-squares fitted to the relationship

$$\sigma = Ad^{-m}$$

where σ is the differential stress, A is a constant, d is the grain size and m is a constant. The constant m has been observed to vary with the mechanism of recrystallization (Guillope & Poirier 1979), typically for hot-worked metals m has a value of between 0.5 and 0.8 (e.g. Roberts *et al.* 1979), and for grain-boundary migration recrystallization in halite $m = 0.78$ (Guillope & Poirier 1979). The least-squares analysis of the data yields $m = 0.93$. Comparison with previously reported grain size–stress relationships is shown in Figure 18. Note the large range of stresses predicted by the various relationships. The data from this study agree most closely with the relationship proposed by Christie *et al.* (1980), although not in slope. Although there is a paucity of data, which is not of the highest quality due to heterogeneity of the deformed specimens, Figure 18 does illustrate the stress-dependent nature of the grain size formed during the deformation.

The geometry of the ductile shears is reminiscent of brittle faults, particularly the angle to the maximum principal stress direction (*c.* 30°), their en echelon nature and the influence of 'end effects' (see Paterson 1978 for review). It is interesting to note that the 'superplastic crack' model for mylonite zones (Ball 1980), which uses energy balance concepts (Griffith's theory) of fracture mechanics, predicts to an order of magnitude the length of ductile shears observed in deformed samples. As an example, with a 'fault' thickness of 0.5 mm and nominal stress of 228 MPa, the fault length of 8 mm is predicted (cf. Fig. 11).

Hydrolytic weakening

Many important factors have a bearing on the hydrolytic weakening effect, but perhaps the most important are the equilibrium hydroxyl concentration under deformation (test) conditions and the mode of incorporation into the quartz structure of the hydroxyl defect.

The initial hydroxyl concentration in flint is extremely high (60 000–120 000 $H/10^6$ Si or 1–2 wt%) and it is incorporated in the 'gel' form, now known to be a very fine distribution

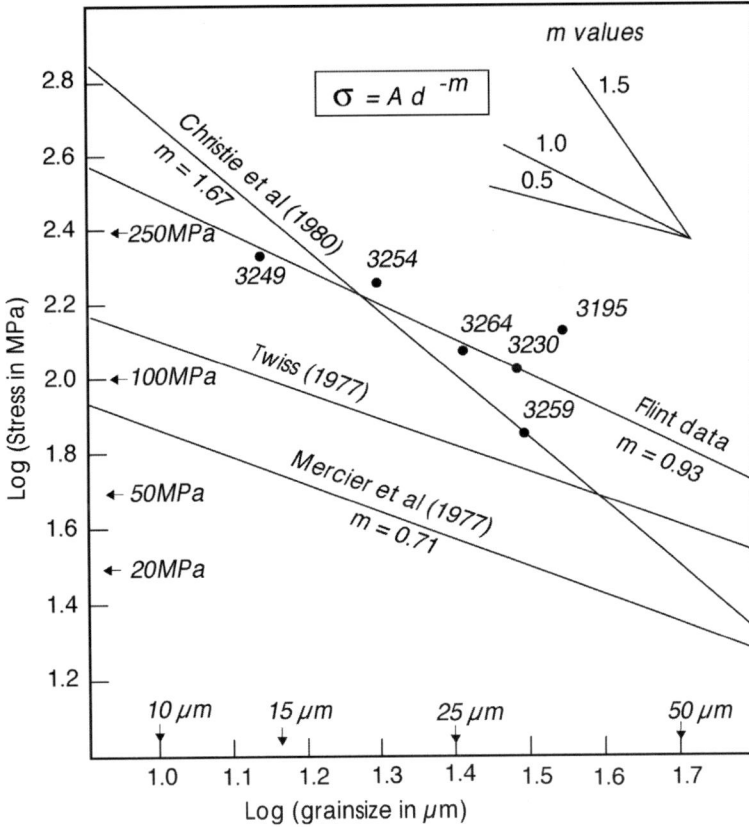

Fig. 18. Grain size v. stress relationship for selected samples compared with previous theoretical and experimental relationships. The numbers adjacent to the data points (black dots) are the run numbers (see Tables 1 and 3).

of molecular water. However, only a fraction of this initial concentration is effective in the deformation, as we will demonstrate in two ways. First, the equilibrium hydroxyl concentration under test conditions is far lower than 1–2 wt%. The amount of effective hydroxyl (and hence probably incorporated in the quartz structure) can be assessed in two ways. First, the temperature dependence of flow stress for flint is similar to that reported for synthetic quartz crystals (Griggs 1967; Hobbs *et al.* 1972) with a marked decrease in strength above the critical weakening temperature (T_c), which is a function of hydroxyl content. The critical weakening temperature appears to be about 600 °C for flint at a strain rate of 10^{-5} s^{-1} (Fig. 4). If the effect of grain boundaries can be ignored for intragranular plasticity, then the T_c value for flint is similar to that of synthetic quartz single crystal W1 (Hobbs *et al.* 1972) that has a hydroxyl content of 925 H/10^6 Si (0.14 wt%) (Hobbs *et al.* 1972; corrected for the factor of 2 error in

the original determination, see Blacic 1975), hydroxyl concentrations well above equilibrium values determined by Cordier & Doukhan (1989) to be approximately 100 H/10^6 Si; hence, the samples are still supersaturated. A second determination can be made using the more direct evidence from infrared absorption of the deformed flint samples, which contain about 1000 H/10^6 Si (Fig. 2) gel-type water, after allowing for some interference from bands associated with the epoxy resin.

The above indications suggest that the vast majority of the hydroxyl present in the starting material is not incorporated into the quartz grains under our test conditions, but remains as a pore fluid. The specimens are weak relative to quartzites deformed at similar conditions (cf. Mainprice & Paterson 1984) and presumably the quartz is weakened by a hydroxyl concentration of only about 1000 H/10^6 Si. The hydroxyl may have been sited preferentially at grain boundaries in the as-received flint, but some must have been in the quartz structure as indicated by the relatively slow strengthening rate with room pressure heat-treatment compared to single crystals (Fig. 8). The water penetration distances in quartzites determined by Post & Tullis (1998) is 15 μm at 700 °C, 1500 MPa for 24 h. Assuming that diffusion rates are linearly related to water fugacity (Giletti & Yund 1984) and the 300 MPa pressure of our experiments, penetration distances would be slightly less than 15 μm on the timescale of our experiments.

At faster strain rates and lower temperatures the constant strain experiments are characterized by; (a) relatively high stresses (>100 MPa); (b) peaked or undulatory stress–strain behaviour; and (c) the presence of displacement zones of larger grain size, typically 10 μm wide. We infer that the displacement zones must be regions of mechanical weakness at high stresses, which is particularly interesting with a view to the local role of hydroxyl. The grain growth rate to produce a 10 μm grain size must be of the order of 10^{-8} m s^{-1}. Growth rates of this order (2×10^{-8} m s^{-1}) have been used to grow hydroxyl-rich synthetic crystals that are known to be weak at room pressure (e.g. Linker & Kirby 1981). The grains within these zones are bounded by grain boundaries of rhombohedral habit plane, a growth orientation known to favour the incorporation of defects and impurities in synthetic crystals (Armington et al. 1980). The preferred orientation within the displacement zones are different from the bulk in that the girdle is stronger than the point maxima. The converse is true in the bulk (Fig. 15). One possible interpretation of the c-axis pole figures is that the point maxima have been produced by $1/3 <$a$>$ (c) slip and the girdle by either [c] (a), [c] (m) or $1/3 <$a$>$ (m) slip. Blacic (1975) and Linker & Kirby (1981) have shown that in, synthetic single crystals, a change from $1/3 <$a$>$ (c) to [c] (a) can be correlated with increasing hydroxyl content or transition from below T_c to above it. Hence, the change in the maximum concentration of the c-axis pole figure from the point maxima in the bulk to the girdle in the recrystallized zones may indicate a change from dominant $1/3 <$a$>$ (c) to [c] (a), [c] (m), or $1/3 <$a$>$ (m) suggesting an increase in hydroxyl concentration in the recrystallized zones. All of the above, growth rate, orientation of grain boundaries and the fabric transition, could be taken to indicate that a chemical softening may be responsible for the formation of the displacement zones. The recent study of flint to high shear strains ($\gamma = 3.3$) in torsion by Schmocker et al. (2003) has identical CPO patterns to those described

here for the displacement zones. Although this is an apparently satisfying interpretation, the shear sense imposed by the experimental deformation geometry does not agree with that derived from c-axis pole figure asymmetry (Fig. 13) using the criteria used in natural (e.g. Bouchez et al. 1983; Law 1990) or experimental (Dell'Angelo & Tullis 1989) quartz fabric studies. Given the abundant microstructural evidence for grain growth we suggest that the fabric is characteristic of grain growth rather than dislocation slip, hence the shear sense would appear to be the inverse of that predicted by the slip hypothesis. A similar situation exists in the non-coaxial experimental deformation of calcite polycrystals, where deformation twinning at low temperature gives a fabric asymmetry in the opposite sense to dislocation glide at high temperature (Schmid et al. 1989).

Twinning, grain growth and petrofabric development

In TEM we observe Brazil law micro-twins in almost every grain. The micro-twins may be of growth origin, a simple accident during the rapid grain-boundary migration (e.g. Kang et al. 1976), such twins are common in materials containing impurities (e.g. Cahn 1953). Alternatively, the twins may be of mechanical origin, being formed by the nucleation and propagation of partial twinning dislocations. We have observed lines of voids in the out-of-contrast images of twin planes, the voids may be radiation-damaged twinning dislocations (Fig. 13a). The relative difficulty in the propagation of the incoherent twin tip of such a twin probably explains why the twins have remained micro-twins of limited development. The nucleation and propagation will be favoured in all orientations where the rhombohedral composition plane is nearly 45° to the principal compressive stress, i.e. orientations with either the c-axis normal or parallel to the compression axis. Such orientations are statistically very frequent in the deformed specimens (Figs 14 and 15). Given that there is no misorientation across the twin plane for the Brazil twin, just a right-handed to left-handed transformation, it appears impossible that they have greatly contributed directly to the fabric pattern. It is well known that twin boundaries have a 'special' structure that results in a very low boundary energy, less than 5% of an arbitrary grain boundary in copper (e.g. Cahn 1983). In the case of the Brazil twin the two individuals are translated with respect to each other by a non-lattice vector such that two of the three silicon atoms and two of the six oxygen atoms per unit cell are brought into register, forming a coincidence site lattice (McLaren 1986). The presence of twins, sometimes misleadingly called 'annealing' twins, in a material undergoing deformation involving extensive boundary migration may lead to the development of grains in a twin orientation relationship with their neighbour at the expense of grains with an arbitrary neighbour orientation. Many grains are observed to have twin orientation grain boundaries, as their boundaries have developed parallel to internal microtwins. We suggest that twinning has imposed a selection rule on grain-boundary migration favouring the growth of grains in a mutual twin orientation, a low-energy configuration. If twin nucleation is a mechanical process, then grains in a favourable orientation for twinning (i.e. c-axis parallel or normal to σ_1) will statistically have a greater chance of growing by creating low-energy twin boundaries, hence contributing to the fabric pattern. The proposed mechanism explains the form of the grains, often square or hexagonal, with straight grain boundaries parallel to rhombohedral planes (twin composition planes). It also explains why high densities of dislocations are not systematically observed adjacent to curved grain boundaries, as might be expected for grain-boundary bulge migration (e.g. Bailey & Hirsch 1962). The mechanism may also explain why in uniaxial compression, for example in the centre of the samples (Fig. 15), the point maximum parallel to σ_1 is more intense than the girdle normal to σ_1. The fabrics measured for flint in axial compression appear to be stronger that those measured from samples deformed in torsion to shear strains of $\gamma = 3.3$, with a maximum concentration of c-axes of only 1.8 times random in torsion (Schmocker et al. 2003) compared with over 2.5 in compression. Such changes in fabric could result from the change in orientation of the local maximum principal compressive stress, to which twinning would be particularly sensitive.

Geological application

The mechanical behaviour flint varies with temperature, strain rate and strain (e.g. Figs 3 and 9), and the traditional 'steady-state' flow-law description (Table 2) is unsatisfactory for the extrapolation of these data to geological strain rates. The constants in the power-law description are similar to previous studies with a low apparent activation energy ($105-117$ kJ mol^{-1}), and a stress exponent between 3 and 5 in the high-

stress regime that is insensitive to strain (see Koch et al. 1989, Paterson & Luan 1990; Gleason & Tullis 1995), but similar to that measured by Luan & Paterson (1992) for synthetic quartz aggregates. In the low-stress regime a more sophisticated model is required due to the strong strain dependence of the rheology. Of the two superplastic models investigated, the Ashby & Verral (1973) constitutive equation more closely describes the data.

The microstructure and fabric observations suggest that for β-quartz aggregates at near-zero effective pressure the fabric development is controlled by grain growth. Gleason et al. (1993) also noted that strain-induced grain-boundary migration plays a major role in fabric development for fine-grained water-rich quartz aggregates at high temperature. We suggest that the nucleation of micro-twins (Brazil law) is favoured by a mechanical origin with the composition plane (rhombohedral) at 45° to the compression direction. Subsequent grain growth favours grains in a low-energy mutual twin orientation, and results in a fabric with a high concentration of c-axes parallel and normal to the compression direction. Photometric fabric determinations have shown that the asymmetry of this grain growth fabric is in the opposite sense to that predicted by a glide mechanism. Hence, caution should be used in the application of fabric asymmetry to β-quartz where grain growth (e.g. Gapais & Barbarin 1986) and glide (e.g. Mainprice et al. 1986) have been observed in naturally deformed quartz in subsolidus granites.

G. Price generously undertook the photometric analysis and J. Shore helped with the X-ray texture goniometry. We would like to thank the many colleagues with whom we have had informal discussions, especially P. Chopra, M. Etheridge, J. FitzGerald, R. Shaw and G. Price. We thank A. Kronenberg for a detailed review and the other anonymous reviewer his constructive comments on the manuscript. We thank L. Burlini for his editoral work. The technical support of G. Horwood, P. Percival and P. Willis created a working environment that was second to none. D. Mainprice would like to thank the Australian National University for financial support (A.N.U. scholarship) during this study.

References

ARMINGTON, A.F., BRUCE, J.A., HALLIBURTON, L.E. & MARKES, M. 1980. Defects induced by seed orientation during quartz growth. *Journal of Crystal Growth*, **49**, 739–742.

AINES, R.D., KIRBY, S.H. & ROSSMAN, G.R. 1984. Hydrogen speciation in synthetic quartz. *Physics and Chemistry of Minerals*, **11**, 204–212.

ASHBY, M.F. 1972. A first report of deformation maps. *Acta Metallurgica*, **15**, 501–511.

ASHBY, M.F. & VERRALL, R.A. 1973. Diffusion accomodated flow and superplasticity. *Acta Metallurgica*, **21**, 149–163.

BAILEY, J.E. & HIRSCH, P.B. 1962. The recrystallization process in some polycrystalline metals. *Proccedings of the Royal Society*, London, **A267**, 11–30.

BAKER, D.W. & WENK, H.-R. 1972. Preferred orientation in a low-symmetry quartz mylonite. *Journal of Geology*, **80**, 81–105.

BAKER, D.W., WENK, H.-R. & CHRISTIE, J.M. 1969. X-ray analysis of preferred orientations in fine-grained quartz aggregates. *Journal of Geology*, **77**, 144–172.

BALL, A. 1980. A theory of geological faults and shear zones. *Tectonophysics*, **61**, T1–T5.

BLACIC, J.D. 1975. Plastic-deformation mechanisms in quartz: the effect of water. *Tectonophysics*, **27**, 271–294.

BOUCHEZ, J.L., LISTER, G.S. & NICOLAS, A. 1983. Fabric asymmetry and shear sense in movement zones. *Geologische Rundschau*, **72**, 401–419.

BRANDON, D.G. 1966. The structure of high-angle grain boundaries. *Acta Metallurgica*, **14**, 1479–1484.

BURTON, B. 1977. *Diffusion Creep of Polycrystalline Materials*. Trans Tech Publications S.A., Zurich, Switzerland.

CAHN, R.W. 1953. Twinned crystals. *Advances in Physics*, **3**, 363–445.

CAHN, R.W. 1983. Recovery and recrystallization. *In*: CAHN, R.W. & HAASEN, P. (eds) *Physical Metallurgy*, 3rd edn. North-Holland, Amsterdam, 1596–1672.

CHOPRA, P.N. & PATERSON, M.S. 1981. The experimental deformation of dunite. *Tectonophysics*, **78**, 453–473.

CHRISTIE, J.M., ORD, A. & KOCH, P.S. 1980. Relationship between recrystallized grain size and flow stress in experimentally deformed quartzite. *American Geophysical Union, Eos*, **61**, 377.

CORDIER, P. & DOUKHAN, J.-C. 1989. Water solubility in quartz and its influence on ductility. *European Journal of Mineralogy*, **1**, 221–237.

CORDIER, P., WEIL, J.A., HOWARTH, D.F. & DOUKHAN, J.-C. 1994. Influence of the (4H)Si deflect on dislocation motion in crystalline quartz. *European Journal of Mineralogy*, **6**, 17–22.

DELL'ANGELO, L. & TULLIS, J. 1989. Fabric development in experimentally sheared quartzites. *Tectonophysics*, **169**, 1–21.

DUNLOP, G.L. & NILSSON, J.-O. 1980. The influence of interfacial structure on the high temperature mechanical behaviour of grain boundaries. *Materals Science and Engineering*, **42**, 273–280.

ETHERIDGE, M.A. & WILKIE, J.C. 1979. Grain size reduction, grain boundary sliding and the flow strength of mylonites. *Tectonophysics*, **58**, 159–178.

EXNER, H.E. 1972. Analysis of grain and particale size distribution in metallic materials. *International Metallurgical Reviews*, **159**(17), 25–42.

FITZGERALD, J.D., BOLAND, J.N., McLAREN, A.C., ORD, A. & HOBBS, B.E. 1991. Microstructures in water-weakened single crystals of quartz. *Journal of Geophysical Research*, **96**, 2139–2155.

GAPAIS, D. & BARBARIN, B. 1986. Quartz fabric transition in a cooling syntectonic granite (Hermitage massif, France). *Tectonophysics*, **125**, 357–370.

GIFKINS, R.C. 1976. Grain boundary sliding and its accomodation during creep and superplasticity. *Metallurgical Transactions, Series A*, **7**, 1225–1232.

GIFKINS, R.C. 1977. The effect of grain size and stress upon grain boundary sliding. *Metallurgical Transactions, Series A*, **8**, 1507–1516.

GILETTI, B.J. & YUND, R.A. 1984. Oxygen diffusion in quartz. *Journal of Geophysical Research*, **89**, 4039–4046.

GLEASON, G.C., TULLIS, J. & HEIDELBACH, F. 1993. The role of dynamic recrystallization in the development of lattice preferred orientation in experimentally deformed quartz aggregates. *Journal of Structural Geology*, **15**, 1145–1168.

GLEASON, G.C. & TULLIS, J. 1995. A flow law for dislocation creep of quartz aggregates determined with the molten salt cell. *Tectonophysics*, **247**, 1–23.

GLEITER, H., MAHAJAN, S. & BACHMANN, J.J. 1980. The generation of lattice dislocations by migrating boundaries. *Acta Metallurgica*, **28**, 1603–1610.

GREEN, H.W., GRIGGS, D.T. & CHRISTIE, J.M. 1970. Syntectonic and annealing recrystallisation of fine-grained quartz aggregates. *In*: PAULITSCH, P. (ed.) *Experimental and Natural Rock Deformation*. Springer, Berlin.

GRIGGS, D.T. 1967. Hydrolytic weakening of quartz and other silicates. *Geophysical Journal of the Royal Astronomical Society*, **14**, 19–32.

GUILLOPE, M. & POIRIER, J.P. 1979. Dynamic recrystallization during creep of single-crystal halite: An experimental study. *Journal of Geophysical Research*, **84**, 5557–5567.

HEAD, A.K., HUMBLE, P., CLAREBROUGH, L.M., MORTON, A.J. & FOREWOOD, C.T. 1973. *Computed Electron Micrographs and Defect Indentification*. North-Holland, Amsterdam.

HEGGIE, M. & JONES, R. 1987. Density functional analysis of the hydrolysis of Si–O bonds in disiloxane Application to hydrolytic weakening in quartz. *Philosophical Magazine Letters*, **55**, 47–51.

HOBBS, B.E., McLAREN, A.C. & PATERSON, M.S. 1972. Plasticity of single crystals of synthetic quartz. *In*: HEARD, H.C., BORG, I.Y., CARTER, N.L. & RALEIGH, C.B. (eds) *Flow and Fracture Rocks, The Griggs Volume*. American Geophysical Union, Geophysical Monograph, **16**, 29–53.

KANG, S.K., BERNSTEIN, I.M. & BAUER, C.L. 1976. *In situ* observations of the recrystallization process in single crystal thin films of gold. *Scripta Metallurgica*, **10**, 693–696.

KARATO, S.-I., PATERSON, M.S. & FITZGERALD, J.D. 1986. Rheology of synthetic olivine aggregates; influence of grain size and water. *Journal of Geophysical Research*, **91**, 8151–8176.

KEKULAWALA, K.R.S.S., PATERSON, M.S. & BOLAND, J.N. 1978. Hydrolytic weakening in quartz. *Tectonophysics*, **46**, T1–T6.

KERN, H. 1977. Preferred orientation of experimentally deformed limestone, marble, quartzite and rock salt at different temperatures and state of stress. *Tectonophysics*, **39**,103–120.

KOCH, P.S., CHRISTIE, J.M., ORD, A. & GEORGE, R.P., JR. 1989. Effect of water on the rheology of experimentally deformed quartzite. *Journal of Geophysical Research*, **94**, 13 975–13 996.

KRONENBERG, A.K. & TULLIS, J. 1984. Flow strengths of quartz aggregates: grain size and pressure effects due to hydrolytic weakening. *Journal of Geophysical Research*, **89**, 4281–4297.

KRONENBERG, A.K., KIRBY, S.H., AINES, R.D. & ROSSMAN, G.R. 1986. Solubility and diffusional uptake of hydrogen in quartz at high pressures: implications for hydrolytic weakening. *Journal of Geophysical Research*, **91**, 12 723–12 744.

LAW, R.D. 1990. Crystallographic fabrics: a selective review of their applications to research in structural geology. *In*: KNIPE, R.J. & RUTTER, E.H. (eds) *Deformation Mechanisms, Rheology and Tectonics*. Geological Society, London, Special Publications, **54**, 335–352.

LINKER, M.F. & KIRBY, S.H. 1981. Anisotropy in the rheology of hydrolytically weakened synthetic quartz crystals. *In*: CARTER, N.L., FRIEDMAN, M., LOGAN J.M. & STEARNS, D.W. (eds) *Mechanical Behavior of Crustal Rocks*. American Geophysical Union, Geophysical Monograph, **24**, 29–48.

LUAN, F.C. & PATERSON, M.S. 1992. Preparation and deformation of synthetic aggregates of quartz. *Journal of Geophysical Research*, **97**, 301–320.

LUTON, M.J. & SELLARS, C.M. 1969. Dynamic recrystallization in nickel and nickel-iron alloys during high temperature deformation. *Acta Metallurgica*, **17**, 1033–1043.

MAINPRICE, D., BOUCHEZ, J.L., BLUMENFELD, P. & TUBIA, J.M. 1986. Dominant c-slip in naturally deformed quartz: implications for dramatic plastic softening at high temperature. *Geology*, **14**, 812–822.

MAINPRICE, D.H. & PATERSON, M.S. 1984. Experimental studies on the role of water in the plasticity of quartzites. *Journal of Geophysical Research*, **89**, 4257–4269.

MASUDA, T. & FUJIMURA, A. 1981. Microstructural development of fine-grained quartz aggregates by syntectonic recrystallization. *Tectonophysics*, **72**, 105–128.

MCLAREN, A.C. 1986. Some speculations on the nature of high-angle grain boundaries in quartz rocks. *In*: HOBBS, B.E. & HEARD, H.C. (eds) *Minerals and Rock Deformation: Laboratory Studies*.

American Geophysical Union, Geophysical Monograph, **36**, 233–245.

MCLAREN, A.C. & PHAKEY, P. 1966. Electron microscope study of Brazil twin boundaries in amethyst quartz. *Physica Status Solidi*, **13**, 413–422.

MCLAREN, A.C., FITZGERALD, J.D. & GERRETSEN, J. 1989. Dislocation nucleation and multiplication in synthetic Quartz: Relevance to water weakening. *Physics and Chemistry of Minerals*, **16**, 465–482.

PATERSON, M.S. 1970. A high pressure, high temperature apparatus for rock deformation. *International Journal of Rock Mechanics and Mining Science*, **7**, 517–526.

PATERSON, M.S. 1978. *Experimental Rock Deformation – The Brittle Field*. Springer, Berlin.

PATERSON, M.S. 1982. The determination of hydroxyl by infrared absorption. *In*: *Quartz, Silicate Glasses and Similar Materials. Bulletin Minéralogie*, **105**, 20–29.

PATERSON, M.S. 1986. The thermodynamics of water in quartz. *Physics and Chemistry of Minerals*, **13**, 245–255.

PATERSON, M.S. & LUAN, F.C. 1990. Quartzite rheology under geological conditions. *In*: KNIPE, R.J. & RUTTER, E.H. (eds) *Deformation Mechanisms, Rheology and Tectonics*. Geological Society, London, Special Publications, **54**, 299–307.

POND, R.C., SMITH, D.A. & SOUTHERDEN, P.W.J. 1978. On the role of grain boundary dislocations in high temperature creep. *Philosophical Magazine*, **A37**, 27–40.

POST, A. & TULLIS, J. 1998. The rate of water penetration in experimentally deformed quartzite: implications for hydrolytic weakening. *Tectonophysics*, **295**, 117–137.

PRICE, G.P. 1973. The photometric method in microstructural analysis. *Journal of American Science*, **273**, 523–537.

PRICE, G.P. 1980. The analysis of quartz c-axes fabrics by the photometric method. *Journal of Geology*, **88**, 181–195.

RAY, C.M.F. & SMITH, D.A. 1980. On the mechanisms of grain boundary migration. *Philosophical Magazine*, **A41**, 477–492.

ROBERTS, W., BODÉN, H. & AHLBLOM, B. 1979. Dynamic recrystallization kinetics. *Metal Science*, **13**, 195–205.

SAH, J.P., RICHARDSON, G.J. & SELLARS, C.M. 1974. Grain size effects during dynamic recrystallisation of nickel. *Metal Science*, **8**, 325–331.

SCHMID, S.M., BOLAND, J.N. & PATERSON, M.S. 1977. Superplastic flow in fine grained limestome. *Tectonophysics*, **43**, 257–291.

SCHMID, S.M., PANOZZO, R. & BAUER, S. 1989. Simple shear experiments on calcite rocks: rheology and microfabric. *Journal of Structural Geology*, **9**, 747–778.

SCHMOCKER, M., BYSTRICKY, M., KUNZE, K., BURLINI, L., STÜNITZ, H. & BURG, J.-P. 2003. Granular flow and Riedel band formation in water-rich quartz aggregates experimentally deformed in torsion. *Journal of Geophysical Research*, **108**, ECV 1–16.

SELLARS, C.M. 1978. Recrystallization of metals during hot deformation. *Philosophical Transations of the Royal Society London*, **135**, 513–516.

WENK, H.R., BAKER, D.W. & GRIGGS, D.T. 1967. X-ray analysis of hot-worked and annealed flint. *Science*, **157**, 1447–1449.

WHITE, S. 1976. The effect of strain on the microstructure fabrics and deformation mechnisms in quartzite. *Philosophical Transations of the Royal Society, London*, **A283**, 69–86.

WIRMARK, G., NILSSON, J.-O. & DUNLOP, G.L. 1981. Sliding at twin boundaries during high-temperature creep. *Philosophical Magazine*, **A43**, 93–101.

High-strain deformation tests on natural gypsum aggregates in torsion

V. BARBERINI [1], L. BURLINI[2], E. H. RUTTER[3] & M. DAPIAGGI[1]

[1]*Università degli Studi di Milano, Dipartimento Scienze della Terra 'A. Desio',*
via Botticelli 23, I-20133 Milano, Italy (e-mail: valentina.barberini@unimi.it)
[2]*Geological Institute, Sonneggstrasse 5, CH-8092 Zürich, Switzerland*
[3]*University of Manchester, Department of Earth Sciences,*
Oxford Road, Manchester M13 9PL, UK

Abstract: Evaporitic minerals, such as gypsum, within sedimentary sequences play an important role in localizing deformation, especially in thrust tectonics, implying that their strength is generally lower than that of other rocks. To study the rheological and microstructural evolution of gypsum with strain, a set of experiments was performed on natural gypsum samples from Volterra (Italy). To reach high shear strain (up to $\gamma = 5$), deformation tests were performed in torsion at 300 MPa confining pressure, at temperatures up to 127 °C, and at shear strain rates between 10^{-3} and $10^{-5}\,s^{-1}$. All samples were studied by optical microscopy, to investigate the evolution of the microstructure with strain, and by X-ray diffraction (XRD) analyses, to determine whether and to what extent gypsum dehydrated during deformation. The shear stress increased with shear strain rate and decreased with temperature. A peak stress was usually reached at γ between 0.5 and 1.5. After the peak, 30–40% of weakening occurred but mechanical steady-state conditions were never reached. The microstructure evolved from a plastic deformation microstructure, where grains changed shape according to the bulk strain imposed, into a recrystallization-dominated microstructure, where grains were more equant. The shear stress sensitivity to strain rate increases with progressive strain, thus a meaningful constitutive flow law can only be determined from experiments in which steady-state flow is eventually reached. These results imply that gypsum in nature will flow plastically at shear stress levels lower than those expected from previous experimental studies due to the strain weakening associated with dynamic recrystallization, which can occur at temperatures even lower than gypsum dehydration.

Gypsum, together with halite and anhydrite, are the main minerals forming evaporitic rocks. When interlayered within sedimentary sequences, they can play an important role in localizing deformation. Evaporitic levels have often been found along the zones of movement of overthrust faults or at the lower boundaries of nappes (e.g. Heard & Rubey 1966; Laubscher 1975; Malavielle & Ritz 1989). The worldwide occurrence of deformation that involves evaporites sequences in compressive settings is reviewed by Warren (1999).

The deformability of evaporite rocks and minerals is the result of:

- an intrinsically low resistance to deformation by intracrystalline plasticity compared to carbonates and silicate rocks (Heard & Rubey 1966);
- the possibility of the development of excess pore fluid pressures caused by dehydration reactions (such as gypsum → anhydrite + water). This causes lowering of the resistance to brittle failure as a result of decrease in effective mean stress (Heard & Rubey 1966);
- because many evaporite minerals can display dynamic recrystallization or growth of new phases at rather low temperatures, it is possible for the growth of transiently fine-grained reaction products to favour flow by diffusion-accommodated grain-boundary sliding (Brodie & Rutter 1985);
- evaporite minerals can be relatively highly soluble in groundwater, and hence may

From: BRUHN, D. & BURLINI, L. (eds) 2005. *High-Strain Zones: Structure and Physical Properties.*
Geological Society, London, Special Publications, **245**, 277–290.
0305-8719/05/$15.00 © The Geological Society of London 2005.

deform by water-assisted diffusive mass-transfer processes (e.g. Peach & Spiers 1996; de Meer & Spiers 1995, 1999).

A good knowledge of the mechanical behaviour and long-term stability of evaporites is of great importance in several fields. In the oil industry, evaporites often form the cap rock of oil and gas reservoirs, and rock salt formations are important in nuclear waste disposal (Schulze et al. 2001) and hydrocarbon storage. To understand how different factors, such as confining pressure, temperature, differential stress, strain rate and dehydration reactions, influence the behaviour of evaporites, it is essential to perform experimental deformation tests.

Many experimental studies have been performed on halite (e.g. Carter & Heard 1970; Chester 1988; Peach & Spiers 1996; Popp et al. 2001; Schulze et al. 2001), anhydrite (e.g. Dell'Angelo & Olgaard 1995; Heidelbach et al. 2001) and gypsum (e.g. Heard & Rubey 1966; Murrel & Ismail 1976; Ko et al. 1995, 1997; Olgaard et al. 1995; de Meer & Spiers 1995, 1999; Stretton 1996; de Meer et al. 2000). In most cases these tests were performed under coaxial deformation conditions under standard 'triaxial' loading configurations, or to study hydrostatic compaction, and generally up to only a small amount of strain. However, many geological deformations are non-coaxial (e.g. simple shear) and show evidence of large amounts of strain. The significant strain dependence observed in the behaviour of some relevant geological material deformed up to large shear strains, such as calcite (e.g. Pieri et al. 2001; Barnhoorn et al. 2004), olivine (e.g. Bystricky et al. 2000) and anhydrite (e.g. Heidelbach et al. 2001), have shown the importance of high strain experiments. To create such conditions in experimental rock deformation tests, there are two main techniques that can be used. The first is the saw-cut (direct shear) assembly employed in triaxial testing machines and the second is the torsion technique. The latter was used in rock mechanical studies by Handin et al. (1960), but has recently been further developed for elevated temperature testing (e.g. Paterson & Olgaard 2000). A few deformation experiments in simple shear were carried out on 'synthetic' gypsum rock using the saw-cut assembly (Olgaard et al. 1986; Panozzo Heilbronner & Olgaard 1987), but torsion experiments on gypsum have never previously been reported. Here we report a set of preliminary torsion experiments on natural gypsum rock that provide a pointer for further high strain studies on this material.

Starting material

Specimens were cored from the same block of natural alabaster from the Volterra Basin (Tuscany, Italy), and with the same orientation used by Ko et al. (1995, 1997), Olgaard et al. (1995) and by Stretton (1996) in their experimental work under coaxial strain conditions.

The Volterra Basin is the largest among the N–S elongated extensional basins that opened along with the Tyrrhenian Basin beginning in the late Tortonian on the back of the E-migrating Apennine thrust belt, where Messinian evaporites were deposited (Testa & Lugli 2000). Within this basin, different gypsum facies have been described (selenite gypsum, gypsarenites and gypsiferous arenites, gypsum laminites, and massive and nodular microcrystalline gypsum; Testa 1996). The microcrystalline variety, derived from rehydration of secondary anhydrite (Testa 1996), has been chosen among the others because it is fine grained (80–100 μm) and contains less than 1% impurity phases (Ko et al. 1995). Moreover, the microstructure is very homogeneous (Fig. 1), with no shape or lattice preferred orientation. Grains are often subhedral and form nearly equigranular aggregates. Total porosity is of the order of 0.1%.

Experimental methods

Deformation apparatus

Torsion experiments were carried out in a high-pressure–high-temperature gas-medium apparatus at ETH, Zürich. The apparatus is internally heated and equipped with a torsion loading system in addition to the standard axial loading system (Paterson Instruments Ltd, Canberra). The testing column is positioned in a vertically oriented pressure vessel. External actuators supply the axial load and the torque that are transmitted to the specimen assembly (Paterson & Olgaard 2000).

The resolution of the axial load and the torque are, respectively, ± 100 N and ± 1 N m (Pieri et al. 2001). These loads correspond, for a typical 10 mm-diameter cylindrical gypsum sample, to a resolution of ± 1.27 MPa for the axial stress σ_1 and of ± 4 MPa for the shear stress τ. The furnace was carefully calibrated so that the temperature measurement uncertainty was less than ± 1 °C.

Sample preparation and specimen assembly

Torsion experiments were performed on precision-ground cylindrical specimens with

Fig. 1. Microphotographs (crossed polars) showing microcrystalline gypsum from Volterra (Tuscany, Italy), used as starting material in the torsion experiments.

diameters between 9.71 and 9.96 mm, and with lengths between 3.47 and 10.15 mm. The specimens were assembled within the torsion column as schematically drawn in Figure 2. The whole torsion column is inserted in a 0.25 mm wall thickness non-annealed metal (iron or copper) jacket with a 15 mm inner diameter. Non-annealed jackets were used instead of annealed ones because although the annealed ones are softer, the effects of strain-hardening are greater at the high strains imposed. Moreover, by using non-annealed jackets the variation in jacket strength throughout the test was minimized. Because the specimen diameter was about 5 mm smaller than that of the pieces that make frictional contact with the specimen, the jacket had to be spun down in its central section on a lathe until it

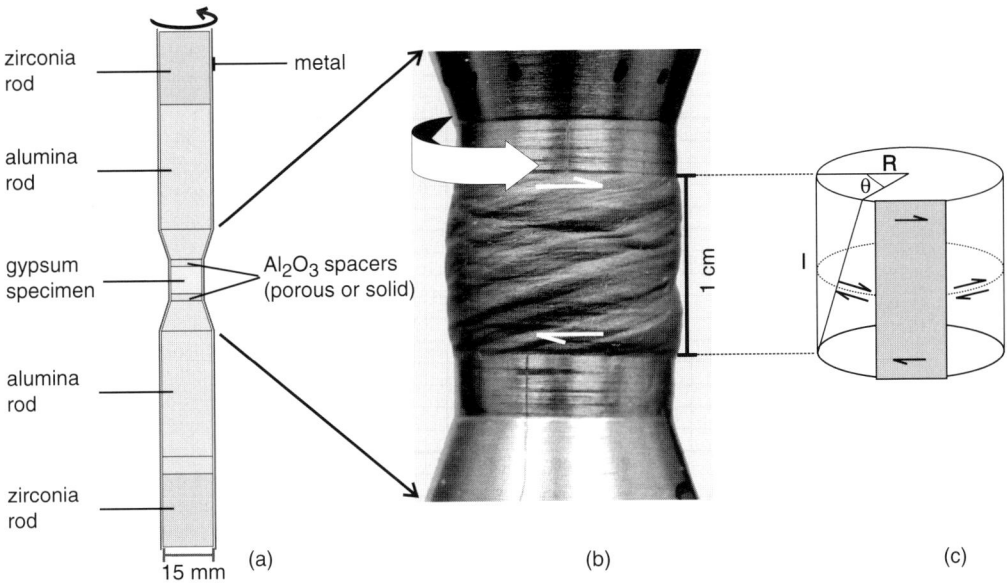

Fig. 2. Set-up of the torsion experiment. (**a**) The torsion column, comprising the specimen between two series of alumina and zirconia end-pieces, is inserted in a metal jacket. (**b**) Photograph of a deformed sample, sense of shear is indicated. The rolling traces on the jacket, which were vertically oriented before the test, serve as passive strain markers. (**c**) Schematic drawing of a torsion sample, where R is the radius, l is the length of the cylindrical specimen and θ is the twist angle. The position of a thin section plane in a specimen deformed in torsion is shown (grey area).

fitted snugly around the sample. The jacket and the specimen assembly were sealed with O-rings against the steel end-pieces outside the furnace at the top and bottom of the assembly.

Experimental conditions

Torsion experiments (Table 1) were conducted at temperatures of 70 ± 1, 90 ± 1 and $127 \pm 1\ °C$, and to twist angles ϑ between $46°$ and $389°$. Because of the low temperatures used ($70–127\ °C$), the strength of gypsum + jacket is quite high, therefore it is necessary to deform at high confining pressure (300 MPa) to avoid frictional sliding at the ends of the specimen cylinder. The angular displacement rate was chosen for each test depending on the desired shear strain rate on the outer surface of the sample, and on its radius and length (see equation 2). Constant shear strain-rate tests were performed either under undrained or totally drained conditions at shear strain rates between 4×10^{-5} and $6 \times 10^{-4}\ s^{-1}$. In addition, some twist-rate-stepping tests (involving a strain-rate step from 10^{-5} to $6 \times 10^{-4}\ s^{-1}$) were carried out to determine the value of the stress exponent n.

Determination of jacket strength

To determine the strength of iron and copper jackets two calibration tests were performed. Two nylon specimens (negligible shear strength) of 10 mm diameter were assembled as previously described, one jacketed with iron and one with copper. Each column was deformed in torsion under conditions similar to those used for deformation tests on gypsum, i.e. at a confining pressure of 300 MPa and $90 \pm 1\ °C$ (copper), and $90 \pm 1\ °C$ and $127 \pm 1\ °C$ (iron). Also, different shear strain rates ($5 \times 10^{-5}–2 \times 10^{-4}\ s^{-1}$ for iron and $10^{-4}–10^{-3}\ s^{-1}$ for copper) were used. The iron strength was higher than the copper strength (Fig. 3), therefore copper jackets were used in most tests, even though copper showed some strain hardening. These data were used to correct the stress–strain data for the gypsum experiments, subtracting the jacket strength at the corresponding strain and temperature.

Rheological analysis of the torsion test

The approach to the deformation of a uniform (right circular) cylinder in torsion is based on the assumption that strain is homogeneous along the cylinder length and constitutes locally simple shear (e.g. Handin *et al.* 1960; Paterson & Olgaard 2000). The relation between measured twist angle and torque, and shear strain and shear stress are as follows (Paterson & Olgaard 2000). The shear strain is given by:

$$\gamma_r = \frac{r}{l}\vartheta \tag{1}$$

where r is any radius ($0–R$), l is cylinder length and ϑ is twist angle (radians), and analogously for the strain rate:

$$\dot{\gamma}_r = \frac{d\gamma_r}{dt} = \frac{r}{l}\dot{\vartheta}. \tag{2}$$

It is a reasonable assumption for most relevant materials that the solid cylinder deforms with power-law creep rheology, with constant parameters anywhere in the sample. Then, the shear stress at any radius is:

$$\tau_r = \left(\frac{r}{R}\right)^{(1/n)}\tau_R \tag{3}$$

and the shear stress at maximum radius $r = R$ relates to the torque M applied to the cylinder by:

$$\tau_R = \frac{(3 + 1/n)}{(2\pi R^3)}M. \tag{4}$$

The stress exponent n can be experimentally determined from twist-rate-stepping tests because the slope of the line in a log–log plot of ϑ v. M is given by:

$$\log \vartheta = n \log M + \text{constant}. \tag{5}$$

The applied shear stress within the sample (Pieri *et al.* 2001) increases non-linearly with radius according to equation (3) and, for high exponent ($n \approx 10$), the shear stress remains nearly constant for $r > R/2$. Shear strain, however, varies linearly within the sample radius, from $\gamma = 0$ at the cylinder centre to $\gamma = \gamma_{max}$ at the outer cylinder surface.

Microstructural studies on
torsion specimens

Because of the strain gradient occurring along the radius, microstructural studies from thin sections of samples deformed in torsion are more complex than for those deformed in coaxial compression deformation tests. Thin sections were cut parallel to the cylinder axis close to the outer surface (Fig. 2). The shear direction is approximately parallel to this section. Shear strain, shear strain rate and shear stress are nearly constant across such sections. Thin sections (20 μm thick) of the deformed specimens were prepared using ethyl alcohol, to avoid rehydrating any bassanite formed during experiments.

Table 1. *Test details*

Run No.	Temperature (°C)	Sample No.	Diameter $2R$ (mm)	Length l (mm)	Total twist ϑ (rad)	Shear strain γ	Shear strain rate (s^{-1})	Drained (D) undrained (U)	Gypsum (G) bassanite (B)	Remarks
PO382	70	V7	9.72	8.65	0.87	0.49	5×10^{-5}	U	G	Jacket failure
PO383	90	V6	9.78	8.34	1.91	1.12	Up to 1×10^{-4}	U	G+B	Stepping, jacket failure
PO384	70–90	Nylon	9.99	7.45	1.09	0.73	Up to 2×10^{-4}	–	–	Calibration of iron jacket strength – stepping
PO385	127	V4	9.70	9.92	0.73	0.36	5×10^{-5}	D	B+G	Jacket failure
PO392	90	V8	9.91	7.66	1.84	1.19	1×10^{-4}	U	G+B	Jacket failure
PO393	90	V9	9.90	10.15	5.10	2.49	1×10^{-4}	U	G+B	Jacket failure
PO394	90	Nylon	9.96	9.14	5.26	2.87	Up to 1×10^{-3}	–	–	Calibration of copper jacket strength – stepping
PO395	90	V10	9.96	8.31	6.79	4.07	2×10^{-4}	D	G	Jacket failure
PO396	90	V11	9.92	9.23	5.39	2.89	6×10^{-4}	D	G	Jacket failure
PO403	90	V13	9.83	7.10	2.58	1.78	1×10^{-4}	D	G+B	Jacket failure
PO404	90	V14	9.85	8.80	4.67	2.61	2×10^{-4}	D	G	Jacket failure
PO405	70	V15	9.79	6.83	2.51	1.80	1×10^{-4}	D	G	–
PO455	70	V16	9.71	7.25	6.15	4.11	1×10^{-4}	D	G	Jacket failure
PO456	70	V18	9.81	6.28	4.83	2.59	Up to 6×10^{-4}	D	G	Stepping, jacket failure
PO472	70	V20	9.82	3.47	3.40	4.82	1×10^{-4}	D	G	Jacket failure

Experimental conditions and specimen dimensions for the performed tests. All samples were deformed at 300 MPa confining pressure. Details of the experiments for jacket strength calibration are included.

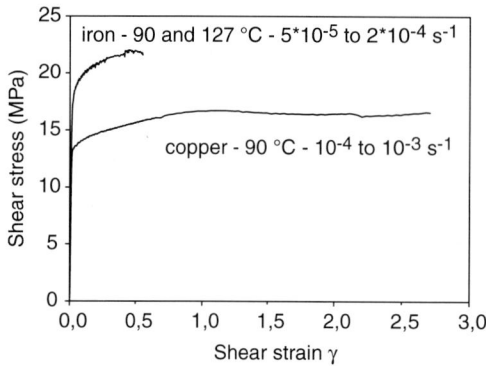

Fig. 3. Shear stress v. shear strain curves for iron and copper jackets, obtained by torsion-testing jacketed nylon specimens. Nylon strength is negligible compared to copper or iron, therefore the torque is entirely supported by the jacket.

Microstructural information is presented in the standard reference frame commonly used for observations of naturally deformed mylonites, with the section cut normal to the foliation containing the lineation. Descriptions of observed microstructural features reported in this paper are based on criteria given by Passchier & Trouw (1996).

X-ray diffraction analyses

In the laboratory dehydration of gypsum ($CaSO_4 \cdot 2H_2O$), bassanite ($CaSO_4 \cdot \frac{1}{2}H_2O$) is generally produced as an intermediate phase prior to complete dehydration to anhydrite. X-ray diffraction (XRD) analyses were used to single out gypsum and bassanite in all samples. X-ray diffraction analyses on both the starting material and the deformed samples were carried out using the Philips X'Pert diffractometer at the Department of Earth Sciences (University of Milano).

The investigated 2ϑ range ($2°-30°$), although small, allowed discrimination between gypsum and bassanite, as at least three peaks from each phase were present. Moreover, the diffraction pattern acquisition was performed, for each specimen, directly on the surface from which the thin section was cut, with no need to risk damage to the specimen from grinding.

Experimental results

Shear stress v. strain data

Deformation was usually homogeneous but not always. In some tests several shear bands

developed, although they were homogeneously distributed throughout the sample. At 70 °C (Fig. 4a) peak shear stresses were reached at γ between 1 and 1.5, at 90 °C (Fig. 4b) at γ below 0.5 for the slower tests, and at γ between 0.5 and 1 for the faster ones. At 127 °C (Fig. 4c) peak shear stress was reached at $\gamma = 0.3$. After the peaks, various amounts of strain-dependent weakening (up to 40%) occurred at all temperatures. Moreover, mechanical steady-state conditions (constant shear stress at constant twist rate) were never reached. Most often, jacket failure ended the experiment prematurely owing to the limited ductility of the metal at low temperatures.

The torque measured during each experiment was converted into shear stress at the outer cylinder radius using equation (4), with a stress exponent obtained from strain-rate-stepping tests.

Power-law creep exponent

Two strain-rate-stepping tests were performed to determine the value of the stress exponent n at temperatures of 70 and 90 °C, and after different amounts of strain. The specimen was twisted at a constant angular displacement rate until the torque reached an almost constant value. The twist-rate was then decreased by approximately one order of magnitude and kept constant until the torque value reached a new constant value, then the twist rate was increased in several small increments, extending above the initial rate. The twist-rate values and the corresponding torques measured were plotted in a log–log diagram (Fig. 5). The slope of the linear regression of the data gives the stress exponent n according to equation (5).

n values of 9.6 and 3.9 were determined at 70 °C for $\gamma = 1.2$ and 3.5, respectively, whereas $n = 12$ was determined at 90 °C for $\gamma = 1$. Therefore, the n value apparently decreases with strain, probably in response to a change in dominant deformation mechanism, but it does not seem to be very sensitive to temperature for a given twist rate.

Microstructural features

A similar microstructural evolution was observed with strain at 70 and 90 °C (Figs 6 and 7). After $\gamma = 0.5$ (Fig. 6a & b) undulose extinction and kink bands formed within the few largest grains. The formation of undulose extinction and kink bands affected an increasing number of grains of different size after $\gamma = 1$ (Fig. 7a & b). Also, a shape preferred orientation (SPO) of grains consistent with the shear sense

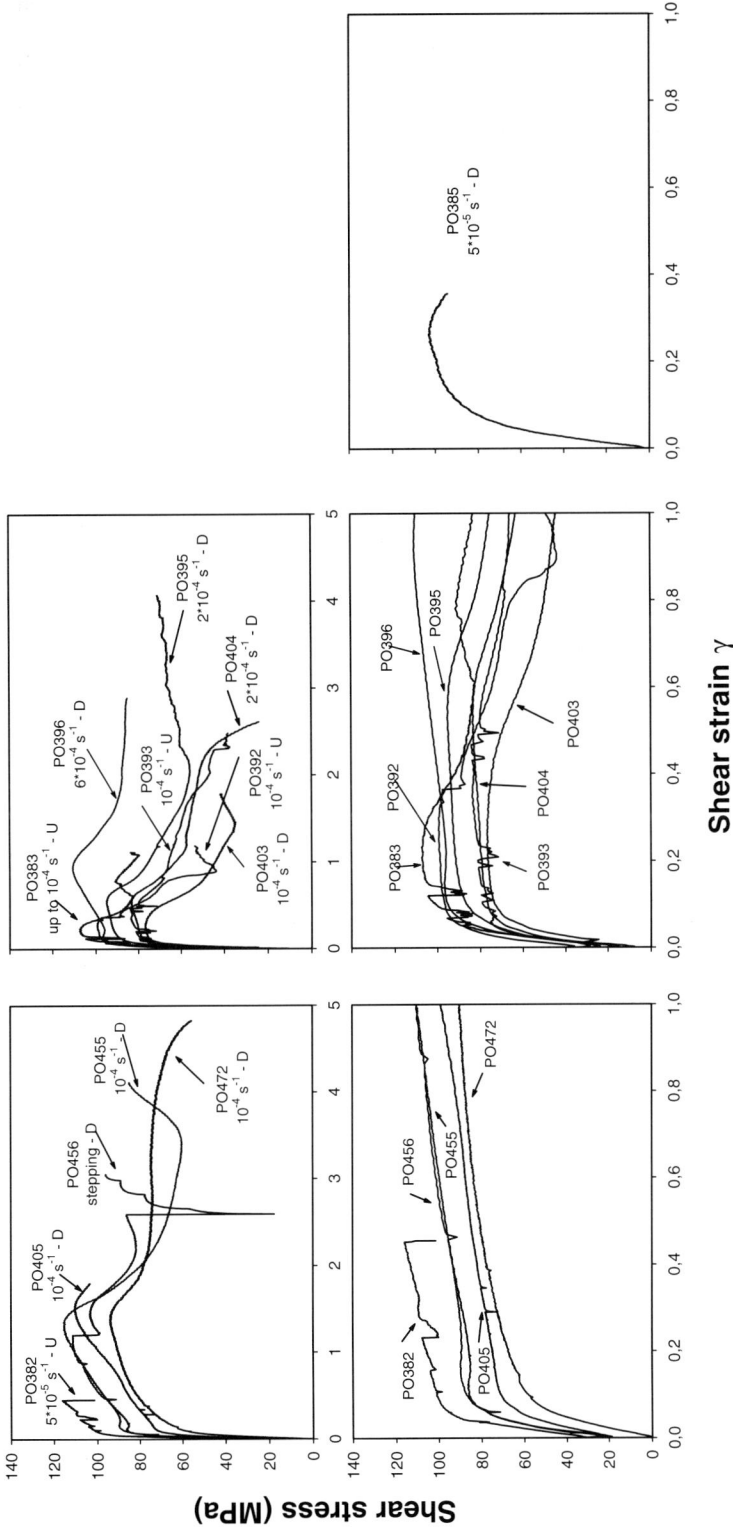

Fig. 4. Shear stress *v.* shear strain graphs obtained from torsion testing on gypsum specimens. All data are corrected for jacket strength. Shear stress is calculated using the stress exponent value from strain-rate stepping tests. All curves are characterized by a post-yield stress maximum followed by continuous strain softening. (**a**) Tests at 70 °C. (**b**) Tests at 90 °C. (**c**) Test at 127 °C. The bottom part of each graph represents a magnification of the top part. Shear strain rate and indication of drained (D) or undrained (U) conditions are reported for each curve.

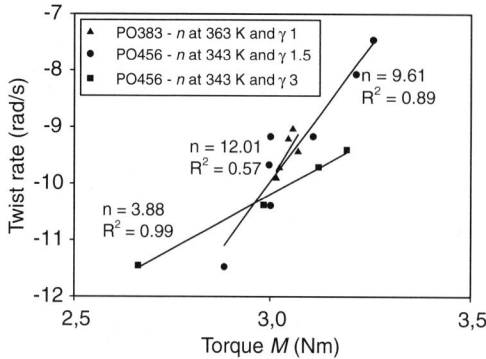

Fig. 5. Results from strain-rate stepping tests. Twist rate v. torque in a log−log plot. The slopes of the fitted lines give the stress exponent *n* for the respective temperature and shear strain.

started to develop. Attempts to measure lattice preferred orientation (LPO) by X-ray goniometry were not successful, but flat-stage optical microscopy showed that an LPO developed and progressively strengthened with increasing strain. After $\gamma = 1.7$ (Figs 6c & d and 7c & d) SPO and LPO became strongly developed and most grains showed undulose extinction and kink bands.

At $\gamma > 2.5$ (Figs 6e & f and 7e & f), shear bands and subgrain boundaries were observed in some of the grains in addition to undulose extinction and kink bands. Moreover, several small and undeformed new grains began to appear. At $\gamma > 3.5$ (Figs 6g & h and 7g & h), although deformed grains were still present, a significant amount of new grains had formed by dynamic recrystallization.

No significant differences were observed between the microstructures obtained at 70 and 90 °C, except that at 90 °C dynamic recrystallization with concomitant grain size reduction started to develop at lower γ values. At 127 °C, however, the microstructure produced was quite different because the gypsum has been almost wholly replaced by fibrous bassanite (Fig. 8a & b).

Shear cataclastic bands in Riedel orientation formed within the sample deformed under undrained conditions (Fig. 9a & b). These are probably caused by the pressure excess resulting from gypsum dehydration. Under drained conditions the water escaped from the samples and only shear bands developed (Fig. 9c & d), with the characteristic bending of the tips of the grains.

Gypsum dehydration

The results from XRD analyses on the specimens deformed in torsion were as follows:

(a) at 70 °C, the lowest temperature investigated, only gypsum was present;

(b) at 127 °C, the highest temperature investigated, bassanite was dominant but a small amount of gypsum was still present;

(c) at 90 °C, the temperature at which the widest shear strain rate range was applied, bassanite formed when the shear strain rate was below or equal to $10^{-4}\,\mathrm{s}^{-1}$, whereas only gypsum was present for higher shear strain rates.

Thus, both temperature and shear strain rate influence the dehydration of gypsum. Probably more important than strain rate is the time at conditions. Slowest strain rates imply longer experimental time for reaching the same amount of strain. Stretton (1996) has already pointed out that the amount of time spent at a temperature influences the amount of gypsum dehydration, as the reaction progressively accelerates over the range 70−120 °C. Thus, from the observation in (c) above we cannot separate the effects of time at temperature from strain rate.

Discussion of experimental results

Deformation mechanisms

The occurrence of intragranular kink bands, undulose extinction, development of SPO and LPO, and, at high strains, dynamic recrystallization, point to the importance of deformation by intracrystalline plasticity at all experimental conditions studied. The peak shear stress values recorded ranged up to a maximum of 100 MPa at the lowest temperature and highest strain rate used. This value is substantially smaller than the confining pressure employed (300 MPa), and strongly suggests that deformation was dominated by intracrystalline plasticity, without significant contribution from cataclasis and associated dilatancy. The near-zero starting porosity precludes any significant strain associated with compaction. Stretton (1996) measured only small dilatant volume changes associated with axisymmetric shortening deformation at the same confining pressure, and a low sensitivity to confining pressure changes. The low crystallographic symmetry of gypsum means that there are few available slip systems, hence the formation of kink bands to help accommodate strain. A small amount of dilatation is probably required to accommodate intergranular strain

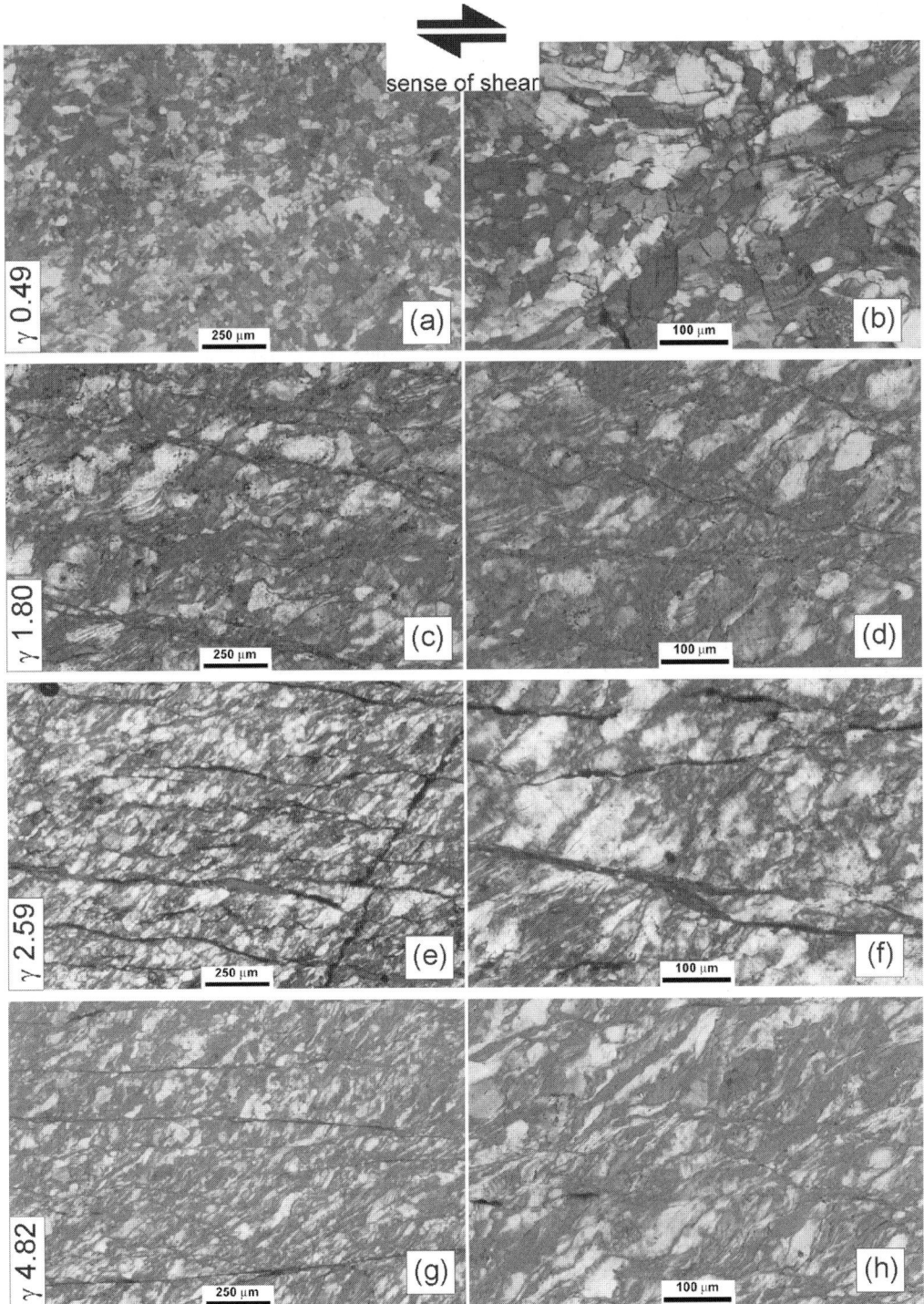

Fig. 6. Microstructures at 70 °C (optical microphotographs between crossed polars with gypsum plate). For increasing shear strain values, development of LPO and shape fabric can be observed. (**a**) & (**b**) Run PO382; (**c**) & (**d**) PO405; (**e**) & (**f**) PO456; (**g**) & (**h**) PO472.

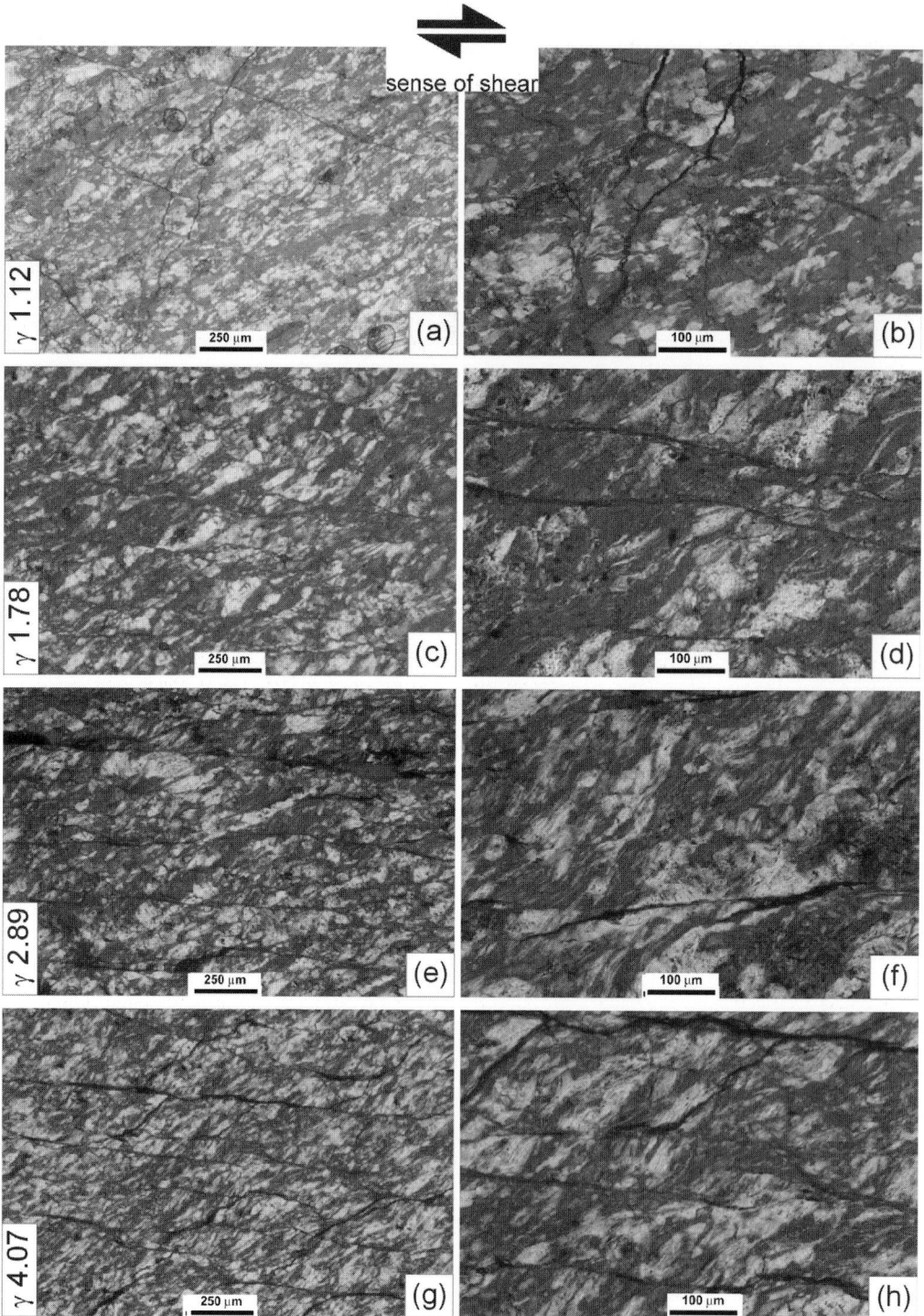

Fig. 7. Microstructures at 90 °C (microphotographs between crossed polars with gypsum plate). For increasing shear strain values, development of LPO and shape fabric can be observed. (**a**) & (**b**) Run PO383; (**c**) & (**d**) PO403; (**e**) & (**f**) PO396; (**g**) & (**h**) PO395.

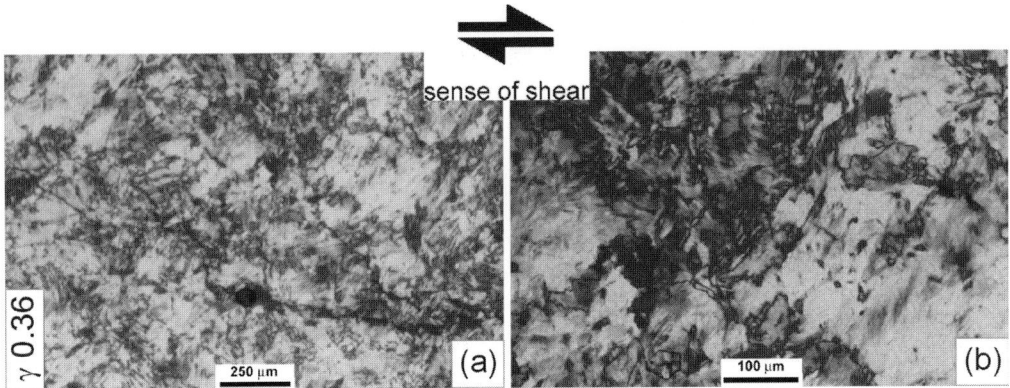

Fig. 8. Microstructures at 130 °C (run PO385; microphotographs between crossed polars with gypsum plate). Fibrous bassanite aggregates replacing gypsum can be observed.

incompatibilities and the opening of small voids in the hinges of kink bands. Stretton (1996) also reported a reduction in confining pressure sensitivity once dynamic recrystallization began to be important, thus being dynamic-recrystallization-efficient in reducing dilatancy.

The small number of experiments reported here do not allow the determination of a constitutive flow law for gypsum. This is because no constant value of the stress exponent n could be determined. It seems likely that the n value decreases with increasing strain. This finding is consistent with the observation that there is continuous strain softening up to high shear strains, implying that the dominant deformation mechanism is evolving over that strain interval, so that no steady-state flow can be established. This is in turn consistent with the observation that strain- and/or time-dependent dynamic recrystallization develops progressively over the duration of our experiments. A truly steady-state flow stress would not be expected until a sufficient combination of high strains and low strain rate were to be established, perhaps through the mechanism of intracrystalline plasticity accommodated by cyclic dynamic recrystallization. Stretton (1996) found for the regime of intracrystalline plastic flow without dynamic recrystallization that the flow stress is not very sensitive to large changes in strain rate. The present study points to a much greater sensitivity of flow stress to strain rate developing at high strains and low strain rates, and hence that a flow law for intracrystalline plasticity of gypsum rock that might be validly extrapolated to natural conditions must be determined wholly in the regime of plastic flow accompanied by dynamic recrystallization and in absence of

dehydration reaction. In presence of the gypsum dehydration to bassanite, the rheological behaviour of bassanite should also be taken into account.

Geological implications

Following the work of Heard & Rubey (1966) and Ko *et al.* (1995, 1997), the commonly cited association between gypsiferous horizons and tectonic detachment has tended to be interpreted in terms of transient weakening and embrittlement associated with dehydration of gypsum to anhydrite under undrained conditions. On the influence of the formation of bassanite during deformation we must recall that shear stress drops were recorded episodically during undrained experiments (Fig. 4). We know from XRD results that some gypsum underwent dehydration with formation of bassanite and water in these experiments. One possible interpretation is that the water formed during dehydration could not leave the sample causing a pore fluid pressure increase that resulted in an abrupt lowering of the effective stresses (with consequent shear stress drop). Interestingly, the numbers of shear bands observed on the outer jacket of the sample corresponded with the numbers of shear stress drops in the stress–strain curve. Conversely, under drained conditions the water eventually formed could leave the sample and, in fact, shear stress drops did not occur.

Although there is dispute regarding the temperature at which dehydration of gypsum rocks takes place in nature, it certainly lies between 40 and 90 °C. This corresponds to a depth range of between 1 and 3, or, perhaps, 4 km. This is not what one would associate with the

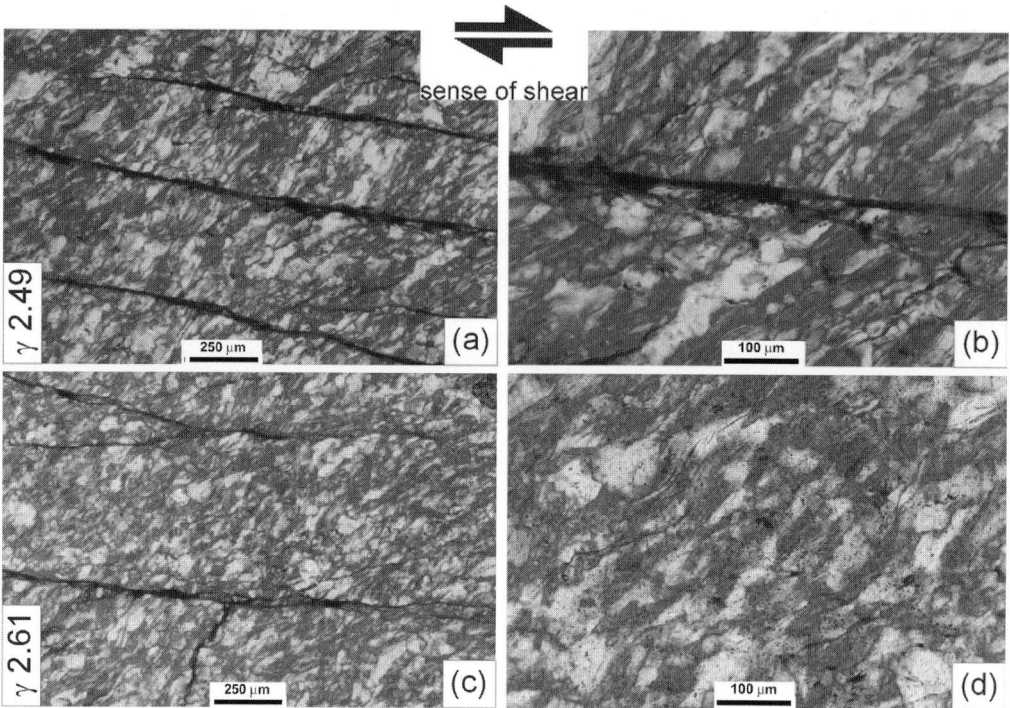

Fig. 9. Microstructures observed within specimens deformed under (**a**) & (**b**) undrained (run PO393) and (**c**) & (**d**) drained (run PO404) conditions (microphotographs between crossed polars with gypsum plate). A strong shape fabric can be observed. Cataclastic shear bands in Riedel orientation formed under undrained conditions. Under drained conditions minor shear bands in Riedel orientation are only visible at high magnification. Note that the shear strain in the two samples is comparable.

basal zone of a large overthrust, but it could be important to relatively smaller scale thrust tectonics in the external part of a foreland fold-and-thrust belt such as the Jura, or in hydro-carbon cap rocks where plastic flow would help maintain its integrity as a pressure seal. The present work shows that gypsum rocks can be plastically weak, even without invoking further transient weakening arising from dehydration. The high confining pressures used in these experiments, in order to be able to measure the plastic flow strength without slippage at the ends of the samples, means that the flow stresses measured represent upper bounds on those that could develop in nature, but which could conceivably be developed in a thrust belt wedge. More importantly, the possibility of low-temperature dynamic recrystallization, with a low stress exponent, implies that gypsum rocks flowing plastically at temperatures just below those required for dehydration may be very weak indeed. This conclusion points to the future importance of: (a) determining experimentally the flow-law parameters exclusively in the

regime of plasticity plus dynamic recrystalliza-tion; and (b) seeking to compare the microstruc-tures and LPOs of naturally and experimentally deformed gypsum rocks.

Conclusions

We have performed a preliminary set of experi-ments on natural, polycrystalline gypsum to high strains in torsion, at 70, 90 and 127 °C, and at a confining pressure of 300 MPa and at shear strain rates ranging between 4×10^{-5} and $6 \times 10^{-4} \, s^{-1}$. At 130 °C the rock dehy-drated almost completely to bassanite. At the lower temperatures post-yield behaviour was dominated by progressive strain softening, so that a mechanical steady state was never achieved. At low strains deformation micro-structures were dominated by intracrystalline plasticity with kink-band formation, and devel-opment of SPO and LPO. At progressively higher strains dynamic recrystallization with grain size reduction began to overprint the micro-structure. It is inferred that this led to a progressive change in dominant deformation mechanism to dynamic recrystallization-accommodated plastic flow, causing progressive, strain-dependent weakening and an increase in the sensitivity of strength to deformation rate. This, together with the occurrence of the dehydration reaction gypsum → bassanite + water, meant that flow-law parameters could not be obtained from these experiments.

Thus, gypsum in nature will flow plastically at shear-stress levels lower than those expected from previous experimental studies due to the strain weakening associated with dynamic recrystallization, which can occur at temperatures even lower than gypsum dehydration. These results point to the potential importance of the inherent weakness of gypsum rocks at high strains and natural strain rates, when deforming by plastic processes, without invoking weakening and embrittlement through dehydration reactions and transient elevation of pore fluid pressure.

A. Barnhoorn, M. Bystricky and T. Hirose (Rock Deformation Laboratory, Zürich) are thanked for teaching and assistance with the torsion rig. U. Gerber, R. Hofmann and F. Pirovino (ETH technical staff) are thanked for their excellent work. D. Olgaard and C. Spiers are gratefully acknowledged for their very constructive reviews as well as D. Bruhn for his encouragement and useful suggestions. V. Barberini is grateful to A. Boriani for support during her PhD. This work was supported by Italian MIUR COFIN 2001 grant and by ETH grants (Prof. Burg's Berufungskredit # 0-42073-93/3392 and ETH project # 01066/41-2704 5).

References

BARNHOORN, A., BYSTRICKY, M., BURLINI, L. & KUNZE, K. 2004. The role of recrystallisation o the deformation behaviour of calcite rocks: large strain torsion experiments on Carrara marble. *Journal of Structural Geology*, **26**, 885–903.

BRODIE, K.H. & RUTTER, E.H. 1985. The role of transiently fine-grained reaction products in syntectonic metamorphism: natural and experimental examples. *Canadian Journal of Earth Sciences*, **24**, 556–564.

BYSTRICKY, M., KUNZE, K., BURLINI, L. & BURG, J.-P. 2000. High shear strain of olivine aggregates: rheological and seismic consequences. *Science*, **290**, 1564–1567.

CARTER, N.L. & HEARD, H.C. 1970. Temperature and rate dependent deformation halite. *American Journal of Science*, **269**, 193–196.

CHESTER, F.M. 1988. The brittle–ductile transition in a deformation-mechanism map for halite. *Tectonophysics*, **154**, 125–136.

DELL'ANGELO, L.N. & OLGAARD, D.L. 1995. Deformation of fine-grained anhydrite: evidence for dislocation and diffusion creep. *Journal of Geophysical Research*, **100**, 15 425—15 440.

DE MEER, S. & SPIERS, C.J. 1995. Creep of wet gypsum aggregates under hydrostatic loading conditions. *Tectonophysics*, **245**, 171–183.

DE MEER, S. & SPIERS, C.J. 1999. Influence of pore-fluid salinity on pressure solution creep in gypsum. *Tectonophysics*, **308**, 311–330.

DE MEER, S., SPIERS, C.J. & PEACH, C.J. 2000. Kinetics of precipitation of gypsum and implications for pressure-solution creep. *Journal of the Geological Society, London*, **157**, 269–281.

HANDIN, J., HIGGS, D.V. & O'BRIEN, J.K. 1960. Torsion of Yule marble under confining pressure. *In*: GRIGGS, D. & HANDIN, J. (eds) *Rock Deformation*. Geological Society of America Memoir, **79**, 245–274.

HEARD, H.C. & RUBEY, W.W. 1966. Tectonic implications of gypsum dehydration. *Geological Society of America Bulletin*, **77**, 741–760.

HEIDELBACH, F., STRETTON, I.C. & KUNZE, K. 2001. Texture development of polycrystalline anhydrite experimentally deformed in torsion. *International Journal of Earth Sciences*, **90**, 118–126.

KO, S.C., OLGAARD, D.L. & BRIEGEL, U. 1995. The transition from weakening to strengthening in dehydrating gypsum: evolution of excess pore pressure. *Geophysical Research Letters*, **22**, 1009–1012.

KO, S.C., OLGAARD, D.L. & WONG, T.-F. 1997. Generation and maintenance of pore pressure excess in a dehydrating system: 1) Experimental and microstructural observations. *Journal of Geophysical Research*, **102**, 825–839.

LAUBSCHER, H.P. 1975. Viscous components of Jura folding. *Tectonophysics*, **27**, 239–254.

MALAVIELLE, J. & RITZ, J.F. 1989. Mylonitic deformation of evaporites in decollements; examples from the Southern Alps, France. *Journal of Structural Geology*, **11**, 583–590.

MURRELL, S.A.F. & ISMAIL, I.A.H. 1976. The effect of decomposition of hydrous minerals on the mechanical properties of rocks at high pressures and temperatures. *Tectonophysics*, **31**, 207–258.

OLGAARD, D.L., HEILBRONNER, R., BRIEGEL, U. & BLENDELL, J.E. 1986. Hot pressing methods for four synthetic rocks: calcite, quartz, anhydrite and gypsum. *Eos Transactions, American Geophysical Union*, Fall Meeting, San Francisco, **67**, 44, 1207.

OLGAARD, D.L., KO, S.C. & WONG, T.-F. 1995. Deformation and pore pressure in dehydrating gypsum under transiently drained conditions. *Tectonophysics*, **245**, 237–248.

PANOZZO, HEILBRONNER, R. & OLGAARD, D.L. 1987. Experimental shear deformation of synthetic gypsum rock. *Terra Cognita*, **7**, 64–65.

PASSCHIER, C.W. & TROUW, R.A.J. 1996. *Microtectonics*. Springer, Berlin.

PATERSON, M.S. & OLGAARD, D.L. 2000. Rock deformation tests to large shear strains in torsion. *Journal of Structural Geology*, **22**, 1341–1358.

PEACH, C.J. & SPIERS, C.J. 1996. Influence of crystal plastic deformation on dilatancy and permeability development in synthetic salt rock. *Tectonophysics*, **256**, 101–128.

PIERI, M., BURLINI, L., KUNZE, K., STRETTON, I.C. & OLGAARD, D.L. 2001. Rheological and micro-structural evolution of Carrara marble wilth high shear strain: results from high temperature torsion experiments. *Journal of Structural Geology*, **23**, 1393–1413.

POPP, T., KERN, H. & SCHULZE, O. 2001. Evolution of dilatancy and permeability in rock salt during hydrostatic compaction and triaxial deformation. *Journal of Geophysical Research, Solid Earth*, **106**, 4061–4078.

SCHULZE, O., POPP, T. & KERN, H. 2001. Development of damage and permeability in deforming rock salt. *Engineering Geology*, **61**, 163–180.

STRETTON, I.C. 1996. *An experimental investigation of the deformation properties of gypsum*. PhD thesis, University of Manchester.

TESTA, G. 1996. *Il bacino evaporitico messiniano di Volterra: evoluzione sedimentaria, tettonica e diagenetica*. PhD thesis, University of Pisa.

TESTA, G. & LUGLI, S. 2000. Gypsum–anhydrite transformations in Messinian evaporites of central Tuscany (Italy). *Sedimentary Geology*, **130**, 249–268.

WARREN, J. 1999. *Evaporites: Their Evolution and Economics*. Blackwell Science, Oxford.

The effect of grain size and melt distributions on the rheology of partially molten olivine aggregates

DAVID BRUHN[1], DAVID L. KOHLSTEDT & KYONG-HO LEE

Department of Geology & Geophysics, University of Minnesota,
Minneapolis, MN 55455, USA

[1]*Present address: GFZ Potsdam, Section 5.2, 14473 Potsdam, Germany*
(e-mail: dbruhn@gfz-potsdam.de)

Abstract: Grain size distribution affects the distribution of melt in partially molten rocks, as melt is partitioned into the finer grained regions. As the creep strength of partially molten rocks depends on melt fraction, grain size distribution affects rheology in two ways: in the diffusion-creep regime the strength of the rock decreases with decreasing grain size; and by affecting the melt distribution, grain size distribution affects rock strength in the dislocation and the diffusion-creep regimes. To investigate this effect on partially molten dunites, we deformed synthetic aggregates of olivine + mid-ocean ridge basalt (MORB) with varying grain size distributions and melt fractions. We found that samples with bimodal grain size distributions did not obey the flow law derived for samples with a uniform grain size distribution. In the samples with bimodal grain size distributions the finer grained parts are enriched in melt, reducing the relative melt fraction in the coarser grained parts. The stress exponent is close to 1 for lower stresses for samples with non-uniform grain size distributions, while it is close to 3 for samples with a uniform grain size distribution. Electron backscatter diffraction (EBSD) analysis shows crystallographic preferred orientation (CPO) well developed in all samples. Microstructural investigation shows that partitioning of the liquid out of the coarser grained part of the aggregate into the finer grained part leads to a localization of the strain into the weaker, finer grained and melt-enriched part. By treating the two parts as mechanically different phases that deform at a uniform stress, we are able to approximate the rheological behaviour of the aggregates with bimodal grain size distribution.

Grain size is an important factor in the rheology of rocks. As one example, viscosity increases as grain size cubed if deformation is dominated by grain-boundary diffusion accommodated by grain-boundary sliding (Coble 1963; Raj & Ashby 1971), often referred to as *Coble* creep. The determination of the exact effect of grain size on physical properties is often complicated by the fact that a material rarely contains a single grain size, but rather is characterized by a distribution of grain sizes. The term 'grain size' is commonly used synonymously for the average or mean grain size. However, different grain size distributions can have the same mean value. For physical properties that are a function of the grain size, consideration of grain size distribution can be critical for the understanding of material behaviour (e.g. Ghosh & Raj 1981, 1986; Raj & Ghosh 1981; Heilbronner & Bruhn 1998). For example, near the transition from grain size insensitive dislocation creep to grain size sensitive diffusion creep, the fine-grained parts of the rock deform by diffusion creep and the coarser-grained regions deform by dislocation creep. The transition region can extend over several orders of magnitude in stress–strain rate, depending on the grain size distribution (e.g. Freeman & Ferguson 1986; Bruhn *et al.* 1999). Therefore comparison of rheological data from different samples using mean grain sizes is of limited use unless grain size distributions are similar from one sample to the next.

In a rock that contains a fluid such as water or melt, the grain size distribution affects the distribution of the liquid. In experiments using two juxtaposed fluid-bearing rocks with different grain sizes, Wark & Watson (1999) showed that the liquid partitions preferentially into the finer-grained rock. Hence, grain size distribution affects creep rate, $\dot{\varepsilon}$, in two ways. First, in the diffusion-creep regime, grain size, d, enters directly into the creep equation as for Coble creep where $\dot{\varepsilon} \propto 1/d^3$. Second, in both the diffusion-creep and the dislocation-creep regimes, strain rate

From: BRUHN, D. & BURLINI, L. (eds) 2005. *High-Strain Zones: Structure and Physical Properties.*
Geological Society, London, Special Publications, **245**, 291–302.
0305-8719/05/$15.00 © The Geological Society of London 2005.

increases with increasing melt fraction, ϕ (Hirth & Kohlstedt 1995*a, b*). Therefore, grain size distribution indirectly influences strain rate through its effect on melt distribution, such that $\dot{\varepsilon} = \dot{\varepsilon}(\phi(d))$.

In this paper we present the results of an experimental investigation of the effects of grain size and melt distributions on the deformation behaviour of partially molten aggregates of olivine plus basalt. By systematically varying grain size distribution and melt fraction, we demonstrate that the effect of grain size distribution on viscosity can be dramatic and that differences in flow behaviour between two samples cannot be explained by differences in average grain sizes. Partitioning of the liquid out of the coarser grained part of the aggregate into the finer grained part leads to a localization of the strain into the weaker, finer grained and melt-enriched part. The rheological behaviour can be approximated by treating the two parts as mechanically different phases that deform at a uniform stress. Such an approach describes the observed behaviour of the aggregates with bimodal grain size distributions very well.

Sample preparation

Three series of samples were prepared from olivine plus mid-ocean ridge basalt (MORB) by hot isostatic pressing mineral powders with different particle size distributions and varying amounts of MORB: (A) fine-grained olivine with a narrow, uniform grain size distribution; (B) fine-grained olivine with a broader, quasi-bimodal grain size distribution; and (C) coarser grained olivine with an even broader, quasi-bimodal grain size distribution.

Olivine powders were obtained by grinding gem-quality single crystals from San Carlos, Arizona. Different particle size distributions were obtained by using different processing techniques. For the two fine-grained series (A and B), olivine crystals were first crushed and then ground in a fluid energy mill. For series A, the finest grained powders with the most uniform particle size distribution were collected from the dust filter of the mill. For series B, powders with a particle size of $<10 \, \mu m$ were produced by the fluid energy mill and subsequently sorted by Stokes' settling in water. For series C, powders were simply crushed and then sieved to obtain initial particle sizes of $45–75 \, \mu m$. The MORB powder used for the olivine + melt samples is from the same olivine tholeiite used in previous studies (Cooper & Kohlstedt 1984; Riley & Kohlstedt 1991; Daines & Kohlstedt 1997; Hirth & Kohlstedt

1995*a, b*). MORB powders with an average particle size of $8 \, \mu m$ and olivine powders were mechanically mixed. The amount of MORB added ranged from 2 to 10 vol% in the two fine-grained olivine series and from 4 to 6 vol% MORB in the coarser grained series. All powders were dried at $1100 \, ^{\circ}C$ at a controlled oxygen fugacity equivalent to the Ni–NiO phase boundary for 12 h. The dry powders were then stored under vacuum at $140 \, ^{\circ}C$ until they were used for sample preparation.

All the samples were fabricated individually by uniaxially cold-pressing the powders or powder mixtures, with an applied load of 2000 kg, into nickel cans 26 mm in length, 10.0 mm in inner diameter and 11.6 mm in outer diameter. The cold-pressed samples were then isostatically hot-pressed at $1250 \, ^{\circ}C$ and 300 MPa for approximately 3 h using an internally heated gas-confining medium apparatus. During hot-pressing, as well as during subsequent deformation experiments, the oxygen fugacity was buffered near the Ni–NiO boundary by oxidation of the Ni can. After hot-pressing, jackets were dissolved in aqua regia. The porosity of our samples was generally $<1 \, vol\%$. A small piece of every hot-pressed sample was retained for microscopic analysis. Finally, the resulting cylindrical samples with a length of $14–18 \, mm$ and a diameter of $8–9 \, mm$ were cut, dried and jacketed for deformation experiments.

Examples of the microstructures resulting from our deformation experiments are shown in Figure 1. After hot-pressing, the grains in samples of series A were generally $<15 \, \mu m$ with an average of $4–5 \, \mu m$. In series B, the maximum grain size is about $25 \, \mu m$ and the average grain size is $6 \, \mu m$. In series C, the maximum grain size is $<75 \, \mu m$ with an average grain size of $7 \, \mu m$. All grain sizes are given as mean intercept lengths without any correction factor – Hirth & Kohlstedt (1995*a, b*) used a factor of 1.5 for their average grain sizes. Grain sizes and grain size distributions do not change significantly during deformation, as the melt pins grain boundaries preventing grain-boundary migration, as shown by Hirth & Kohlstedt (1995*a*).

Apparatus and experimental procedures

Our deformation experiments were performed in an internally heated servo-controlled triaxial deformation apparatus with internal load cell (Paterson 1990). The temperature varied by not more than $\pm 1 \, ^{\circ}C$ along the sample length and was held within $\pm 2 \, ^{\circ}C$ of the set-point over the duration of the experiment. The confining

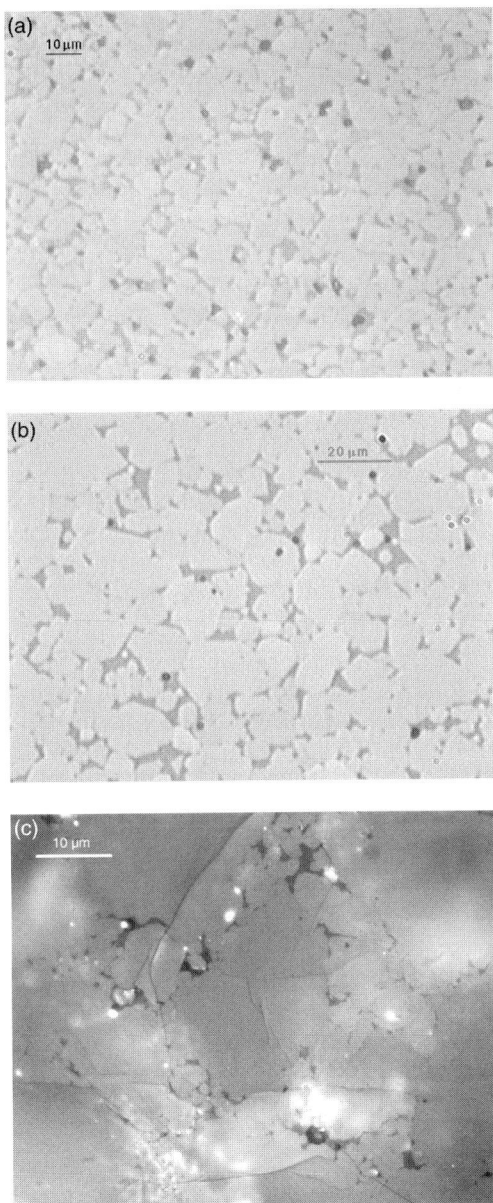

Fig. 1. Photomicrographs of olivine + MORB aggregates. Light grey, olivine; dark grey, MORB; black, pores/holes from polishing. (**a**) Sample of series A, with 10 vol% MORB. (**b**) Sample of series B with 10 vol% MORB. (**c**) Sample of series C with 4 vol% MORB. Note the concentration of melt in fine-grained areas of the sample.

pressure was kept constant within ± 5 MPa. The precision of the load measurements is 0.05 kN (corresponding to a stress resolution of < 1 MPa on our samples). For all experiments,

samples were enclosed in Ni sleeves approximately 0.5 mm thick. This sample assembly along with the alumina and zirconia pistons were jacketed with a thin-walled iron tube.

All deformation tests were performed as constant load 'creep' tests at loads corresponding to differential stresses (σ) of 100–600 MPa. Between three and six constant load tests were performed per sample to determine the stress exponent, n, in the power-law relation $\dot{\varepsilon} \propto \sigma^n$. The temperature was 1200 °C, and the confining pressure was 200 MPa. At these conditions, the MORB is molten such that vol% MORB $\approx$ vol% melt.

At the end of the last step of a deformation experiment, the sample was cooled quickly (1 °C s^{-1}) under load to preserve the deformation-induced distribution of melt. After the temperature dropped to below 1100 °C (the solidus temperature for the MORB is approximately 1150 °C), the load was removed and the samples were cooled to room temperature at a rate of $c.$ 0.5 °C s^{-1}.

After the experiment, the iron and nickel jackets were dissolved in aqua regia. Samples were cut longitudinally and polished for optical microscopy, as well as for scanning electron and transmission electron microscopy (SEM and TEM). The volume percentage of melt was determined using binary digital images, as outlined by Daines & Kohlstedt (1997). Grain sizes and grain size distributions were determined on optical micrographs obtained in reflected and transmitted light by the linear intercept method using the public domain software NIH Image. Individual grain intercept lengths were measured along a square grid placed over the images with lines parallel and perpendicular to the shortening direction. Analyses of crystallographic preferred orientations (CPO) by electron backscatter diffraction (EBSD) were carried out on a JEOL 840 SEM customized with HKL's EBSD system, following the procedure outlined by Lee *et al.* (2002). All EBSD patterns were manually indexed using HKL's Channel14 software package.

Results

Experimental data

The experimental results are listed in Table 1. In Figure 2, flow laws derived from our experimental data are plotted and compared to the individual data points. In the flow laws for both the dislocation-creep regime and the diffusion-creep regime, we incorporated the exponential dependence of strain rate ($\dot{\varepsilon}$) on melt fraction

Table 1. *Experimental data*

Experiment	ϕ (vol%)	$\dot{\varepsilon}$ (s^{-1})	ε (%)	σ (MPa)	d^* (µm)	$d^\dagger$ (µm)
(a) Uniform grain size distribution						
PI-749	5	1.76×10^{-5}	1.0	109.1	4.9	
		2.41×10^{-5}	1.0	139.3		
		3.03×10^{-5}	1.0	165.2		
		4.64×10^{-5}	0.9	202.3		
		6.38×10^{-5}	0.9	238.4		
		1.62×10^{-4}	0.8	292.9		
PI-754	10	1.60×10^{-4}	1.9	94.6	4.3	4.1
		1.37×10^{-4}	1.4	115.7		
		1.49×10^{-4}	1.3	137.5		
		2.50×10^{-4}	1.5	164.8		
		2.75×10^{-4}	1.5	190.4		
		3.34×10^{-4}	2.2	214.2		
		6.05×10^{-4}	1.9	238.3		
		6.79×10^{-4}	3.7	267.8		
(b) Bimodal, fine-grained						
PI-419	10	6.44×10^{-5}	2.9	134.6	6.0	4.6
		4.01×10^{-5}	1.4	132.8		
		5.62×10^{-5}	5.9	168.1		
		1.73×10^{-4}	4.3	170.5		
PI-424	10	9.95×10^{-5}	9.3	170.3		
		7.35×10^{-5}	6.8	184.5		
		7.05×10^{-5}	2.5	194.3		
		7.80×10^{-5}	2.1	206.4		
PI-427	5	5.16×10^{-5}	6.8	189.5	5.7	4.8
		6.20×10^{-5}	2.1	219.2		
		8.07×10^{-5}	1.7	272.5		
		8.71×10^{-5}	0.8	285.4		
		8.21×10^{-5}	1.8	298.0		
		8.80×10^{-5}	2.2	319.1		
PI-435	3	4.15×10^{-5}	1.4	218.9	6.1	4.9
		4.11×10^{-5}	2.7	245.8		
		4.24×10^{-5}	3.5	280.0		
		4.87×10^{-5}	2.0	314.1		
		6.62×10^{-5}	2.5	371.5		
		8.30×10^{-5}	1.3	413.7		
PI-438	2	2.90×10^{-5}	2.8	214.6		
		3.55×10^{-5}	1.1	252.8		
		4.73×10^{-5}	1.3	291.2		
		5.65×10^{-5}	1.4	329.7		
		6.39×10^{-5}	1.2	366.8		
		7.46×10^{-5}	1.4	404.1		
(c) Bimodal, coarse-grained						
PI-479	4	8.04×10^{-5}	2.5	303.9	7.0	4.5
		1.19×10^{-4}	2.5	370.9		
		1.50×10^{-4}	2.6	415.9		
		1.90×10^{-4}	2.6	458.4		
		2.27×10^{-4}	2.7	489.7		
		2.75×10^{-4}	2.8	518.5		
PI-499	6	3.71×10^{-4}	4.5	387.1		
		4.23×10^{-4}	3.6	457.4		
		4.93×10^{-4}	3.8	519.9		
		6.09×10^{-4}	2.1	561.3		
		7.78×10^{-4}	1.9	583.0		

*Average grain size.
†Average grain size of fraction of grains <8µm.

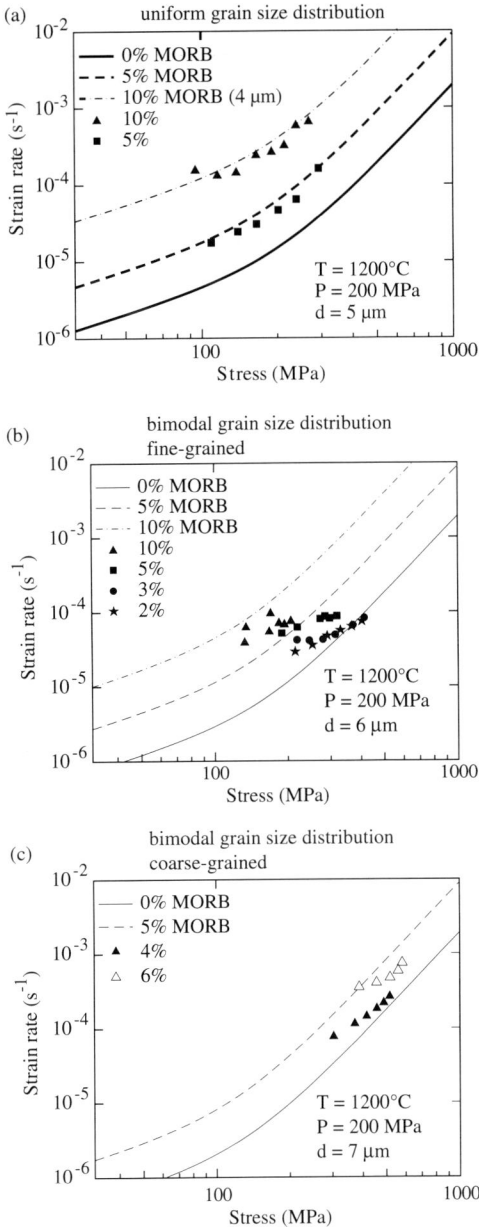

Fig. 2. Comparison of experimental data with constitutive equation for olivine + MORB. Input parameters are given in the bottom right. (**a**) Data for samples with uniform grain size distribution, series A. 10%, sample PI-754; 5%, sample PI-749. (**b**) Data for samples of series B; 10%, samples PI-419 and PI-424; 5%, sample PI-427; 3%, sample PI-435; 2%, sample PI-438. (**c**) Data for samples of series C; 6%, sample PI-499; 4%, sample PI-479.

(ϕ) noted by Kelemen *et al.* (1997) and Mei *et al.* (2002) in the flow law

$$\dot{\varepsilon} = A_{diff} \frac{\sigma}{d^3} \exp\left(\alpha_{diff}\phi\right) + A_{disl}\sigma^{3.5}\exp\left(\alpha_{disl}\phi\right).$$

(1)

Values of $\alpha_{diff} = 26$ for diffusion creep and $\alpha_{disl} = 31$ for dislocation creep from Mei *et al.* (2002) were used in this comparison.

In Figure 2a, the slope of a best-fit line through the data points for the experiments on samples with the uniform grain size distribution (series A) suggests that deformation was accommodated by a combination of diffusion creep and dislocation creep, as the stress exponent n increases from a value of <2 at lower stresses to a value of ≈ 3 at stresses greater than 200 MPa.

The data for experiments on the samples with bimodal grain size distributions (series B and C) are not in good agreement with the constitutive equation derived for the samples with a uniform grain size distribution (series A). Both the fine-grained (Fig. 2b) and the coarse-grained (Fig. 2c) samples are stronger than predicted by the flow laws, and they are stronger than the samples with a uniform grain size distribution at the higher stresses. Moreover, the best-fit line through the data for the fine-grained samples (Fig. 2b) has a slope of 1 for the higher melt contents (5 and 10 vol%), suggesting that deformation occurred in the diffusion-creep regime. For samples with smaller melt contents (2 and 3 vol%), the slope increases with increasing stress, suggesting a transition from diffusion creep to dislocation creep. Similarly, the coarser grained sample (Fig. 2c) with a melt volume of 6% deformed in the transition region between the diffusion and dislocation-creep regimes with a stress exponent of $n \approx 3$ at higher stresses, and deformation of the sample with 4% melt was mainly in the dislocation-creep regime ($n = 3.8$).

Grain size distributions

Examples of the grain size distributions for all three sample types are shown in Figure 3. In these histograms, the percentage of the total intercept length measured is shown for each grain size fraction instead of the commonly used frequency plots, where the number of grains per grain size fraction is used. The intercept lengths of the grains are added for each grain size class and then normalized by the sum of the intercept lengths of all the measured grains to give an impression of the relative volume fractions of the different grain size classes. A small, statistically insignificant

(a)

(b)

(c)

(d)

(e)

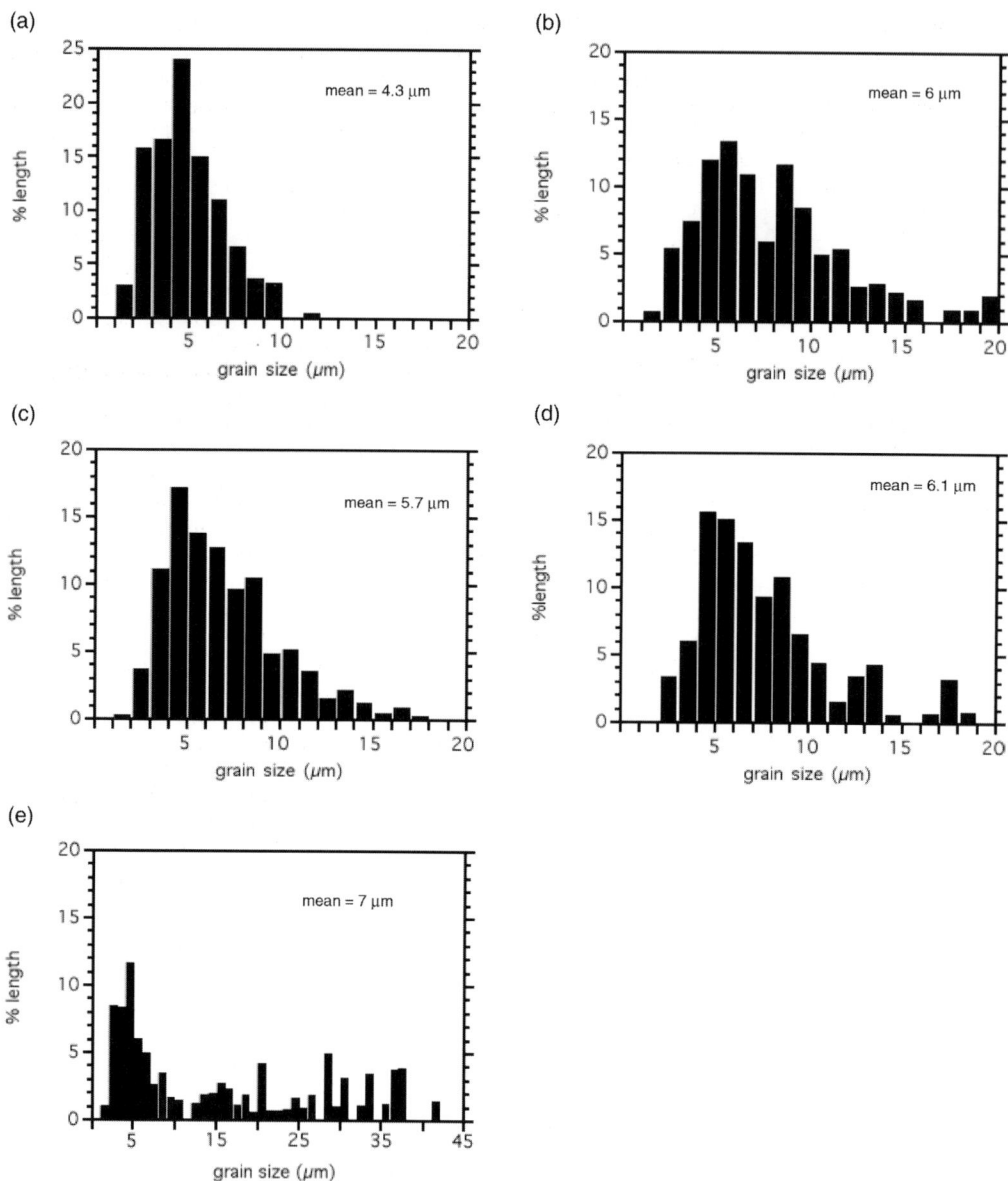

Fig. 3. Grain size distributions representing the sum of all intercept lengths measured per size class. Mean grain sizes are given in the top right corner. (**a**) Series A, 10 vol% MORB. (**b**) Series B, 10 vol% MORB. (**c**) Series B, 5 vol% MORB. (**d**) Series B, 3 vol% MORB. (**e**) Series C, 4 vol% MORB.

number of large grains would not show up in a frequency plot, even though the large grains may occupy a relatively large area of the sample image. For the interpretation of the rheology of materials with a bimodal grain size distribution, the volumetric distribution plot allows quantification of the effect of the grain size distribution on the deformation behaviour.

The main difference between the three distributions is their width; that is, the maximum grain size and the number of large grains increases from series A to C. The grain size distribution in Figure 3a is for a sample of series A; the distribution is narrow and uniform, with only one peak close to the mean value of 4 μm. For series B, three examples for samples with different melt fractions are shown in Figure 3b–d. The mean values and the distributions are similar, with a peak close to the mean value of 6 μm, and a second peak between 8 and 9 μm. The peak at smaller grain size fraction is asymmetric, with a 'tail' towards the larger grain sizes. The distributions for the coarser grained samples of series C are polymodal, with a peak in the small grain size fraction below the mean value of 7 μm and several smaller peaks at larger grain sizes.

Crystallographic preferred orientation

The CPO of the olivine grains for four samples with different melt fractions was measured using the EBSD technique. One undeformed sample containing 3% melt was analysed for comparison. Usually 300–600 points were measured along square grids in 10 μm intervals. EBSD patterns for only 25–55% of all the points measured in the deformed samples were of high enough quality to be analysed.

From the EBSD patterns analysed, pole figures were derived for (100), (010) and (001) of forsterite. For the undeformed sample (Fig. 4a), the pole figures show a random crystallographic distribution, demonstrating that there was no initial CPO introduced during sample preparation. In all of the deformed samples (Fig. 4b–e), a clear CPO exists with [010] forming a point maximum aligned parallel to the orientation of the maximum principal stress σ_1. The a-axes and c-axes form girdles in the plane normal to the compression direction. This CPO is consistent with slip on the (010)[100] system. In the samples with bimodal grain size distributions, EBSD patterns taken from neighbouring points were often almost identical, suggesting that the two points were in the same grain.

Discussion

The flow laws that fit the data for samples with a uniform grain size distribution do not describe the experimental data for the samples with bimodal grain size distributions. At high stresses the finest grained samples of series A were stronger than the samples of series B (fine-grained, bimodal). This result is somewhat surprising because from the normal flow strength–grain size relation (Kohlstedt & Wang 2001), the coarser grained samples should be stronger than the finer grained samples. In addition, the samples of series A deformed in the dislocation-creep regime, while the lower stress exponents derived for the samples with bimodal grain size distributions of series B and C suggest deformation was dominated by diffusion creep and the transition to the dislocation-creep regime occurs at much higher stresses than it does for the finer grained samples with the uniform grain size distribution. Thus, the three data sets cannot be explained by the same flow law, if only the difference in mean grain size is considered. In addition, there is a clear CPO in the bimodal samples, indicative of dislocation creep, and yet the stress exponent is 1, suggestive of diffusion creep, at least for the finer grained bimodal samples of series B.

The flow law derived from data by Hirth & Kohlstedt (1995a, b) underestimates the strength of our fine-grained samples with a uniform grain size distribution of series A (fine-grained, uniform) by a factor of approximately 3. There are two possible reasons for this discrepancy. One involves the chemistry of the melt, as Hirth & Kohlstedt (1995a, b) added c. 5vol% enstatite to their olivine + MORB samples. The difference in melt chemistry may affect the grain-boundary diffusion rates, as suggested by the experiments of Hirth & Kohlstedt (1995a). An alternative reason involves differences in flow strength associated with grain size distributions. The grain size distributions described by Hirth & Kohlstedt (1995b) are much broader than for our series A and series B samples, and also less uniform. Even though the grain size factor is incorporated in the flow law, possible differences in grain size distributions between our samples and those of Hirth & Kohlstedt (1995a, b) are not accounted for.

In partially molten samples the analysis of the effect of grain size distribution on creep behaviour is complicated by the distribution of the melt, which is influenced by local differences in grain size (Wark & Watson 2000). The local melt fraction also has an effect on the deformation behaviour of partially molten rocks, as melt provides a rapid diffusion path along wetted grain boundaries resulting in deformation rates that are locally higher than average (Hirth & Kohlstedt 1995a).

An example of the effect of grain size distribution on melt distribution is apparent in Figure 1c. Melt is partitioned into the finer grained parts of a bimodal aggregate, while grain boundaries between large grains contain no melt. These samples consist of two parts, a fine-grained region with a relatively high melt fraction and a coarse-grained region with a relatively small melt fraction. Since the strength of olivine + MORB aggregates decreases with increasing melt fraction (Hirth & Kohlstedt 1995a, b), the finer grained, melt-enriched part

D. BRUHN *ET AL.*

100   **010**

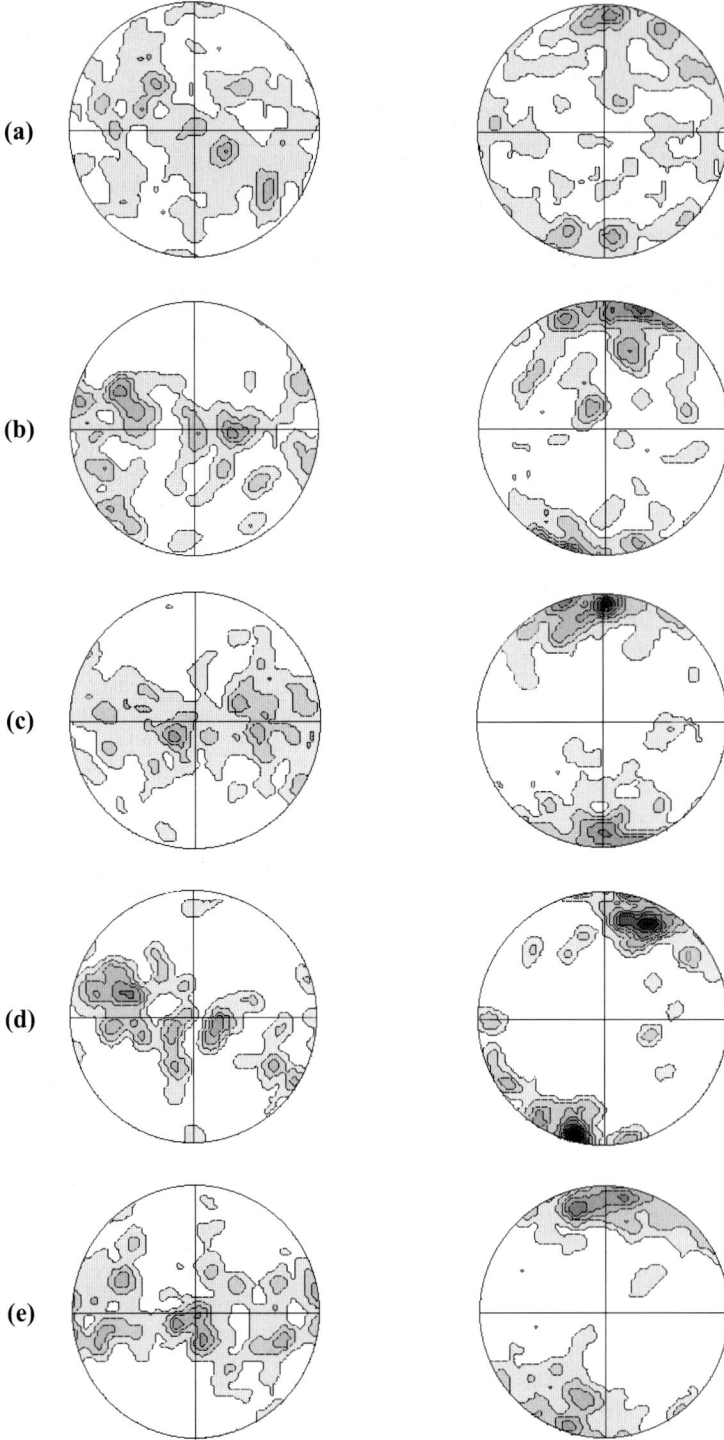

is weaker than the coarser grained melt-free part of the aggregate. As a consequence, strain is partitioned into the weaker part of the sample, and the fine-grained melt-enriched part accommodates most of the deformation of the aggregate. To explain the rheological behaviour of these samples, the two parts were treated as two mechanically different phases, and the physical properties of these phases were combined to calculate the behaviour of the bulk rock.

The deformation behaviour of two-phase materials is often described by upper and lower bounds (e.g., Tullis *et al.* 1991). Two rigorous bounds are presented by the assumptions that both phases deform to uniform strain (Voigt 1928) or at uniform stress (Reuss 1929). Even though these conditions are an oversimplification of the deformation processes in a two-phase material, they do present end-member cases and as such serve as a useful reference frame that can be described by simple mathematical formulations. While these end-member conditions were originally formulated to describe the elastic strength of two-phase materials, the approach can be adapted to plastic flow behaviour if there are no interactions other than mechanical between the phases. The limitations of such an approach are discussed in more detail elsewhere (e.g. Tullis *et al.* 1991; Bruhn *et al.* 1999).

If the weaker phase in a material consisting of two mechanically different phases is interconnected, deformation is strongly partitioned into that weaker phase (e.g. Handy 1990). Such a scenario does not correspond with the uniform strain condition but is best described by the uniform stress condition, which assumes strain localization into the weaker phase at a uniform stress (Reuss 1929). To evaluate the effect of the partitioning of melt and strain into the finer-grained parts of our sample on strength we compare our data to the uniform stress case.

If the stress is uniform throughout the aggregate

$$\sigma_t = \sigma_1 = \sigma_2 \qquad (2a)$$

then

$$\dot{\varepsilon}_t = V_1 \dot{\varepsilon}_1 + V_2 \dot{\varepsilon}_2 \qquad (2b)$$

where the subscripts 1 and 2 refer to the two distinct regions in the sample, and the subscript t

refers to the bulk sample. For the partially molten olivine aggregates with bimodal grain size distributions, $\dot{\varepsilon}$ as a function of σ was calculated for the uniform stress condition using equation (2b), with the flow law parameters given in equation (1). The pre-exponential factor A is 5×10^{-24} and 6×10^{-14} for the diffusion creep and for the dislocation creep parts of the equation respectively. We used a combined flow law incorporating a term for diffusion creep and for dislocation creep for both the fine-grained and coarse-grained parts of the rock analogous to Bai *et al.* (1991). The only difference between the two parts is the grain size used for each part. To simplify the analysis, we assume that all the melt partitioned into the finer grained regions of the samples.

If the flow-law parameters are fixed, the flow strength of the aggregates depends on the volume fraction of each of the two parts, as well as on the grain size and melt fraction in the fine-grained part. Because the flow law for dislocation creep is not a function of grain size and as the part of the aggregate deforming by dislocation creep is nearly melt-free, grain size, volume fraction and melt fraction only affect the aggregate strength through the diffusion creep flow law. In the diffusion-creep regime, sample strength decreases with decreasing grain size as $1/d^3$ and exponentially with increasing melt fraction (equation 1). The effect of melt fraction and volume fraction of the two parts are interdependent. The overall melt fraction (ϕ_t) in our samples is fixed. However, if all the melt is partitioned into the fine-grained parts of a sample with a bimodal grain size distribution, the local melt fraction in the fine-grained part depends on the volume fraction of the fine-grained part in the aggregate.

If the sample volume $V_t = V_c + V_f + V_m$, then the local melt fraction ϕ_1 is

$$\phi_1 = \frac{V_m}{V_m + V_f} \qquad (3)$$

where V_c, V_f and V_m are the volumes of the coarse-grained and the fine-grained parts of the rock and of the melt, respectively. Thus, an overall melt fraction of 0.1 corresponds to a local melt fraction of 0.12 if the ratio $V_f : V_c = 4:1$, and the local melt fraction is 0.18 if $V_f : V_c = 1:1$. Thus, the effect of increasing

Fig. 4. Pole figure contour plots of EBSD patterns of olivine *a*- and *b*-axes, showing contours of density of poles in the lower-hemisphere, equal-area stereographic projection. In the deformed samples, the orientation of the maximum principal stress σ_1 is vertical. (**a**) Undeformed sample of series B, 3% MORB. (**b**) Sample of series A, 10 vol% MORB. Strain *x*%. (**c**) Sample of series B, 10% MORB. (**d**) Sample of series B, 5% MORB. (**e**) Sample of series B, 2% MORB.

the local melt fraction on decreasing the strength of the aggregate is offset by that of decreasing the volume fraction of the weaker phase. Figure 5a shows how stress is affected by the change in the ratio $V_f{:}V_c$ and by the corresponding change in local melt fraction. Generally, an increase in local melt fraction decreases the strength, despite the decrease in V_f.

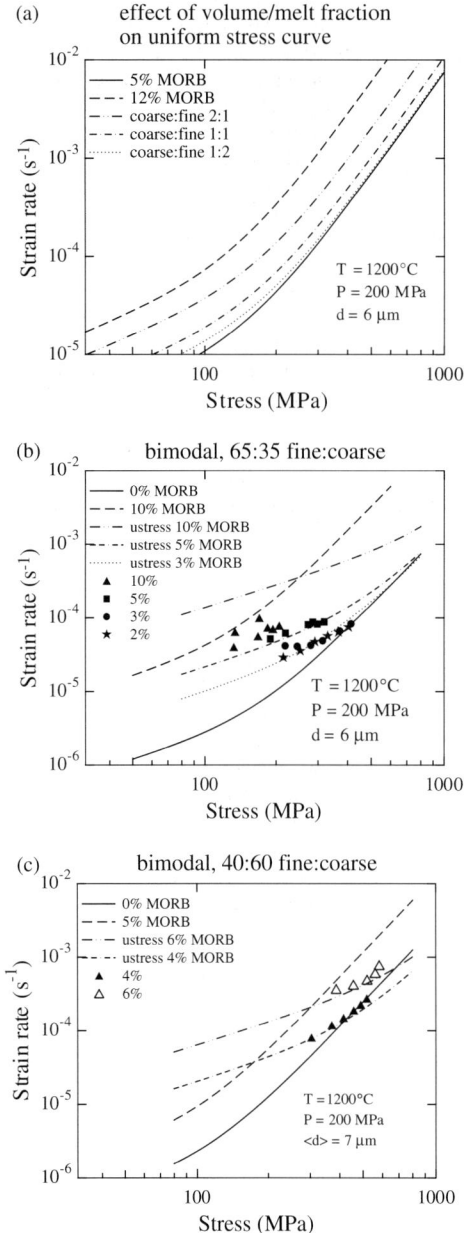

(a) effect of volume/melt fraction on uniform stress curve

5% MORB
12% MORB
coarse:fine 2:1
coarse:fine 1:1
coarse:fine 1:2

$T = 1200\,^{\circ}C$
$P = 200$ MPa
$d = 6\ \mu m$

Strain rate (s^{-1})
Stress (MPa)

(b) bimodal, 65:35 fine:coarse

0% MORB
10% MORB
ustress 10% MORB
ustress 5% MORB
ustress 3% MORB
10%
5%
3%
2%

$T = 1200\,^{\circ}C$
$P = 200$ MPa
$d = 6\ \mu m$

Strain rate (s^{-1})
Stress (MPa)

(c) bimodal, 40:60 fine:coarse

0% MORB
5% MORB
ustress 6% MORB
ustress 4% MORB
4%
6%

$T = 1200\,^{\circ}C$
$P = 200$ MPa
$<d> = 7\ \mu m$

Strain rate (s^{-1})
Stress (MPa)

There is no unique combination of grain size and $V_f{:}V_c$ that describes any particular stress–strain-rate conditions, as the effect of a change in grain size can be offset by a change in $V_f{:}V_c$. For our samples, the relative volume fractions of each part of the aggregate can be established and the average grain sizes for each part of the aggregate can be determined from the volumetric grain size distributions in Figure 3. The relative volume fractions can be estimated from the histograms, assuming that the grain size classes forming the second peak constitute the coarser grained part of the sample. Even though such an assumption is an oversimplification, it does present a first-hand approximation of the relative volumetric distributions. For all of the samples with a bimodal grain size distribution, the second peak is formed by all grain sizes larger than 8 μm. The average grain sizes of the fine-grained parts can be calculated for the grain sizes constituting the first peak in the histograms (Fig. 3). The volume fraction of grains <8 μm and the average grain size of those finer-grained parts of the samples are listed in Table 1.

Using the values in Table 1, the uniform stress case is compared to our experimental data in Figure 5b & c. For the data of series B, the finer grained bimodal samples, the uniform stress curve is calculated with 65 vol% of all grains deforming by diffusion creep, which corresponds to the fraction of small grains (<8 μm) in the distribution histograms. That fraction of small grains has an average grain size of 4.6–4.9 μm (Table 1). The uniform stress curves fit our data reasonably well at lower melt contents, but still underestimate the strength at 10% melt.

For the data of series C, the coarser-grained bimodal samples, the uniform stress curve is calculated with 40 vol% of all grains deforming by diffusion creep, which corresponds to the

Fig. 5. Comparison of uniform stress case calculated from flow laws of olivine + MORB in dislocation creep and in diffusion creep with flow laws and experimental data. (**a**) shows the effect of a change in the ratio of diffusion creep and dislocation creep for a total melt content of 10%, assuming all the melt is in the finer grained part region deforming by diffusion creep. (**b**) Uniform stress case compared to data of series B, using the 65:35 ratio of diffusion creep:dislocation creep derived from grain size distribution histograms in Figure 3b–d; data set identical to Figure 2b. (**c**) Uniform stress case compared to data form series C, using the 40:60 ratio of diffusion creep:dislocation creep derived from the grain size distribution histograms in Figure 3e; data set identical to Figure 2c.

fraction of small grains (<8 μm) in the distribution histograms. That fraction of small grains has an average grain size of 4.5 μm. For both aggregates, PI-479 (4% MORB) and PI-499 (6% MORB), the uniform stress curve fits the experimental data very well (Fig. 5c).

The approach applied above, using the uniform stress condition to calculate the strength of our partially molten olivine aggregates with bimodal grain size distributions, qualitatively describes the rheological behaviour we observe. Deformation of samples with a bimodal grain size and corresponding melt distribution yields $n = 1$ at conditions for which samples with a uniform grain size distribution deform in the dislocation-creep regime ($n \approx 3-4$). In addition, the transition from diffusion creep to dislocation creep occurs at higher stresses than predicted by the constitutive equation derived for samples with a uniform grain size distribution.

Calculation of the uniform strain rate bound to match our experimental data by a procedure analogous to the one applied for the uniform stress bound cannot reproduce our data. The calculated bound is more than an order of magnitude weaker than the observed sample strengths. A good match of the uniform strain rate bound with our data would have been surprising, as it would imply no strain partitioning between coarser grained and the fine-grained parts of a sample.

Two assumptions in the calculation of the uniform stress condition represent end-member cases: (1) all the melt is partitioned into the finer grained part of the aggregates; and (2) the stress is uniform throughout the aggregate. In reality, stresses are probably not completely homogeneously distributed. Also, some melt remains in the coarser grained parts of the samples, especially in the finer grained bimodal samples of series B, for which the differences in grain size are not as large as in series C. Nonetheless, this end-member model illustrates the consequence of deforming partially molten samples with bimodal grain size distributions where strain is strongly partitioned into the weaker parts of the aggregates: the rheology is dominated by the finer grained parts, while the coarse-grained parts only affect the strength. Thus, aggregates can be stronger than predicted by a flow law for aggregates with uniform grain size distribution and still deform in the diffusion-creep regime, as we observe in our samples.

The analysis presented above to describe the rheology of aggregates in terms of two mechanically distinct parts in our aggregates can also be used to explain the apparent discrepancy between the measured value for the stress exponent ($n = 1$) and the observed CPO. Theoretically, it is possible that the last deformation step of an experiment was mainly in the dislocation-creep regime and that the CPO may not be representative of the lower stress deformation steps that define the $n = 1$ part of the mechanical data, as the tests were run at progressively higher stresses. However, as the strains for the individual deformation steps were rather small (Table 1), fabric is more likely to have developed successively during all deformation steps. The coarser grained part of the rock, deformed in the dislocation-creep regime, may develop a CPO, while the finer grained part may lack a strong fabric while controlling the dominant deformation mechanism leading to $n = 1$. Because some of the EBSD patterns of neighbouring points (spaced 10 μm apart) were identical, as pointed out above, suggesting they were measured in the same grain, the grains must have been >10 μm, which is relatively large for our samples. If the finer grained part is enriched in melt, the chance of measurements in melt pockets or on grain boundaries is higher than in the coarser grained part of the sample; thus, fewer EBSD measurements in the finer grained parts yielded patterns that could be analysed successfully than in the coarse-grained parts. As a result, the pole figures in Figure 4 may over-represent the mean CPO of the sample.

Implications

Results for our samples with bimodal grain size distribution have significant implications for the rheology of the mantle and for the rate of segregation and localization of melt beneath mid-ocean ridges. Our experimental results suggest that deformation will be localized into weaker parts of the mantle that are finer grained and have a higher local melt fraction than the average melt fraction of the region. If recrystallization leads to a non-uniform grain size distribution as observed by Lee et al. (2002), melt is partitioned into the finer grained domains, inhibiting grain growth and weakening that domain. Localization of the deformation into that weaker area will lead to the formation of a shear zone, which has a lower viscosity than the average viscosity estimated for the mantle. Such a shear zone would also serve to localize melt flow and provide a high-permeability path for melt extraction from the mantle. A natural example of non-uniform grain size distributions in a super-solidus shear zone within the Oman ophiolite is described by Dijkstra et al. (2002).

With increasing strain, there is no change in the modal grain size, but the grain size distribution narrows as the volume fraction of fine material increases by the recrystallization of coarse grains.

If the grain size distribution is non-uniform and our experimental results can be extrapolated to mantle conditions, the presence of seismic anisotropy due to CPO of olivine does not necessarily imply a non-Newtonian viscosity of the mantle. Our experiments demonstrate that it is possible to have a CPO even for small strains, even if the overall deformation is in the diffusion-creep regime and stress exponents are close to 1.

We are grateful for the scientific and technical support by M.E. Zimmerman. Detailed and constructive reviews by A. Kronenberg and by M. Drury helped to clarify the manuscript. L. Burlini's editorial comments and his patience are greatly appreciated. Financial support for this research was provided through a Forschungsstipendium by the Deutsche Forschungsgemeinschaft (to D. Bruhn) and through grants EAR-0126277 and OCE-0327143 from the NSF.

References

BAI, Q., MACKWELL, S.J. & KOHLSTEDT, D.L. 1991. High-temperature creep of olivine single crystals; 1, Mechanical results for buffered samples. *Journal of Geophysical Research*, **96**, 2441–2463.

BRUHN, D.F., OLGAARD, D.L. & DELL'ANGELO, L.N. 1999. Evidence for enhanced deformation in two-phase rocks: Experiments on the rheology of calcite-anhydrite aggregates. *Journal of Geophysical Research*, **104**, 707–724.

COBLE, R.L. 1963. Model for boundary diffusion-controlled creep in polycrystalline materials. *Journal of Applied Physics*, **34**, 1679–1682.

COOPER, R.F. & KOHLSTEDT, D.L. 1984. Solution-precipitation enhanced diffusional creep of partially molten olivine-basalt aggregates during hot-pressing. *Tectonophysics*, **107**, 207–233.

DAINES, M.J. & KOHLSTEDT, D.L. 1997. Influence of deformation on melt topology in peridotites. *Journal of Geophysical Research*, **102**, 10 257–10 271.

DIJKSTRA, A.H., DRURY M.R. & FRIJHOF, R. 2002. Microstructures and lattice fabrics in the Hilti mantle section (Oman ophiolite): Evidence for shear localization and melt weakening in the crust–mantle transition zone. *Journal of Geophysical Research*, **107**, 2270–2287.

FREEMAN, B. & FERGUSON, C.C. 1986. Deformation mechanism maps and micromechanics of rocks with distributed grain sizes. *Journal of Geophysical Research*, **91**, 3849–3860.

GHOSH, A.K. & RAJ, R. 1981. Grain size distribution effects in superplasticity. *Acta Metallurgica*, **29**, 607–616.

HANDY, M.R. 1990. The solid-state flow of polymineralic rocks. *Journal of Geophysical Research*, **95**, 8647–8661.

HEILBRONNER, R. & BRUHN, D.F. 1998. The influence of three-dimensional grain-size distributions on the rheology of polyphase rocks. *Journal of Structural Geology*, **20**, 695–705.

HIRTH, G. & KOHLSTEDT, D.L. 1995a. Experimental constraints on the dynamics of the partially molten upper mantle: Deformation in the diffusion creep regime. *Journal of Geophysical Research*, **100**, 1981–2001.

HIRTH, G. & KOHLSTEDT, D.L. 1995b. Experimental constraints on the dynamics of the partially molten upper mantle 2. Deformation in the dislocation creep regime. *Journal of Geophysical Research*, **100**, 15 441–15 449.

KELEMEN, P.B., HIRTH, G., SHIMIZU, N. SPIEGELMAN, M. & DICK, H.J.B. 1997. A review of melt migration in the adiabatically upwelling mantle beneath oceanic spreading ridges. *Philosophical Transactions of the Royal Society, London*, **A355**, 283–318.

KOHLSTEDT, D.L. & WANG, Z. 2001. Grain boundary sliding accommodated by dislocation creep in dunite. (Abstract.) *Transactions of the American Geophysical Union, EOS*, **82**, F1137.

LEE, K.-H., JIANG, Z. & KARATO, S.-I. 2002. A scanning electron microscopy study of the effects of dynamic recrystallization on lattice preferred orientation in olivine. *Tectonophysics*, **351**, 331–341.

MEI, S., BAI, W., HIRAGA, T. & KOHLSTEDT, D.L. 2002. Influence of melt on the creep behaviour of olivine-basalt aggregates under hydrous conditions. *Earth and Planetary Science Letters*, **201**, 491–507.

RAJ, R. & GHOSH, A.K. 1981. Micromechanical modeling using distributed parameters. *Acta Metallurgica*, **29**, 283–292.

REUSS, A. 1929. Berechnung der Fließgrenze von Mischkristallen auf Grund den Konstanten des Einkristalls. *Zeitschrift für Angewandte Mathematik und Mechanik*, **9**, 49–58.

RILEY, G.N. JR. & KOHLSTEDT, D.L. 1991. Kinetics of melt migration in upper mantle-type rocks. *Earth and Planetary Science Letters*, **105**, 500–521.

TULLIS, T.E., HOROWITZ, F.G. & TULLIS, J. 1991. Flow laws of polyphase aggregates from end-member flow laws. *Journal of Geophysical Research*, **96**, 8081–8096.

VOIGT, W. 1928. *Lehrbuch der Kristallphysik*. B.G. Teubner, Leipzig (Germany).

WARK, D.A. & WATSON, E.B. 2000. Effect of grain size on the distribution and transport of deep-seated fluids and melts. *Geophysical Research Letters*, **27**, 2029–2032.

Influence of phyllosilicates on fault strength in the brittle–ductile transition: insights from rock analogue experiments

A. R. NIEMEIJER & C. J. SPIERS

HPT Laboratory, Faculty of Geosciences, Utrecht University, PO Box 80.021,
3508 TA, Utrecht, the Netherlands (e-mail: niemeyer@geo.uu.nl)

Abstract: Despite the fact that phyllosilicates are ubiquitous in mature fault and shear zones, little is known about the strength of phyllosilicate-bearing fault rocks under brittle–ductile transitional conditions where cataclasis and solution-transfer processes are active. In this study we explored steady-state strength behaviour of a simulated fault rock, consisting of muscovite and halite, using brine as pore fluid. Samples were deformed in a rotary shear apparatus under conditions where cataclasis and solution transfer are known to dominate the deformation behaviour of the halite. It was found that the steady-state strength of these mixtures is dependent on normal stress and sliding velocity. At low velocities ($<0.5 \mu m \, s^{-1}$) the strength increases with velocity and normal stress, and a strong foliation develops. Comparison with previous microphysical models shows that this is a result of the serial operation of pressure solution in the halite grains accommodating frictional sliding over the phyllosilicate foliation. At high velocities ($>1 \mu m \, s^{-1}$), velocity-weakening frictional behaviour occurs along with the development of a structureless cataclastic microstructure. Revision of previous models for the low-velocity behaviour results in a physically realistic description that fits our data well. This is extended to include the possibility of plastic flow in the phyllosilicates and applied to predict steady-state strength profiles for continental fault zones containing foliated quartz–mica fault rocks. The results predict a significant reduction of strength at mid-crustal depths and may have important implications for crustal dynamics and seismogenesis.

Classical models for the steady-state strength of the crust consist of a two-mechanism brittle–ductile strength profile, based on Byerlee's law plus a dislocation creep law for quartz (e.g. Sibson 1977; Schmid & Handy 1991; Scholz 2002). However, grain size sensitive processes, such as pressure solution, and the production of weak phyllosilicates known to be important under mid-crustal, brittle–ductile transitional conditions (e.g. Rutter & Mainprice 1979; Passchier & Trouw 1996; Imber *et al.* 2001; Holdsworth *et al.* 2001) are neglected, as are the effects of phyllosilicate foliation development. Such processes have long been anticipated to lead to some form of hybrid frictional–viscous rheological behaviour in the brittle–ductile transition (Sibson 1977; Rutter & Mainprice 1979; Lehner & Bataille 1984/85; Schmid & Handy 1991; Wintsch *et al.* 1995; Handy *et al.* 1999). The steady-state stress levels at which mid-crustal fault rocks deform may, therefore, be much lower than those predicted using a classical two-mechanism strength profile (see Fig. 1) (Sibson 1977; Byerlee 1978; Schmid & Handy

1991; Scholz 2002). Moreover, phyllosilicate foliation development and processes such as pressure solution can be expected to play an important role in controlling transient healing, cementation and strength recovery of fault rocks, thus influencing the rate- and state-dependent frictional and seismogenic behaviour (Fredrich & Evans 1992; Beeler *et al.* 1994; Karner *et al.* 1997; Bos & Spiers 2000, 2002a; Beeler & Hickman 2001; Saffer & Marone, 2003).

Numerous authors have considered the possible weakening effects of phyllosilicates, foliation development, pressure solution and cataclasis within faults and shear zones in the brittle–ductile transition, using both theoretical and experimental approaches (Rutter & Mainprice 1979; Lehner & Bataille 1984/85; Logan & Rauenzahn 1987; Kronenberg *et al.* 1990; Shea & Kronenberg 1992, 1993; Mares & Kronenberg, 1993; Chester 1995; Blanpied *et al.* 1998; Gueydan *et al.* 2001, 2003). Such studies have produced a consensus that fault rocks containing a contiguous and

From: BRUHN, D. & BURLINI, L. (eds) 2005. *High-Strain Zones: Structure and Physical Properties.*
Geological Society, London, Special Publications, **245**, 303–327.
0305-8719/05/$15.00 © The Geological Society of London 2005.

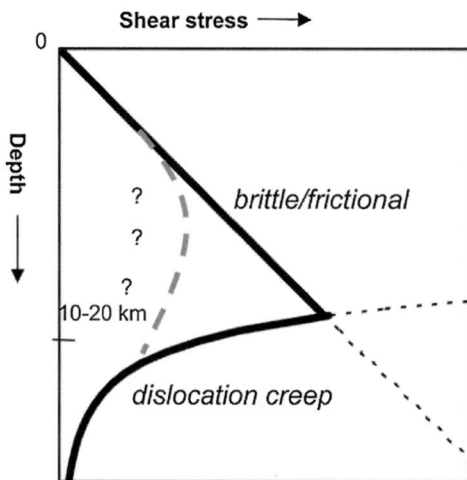

Fig. 1. Schematic of classical crustal strength profile, showing brittle–frictional behaviour dominating at upper crustal levels, and dislocation creep determining crustal strength at deeper levels (solid lines). The dashed line represents the widely accepted effects of fluid-assisted deformation mechanisms on crustal strength (After Bos *et al.* 2000*b*.)

well-developed phyllosilicate (mica) foliation can potentially be as weak as the frictional strength of the phyllosilicate, at high crustal levels (Logan & Rauenzahn 1987; Shea & Kronenberg 1992), or as weak as the crystal-plastic flow strength of the phyllosilicate basal plane, at deeper levels (Hickman *et al.* 1995; Wintsch *et al.* 1995). However, the time and technical limitations of experiments on silicate fault rocks (e.g. Blanpied *et al.* 1991, 1995; Chester & Higgs 1992; Chester 1994; Kanagawa *et al.* 2000) have precluded systematic studies of large strain sliding behaviour, with associated foliation development, under hydrothermal, brittle–ductile conditions where the relevant processes are active. Large strain, steady-state rheological laws for faults incorporating the effects of cataclasis, pressure solution and foliation development in the brittle–ductile transition are therefore not available.

Recently, Bos and co-workers (Bos *et al.* 2000*a*, *b*; Bos & Spiers 2001) performed ultra-high strain rotary shear experiments on simulated fault gouges consisting of mixtures of halite (rock salt) and kaolinite. The experiments were carried out under room temperature conditions where pressure solution and cataclasis are known to dominate over dislocation creep in the halite, thus modelling the brittle–ductile transition. In wet samples with >10 wt% clay

brittle failure was followed by strain weakening towards a steady-state shear strength that was dependent on both sliding rate and normal stress (frictional–viscous flow). Strongly foliated microstructures were produced in these experiments, closely ressembling natural mylonite or phyllonite microstructures, without the operation of dislocation creep. Bos & Spiers (2002*b*) developed a microphysical model to explain the observed steady-state behaviour, based on the steady-state microstructure and corresponding mechanical analogue diagram of Figure 2. In this model, the shear strength of the gouge is determined by the combined resistance to shear offered by frictional sliding on the phyllosilicate foliae, pressure solution in the halite and dilatation on the foliation (work against the normal stress). The model accordingly predicts three velocity regimes, namely: (1) a low-velocity regime, where pressure solution is so easy that the strength of the gouge is determined by sliding friction on the foliation; (2) an intermediate velocity regime, where the strength of the gouge is determined by accommodation through pressure solution; and (3) a high-velocity regime, where pressure solution is too slow to accommodate geometric incompatibilities, so that dilatation occurs. Bos & Spiers (2002*b*) reported good agreement between their experimental data and model, and went on to apply the model to predict crustal strength profiles for quartz–mica fault rocks. These showed a major weakening (2–5 times) of crustal fault zones around the brittle–ductile transition (5–15 km, see Bos & Spiers 2002*b*), in qualitative agreement with inferences drawn from numerous geological and geophysical studies (Lachenbruch & Sass 1980; Schwarz & Stöckhert 1996; Imber *et al.* 2001; Stewart *et al.* 2000; Zoback 2000; Townend & Zoback 2001).

However, there are several aspects of the Bos–Spiers model that have not been tested or are not physically realistic. First, the model is based on experiments in which the phyllosilicate phase (ultrafine kaolinite) was unrealistically fine in relation to the halite grains and to natural fault-rock microstructures (e.g Imber *et al.* 2001). Second, the model has not been tested in the low-velocity regime (Regime 1), nor adequately at high velocities (Regime 3) where it must eventually break down due to fault-rock failure. Third, the model employs an unnecessary and physically unrealistic approximation to couple the mechanical effects of pressure solution and dilatation, and deals only with a single-valued grain size. Finally, the model is restricted to frictional behaviour of the phyllosilicate foliae, whereas under the conditions of the

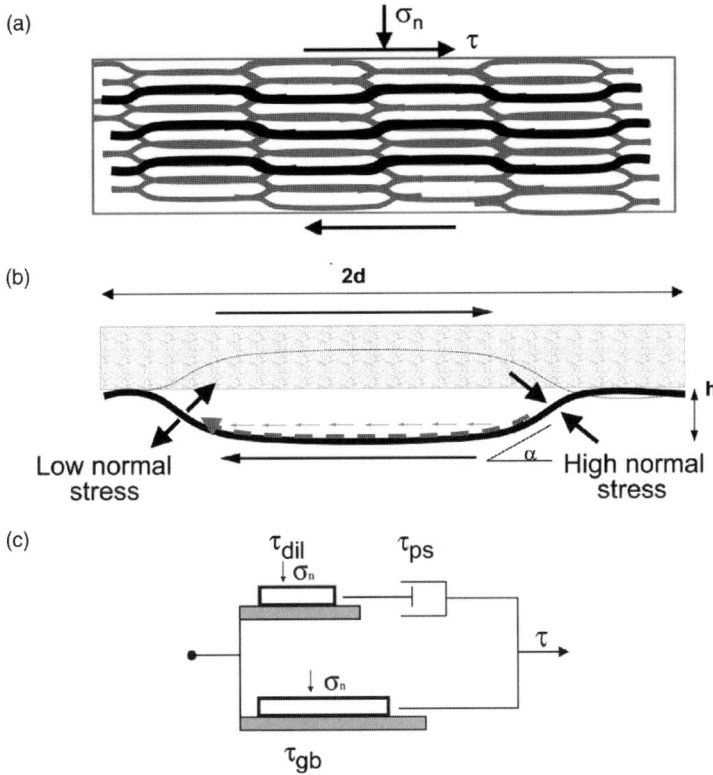

Fig. 2. (**a**) Schematic diagram of the model microstructure proposed by Bos & Spiers (2002*b*), showing contiguous, anastomosing network of phyllosilicates surrounding elongate grains of a soluble solid. The shear strength of the gouge is determined by the combined resistance to shear offered by frictional sliding on the phyllosilicate foliae, pressure solution in the halite and dilatation on the foliation (work done against the normal stress). (**b**) Schematic drawing of representative grain element of the matrix, showing an active sliding surface in black. Shear sense is right-lateral. The diffusive mass flux from source to sink regions is indicated by a dashed arrow. The foliation waves have amplitude h, the grains have long axis d. The leading edge of the grain is inclined at angle α to the horizontal (after Bos & Spiers 2002*b*). (**c**) Mechanical analogue diagram for shear deformation of the model microstructure, assuming zero porosity. τ_{gb} is the shear stress contribution offered by frictional sliding on the phyllosilicate foliae (Regime 1), τ_{ps} is the shear stress contribution offered by pressure solution of the soluble solid (Regime 2), τ_{dil} is the shear contribution offered by dilatation on the foliation (work against the normal stress, Regime 3) and σ_n is the normal stress on the gouge.

brittle–ductile transition phyllosilicates may more easily deform by dislocation-creep processes (see Kronenberg *et al.* 1990; Shea & Kronenberg, 1992, 1993; Mares & Kronenberg, 1993).

In this paper we present experimental results on synthetic halite–muscovite mixtures, obtained using a more realistic phyllosilicate grain size ($d_{med} = c.$ 13 μm). We explore a wider range of sliding velocities (0.001– 13 μm s^{-1}) than Bos & Spiers (2002*b*) accessing both low- and high-velocity regimes, and thus testing their model across regimes 1–3. In addition, we re-formulate the Bos–Spiers model to incorporate proper mechanical coupling of viscous (pressure solution) and frictional

(dilatation) processes, and to allow for crystal-plastic flow in the phyllosilicates. We also consider the effects of a distributed grain size. Finally, we extrapolate our improved model to natural conditions to construct crustal strength profiles and we consider their implications for mature crustal fault zones.

Experimental method

Apparatus

The present experiments were conducted at room temperature on simulated (NaCl–muscovite) fault gouge using the rotary shear apparatus, previously described in detail by Bos *et al.* (2000*a, b*)

(a)

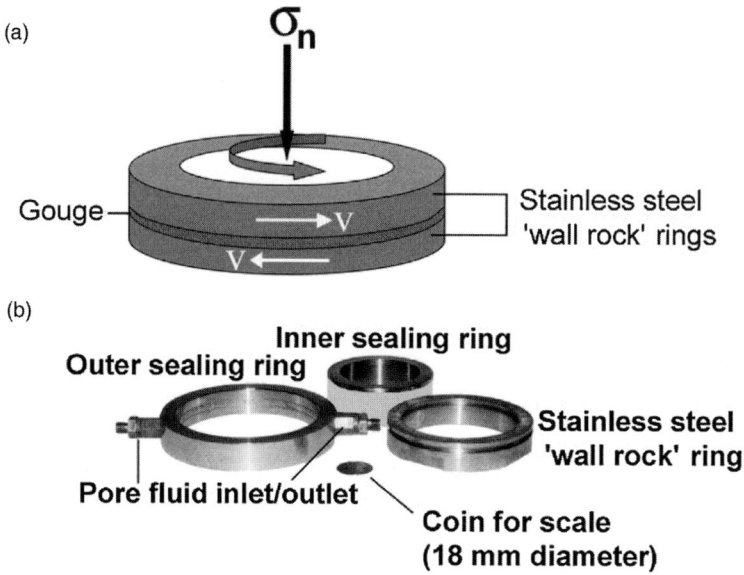

(b)

Fig. 3. (**a**) Schematic diagram of the ring-shear configuration used in this study. Granular fault gouge is sheared between annular stainless steel wallrock rings at controlled velocity *v* and normal stress σ_n. (**b**) Photograph of the wallrock rings and the sealing rings that constitute the sample assembly.

with the annular sample assembly shown in Figure 3a. This assembly consists of two, toothed, stainless steel rings (tooth height is *c.* 0.1 mm), sandwiching a layer of synthetic halite–muscovite gouge. When located in the shear apparatus the rings are rigidly gripped between two cylindrical forcing blocks. The lower forcing block is rotated at controlled angular velocity by a servo-controlled drive motor plus gearbox, while the upper block is maintained stationary, thus leading to shear on the synthetic fault. Sliding velocity is controlled to within 0.001 μm s^{-1}. Normal stress is applied and servo-controlled to within 0.01 MPa using an Instron 1362 loading frame and measured using a 100 kN Instron load cell. Shear stress on the fault is measured with a resolution of approximately 10 kPa, using a torque gauge that provides the couple necessary to hold the upper forcing block stationary. Displacement normal to the fault surface (i.e. gouge compaction/dilatation) is measured using a Linear Variable Differential Transformer (LVDT, 1 mm full scale, 0.01% resolution) located inside the upper forcing block. Rotary displacement is measured using a potentiometer geared to the rotation of the lower forcing block. The synthetic fault assembly is sealed using a plastic inner ring and a stainless steel outer ring, both fitted with O-rings (see Fig. 3b). Pore fluid is introduced through the outer sealing ring. All experiments were conducted at room temperature and atmospheric pore fluid pressure (drained conditions).

Sample material

Synthetic fault gouge material was prepared by mixing analytical-grade granular halite with fine-grained muscovite in the appropriate proportions (0, 10, 20, 30, 50, 80 and 100 wt% muscovite, see Table 1). Particle size analysis, performed using a Malvern particle sizer, showed that the halite had a median grain size of 104 μm, with 90% of the grains in the range 60–110 μm. The muscovite was a commercially available natural muscovite mined in Aspang, Austria, and was supplied by Internatio B.V. The median grain size of this material was 13 μm (equivalent spherical diameter), with 90% of the grains in the range 3–50 μm. Sodium chloride solution (brine), saturated with respect to the halite, was used as the pore fluid, except for one control experiment in which silicone oil was used.

Testing procedure

Samples were prepared for testing by sandwiching a quantity of uniformly distributed gouge (usually about 8 g) between the two stainless

Table 1. List of experiments performed plus corresponding experimental conditions. Note that σ_n is the normal stress on the fault and v is the applied sliding velocity.

Sample code	Composition Muscovite: Halite	Dry run-in displacement (mm)	Fluid	σ_n stepping sequence (MPa)	Velocity-stepping sequence (μm s^{-1})	Total shear displacement*	Final gouge thickness (μm)
Hal1[†]	0%:100%	0	brine	1, 0.5, 2, 3, 1	1	29.38	n.d.
Mus1	100%:0%	69	brine	5, 7, 9, 5, 3, 1	1, 10, 0.1, 1, 0.03, 1	40.01	1028
Mus2	10%:90%	59	brine	5, 9, 1, 5	1, 10, 0.1, 1, 0.03, 1	44.18	1470
Mus3	20%:80%	49	brine	5	1, 0.8, 0.5, 1, 0.1, 0.03, 0.3, 1, 0.01, 0.003, 0.008, 0.3, 1	44.96	1318
Mus4	20%:80%	61	brine	5	1, 3, 5, 7, 9, 11, 13, 11, 9, 7, 5, 3, 1, 5, 10	171.04	975
Mus5	20%:80%	50	brine	5	0.03	35.35	967
Mus6	20%:80%	50	brine	5	1	30.10	1142
Mus7	20%:80%	50	brine	5	5	30.11	1135
Mus8	20%:80%	53	brine	5	13	30.05	1295
Mus9	20%:80%	0	brine	5, 4, 3, 2, 1	1	32.43	1212
Mus10	20%:80%	0	brine	5, 4, 3, 2, 1	5	38.87	1361
Mus11	20%:80%	0	brine	5, 4, 3, 2, 1, 2, 3, 4, 5	10	60.14	1376
Mus12	20%:80%	0	silicone oil	5	1, 10, 0.1, 1, 0.03, 1, 3, 5, 1, 0.1	44.62	1007
Mus13	20%:80%	0	none	5	1, 10, 0.1	38.05	n.d.
Mus14	30%:70%	45	brine	5	1, 10, 0.1, 1, 0.03, 1, 3, 5, 10, 1	33.72	n.d.
Mus15	30%:70%	66	brine	5	1, 10, 0.1, 1, 0.03, 1, 10, 1, 3, 1, 0.01	50.53	997
Mus16	20%:80%	0	none	5	1, 10, 0.1, 1, 0.03, 1, 10, 1, 0.1	38.46	1555
Mus17	20%:80%	0	brine	5	1, 0.001	8.27	880
Mus20	0%:100%	54.8	brine	5	1, 10, 0.1, 1, 0.03, 1, 3, 5, 7, 9, 11, 13, 11, 1, 0.01, 0.003	106.51	1100
Mus21	80%:20%	50.3	brine	5	1, 10, 0.1, 1, 0.03, 1, 0.00	30.91	793
Mus22	50%:50%	55.5	brine	5	1, 10, 0.1, 1, 0.03, 1, 0.01	47.05	955
Mus23	20%:80%	49.5	brine	5	0.1	30.3	1026
Mus24	20%:80%	51.8	none	5	1	–	1568
Mus25	20%:80%	0	brine	5, 4, 3, 2, 1	0.03	30	1300

*Total shear displacement after the dry run-in stage.
[†]Hal1, Data from Bos et al. (2000b).
n.d., not determined.

steel wallrock rings shown in Figure 3, with the two sealing rings in place. This yielded a starting gouge thickness of approximately 1.5–2.0 mm. The sample assembly was then located in the rotary shear machine and subjected to a fixed normal stress of 1 MPa under 'room dry' conditions for 10 min, by which point all compaction had ceased. Most samples were dry-sheared to a displacement of approximately 50 mm at a sliding velocity of 1 μm s^{-1} and a normal stress of 5 MPa, to produce a well-controlled microstructure for the wet stage of the experiments. The samples were then unloaded and the thickness of the gouge was determined. The assemblies were then connected to the pore fluid system, evacuated and saturated with the appropriate pore fluid. They were then reloaded to the prescribed normal stress (σ_n) and sheared at a chosen constant sliding velocity (v). Both normal stress and velocity stepping techniques were used to obtain the shear stress as a function of σ_n and v for individual samples (i.e. fixed mica content), after achieving steady-state sliding behaviour. The experiments were terminated by halting the rotary drive motor and unloading the sample. The pore fluid system was then disconnected from the sample assembly and residual brine was flushed out using dry hexane. Finally, the thickness of the sample assembly was re-measured, the sealing rings were removed and the gouge was carefully detached from the stainless steel rings. After macroscopic inspection, the samples were dried, impregnated with blue-stained epoxy resin and sections were cut for optical microscopy study.

In this way, the sliding behaviour of the synthetic halite–muscovite gouge was systematically investigated, varying normal stresses in the range 1–9 MPa, sliding velocities in the range 0.001–13 μm s^{-1} and muscovite content in the range 0–100 wt%. We also performed a number of experiments at constant sliding velocity and constant normal stress to isolate the influence of displacement on sliding behaviour. Note that the dry shearing stage was omitted from a total of eight experiments to be able to directly compare the results with data obtained by Bos *et al.* (2000*a*); this has little influence on the results for steady-state strengths. All experiments and corresponding experimental conditions are listed in Table 1.

Results

Dry run-in stage

Shear stress and compaction v. displacement curves for the dry run-in stage of samples

spanning the full range of muscovite content investigated (0–100 wt%, Mus1, Mus2, Mus4, Mus15 and Mus20–22) are presented in Figure 4a & b. Recall that all dry shearing was performed at a sliding velocity of 1 μm s^{-1} and a normal stress of 5 MPa. The pure muscovite sample and the 80 wt% muscovite sample (Mus1 and 21) showed initial quasi-elastic loading followed by 'yield' and then stable sliding with gradual slip hardening. At approximately 60 mm of displacement the strength of the pure muscovite gouge reached a steady value of *c.* 2.3 MPa. The sample with 80 wt% muscovite (Mus21) is slightly stronger, reaching a steady value of approximately 2.6 MPa at *c.* 50 mm of displacement.

Salt–mica samples with less than 80 wt% muscovite showed an initial quasi-elastic stress increase, followed at about 2–3 mm displacement by a peak strength and then marked slip weakening up to 15–35 mm displacement (see Mus2, Mus4, Mus15 and Mus22, Fig. 4a). Beyond about 15–35 mm of displacement a transition from stable sliding to stick–slip behaviour occurred. No systematic correlation was found between the amount of muscovite and the displacement at which the transition to stick–slip behaviour occurred, although the sample with 50 wt% muscovite (Mus22) showed stable sliding up to 35 mm of displacement. Beyond approximately 35 mm displacement, the peak strength of salt–mica samples with less than 80 wt% muscovite lay in the range 2–2.6 MPa. In contrast to these halite–mica samples, the single pure salt sample tested (Mus20) showed only a minor stress drop after initial loading, followed by slip hardening beyond *c.* 3 mm displacement to result in a final shear stress of about 4.6 MPa at 50 mm displacement.

All samples showed strong initial compaction normal to the fault in the first 5–10 mm of dry displacement, except for some minor dilatation just before the peak stress was reached. Beyond approximately 5–10 mm of displacement, compaction proceeded at a lower, but still significant, rate. The muscovite-rich samples exhibited particularly large compaction up to around 50 mm of displacement. The pure salt control sample showed continuous compaction, except for a strong dilatation event at approximately 3 mm displacement, just prior to the peak shear strength.

Wet stage: stress–displacement behaviour

In Figure 5a & b, typical shear stress and compaction v. displacement curves are presented for halite–muscovite samples deformed wet at

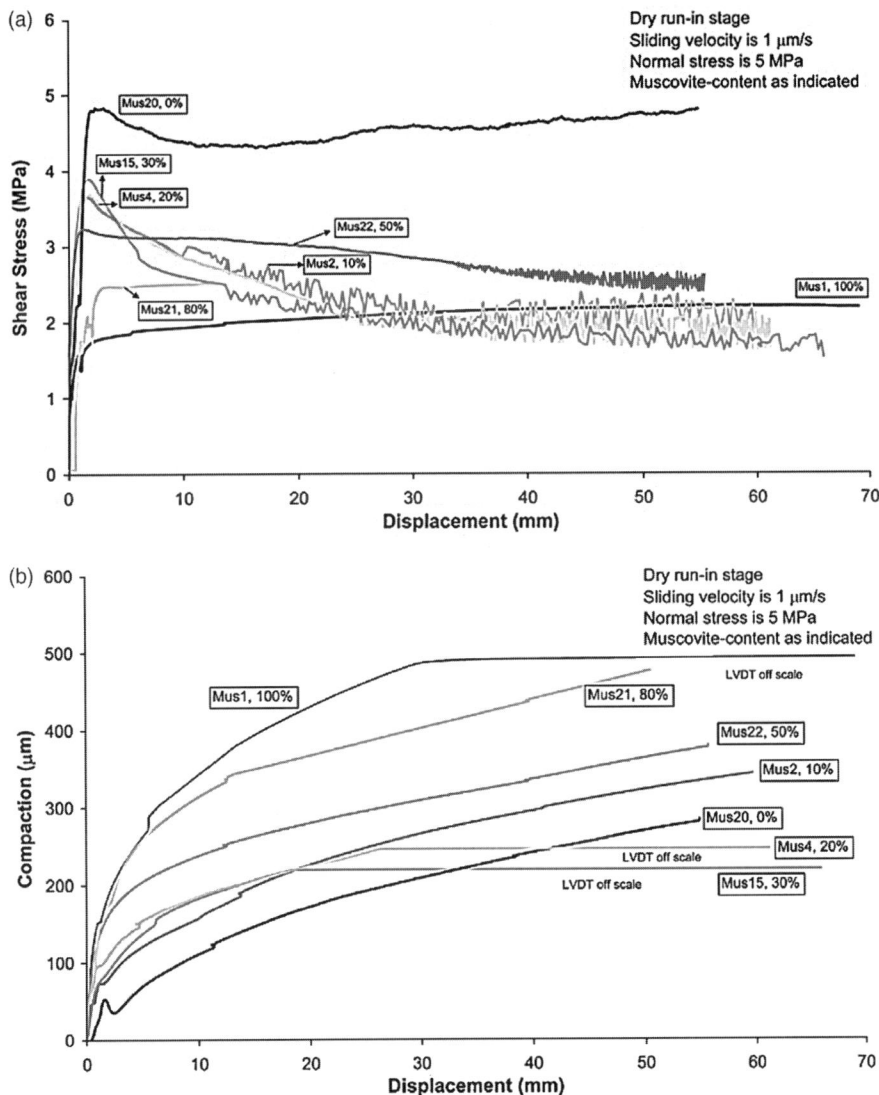

Fig. 4. Shear stress and volume change v. displacement data for the dry run-in stage for samples Mus1, Mus4, Mus15, Mus20, Mus21 and Mus22. (**a**) Stress v. displacement curves. (**b**) Compaction v. displacement curves.

fixed sliding velocity and normal stress. The samples depicted all contained 20 wt% muscovite (Mus5, Mus6–Mus8 and Mus23). Samples deformed at low velocities ($<1 \mu m\ s^{-1}$) show smooth yielding behaviour followed by stable steady-state sliding. Samples deformed at sliding velocities of 1 $\mu m\ s^{-1}$ and above consistently show an initial stress peak followed by rapid weakening. After displacements of approximately 10 mm steady-state sliding was achieved, except at the highest sliding velocity investigated (13 $\mu m\ s^{-1}$), where stick–slip

behaviour was obtained beyond about 4 mm of displacement. The peak stress of such stick–slip events remained roughly constant, but the amplitude of the cycles increased (i.e. the minimum stress decreased). All samples compacted rapidly during loading to the peak stress or smooth yield point (Figure 5b), beyond which compaction continued in a steady fashion. Samples showing a peak stress followed by slip weakening generally displayed minor dilatation just prior to the stress peak. Total compaction was highest in the slowest experiments.

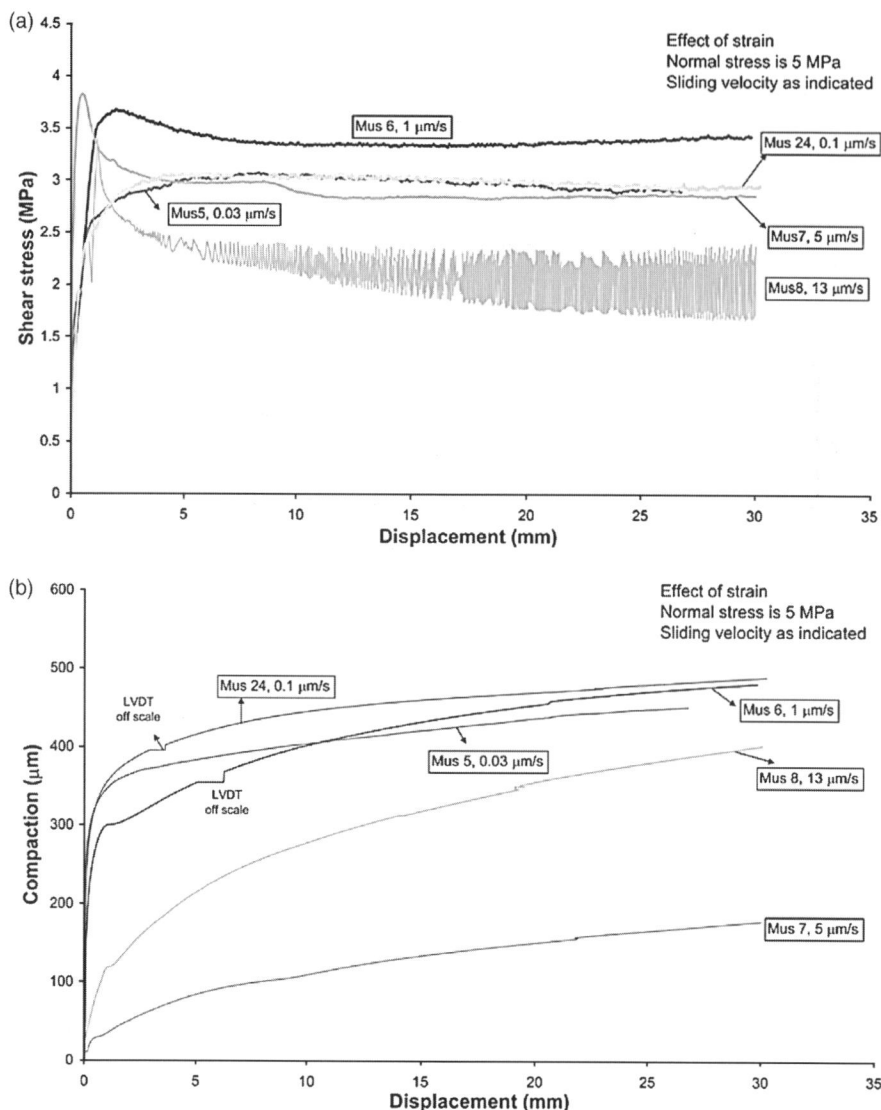

Fig. 5. Shear stress and volume change v. displacement data for samples containing 20 wt% muscovite deformed at fixed sliding velocity and normal stress (Mus5, Mus6, Mus7, Mus8 and Mus24). (**a**) Stress v. displacement curves. (**b**) Compaction v. displacement curves.

Wet stage: effect of normal stress

In Figure 6a we show representative mechanical data obtained from one of the four normal stress-stepping experiments (Mus10) performed under wet conditions on samples with fixed mica content (20 wt%). All such experiments showed significant slip weakening in the first 5–10 mm of displacement after the initial peak stress, then reaching a steady-state strength. Subsequent downwards (or upwards) normal stress stepping

led to an immediate decrease (or increase) in shear stress and to transient dilatation (or increased compaction) of the samples. Typical shear stress v. normal stress data obtained for samples containing 20 wt% muscovite, at fixed sliding velocities of 0.03, 1, 5 and 10 μm s^{-1} (Mus9–Mus11 and Mus25), are presented in Figure 6b. For comparison, similar data are added for pure halite and pure muscovite samples sheared at 1 μm s^{-1} (Hal1 and Mus1,

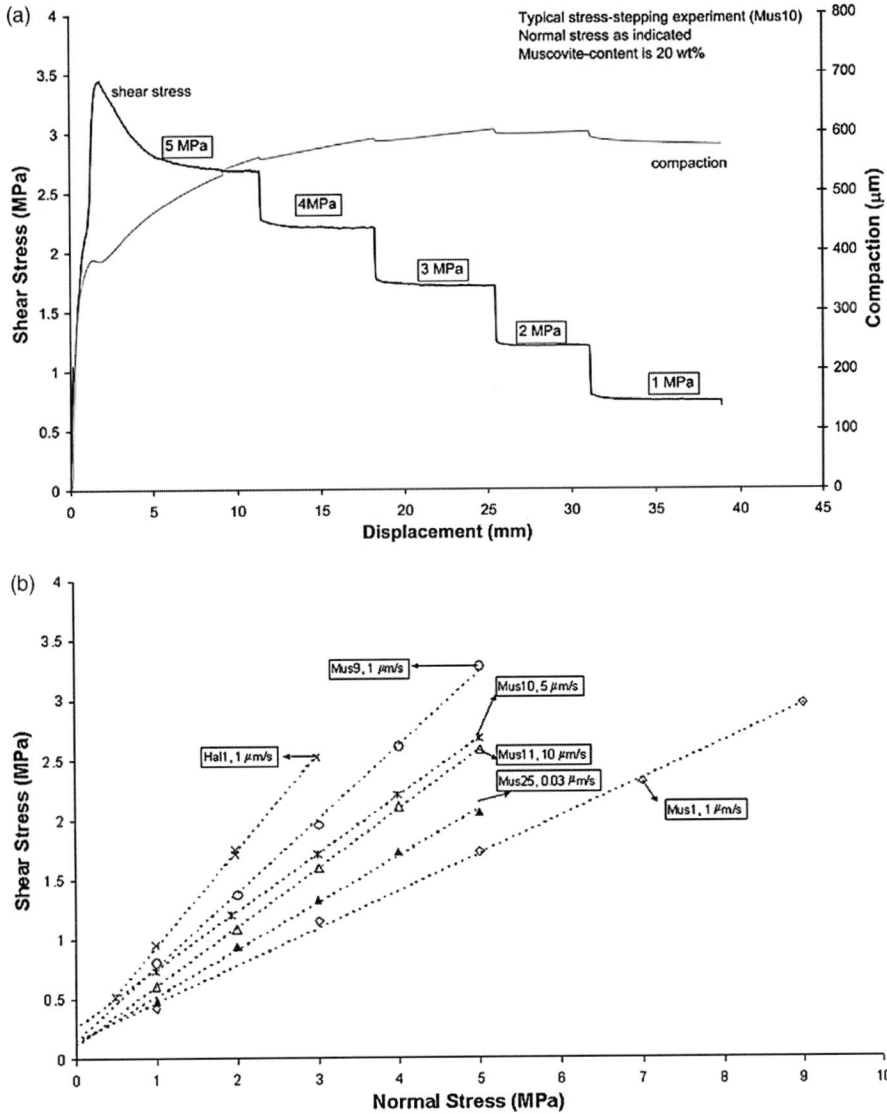

Fig. 6. Typical data obtained in normal stress-stepping experiments. (**a**) Shear stress and compaction strain v. sliding displacement for a 20 wt% muscovite sample (Mus10), subjected to normal stress stepping at 5 μm s^{-1}. (**b**) Steady-state shear stress v. normal stress data that can be described by a Coulomb-type criterion with a friction coefficient and cohesion.

respectively). We plot only data obtained for stable, steady-state sliding. Note that each sample shows a linear relation between shear stress and normal stress, which can be accurately described by a Coulomb-type criterion written

$$\tau = \mu^{ss}\sigma_n + C \qquad (1)$$

where μ^{ss} is a steady-state friction coefficient and C is the cohesion (in MPa). The friction

coefficient and cohesion values obtained are given in Table 2. Note from Figure 6b that the pure halite sample (Hal1) shows the highest strength with a coefficient of friction of approximately 0.8 (Table 2), while the pure muscovite sample is the weakest, with a coefficient of friction of c. 0.3 (Table 2). The samples containing 20 wt% muscovite show intermediate strength. Note also that the friction coefficient of the 20% muscovite–halite mixture decreases with

Table 2. *Strength parameters determined for samples Hal1, Mus1, Mus9, Mus10, Mus11 and Mus25*

Composition	Sliding velocity (μm s^{-1})	Friction coefficient, μ	Cohesion (MPa)	Experiment
100% halite	1	0.802	0.137	Hal1
80% halite, 20% muscovite	1	0.619	0.149	Mus9
80% halite, 20% muscovite	5	0.486	0.261	Mus10
80% halite, 20% muscovite	10	0.496	0.111	Mus11
100% muscovite	1	0.312	0.160	Mus1

increasing sliding velocity for velocities from 1 μm s^{-1} (refer to Fig. 6b and Table 2).

Wet stage: effect of sliding velocity

We present data from a typical velocity-stepping experiment (Mus2) in Figure 7a. Upon a change in sliding velocity, an instantaneous effect on the strength was observed, in all such experiments, followed by a gradual approach to a new steady-state strength (cf. conventional rate- and state-dependent friction experiments: Dieterich, 1979; Ruina 1983; Marone 1998; Scholz 1998). A representative set of data on steady-state shear strength v. sliding velocity obtained in our experiments is presented in Figure 7b. These data again show that the pure halite and pure muscovite gouges are the strongest and weakest materials tested, with mixed compositions being intermediate in strength. Moreover, the steady-state shear strengths of pure halite and pure muscovite gouge are more or less velocity independent. However, mixed muscovite–halite gouges show clear rate effects. For samples containing 10–80 wt% muscovite, two types of behaviour can tentatively be identified. At sliding velocities below about 1 μm s^{-1}, the steady-state shear strength increases with sliding velocity, i.e. the samples are velocity-strengthening. At higher velocities, the samples become weaker with increasing velocity, i.e. the samples are velocity-weakening.

Wet stage: effect of muscovite content

The effect of muscovite content on steady-state shear stress is plotted in Figure 8, by combining shear stress data obtained from different samples at fixed sliding velocities and at a fixed normal stress of 5 MPa. The curves for samples deformed at low and high velocities show a systematic decrease of steady-state shear stress supported with increasing muscovite content, notably up to a muscovite content of 30 wt%. Above 30 wt% muscovite, the steady-state shear strength of the mixed samples is similar

to that of the pure muscovite (see also Fig. 7b). Samples deformed at intermediate sliding velocities of approximately 1 μm s^{-1} show a more continuous decrease in strength with muscovite content (Fig. 8).

Microstructural observations

The microstructure developed in the dry run-in stage of our experiments was investigated using the sample obtained from experiment Mus24 (20 wt% muscovite), which was terminated directly after the dry run-in. Its microstructure is illustrated in Figure 9a & b. The gouge shows numerous halite clasts (30 vol%) that are little affected by the deformation, as evidenced by their resemblance to the starting halite grains, and by the absence of intragranular fractures. These are embedded in a fine-grained mixture of halite and muscovite. The halite grains in the matrix are around 10 μm in size and therefore significantly finer than the starting grain size fraction (60–110 μm). The muscovite grains appear little affected by deformation and their size is similar to the starting grain size fraction. Some regions of the gouge (c. 20 vol.%, especially near the edges of the gouge) are relatively enriched in muscovite and contain fewer halite clasts than other regions. In plane polarized light (Fig. 9a), the dry sheared gouge appears structureless with no clear foliation. However, with crossed polarizers the muscovite flakes are seen to define a weak, anastomosing foliation orientated at approximately 30° to the fault-zone boundaries (Fig. 9b). Some halite clasts are crudely elongated within this foliation. A sharp, continuous shear band was observed parallel and adjacent to the lower shear-zone boundary of sample Mus24 (Y-shear, see arrows in Fig. 9b). No such feature was present near the upper boundary. This boundary-parallel Y-shear mainly consists of fine, aligned muscovite flakes and is about 5–15 μm thick.

The wet-deformed samples show little or no evidence for boundary-parallel faults (Y-shears). Instead, these samples show a

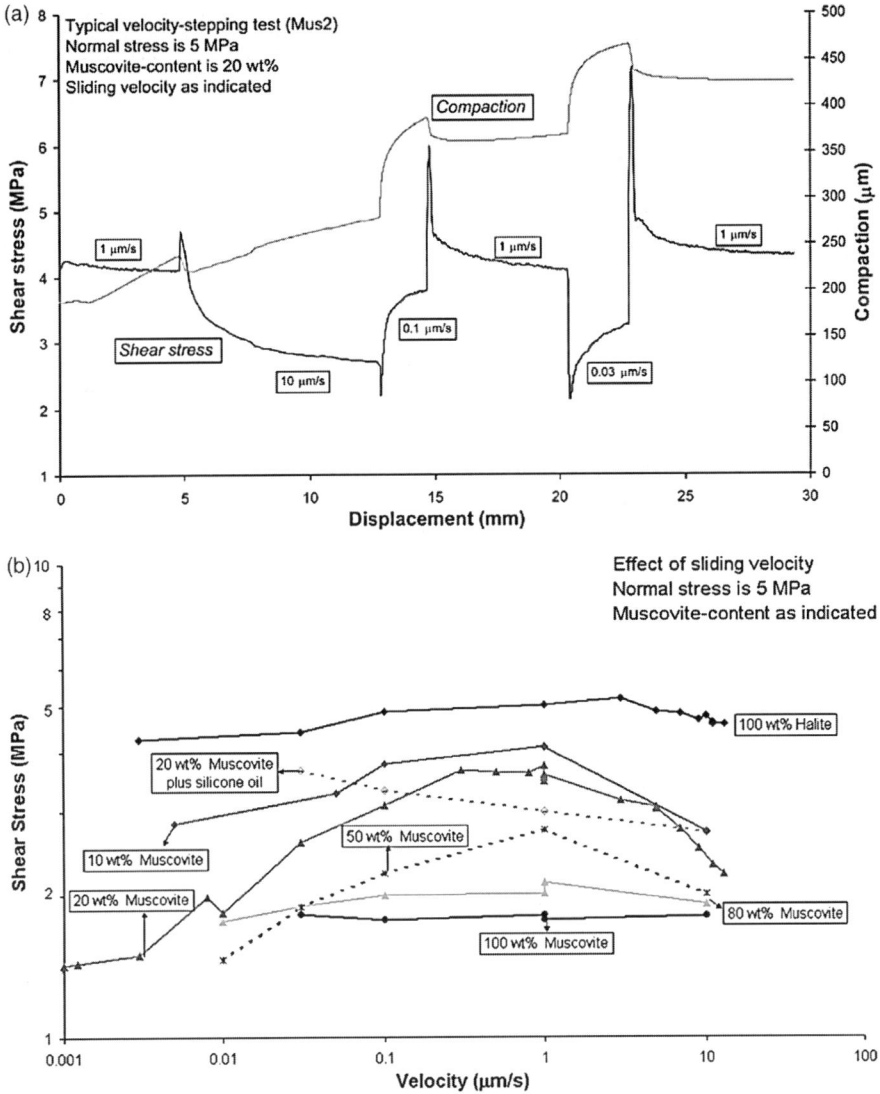

Fig. 7. Effect of sliding velocity on steady-state shear strength of gouges. (**a**) Shear stress and compaction strain v. sliding displacement for a 20 wt% muscovite sample (Mus2), subjected to velocity stepping at a normal stress of 5 MPa. (**b**) Steady-state shear stress v. sliding velocity plot using representative data from velocity-stepping experiments.

complex range of microstructures that depend on wet sliding velocity. Figure 10a–d illustrates the microstructures developed in four samples deformed to approximately 30 mm of wet displacement at sliding velocities of 0.03, 0.1, 5 and 13 μm s^{-1} (Mus5, Mus23, Mus7 and Mus8, respectively) after the dry run-in stage. This range of velocities covers both the velocity-strengthening and the velocity-weakening regimes of Figure 7b. Unlike the dry-sheared

sample, the sample that deformed wet at the lowest sliding velocity (Mus5, Fig. 10a) shows a strongly foliated microstructure with only a few (c. 10 vol%) asymmetric sigmoidal halite clasts and has the general appearance of an S–C mylonite. The fine-grained matrix consists of an intercalation of muscovite (c. 10 μm) and halite grains (c. 15–30 μm, see Fig. 11a), in which the muscovite flakes together with elongated halite grains define the main foliation

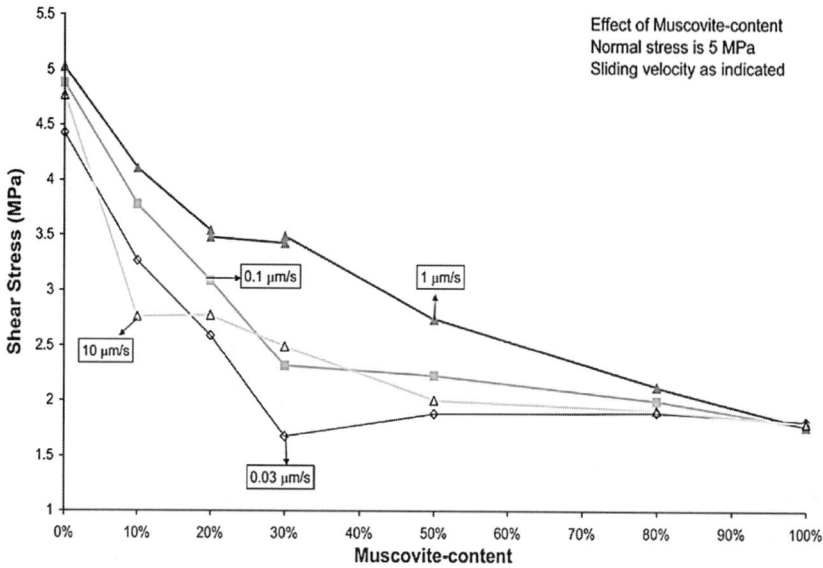

Fig. 8. Effect of muscovite content on steady-state shear strengths of gouges at different sliding velocities. Normal stress is 5 MPa in all experiments.

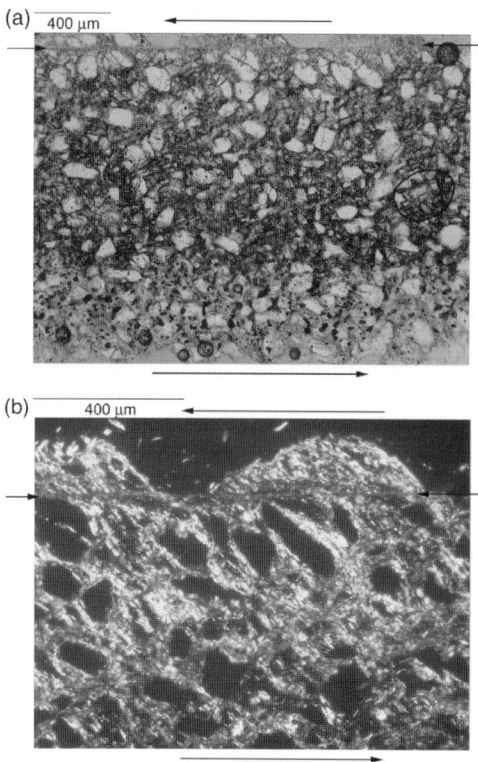

(a) 400 μm

(b) 400 μm

at an angle of approximately 25° to the shear-zone boundary. With increasing velocity (i.e. going from Fig. 10a to d), the number and volumetric fraction of halite clasts increases, and the gouge becomes less well foliated and more chaotic. The shape of the halite clasts progressively changes from sigmoidal forms at 0.03 and 0.1 μm s^{-1} (Fig. 10a & b) to grains with a much more blocky and angular appearance at velocities of 5 and 13 μm s^{-1}. Indeed, the microstructure evolves from one with a foliated mylonitic character at low sliding velocities ($\leq$0.1 μm s^{-1}) to an almost structureless cataclastic microstructure at high velocities ($\geq$5 μm s^{-1}). The sample deformed at the highest sliding velocity (Mus8, Fig. 10d) shows almost no foliation, but is characterized by angular halite clasts embedded in a fine-grained

Fig. 9. Optical micrographs showing microstructures developed in the initial dry run-in stage of deformation, sample Mus24 (20 wt% muscovite, shear strain is approximately 40) (**a**) Plane polarized light: note the chaotic structure with halite clasts embedded in a fine halite–muscovite matrix. A boundary-parallel Y-shear is evident (small arrows). Shear sense as indicated. (**b**) Crossed polarized light: note the weak foliation and Y-shear developed at the boundary of the gouge (small arrows). Shear sense as indicated.

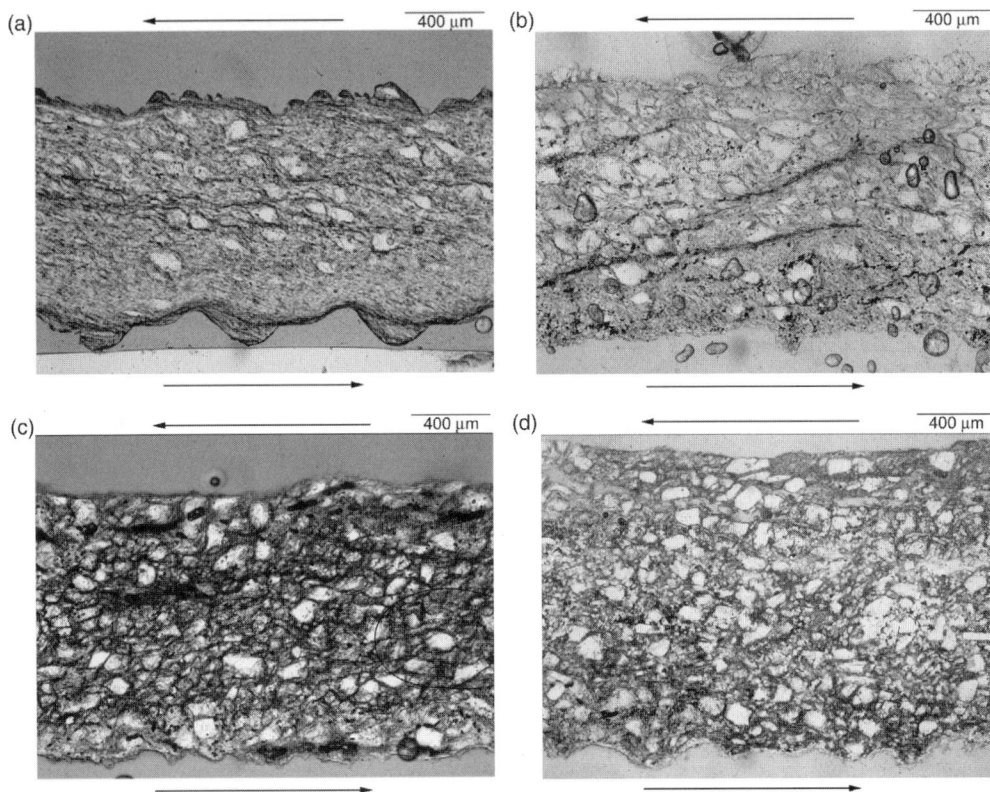

Fig. 10. Microstructural development with increasing sliding velocity in samples deformed wet at velocities of (**a**) 0.03 (Mus5), (**b**) 0.1 (Mus24), (**c**) 5 (Mus7) and (**d**) 13 μm s^{-1} (Mus8). Muscovite content is 20 wt%, normal stress is 5 MPa and total (wet) shear strain is *c*. 25 (see also Table 1). Note progression from a mylonitic structure at low velocity to a cataclastic structure at high velocity. Shear sense as indicated.

structureless matrix. The matrix consists of a poorly sorted mixture of fine halite with muscovite flakes coating all the grain boundaries (Fig. 11b).

For samples Mus5 and Mus23 (constant sliding velocity tests) we determined halite grain sizes using manual tracings and image analysis. It was found that the halite grain size was more or less log-normally distributed in these samples with an average grain size of 20 μm and a relative standard deviation of 0.8.

Discussion

The aim of the present experiments was to test the microphysical model for steady-state frictional–viscous flow of foliated fault rock, developed by Bos & Spiers (2002*b*), for sliding velocities covering all behavioural regimes predicted by the model and for a realistic phyllosilicate grain size. In the following we compare

our experimental data and microstructural observations with those of Bos and co-workers (Bos *et al.* 2000*a, b*; Bos & Spiers 2001, 2002*b*) and with the Bos–Spiers model. In subsequent sections, we go on to reformulate part of the model and extend it to account for effects of crystal–plastic flow within the phyllosilicate foliation, in addition to the mechanisms of frictional slip, pressure solution and dilatation already incorporated. Finally, we apply the new model to predict steady-state strength profiles in crustal faults and we examine the implications thereof.

Dry sliding behaviour

Samples deformed in the dry run-in stage of our experiments showed marked slip weakening after the initial peak strength (Fig. 4a), plus an overall tendency for decreasing strength with increasing muscovite content. Those with an

(a)

(b)

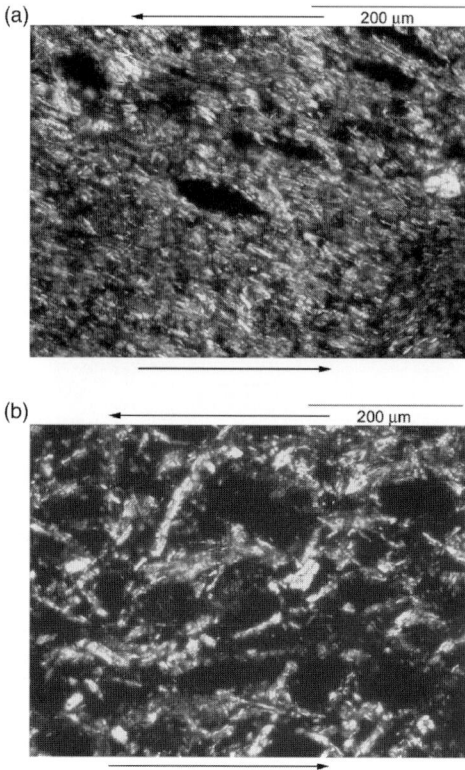

Fig. 11. Optical micrographs showing matrix microstructure in samples deformed at (**a**) 0.03 μm s^{-1} (Mus5) and (**b**) 13 μm s^{-1} (Mus8). Note the change from a foliated to a structureless matrix. Shear sense as indicated.

intermediate muscovite content (10–50 wt%) exhibited stick–slip behaviour after displacements of approximately 15 mm, but showed little effect of muscovite content on peak stick–slip strength. This behaviour is closely similar to that reported by Bos *et al.* (2000*a*) for halite plus kaolinite gouges, indicating that the phyllosilicate grain size and type does not strongly influence dry slip. Our microstructural observations (Fig. 9) demonstrate significant grain size reduction by cataclasis in the halite during the dry run-in stage plus the development of a weak foliation oblique to the shear plane. The stick–slip behaviour of intermediate muscovite content (10–50 wt%) samples, for displacements beyond 15 mm displacement, corresponds to the presence of a through-going boundary-parallel shear (*Y*-shear, cf. Mus24, Fig. 9). Localized slip on a *Y*-shear is often associated with onset of unstable (stick–slip) sliding (Chester & Logan 1990; Bos *et al.* 2000*b*), so we infer

that slip on the boundary-parallel shear occurred after about 15 mm of displacement. The fact that the peak strength of the mixtures during stick–slip did not depend on the muscovite content was probably due to the fact that slip was localized on a muscovite-enriched boundary-parallel shear. Stable sliding prior to stick–slip behaviour may have involved more distributed deformation, presumably by cataclasis, grain size reduction and redistribution of material, leading to the formation of the observed foliation or anastomosing muscovite network. Despite the fact that most deformation in the dry run-in stage of samples with intermediate muscovite content (10–50 wt%) was localized on a boundary-parallel shear, the anastomosing network of muscovite flakes was therefore present at the start of the wet stage of the experiments.

Wet sliding behaviour

From the wet runs on pure halite and pure muscovite, and from the single test on material consisting of 20% muscovite with halite plus silicone oil, it is evident that the steady-state shear strength of these samples show little or no dependence on sliding velocity (Fig. 7b). This suggests purely frictional behaviour as reported by Bos *et al.* (2000*a*) for pure salt and pure kaolinite, with a coefficient of friction ($\mu = \tau/\sigma_n$) of approximately 0.8 for halite and 0.3 for muscovite (cf. 0.2 for kaolinite: Bos *et al.* 2000*a*).

In contrast, all salt–muscovite plus brine samples showed a pronounced velocity strengthening up to *c*. 0.3 μm s^{-1}, with a broad maximum in strength at velocities of approximately 0.3–1 μm s^{-1} and velocity weakening at higher velocities (Fig. 7b). In broad terms then, the data obtained for muscovite and halite samples show two regimes of behaviour: a low-rate, velocity-strengthening regime; and a high-rate, velocity-weakening regime. Note that across the entire range of velocities investigated, the steady-state strength of most halite–muscovite samples is intermediate between that of pure halite and pure muscovite, with a clear decrease in strength as muscovite content increases (Figs 7b and 8).

Halite–muscovite samples deformed in the low-rate, velocity-strengthening regime show no slip weakening in the shear stress v. displacement curves (Fig. 5a). However, their steady-state strength is strongly dependent on both normal stress (σ_n) and in a highly non-linear way, on sliding velocity (see Figs 6b and 7b). This 'frictional–viscous' behaviour, and the strongly foliated mylonitic microstructure

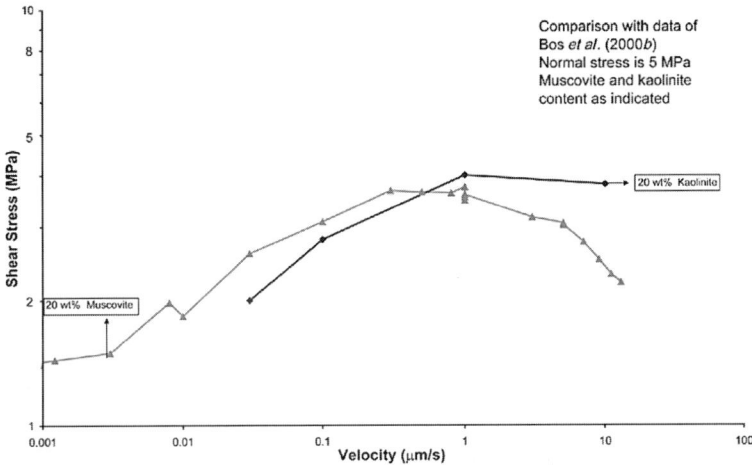

Fig. 12. Comparison of our experimental data on muscovite–halite mixtures with date of Bos *et al.* (2000*a*) on kaolinite–halite mixtures.

that develops in this regime, are closely similar to those found by Bos *et al.* (2000*a*) in halite–kaolinite gouge at similar velocities and normal stresses (Fig. 12). Such behaviour does not occur in our pure, wet halite or pure, wet muscovite samples, nor in the oil-flooded sample in which pressure solution is not possible. The observed frictional–viscous behaviour therefore seems to require the presence of a foliated, halite-plus-muscovite microstructure as well as conditions favourable for pressure solution of the halite. As for halite–kaolinite gouges (Bos *et al.* 2000*a*), we infer that the mechanism of deformation in the low-velocity regime must be one of frictional sliding on/in the anastomosing muscovite foliae, with accommodation of ensuing geometric misfits by pressure solution of the intervening halite.

One important feature of our results for the halite–muscovite gouges, seen at velocities below those investigated by Bos *et al.* (2000*a*, i.e. below 0.03 μm s^{-1}), is a clear change in the sensitivity of steady-state strength to sliding velocity. At the lowest velocities reached, our data show strengths slightly below that of pure muscovite, with almost no dependence on sliding velocity (Figs 7b and 12). This decrease in the sensitivity of steady-state strength to velocity towards very low sliding velocities was not observed by Bos *et al.* (2000*a*, *b*), but is predicted by their microphysical model for pressure-solution-accommodated slip in a foliated fault rock. The Bos–Spiers model predicts rate-independent sliding behaviour at the lowest velocities (Regime 1). Here, pressure-solution

accommodation is predicted to be so easy that frictional sliding on the foliation controls the strength. As the sliding velocity increases, however, pressure-solution accommodation begins to contribute to the bulk strength leading to truly frictional–viscous behaviour (Regime 2). The present results therefore seem to confirm the validity of regimes 1 and 2 of the Bos–Spiers model for the lower sliding velocities accessed here.

At velocities of 0.3–1 μm s^{-1} our wet halite–muscovite samples show a broad maximum in steady-state strength, and at velocities above 1 μm s^{-1} wet samples show strong velocity weakening. In this range of velocities slip weakening and, finally, stick–slip becomes important (Fig. 5a), and the samples are characterized by a structureless microstructure, rather than a foliated one, resembling that of a cataclasite (Fig. 10c & d). At sliding velocities above about 1 μm s^{-1} the steady-state strength of the samples remains linearly related to normal stress, but the coefficient of friction ($\mu = \tau/\sigma_n$) decreases with increasing sliding velocity. The data presented by Bos *et al.* (2000*a*, *b*) for wet halite–kaolinite gouges at velocities of 0.3–10 μm s^{-1} are significantly different from ours in that they show an increase in strength to a more or less constant value at 1–10 μm s^{-1} with no significant velocity weakening (Fig. 12). This strength plateau was interpreted by Bos *et al.* (2000*a*, *b*) to correspond to the onset of dilatational processes within the gouge, as slip on the anastomosing foliation becomes too fast to be accommodated by pressure solution in the halite. Such behaviour is described by Regime 3 of the Bos–Spiers model. It seems reasonable to suppose that at sufficiently high velocities, deformation of the gouge will indeed occur by a cataclastic process, involving disruption of any foliation produced during the dry run-in or earlier slow wet sliding, grain rotation, grain sliding and microfracturing. That such processes operated in our experiments is also supported by our microstructural observations that show a chaotic microstructure at the highest sliding velocity. However, purely cataclastic deformation is not expected to show a rate dependence, whereas we found a very strong, inverse dependence of steady-state shear strength on velocity (Fig. 7b). This could be explained by severe grain size reduction in the halite, enhancing pressure-solution rates and making it again possible for pressure solution to operate at a significant rate, accomodating slip at geometric obstacles. However, as the velocity-weakening behaviour that we observed is reversible with velocity-stepping direction, we believe that this is not likely. Another possible explanation is that deformation becomes localized in weak zones in the gouge at high velocities. However, we found little microstructural evidence for Y-shears or other signs of localized deformation in the high-velocity samples examined (although one sample did show stick–slip, which is suggestive of localization). Clearly, more data are required to explain this velocity-weakening effect. We will adress this in a future paper incorporating the results of slide–hold–slide experiments in this regime.

Revised model for frictional–viscous flow of foliated fault rock

Viscous-dilatational coupling

We have argued above that the Bos–Spiers model for regimes 1 and 2 is consistent with the behaviour seen at low velocities in our experiments ($<0.3 \ \mu\mathrm{m \ s}^{-1}$). The original model is based on the foliated steady-state microstructure shown in Figure 2, and on the mechanical analogue diagram of Figure 2c. An expression for the resistance to steady-state slip on phyllosilicate foliae was proposed by Bos & Spiers (2002b), with pressure-solution accommodation at low rates and dilatation at high rates:

$$\tau = \tau_{\mathrm{gb}} + \left(\frac{1}{\tau_{\mathrm{ps}}} + \frac{1}{\tau_{\mathrm{dil}}} \right)^{-1} \qquad (2)$$

where τ is the total shear stress supported by the shear zone, τ_{gb} is the frictional shear stress

generated by sliding on/within the phyllosilicate foliae, τ_{ps} is the shear stress offered by pressure solution (dissolution, diffusion and precipitation) of the intervening soluble grains and τ_{dil} is the shear stress contribution due to dilatational accommodation of slip on the foliation (modelled as work against the normal stress). The bracketed term $\tau \equiv (1/\tau_{\mathrm{ps}} + 1/\tau_{\mathrm{dil}})^{-1}$, on the right-hand side of the equation, is an approximation for the stress (τ_{u}) supported by the upper branch in the mechanical analogue diagram of Figure 1c. This approximation, $\tau_{\mathrm{u}} \simeq \tau$, was employed by Bos & Spiers (2002b) to smooth out the abrupt transition from a pressure-solution-controlled strength of the upper branch of $\tau_{\mathrm{u}} \simeq \tau_{\mathrm{gb}}$ when $\tau_{\mathrm{ps}} \simeq \tau_{\mathrm{dil}}$ (Regime 2), to a dilatation controlled strength at the limiting value of $\tau_{\mathrm{u}} = \tau_{\mathrm{dil}}$ when $\tau_{\mathrm{ps}} \geq \tau_{\mathrm{dil}}$ (Regime 3). The approximation is accurate for $\tau_{\mathrm{ps}} \gg \tau_{\mathrm{dil}}$ and for $\tau_{\mathrm{dil}} \gg \tau_{\mathrm{ps}}$, but is physically unfounded and inaccurate by a factor of 2 when $\tau_{\mathrm{ps}} = \tau_{\mathrm{dil}}$, as at this point $\tau = \tau_{\mathrm{dil}}/2$ whereas $\tau_{\mathrm{u}} = \tau_{\mathrm{dil}}$. The transition from pressure-solution-controlled strength ($\tau_{\mathrm{u}} = \tau_{\mathrm{ps}}$) to dilatation-controlled strength at the limiting value of $\tau_{\mathrm{u}} = \tau_{\mathrm{dil}}$, in the region $\tau_{\mathrm{ps}} \geq \tau_{\mathrm{dil}}$, is better described by re-defining the bracketed term in equation (2). Then, our model for frictional–viscous flow now takes the form

$$\tau = \tau_{\mathrm{gb}} + \{(\tau_{\mathrm{ps}})^{-n} + (\tau_{\mathrm{dil}})^{-n}\}^{-1/n}. \qquad (3)$$

For large values of the exponent n (e.g. $n = 20$), this equation faithfully reproduces the sharp transition from pressure-solution to dilatation-controlled strength, while still allowing the strength behaviour of the system to be represented using a simple continuous function (equation 3).

In Figure 13 we plot equation (3) as a function of sliding velocity for n values of 1 (which is equivalent to the original model of Bos & Spiers 2002b), as described by equation (2), and 20, and we compare the results with our experimental data on strength v. velocity for samples deformed at velocities below $1 \ \mu\mathrm{m \ s}^{-1}$ (i.e. those showing a foliation). In plotting equation (3), we used parameter values to describe pressure solution in halite taken from Spiers et al. (1990). The grain size of the intervening halite grains was taken to be 30 $\mu\mathrm{m}$ on the basis of our microstructural observations (Fig. 10). For the coefficient of frictional sliding on the phyllosilicate foliation we took a value of 0.31 obtained for pure muscovite plus brine from our own tests. Based on our microstructural observations, we took a value of $25°$ for the angle of dilatation. Increasing the value

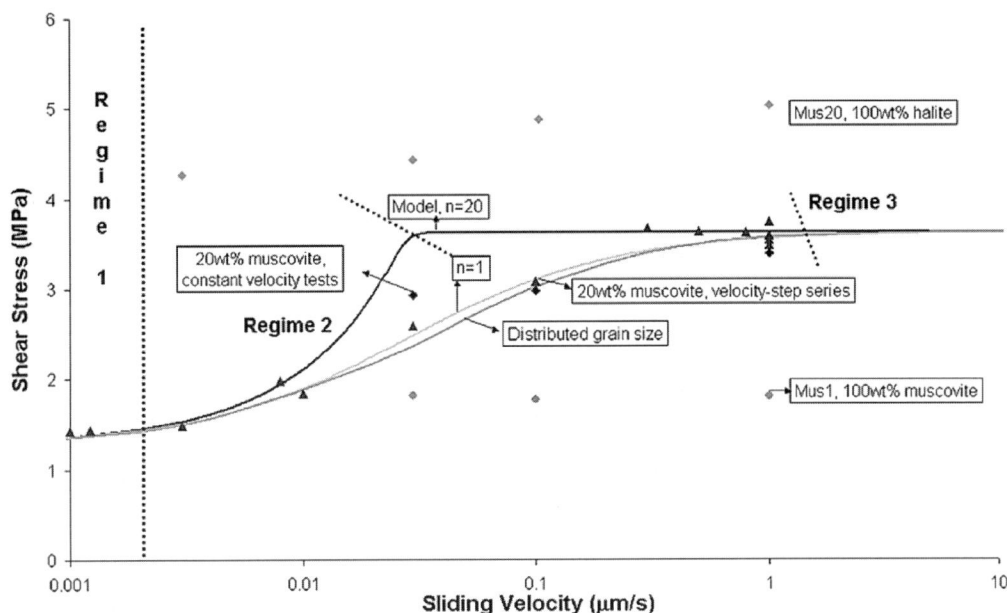

Fig. 13. Comparison of representative experimental data with the different microphysical models. The halite grain size used in the model calculation is 30 μm and the angle of dilatation used is 25°. For the distributed grain size simulation a log-normal distribution was used with a mean grain size of 20 μm and a relative standard deviation of 1. Dilatation angle in this model is 25°.

of n from 1 to 20 clearly demonstrates the much sharper change from Regime 2 (pressure solution dominated) to Regime 3 (dilatation dominated). The revised model with $n = 20$ fits the experimental data at the lowest and highest velocities, but at intermediate velocities (0.02–0.2 μm s^{-1}) the revised model ($n = 20$) predicts higher values for the steady-state shear strength, and a sharper increase in strength than we observed. However, the model is derived for a single-valued grain size of the soluble phase (Bos & Spiers 2002b). Incorporating a grain size distribution into the revised model requires equation (3) to be solved for each incremental grain size class and their stress contributions to be volume-averaged accordingly. Assuming a homogeneous deformation rate in the soluble phase, this can be performed numerically following the method described by Freeman & Ferguson (1981) and Ter Heege et al. (2004). We performed such a calculation, taking $n = 20$, for a population of 1000 grains, randomly generated via a log-normal distribution consistent with that seen in samples Mus5 and Mus23. The result of this simulation is compared with our single-valued grain size calculation ($n = 20$) and experimental data in Figure 13. The distributed grain size leads to a smoother

transition between Regime 2 and Regime 3, bringing the model and experimental data into good agreement over the entire range of velocities below approximately 1 μm s^{-1}. We accordingly infer that the discrepancy between our revised model ($n = 20$), for a single-valued grain size, and our experimental data may, indeed, be due to the effects of the distributed grain size of our samples. The implication is that our revised model applied for a single-valued grain size offers a realistic prediction of steady-state frictional–viscous behaviour of foliated fault gouge, but that the predicted transition from Regime 2 to Regime 3 is too sharp. The excellent fit of the original Bos–Spiers model (equation 2, or 3 with $n = 1$) to data obtained in Regime 2 at velocities of 0.02–0.2 μm s^{-1} (Fig. 13) may thus have been largely fortuitous.

Incorporation of phyllosilicate plasticity

To apply our re-formulated microphysical model (equation 3, $n = 20$) to faults cutting the middle and lower crust, we need to include the possibility of plastic flow in the phyllosilicate foliae, operating concurrently with frictional slip. This can easily be incorporated in the model using the following equation describing the mechanical

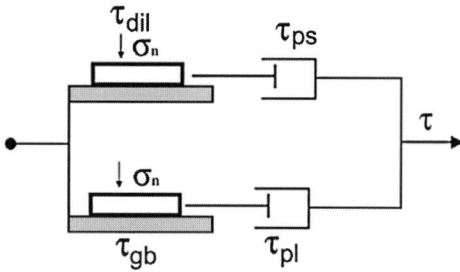

Fig. 14. Mechanical analogue diagram for the postulated shear-deformation process at zero porosity including plastic flow of the phyllosilicate foliae. τ_{pl} is the shear contribution offered by plastic flow in the phyllosilicate foliae (Regime A), τ_{gb} is the shear stress contribution offered by frictional sliding on the phyllosilicate foliae (Regime B), τ_{ps} is the shear stress contribution offered by pressure solution of the soluble solid (Regime C), τ_{dil} is the shear contribution offered by dilatation on the foliation (work against the normal stress, Regime D) and σ_n is the normal stress on the gouge.

analogue diagram of Figure 14:

$$\tau = \{(\tau_{gb})^{-m} + (\tau_{pl})^{-m}\}^{-1/m}$$
$$+ \{(\tau_{ps})^{-n} + (\tau_{dil})^{-n}\}^{-1/n}. \quad (4)$$

Here τ_{gb}, τ_{ps}, τ_{dil} and n are the same as in equation (3), τ_{pl} is the shear strength contribution offered by plastic flow in the phyllosilicate foliae and m ($=20$) is an exponent, like n, that allows for a sharp, but continuous, transition from frictional slip to plastic flow in the foliation. To our knowledge, the only phyllosilicates for which a flow equation has been determined are biotite single crystals (Kronenberg *et al.* 1990), 75 wt% biotite schist (Shea & Kronenberg 1992) and muscovite single crystals (Mares & Kronenberg 1993). As there are large uncertainties in the flow behaviour determined for muscovite (Mares & Kronenberg 1993), and since our model is concerned with plastic flow in pure phyllosilicate foliae, we used the following equation determined for compressive deformation of biotite at 45° to (001) at temperatures of 20–400° (Kronenberg *et al.* 1990) as a first step to incorporating phyllosilicate plasticity in our frictional–viscous flow model:

$$\dot{\varepsilon} = C \exp(\alpha \sigma_d) \exp(-Q/RT). \quad (5)$$

In this equation $\dot{\varepsilon}$ is axial strain rate (s^{-1}), σ_d is applied differential stress ($\sigma_1 - \sigma_3$), T is absolute temperature, R is the gas constant, Q is an apparent activation energy, and C and α are empirical

constants. Rearranging equation (5) and converting it for simple shear by dislocation slip on (001) leads to the relation:

$$\tau_{pl} = \frac{1}{\alpha} \log\left\{\frac{3^{1/2}\dot{\gamma}_{pl}}{C \exp(-Q/RT)}\right\}. \quad (6)$$

Note that the shear strain rate, $\dot{\gamma}_{pl}$, in this equation is the shear strain rate within the phyllosilicate foliae, defined as $\dot{\gamma}_{pl} = \gamma/R$, where $\dot{\gamma}$ is the bulk shear strain rate of the gouge and R is ratio of the thickness of the phyllosilicate foliae to the total gouge thickness.

Given appropriate constitutive parameters allowing computation of τ_{gb}, τ_{pl}, τ_{ps} and τ_{dil}, equation (4) allows the steady-state strength of a foliated quartz–mica or other similar fault rock to be modelled for deformation by plastic and/or frictional slip on the anastomosing phyllosilicate foliation, with accommodation by pressure solution of the intervening mineral phase or by dilatation. From inspection of Figure 14 and the new model embodied by equation (4), it is evident that the model predicts four possible regimes of frictional and/or viscous behaviour with increasing strain rate, in which strength is controlled, respectively, by: (A) plastic flow in the phyllosilicate foliae; (B) frictional slip in the phyllosilicate foliae; (C) pressure solution of the soluble mineral phase; and (D) dilatational cataclasis. We will refer to these as regimes A–D hereforth. Note that the model addresses steady-state velocity-strengthening behaviour only, and does not include velocity-weakening behaviour of the type seen in our experiments at rapid sliding velocities where cataclastic flow destroys the foliation seen at low velocities.

Strength profiles for phyllosilicate-bearing mature crustal fault zones

Model application and predictions

We now construct crustal fault strength profiles using equation (4) and the appropriate constitutive equations for each process involved, assuming a foliated quartz–mica fault rock. Following Bos & Spiers (2002b), we apply the revised model to four different tectonic settings: (1) a transcrustal strike-slip fault zone (such as the San Andreas fault zone); (2) a normal fault zone accommodating crustal extension; (3) a thrust fault zone deforming under high-pressure, low-temperature conditions typical of a subduction setting; and (4) a thrust fault zone deforming under intermediate conditions typical of continent–continent collision zones. The geothermal gradients used in these models are 25, 35, 15 and 25 °C km^{-1}, respectively. Crustal density is taken to be 2.75 g cm^{-3} and pore pressure is assumed hydrostatic at all depths. At upper crustal levels, we will assume pure frictional behaviour with an average friction coefficient of 0.75 (Byerlee 1978). The minimum shear stress to activate frictional sliding is given by

$$\tau_{dil} = 1/2(\sigma_1 - \sigma_3)\sin[\tan^{-1}(1/\mu)] \quad (7)$$

where $\sigma_1 - \sigma_3$ is the minimum differential stress to activate sliding, which is different for each tectonic setting (Sibson 1977; see also Bos & Spiers, 2002b). Turning to the constitutive behaviour of the fault rock (equation 4) for frictional sliding in/on the phyllosilicate foliae ($\tau_{gb} = \mu\sigma_n$), we used the friction coefficient (μ) of muscovite as determined in this study, which is similar to that determined for biotite by Kronenberg et al. (1990). For the plasticity of the phyllosilicate at stress τ_{pl} we used equation (6) and the parameter values for single crystal biotite given by Kronenberg et al. (1990). We took the ratio R of equation (6) as 0.1. Although Bos & Spiers (2002b) used a diffusion-controlled pressure-solution equation for quartz in their study, recent data suggest that interface reaction is likely to be rate-limiting in quartz (Renard et al. 1997, 1999, 2000b, 2001; Niemeijer & Spiers 2002; Niemeijer et al. 2002; Alcantar et al. 2003). We therefore took dissolution/precipitation as the rate-controlling process in our calculation of τ_{ps} in equation (4) using

$$\tau_{ps} = \frac{ARTd}{B^2 k_{s,p}\Omega_s}\gamma. \quad (8)$$

Here A is a constant reflecting the proportion of grain contacts actively dissolving at any instant of time (taken as 1), B is the grain aspect ratio (taken as 5), d is the grain size (taken as 50 μm), Ω_s is the molar volume of quartz and $k_{s,p}$ is the phenomenological rate coefficient for dissolution (subscript s) or precipitation (subscript p). We used an empirical equation for k_s as a function of temperature determined by Rimstidt & Barnes (1980). To describe crystal–plastic flow at greater depth we use the creep equation for wet quartz of Luan & Paterson (1992), rewritten for the case of simple shear. In Figure 15, we show crustal strength profiles resulting from equation (4) for the four different tectonic regimes. The profiles have been constructed using strain rates of 10^{-10} and 10^{-12} s^{-1}, equivalent to a fault zone of 10–1000 m

wide sliding at 30 mm year^{-1}. These rates were chosen by reasoning that the narrow, upper crustal portions of major faults experience higher average strain rates (e.g. 10^{-10} s^{-1}) than the broader mid-crustal reaches (e.g. 10^{-12} s^{-1}: Sibson 1977; Bos & Spiers 2002b). Note in Figure 15 that Byerlee's law plus the dislocation-creep equation used for pure wet quartz for the strain rates shown (shaded lines) together represent a 'classical' two-mechanism crustal strength profile, with the brittle–ductile transition located at depths of 8–15 km and peak crustal strengths at these depths of 50–200 MPa. By contrast, our model for foliated (i.e. high strain) phyllosilicate-bearing fault zones predicts a significant truncation of the brittle–ductile transition in all tectonic regimes, reducing the peak crustal strengths to values between 15 and 90 MPa, and reducing mid-crustal load-bearing capacity by 50–70%. For a fault zone in a strike-slip setting with a geothermal gradient of 25 °C km^{-1}, such as the San Andreas Fault, the peak strength of 30 MPa predicted by our frictional–viscous model occurs at the fully ductile transition from Regime B (frictional sliding with pressure-solution accommodation) to dislocation creep of quartz at a depth of approximately 14 km, assuming a strain rate there of 10^{-12} s^{-1} due to widening of the shear zone with depth (Sibson 1977). The predicted brittle–ductile transition (i.e. the transition from cataclastic Regime D to Regime C in our model), which is much sharper than given by the earlier model of Bos & Spiers (2002b), occurs at a depth of approximately 3.5 km and has a peak strength of about 20 MPa. This marks the predicted transition from seismic to aseismic slip, as the deformation behaviour below this depth is dominated by velocity-strengthening viscous processes (regimes A–C). Note, however, that the brittle–ductile transition is likely to be less sharp than predicted due to the effects of distributed grain sizes, as discussed earlier.

The extension of our model to incorporate the possibility of plastic flow of the phyllosilicates does not significantly modify the predicted crustal-strength profiles compared with the earlier work of Bos & Spiers (2002b). The only tectonic regime where this has a clear effect is the high-pressure, low-temperature (subduction) setting (Fig. 15c). Here, the inclusion of the plasticity of biotite reduces the predicted peak strength at the transition to homogeneous plastic flow of wet quartz from approximately 120 to 90 MPa. It should be noted, however, that the experimental flow law for biotite (Kronenberg et al. 1990) is less well constrained than experimental flow laws for quartz. Clearly, more data are needed on plastic flow in phyllosilicates to evaluate their role in determining the rheology of foliated fault rock.

Pressure-solution-controlled slip on the foliation in high strain phyllosilicate-bearing fault zones (Regime C behaviour) is relatively unimportant in our predicted strength profiles. The zone where crustal strength is dominated by pressure solution is relatively narrow (c. 2–5 km). We have to bear in mind,

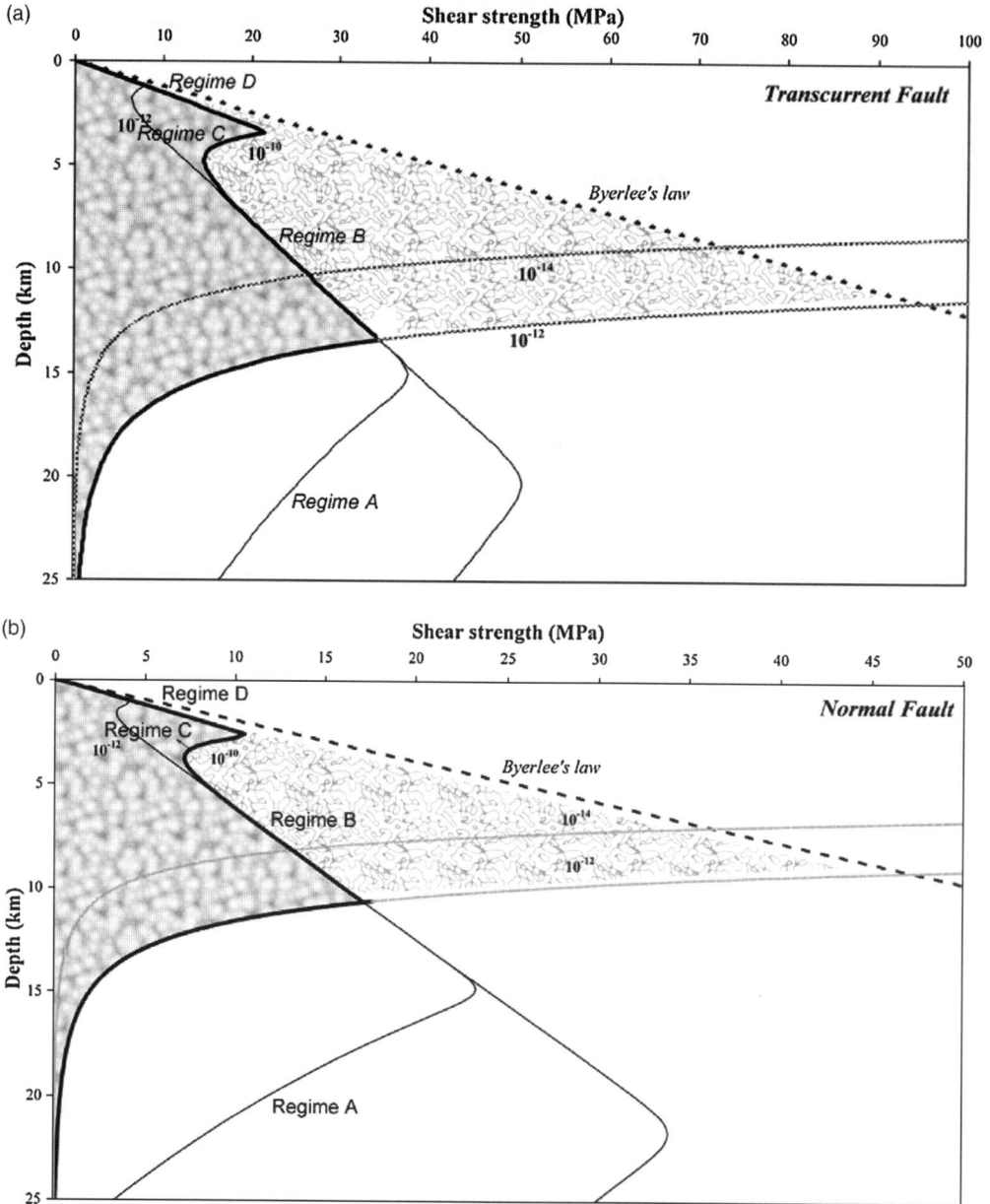

Fig. 15. Crustal strength profiles for four different tectonic regimes. Geothermal gradients used are 25, 35, 15 and 25 °C km^{-1} for cases A, B, C and D, respectively. The grain size used is 50 μm in all cases. Regime A, plastic flow in the phyllosilicate foliae. Regime B, frictional sliding in/over the phyllosilicate foliae. Regime C, pressure-solution-controlled strength. Regime D, sliding accommodated by dilatation.

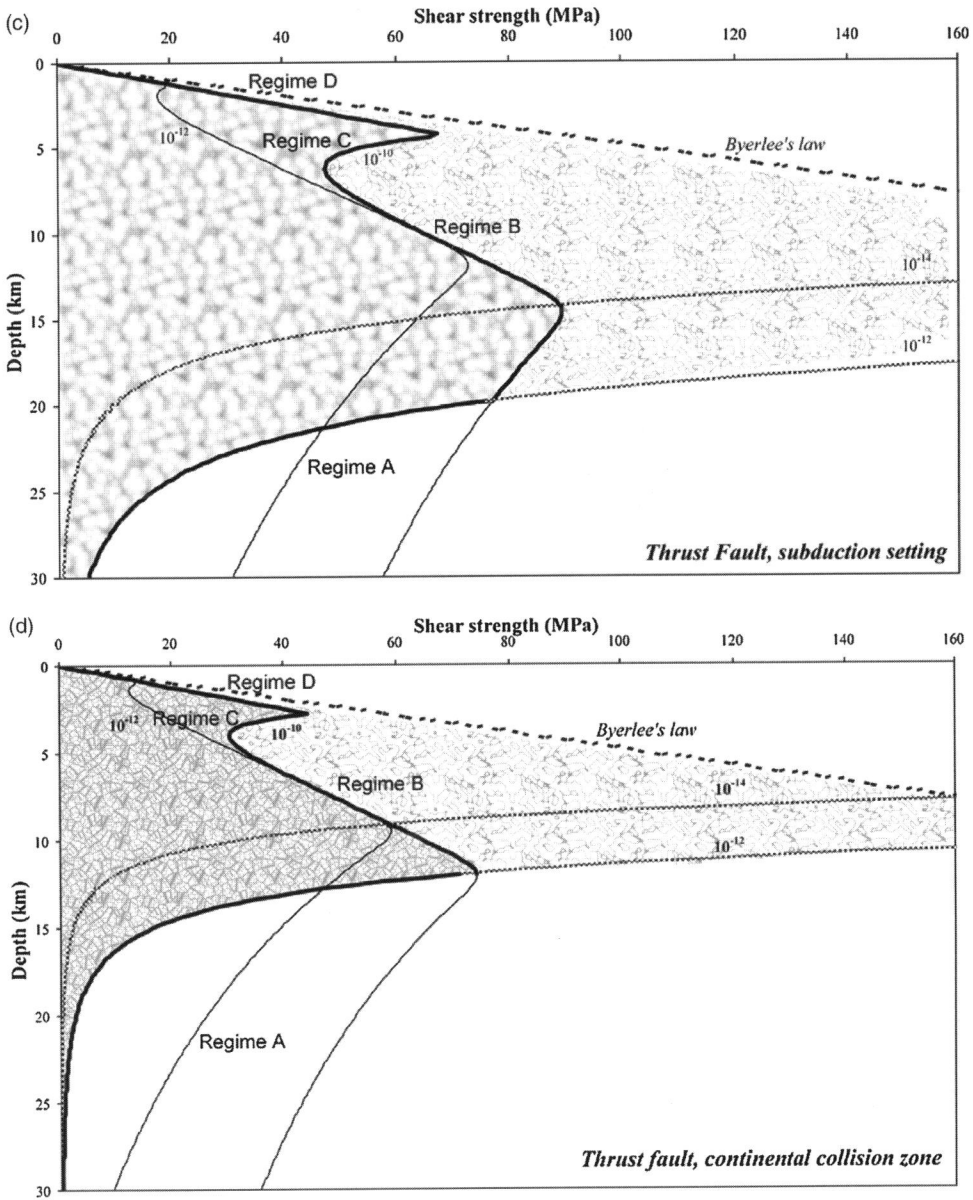

Fig. 15. *Continued.*

however, that the parameters describing pressure solution are relatively poorly constrained (e.g. see Bos & Spiers 2002*b*; Renard *et al.* 2000*a*; Niemeijer & Spiers 2002; Niemeijer *et al.* 2002) and that the use of different estimates for reaction control can result in differences in the predicted strength τ_{ps} of perhaps one order of magnitude. In addition, variations in grain size (<50 μm) and/or strain rate will influence the depth range over which fault slip is controlled by pressure-solution processes.

Further considerations and implications

Our improved microphysical model for frictional–viscous flow of high-strain (mature) phyllosilicate-bearing fault zones predicts significant weakening compared to 'classical' two-mechanism strength profiles. While the predicted brittle–ductile transition and the transition to quartz plasticity are presumably too sharp, the predicted average shear strength is roughly consistent with the low values for the San Andreas Fault inferred from heat-flow measurements (Lachenbruch & Sass 1980) and stress orientation data (Zoback 2000; Townend & Zoback 2001). Moreover, the predicted brittle–ductile transition depth of 3–5 km is consistent with the conclusion drawn from recent field studies of phyllosilicate-bearing mylonite zones, that progressive localization of deformation into such zones leads to a shallowing of the brittle–ductile transition from 10–15 to approximately 5 km (Imber et al. 2001; Stewart et al. 2000). We are suggesting, therefore, that localization of continental fault zones may involve progressive weakening from the intact two-mechanism strength profile down to one described by our model, as the core of the fault zone is gradually converted into (overprinted by) a foliated, phyllosilicate-bearing mylonite at depths of 5–15 km and into a cataclastic equivalent at shallower levels. This does not exclude other mechanisms of fault-zone weakening, such as fluid-overpressuring effects (Hickman et al. 1995; Rutter et al. 2001), but it does seem consistent with field observations, at least on long-lived reactivated fault zones (Imber et al. 2001; Stewart et al. 2000; Holdsworth et al. 2001).

Despite this consistency with field observations, our predicted strength profiles display one major inconsistency with seismological observations on the depth distribution of seismicity on mature faults (e.g. Scholz 2002). Our steady-state model would suggest that crustal seismogenesis should be limited to the upper 5 km, with frictional–viscous flow precluding seismicity at greater depths, whereas the observed distribution of seismicity typically extends to depths of 10–15 km (Sibson 1977; Scholz 2002). This problem requires detailed examination. Preliminary, however, we suggest that the discrepancy lies in the complex anastomosing internal structure and compositional heterogeneity of (mature) fault zones. This complexity will inevitably lead to the development of internal stress concentrations around geometric irregularities or 'rigid' lenses of more intact or stronger rock, as fault motion proceeds, producing local brittle failure. A mature fault zone, consisting of an anastomosing three-dimensional network of weak, aseismically creeping mylonitic–phyllonitic strands with intervening strong lenses and by locked portions of widely varying length scale, would accordingly produce seismic events of varying magnitude and depth

right down to the classical brittle–ductile transition.

A similar argument has recently been put forward by Faulkner et al. (2003) who studied the Carboneras fault in SE Spain. This ancient, large displacement (>40 km) strike-slip fault is approximately 1 km wide and is composed of continuous strands of phyllosilicate-rich fault gouge that bound lenses of variably broken-up protolith. The gouge strands are no wider than 5 m and link up to form an anastomosing network with a width of about 1 km. Faulkner et al. (2003) suggest that these strands are capable of accommodating large amounts of strain through aseismic creep, while the intervening lenses of (more or less) intact protolith are the possible sites of stress build-up and of large seismic moment earthquakes. Notably, several geophysical observations (see Faulkner et al. 2003 and references therein) suggest that the San Andreas fault zone around Parkfield with creeping and 'locked' segments has a similar anastomosing fault structure at depth.

Conclusions

Room-temperature rotary shear experiments have been performed on simulated fault gouge consisting of halite–muscovite mixtures at sliding velocities of 0.001–13 μm s^{-1} and normal stresses of 1–9 MPa. The aim was to test the model for steady-state frictional–viscous flow of foliated phyllosilicate-bearing fault rock developed by Bos & Spiers (2002b), over a wide range of sliding velocities and using a realistic phyllosilicate grain size. A unnecessary and physically unrealistic approximation in the model has been removed and the model has been extended to include the possibility of plastic flow in the phyllosilicates. The revised model was then applied to construct strength profiles for mature phyllosilicate-bearing continental fault zones. The following conclusions can be drawn:

- At sliding velocities of less than c. 0.5 μm s^{-1}, halite–muscovite mixtures develop an anastomosing muscovite foliation and deform by a combination of frictional sliding in/on the muscovite foliation with accommodation by pressure solution of the intervening halite grains. The general behaviour is similar to that of halite–kaolinite mixtures implying that the grain size and type of phyllosilicate has little effect on the sliding behaviour.
- At low velocities (<0.5 μm s^{-1}) our revised model describes the observed behaviour well, assuming frictional behaviour in the

phyllosilicate foliation only. At the lowest velocities, the strength of the synthetic fault gouge is purely frictional and consistent with purely frictional sliding in/on the foliation. This confirms the predictions of our model at velocities that had not been explored previously.

- At velocities higher than 0.5 μm s^{-1}, a gradual transition to velocity-weakening behaviour and cataclastic microstructures was observed in our halite–muscovite samples, whereas halite–kaolinite samples show constant strength. A transition to dilatational/cataclastic flow at constant shear stress is predicted by our model, but the velocity weakening we observed is not. More research is necessary to determine the operative deformation mechanisms and to evaluate the possible implications of such behaviour for seismogenesis.

- Extending our model to foliated, phyllosilicate-bearing quartz–mica fault rocks in mature continental fault zones predicts a significant truncation of classical two-mechanism crustal strength profiles, as well as a strength reduction of the upper crust of 50–70%. The strength–depth profiles predicted by our model suggest that frictional slip and plastic deformation on and within the phyllosilicate foliation will be the main factors controlling the strength of phyllosilicate-bearing fault rocks. While pressure solution is important in accommodating this process, it may be relatively unimportant in controlling strength. Flow laws for phyllosilicates need to be better constrained for future modelling.

- Our revised model predicts the possibility of seismogenic slip on continental phyllosilicate-bearing fault to a depth of only 3–5 km, whereas seismicity is observed to depths of 5–15 km. We suggest that in such faults, seismicity in this depth range may be triggered at geometric irregularities or at 'rigid' lenses of intact or stronger rock within the anastomosing internal structure of the heterogeneous fault zone.

We thank R. Holdsworth, T. Shimamoto, T. Hirono and A. Kellermann-Slotemaker for discussions. C. Peach, E. de Graaff, G. Kastelein and P. van Krieken are thanked for help in the laboratory and technical support. We also thank D. Moore and A. Kronenberg for their constructive reviews that helped improve this paper.

References

ALCANTAR, N., ISRAELACHVILI, J. & BOLES, J. 2003. Forces and ionic transport between mica surfaces: implications for pressure solution. *Geochimica et Cosmochimica Acta*, **67**, 1289–1304.

BEELER, N.M. & HICKMAN, S.H. 2001. Earthquake stress drop and laboratory-inferred interseismic strength recovery. *Journal of Geophysical Research*, **106**, 30 701–30 713.

BEELER, N.M., TULLIS, T.E. & WEEKS, J.D. 1994. The roles of time and displacement in the evolution effect in rock friction. *Geophysical Research Letters*, **21**, 1987–1990.

BLANPIED, M.L., LOCKNER, D.A. & BYERLEE, J.D. 1991. Fault stability inferred from granite sliding experiments at hydrothermal conditions. *Geophysical Research Letters*, **18**, 609–612.

BLANPIED, M.L., LOCKNER, D.A. & BYERLEE, J.D. 1995. Frictional slip of granite at hydrothermal conditions. *Journal of Geophysical Research*, **100**, 13 045–13 064.

BLANPIED, M.L., MARONE, C.J., LOCKNER, D.A., BYERLEE, J.D. & KING, D.P. 1998. Quantitative measure of the variation in fault rheology due to fluid–rock interactions. *Journal of Geophysical Research*, **103**, 9691–9712.

BOS, B. & SPIERS, C.J. 2000. Effect of phyllosilicates on fluid-assisted healing of gouge-bearing faults. *Earth and Planetary Science Letters*, **184**, 199–210.

BOS, B. & SPIERS, C.J. 2001. Experimental investigation into the microstructural and mechanical evolution of phyllosilicate-bearing fault rock under conditions favouring pressure solution. *Journal of Structural Geology*, **23**, 1187–2002.

BOS, B. & SPIERS, C.J. 2002a. Fluid-assisted healing processes in gouge-bearing faults: insights from experiments on a rock analogue system. *Pure and Applied Geophysics*, **159**, 2537–2566.

BOS, B. & SPIERS, C.J. 2002b. Frictional-viscous flow of phyllosilicate-bearing fault rock: microphysical model and implications for crustal strength profiles. *Journal of Geophysical Research*, **107**, 1–13, doi: 10.1029/2001JB000301.

BOS, B., PEACH, C.J. & SPIERS, C.J. 2000a. Frictional–viscous flow of simulated fault gouge caused by the combined effects of phyllosilicates and pressure solution. *Tectonophysics*, **327**, 173–194.

BOS, B., PEACH, C.J. & SPIERS, C.J. 2000b. Slip behavior of simulated gouge-bearing faults under conditions favoring pressure solution. *Journal of Geophysical Research*, **105**, 16 669–16 718.

BYERLEE, J. 1978. Friction of rocks. *Pure and Applied Geophysics*, **116**, 615–627.

CHESTER, F.M. 1994. Effects of temperature on friction: Constitutive equations and experiments with quartz gouge. *Journal of Geophysical Research*, **99**, 7247–7261.

CHESTER, F.M. 1995. A rheologic model for wet crust applied to strike-slip faults. *Journal of Geophysical Research*, **100**, 13 033–13 044.

CHESTER, F.M. & LOGAN, J.M. 1990. Frictional faulting in polycrystalline halite: correlation of microstructure, mechanisms of slip, and constitutive behaviour. *In*: DUBA, A.G., DURHAM, W.B., HANDIN, J.W. & WANG, H.F. (eds) *The Brittle-Ductile Transition in Rocks: The Heard Volume*. American Geophysical Union, Geophysical Monograph, **56**, 49–65.

CHESTER, F.M. & HIGGS, N.G. 1992. Multimechanism friction constitutive model for ultrafine quartz gouge at hypocentral conditions. *Journal of Geophysical Research*, **97**, 1859–1870.

DIETERICH, J.H. 1979. Modeling of rock friction 1. Experimental results and constitutive equations. *Journal of Geophysical Research*, **84**, 2162–2168.

FAULKNER, D.R., LEWIS, A.C. & RUTTER, E.H. 2003. On the internal structure and mechanics of large strike-slip fault zones: field observations of the Carboneras fault in southeastern Spain. *Tectonophysics*, **367**, 235–251.

FREDRICH, J.T. & EVANS, B. 1992. Strength recovery along simulated faults by solution-transfer: Implications for diagenesis, lithification, and strength recovery. *In*: TILLERSON, J.R. & WAWERSIK, W.R. (eds) *Rock Mechanics*, Proceedings of the 33rd US Symposium. AA Balkema, Rotterdam, 121–130.

FREEMAN, B. & FERGUSON, C.C. 1981. Deformation mechanism maps and micromechanics of rocks with distributed grain sizes. *Journal of Geophysical Research*, **91**, 3849–3860.

GUEYDAN, F., LEROY, Y.M. & JOLIVET, L. 2001. Grain-size-sensitive flow and shear-stress enhancement at the brittle-ductile transition of the continental crust. *International Journal of Earth Sciences (Geologische Rundschau)*, **90**, 181–196.

GUEYDAN, F., LEROY, Y.M., JOLIVET, L. & AGARD, P. 2003. Analysis of continental midcrustal strain localization induced by microfracturing and reaction-softening. *Journal of Geophysical Research*, **107**, 1–16, doi: 10.1029/2001JB000611.

HANDY, M.R., WISSING, S.B. & STREIT, L.E. 1999. Frictional-viscous flow in mylonite with varied bimineralic composition and its effect on lithospheric strength. *Tectonophysics*, **303**, 175–191.

HICKMAN, S., SIBSON, R. & BRUHN, R. 1995. Introduction to special section: Mechanical involvement of fluids in faulting. *Journal of Geophysical Research*, **100**, 12 831–12 840.

HOLDSWORTH, R.E., STEWART, M., IMBER, J. & STRACHAN, R.A. 2001. The structure and rheological evolution of reactivated continental fault zones: a review and a case study. *In*: MILLER, J.A., HOLDSWORTH, R.E., BUICK, I.S. and HANDY, M.R. (eds) *Continental Reactivation and Reworking*. Geological Society, London, Special Publications, **184**, 115–137.

IMBER, J., HOLDSWORTH, R.E., BUTLER, C.A. & STRACHAN, R.A. 2001. A reappraisal of the Sibson–Scholz fault-zone model: the nature of the frictional to viscous ('brittle–ductile') transition along a long-lived crustal-scale fault, Outer Hebrides, Scotland. *Tectonics*, **20**, 601–624.

KANAGAWA, K., COX, S.H. & ZHANG, S. 2000. Effects of dissolution–precipitation processes on the strength and mechanical behavior of quartz gouge at high-temperature hydrothermal conditions. *Journal of Geophysical Research*, **105**, 11 115–11 126.

KARNER, S.L., MARONE, C. & EVANS, B., 1997. Laboratory study of fault healing and lithification in simulated fault gouge under hydrothermal conditions. *Tectonophysics*, **277**, 41–55.

KRONENBERG, A.K., KIRBY, S.H. & PINKSTON, J. 1990. Basal slip and mechanical anisotropy of biotite. *Journal of Geophysical Research*, **95**, 19 257–19 278.

LACHENBRUCH, A.H. & SASS, J.H. 1980. Heat flow and energetics of the San Andreas fault zone. *Journal of Geophysical Research*, **85**, 6185–6222.

LEHNER, F.K. & BATAILLE, J. 1984/85. Nonequilibrium thermodynamics of pressure solution. *Pure and Applied Geophysics*, **122**, 53–85.

LOGAN, J.M. & RAUENZAHN, K.A. 1987. Frictional dependence of gouge mixtures of quartz and montmorillonite on velocity, composition and fabric. *Tectonophysics*, **144**, 87–108.

LUAN, F.C. & PATERSON, M.S. 1992. Preparation and deformation of synthetic aggregates of quartz. *Journal of Geophysical Research*, **97**, 301–320.

MARES, V.M. & KRONENBERG, A.K. 1993. Experimental deformation of muscovite. *Journal of Structural Geology*, **15**, 1061–1075.

MARONE, C. 1998. Laboratory-derived friction laws and their application to seismic faulting. *Annual Review of Earth and Planetary Science Letters*, **26**, 643–696.

NIEMEIJER, A.R. & SPIERS, C.J. 2002. Compaction creep of quartz-muscovite mixtures at 500 °C: Preliminary results on the influence of muscovite on pressure solution. *In*: DE MEER, S., DRURY, M.R., DE BRESSER, J.H.P. & PENNOCK, G.M. (eds) *Deformation Mechanisms, Rheology and Tectonics: Current Status and Future Perspectives*. Geological Society, London, Special Publications, **200**, 61–72.

NIEMEIJER, A.R., SPIERS, C.J. & BOS, B. 2002. Compaction creep of quartz sand at 400–600 °C: experimental evidence for dissolution-controlled pressure solution. *Earth and Planetary Science Letters*, **195**, 261–275.

PASSCHIER, C.W. & TROUW, R.A.J. 1996. *Microtectonics*. Springer, New York.

RENARD, F., ORTOLEVA, P. & GRATIER, J.-P. 1997. Pressure solution in sandstones: influence of clays and dependence on temperature and stress. *Tectonophysics*, **280**, 257–266.

RENARD, F., PARK, A., ORTOLEVA, P. & GRATIER, J.-P. 1999. An integrated model for transition pressure solution in sandstones. *Tectonophysics*, **312**, 97–115.

RENARD, F., GRATIER, J.P. & BROSSE, É. 2000a. The different processes involved in the mechanism of pressure solution in quartz-rich rocks and their interactions. *In*: WORDEN, R. & MORAD, S. (eds) *Quartz Cement in Oil-Field Sandstones*. International Association of Sedimentology, Special Publications, **29**, 67–78.

RENARD, F., GRATIER, J.-P. & JAMTVEIT, B. 2000b. Kinetics of crack-sealing, intergranular pressure solution, and compaction around active faults. *Journal of Structural Geology*, **22**, 1395–1407.

RENARD, F., DYSTHE, D., FEDER, J., BJØRLYKKE, K. & JAMTVEIT, B. 2001. Enhanced pressure solution creep rates induced by clay particles: Experimental evidence in salt aggregates. *Geophysical Research Letters*, **28**, 1295–1298.

RIMSTIDT, J.D. & BARNES, H.L. 1980. The kinetics of silica-water reactions. *Geochimica et Cosmochimica Acta*, **44**, 1683–1699.

RUINA, A. 1983. Slip instability and state variable friction laws. *Journal of Geophysical Research*, **88**, 10 359–10 370.

RUTTER, E.H. & MAINPRICE, D.H. 1979. On the possibility of slow fault slip controlled by a diffusive mass transfer process. *Gerlands Beitrage Geophysik*, **88**, 154–162.

RUTTER, E.H., HOLDSWORTH, R.E. & KNIPE, R.J. 2001. The nature and tectonic significance of fault zone weakening: an introduction. *In*: HOLDSWORTH, R.E., STRACHAN, R.A., MAGLOUGHLIN, J.F. & KNIPE, R.J. (eds) *The Nature and Tectonic Significance of Fault Zone Weakening*. Geological Society, London, Special Publications, **186**, 1–12.

SAFFER, D.M. & MARONE, C. 2003. Comparison of smectite- and illite-rich gouge frictional properties: application to the updip limit of the seismogenic zone along subduction megathrusts. *Earth and Planetary Science Letters*, **215**, 219–235.

SCHMID, S.M. & HANDY, M.R. 1991. Towards a genetic classification of fault rocks: geological usage and tectonophysical implications. In: MULLER, D.W. McKENZIE, J.A. & WEISSERT, H. (eds) *Controversies in Modern Geology*. Academic Press, London, 339–361.

SCHOLZ, C.H. 1998. Earthquakes and friction laws. *Nature*, **391**, 37–42.

SCHOLZ, C.H. 2002. *The Mechanics of Earthquakes and Faults*. Cambridge University Press, Cambridge.

SCHWARZ, S. & STÖCKHERT, B. 1996. Pressure solution in siliciclastic HP–LT metamorphic rocks – constraints on the state of stress in deep levels of accretionary complexes. *Tectonophysics*, **255**, 203–209.

SHEA, W.T. & KRONENBERG, A.K. 1992. Rheology and deformation mechanisms of an isotropic mica schist. *Journal of Geophysical Research*, **97**, 15 201–15 237.

SHEA, W.T. & KRONENBERG, A.K. 1993. Strength and anisotropy of foliated rocks with varied mica contents. *Journal of Structural Geology*, **15**, 1097–1121.

SIBSON, R.H. 1977. Fault rocks and fault mechanisms. *Journal of the Geological Society, London*, **133**, 191–213.

SPIERS, C.J., SCHUTJENS, P.M.T.M., BRZESOWSKY, R.H., PEACH, C.J., LIEZENBERG, J.L. & ZWART, H.J. 1990. Experimental determination of constitutive parameters governing creep of rocksalt by pressure solution. *In*: KNIPE, R.J. & RUTTER, E.H. (eds) *Deformation Mechanisms, Rheology and Tectonics*. London, Special Publications, **54**, 215–227.

STEWART, M., HOLDSWORTH, R.E. & STRACHAN, R.A. 2000. Deformation processes and weakening mechanisms within the frictional–viscous transition zone of major crustal-scale faults: insights from the Great Glen Fault Zone, Scotland. *Journal of Structural Geology*, **22**, 543–560.

TER HEEGE, J.H., DE BRESSER, J.H.P. & SPIERS, C.J. 2004. Composite flow laws for crystalline materials with log-normally distributed grain size: Theory and application to olivine. *Journal of Structural Geology*, **26**, 1693–1705.

TOWNEND, J. & ZOBACK, M. 2001. Implications of earth quake focal mechanisms for the frictional strength of the San Andreas fault system. *In*: HOLDSWORTH, R.E. STRACHAN, R.A., MAGLOUGHLIN, J.F. & KNIPE, R.J. (eds) *The Nature and Tectonic Significance of Fault Zone Weakening*. Geological Society, London, Special Publications. **186**, 13–22.

WINTSCH, R.P., CHRISTOFFERSEN, R. & KRONENBERG, A.K. 1995. Fluid-rock reaction weakening of fault zones. *Journal of Geophysical Research*, **100**, 13 021–13 032.

ZOBACK, M.D. 2000. Strength of the San Andreas. *Nature*, **405**, 31–32.

Creep constitutive laws for rocks with evolving structure

BRIAN EVANS

Department of Earth, Atmospheric, and Planetary Sciences, 54-718,
Massachusetts Institute of Technology, Cambridge, MA 02139, USA
(e-mail: brievans@mit.edu)

Abstract: Zones of localized strain are common geological features at all spatial scales, are persistent in time and have important implications for rock strength. Technical advances now allow mechanical tests to be carried out to high strain, and, thus, there is both need and the opportunity to formulate constitutive laws for rocks as their structure (i.e. fabric, texture and mineralogy) changes. This paper reviews some attempts to describe quantitatively the mechanical behaviour of calcite rocks deformed to large strains. To include the effect of evolving structure on strength, three or more interlinked laws are needed: an evolution equation describing the rate of change for each structural variable; a kinetic equation relating mechanical and thermodynamic loading and the structural variables to the rate of strain; and a kinematic equation involving a time integral of inelastic strain rate. Structure variables may be explicitly identified or implicitly determined without identification. Appropriate explicit state variables might include aspects of the dislocation microstructure or the grain size, as are common in studies of plasticity in metals. But because natural tectonic situations are more complex, a much broader class of state variables will be needed. Among these additional variables might be crystal lattice preferred orientation, progress variables for metamorphic reactions, solid–solution chemistry, porosity and pore fluid fugacity.

To describe the production of field structures or to solve the mechanics of natural deformation requires accurate constitutive laws for rocks (Ranalli 1987). Typical formulations relate external thermodynamic variables, including stress, temperature, chemical activity and fluid pressure (or fugacity), to strain rate through material parameters determined by mineralogy and the rate-limiting process (Table 1) (see Poirier 1985; Evans & Kohlstedt 1995). The most commonly used laws assume that rock strength does not evolve, i.e. rapidly achieves steady state. However, there are important exceptions (e.g. Lerner & Kohlstedt 1981; Stone 1991; Covey-Crump 1994b, 1998; Rutter 1999, 2001); these are discussed below.

In part, the motivation to use steady-state descriptions arises from the desire to formulate simple relations containing only external thermodynamic variables. For at least one major tectonic process, creep in the upper mantle, this approach is certainly justified (Kohlstedt *et al.* 1995; Hirth 2002). The homologous temperatures (ratio of deformation temperature to melting temperature) at which peridotites are deformed in the mantle are much higher than those of rocks deformed in the crust, suggesting that diffusion and recovery in the mantle might

be rapid, and that steady-state is attained (Ashby & Verall 1973). Indeed, dislocation microstructures in naturally deformed mantle rocks confirm that climb and other recovery mechanisms have operated extensively (Nicolas *et al.* 1971, 1973; Drury *et al.* 1990, 1991).

In distinct counterpoint to rocks deformed at high temperatures, field observations of exhumed crustal shear zones often indicate severe strain gradients, accompanied by dramatic changes in rock structure that occur over spatial scales from kilometres to millimetres (Nicolas & Bouchez 1977; Burg & Laurent 1978; Lister & Williams 1979; Ramsay 1980; Means 1984; White *et al.* 1986; Vissers *et al.* 1997; Kenkmann & Dresen 1998; Bestmann *et al.* 2000). In this paper, rock structure or, more loosely, structure, is meant to include all elements of fabric, texture, microstructure, mineralogy and mineral distribution that are necessary to describe the rock fully.

In addition to field observations, mechanical analyses of strength instabilities in rocks (Rudnicki & Rice 1975; Poirier 1980; Rice & Ruina 1983; Ruina 1985; Hobbs *et al.* 1990; Handy 1994; Montesi & Zuber 2002; Montesi & Hirth 2003) and large strain experiments (Drury *et al.* 1985; Schmid *et al.* 1987; Tullis *et al.* 1990, Rutter *et al.* 1994; Rutter 1998, 1999;

From: BRUHN, D. & BURLINI, L. (eds) 2005. *High-Strain Zones: Structure and Physical Properties.*
Geological Society, London, Special Publications, **245**, 329–346.
0305-8719/05/$15.00   © The Geological Society of London 2005.

Table 1. *Steady-state constitutive laws for creep processes (after Evans & Kohlstedt 1995)*

Mechanism	Constitutive law	Remarks
Harper–Dorn creep	$$\dot{\varepsilon} = A_{H-D} \frac{\Omega \rho_m}{\lambda} \frac{D\mu b}{kT} \left(\frac{\sigma}{\mu}\right)$$	Low stress, low dislocation density
Viscous dislocation glide	$$\dot{\varepsilon} = \frac{1}{8 c_o \varepsilon_a^2} \left(\frac{kT}{\mu b^3}\right) \frac{D_{sol}}{b^2} \left(\frac{\sigma}{\mu}\right)^3$$	Glide controlled; dislocations dragging solute atmosphere. High temperature, intermediate stress
High and low temperature diffusion creep	$$\dot{\varepsilon} = A_{creep} D_{eff} \frac{\mu b}{kT} \left(\frac{\sigma}{\mu}\right)^{n'}$$ where $D_{eff} = D_v + 10 \frac{a_c}{b^2} \left(\frac{\sigma}{\mu}\right)^2 D_c$	Effective diffusion dominated by lattice diffusion (high temperature) or pipe diffusion (low temperature)
Recovery-controlled dislocation creep	$$\dot{\varepsilon} = \frac{A_{rec}}{b^{3.5}} \frac{D}{M^{0.5}} \left(\frac{\sigma}{\mu}\right)^{4.5} \frac{\mu \Omega}{kT}$$	Creep rate controlled by escape rate of pile ups. High temperature, Intermediate stress
Cross-slip controlled dislocation creep	$$\dot{\varepsilon} = A_{xs} \left(\frac{\sigma}{\mu}\right)^2 \exp - \left[\frac{Q_{xs}\left[\left(1 - \frac{\alpha_{xs} b \sigma}{\gamma}\right)\right]}{RT}\right]$$	Localized obstacles require cross-slip. High temperature, intermediate stress
Glide-controlled dislocation creep	$$\dot{\varepsilon} = 12 v a M^{-0.5} \left(\frac{\sigma}{\mu}\right)^{2.5} \exp - \frac{\Delta G_{glide}}{kT} \left[1 - \frac{\pi \sigma}{2\tau}\right]$$	Stress-activated. Low temperature, high stress
Barrier-controlled dislocation creep	$$\dot{\varepsilon} = A_{barrier} \exp\left[-\frac{\Delta G_{barrier}}{kT}\left(1 - \frac{\sigma}{\tau}\right)^{p'}\right]^{q'}$$	Obstacle dominated; stress-activated. Low temperature; high stress

Additional symbols not defined in text: $\dot{\varepsilon}$, inelastic shear strain rate (usually excludes volumetric strain); σ is the differential stress; Ω, molecular volume; ρ_m, mobile dislocation density; D, diffusion coefficient for non-specified, rate-limiting species; μ, G, shear or rigidity modulus; b, Burgers vector; k, Boltzman constant; T, absolute temperature; D_{sol}, diffusion coefficient of solid solute; c_o, solute concentration; ε_a, elastic misfit strain of impurity; D_{eff}, effective diffusion coefficient; D_v, lattice diffusion coefficient; D_c, pipe diffusion coefficient; v, frequency of vibration of dislocation core segments; ΔG, free energy of activation for particular process (glide, barrier); M, density of dislocation cores; Q_{xs}, activation enthalpy for cross-slip; α_{xs}, apparent activation area for cross-slip event; γ, specific free energy for stacking fault; τ, lattice friction stress at 0 K; p', exponent for barrier-controlled glide; q', exponent in piezometer or for barrier-controlled glide. The remainder of the symbols are materials or process constants (see Evans & Kohlstedt 1995 for more details).

Bystricky *et al.* 2000; Pieri *et al.* 2001*a*; Ter Heege *et al.* 2002; Rybacki *et al.* 2003; Schmocker *et al.* 2003) suggest that the evolution of rock structure may substantially change rock strength. Thus, it is important to develop constitutive laws that include and predict changes in structure variables as they occur during deformation.

The purpose of this paper is to review progress in the quest for creep-constitutive laws that include structure evolution at large strains and to suggest directions for new research. A related problem, not discussed here, is transient creep at very low strains ($\ll 5\%$). Such behaviour might be important, for example, in glacial rebound studies. Even with this restriction, the reader may find this note inadequate in length, lacking in specificity, incomplete in review or simplistic in depth. All these objections are arguably valid; I presume only to present my opinion of a possible and productive way forward. Actually, the journey to a complete mechanical description

has scarcely begun, and the reader should view this note only as a signpost along the road.

Rock structure and strength during creep

Much discussion in understanding localization during creep deformation of rocks and in interpreting the mechanics of ductile shear zones has centred on a putative transition between two steady-state creep mechanisms: dislocation creep and grain-boundary sliding accommodated by diffusion. Such a transition is often inferred from the commonly observed correlation of reduced grain size with increased strain (Fig. 1). That inference has been questioned based on mechanistic arguments (de Bresser *et al.* 1998, 2001) and on the results of experiments carried out in torsion to very large strains (Pieri *et al.* 2001*a*, *b*).

An alternative hypothesis to explain localization within shear zones is that the rocks deformed by a single mechanism that contained

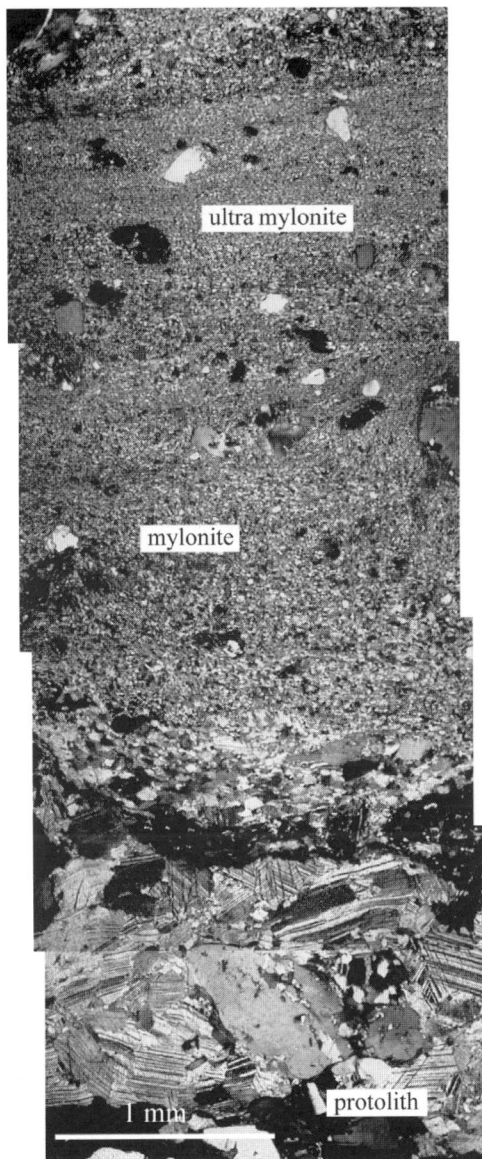

Fig. 1. Mosaic of optical photomicrographs of the microstructure of a marble from the Val Vogna shear zone (Sesia zone, Western Alps) (B. Stöckhert pers. comm. 2004). The transition from protolith (bottom of photomicrograph) to mylonite occurs in a few centimetres. Grain sizes are reduced from 1 to 20 μm.

a strain-weakening term (Rutter 1999). Indeed, large strain deformation tests in monomineralic rocks, such as marbles, limestones and quartzites, often exhibit work-hardening and work-softening behaviours that are qualitatively similar to those observed during metal plasticity

(Rutter 1999; Bystricky *et al.* 2000; Zhang *et al.* 2000, Pieri *et al.* 2001*a*). This behaviour has often been explained qualitatively by appealing to changes in rock structure involving grain growth, recrystallization, recovery and texture formation. But, in addition, natural deformation often involves changes in mineralogy due to metamorphic reactions (Rutter *et al.* 1995; Selverstone & Hyatt 2003), changes in the pore structure, changes in the secondary mineral dispersion and grain sizes (Handy 1994; Herwegh & Kunze 2002), and influx and egress of fluids (Selverstone *et al.* 1991). All of these variables, many of which are not included in the metal plasticity treatments, can potentially cause significant changes in the mechanical properties of rocks. Thus, it seems critical to develop an array of constitutive relations that directly include a broad variety of modifications to rock structure.

State-variable approach

In order to accommodate evolving structure into a flow law, it is necessary to include in the constitutive equation one or more internal state variables (ζ_i) that provide a macroscopic measure of the elements of structure that are pertinent to rock strength. The identities of these variables are discussed below. I follow the general philosophy given by Stouffer & Dame (1996, pp. 267–272), although other treatments might do as well. The state-variable approach has also been widely used by Earth scientists to describe brittle sliding along faults (Dietrich 1978; Rice & Ruina 1983; Ruina 1985), but in this paper the focus is on creep processes.

Three general equations are needed: a state or evolution equation (1) that prescribes the rate at which the state variable changes, a kinetic equation (2) that relates the strain rate to the state variables and the applied thermodynamic variables, and a kinematic equation (3) that prescribes the total strain resulting from a given deformation path. In general terms, these equations are:

$$\dot{\zeta}_m = \Sigma(\dot{\varepsilon}'_{ij}, \sigma_{kl}, T, P, f_p, \zeta_r, \dots) \quad \text{evolution} \quad (1)$$

$$\dot{\varepsilon}'_{ij} = E(\sigma_{kl}, T, P, f_p, \zeta_r, \dots) \quad \text{kinetic} \quad (2)$$

$$\varepsilon_{ij} = \varepsilon^{EL}_{ij} + \oint \dot{\varepsilon}'_{ij} dt \quad \text{kinematic} \quad (3)$$

The *evolution* equation (1) expresses the rate of change of the *m*th state variable ($\dot{\zeta}_m$), as a function (Σ) of the inelastic strain rate ($\dot{\varepsilon}'_{ij}$), the stress tensor (σ_{kl}), the absolute temperature (T), the mean stress (P), the chemical fugacity (or activity) of the pth chemical component (f_p) and the other state variables (ζ_r). When more than one state variable is required, there will be an evolution equation for each. The *kinetic* equation (2) expresses $\dot{\varepsilon}'_{ij}$ as a function (E) of the external thermodynamic state variables, σ_{kl}, T, P, and f_p, and the internal state variable, ζ_r. Finally, the *kinematic* equation (3) gives the total strain as the sum of the instantaneous elastic strain (ε^{EL}_{ij}), which includes reversible strain due to mechanical and thermal loading, and the path integral of $\dot{\varepsilon}'_{ij}$ over some prescribed time interval. Notice that any or all of the state variables and the external thermodynamic variables might change as the rocks are deformed. Finally, it may be necessary, in some cases, to formulate the kinetic equation so as to include two or more processes that operate simultaneously.

Identity of the internal state variables

For creep deformation occurring in complex rock masses, many internal state variables can be imagined (Poirier 1980; Hobbs *et al.* 1990), and the form of equations (1) and (2), and the relative pertinence of each candidate variable, will depend on the mineralogy and petrology of the rocks, the temperature, pressure and stress state (cf. Ashby 1972; Korhonen *et al.* 1987).

The choice of the equation and the variables should be made on a physical and mechanistic basis (Argon 1975); these selections should be informed by observations of the microstructure of the deformed rock when possible (Hirth & Tullis 1992; Tullis 2002). In the best of all possible worlds, constitutive models would predict the mechanical behaviour of polyminerallic rocks with volatiles present (Tullis 2002), subjected to multiaxial stress states (Paterson 2001), over a broad range of thermodynamic conditions (Ashby 1972). These models would include metamorphic reactions (Rutter *et al.* 1982; Brodie & Rutter 1987; Stunitz & Tullis 2001). The reader may have noticed that we do not live in the best of all worlds (Gibbs 1878).

Candidate state variables can be arranged into three categories: external thermodynamic variables; implicit internal variables; and explicit internal variables. See Stouffer & Dame (1996) and Table 2. Implicit state variables provide a description of macroscopic mechanical properties, but are determined empirically using specific mechanical tests. Examples include the hardness parameter (Hart 1970, 1976; Korhonen *et al.* 1987; Covey-Crump 1998, 2001) and the internal friction variable of the rate-state friction law (Dietrich 1978; Ruina 1985; Tullis 1986; Tullis & Weeks 1986; Marone 1998). The facts that a detailed description of the implicit structure variable is not necessary and that the variable is determined empirically are both strengths and weaknesses of this approach (Stone 1991).

The implicit variable approach is most accurate when tests to determine the variable and its evolution are performed under conditions nearly the same as those for which the constitutive law is to be applied. However, unless the implicit state variable is identified and placed into micromechanical context within the evolution and kinetic equations, it remains a fitting parameter and cannot be extrapolated with confidence to different loading and thermodynamic conditions. Because this extrapolation is a necessary condition for application of laboratory data to natural deformation (Paterson 2001), it is a significant disadvantage.

Explicit internal state variables require a detailed description of some aspect of the microstructure. For naturally deformed rocks, explicit variables have the advantage that the strength at a prior point of the deformation history may sometimes be estimated by observing a relict structure. Influenced by Argon (1975), Stone (1991), Stone *et al.* (2004) and others, I depart from Stouffer & Dame (1996) and argue that explicit internal state variables, which are based on physical mechanisms and are potentially observable, provide a more predictive approach.

Whether explicit or implicit, the choice of the variables to be used should be governed by evaluation of the mechanism to be described.

Table 2. *Some possible state variables*

External thermodynamic	Implicit internal	Explicit internal
Lithostatic pressure	Internal friction variable	Subgrain (dislocation) structure
Lithostatic deviatoric stress	Hardness	Grain Size (grain size distribution)
Pore pressure		Crystal orientation distribution
Temperature		Pore structure
Chemical fugacity (or activity)		Reaction progress

For example, during diffusion creep, average grain size and, perhaps, grain size distribution arise naturally as explicit state variables (Ashby 1972; de Bresser *et al.* 2001; Ter Heege *et al.* 2002). But for both variable classes, three other principles should also be met: parsimony, objectivity and avoidance of dependence on history (Ruina 1985; Stouffer & Dame 1996). The principle of parsimony, an example of Occam's razor, suggests that the best description is one that uses a minimum number of variables, or to use a phrase attributed to Einstein (Hodge 1985), one that is 'as simple as possible and no simpler'. Objectivity requires that the constitutive description be unchanged by transformations in the spatial and temporal reference frames. In other words, the strength of the material should not depend on the choice of origin or the starting time. A dependence of strength on history is avoided if the internal variables are path-independent state variables in the thermodynamic sense. If this principle is not obeyed, geologists face the unpleasant task of determining the strength of each rock separately, as each has a unique history. If dependence on history is to be avoided, and if the principle of objectivity is to be observed, then strain should not be adopted as a state variable. Notice that in the formulation given above strain is a dependent variable that is the net result of the integrated deformation history (equation 3).

Explicit state-variable approach applied to naturally deformed rocks

Owing to the complexity of nature, the explicit state variables appropriate for describing the deformation of rocks will be broader in class and greater in number than those traditionally considered by metallurgists. But, formalism and complexity notwithstanding, equations (1)–(3) have simple interpretations that coincide with common geological pedagogy. For example, equation (1) prescribes the kinetic evolution of some common petrological structures, including nucleation and growth of new minerals, changes in composition of phases, mineral dispersion, grain size and its distribution, crystal lattice preferred orientation (CPO), or other elements of rock fabric, texture or microstructure.

For steady-state deformation, the kinetic equation is known as the constitutive equation, and the evolution equation is, by definition, unimportant. As reiterated by many authors (e.g. Paterson 1987; Ranalli 1987; Kohlstedt *et al.* 1995; Paterson 2001), it is important for the Earth scientist to use descriptions of as many rheological processes as possible to determine which constitutive law is most important for a given geological situation. Deformation mechanism maps are a convenient tool for this determination and have been used to consider steady-state mechanisms for some monomineralic rocks (Ashby 1972; Frost & Ashby 1982). Similar maps could be developed for the rheology of rocks with evolving structure, but, due to the larger number of variables, they would need to be multidimensional (cf. Carter & Tsenn 1987).

The last, or kinematic, equation emphasizes the fact that total strain is not a thermodynamic state variable, but rather a path-dependent result of deformation history. In fact, a significant part of the task assigned to the field structural geologist involves parsing the structural elements in the rock and unravelling the past deformation history (Ramsay & Huber 1983). It follows that using strain as a state variable would require an *a priori* specification of the history of all the other thermodynamic variables.

In the following sections are brief discussions of selected experiments involving the creep of calcite rocks with evolving microstructure and of the attempt to formulate constitutive relations to describe that behaviour. These discussions are only a few examples of a very broad class of situations that might occur during natural deformation of both silicate and non-silicate rocks. To describe a complete inventory of constitutive laws for rocks with evolving structure will require a vigorous and sustained effort on the part of a community of scientists, including practitioners of solid mechanics, laboratory experimentalists and field geologists.

Strength evolution in calcite rocks: a case example

Strain hardening up to a peak stress, followed by weakening, is a common feature in mechanical tests of many rock types, deforming by both cataclastic and crystal-plastic mechanisms (Heard 1960; Paterson 1978; Tullis & Yund 1985, 1992; Evans *et al.* 1990; Rutter 1998, 1999). In many silicate rocks it is difficult to separate the contributions of the various brittle and crystal-plastic processes (for a discussion see Tullis & Yund 1992), rendering formulation of a constitutive law difficult. Here we are interested in creep behaviour without the additional complication of cataclastic mechanisms. Deformation tests in dense calcite rocks exhibit the full transition from brittle failure to semi-brittle behaviour, and then to non-dilatant flow over pressures and strain rates that are experimentally accessible, even at room temperature (Heard 1960; Fischer & Paterson 1989; Fredrich *et al.* 1989). Thus, an extensive data set exists that includes useful information on the effect of evolving structure on the creep strength of limestone and marble. For that reason, and for the sake of brevity, the discussion below uses the deformation of limestones and marbles as a case study. A summary of the evolutionary constitutive equations is given Table 3.

Single crystals of calcite oriented for multiple slip (de Bresser & Spiers 1993, 1997) and polycrystals of marble and limestone (Casey *et al.* 1998; Covey-Crump 1998, 2001; de Bresser *et al.* 1998; Pieri *et al.* 2001a; Ter Heege *et al.* 2002; Rybacki *et al.* 2003; Barnhoorn *et al.* 2004) typically exhibit changes in microstructure and in strength, in some cases to very large strains, but the details of the mechanical behaviour differ significantly depending on the test configuration, temperature and pressure (Fig. 2). Test configurations are discussed by Tullis & Tullis (1986), Paterson & Olgaard (2000) and Mackwell & Paterson (2002). The slope of the stress–strain curve, i.e. the hardening coefficient ($h = d\sigma/d\varepsilon$ or, alternatively, $h = d\ln\sigma/d\varepsilon$, where σ is the differential stress and ε the strain), is often used as a convenient macroscopic measure of the evolution of strength. When the coefficient is zero, the structure variables do not change and recovery processes balance any hardening processes.

Below 800–900 K, stress–strain curves for a variety of marbles and limestones tested in conventional triaxial compression tests often indicate continuous strain hardening up to the maximum strains achieved (usually about 0.2) (Heard & Raleigh 1972; Covey-Crump 1994b). At lower temperatures, especially when effective pressure is also low, a brief hardening period is followed by rapid load drop. Often the peak stress and hardening rate are pressure-dependent (Heard 1960; Fischer & Paterson 1989; Fredrich *et al.* 1989). Such behaviour is thought to be semi-brittle and to involve cataclastic processes as well as dislocation creep (for a review see Evans *et al.* 1990; Wong *et al.* 1997; Baud *et al.* 2000). When cataclastic behaviour can be ruled out, the qualitative aspects of the stress–strain curves for both single crystals (de Bresser & Spiers 1990, 1993, 1997) and polycrystalline aggregates of calcite (Covey-Crump 1998, 2001; Rutter 1999) are reminiscent of low-temperature creep behaviour in metals.

Two recent studies in a conventional triaxial machine focused on the correlations among grain size, grain size distribution and strength (de Bresser 2002; Ter Heege *et al.* 2002). Their compression tests of Carrara marble at strain rates of $3 \times 10^{-6} - 5 \times 10^{-4} \, s^{-1}$, at temperatures between 970 and 1180 K, and confining pressures of 150–300 MPa, indicate stresses increase up to natural strains of 0.1–0.2, followed by strain softening that occurs over a substantial period of strain. The total weakening was $\leq 25\%$, with larger values occurring at higher temperatures. Steady-state behaviour was not achieved in the compression tests even at natural strains as high as 0.9. Detailed measurements of grain size and grain size distribution indicated that grain size reductions begin at low strain (0.1) and are significant at strains of about 0.4.

Torsion testing may now be carried out on rocks at high temperature and pressure. Such tests allow deformation to natural strains of 5.0 and higher (Paterson & Olgaard 2000), and can produce microstructures similar to those observed in natural shear zones (Casey *et al.* 1998). Experimental details of the torsion test and conversion from torque and twist angle to stress and strain are given by Paterson & Olgaard (2000) and Mackwell & Paterson (2002). A recent suite of torsion tests were performed on Carrara marble at strain rates of $6 \times 10^{-5} - 3 \times 10^{-3} \, s^{-1}$, temperatures of between 750 and 1200 K, confining pressures of 300 MPa, up to strains as large as 50 (Pieri *et al.* 2001a, b; Barnhoorn *et al.* 2004). Plots of torque v. shear strain for these experiments indicate that the samples bear increasing torque up to natural strains of 1.0–2.0, followed by weakening up to strains of 5.0. Apparent steady state follows at very high strains. The peak torque occurs at smaller twists for higher temperatures; and the difference between the peak stress and the stress at very high strains is between 5 and 25%, i.e. similar in magnitude to the compression experiments, but occurring over a much larger strain increment.

Steady state or evolving state?

The mechanical-creep behaviour of calcite rock is often divided into three major regimes (Rutter 1974; Schmid *et al.* 1977, 1980; de Bresser & Spiers 1990, 1993; Walker *et al.* 1990). There are a number of ways to characterize the three different regimes, but one way is to calculate the apparent sensitivity of strain rate to stress, $n^* = \ln\dot{\varepsilon}/\ln\sigma$, or its inverse, $1/n^*$ (Fig. 3a). At low temperatures and high stress $1/n^*$ for calcite rocks is quite small (<0.1). At intermediate temperatures and stresses $1/n^*$

Table 3. *Constitutive equations for evolving state*

Constant law	State variable	Kinetic	State evolution equation	References
Hart	Hardness, σ^*	$\sigma = G\left(\dfrac{\dot\varepsilon}{a^*}\right)^{1/M} + \sigma^* \exp-\left(\dfrac{\dot\varepsilon^*}{\dot\varepsilon}\right)^\lambda$	$\dfrac{d\ln\sigma^*}{d\varepsilon} = \gamma^*(\sigma^*)\exp-\left(\dfrac{\dot\alpha^*}{\dot\varepsilon}\right)^\lambda$	1
Strain weakening 1	Strain, ε	$\dot\varepsilon = \mathscr{U}(\varepsilon)\left(\dfrac{\sigma}{\mu}\right)^n \exp\left(\dfrac{-Q_{plc}}{RT}\right)$	$\mathscr{U}(\varepsilon) = A_{peak}\left[W^{-n} + (1 - W^{-n})\exp\{-a(\varepsilon - \varepsilon_{crit})^r\}\right]$	2
Strain weakening 2	Strain, ε	$\sigma = k'\left(\dfrac{\varepsilon}{\varepsilon_0}\right)^r \dot\varepsilon^{1/n}$	$\varepsilon_0 \propto \dot\varepsilon \exp(H/RT)$	3
Stone/Hart	Subgrain size, δ	$\sigma = \int_0^{\delta_0}\left(\dfrac{\delta^q\dot\varepsilon}{B}\right)^{1/n} f(\delta)d\delta + \int_{\delta_0}^\infty \dfrac{A}{\delta^r} f(\delta)d\delta$	$\delta_0 = \left(\dfrac{BA^n}{\dot\varepsilon}\right)^{1/\Lambda}$	4
Diffusion	Grain size, d	$\dot\varepsilon = A_{diff}\dfrac{\sigma}{d^3}\exp-\left(\dfrac{Q_{diff}}{RT}\right)$	$\left.\dfrac{\partial d}{\partial t}\right|_{gg} = M\left(T, P, f_i, \phi, \ell_\phi, \dfrac{1}{d}, \cdots\right)\dfrac{\gamma}{d}$	5
Composite power law	Grain size, d	$\dot\varepsilon = \dot\varepsilon_{xs}V_{xs} + \dot\varepsilon_{diff}V_{diff}$	See Table 4	6
Peierls	Grain size (?), d	$\dot\varepsilon = A_{pei}\left(\dfrac{\sigma}{\mu}\right)^2 \exp-\dfrac{Q}{RT}\left(f\left(\sigma, \dfrac{1}{d}\right)\right)$	State variable not identified; If grain size, see above	7
Power-law creep porous solid	Mean stress, P; Triaxiality, x; Porosity, ϕ	$\dot\varepsilon_{axial}^{pore} = \dot\varepsilon(T, P, \sigma, \ldots)[1 + \phi f(x)]$	$\dot\phi = \phi\dot\varepsilon_0(T, S_{ij}, \bar\sigma, \dot\varepsilon_{ij}, \cdots)$	8

References: 1, Hart 1970, 1976; Korhonen *et al.* 1987; Covey-Crump 1998. 2, Barnhoorn *et al.* 2004. 3, Rutter 1999. 4, Stone 2001. 6, de Bresser *et al.* 2001. Montesi & Hirth 2003. 7, Renner *et al.* 2002. 8, Xias & Evans 2003.

Additional symbols not defined in text: W, ratio of strength at large natural strain to peak strength; ε_{crit}, critical strain to peak strength; f, process-dependent parameter in Hart formulation; Λ, exponent in Hart formulation; $f(\delta)$, distribution of sub-grain size (δ); $\Lambda = 1/(nr + q)$, (Stone formulation); M, grain boundary mobility; $\bar\sigma$, mean stress, S_{ij}, deviatoric stress tensor; γ, specific free energy of boundary; ϕ, porosity, l_ϕ, representative length scale of pore.

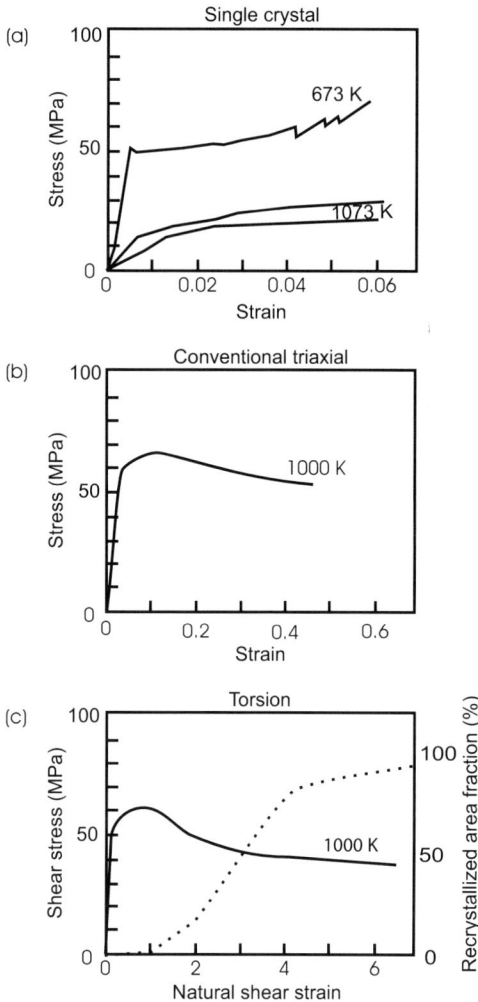

(a) Single crystal

(b) Conventional triaxial

(c) Torsion

Fig. 2. Schematic of mechanical behaviour of single crystals, polycrystalline aggregates in conventional triaxial configuration and polycrystalline aggregates in torsion. Notice that the scale along the strain axes increases by a factor of 10 between each graph. (**a**) Strain hardening curves for calcite single crystals compressed along [$22\bar{4}3$] at 673 and 1073 K, and a strain rate of $3 \times 10^{-5}\,\text{s}^{-1}$ (de Bresser 1991). (**b**) Carrara marble deformed in conventional triaxial apparatus at 1000 K to strains of 0.4 at a strain rate of about 3×10^{-6} s^{-1}. (Ter Heege *et al.* 2002). (**c**) Schematic of shear stress v. shear strain during torsion of Carrara marble at temperatures at about 1000 K and a strain rate of $3 \times 10^{-4}\,\text{s}^{-1}$. Dotted line is the areal fraction of recrystallized grains (Pieri *et al.* 2001a).

might be between 0.1 and 0.5; and at the lowest stresses and highest temperatures it might be nearly 1. A variety of mechanistic interpretations of these regimes have been given (see Table 1).

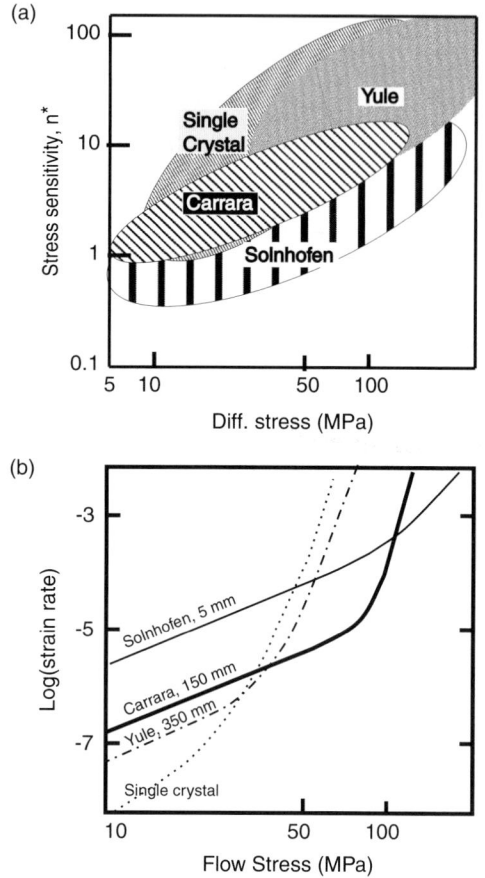

Fig. 3. (**a**) Schematic of apparent stress sensitivity of strain rate, n^*, for Solnhofen limestone, Carrara marble, Yule marble and single crystals. The values are calculated for a small differential stress interval and are plotted against the average stress value. Notice that n^* monotonically increases over the entire stress range. (**b**) Strain rate v. differential stress for Solnhofen, Carrara, Yule and single crystals at 973 K. At low stresses, finer grained materials creep faster than coarse-grained rocks. The relationship reverses as stress increases. See Renner & Evans (2002) for a discussion and references.

In the low-stress–high-temperature regime, where diffusion creep and grain-boundary sliding seem to be active (Schmid 1976; Schmid *et al.* 1977; Walker *et al.* 1990; Herwegh *et al.* 2003), the constitutive law is commonly given as:

$$\dot{\varepsilon}_{\text{diff}} = A_{\text{diff}}\frac{\sigma}{d^m}\exp\left(-\frac{Q_{\text{diff}}}{RT}\right) \qquad (4)$$

where $\dot{\varepsilon}_{\text{diff}}$ is the strain rate, A_{diff} may depend on the chemical fugacity of major or impurity components, σ is the differential stress, d is the grain size, m is an index of the sensitivity of strain rate to grain size and is 2 or 3 depending on the dominant diffusion path, Q_{diff} is the activation energy for lattice or grain-boundary diffusion, R is the gas constant and T is the absolute temperature.

In the intermediate regime, the most commonly used flow law is the power-law creep equation:

$$\dot{\varepsilon}_{plc} = A_{plc}\sigma^n \exp\left(-\frac{Q_{plc}}{RT}\right) \qquad (5)$$

where A_{plc} is a pre-exponential factor perhaps containing a dependence on chemical fugacity and temperature, n is often called the stress exponent and Q_{plc} is the activation enthalpy for creep (Kohlstedt et al. 1995).

Alternative descriptions for deformation in this regime also exist. For example, de Bresser and co-workers (de Bresser & Spiers 1993; de Bresser 2002; Ter Heege et al. 2002) have suggested that deformation might be limited by cross-slip, for which:

$$\dot{\varepsilon}_{xs} = A_{xs}\sigma^2 \exp\left[-\frac{Q_{xs}}{RT}(1 - c_{xs}\sigma)\right] \qquad (6)$$

where A_{xs}, Q_{xs} and c_{xs} are constants simplified from those given in Table 1.

At the highest stresses, the mechanism might change again (Frost & Ashby 1982; Carter & Tsenn 1987). Constitutive laws for this mechanism, often called power-law breakdown, vary in form, but are usually similar to equation (6) in that they contain an exponential term that is thermally activated, but is also a function of stress (see Table 1).

If diffusion flow (equation 4) or power-law creep (equation 5) is valid, then there will be regions in temperature and stress space where n and Q_{plc} are relatively constant. Equations (5) and (6), as written here, are steady-state descriptions, At higher temperatures, grain-boundary sliding accommodated by dislocation creep is sometimes inferred to occur. This composite mechanism is grain-size dependent. If grain size changed, both this latter mechanism and diffusion creep, equation (4), would become kinetic equations of a state-variable formulation, with grain size as the state variable.

Power-law creep descriptions alone do not adequately describe the conventional triaxial mechanical data for calcite rocks at any temperature (Schmid et al. 1980; de Bresser & Spiers 1990, 1993; Walker et al. 1990; Covey-Crump 2001; de Bresser 2002; Renner & Evans 2002). For example, if n^* is computed for small increments of stress at constant temperatures above 700 K, then n^* is found to vary with increasing stress from 1 to nearly 100 (Renner & Evans 2002) (Fig. 3a). The temperature sensitivity of log strain rate, i.e. $Q' = R\ln(\dot{\varepsilon}_1/\dot{\varepsilon}_2)/(1/T_2 - 1/T_1)|_{\sigma,d}$, also varies with stress (Fig. 3b). Subscripts 1, 2 denote two conditions with constant stress but differing temperatures; other symbols are as before. For temperatures below 700 K, where steady state does not occur within the strain range measured, Covey-Crump (2001) computed n^* at constant structure using the Hart formulation (see below) and concluded that the power-law parameters varied with strain and temperature. Finally, and most tellingly, the high strain torsion experiments indicate an evolution of structure and of strength up to shear strains as large as ≈ 30 at 1000 K (Pieri et al. 2001a, b; Rybacki et al. 2003; Barnhoorn et al. 2004).

Implicit internal state variables

Covey-Crump (1994a, b, 1998, 2001) was first to use an internal state variable theory to describe the creep of calcite. By using a testing technique involving interrupted constant strain-rate tests in a conventional triaxial compression machine, he applied Hart's (1976) formulation of mechanical equation of state. In its simplest form, Hart's kinetic equation expresses the idea that the deformation resistance of the aggregate can be represented by an array of mechanical processes arranged in either sequential order or parallel branches, as motivated by physical arguments about the processes. Covey-Crump (1994b) suggests that glide resistance might be represented by a viscous element (i.e. high dependence of strength on strain rate), while the passage of a dislocation through a dislocation tangle or other concentrated defect might be represented by a plastic element.

To discuss low-temperature (<700 K) creep in calcite, Covey-Crump (1994b) (see also Table 3) used a two-element configuration with plastic and viscous elements in parallel; thus, anelastic and elastic contributions are assumed to be negligible. For this case, the kinetic equation expresses creep resistance as a function of total strain rate ($\dot{\varepsilon}$), and an implicit internal state variable (σ^*) often called hardness:

$$\sigma = G\left(\frac{\dot{\varepsilon}}{\dot{a}^*}\right)^{1/M} + \sigma^* \exp-\left(\frac{\dot{\varepsilon}^*}{\dot{\varepsilon}}\right)^\lambda \qquad (7)$$

where G is the rigidity modulus, $\dot{a}^*$ is a rate parameter that depends on T and σ^*, M may depend on T, λ is a material constant and $\dot{\varepsilon}^*$ is a scaling parameter that depends on T and σ^* in a way that is reminiscent of power-law creep (equation 5):

$$\dot{\varepsilon}^* = f_0 \left(\frac{\sigma^*}{G}\right)^m \exp\left(-\frac{H}{RT}\right) \qquad (8)$$

where f_0 and m are material (process)-dependent parameters. The first term in the kinetic equation (7) has a larger dependence on strain rate and describes a non-linear friction element, while the second term is relatively independent of rate and describes a process that is plastic.

The evolution equation for the state variable can be written as:

$$\frac{d \ln \sigma^*}{d\varepsilon} = \gamma^*(\sigma^*) \exp -\left(\frac{\dot{\alpha}^*}{\dot{\varepsilon}}\right)^\Lambda \qquad (9)$$

where $\gamma^*(\sigma^*)$ and $\dot{\alpha}^*$ may be functions of σ^* and T, and are determined by experiment.

As temperature increases, recovery processes including dislocation cross-slip, dislocation climb, twin-boundary migration, grain-boundary migration and dynamic recrystallization increase in activity. To describe mechanical tests carried out at 673 K, Covey-Crump (1998) suggested that the first term of equation (7) should be neglected, and that the evolution equation for temperatures between 700 and 900 K should have an additional recovery term ($\mathfrak{R}$) that expresses the rate of change of σ^* due to processes that are nominally independent of deformation. A measure of recovery rate can be glimpsed by measuring static recrystallization after deformation (Covey-Crump 1994a).

The application of the Hart formulation to low- and intermediate-temperature regimes of the deformation of calcite rocks is notable in that it recognizes the importance of evolving state to rock strength. Although based at least in part on micromechanical grounds, this approach suffers from the fact that the identity of the implicit internal state variable is cloaked. The formulation also contains many material and process variables that may be fitted to the experimental data. Thus, although data fit is often quite good, confidence in extrapolation would be increased if the parameters were guided by micromechanical arguments.

Explicit internal state variables

What variable(s) are responsible for strain hardening and weakening? In the intermediate-temperature creep regime (above 700 K for Carrara marble), during the hardening portion of the curve, there seems to be a clear correlation between strength and initial grain size (Renner & Evans 2002; Renner et al. 2002): rocks with finer grain size are stronger. The sense of this correlation is exactly opposite to that observed in torsion after the peak strength has been reached. The onset of weakening is correlated with the appearance of dynamically recrystallized grains and the production of CPO in all configurations (Schmid et al. 1987; Pieri et al. 2001a, b; de Bresser 2002; Ter Heege et al. 2002; Barnhoorn et al. 2004). Interestingly, peak stresses were observed at much lower strains in triaxial compression than in torsion. Notice also that the size of the recrystallized grains agreed with a grain size piezometer for subgrain migration (Rutter 1995) in both torsion and conventional triaxial loading. The causes of the transition from hardening to weakening, and of the difference in the weakening behaviour for different testing configurations, although important, are not well explained.

From the point of view of the state-variable approach, the hardening and weakening in the calcite experiments may be due to relative changes in the influence of competing terms in the kinetic equation, for example between dislocation creep and grain-boundary sliding or diffusion creep (Schmid et al. 1987; Walker et al. 1990; Rutter 1995; Ter Heege et al. 2002). Alternatively, there may be changes in the evolution equation. For example, do the rates of recovery vary as peak stress is approached, perhaps due to boundary formation and migration during recrystallization? (See Rutter 1998; Ter Heege et al. 2002; Barnhoorn et al. 2004.) Or could the rate of production of CPO cause a geometric softening due to a change in the Taylor factor? (See Schmid et al. 1987; Pieri et al. 2001a; Ter Heege et al. 2002).

Although most of the discussion of experiments in dominantly monomineralic calcite rocks has been focused on grain size changes, other elements of the structure will also change during deformation of marble and limestone. Other important variables might include, for example, the dislocation structure (Stone 1991), CPO (Schmid et al. 1987), second-phase distribution (Rybacki et al. 2003) and porosity (Xiao & Evans 2003).

Dislocation structure. The hardness parameter, σ^*, used in the Hart equation of state represents an implicit, unspecified aspect of the dislocation microstructure. Recently, Stone (1991) and Stone et al. (2004) have proposed that the subgrain diameter (δ) can be used as an explicit descriptor of dislocation structure. Stone's approach assumes that dislocation glide and subgrain-boundary migration operate as independent, parallel processes. In this formulation, glide stress is assumed to be strain-rate independent (i.e. plastic), but inversely dependent on δ. The strain rate due to subgrain migration is assumed to be linearly viscous and inversely dependent on δ^2 or δ^3. If statistical variations in subgrain size distribution are incorporated into the kinetic equation, then a scaling parameter similar to $\dot{\varepsilon}^*$ arises from the theory (see also Table 3).

Using constant-stress creep and load-relaxation tests on single crystals of halite, Stone *et al.* (2004) were able to correlate the creep resistance in halite single crystals to the dislocation subgrain structure to demonstrate that the subgrain structures were similar in shape, and to show that strength curves could be scaled to constant structure (δ) using $\dot{\varepsilon}^*$. If the results can be applied to calcite, the state variable pertinent to the hardening part of stress–strain curves in the intermediate-temperature regime might be characteristic of the dislocation structure. If that variable correlated with grain size at small strains, but not at large strains (post-peak stress), then the experimental results could be rationalized. Perhaps subgrain migration is less important as dynamic recrystallization occurs, or perhaps the dislocation subgrain size during creep is altered by recovery due to recrystallization.

Grain size evolution during diffusion creep. In a fine-grained monomineralllic aggregate, deforming by diffusion creep, the grain size of the aggregate often increases during deformation. Then equation (4) may be cast as an evolving constitutive law. If grain growth during diffusion creep is analogous to that under static conditions, as was suggested for olivine (Karato 1989), then the evolution equation might be a boundary-mobility equation similar to:

$$\left.\frac{\partial d}{\partial t}\right|_{gg} = M_{GG}\frac{\gamma}{d} \qquad (10)$$

where γ is the specific grain boundary energy and $M_{GG} = M(T, P, f_i, \phi, \ell_\phi, 1/d, \ldots)$ is the mobility of the grain boundary. Grain-boundary mobility is a function of temperature, pressure, total porosity (ϕ), average pore size (ℓ_ϕ), solid–solution impurities (a_i) and grain-boundary curvature, and, hence, grain size (d) (for a review of static grain growth, see Evans *et al.* 2001). Assuming that mobility is a function of $1/d$ to some power, $p - 1$, the evolution

equation for grain growth during diffusion creep can be recast as:

$$\left.\frac{\partial d}{\partial t}\right|_{gg} = M(T, P, a_i, \phi, \ell_\phi)\frac{\gamma}{d^{p-1}}. \qquad (11)$$

Solid–solution impurities might also affect the major element defect concentrations or mobility. Thus, impurity concentration could act as a second internal variable in two ways, either changing the diffusion coefficient in the kinetic equation or by altering the grain-boundary mobility in the evolution equation. The exact effects can be complex; and diffusion rate and grain-boundary mobility can be affected in different ways. For example, trace amounts of manganese accelerate creep of synthetic marbles at low stresses (Freund *et al.* 2004) perhaps due to an impurity effect on diffusion rate (equation 4). In another study (Herwegh *et al.* 2003) larger amounts of magnesium did not seem to affect the diffusion-creep rate, provided that materials of equivalent grain size were considered. The dissolved magnesium did reduce grain growth rates, probably by reducing grain-boundary mobility in equation (11).

Grain size evolution. When a coarse-grained material is deformed by dislocation creep to sufficiently high strain, the grain size and its distribution evolves to a value that often can be correlated to the deformation conditions. Average recrystallized grain sizes produced during dislocation creep are often found to inversely correlate with the creep strength (Twiss 1977; Drury *et al.* 1985; Derby 1991; Rutter 1995; de Bresser *et al.* 1998):

$$d_{ssrx} = C_{pz}\sigma^{-q} \qquad (12)$$

where the exponent q is about 1. Because this piezometric relation can be used with observations of grain size to infer stresses during natural deformation (e.g. White 1979; Kennedy & White 2001) attention has been paid to determining whether C_{pz} and q depend on the recrystallization mechanism (Poirier & Guillopé 1979; Rutter 1995; Post & Tullis 1999), temperature, strain rate (de Bresser *et al.* 1998) or some other variables (Etheridge & Wilkie 1981).

When the grain size from the dislocation-creep piezometer is predicted to be so small that the strain rate due to diffusion creep would be larger than that owing to dislocation creep, then the average grain size and the distribution of grain sizes may increase back to values that would balance the dislocation and

Table 4. *Grain size evolution laws*

	Grain growth		Grain reduction	Reference
Field boundary	$\dfrac{\partial d}{\partial t}\bigg|_{gg} = M\dfrac{\gamma}{d^{p-1}}$		$\dot{d} = -\dfrac{\dot{\varepsilon}_{disl}d}{\varepsilon_c}$	1
Continuous Recrystallization	$\dfrac{\partial d}{\partial t}\bigg|_{gg} = M\dfrac{\gamma}{d^{p-1}} - \dfrac{\dot{\varepsilon}_{disl}d}{\varepsilon_c}$			2
Total Rate Relaxation	$\dot{d} = -\dfrac{\dot{\varepsilon}_{total}(d_{ssrx} - d)}{\varepsilon_c},$	where $d_{ssrx} = C_{pz}\sigma^{-q}$		3
Dislocation Rate Relaxation	$\dot{d} = -\dfrac{\dot{\varepsilon}_{disl}(d_{ssrx} - d)}{\varepsilon_c},$	where $d_{ssrx} = C_{pz}\sigma^{-q}$		4

For a discussion of these evolution equations, see Montesi & Hirth (2003). References are as follows. 1. de Bresser *et al.* 1998, 2001. 2, Hall & Parmentier 2003. 3, Kameyama *et al.* 1997; Braun *et al.* 1999. 4, Montesi & Zuber 2002.

diffusion-creep rates at a critical ratio (de Bresser *et al.* 2001). Under this hypothesis, the grain size would approach a steady-state value given by a relation similar to equation (12), but where C_{pz} would be temperature-dependent, and $q = (n - 1)/m$, where m and n are given in equations (4) and (5).

The grain size evolution equations can be written in several ways, depending on the assumptions regarding the steady-state value and the mechanisms of reduction or grain growth (for a summary see Table 4) (see Montesi & Hirth 2003). If the growth process is separate and exclusive of the refinement process, then the evolution equation for grain growth might be similar to equation (11). One possibility for rate of grain size refinement would set the rate $\dot{d}$ as proportional to the product of the dislocation strain rate, $\dot{\varepsilon}_{disl}$, and the difference between a steady-state grain size and the current grain size, i.e. $(d_{ssrx} - d)$, and inversely proportional to a critical strain, ε_c:

$$\dot{d} = -\frac{\dot{\varepsilon}_{disl}(d_{ssrx} - d)}{\varepsilon_c}.$$

Other variations have been posited: for a discussion see Kameyama *et al.* (1997), Braun *et al.* (1999), Montesi & Zuber (2002), Hall & Parmentier (2003) and Montesi & Hirth (2003).

Grain size evolution and mixed mechanisms. Typically a range of grain sizes exists in a polycrystalline aggregate, and at high temperatures and low strain rates it is possible that both dislocation processes and diffusion creep occur (i.e. under conditions where strain rates predicted by equations (4) and (5) are nearly equal) (Ter Heege *et al.* 2002). Assume the grain size distribution of the solid is divided into discrete classes of grain size (d_i) with total volume fraction of each class (v_i). If the two

mechanisms are independent, then within each class, the strain rate ($\dot{\varepsilon}^i$) will be the sum of equations (4) and (5):

$$\dot{\varepsilon}^i = \dot{\varepsilon}^i_{disl} + \dot{\varepsilon}^i_{diff}$$

where $\dot{\varepsilon}^i_{disl}$ and $\dot{\varepsilon}^i_{diff}$ are the strain rates for each class due to dislocation and diffusion creep, respectively. Of course, the relative contribution of each process will depend on the value of d_i. For the entire aggregate, the strain rate might be approximated by the volume average of the strain rates within each grain size class, which would be:

$$\dot{\varepsilon}^{agg} = \sum_i v_i\dot{\varepsilon}^i \qquad (13)$$

assuming constant stress on each grain. A similar volume average of stress can be computed assuming that strain rate is constant in each grain.

Crystal lattice preferred orientation. Given the potentially large anisotropy of creep strength in minerals, it likely that the strength of a textured calcite polycrystal could vary in strength from that of one with random CPO (Dawson 2002; Wenk 2002). From the point of view of the formalism used here, some important aspect of the CPO is adopted as a state variable: e.g. a probability distribution function that describes the lattice orientation of all the crystals. The evolution equation would then be a prediction of the rate of change of the distribution function. At each instant, the kinetic equation would involve a calculation of the total creep rate of the aggregate given the thermodynamic conditions and the current probability distribution function. The calculation involves a volume average of the mechanical behaviour of each crystal over the entire ensemble. Bounds are obtained by assuming that each crystal experiences the same strain rate (Taylor approximation) or the same stress (Sachs approximation). Improved estimates can be obtained by self-consistent average schemes (Wenk *et al.* 1997) or small-scale finite-element calculations (Lebensohn *et al.* 1998; Dawson 2002). All techniques depend on having detailed information on plastic flow strength for single crystals and the way it evolves during deformation. The calculations become particularly challenging when deformation is highly heterogeneous, when recrystallization and grain-boundary sliding occur (Wenk *et al.* 1997), when hardening rates are also anisotropic, or when second phases with different strength are present, e.g. most rocks (Wenk 2002).

Wenk (2002) reviewed the difficulties in matching the observations of texture production in the high strain experiments (Pieri *et al.* 2001*b*) with predictions using polycrystal plasticity theory (Lebensohn *et al.* 1998). The CPO production seen in the experiments can be predicted if slip on $\{10\bar{1}4\}\langle11\bar{2}0\rangle$ is the dominant slip system. However, neither single crystal tests nor transmission electron microscopy (TEM) observations seem to support this contention (de Bresser & Barber pers. comm. cited in Wenk 2002). Similarly, in the converse problem, the relative importance of CPO production (Barnhoorn *et al.* 2004) and recrystallization (Ter Heege *et al.* 2002) on the creep strength of marble also seems to be in question.

Conclusions and suggestions

Some general conclusions emerge from considering the previous experiments and analyses. A couple of suggestions for further progress are also given.

- Many field observations, laboratory experiments and mechanical analyses show that evolution of structure is common during natural deformation of rocks.
- The state-variable approaches are notable in recognizing that constitutive equations must incorporate evolving structure and are capable of fitting detailed experimental data over ranges of stress and temperature space. If strain is eschewed as a state variable, it is conveniently represented as a history-dependent integral of the deformation rates. As with all constitutive relations, the kinetic and evolution equations must be designed to include all processes that occur under the actual conditions of deformation.
- Implicit state-variable formulations are capable of describing detailed mechanical

behaviour over limited ranges of (σ, P_{lith}, P_{fluid}, T, μ_i) space. Because of their semi-empirical nature, large extrapolations in conditions may reduce the confidence that all processes have been correctly identified or described. To improve the confidence of the extrapolation from laboratory conditions to conditions in the Earth, it is desirable 'in my humble, but nevertheless correct opinion' (M. Batzle, numerous pers. comms, 1978–2002) to formulate the state variables as explicit internal state variables, if at all possible. They should be incorporated within the kinetic and evolution equations by expressions that describe micromechanical processes which can be justified to operate under natural conditions. Such justification could come from microstructural observations of naturally deformed rocks, from rate arguments or, preferably, from both.

- In general, a large number of explicit internal state variables can be imagined for natural processes. These variables included in the description of low-temperature creep of metals are probably pertinent in many situations: grain size, dislocation structure and solid dispersions. But, because of the complexity of natural rocks, a broader range of parameters will need accounting, including, for example, individual phase composition: average grain size, grain shape or distributions, fugacity of fluids, activity of chemical components, and volume or mass proportion for each phase. Although not discussed here, it is likely that changes in mineralogy and petrology are capable of producing profound changes in rock strength.
- If the kinetic and evolution equations are formulated with guidance from specific physical, chemical and mechanistic grounds, and if observable explicit state variables are developed, it is possible, at least in principle, to use current and relict microstructures to estimate rock strength at certain points in the deformation history.
- For the specific case of creep deformation of calcite rocks at moderate pressures without fluids present:
  - both conventional triaxial tests and large strain torsion tests indicate that mechanical strength evolves with structure during deformation, reaching a peak stress after a period of work hardening, and then work weakening over another, longer period. However, the details of the strength evolution are different between the two types of test. In both cases, the

peak stress is correlated with the appearance of newly recrystallized grains;

- in the dislocation regime, flow laws for steady-state power-law creep, or for steady-state creep controlled by cross-slip, do not accurately represent the mechanical data for calcite rocks over any temperature or stress range for which data are currently available;
- the exact identity of the pertinent state variable(s) for the intermediate temperature creep regime still remain(s) in doubt. In the work-hardening portion of deformation at fixed temperature and stress, increasing strain rate correlates with increasing grain size. In the work-softening portion of the curve, strength decreases as the proportion of fine, recrystallized grains increases. It is possible that the pertinent state variable has not been identified, that multiple state variables are involved or that two competing processes operate;
- a composite flow law incorporating diffusion flow and steady-state, recovery-controlled, dislocation mechanisms can describe the transition between the linearly viscous creep and non-linear viscous creep at low stresses, high temperatures and fine grain sizes, but, as mentioned above, steady-state descriptions of deformation controlled by cross-slip or power-law creep fail at higher stresses;
- current explicit state variable descriptions do not include the effect of crystal lattice preferred orientation. Some suggestions for evolution of grain size have been made, but, at any rates, they have yet to be validated by careful experimentation. Other important parameters that need more examination are the effect of second phases with different strength, the effect of porosity and the influence of metamorphic reactions.

M. Herwegh, J. Renner and X. Xiao contributed insight, observations and data that helped inform my view. D. Kohlstedt, M. Paterson and E. Rybacki provided rapid and constructive reviews. The editor, D. Bruhn, showed remarkable patience. Funding was provided by NSF EAR 0125669.

References

ARGON, A.S. 1975. Physical basis of constitutive equations for inelastic deformation. *In*: ARGON, A.S. (ed.) *Constitutive Equations in Plasticity*. MIT Press, Cambridge, MA, 1–22.

ASHBY, M.F. 1972. A first report on deformation mechanism maps. *Acta Metallurgica*, **29**, 293–299.

ASHBY, M.F. & VERALL, R.A. 1973. Diffusion accomodated flow and superplasticity. *Acta Metallurgica*, **21**, 149–163.

BARNHOORN, A., BYSTRICKY, J., BURLINI, L. & KUNZE, K. 2004. The role of recrystallisation on the deformation behaviour of calcite rocks: large strain torsion experiments on Carrara marble. *Journal of Structural Geology*, **26**, 885–904.

BAUD, P., SCHUBNEL, A. & WONG, T.-F. 2000. Dilatancy, compaction, and failure mode in Solnhofen limestone. *Journal of Geophysical Research – Solid Earth*, **105**, 19 289–19 303.

BESTMANN, M., KUNZE, K. & MATTHEWS, A. 2000. Evolution of a calcite marble shear zone complex on Thassos Island, Greece: microstructural and textural fabrics and their kinematic significance. *Journal of Structural Geology*, **22**, 1789–1807.

BRAUN, J., CHERY, J., POLIAKOV, A., MAINPRICE, D., VAUCHEZ, A., TOMASSI, A. & DAIGNIERES, M. 1999. A simple parameterization of strain localization in the ductile regime due to grain-size reduction: a case study for olivine. *Journal of Geophysical Research – Solid Earth*, **104**, 25 167–25 181.

BRODIE, K.H. & RUTTER, E.H. 1987. The role of transiently fine-grained reaction products in syntectonic metamorphism; natural and experimental examples. *Canadian Journal of Earth Sciences*, **24**, 556–564.

BURG, J.P. & LAURENT, P. 1978. Strain analysis of a shear zone in a granodiorite. *Tectonophysics*, **47**, 15–42.

BYSTRICKY, M., KUNZE, K., BURLINI, L. & BURG, J.P. 2000. High shear strain of olivine aggregates: Rheological and seismic consequences. *Science*, **290**, 1564–1567.

CARTER, N. & TSENN, M.C. 1987. Flow properties of continental lithosphere. *Tectonophysics*, **136**, 27–63.

CASEY, M., KUNZE, K. & OLGAARD, D.L. 1998. Texture of Solnhofen limestone deformed to high strains in torsion. *Journal of Structural Geology*, **20**, 255–267.

COVEY-CRUMP, S.J. 1994a. The high temperature static recovery and recrystallization behaviour of cold-worked Carrara marble. *Journal of Structural Geology*, **19**, 225–241.

COVEY-CRUMP, S.J. 1994b. The application of Hart's state variable description of inelastic deformation to Carrara marble at $T < 450°C$. *Journal of Geophysical Research – Solid Earth*, **99**, 19 793–19 808.

COVEY-CRUMP, S.J. 1998. Evolution of mechanical state in Carrara Marble during deformation at 400 °C to 700 °C. *Journal of Geophysical Research – Solid Earth*, **103**, 29 781–29 794.

COVEY-CRUMP, S.J. 2001. Variation of the exponential and power law creep parameters with strain for Carrara Marble deformed at 120 degrees to 400 degrees C. *Geophysical Research Letters*, **28**, 2301–2304.

DAWSON, P.R. 2002. Modeling deformation of polycrystalline rocks. *In*: KARATO, S. & WENK, H.-R. (eds) *Plastic Deformation of Minerals and Rocks. Reviews in Mineralogy and Geochemistry*, **51**, 331–352.

DE BRESSER, J.H.P. 1991. Intracrystalline deformation of calcite. *Geologica Ultraiectina*, **79**, 191.

DE BRESSER, J.H.P. 2002. On the mechanism of dislocation creep of calcite at high temperature: Inferences from experimentally measured pressure sensitivity and strain rate sensitivity of flow stress. *Journal of Geophysical Research – Solid Earth*, **107**, 2337, doi: 10.1029/B122002JB001812.

DE BRESSER, J.H.P. & SPIERS, C.J. 1990. High-temperature deformation of calcite single crystals by r^+ and f^+ slip. *In*: KNIPE, R.J. & RUTTER, E.H. (eds) *Deformation Mechanisms, Rheology and Tectonics*. Geological Society, London, Special Publications, **54**, 285–298.

DE BRESSER, J.H.P. & SPIERS, C.J. 1993. Slip systems in calcite single crystals deformed at 300–800 °C. *Journal of Geophysical Research – Solid Earth*, **98**, 6397–6409.

DE BRESSER, J.H.P. & SPIERS, C.J. 1997. Strength characteristics of the r, f, and c slip systems in calcite. *Tectonophysics*, **272**, 1–23.

DE BRESSER, J.H.P., PEACH, C.J., REIJS, J.P.J. & SPIERS, C.J. 1998. On dynamic recrystallization during solid state flow: Effects of stress and temperature. *Geophysical Research Letters*, **25**, 3457–3460.

DE BRESSER, J.H.P., TER HEEGE, J.H. & SPIERS, C.J. 2001. Grain size reduction by dynamic recrystallization: can it result in major theological weakening? *International Journal of Earth Sciences*, **90**, 28–45.

DERBY, B. 1991. The dependence of grain-size on stress during dynamic recrystallization. *Acta Metallurgica et Materialia*, **39**, 955–962.

DIETRICH, J.H. 1978. Modeling of rock friction, 1, experimental results and constitutive equations. *Journal of Geophysical Research*, **84**, 2161–2168.

DRURY, M.R., HUMPHREYS, F.J. & WHITE, S. 1985. Large strain deformation studies using polycrystalline magnesium as a rock analogue. II. Dynamic recrystallization mechanism at high temperatures. *Physics of Earth and Planetary Interiors*, **40**, 208–222.

DRURY, M.R., HOOGERDUIJN STRATING, E.H., VISSERS, R.L.M. & VAN WAMEL, W.A. 1990. Shear zone structures and microstructures in mantle peridotites from the Voltri Massif, Ligurian Alps, N.W. Italy. *Geologie en Mijnbouw*, **69**, 3–17.

DRURY, M.R., VISSERS, R.L.M., VAN DER WAL, D. & HOOGERDUIJN STRATING, E.H. 1991. Shear localisation in upper mantle peridotites. *Pure and Applied Geophysics*, **137**, 439–460.

ETHERIDGE, M.A. & WILKIE, J.C. 1981. An assessment of dynamically recrystallized grain-size as a paleopiezometer in quartz-bearing mylonite zones. *Tectonophysics*, **78**, 475–508.

EVANS, B. & KOHLSTEDT, D.L. 1995. Rheology of rocks. *In*: AHRENS, T.J. (eds) *Rock Physics and Phase Relations, Handbook of Physical Properties of Rocks*. American Geophysical Union, Washington, DC, 148–165.

EVANS, B., FREDRICH, J. & WONG, T.-F. 1990. The brittle-ductile transition in rocks: Recent experimental and theoretical progress. *In*: DUBA, A., DURHAM, W.B., HANDIN, J.W. & WANG, H. (eds) *The Brittle–Ductile Transition in Rocks, The Heard Volume*. American Geophysical Union, Geophysical Monograph, **56**, 1–21.

EVANS, B., RENNER, J. & HIRTH, G. 2001. A few remarks on the kinetics of static grain growth in rocks. *International Journal of Earth Sciences*, **90**, 88–103.

FISCHER, G.J. & PATERSON, M.S. 1989. Dilatancy during rock deformation at high temperatures and pressures. *Journal of Geophysical Research – Solid Earth and Planets*, **94**, 17 607–17 617.

FREDRICH, J.T., EVANS, B. & WONG, T.-F. 1989 Micromechanics of the brittle to plastic transition in Carrara Marble. *Journal of Geophysical Research – Solid Earth and Planets*, **94**, 4129–4145.

FREUND, D., WANG, Z.-C., RYBACKI, E. & DRESEN, G. 2004. High-temperature creep of synthetic calcite aggregates: influence of Mn-content. *Earth and Planetary Science Letters*, **226**, 433–448.

FROST, H.J. & ASHBY, M.F. 1982. *Deformation–Mechanism Maps: The Plasticity and Creep of Metals and Ceramics*. Pergamon, Oxford.

GIBBS, J.W. 1878. On the equilibrium of heterogeneous substances. (Abstract.) *American Journal of Science, Series 3*, **16**, 441–458.

HALL, C.E. & PARMENTIER, E.M. 2003. Influence of grain size evolution on convective instability. *Geochemistry Geophysics Geosystems*, **4**, 10.1029/2002GC000308.

HANDY, M.R. 1994. Flow laws for rocks containing two non-linear viscous phases: a phenomenological approach. *Journal of Structural Geology*, **16**, 287–301.

HART, E.W. 1970. A phenomenological theory for plastic deformation of polycrystalline metals. *Acta Metallurgica*, **18**, 599–610.

HART, E.W. 1976. Constitutive relations for the nonelastic deformation of metals. *Journal of Engineering Material Technology*, **98**, 193–202.

HEARD, H.C. 1960. Transitions from brittle to ductile flow in Solnhofen limestone as a function of temperature, confining pressure, and interstitial fluid pressure. *In*: GRIGGS, D. & HANDIN, J. (eds) *Rock Deformation (A Symposium)*. Geological Society of America Memoirs, **79**, 133–192.

HEARD, H.C. & RALEIGH, C.B. 1972 *Steady-state flow in marble at 500 degrees to 800 degrees C*. Geological Society of America Bulletin, **83**, 935–956.

HERWEGH, M. & KUNZE, K. 2002. The influence of nano-scale second-phase particles on deformation of fine-grained calcite mylonites. *Journal Structural Geology*, **24**, 1463–1478.

HERWEGH, M., XIAO, X.H. & EVANS, B. 2003. The effect of dissolved magnesium on diffusion creep in calcite. *Earth and Planetary Science Letters*, **212**, 457–470.

HIRTH, G. 2002. Laboratory constraints on the rheology of the upper mantle. *In*: KARATO, S.-I. & WENK, H.-R. (eds) *Plastic Deformation of Minerals and Rocks. Reviews in Mineralogy and Geochemistry*, **51**, 97–120.

HIRTH, G. & TULLIS, J. 1992. Dislocation creep regimes in quartz aggregates. *Journal of Structural Geology*, **14**, 145–159.

HOBBS, B.E., MUHLHAUS, H.-B. & ORD, A. 1990. Instability, softening, and localization of deformation. *In*: KNIPE, R.J. & RUTTER, E.H. (eds) *Deformation, Mechanisms, Rheology and Tectonics*. Geological Society, London, Special Publications, **54**, 143–173.

HODGE, P.G., JR. 1985. William Prager: personal recollections. *In*: BAZANT, Z. (ed.) *Mechanics of Geomaterials*. Wiley, London, 13–18.

KAMEYAMA, M., YUEN, D.A. & FUJIMOTO, H. 1997. The interaction of viscous heating with grain-size dependent rheology in the formation of localized slip zones. *Geophysical Research Letters*, **24**, 2523–2526.

KARATO, S. 1989. Grain-growth kinetics in olivine aggregates. *Tectonophysics*, **168**, 255–273.

KENKMANN, T. & DRESEN, G. 1998. Stress gradients around porphyroclasts palaeopiezometric estimates and numerical modelling. *Journal of Structural Geology*, **20**, 163–173.

KENNEDY, L.A. & WHITE, J.C. 2001. Low-temperature recrystallization in calcite; mechanisms and consequences. *Geology*, **29**, 1027–1030.

KOHLSTEDT, D.L., EVANS, B. & MACKWELL, S.M. 1995. Strength of the lithosphere: Constraints imposed by laboratory experiments. *Journal of Geophysical Research – Solid Earth*, **100**, 17 587–17 603.

KORHONEN, M.A., HANNULA, S.-P. & LI, C.-Y. 1987. State variable theories based on Hart's formulation. *In*: MILLER, A.K. (eds) *Unified Constitutive Equations for Creep and Plasticity*. Elsevier Applied Science, London, 89–137.

LEBENSOHN, R.A., WENK, H.-R. & TOME, C.N. 1998. Modelling deformation and recrystallization textures in calcite. *Acta Materialia*, **46**, 2683–2693.

LERNER, I. & KOHLSTEDT, D.L. 1981. Effect of gamma radiation on plastic flow of NaCl. *Journal of the American Ceramic Society*, **64**, 105–108.

LISTER, G.S. & WILLIAMS, P.F. 1979. Fabric development in shear zones: theoretical controls and observed phenomena. *Journal of Structural Geology*, **1**, 283–297.

MACKWELL, S.J. & PATERSON, M.S. 2002. New developments in deformation studies: High-strain deformation. *In*: KARATO, S. & WENK, H.-R. (eds) *Plastic Deformation of Minerals and Rocks. Reviews in Mineralogy and Geochemistry*, **51**, 1–19.

MARONE, C.J. 1998. Laboratory-derived friction laws and their application to seismic faulting. *Annual Reviews of Earth and Planetary Sciences*, **26**, 643–696.

MEANS, W.D. 1984. Shear zones of type I and II and their significance for reconstruction of rock history. *Geological Society of America, Abstracts with Programs*, **16**, 50.

MONTESI, L. & ZUBER, M. 2002. A unified description of localization for application to large-scale tectonics. *Journal of Geophysical Research – Solid Earth*, **107**, 2045, doi: 10.1029/2001JB000465.

MONTESI, L.G.J. & HIRTH, G. 2003. Grain size evolution and the rheology of ductile shear zones: from laboratory experiments to postseismic creep. *Earth and Planetary Science Letters*, **211**, 97–110.

NICOLAS, A. & BOUCHEZ, J.L. 1977. Geological aspects of deformation in continental shear zones. *Tectonophysics* 42, 55–73.

NICOLAS, A., BOUCHEZ, J.L., BOUDIER, F. & MERCIER, J.C. 1971. Textures, structures and fabrics due to solid state flow in some European lherzolites. *Tectonophysics*, **12**, 55–86.

NICOLAS, A., BOUDIER, F. & BOULLIER, A.M. 1973. Mechanisms of flow in naturally and experimentally deformed peridotites. *American Journal of Science*, **273**, 853–876.

PATERSON, M.S. 1978. *Experimental Rock Deformation – The Brittle Field*. Springer, New York.

PATERSON, M.S. 1987. Problems in the extrapolation of laboratory rheological data. *Tectonophysics*, **133**, 33–43.

PATERSON, M.S. 2001. Relating experimental and geological rheology. *International Journal of Earth Science (Geologisches Rundschau)*, **90**, 157–167.

PATERSON, M.S. & OLGAARD, D.L. 2000. Rock deformation tests to large shear strains in torsion. *Journal of Structural Geology*, **22**, 1341–1358.

PIERI, M., BURLINI, L., KUNZE, K., STRETTON, I. & OLGAARD, D.L. 2001a. Rheological and microstructural evolution of Carrara Marble with high shear strain; results from high temperature torsion experiments. *Journal of Structural Geology*, **23**, 1393–1413.

PIERI, M., KUNZE, K., BURLINI, L., STRETTON, I., OLGAARD, D.L., BURG, J.-P. & WENK, H.-R. 2001b. Texture development of calcite by deformation and dynamic recrystallization at 1000 K during torsion experiments of marble to large strains. *Tectonophysics*, **330**, 119–140.

POIRIER, J.P. 1980. Shear localization and shear instability in materials in the ductile field. *Journal of Structural Geology*, **2**, 135–142.

POIRIER, J.P. 1985. *Creep of Crystals: High-temperature Deformation Processes in Metals, Ceramics, and Minerals*. Cambridge University Press, Cambridge.

POIRIER, J.P. & GUILLOPÉ, M. 1979. Deformation induced recrystallization of minerals. *Bulletin de la Société minéralogique de France*, **102**, 67–74.

POST, A. & TULLIS, J. 1999. A recrystallized grain size piezometer for experimentally deformed feldspar aggregates. *Tectonophysics*, **303**, 159–173.

RAMSAY, J.G. 1980. Shear zone geometry: A review. *Journal of Structural Geology*, **2**, 83–89.

RAMSAY, J.G. & HUBER, M.I. 1983. *The Techniques of Modern Structural Geology, Volume 1: Strain Analysis.* Academic Press, London.

RANALLI, G. 1987. *Rheology of the Earth.* Allen & Unwin, Boston, MA.

RENNER, J. & EVANS, B. 2002. Do calcite rocks obey the power law creep equation? *In:* DE MEER, S., DRURY, M.R., DE BRESSER, J.H.P. & PENNOCK, G. (eds) *Deformation Mechanisms, Rheology, and Tectonics: Current Status and Future Perspectives.* Geological Society, London, Special Publications, **200**, 309–329.

RENNER, R., EVANS, B. & SIDDIQI, G. 2002. Dislocation creep of calcite. *Journal of Geophysical Research – Solid Earth*, **107**, 2364, doi: 10.1029/2001JB001680.

RICE, J.R. & RUINA, A.L. 1983. Stability of steady frictional slipping. *Journal of Applied Mechanics, Transactions of the ASME*, **50**, 343–349.

RUDNICKI, J.W. & RICE, J.R. 1975. Conditions for the localization of deformation in pressure sensitive dilatant materials. *Journal of the Mechanics and Physics of Solids*, **23**, 271–394.

RUINA, A.L. 1985. Constitutive relations for frictional slip. *In:* BAZANT, Z.P. (eds) *Mechanics of Geomaterials: Rocks, Concrete, Soils.* Wiley, London, 169–188.

RUTTER, E.H. 1974. The influence of temperature, strain rate and interstitial water in the experimental deformation of calcite rocks. *Tectonophysics*, **22**, 311–334.

RUTTER, E.H. 1995. Experimental study of the influence of stress, temperature, and strain on the dynamic recrystallization of Carrara Marble. *Journal of Geophysical Research – Solid Earth*, **100**, 24 651–24 663.

RUTTER, E.H. 1998. Use of extension testing to investigate the influence of finite strain on the rheological behaviour of marble. *Journal of Structural Geology*, **20**, 243–254.

RUTTER, E.H. 1999. On the relationship between the formation of shear zones and the form of the flow law for rocks undergoing dynamic recrystallization. *Tectonophysics*, **303**, 147–158.

RUTTER, E.H., PEACH, C.J., WHITE, S.H., JOHNSON, D. & BROWN, M. 1982. On the inter-relationships between deformation and metamorphism, an experimental study of the syntectonic hydration of basalt. *Journal of the Geological Society, London*, **139**, 366.

RUTTER, E.H., CASEY, M. & BURLINI, L. 1994. Preferred crystallographic orientation development during the plastic and superplastic flow of calcite rocks. *Journal of Structural Geology*, **16**, 1431–1446.

RUTTER, E.H., BRODIE, K.H. & BOYLE, A.P. 1995. Mechanistic interactions between deformation and metamorphism. *Geological Journal*, **30**, 227–240.

RUTTER, E.H., HOLDSWORTH, R.E. & KNIPE, R.J. 2001. The nature and significance of fault-zone weakening; an introduction. *In:* HOLDSWORTH, R.E., STRACHAN, R.A., MAGLOUGHLIN, J.F. & KNIPE, R.J. (eds) *The Nature and Tectonic Significance of Fault Zone Weakening.* Geological Society, London, Special Publications, **186**, 1–11.

RYBACKI, E., PATERSON, M.S., WIRTH, R. & DRESEN, G. 2003. Rheology of calcite–quartz aggregates deformed to large strain in torsion. *Journal of Geophysical Research – Solid Earth*, **108**, 2089, doi: 10.1029/2001JB0006942174.

SCHMID, S.M. 1976. Rheological evidence for changes in the deformation mechanism of Solnhofen limestone towards low stress. *Tectonophysics*, **31**, T21–T28.

SCHMID, S.M., BOLAND, J.M. & PATERSON, M.S. 1977. Superplastic flow in fine grained limestone. *Tectonophysics*, **43**, 257–291.

SCHMID, S.M., PATERSON, M.S. & BOLAND, J.N. 1980. High temperature flow and dynamic recrystallization in Carrara Marble. *Tectonophysics*, **65**, 245–280

SCHMID, S.M., PANOZZO, R. & BAUER, S. 1987. Simple shear experiments on calcite rocks: Rheology and microfabric. *Journal of Structural Geology*, **9**, 747–778.

SCHMOCKER, M., BYSTRICKY, M., KUNZE, K., BURLINI, L., STUNITZ, H. & BURG, J.P. 2003. Granular flow and Riedel band formation in water-rich quartz aggregates experimentally deformed in torsion. *Journal of Geophysical Research – Solid Earth*, **108**, 2242, doi: 10.1029/2002JB001958.

SELVERSTONE, J. & HYATT, J. 2003. Chemical and physical responses to deformation in micaceous quartzites from the Tauern Window, Eastern Alps. *Journal of Metamorphic Geology*, **21**, 335–345.

SELVERSTONE, J., MORTEANI, G. & STAUDE, J.M. 1991. Fluid channelling during ductile shearing; transformation of granodiorite into aluminous schist in the Tauern Window, Eastern Alps. *Journal of Metamorphic Geology*, **9**, 419–431.

STONE, D.S. 1991. Scaling laws in dislocation creep. *Acta Metallurgica et Materialia*, **39**, 599–608.

STONE, D.S., PLOOKPHOL, T. & COOPER, R.F. 2004. Simlarity and scaling in creep and load relaxation of single crystal halite (NaCl). *Journal of Geophysical Research – Solid Earth*, **109**, 10.1029/2004JB003064.

STOUFFER, D.D. & DAME, L.T. 1996. *Inelastic Deformation of Metals: Models, Mechanical Properties and Metallurgy.* Wiley, New York.

STUNITZ, H. & TULLIS, J. 2001. Weakening and strain localization produced by syn-deformational reaction of plagioclase. *International Journal of Earth Science (Geologisches Rundschau)*, **90**, 136–148.

TER HEEGE, J.H., DE BRESSER, J.H.P. & SPIERS, C.J. 2002. The influence of dynamic recrystallization on the grain size distribution and rheological behaviour of Carrara marble deformed in axial compression. *In:* DE MEER, S., DRURY, M., DE BRESSER, J.H.P. & PENNOCK, G.M. (eds) *Deformation Mechanisms, Rheology, and Tectonics: Current Status and Future Perspectives.* Geological Society, London, Special Publications, **200**, 331–353.

TULLIS, J. 2002. Deformation of granitic rocks: Experimental studies and natural examples. *In*: KARATO, S. & WENK, H.-R. (eds) *Plastic Deformation of Minerals and Rocks. Reviews in Mineralogy and Geochemistry*, **51**, 53–95.

TULLIS, J. & YUND, R.A. 1985. Dynamic recrystallization of feldspar; a mechanism for ductile shear zone formation. *Geology*, **13**, 238–241.

TULLIS, J. & YUND, R.A. 1992. The brittle–ductile transition in feldspar aggregates: An experimental study. *In*: EVANS, B. & WONG, T.-F. (eds) *Fault Mechanics and Transport Properties of Rocks*. Academic Press, San Diego, CA, 89–117.

TULLIS, J., DELL'ANGELO, L. & YUND, R.A. 1990. Ductile shear zones from brittle precursors in feldspathic rocks; the role of dynamic recrystallization. *In*: DUBA, A.G., DURHAM, W.B., HANDIN, J.W. & WANG, H.F. (eds) *The Brittle–Ductile Transition in Rocks: The Heard Volume*. American Geophysical Union, Geophysical Monograph, **56**, 67–82.

TULLIS, T.E. 1986. Friction and faulting, Editor's note. *Pure and Applied Geophysics*, **124**, 375–382.

TULLIS, T.E. & TULLIS, J. 1986. Experimental rock deformation techniques. *In*: HOBBS, B.E. & HEARD, H.C. (eds) *Mineral and Rock Deformation; Laboratory Studies; the Paterson Volume*. American Geophysical Union, Geophysical Monograph **36**, 297–324.

TULLIS, T.E. & WEEKS, J.D. 1986. Constitutive behavior and stability of frictional sliding of granite. *Pure and Applied Geophysics*, **124**, 383–414.

TWISS, R.J. 1977. Theory and applicability of a recrystallized grain size paleopiezometer. *Pure and Applied Geophysics*, **115**, 227–244.

VISSERS, R.L.M., DRURY, M.R., NEWMAN, J. & FLIERVOET, T.F. 1997. Mylonitic deformation in upper mantle peridotites of the North Pyrenean Zone (France); implications for strength and strain localization in the lithosphere. *Tectonophysics*, **279**, 303–325.

WALKER, A.N., RUTTER, E.H. & BRODIE, K.H. 1990. Experimental study of grain-size sensitive flow of synthetic, hot-pressed calcite rocks. *In*: KNIPE, R.J. & RUTTER, E.H. (eds) *Deformation Mechanisms, Rheology and Tectonics*. Geological Society, London, Special Publications, **54**, 259–284.

WENK, H.R. 2002. Texture and anisotropy. *In*: KARATO, S. & WENK, H.-R. (eds) *Plastic Deformation of Minerals and Rocks. Reviews in Mineralogy and Geochemistry*, **51**, 291–329.

WENK, H.-R., CANOVA, G., BRECHET, Y. & FLANDIN, L. 1997. A deformation-based model for recrystallization of anisotropic materials. *Acta Materialia*, **45**, 3283–3296.

WHITE, S. 1979. Difficulties associated with paleostress estimates. *Bulletin de la Société minéralogique de France*, **102**, 210–215.

WHITE, S.H., BRETAN, P.G. & RUTTER, E.H. 1986. Fault-zone reactivation – kinematics and mechanisms. *Philosophical Transactions of the Royal Society of London*, **A317**, 81–97.

WONG, T.-F., DAVID, C. & ZHU, W. 1997. The transition from brittle faulting to cataclastic flow in porous sandstones: Mechanical deformation. *Journal of Geophysical Research – Solid Earth*, **102**, 3009–3026.

XIAO, X.H. & EVANS, B. 2003. Shear-enhanced compaction during non-linear viscous creep of porous calcite-quartz aggregates. *Earth and Planetary Science Letters*, **216**, 725–740.

ZHANG, S.Q., KARATO, S., GERALD, J.F., FAUL, U.H. & ZHOU, Y. 2000. Simple shear deformation of olivine aggregates. *Tectonophysics*, **316**, 133–152.

Initiation of basement thrust detachments by fault-zone reaction weakening

CHRISTOPHER A. J. WIBBERLEY

Géosciences Azur, Université de Nice Sophia-Antipolis, 250 rue A. Einstein, 06560 Valbonne, France (e-mail: wibbs@geoazur.unice.fr)

Abstract: This paper examines how crystalline basement thrust sheets can detach in foreland thrust belts, in terms of the deformation mechanisms and rheological evolution of the detachment fault zones. Basement thrust fault zones of the Moine Thrust Belt and the external Western Alps show relatively narrow thrust zones considering the large displacements accommodated. Microscopic examination of fault rocks from these high strain thrust zones show that syntectonic alteration of fractured feldspars to white mica of strong preferred orientation generated ultramylonites deforming by diffusion creep and other viscous deformation mechanisms, similar to documented basement thrust zones in North America. Motivated by these observations coupled with other published examinations of foreland basement thrust zones, and recent developments in crustal hydrology, a conceptual model is proposed to explain basement detachment formation and evolution. Meteoric fluid that percolated into a previously fractured upper crust is drawn into developing fault zones by dilatancy pumping during the early stages of thrust-related deformation. The generation of cataclastic fault rocks with fresh fracture surfaces by microfracturing enhances the rate of fluid–rock interaction. Syntectonic alteration causes a deformation-mechanism transition to phyllosilicate-dominated ductile fault-rock rheologies, resulting in a large ductility contrast between host rock and fault zone that inhibits growth of the zone into the wall rock and weakens the thrust. Deformation becomes focused into these weakened early thrust zones so that they become zones of high strain, preventing the development of other newer fault zones elsewhere. This model explains the detachment and continued sliding of basement thrust sheets on narrow mica-rich zones of high strain in foreland thrust belts, and suggests that reaction weakening of the basement is important in decreasing the strength of the foreland crust in orogenic wedge evolution.

Foreland thrust belts are characterized by thrust faults that typically form as series of bedding-parallel flats or décollements linked by ramps, forming a staircase geometry dipping down towards the hinterland (Boyer & Elliott 1982). These décollements detach the thrust sheets at their bases from the underlying footwall. The formation of detachments in this way is relatively easy where a weak stratigraphic layer is present (e.g. gypsum or anhydrite) that can be exploited by a fault as it propagates. However, many foreland thrust belts such as the Moine Thrust Belt (NW Scotland) and the external Western Alps contain crystalline basement thrust sheets emplaced along a series of flats and ramps (e.g. Elliott & Johnson 1980), yet the initiation of the underlying detachment in the basement (such as the one shown in Fig. 1a) cannot have exploited similar low-angle stratigraphic weaknesses (Hatcher & Williams 1986). Deformation in such 'thin-skinned' cases is typically localized close to the thrust surfaces with little penetrative strain within the basement (Erslev & Rogers 1993). In other cases basement is involved in shortening throughout the crust (thick-skinned behaviour) with deformation being accommodated by horizontal flattening (e.g. Deremond 1979), or shear on steep reverse ductile shear zones (Stahl 1992). The initiation of low-angle detachments in the basement can therefore dictate the entire behaviour of the crust, yet little is known about how they initiate, beyond that they often occur around the 'brittle–ductile transition' in the crust (Armstrong & Dick 1974).

Basement deformation mechanisms typical of generally low-pressure and temperature grades have been documented in foreland thrust belts, such as brittle fracture and cataclasis (e.g. Wojtal & Mitra 1986; Evans 1990*a*, 1993). Associated with these deformation processes are the syntectonic alteration of feldspars to clays (e.g. Evans 1993) or to micas (Yonkee & Mitra

From: BRUHN, D. & BURLINI, L. (eds) 2005. *High-Strain Zones: Structure and Physical Properties.*
Geological Society, London, Special Publications, **245**, 347–372.
0305-8719/05/$15.00 © The Geological Society of London 2005.

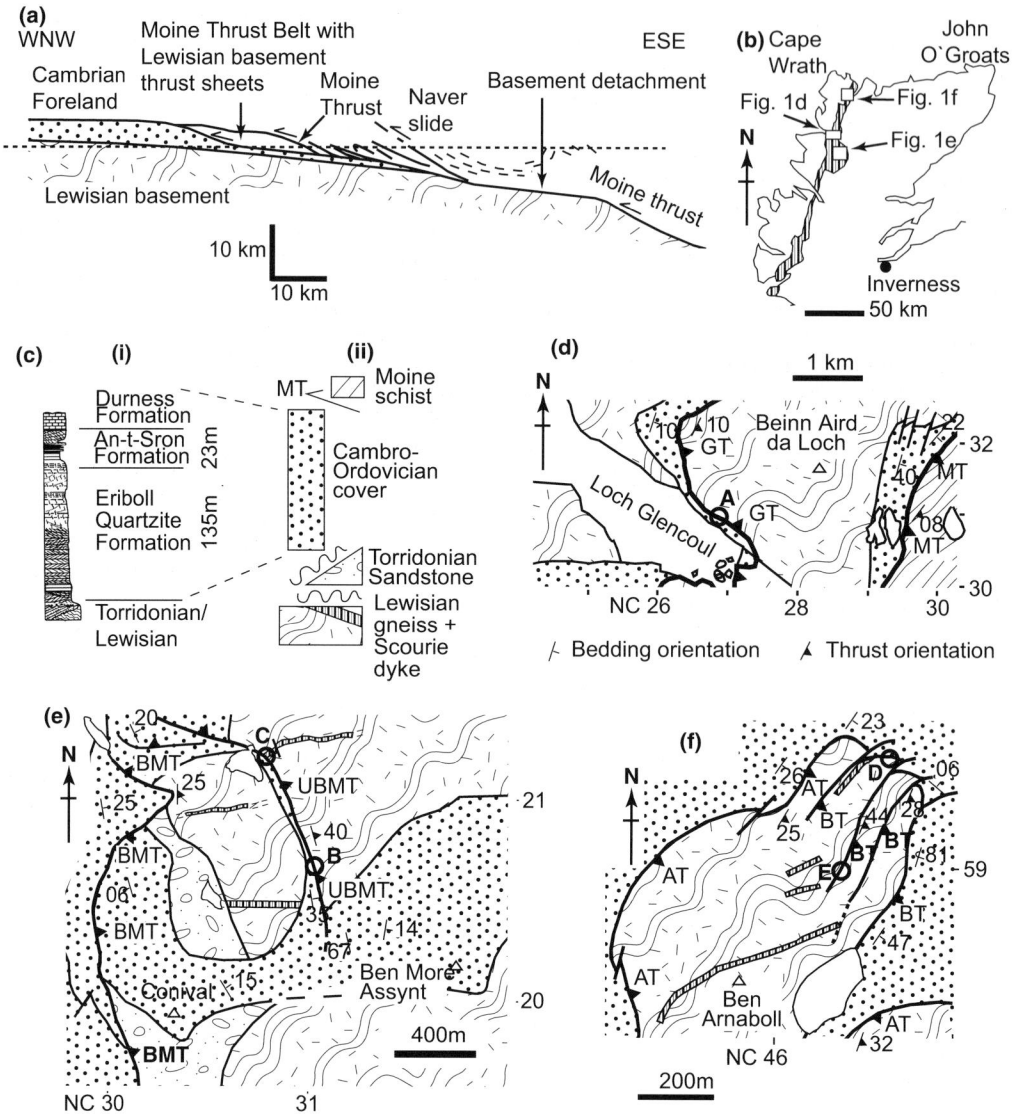

Fig. 1. Geological setting of the study areas in the Moine Thrust Belt. (**a**) Schematic cross-section through northern Scotland. Note the footwall flat in the basement Lewisian. Modified from Coward (1988). (**b**) Locality map of north Scotland, and the Moine Thrust Belt (shaded region). (**c**) Summary of foreland stratigraphy involved in the Moine Thrust Belt. (i) Stratigraphic column from McKie (1989) with thicknesses measured by the author in the Assynt district. (ii) Sketch column and key showing the structural relationships of the units as used in this paper. (**d**) Map of the Glencoul area, based on mapping by Peach et al. (1907) and Wibberley (1995). (**e**) Map of the Ben More Assynt area, from Wibberley (1995). (**f**) Map of the basement structure around Ben Arnaboll, based on Coward (1984) and Wibberley (1995). The Arnaboll Thrust (AT) is cut by several breaching thrusts (BT).

1993) during fluid–rock interaction during thrusting. Previous work on basement thrust-zone evolution suggests that a transition can occur from feldspar- and quartz-dominated rheologies to phyllosilicate-dominated rheologies due to fluid-assisted alteration of the feldspars to mica and other fine-grained reaction products, notably quartz (but in some cases also epidote and/or calcite) (e.g. White et al. 1982; Evans 1990a; Mitra 1992). In many cases the phyllosilicate forms a fine-grained matrix with strong grain shape and crystallographic preferred orientations that, along with the production of new strain-free quartz grains, promotes dislocation and diffusion-creep deformation mechanisms ('reaction-enhanced ductility'; White & Knipe 1978). For shallow- to mid-crustal fault zones that initially suffered cataclastic deformation of quartz and feldspar, this could cause a shallowing of the frictional–viscous transition zone in the crust and a switch to a viscous phyllosilicate-dominated rheology, resulting in significant weakening (Janecke & Evans 1988; Schmid & Handy 1991; Wintsch et al. 1995; Imber et al. 2001). Detailed microstructural observations suggest that the trigger for such 'reaction weakening' is the intra-granular fracturing of the feldspars (Janecke & Evans 1988; Wibberley & McCaig 2000; Gueydan et al. 2003), which serves to locally increase the permeability to allow fluid access to the reactants, and increases the reactant surface area and therefore the rate of reaction. For deeper crustal fault zones, syntectonic alteration of feldspars would therefore not be expected to occur at high rates in shear zones at higher temperatures than the brittle–crystal-plastic transition for feldspar. Nevertheless, the production of fine-grained reaction products and products of dynamic recrystallization may induce weakening by promoting diffusion creep, particularly in polymineralic aggregates where diffusion is enhanced along non-like grain boundaries (Wheeler 1992; Stünitz & FitzGerald 1993; Bruhn et al. 1999). Hence, a range of different weakening mechanisms may operate at different levels in the crust, depending on the initial deformation mechanisms, grain size and mineralogy of the fault rock.

The reactions that can occur in fault zones depend heavily on the temperature, fluid pressure and composition, with mid-greenschist conditions and a weakly acidic (e.g. meteoric) fluid generally favouring alteration of feldspars to white mica (e.g. Hemley & Jones 1964; Wintsch et al. 1995). At shallower conditions, clays are typical reaction products of feldspar breakdown in basement fault zones (e.g. Evans 1993). At higher metamorphic grades, felspathization of mica or albitization of orthoclase may occur as evidenced in higher temperature greenschist fault rocks (e.g. Wintsch & Knipe 1985; McCaig 1987), so that reaction hardening may also occur. In some circumstances, the direction of a reaction may change through time and vary locally within the shear zone, causing local partitioning of strain that oscillates with the reaction (Knipe & Wintsch 1985). Initiation of a weak detachment in the

basement is therefore very dependent on depth, temperature (and therefore geothermal gradient), and fluid composition and supply. Even after initiation, the burial and exhumation history of the thrust during its activity will control both the geochemical reactions and the deformation mechanisms occurring, and the interplay between them. This potentially complex history through which basement detachment zones may have passed hinders our understanding of the initiation and operation of basement detachments during foreland thrust belt evolution.

This paper investigates the mechanism of basement detachment by studying several examples of basement thrust fault zones in the Moine Thrust Belt and the external Western Alps, and also by drawing on previously published examples from North America. The results of field and microscopic examinations of the thrust fault rocks are presented, and previously presented geochemical analyses are drawn on to link microstructural and chemical evolutions. An important characteristic that all these faults have in common is a white mica-rich (usually muscovite) ultramylonitic matrix, despite the lack of phyllosilicates in the protoliths, and these observations motivate a focus on syntectonic reactions affecting the deformation mechanisms and consequent basement fault-zone rheology. The role that fluid–rock interaction may play in fault-zone growth and fault-network development in foreland crystalline basement is discussed in the light of this, from which a general conceptual model is proposed for the initiation of low-angle basement detachment zones in the foreland during orogenesis.

Examples of basement thrusting in the Moine Thrust Belt

The Moine Thrust Belt (Fig. 1) is a narrow belt of Caledonian deformation between the Moine Series to the ESE and the Lewisian basement and Cambro-Ordovician foreland succession to the WNW (Elliott & Johnson 1980; McClay & Coward 1981). The Moine Thrust Belt was generated during the collision of Baltica with Laurentia by WNW-directed imbricate thrusting of the Laurentian foreland succession in the footwall of the Moine Thrust (Soper & Hutton 1984). Apart from the Moine Thrust itself, the large-displacement thrusts in the northern Moine Thrust Belt are the Glencoul and Ben More thrusts in the Assynt region (Fig. 1d & e), and the Arnaboll Thrust in the Eriboll region (Fig. 1f). In all of these cases Lewisian basement is involved in the thrusting and balanced cross-sections suggest that a detachment level to the foreland thrust belt was generated within the Lewisian basement (Elliott & Johnson 1980). The Lewisian basement is a gneissic rock with a cm-scale banding defined by segregation into leucocratic and mafic layers containing quartz, orthoclase, plagioclase, pyroxenes and amphiboles. The long and complicated metamorphic and deformation history generating and affecting the Lewisian gneiss gave rise to a very heterogeneous crystalline basement, with both structural and compositional features becoming potentially important factors in controlling Caledonian basement deformation. Amongst the most important of these structural heterogeneities are 1–50 m-wide WNW-striking amphibolite Scourie dykes. However, the latest pre-Caledonian deformation to affect the Lewisian basement in the region is thought to be crustal extension as evidenced by Torridonian syndepositional normal faults of late Precambrian age (Soper & England 1995). This section presents data from three different areas in the Moine Thrust Belt (Fig. 1b). First, observations from optical microscope work on fault rocks of the Glencoul Thrust are presented as an example of a large-displacement basement foreland thrust (order of displacement of tens of km). Although published work on the microstructural evolution of the sedimentary cover rocks has constrained the metamorphic conditions of thrust-sheet emplacement (Knipe 1990), no detailed work of basement Glencoul Thrust fault rocks has previously been presented. Secondly, work from a previously undescribed fault, named by Wibberley (1995) as the Upper Ben More Thrust, is presented as an illustration of a smaller displacement basement thrust (displacement approximately 100–200 m) operating at a similar metamorphic grade. The Upper Ben More Thrust has anisotropic gneiss and isotropic mafic dyke rocks exposed on its hanging-wall side, allowing for an examination of the effect of anisotropy on strain-localization processes during the evolution of the fault. Thirdly, two examples of basement thrust faulting are presented from the Arnaboll Thrust sheet further north in the Moine Thrust Belt: the Arnaboll Thrust, although previously described (White et al. 1982) provides an interesting comparison of deformation at slightly higher temperature–slower strain rates; and one of a series of later breaching thrusts provides an example of shallower grade deformation on a small fault (displacement approximately 10 m) whose structure has not been destroyed by further movement.

The Glencoul Thrust

The Glencoul Thrust emplaced Lewisian basement on top of Cambro-Ordovician stratigraphy with an estimated displacement of 19–35 km (Ramsay 1969; Coward et al. 1980; Elliott & Johnson 1980). Conditions under which this deformation occurred are considered to be lower greenschist facies (250–350 °C, 5–9 km depth, e.g. Knipe 1990), resulting from burial by structurally higher units such as the Moine Series. Most of the juxtaposition history of the hanging-wall basement was on a footwall flat of Eriboll Quartzites, although initially the basement moved along a detachment and up a ramp within the basement (Elliott & Johnson 1980) (Fig. 1a).

On the NE side of Loch Glencoul (locality A, Fig. 1d), the Glencoul Thrust is defined by a 2 m-wide zone of phyllosilicate-rich mylonite at the base of the overthrusted Lewisian gneiss, in direct contact with Ordovician Durness Limestone in the footwall. Cross-sections suggest that the hanging wall at this locality moved several hundred metres on a footwall flat in the Durness Limestone during the latest part of its emplacement history. Three metres above the thrust contact the original Lewisian gneissic foliation is visible, and the gneiss shows little evidence of deformation associated with the thrusting. Thin section examination shows that the Lewisian gneissic microstructure is intact, although the feldspar grains have undergone widespread internal patchy alteration to very-fine-grained mica (Fig. 2a). Fifty centimeters above the thrust contact, the quartz underwent severe fracturing and cataclastic grain size reduction (Fig. 2b). The feldspars (orthoclase and plagioclase $An_{25}-An_{30}$) have been extensively replaced by fine-grained mica, which has a mesh-like microfabric with little preferred orientation, controlled by the arrangement of the original dilatant microfracture network (Fig. 2c).

A thin section across the thrust contact (Fig. 2d) shows white mica-rich ultramylonite in the hanging wall with a striking lack of any fractured quartz or feldspar. Instead, it has a very strong fabric defined by the shape preferred orientation (SPO) of fine-grained, secondary white mica. The absence of any visible feldspar in thin section suggests that syntectonic replacement of feldspar by mica and possibly other phases was extremely important. Other new phases present are highly elongate domains of fine-grained quartz aligned parallel to the fabric and thrust contact, and calcite with the white mica in the fine-grained matrix. At the thrust

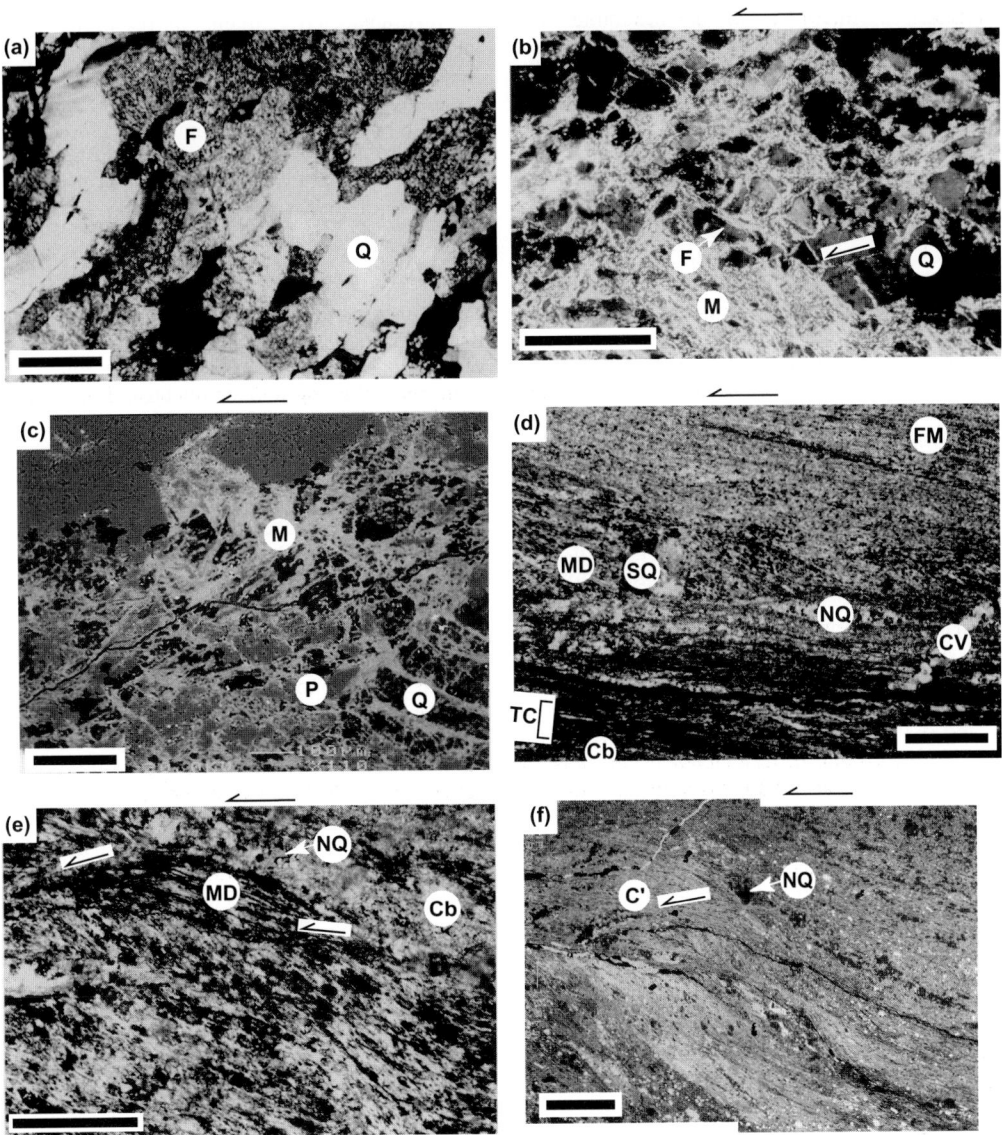

Fig. 2. Photomicrographs (optical unless stated) of Lewisian basement samples from the Glencoul thrust zone at locality A (see Fig. 1d). All scale bars represent 0.25 mm. All sections are vertical, orientated WNW (left)–ESE (right). For (b)–(f), the sense of shear on the Glencoul Thrust is indicated by the top-to-left arrows, with smaller shear arrows indicating individual minor shear bands. (**a**) Lewisian gneiss, 3 m above the thrust contact, with unfractured quartz ('Q'), and undeformed but altered plagioclase feldspar ('F'). (**b**) Deformed Lewisian gneiss, 50 cm above the thrust contact, showing fractured quartz ('Q'), and a mesh of fine-grained white mica ('M') that replaced most of the feldspar ('F'). (**c**) Backscattered scanning electron micrograph of the deformed Lewisian gneiss shown in (b), showing details of the mesh fabric of new muscovite ('M') in a cataclastic matrix of quartz ('Q') and relict plagioclase grains ('P'). (**d**) Ultramylonite from the thrust contact with the thrust contact zone ('TC'), above which the ultramylonite, with its ultrafine matrix ('FM'), is derived from the Lewisian basement in the hanging wall, and below which the ultrafine carbonate material ('Cb') is derived from Durness Limestone footwall. Note the domains of mica ('MD') either side of a square-shaped domain of new calcite and quartz ('SQ'), and the domain of newly crystallized quartz ('NQ'), cross-cut by a late carbonate vein ('CV'). (**e**) Magnification of part of the basement-derived ultramylonitic fabric from the sample shown in (d), showing how the fine-grained zones are made up of domains of mica ('MD'), carbonate ('Cb') and newly crystallized quartz ('NQ'). (**f**) Backscattered scanning electron micrograph of the ultramylonite at the thrust contact showing late C'-slip planes ('C'') in the aligned mica domains with new quartz ('NQ') precipitated in dilational sites.

contact, basement-derived ultramylonite and footwall limestone-derived carbonate–ultramylonite define a narrow zone approximately 0.25 mm thick (Fig. 2d). Calcite exists in the fine-grained ultramylonitic matrix above this contact, as well as in the carbonate–mylonite below, but only close (1–2 mm) to the contact suggesting its presence is due to local diffusion of carbonate from the footwall limestone during the latest stages of emplacement. Figure 2d also shows domains of aligned, fine-grained white mica growth arranged symmetrically around a square-shaped carbonate and quartz domain (possibly formed by replacement of a cubic non-silicate grain). This is interpreted to represent the product of aligned growth of new mica grains in a strain shadow around a rigid cubic grain (e.g. Passchier & Trouw 1998), followed by replacement of the cubic mineral by calcite and quartz. The interpretation that the aligned white mica formed by growth of new grains during shear deformation is significant because it suggests that the alignment fabric of strong mica preferred orientation was formed directly from the production of mica by metamorphic reaction in the presence of a fluid during deformation, and not by the mechanical rotation of the mica grains during shear after the transformation (although the latter cannot be ruled out as an additional process). Deformation was apparently further localized into microshears that deflect the fabric of the secondary mica both in C-bands, approximately parallel to the thrust contact, and in low angle C'-bands (Fig. 2e). These shear bands show concentrations of opaque material suggesting that pressure solution accompanied the localization of deformation. The presence of quartz in dilational sites related to slip on these microshears (Fig. 2f) suggests that the growth of new grains continued during the localization of shear strain, and such slip localization was similarly evidenced by microstructural examination of

phyllosilicate-rich domains in mylonites from the structurally higher Moine Thrust (Knipe & Wintsch 1985). Fracturing and carbonate vein formation also occurred in the mylonite and footwall limestone, cross-cutting the aligned mica fabric and being therefore late (although they could be episodic with other deformation mechanisms in the later stages of fault movement).

The Upper Ben More Thrust

The Ben More thrust sheet lies structurally above the Glencoul thrust sheet in the imbricate sequence at Assynt (Bailey 1935; Elliott & Johnson 1980; Kim 1981). Close to the Ben More Thrust, a second, structurally higher thrust with a displacement of approximately 100–200 m has been identified and named the Upper Ben More Thrust (Wibberley 1995) (see also Fig. 1e). The outcrop fabrics and microstructures of the Upper Ben More Thrust are important in this study for two reasons: (1) to provide an example of fault-zone structure and textural–mineralogical evolution of a relatively low-displacement basement thrust as a comparison to the Glencoul Thrust, operative under similar conditions; and (2) by comparing parts of the thrust affecting Lewisian gneiss with those affecting isotropic amphibolite dykes, the effect of basement anisotropy on strain localization in the thrust zones is assessed.

The Lewisian gneiss adjacent to the Upper Ben More Thrust in the hanging wall has developed a finely spaced fabric, and become deformed and altered just above the thrust contact. These fabrics, and the thrust surface, strike parallel to the gneissic foliation of the Lewisian basement in this area, but dip less steeply to the east. However, the dip of the gneissic foliation in the hanging wall decreases on approaching the thrust surface, suggesting that some rotation of the gneissic foliation occurred during thrusting. Coarse-grained amphibolite

---

Fig. 3. Optical photomicrographs and outcrop sketches comparing the fault rocks at the Upper Ben More Thrust from both Lewisian gneiss and amphibolite Scourie dyke rock. In (a) and (b), the thrust orientation and sense of shear are marked by the top-to-left arrows, and the thin section orientations marked by the orientated arrows. Scale bars represent 0.2 mm. In (c) and (d), the thrust sense is top-towards-viewer. (**a**) Mica-rich cataclasite from the Upper Ben More thrust zone, taken 60 cm above the thrust contact at locality B (see Fig. 1e). Fragments of slightly altered plagioclase ('P'), and lenses of Lewisian quartz ('Q'), are isolated by a through-going network of anastomosing fine-grained secondary white mica zones ('AM'). (**b**) Ultramylonite derived from the amphibolite Scourie dyke, taken from 10 cm above the thrust contact at locality C (see Fig. 1e), showing a compositional banding of secondary chlorite ('C') and mica ('M'). (**c**) Outcrop sketches showing the variation in deformation fabrics moving towards the Upper Ben More Thrust from the hanging wall Lewisian gneiss, at locality B, with distance from the thrust contact for each sketch. Note that the schistose fabric in the mafic bands parallels the Lewisian gneissic foliation, which gets progressively destroyed closer to the thrust contact. (**d**) Outcrop sketches of deformation fabrics in the amphibolite Scourie dyke above the Upper Ben More Thrust at locality C.

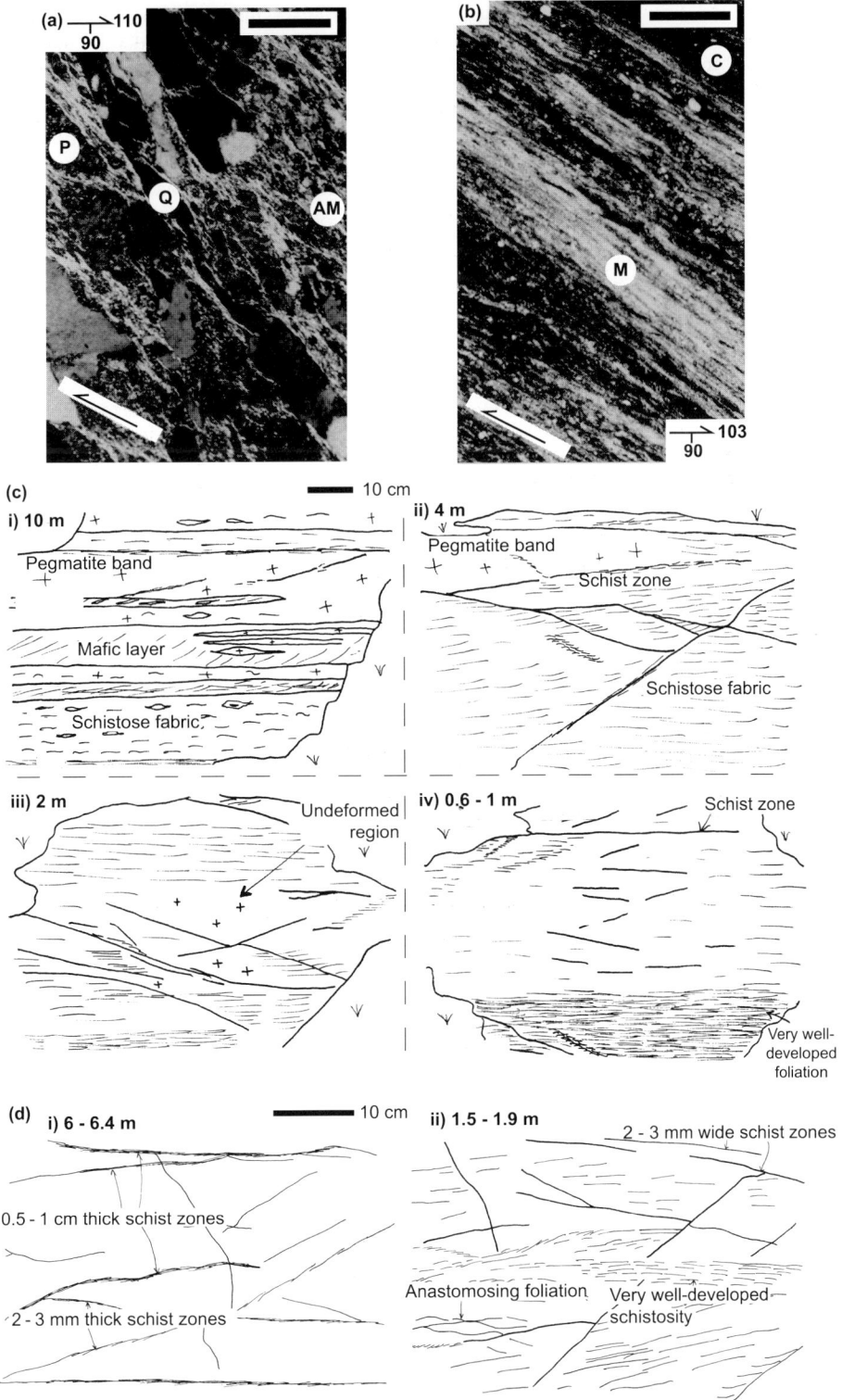

Scourie dykes, ranging in thickness from tens of centimetres to metres, cut the Lewisian gneiss in this area. Away from the thrust zones they are isotropic, having a granular texture. Such an amphibolite dyke is truncated by the Upper Ben More Thrust on its hanging-wall side (locality C on Fig. 1e). In order to address point (1) above, thrust fault rocks were examined from the two basement lithologies. In both cases, significant replacement of coarse feldspar and/or amphibole by aligned fine-grained phyllosilicates of strong preferred orientation is evident. In the case of cataclastic thrust fault rocks derived from Lewisian gneiss 60 cm above the thrust contact (locality B on Fig. 1e), the plagioclase is heavily altered to aligned fine-grained mica of which the strong preferred orientation defines the foliation (Fig. 3a). Quartz grains have been fractured, yet in places retain their elongate Lewisian form in a direction parallel to the mica fabric. This supports field evidence that the mica preferred orientation fabric parallels the earlier gneissic foliation which it has overprinted. Some newly/recrystallized quartz with grain sizes of 0.02–0.1 mm is visible in the sample. Thrust fault rocks derived from the amphibolite Scourie dyke at locality C (Fig. 1e) are ultramylonites composed of bands of fine-grained white mica and chlorite alternating with bands of quartz (e.g. Fig. 3b). No relicts of original feldspar or amphibole grains were observed. The compositional banding and strong preferred orientation of phyllosilicate grains within each band define the foliation.

The degree of development of foliation in the hanging wall with varying proximity to the thrust contact was studied by detailed mapping and 'logging' of fabrics in order to address point (2) above. In the Lewisian gneiss 10 m away from the thrust at locality B, localized anastomosing bands of schistosity a few millimetres thick are parallel or at low angles to the original gneissic foliation (Fig. 3ci). These become better connected and more frequent closer to the thrust (Fig. 3cii) until the schistosity comprises the bulk of the rock, anastomosing around pods of undeformed material (Fig. 3ciii). A finely spaced schistosity also present parallel to the gneissic foliation is concentrated in the mafic bands, suggesting preferential parallel overprinting of the gneissic foliation, 'pseudoreactivation', in these mafic bands. From the thrust contact up to 60 cm away, the basement is very well foliated and altered to form the thrust-fault zone (of which the edge of this zone was sampled in Fig. 3a). Fabrics in the isotropic amphibolite Scourie dyke (Fig. 3d) are present in a narrower zone adjacent to the thrust contact. More than 4 m from the thrust (Fig. 3di) the amphibolite dyke contains widely spaced schist bands that decrease in density further away from the thrust (Fig. 4). These schist bands become much more closely spaced in a denser anastomosing network approaching the thrust (e.g. Fig. 3dii) until approximately 1.5 m away a thrust-parallel mylonitic foliation starts to dominate. This mylonitic foliation increases in intensity towards the thrust until it defines an ultramylonite zone 30 cm thick immediately above the thrust contact. Structural logging data of foliation parting frequencies for the gneiss and the dyke (Fig. 4) show that deformation was more

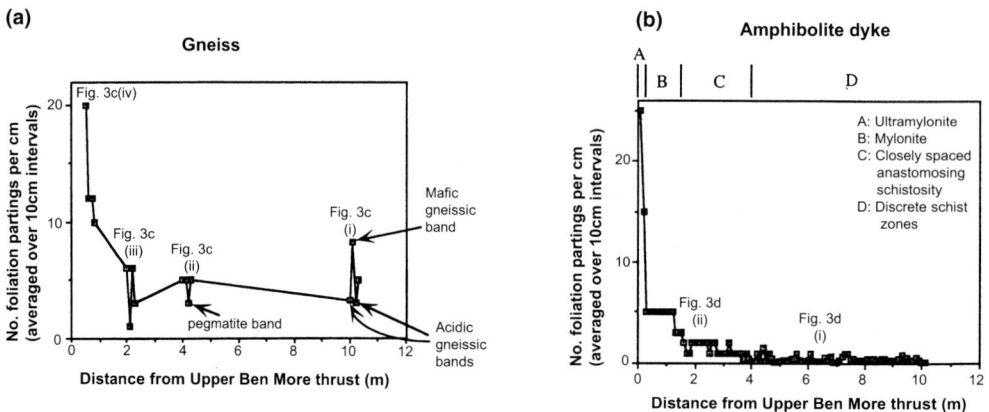

Fig. 4. Variation in fabric intensity in the hanging wall away from the Upper Ben More Thrust, for (**a**) Lewisian gneiss and (**b**) amphibolite Scourie dyke.

localized close to the thrust in the isotropic amphibolite dyke than in the anisotropic gneiss. Given that the gneissic fabric had been rotated parallel to the thrust surface, this suggests that this pre-existing anisotropy promoted a wider distribution of the deformation, and contributed to the increased deformability of the wall rock relative to the isotropic dyke.

The Arnaboll thrust sheet

The Arnaboll thrust sheet is the most northerly basement thrust sheet of the Moine Thrust Belt exposed on the mainland of NW Scotland. The Moine Thrust Belt is much narrower here (around 2 km) than at Assynt, and the basement thrust sheets occur only in the structurally highest parts, such as the Arnaboll thrust sheet (Coward 1984). They overlie thrust sheets composed of only Cambro-Ordovician cover rocks, suggesting that the region displays an overall shallowing in origin of the thrust sheets towards the foreland. This relates to observations by Butler (1982) that the sequence of thrusts several kilometres to the south shows a transition from dominantly ductile fabrics in the structurally higher thrusts to dominantly cataclastic textures in the lower thrusts. The Arnaboll thrust sheet lies on a mylonite zone that is 0.5 m thick where pegmatitic material in the hanging wall is in contact with the thrust zone, and 1–4 m thick where anisotropic gneiss is in contact with the thrust zone, with sheared Eriboll Quartzite (Fig. 1c) in the footwall flat underneath. Within this zone, the Lewisian basement has undergone significant grain size reduction and has developed a mylonitic fabric parallel to the thrust contact (Lapworth 1883). Mineral stretching lineations indicate a west to WNW transport direction for the Arnaboll Thrust (Wibberley 1995).

Samples were taken from the Arnaboll Thrust fault-rock zone (locality D on Fig. 1f) where intermediate composition gneiss is deformed by progressive mylonitization from a blastomylonite approximately 2 m above the thrust contact to an ultramylonite at the surface. Arnaboll Thrust mylonites have previously been described in detail by White *et al.* (1982), so only a short summary is presented here. Coarse-grained mylonite 1.25 m above the thrust surface consists of dynamically recrystallized quartz, typically 0.05 mm in size, surrounding coarse feldspar grains. These feldspars show twinning, undulose extinction, kinking and cleavage-plane slip (e.g. Fig. 5a). The feldspars are also veined along cleavage planes by finely recrystallized quartz. Some of the plagioclase grains are partially

altered internally with 'ghost' twinning still easily visible. White mica is often present between coarse feldspar grains, whilst chlorite and epidote exist in mafic bands. These features suggest crystal-plastic deformation of quartz and a combination of brittle and crystal-plastic deformation of feldspars occurred simultaneously (see also FitzGerald & Stünitz 1993). Closer to the thrust, the fault rocks become much finer. A sample taken 25 cm from the thrust contact shows that quartz has undergone dynamic recrystallization to a size of approximately 0.01 mm (Fig. 5b). These grains exist in pod-shaped domains similar in shape and size to the few remaining large quartz grains, surrounded by white mica. The elongate pods and the mica define the foliation, whilst the finely recrystallized quartz grains are aligned slightly oblique to this foliation. A low-angle C'-type shear band of white mica deflects the main quartz–mica foliation. A thin section of the thrust contact between Lewisian basement-derived mylonite and Eriboll Quartzite-derived mylonite shows thrust-parallel domains of fine-grained white mica and finely recrystallized quartz (grain size *c.* 0.01–0.02 mm). However, fine-grained chlorite is also present in the Lewisian basement-derived mylonite, which shows a complete absence of the feldspar present in the protolith Lewisian gneiss. Synemplacement veins cross-cut the tectonic contact, yet are deformed themselves, such as the folded 0.1 mm-thick quartz vein shown in Figure 5c. These observations on the Arnaboll Thrust mylonites are consistent with semi-ductile deformation switching to ductile deformation by reaction weakening of feldspar to mica and chlorite under lower greenschist metamorphic facies conditions, probably around 300–350 °C (White *et al.* 1982).

The Arnaboll Thrust and thrust sheet are cut by a small array of later thrusts, each dipping moderately steeply to the ESE with displacements of approximately 10 m. One of these thrust fault zones is exposed at grid reference NC 4621 5925 (locality E, Fig. 1f), where a crevice represents the thrust surface within a 2 m-wide zone of highly weathered, schistose, dark crystalline rock. Figure 6 shows the change in microstructure from the hanging-wall edge to the centre of the fault zone. Two metres from the centre of the fault, slight alteration of the feldspar appears to be the only modification of the Lewisian microstructure (Fig. 6a). Closer to the centre of the fault, growth of new white mica is localized in seams 0.05–0.1 mm thick, typically spaced 1 mm apart, and is interpreted to result from precipitation from a fluid

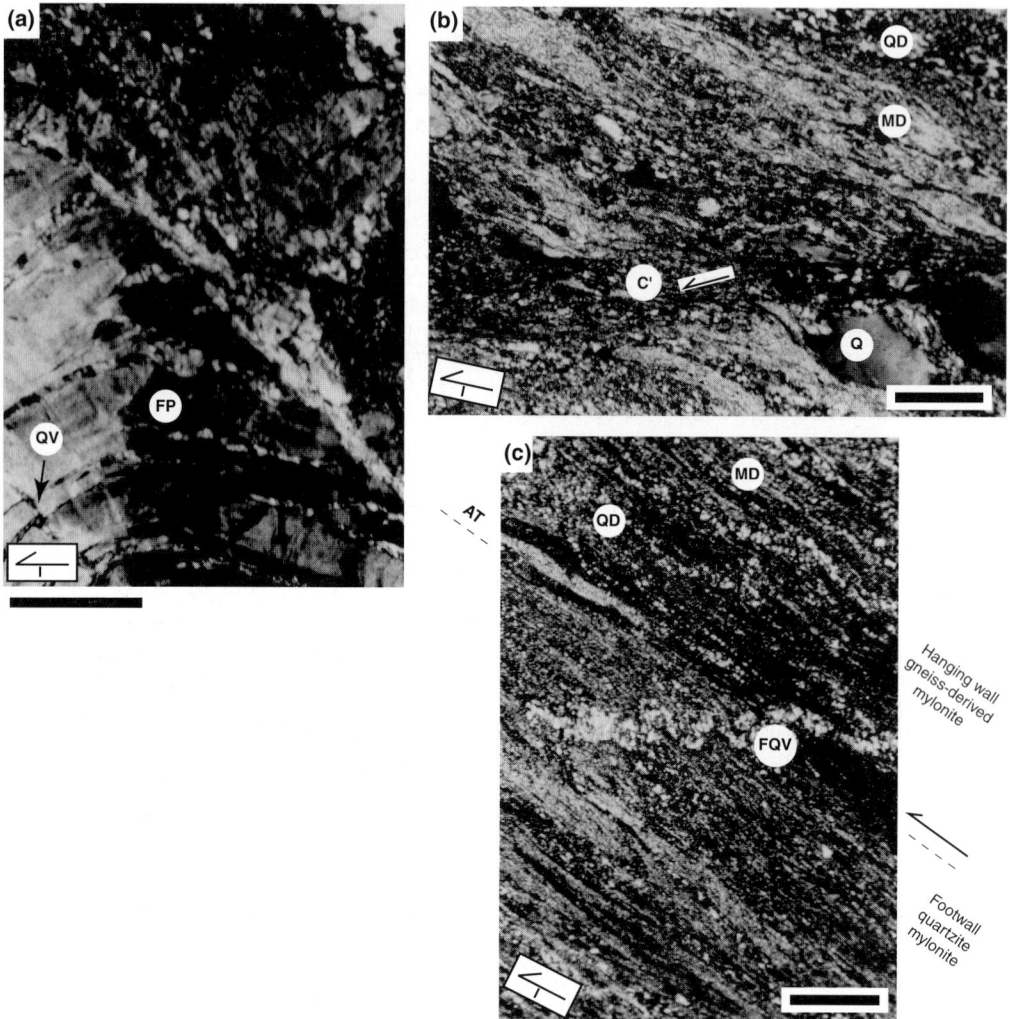

Fig. 5. Optical photomicrographs of samples taken from the Arnaboll Thrust zone at locality D (see Fig. 1f). Orientation arrows indicate the horizontal direction to the WNW for a vertical thin section. (**a**) Kinked feldspar in a protomylonite sample taken 1.25 m above the thrust contact, with slip on cleavage plans in the folded plagioclase ('FP') accompanied by the opening up of perpendicular quartz veins ('QV'). Scale bar represents 0.2 mm. (**b**) Mica-rich mylonite 25 cm above the thrust contact, showing bands of quartz domains ('QD'), mica domains ('MD') and relict Lewisian quartz grains ('Q') deflected by C′ microshear bands ('C″'). Scale bar represents 0.5 mm. (**c**) Ultramylonite at the thrust contact of the Arnaboll Thrust ('AT'), cross-cut by a folded quartz vein (FQV). Scale bar represents 0.5 mm.

in a fracture network (Fig. 6b). These parallel the fabric defined by elongate 0.5 mm quartz grains, which is thought to be originally a Lewisian fabric as it is parallel to the adjacent gneissic foliation. Figure 6c shows that next to the fault crevice, in a zone of very well foliated fault rock approximately 20 cm wide, syntectonic alteration has resulted in a fine mesh-like network of white mica domains. Both the feldspars and quartz grains have undergone

cataclastic grain size reduction, suggesting either lower temperatures or higher strain rates during faulting than those for the Arnaboll mylonite. These breaching thrust zones have accommodated much smaller displacements than the Arnaboll thrust zone itself, yet their microstructural development results in significant fault fabrics, and the overall width of the fault zone is not much smaller than that of the Arnaboll Thrust mylonite zone.

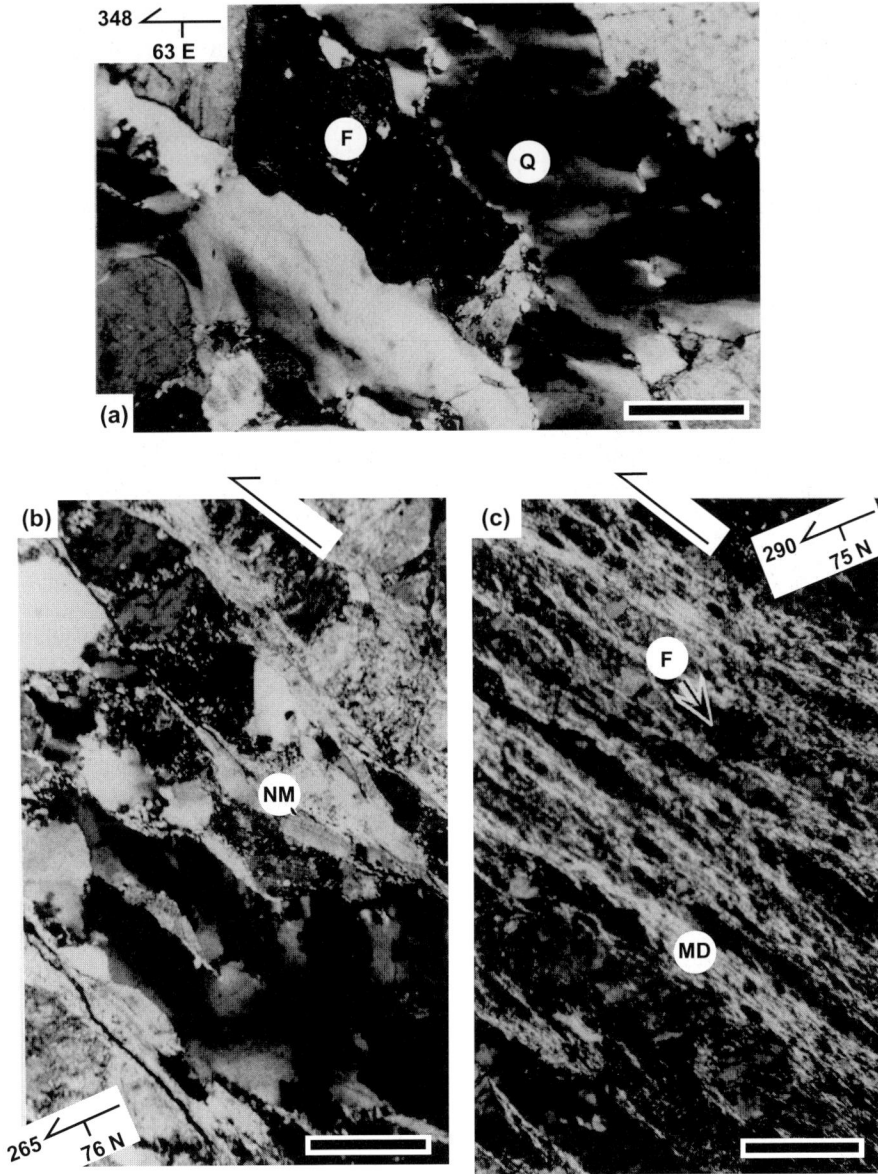

Fig. 6. Optical photomicrographs of samples taken from the breaching thrust zone in the Arnaboll thrust sheet at locality D (see Fig. 1f). Long axes represent 3 mm in each case. In (b) and (c), the orientation and sense of shear of the fault is indicated by the shear arrows, and in all the photographs, the orientation of the thin section is indicated by the numbered arrows. (**a**) Sample of Lewisian gneiss taken from the hanging wall, 2 m from the fault. Neither the Lewisian quartz ('Q'), nor the slightly altered feldspar ('F') are deformed by the faulting. (**b**) Sample of fractured Lewisian gneiss taken from the hanging wall, 50 cm from the fault. Note the growth of new mica in parallel zones ('NM'). (**c**) Phyllosilicate-rich mylonite in the fault zone, containing clasts of broken feldspar ('F') in zones strung out parallel to the aligned fine-grained secondary white mica fabric ('MD').

Example of basement thrusting in the Pelvoux Massif, external Western Alps

The foreland thrust belt of the external Western Alps involves European basement with a variety of pre-Alpine structures exposed in the Alps as External Crystalline Massifs (Fig. 7a). The pre-Alpine structural heterogeneity was strongly influenced by Mesozoic crustal extension, prominently expressed in and around the western Pelvoux, Belledonne, Taillifer and Grandes Rousses massifs by major ESE-dipping normal faults with basement in the footwall and cover in the hanging wall (Fig. 7b) (Barféty & Gidon 1980; Tricart 1984; Gillcrist et al. 1987). These Mesozoic normal faults, and perpendicular subvertical transfer faults, influence Alpine basement–cover deformation patterns in a range of ways, depending upon the geometry of the contacts, the fault dips and the rheological properties of the basement and cover lithologies involved (Gillcrist et al. 1987; Wibberley 1997). Post-dating and in some places reactivating these high-angle structures, a series of thrust zones indicate emplacement of relatively rigid basement thrust sheets on top of one another in an imbricate stack to form the Pelvoux Massif culmination (Beach 1981) (see also Fig. 7b). These basement thrust sheets have been carried, with the Belledonne Massif thrust sheets, on a crustal-scale Alpine sole thrust detachment onto the European foreland (e.g. Beach 1981; Mugnier et al. 1990), with overall displacement estimates of 35–50 km (Butler 1990; Mugnier et al. 1990). Within the framework of wedge models for orogen evolution (e.g. Chapple 1978), the external Alpine basement became incorporated into the front of the wedge relatively late in the evolution of the Alpine Orogen (Pfiffner 1986; Platt 1986; Escher & Beaumont 1997; Schmid & Kissling 2000). Hence, the basement at the present erosion level was buried to relatively shallow depths by the advancing orogenic wedge and experienced peak metamorphism up to lower greenschist facies related to this burial (Frey et al. 1974, 1999). This section presents work carried out around the foreland flank of the Pelvoux Massif. First, it summarizes previously published work from the Ser Barbier thrust fault and associated smaller faults in its hanging wall on the linked microstructural and geochemical evolution of the fault zones (Wibberley 1999; Wibberley & McCaig 2000); second, it presents new mapping data of the outcrop-scale fracture–fault arrays around the thrust, which provide the link between the small-scale observations and the scale of the massif.

Field relationships

A major thrust fault on the west side of the Pelvoux Massif, the Ser Barbier Thrust, emplaced crystalline basement towards the WNW over Mesozoic cover rocks under lower greenschist facies conditions (Grand et al. 1989; Jullien & Goffé 1993). The crystalline basement in this part of the Pelvoux Massif is generally Precambrian granitic gneiss and leuco-granite with a mineral assemblage of quartz + albite + orthoclase (Barféty et al. 1988). The Ser Barbier thrust sheet contains this crystalline basement with Liassic marls juxtaposed against the basement by Mesozoic normal faults, as evidenced by Mesozoic olistolithic blocks within the marls adjacent to the basement fault contacts (Fig. 7c). A prominent transfer fault, the WNW–ESE-striking Vallon Fault, is thought to have been part of the network of Mesozoic faults (Wibberley 1997) by analogy with other parts of the Pelvoux Massif (Coward et al. 1991), and is truncated by the Ser Barbier Thrust (Fig. 7c & d). Fracture patterns in the granitic basement adjacent to this fault contain three prominent sets (Fig. 7e). The earliest fracture set strikes parallel to this fault, and is interpreted to have been part of the Mesozoic fracture population from which the fault initiated. Some of the fractures in this set show evidence of fluid infiltration, such as quartz precipitation, although the timing of this is unclear. The two later sets are low-angle Riedel-type small faults thought to have formed by Alpine dextral movement of the Vallon Fault, and a cluster of fractures around a later breaching thrust that cuts the Vallon Fault. Both of these later sets have evidence of syntectonic alteration along them, either in narrow zones of fault rock or adjacent wall rock microfractures (see the next subsection). In summary, the granitic basement was affected by Mesozoic fracturing and normal faulting before further Alpine fracturing with fluid circulation, fault reactivation and generation of new major basement thrust faults.

Microstructures

The crystalline basement of the Ser Barbier thrust sheet contains fractures and small-scale faults identified as being generated by Alpine movement on the major basement faults. Fracture-tip zones, splays and wall rock microfracture arrays show evidence of fracture dilatancy, cataclasis of quartz and feldspar, and the growth of new white mica (Fig. 8a). This suggests that fracture dilatancy of the granite occurred in the presence of a fluid, with the

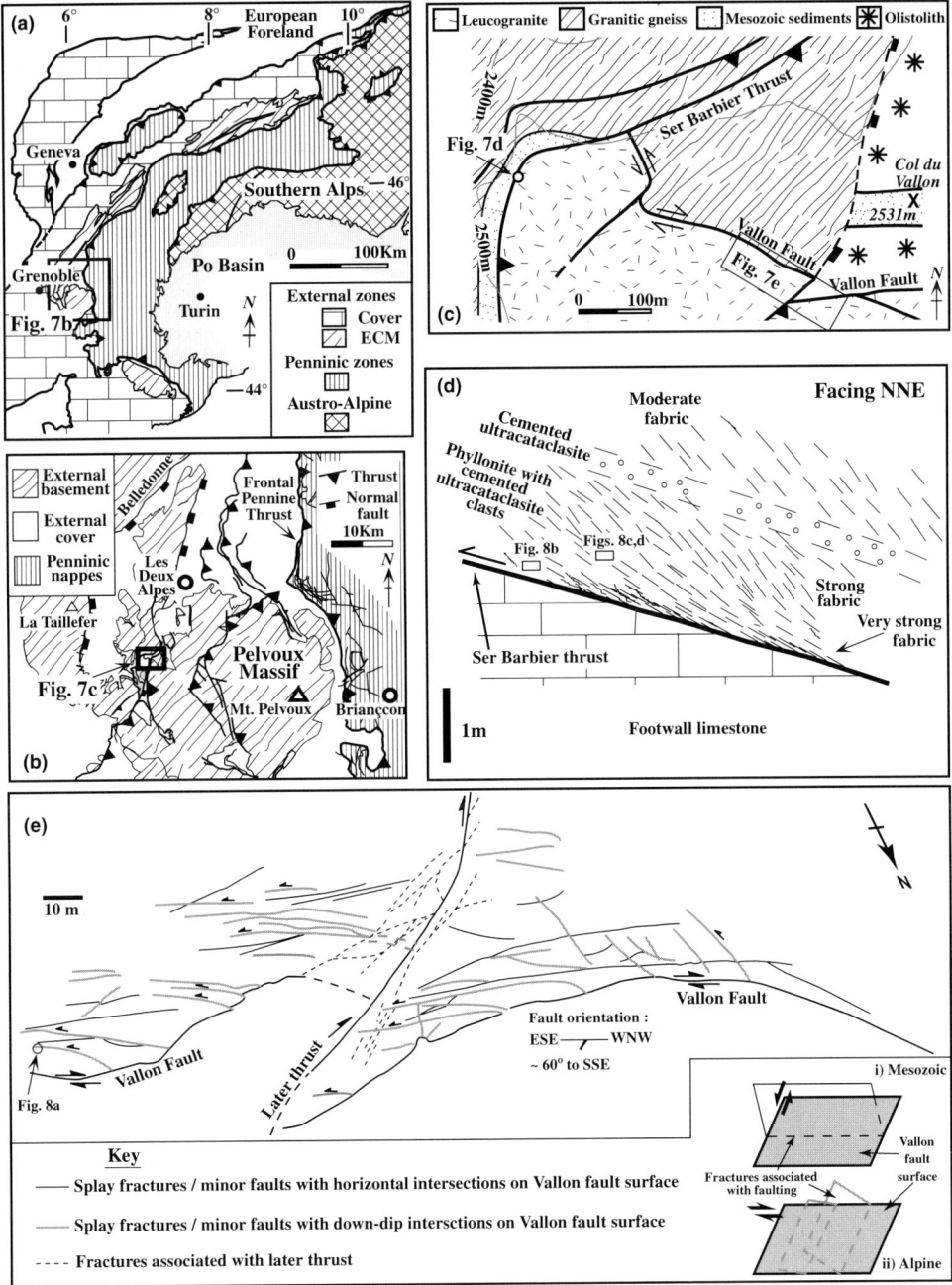

Fig. 7. Structural setting of the Ser Barbier study area in the western Pelvoux Massif, Western Alps. (**a**) The Western Alps. (**b**) The Pelvoux Massif. (**c**) The Ser Barbier thrust sheet (Wibberley 1995). (**d**) Field characterization of the Ser Barbier thrust zone structure. The sketch is of a vertical cliff face. (**e**) Basement fracture map of part of the Ser Barbier thrust sheet. The inset shows an interpretation of the genesis of different fracture patterns in relation to the Vallon Fault, based on detailed field observations and geometric analysis.

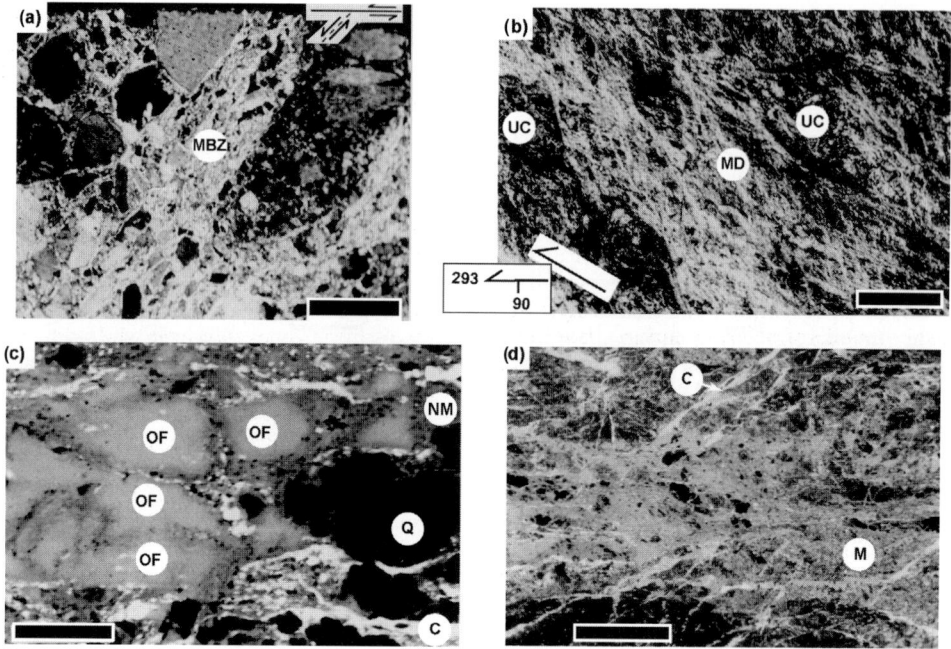

Fig. 8. Microtextures of Alpine Fault rocks from the Ser Barbier Thrust study area. (**a**) Optical photomicrograph of deformation features in the leucogranitic wallrock in the hanging wall to the Ser Barbier Thrust, showing clasts of quartz and feldspar in a microbreccia zone rich in secondary muscovite ('MBZ'). Scale bar represents 1 mm. Shear arrows indicate the microbreccia zone kinematics in relation to a larger (5 cm-wide) Riedel fault. See Figure 7e for location. (**b**) Optical photomicrograph of mylonite 2 cm above the thrust contact (Fig. 7d), with muscovite domains ('MD') enclosing clasts of earlier cemented ultracataclasite ('UC'). The shear arrow indicates the orientation and sense of thrust faulting, and the scale bar represents 1 mm. (**c**) Backscattered scanning electron micrograph magnifying part of the mylonite fabric of the sample shown in (b), showing the corroded fragments of a relict orthoclase feldspar clast ('OF'), surrounded by the secondary muscovite ('NM') alteration product. Quartz ('Q') and a late chlorite seam ('C') are also shown. Scale bar represents 20 μm. (**d**) Backscattered scanning electron micrograph of the muscovite fabric (light grey, 'M') in the sample shown in (b), showing how late localized chlorite seams (white, 'C') give a web-like microtexture to the sample. Scale bar represents 100 μm.

fluid being drawn into the microfracturing zones by dilatancy pumping from the fracture–fault network, consequently triggering precipitation. The Ser Barbier thrust zone is a 2 m-wide zone of black mylonites with a strong, but wavy, foliation containing clasts or bands of earlier quartz-cemented ultracataclasite (Fig. 7d). The mylonitic foliation is defined by a grain SPO of fine-grained muscovite that wraps around or cross-cuts microclasts of the quartz-rich ultracataclasite (Fig. 8b), with the preferred orientation being parallel or slightly inclined to the thrust plane. Corrosion textures of orthoclase feldspar (e.g. Fig. 8c) are visible, with relict fragments surrounded by aligned muscovite, generation of the Ser Barbier thrust zone mylonite by *in situ* fluid-assisted replacement of the feldspars and aligned muscovite growth of strong preferred orientation during deformation. Microprobe and whole-rock geochemical data from these

mylonites and similar fault rocks from the Vallon Fault indicate the replacement of first orthoclase, and later albite, by muscovite (Wibberley & McCaig 2000). Previous microstructural studies suggest that the fate of the silica released in these breakdown reactions seems to be crucial in controlling the microstructural properties and later deformation behaviour of the fault rocks (Wibberley 1999). If the silica is precipitated as quartz in the fault zone, as in the case of the cemented ultracataclasite, further deformation will occur by dilatant fracturing (if the temperature is less than about 300 °C), allowing the influx of further fluid and more alteration of feldspar to muscovite. In the case of the phyllosilicate-rich mylonite, however, the geochemical data suggest silica removal by the fluid flowing through the fault zone, resulting in a muscovite-dominated fault-rock rheology (Wibberley 1999). Fault rocks dominated by fine-grained phyllosilicates typically show marked permeability decreases as the phyllosilicate content increases (Wibberley & Shimamoto 2003), suggesting that permeability decreased as the syntectonic alteration proceeded. A small amount of chlorite precipitation in microbands either parallel to the mica fabric or at higher angles to the thrust surface is localized mainly in the remaining quartz-rich regions, although it also cross-cuts the mica fabric in places (Fig. 8d). This suggests a local increase in permeability within the mylonite, probably by microfracturing during a late lower temperature and/or higher strain-fracturing event (e.g. Lloyd & Knipe 1992).

Discussion

Evidence of syndeformation alteration of feldspar to muscovite has been presented from several important basement thrust zones in the Moine Thrust Belt and the external Western Alps, and also some smaller thrusts and related faults. All of these examples are taken from basement-involved foreland thrust zones where deformation typically occurred at lower greenschist facies metamorphic conditions and microstructural evidence is presented of cataclasis of feldspar and, in most cases (except for the Arnaboll Thrust), quartz in the early stages of deformation. For some of the examples where deformation occurred at around the brittle–crystal-plastic transition for quartz, semi-ductile deformation cannot be ruled out, especially if the quartz switches deformation mechanisms due to strain-rate variations (e.g. Glencoul Thrust: Knipe 1989, 1990). The description of evidence for the generation of phyllosilicate-dominated fault rocks by metamorphic syndeformation alteration in basement thrust zones of similar grade from North America (e.g. Rocky Mountain foreland: Evans 1988, 1990a; Janecke & Evans 1988; O'Hara 1988; Yonkee & Mitra 1993; Goddard & Evans 1995; and the Appalachians at slightly higher metamorphic grades: Mitra 1992; Yonkee & Mitra 1993) suggests that the syntectonic replacement of feldspars by phyllosilicates (mostly white micas) in basement thrust zones is common. This discussion considers the weakening effect of this process (White & Knipe 1978) and implications for fault-zone growth mechanisms in terms of fault-width dependence on displacement. Then a model for basement detachment in shallow mid-crustal foreland thrust belts is proposed, motivated by the observations presented in this paper, considering the likely way in which fluids could infiltrate such detachment zones as

they form. Finally, implications of this model for the rheology of foreland upper crust during horizontal compression and orogenic wedge evolution are discussed.

Reaction and deformation mechanisms

The basement thrust zones studied are zones of extremely high strain inside which microstructural evidence for those mechanisms that predominated during early stages of deformation has been largely destroyed. Examination of the edges of these zones, however, shows the original crystalline basement textures and fabrics partly destroyed by fracturing. The presence of relict fractured clasts and cemented ultracataclasite within the high-strain zones suggests that this fracturing also occurred within the high-strain zones before deformation intensified. Within the high-strain zones, the feldspars are almost completely altered to fine-grained muscovite of strong preferred grain shape orientation, with in some cases the additional presence of quartz and late chlorite and/or calcite. At the edges of the high-strain zones, muscovite precipitated in open fractures and in microfracture mesh-like networks but lacks the preferred orientation observed inside the high-strain zones. This is interpreted as being due to the syndeformational alteration of feldspar to mica inside the zones, with the anisotropic state of stress within the shear zone controlling the preferred orientation of the mica as it forms, along with later deformation possibly modifying the fabric. At the edges of these high-strain zones, however, deformation did not continue after the initial fracturing, so that the mica here is neither strongly aligned nor deformed. Brittle precursor deformation to phyllosilicate-dominated fault-zone rheology has previously been documented from basement fault zones at various scales (e.g. Segall & Pollard 1983; Segall & Simpson 1986; Stel 1986; Janecke & Evans 1988; Stünitz & FitzGerald 1993) and appears to be a very common phenomenon at low- to mid-greenschist grades. Such reactions are usually considered retrograde reactions, because the muscovite stability field in sodium/potassium reactivity space becomes larger at the expense of orthoclase and albite feldspars at lower temperatures (e.g. Hemley & Jones 1964). However, influx of an acidic fluid such as from a meteoric source could move the rock composition into the muscovite stability field, encouraging alteration even if retrogression is not occurring (Wintsch et al. 1995).

The important extent of feldspar-to-mica alteration in the small thrust zones studied (displacements of about 10–100 m) suggests that this alteration probably occurred early on in the development of the large thrusts (displacements of tens of km). The grain scale mechanisms by which the fine-grained mica domains and other fine-grained phases accommodated the continued deformation are extremely difficult to decipher without the aid of more detailed TEM and crystallographic data (e.g. Knipe 1981; White & White 1983; Stel 1986; Stünitz & FitzGerald 1993), but the general impression of the fabrics, and evidence for the precipitation of aligned mica and quartz in dilational sites (e.g. Fig. 2), suggest fluid-enhanced deformation mechanisms such as pressure solution–reprecipitation and diffusional-creep mechanisms are likely to have dominated. The production of new, strain-free muscovite in the Ser Barbier thrust zone was evidenced from TEM work (Wibberley 1999). The alteration of feldspar to fine-grained muscovite of strong preferred orientation at lower greenschist facies is considered to be a process of reaction-enhanced ductility (White & Knipe 1978), because the relatively strong Coulomb-frictional behaviour of feldspar will give way to a viscous rheology due to crystal-plastic and/or fluid-based pressure-solution–reprecipitation deformation mechanisms of white micas (e.g. Janecke & Evans 1988; Schmid & Handy 1991; Imber et al. 1997, 2001; Bos & Spiers 2002). The preferred orientation observed in the phyllosilicate-rich mylonites studied is a grain shape preferred orientation of the mica, but TEM studies of similar fabrics in biotite phyllonites also found a strong crystallographic preferred orientation (CPO) (Goodwin & Wenk 1995). Experimental data on the strength of muscovite under these conditions, although relatively limited, show that deformation could occur at relatively low shear stress by basal plane dislocation glide (Mares & Kronenberg 1993). Arguments based on microstructural information alone also suggest that this type of syntectonic alteration will result in deformation at much lower stresses (White & Knipe 1978; Schmid 1982; Brodie & Rutter 1985; Rubie 1990) due to the following factors:

- the fine grain size of the mica promoting grain size sensitive creep mechanisms;
- the production of new strain-free grains, which can remove the effects of crystal-plastic work hardening;
- the preferred orientation fabric of the mica, which assists non-frictional intergranular slip (superplasticity).

The operation of reaction-enhanced ductility in these thrust zones therefore suggests significant fault weakening.

The fine-grained polymineralic character of the mylonites may permit further weakening mechanisms in the presence of intragranular fluid due to the enhanced diffusion kinetics at boundaries between grains of non-like phases (e.g. Wheeler 1992; Stünitz & FitzGerald 1993; Bruhn et al. 1999). Most of the basement thrust fault rocks examined only showed muscovite as a fine-grained syntectonic reaction product, with two exceptions. (1) The earlier part of the linked reaction-deformation history of the Ser Barbier Thrust generated quartz from the silica released by the feldspar to muscovite reaction. (2) The Glencoul Thrust ultramylonite contains calcite and quartz in addition to muscovite. The calcite is probably present due to local dissolution–reprecipitation of footwall limestone, as the sampling site of the Glencoul Thrust (Fig. 1d) was the only one in this study in which the thrust had climbed up as high in the footwall stratigraphy as the Durness Limestone, and this is the only example that contains calcite in the fine-grained matrix. The quartz may come from two sources, either reprecipitation of original Lewisian quartz or from silica released in the feldspar to muscovite reaction. Either way, its occurrence in dilational sites adjacent to localized muscovite slip zones is common in phyllosilicate-rich mylonites at greenschist grade and suggests the operation of diffusional-creep deformation mechanisms (Knipe & Wintsch 1985) with or without this additional weakening effect.

Fault-zone thickness/displacement ratios

Fault-zone growth mechanisms have been assessed in the past by comparing the thicknesses of large displacement faults with those of smaller faults (Robertson 1983; Scholz 1987). Over a wide range of scales, data typically show a broadly linear increase in thickness with displacement, implying that fault zones grow in thickness in proportion to the accumulation of displacement. Most of these data are from clastic sedimentary rocks (Fig. 9a) (Robertson 1983; Wallace & Morris 1986). However, the constant of proportionality is widely variable, with thickness/displacement ratios being anywhere from 1:10 to 1:1000. Furthermore, complexities such as varying definitions of deformation zone limits (e.g. Blenkinsop 1989) and different host rocks and/or fault rock types (Evans 1990b) make a universal growth path less probable.

Figure 9a compares published data on fault and shear-zone thickness–displacement relationships with data from phyllosilicate-rich and phyllosilicate-poor basement thrust zones, taken from

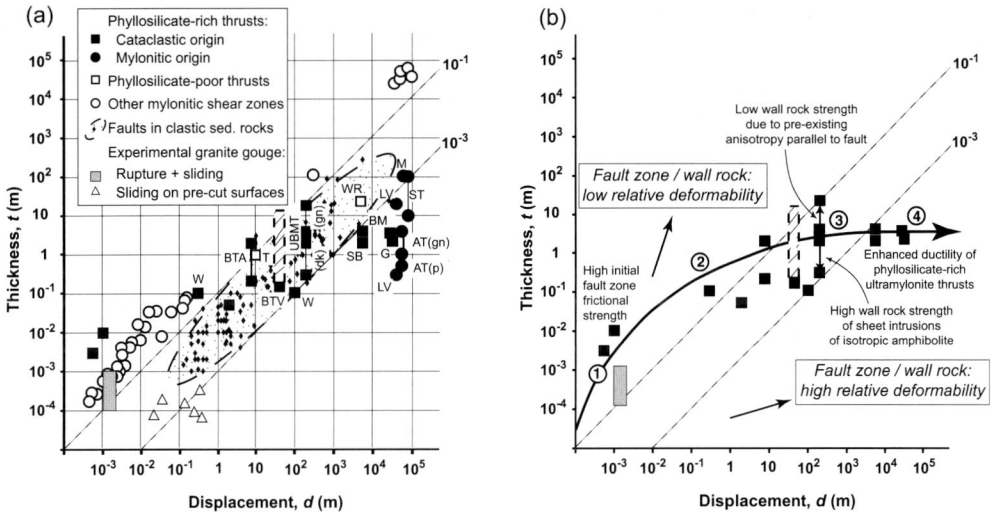

Fig. 9. Fault-zone thickness versus displacement. (**a**) Log–log plot of fault-zone thickness v. total displacment, for various fault rock and lithological types as indicated in the key. Data from this study are the Glencoul (G), Ben More (BM), Upper Ben More (UBMT), Arnaboll (AT) and Ser Barbier (SB) thrusts, plus the breaching thrust in the Arnaboll thrust sheet (BTA), the breaching thrust of the Vallon Fault (BTV) (Fig. 7e) with its fracture damage zone width indicated by the dashed box, and some much smaller thrust faults and microfault zones from the Ser Barbier thrust sheet. Variation in fault-zone thickness is denoted by a vertical line joining two points that mark the maximum and minimum thicknesses, with lithological variations on a single fault shown where appropriate for isotropic amphibolite dyke (dk), Lewisian gneiss (gn) and pegmatite (p). The other named faults are: the Linville Falls (LV) Thrust (Boyer 1978) in the Appalachians; the White Rock (WR) Thrust and Torrey Creek (T) Fault (Mitra 1992) and faults from the Washakie (W) thrust sheet (Evans 1988), from the Wind River Range, Wyoming; the Särv Thrust (ST) in the Swedish Caledonides (Gilotti 1989); the Moine Thrust (M) (Christie 1960). Other mylonite shear zones are from Coward *et al.* (1973), Bak *et al.* (1975) and Mitra (1979). Faults in clastic sedimentary rocks are from Robertson (1983), Wallace & Morris (1986) and unpublished data collected by the author from faults in the Lodève Basin, Southern France. Granitic gouge zones generated in dry experiments by sliding on saw-cut surfaces are from Tullis & Weeks (1986), Yoshioka (1986), Blanpied *et al.* (1987), and Power *et al.* (1988), and gouge zones generated by rupture and continued sliding, shown for a range of data, are from Amitrano & Schmittbuhl (2002). (**b**) Schematic model for growth of phyllosilicate-rich mylonite fault zones developing from cataclastic fault zones with displacement based on the data from thrusts of this kind duplicated from (a).

the thrust zones examined in this paper and from other published examples. The phyllosilicate-rich basement thrust zones are further divided into those that deformed mainly by cataclastic deformation mechanisms prior to alteration, and those that have probably been mylonitic all the way through their deformation history (with the exception of late cataclastic bands), such as the Moine Thrust mylonite. Figure 9a shows that the smallest crystalline basement faults studied (of the type shown in Fig. 8a) show very large thickness-to-displacement ratios of between 1:1 and 10:1. The secondary mica in these zones was precipitated after initial tensile fracturing of the grains, and the early cataclasite generation prior to any alteration was part of a high thickness growth phase due to shearing of wall rock tensile fracture zones during rupture, with probably high fracture permeabilities

associated to them. Similar textures have been observed (without the alteration) in dry rupture experiments of initially intact granites and low-porosity sandstones (e.g. Wibberley *et al.* 2000; Amitrano & Schmittbuhl 2002) that yielded gouge thickness-to-displacement ratios comparable to these small-scale basement faults. The importance of the rupture phase is highlighted by constrasting these data to the low thickness-to-displacement ratios (around 1:1000) of experimentally generated gouge zones from sliding along pre-cut surfaces with no rupture phase (Tullis & Weeks 1986; Yoshioka 1986; Blanpied *et al.* 1987; Power *et al.* 1988). Figure 9a also shows that the high-displacement phyllosilicate-rich thrust zones tend not to grow thicker than 5–10 m if they were of cataclastic origin that suffered reaction-enhanced ductility, so that the examples with the highest displacements (e.g. Glencoul Thrust, Ben More Thrust) have thickness-to-displacement ratios as low as 1:10 000–1:50 000. These contrast with some, but not all, of the thrust zones of mylonitic origin, such as the Moine Thrust, which probably had a wide zone of mylonite before the secondary muscovite was generated.

The difference in thickness-to-displacement ratios of the small and large phyllosilicate-rich thrust zones studied is obvious when these data are separated from the data on other fault zones (Fig. 9b), and there is a clear change from high thickness-to-displacement ratios in the small faults to low thickness-to-displacement ratios in the largest faults. This change in thickness-to-displacement ratio suggests that continued widening of the zone becomes inhibited at a certain displacement. In fact, the active part of the fault zone may become narrower due to strain-softening mechanisms such as the reaction-enhanced ductility, localizing the deformation on a particular part of the fault zone or surface (e.g. Chester & Logan 1986; Hull 1988; Yonkee & Mitra 1993). Thus, fault-zone growth is controlled by the evolution of the relative deformability between host rock and fault zone.

Basement fault-zone initiation and growth mechanisms

The microstructural evidence for reaction-enhanced ductility in the meso-scale faults as well as the large thrusts suggests that this process operated relatively early in the evolution of any one detachment thrust. The generation of low-angle fractures associated with formation of a thrust fault (e.g. Fig. 7e) (see also Wojtal & Mitra 1986) helped provide a fluid pathway to the developing thrust, encouraging the early onset of the reaction-enhanced ductility (similar to small-scale shear zones documented by Segall & Simpson 1986) and thus early initiation of a weak low-angle basement fault.

In the first stage, microcracking associated with rupture leads to nucleation of a microfault (stage 1, Fig. 10) whose thickness increases by the continued development of wall rock fracture damage and progressive incorporation of this damaged material into the developing fault zone, so that the fault zone widens by engulfing the surrounding damage zone towards the limit of the width of the damage zone (stage 2, Fig. 10). This is proposed not just on the basis of microstructures at the edge of the larger thrust fault zones, but also from experimental work, and on the basis of relict microstructures preserved in the centre of some of the fault zones studied, as well as comparison with the smaller fault zones that did not evolve into larger ones (see Means 1995). Continued cataclasis in the centre of the fault zone during this widening phase will reduce the grain size of quartz and feldspar, and provide fresh fracture surface area available for reaction. Cataclastic grain size reduction may result in some initial softening (Goodwin & Wenk 1995) even before the alteration, but the presence of the secondary mica in all the basement thrust fault zones from displacements of 10–100 m upwards suggests that the vast majority of the thrust fault-zone rheology was mica-dominated. Fluid influx into the fracture damage zone from a surrounding pre-existing 'background' fracture network (e.g. Fig. 7e) is encouraged by its high permeability coupled with dilatancy pumping, as suggested by syntectonic micas in the tensile fracturing developing around fault tips (Fig. 8a). Such dilatancy pumping has been previously suggested for thrust faults in other regions (McCaig 1987; Cox & Etheridge 1989;

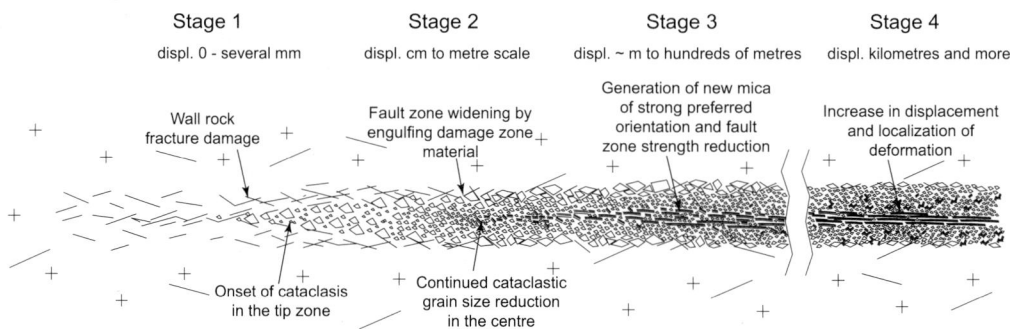

Stage 1	Stage 2	Stage 3	Stage 4
displ. 0 - several mm	displ. cm to metre scale	displ. ~ m to hundreds of metres	displ. kilometres and more

Wall rock fracture damage

Fault zone widening by engulfing damage zone material

Generation of new mica of strong preferred orientation and fault zone strength reduction

Increase in displacement and localization of deformation

Onset of cataclasis in the tip zone

Continued cataclastic grain size reduction in the centre

Fig. 10. Summary of the different stages envisaged in basement thrust fault-zone evolution.

Grant *et al.* 1990; McCaig *et al.* 1990, 2000;) and is probably due to the fracture-dilatant nature of the deformation during rupture and damage zone generation (Brace *et al.* 1966; Evans *et al.* 1997).

The fluid infiltration and the finer grain size in the centre of the fault zone promote alteration of feldspar to fine-grained mica with possibly other phases. The high σ_1/σ_3 stress ratio in the actively deforming part of the fault zone will aid the formation of mica in aligned interconnected domains with a strong preferred orientation (e.g. Bell 1981), so that crystal-plastic and fluid-assisted deformation mechanisms will start to operate at much lower shear stress than frictional deformation mechanisms (Rubie 1990; Wintsch *et al.* 1995). The switch to reaction-enhanced ductility therefore reduces the strength of the fault zone (stage 3, Fig. 10). The bulk strength reduction during reaction-enhanced ductility depends on the degree of interconnectivity of the weak phase and the strength contrast between the relatively strong reactant phase and the weaker product phase (e.g. Handy 1990). Theoretical work on modelling strength in such deforming polyphase aggregates suggests that, given sufficient strength contrasts, only a small percentage of the interconnected weak phase needs to be produced to weaken the aggregate to close to the strength of the pure weak phase (Handy 1994*a, b*). In the light of this, weakening during basement thrust-zone development by reaction-enhanced ductility could occur at relatively low strains, i.e. early on in fault development, inhibiting cataclastic deformation at the edges of the fault zone and limiting the thickness increase of the zone (stage 4, Fig. 10). Finally, continued displacement and possible localization of deformation onto C′ or C-shears within the thrust ultramylonite zone will mean that the overall width of the fault zone will not increase with further displacement, and the 'active' part of the main fault zone may actually episodically decrease and increase within the main fault zone (e.g. Knipe & Wintsch 1985). Evidence of the importance of host-rock properties has been provided by the example of the Upper Ben More Thrust ($d \approx 100$ m) in Figure 3, as highlighted in Figure 9b. In the isotropic, relatively strong amphibolite dyke, the phyllonite zone is relatively narrow (thickness-to-displacement ratio of 1:1000), whereas the deformation is distributed in a much wider zone in the gneiss because of the wall rock weakness due to the heterogeneity of the gneiss (thickness-to-displacement ratio of 1:10–1:100, depending on definition of deformation-zone limits).

Work on the evolution of displacement focusing onto larger faults in a population of normal faults (e.g. Cowie 1998) suggests that a point will come in the development of the fault array when activity will cease on the small and intermediate faults with displacement preferentially focusing on the already large faults (Tchalenko 1970). This 'survival of the largest' behaviour of the fault population was explained by progressive fault weakening to focus the strain on the largest faults at the expense of the small ones (Tchalenko 1970). In the cases of the basement thrust weakening examined here, however, the lack of small- to intermediate-sized basement thrust faults around the large thrusts suggests that weakening is even more extreme and early in relation to the timing of development of the fault population. Microphysical modelling of fluid-assisted deformation mechanisms such as pressure solution and foliation development in crustal faults suggests that this weakening, in comparison to conventional strength profiles through the crust, will be most prominent at depths of 10–15 km, similar to the probable depth of initiation of these basement thrust detachments, and that a crustal strength peak will migrate to shallower depths (Bos & Spiers 2002).

Despite the evidence for early fluid infiltration into the fault zones, and the continued operation of deformation mechanisms requiring pore and grain-boundary fluids throughout the development of the fault zone, no evidence was observed for the flow of large volumes of fluid after the development of the phyllosilicate–ultramylonitic fabric. This contrasts with other studies which have suggested that large fault zones that are weak tend to suffer episodes of reactivation and act as long-lived fluid pathways. An answer to this apparent paradox comes from laboratory measurements of fault-rock permeabilities, which have shown a strong tendency for permeability to decrease (by a few orders of magnitude) as phyllosilicate content increases (Wibberley & Shimamoto 2003). Thus, the longevity of certain other faults as fluid-flow pathways is probably due to brittle reactivation and consequent fracture dilatancy, something that appears not to have happened on the thrusts examined due to the later tectonic histories in the regions of these thrusts and their low-angle geometries. It is therefore suggested that the basement thrusts examined were anomalously weak only during their thrust activity due to the mechanics of the fluid-assisted deformation mechanisms operating. After exhumation to shallower depths (in the brittle field), however, they have neither a low enough friction coefficient nor sufficient permeability for pore fluid infiltration to allow for weak fault behaviour.

Model for foreland basement thrust detachment

The data presented in Figure 9 suggest that the development of an ultramylonitic fabric of preferred orientation of mica from syndeformation alteration of feldspar promotes strain localization in foreland crystalline basement thrust zones. The common occurrence of phyllosilicate-rich ultramylonites in these thrust zones, and lack of a continuous size distribution of small- to intermediate-scale thrust faults around these large basement thrusts, suggest that there is an important link between the weakening effect due to reaction-enhanced ductility and initial basement detachment in the foreland thrust belt. Indeed, strain localization induced by reaction-weakening processes has recently also been suggested as a mechanism for generating low-angle normal fault detachments in the mid-crust (Gueydan et al. 2003) and the upper mantle (Handy & Stünitz 2002). This subsection presents a conceptual model to explain how this phenomenon could occur in foreland thrust belts. The following model is purely conceptual, but different elements are supported by evidence of their general importance from the references cited and those presented in this paper.

Meteoric fluid percolation. The presence of meteoric fluids in microfracture/macrofracture networks at hydrostatic pressure has been evidenced directly from borehole studies even as far down as around 10 km or more in the KTB borehole, Germany (Huenges et al. 1997) and in the Kola superdeep borehole, Russia (Kozlovsky et al. 1987). A wealth of indirect evidence from combined structural and geochemical studies suggests the presence and circulation of meteoric fluids in foreland fracture–thrust systems (e.g. the Canadian Cordillera: Nesbitt & Muehlenbachs 1991; the Sevier Cordilleran thrust belt, USA: Dworkin 1999; the Pannonian Basin, Hungary: Juhász et al. 2002). The presence of a pre-existing high-angle fracture network in the upper crust, possibly formed during an earlier extensional deformation event, facilitates the penetration of a meteoric fluid to the appropriate depths below the basement–cover contact (0.5–5 km) before the onset of orogenesis (stage 1, Fig. 11).

Onset of thrust faulting and dilatancy pumping. The generation of low-angle fracturing

and the initiation of a few small-scale thrust faults with dilatant fracture damage zones at their tips (e.g. Wojtal & Mitra 1986) (see Fig. 8a) accommodates initial basement shortening. Their rapid down-dip linkage to generate a small number of large fractures could achieve a more effective fluid channelization, as suggested by percolation models (Chelidze 1986). Dilatancy pumping of fluid from the pre-existing high-angle fracture network into fracture–dilatant tip damage zones makes these high-permeability regions the foci of fluid infiltration, and the high surface areas of fractured fragments within these fault zones will aid relatively early feldspar-to-mica alteration, resulting in reaction-enhanced ductility (stage 2, Fig. 11), as discussed in the subsection on Basement fault-zone initiation and growth mechanisms (see also Segall & Simpson 1986).

Displacement localization and detachment. Continued, localized deformation focused on these weaker zones, along with further syntectonic alteration and formation of the phyllosilicate-rich ultramylonite, allowing increases of displacement by orders of magnitude. This occurs without such significant increases in the thickness of the ultramylonite zone. Large displacement on these thrust faults detaches the basement by carrying it as relatively intact, coherent thrust sheets that show little evidence of later internal shortening (stages 3 and 4, Fig. 11). Connection by one or more ramps to the sedimentary cover layers allows basement thrust sheets to be emplaced on to stratigraphically higher thrust flats. Continued shortening may be accommodated by displacement on a second basement thrust, initiating at the same décollement level as the first, resulting in foreland propagation of the basement décollement (stage 4, Fig. 11). Hence, detachment and uplift of basement thrust sheets in the foreland thrust belt can be achieved in this way on relatively thin, weak zones at the front of the orogenic wedge.

Data on pressure and temperature during deformation often indicate higher depths of deformation than suggested by the thickness of basement and cover stratigraphy in the thrust sheet (e.g Frey et al. 1974, 1999; Knipe 1990); hence, loading by hinterland material must have been occurring during this process or before the onset of foreland shortening (Fig. 11). This can be related to overall models for orogenic wedge evolution and pressure, temperature and deformation histories (e.g. Platt 1986; Escher & Beaumont 1997; Schmid and Kissling 2000). Within the framework of

Stage 1 : Fluid percolation into fractured basement prior to foreland shortening

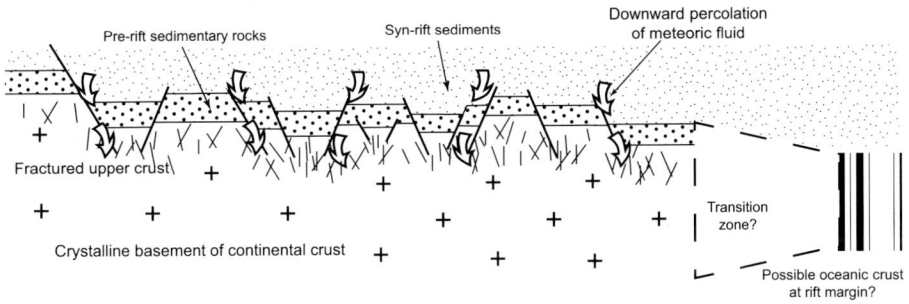

Stage 2 : Onset of foreland shortening by low-angle fracturing with dilatancy pumping

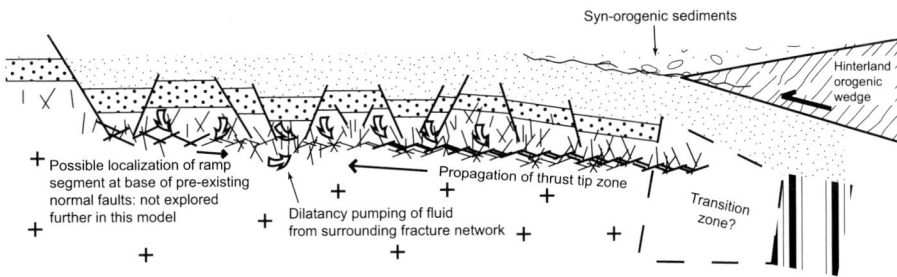

Stage 3 : Strain localization onto narrow basement thrusts by reaction weakening

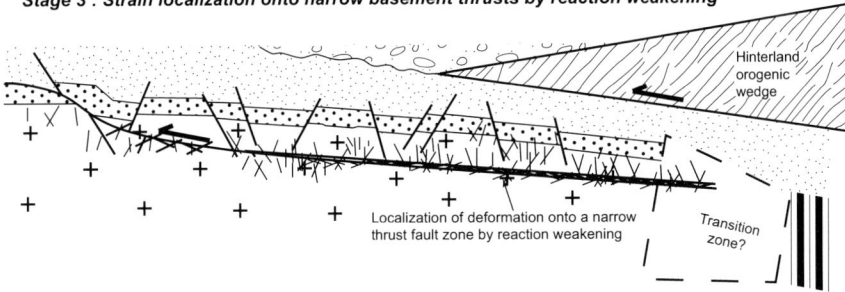

Stage 4 : Basement detachment and overall forelandward propagation of the décollement front

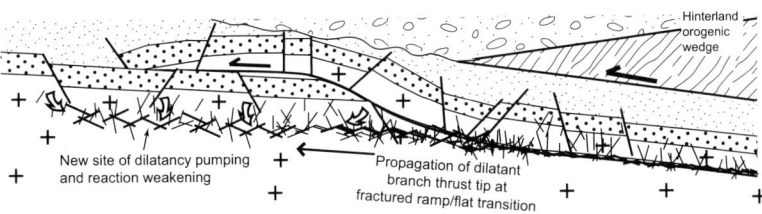

Fig. 11. Model for the detachment of basement in foreland thrust belts. Stage 1 – initial fracturing and meteoric fluid percolation down into the basement. Stage 2 – low-angle dilatant fracturing in damage zones around proto-thrust faults, and fluid influx by dilatancy pumping. Stage 3 – reaction weakening and formation of a detachment. Stage 4 – propagation of the décollement into the foreland as new imbricates form.

such models, foreland crystalline basement shortening occurs as frontal accretion and/or basal accretion at the front of the wedge, thus lengthening the orogenic wedge. The reaction-weakening processes suggested here to occur during formation of the décollement in the foreland basement would result in a much weaker new wedge décollement than predicted by Byerlee friction laws. As noted by Davis *et al.* (1983), orogenic wedges that have a décollement passing down through the brittle–ductile transition appear to have lower critical tapers, predicted to be the case for a weaker décollement. In the case of foreland crystalline basement, however, reaction weakening (or reaction-enhanced ductility) causes such an effect at potentially shallower depths. This highlights the importance of the evolving foreland strength as an additional parameter in wedge mechanics models.

Conclusions

Examples of key foreland basement thrust faults from the Moine Thrust Belt and the Pelvoux Massif, external Western Alps, show widespread evidence for the intense syntectonic alteration of feldspar to muscovite of strong preferred orientation, resulting in the generation of weak phyllosilicate-rich ultramylonite zones during thrusting. These ultramylonite zones rarely achieve thicknesses greater than a few metres, despite accommodating displacements of up to the order of 50 km. Smaller basement thrust fault zones also show evidence for this process, suggesting it occurred early on in the displacement history of the fault zones. Dilatant fracture zones immediately adjacent to the ultramylonite zones contain new mica of no or weak preferred orientation, suggesting that original fracture dilatant thrust zones were the sites of fluid infiltration, with reaction weakening concentrating in the centre of the fault zones, promoting strain localization and the cessation of deformation at the edges (stage 3 above). The initial state of the wall rock and accumulation of fracture damage around the proto-thrust fault is shown to be an important influence on the distribution and thickness of phyllosilicate-rich ultramylonite thrust zones, so that it is the evolving strength contrast between wall rock and thrust fault zone that controls strain localization with displacement.

The common occurrence of thin phyllosilicate-rich ultramylonite thrust fault zones in crystalline basement thrust belts suggests a link between their generation and basement detachment. The key to a model for basement thrust-sheet detachment by reaction weakening is that this process occurs very early on in the deformation history, (i.e. prior to detachment initiation). In order to trigger this, early fluid percolation into a pre-existing high-angle fracture–fault network is invoked, and fracture dilatancy pumping into a high-permeability proto-thrust fault damage zone focused within the low-angle deformation zone. Reaction weakening in this early proto-thrust zone localizes the deformation onto a narrow shear zone, and inhibits the generation of new faults in the adjacent rock that would otherwise distribute the strain. The weak basement thrust can then accommodate large displacement and act as a décollement, detaching and transporting the basement as an intact, coherent thrust sheet. Where the initial basement thrust ramps up into the cover, propagation of similar basement thrusts will encourage migration of the basement décollement into the foreland. This has consequences for the bulk strength of the under-riding basement in orogenic wedge mechanics models, by facilitating frontal accretion of relatively rigid material and by providing a weaker, viscous décollement.

Most of the data collection for this work was carried out whilst the author was conducting a PhD sponsored by NERC under the supervision of R.J. Knipe and R.W.H. Butler, who are thanked for their input. The manuscript has benefited greatly from earlier feedback from R.P. Wintsch and L. B. Goodwin, and thorough reviews from M.R. Handy, R.P. Wintsch and J. Imber.

References

AMITRANO, D. & SCHMITTBUHL, J. 2002. Fracture roughness and gouge distribution of a granite shear band. *Journal of Geophysical Research*, **107**, 2357 doi:10.1029/2002JB001761.

ARMSTRONG, R.L. & DICK, H.J.B. 1974. A model for the development of thin overthrust sheets of crystalline rock. *Geology*, **3**, 35–40.

BAILEY, E.B. 1935. The Glencoul nappe and the Assynt culmination. *Geological Magazine*, **72**, 151–165.

BAK, J., SORENSEN, K., GROCOTT, J., KORSTGARD, J., NASH, D. & WATTERSON, J. 1975. Tectonic implications of Precambrian shear belts in western Greenland. *Nature*, **254**, 566–569.

BARFÉTY, J.-C., MONTJUVENT, G., PÉCHER, A. & CARME, F. 1988. *Carte géologique de la France à 1/50 000. Feuille La Mure.* Bureau de Recherches Géologiques et Minières (BRGM): Orléans.

BARFÉTY, J.-C. & GIDON, M. 1980. Fonctionnement synsédimentaire liasique d'accidents du socle dans la partie occidentale du massif du Pelvoux (région de Venosc, Isère). *Bulletin du BRGM*, Series 2, 1(1), 11–20.

BEACH, A. 1981. Thrust tectonics and crustal shortening in the external French Alps based on a seismic cross-section. *Tectonophysics*, **79**, T1–T6.

BELL, T.H. 1981. Foliation development: the contribution, geometry and significance of progressive, bulk inhomogeneous shortening. *Tectonophysics*, **75**, 273–296.

BLANPIED, M.L., TULLIS, T.E. & WEEKS, J.D. 1987. Frictional behaviour of granite at low and high sliding velocities. *Geophysical Research Letters*, **14**, 554–557.

BLENKINSOP, T.G. 1989. Thickness-displacement relationships for deformation zones: discussion. *Journal of Structural Geology*, **11**, 1051–1054.

BOS B. & SPIERS, C.J. 2002. Frictional-viscous flow of phyllosilicate-bearing fault rock: Microphysical model and implications for crustal strength profiles. *Journal of Geophysical Research*, **107**, 2028, doi:10.1029/2001JB000301.

BOYER, S.E. 1978. *Structure and origin of the Grandfather Mountain window*. PhD thesis, John Hopkins University, Baltimore, MD.

BOYER, S.E. & ELLIOTT, D. 1982. Thrust systems. *AAPG Bulletin*, **66**, 1196–1230.

BRACE, W.F., PAULDING, J.B.W. & SCHOLZ, C. 1966. Dilatancy in the fracture of crystalline rocks. *Journal of Geophysical Research*, **71**, 3939–3953.

BRODIE, K.H. & RUTTER, E.H. 1985. On the relationship between deformation and metamorphism with special reference to the behaviour of basic rocks. *In*: THOMPSON, A.B. & RUBIE, D.C. (eds) *Advances in Physical Geochemistry*, Volume 4. Springer, Berlin, 138–179.

BRUHN, D.F., OLGAARD, D.L. & DELL'ANGELO, L.N. 1999. Evidence for enhanced deformation in two-phase rocks: Experiments on the rheology of calcite-anhydrite aggregates. *Journal of Geophysical Research*, **104**, 707–724.

BUTLER, R.W.H. 1982. A structural analysis of the Moine thrust zone between Loch Eriboll and Foinaven, N.W. Scotland. *Journal of Structural Geology*, **4**, 19–29.

BUTLER, R.W.H. 1990. Balancing cross-sections on a crustal scale: a view from the Western Alps. *In: The European Geotraverse 6th Workshop*, 157–164.

CHAPPLE, W.M. 1978. Mechanics of thin-skinned fold-and-thrust belts. *Bulletin of the Geological Society of America*, **89**, 1189–1198.

CHELIDZE, T.L. 1986. Percolation theory as a tool for imitation of fracture process in rocks. *Pure and Applied Geophysics*, **124**, 731–748.

CHESTER, F.M. & LOGAN, J.M. 1986. Implications for mechanical properties of brittle faults from observations of the Punchbowl fault zone, California. *Pure and Applied Geophysics*, **124**, 77–106.

CHRISTIE, J.M. 1960. Mylonitic rocks in the Moine thrust zone in the Assynt region, North-west Scotland. *Transactions of the Edinburgh Geological Society*, **18**, 79–93.

COWARD, M.P. 1984. A geometrical study of the Arnaboll and Heilam thrust sheets, N.W. of Ben Arnaboll, Sutherland. *Scottish Journal of Geology*, **20**, 87–106.

COWARD, M.P. 1988. The Moine thrust and the Scottish Caledonides. *In*: MITRA, G. & WOJTAL, S.W. (eds) *Geometries and Mechanisms of Thrusting*. Geological Society of America, Special Paper, **222**, 1–16.

COWARD, M.P., GRAHAM, R.H., JAMES, P.R. & WAKEFIELD, J. 1973. A structural interpretation of the northern margin of the Limpopo orogenic belt, southern Africa. *Philosophical Transactions of the Royal Society of London*, **A273**, 487–491.

COWARD, M.P., GILCRIST, R. & TRUDGILL, B. 1991. Extensional structures and their tectonic inversion in the W. Alps. *In*: ROBERTS, A.M., YIELDING, G. & FREEMAN, B. (eds) *The Geometry of Normal Faults*. Geolical Society, London, Special Publications, **56**, 93–112.

COWARD, M.P., KIM, J.H. & PARKE, J. 1980. A correlation of Lewisian structures and their displacement across the lower thrusts of the Moine thrust zone, N.W. Scotland. *Proceedings of the Geological Association of London*, **91**, 327–337.

COWIE, P.A. 1998. A healing–reloading feedback control on the growth rate of seismogenic faults. *Journal of Structural Geology*, **20**, 1075–1087.

COX, S.F. & ETHERIDGE, M.A. 1989. Coupled grain-scale dilatancy and mass transfer during deformation at high fluid pressures: examples from Mount Lyell, Tasmania. *Journal of Structural Geology*, **11**, 147–162.

DAVIS, D., SUPPE, J. & DAHLEN, F.A. 1983. Mechanics of fold-and-thrust belts and accretionary wedges. *Journal of Geophysical Research*, **88**, 1153–1172.

DEREMOND, J. 1979. *Deformation et desplacement des nappes: Example de la Nappe de Gavarnie (Pyrénées Centrales)*. PhD thesis, Toulouse University.

DWORKIN, S.I. 1999. Geochemical constraints on the origin of thrust fault fluids. *Geophysical Research Letters*, **26**, 3665–3668.

ELLIOTT, D. & JOHNSON, M.R.W. 1980. Structural evolution in the northern part of the Moine thrust belt, N.W. Scotland. *Transactions of the Royal Society of Edinburgh: Earth Sciences*, **71**, 69–96.

ERSLEV, E.A. & ROGERS, J.L. 1993. Basement-cover geometry of Laramide fault-propagation folds. *In*: SCHMIDT, C.J., CHASE, R.B. & ERSLEV, E.A. (eds) *Laramide Basement Deformation in the Rocky Mountain Foreland of the Western United States*. Geology Society of America, Special Paper, **280**, 125–146.

ESCHER, A. & BEAUMONT, C. 1997. Formation, burial and exhumation of basement nappes at crustal scale: a geometric model based on the Western Swiss–Italian Alps. *Journal of Structural Geology* 19, 955–974.

EVANS, J.P. 1988. Deformation mechanisms is granitic rocks at shallow crustal levels. *Journal of Structural Geology*, **10**, 437–443.

EVANS, J.P. 1990a. Textures, deformation mechanisms, and the role of fluids in the cataclastic deformation of granitic rocks. *In*: KNIPE, R.J. & RUTTER, E.H. (eds) *Deformation Mechanisms, Rheology and Tectonics*. Geological Society, London, Special Publications, **54**, 29–39.

EVANS, J.P. 1990b. Thickness–displacement relationships for fault zones. *Journal of Structural Geology*, **12**, 1061–1065.

EVANS, J.P. 1993. Deformation mechanisms and kinematics of a crystalline-cored thrust sheet: The EA thrust system. *In*: SCHMIDT, C.J., CHASE, R.B. & ERSLEV, E.A. (eds) *Laramide Basement Deformation in the Rocky Mountain Foreland of the Western United States.* Geological Society of America, Special Paper, **280**, 147–161.

EVANS, J.P., FORSTER, C.B. & GODDARD, J.V. 1997. Permeability of fault-related rocks, and implications for hydraulic structure of fault zones. *Journal of Structural Geology*, **19**, 1393–1404.

FITZGERALD, J.D. & STÜNITZ, H. 1993. Deformation of granitoids at low metamorphic grade. I: Reactions and grain size reduction. *Tectonophysics*, **221**, 269–297.

FREY, M., HUNZEKER, J.C., FRANK, W., BOCQUET, J., DAL PIAZ, G.V., JÄGER, E. & NIGGLI, E. 1974. Alpine Metamorphism of the Alps. *Schweizerische Mineralogische und Petrographische Mitteilungen*, **51**, 247–289.

FREY, M., DESMONS, J. & NEUBAUER, F. 1999. Metamorphic maps of the Alps: Map of Alpine metamorphism. *Schweizerische Mineralogische und Petrographische Mitteilungen*, **79**,1–189.

GILLCRIST, R., COWARD, M. & MUGNIER, J.-L. 1987. Structrual inversion and its controls: examples from the Alpine foreland and the French Alps. *Geodynamica Acta*, **1**, 5–34.

GILOTTI, J.A. 1989. Reaction progress during mylonitisation of basaltic dikes along the Särv thrust, Swedish Caledonides. *Contributions to Mineralogy and Petrology*, **101**, 30–45.

GODDARD, J.V. & EVANS, J.P. 1995. Chemical changes and fluid-rock interaction in faults of crystalline thrust sheets, northwestern Wyoming, U.S.A. *Journal of Structural Geology*, **17**, 553–547.

GOODWIN, L.B. & WENK, H.-R. 1995. Development of phyllonite from granodiorite: Mechanisms of grain-size reduction in the Santa Rosa mylonite zone, California. *Journal of Structural Geology*, **17**, 689–707.

GRAND, T., MÉNARD, G. & BONHOMME, M.G. 1989. Tectonic implications of Upper Oligocene–Lower Miocene K–Ar ages of fine fractions from the metasedimentary cover around Bourg d'Oisans (French Western Alps). *Terra Abstracts*, **1**, 372–373.

GRANT, N.T., BANKS, D.A., McCAIG, A.M. & YARDLEY, B.W.D. 1990. Chemistry, source, and behaviour of fluids involved in Alpine thrusting of the Central Pyrenees. *Journal of Geophysical Research*, **95**, 9123–9131.

GUEYDAN, F., LEROY, Y.M., JOLIVET, L. & AGARD, P. 2003. Analysis of continental midcrustal strain localization induced by microfracturing and reaction-softening. *Journal of Geophysical Research*, **108**, 2064, doi:10.1029/2001JB000611.

HANDY, M.R. 1990. The solid-state flow of polymineralic rocks. *Journal of Geophysical Research*, **95**, 8647–8661.

HANDY, M.R. 1994a. Flow laws for rocks containing two non-linear viscous phases: a phenomenological approach. *Journal of Structural Geology*, **16**, 287–301.

HANDY, M.R. 1994b. Correction to: 'Flow laws for rocks containing two non-linear viscous phases: a phenomenological approach'. *Journal of Structural Geology*, **16**, 1727.

HANDY, M.R. & STÜNITZ, H. 2002. Strain localization by fracturing and reaction weakening – a mechanism for initiating exhumation of subcontinental mantle beneath rifted margins. *In*: DE MEER, S., DRURY, M.R., DE BRESSER, J.H.P. & PENNOCK, G.M. (eds) *Deformation Mechanisms, Rheology and Tectonics: Current Status and Future Perspectives:* Geological Society, London, Special Publications, **200**, 387–407.

HATCHER, R.D., JR. & WILLIAMS, R.T. 1986. Mechanical model for single thrust sheets Part I: Taxonomy of crystalline thrust sheets and their relationships to the mechanical behaviour of orogenic belts. *Geological Society of America Bulletin*, **97**, 975–985.

HEMLEY, J.J. & JONES, W.R. 1964. Chemical aspects of hydrothermal alteration with emphasis on hydrogen metasomatism. *Economic Geology*, **59**, 538–569.

HUENGES, E., ERZINGER, J., KÜCK, J., ENGESER, B. & KESSELS, W. 1997. The permeable crust: Geohydraulic properties down to 9101 m depth. *Journal of Geophysical Research*, **102**, 18 255–18 265.

HULL, J. 1988. Thickness-displacement relationships for fault zones. *Journal of Structural Geology*, **10**, 431–435.

IMBER, J., HOLDSWORTH, R.E. & BUTLER, C.A. 2001. A reappraisal of the Sibson–Scholz fault zone model: The nature of the frictional-to-viscous ('brittle–ductile') transition along a long-lived, crustal-scale fault, Outer Hebrides, Scotland. *Tectonics*, **20**, 601–624.

IMBER, J., HOLDSWORTH, R.E., BUTLER, C.A. & LLOYD, G.E. 1997. Fault zone weakening processes along the reactivated Outer Hebrides Fault Zone, Scotland. *Journal of the Geological Society, London*, **154**, 105–109.

JANECKE, S.U. & EVANS, J.P. 1988. Feldspar-influenced rock rheologies. *Geology*, **16**, 1064–1067.

JUHÁSZ, A, TÓTH, T.M., RAMSEYER, K. & MATTER, A. 2002. Connected fluid evolution in fractured crystalline basement and overlying sediments, Pannonian Basin, SE Hungary. *Chemical Geology*, **182**, 91–120.

JULLIEN, M. & GOFFÉ, B. 1993. Cookeite and pyrophyllite in the Dauphinois black shales (Isère, France): implications for the conditions of metamorphism in the Alpine external zones. *Schweizerische Mineralogische und Petrographische Mitteilungen*, **73**, 357–363.

KIM, J.H. 1981. *Tectonics of the Moine Thrust Zone, Assynt, Scotland.* PhD thesis, Leeds University.

KNIPE, R.J. 1981. The interaction of deformation and metamorphism in slates. *Tectonophysics*, **78**, 249–272.

KNIPE, R.J. 1989. Deformation mechanisms – recognition from natural tectonites. *Journal of Structural Geology*, **11**, 127–146.

KNIPE, R.J. 1990. Microstructural analysis and tectonic evolution in thrust systems: examples from the Assynt region of the Moine Thrust Zone, Scotland. *In*: BARBER, D.J. & MEREDITH, P.G. (eds) *Deformation Mechanisms in Ceramics, Minerals and Rocks*. Unwin Hyman, London, 228–261.

KNIPE, R.J. & WINTSCH, R.P. 1985. Heterogeneous, deformation, foliation development, and metamorphic processes in a polyphase mylonite. *In*: THOMPSON, A.B. & RUBIE, D.C. (eds) *Advances in Physical Geochemistry*, Volume 4. Springer, Berlin, 180–210.

KOZLOVSKY, YE. A. (ed.) 1987. *The Superdeep Well of the Kola Peninsula*. Springer, Berlin.

LAPWORTH, C. 1883. The Highland controversy in British Geology. *Nature*, **32**, 558–559.

LLOYD, G.E. & KNIPE, R.J. 1992. Deformation mechanisms accommodating faulting of quartzite under upper crustal conditions. *Journal of Structural Geology*, **14**, 127–143.

MARES, V.M. & KRONENBERG, A.K. 1993. Experimental deformation of muscovite. *Journal of Structural Geology*, **15**, 1061–1075.

MCCAIG, A.M. 1987. Deformation and fluid-rock interaction in metasomatic dilatant shear bands. *Tectonophysics*, **135**, 121–132.

MCCAIG, A., WICKHAM, S.M. & TAYLOR, H.P. JR. 1990. Deep fluid circulation in alpine shear zones, Pyrenees, France: field and oxygen isotope studies. *Contributions to Mineralogy and Petrology*, **106**, 41–60.

MCCAIG, A.M., WAYNE, D.M. & ROSENBAUM, J.M. 2000. Fluid expulsion and dilatancy pumping during thrusting in the Pyrenees: Pb and Sr isotope evidence. *Bulletin of the Geological Society of America*, **112**, 1199–1208.

MCCLAY, K.R. & COWARD, M.P. 1981. The Moine Thrust Zone: an overview. *In*: MCCLAY, K.R. & PRICE, N.J. (eds) *Thrust and Nappe Tectonics*. Geological Society, London, Special Publications, **9**, 241–260.

MCKIE, T. 1989. Barrier island to tidal shelf transition in the early Cambrian Eriboll Sandstone. *Scottish Journal of Geology*, **25**, 273–293.

MEANS, W.D. 1995. Shear zones and rock history. *Tectonophysics*, **247**, 157–160.

MITRA, G. 1979. Ductile deformation zones in Blue Ridge basement rocks and estimation of finite strains. *Geological Society of America Bulletin*, **90**, 935–951.

MITRA, G. 1992. Deformation of granitic basement rocks along fault zones at shallow to intermediate crustal levels. *In*: MITRA, S. & FISHER, G.W. (eds) *Structural Geology of Fold and Thrust Belts*. John Hopkins University Press, Baltimore, MD, 123–144.

MUGNIER, J.-L., GUELLEC, S., MÉNARD, G., ROURE, F., TARDT, M. & VAILON, P. 1990. A crustal scale balanced cross-section through the external Alps deduced from the ECORS profile. *In*: ROURE, F., HEITZMANN, P. & POLINO, R. (eds) *Deep Structure of the Alps*. Mémoire de la Société géologique de la France, **156**. Mémoire de la Société géologique de la Suisse, **1**, 203–216.

NESBITT, B.E. & MUEHLENBACHS, K. 1991. Stable isotopic constraints on the nature of the syntectonic fluid regime of the Canadian Cordillera. *Geophysical Research Letters*, **18**, 963–966.

O'HARA, K. 1988. Fluid flow and volume loss during mylonitisation: an origin for phyllonite in an overthrust setting, North Carolina, U.S.A. *Tectonophysics*, **156**, 21–36.

PASSCHIER, C.W. & TROUW, R.A.J. 1998. *Microtectonics*. Springer, Berlin.

PEACH, B.N., HORNE, J., GUNN, W., CLOUGH, C.T. & HINXMAN, L.W. 1907. The geological structure of the north-west Highlands of Scotland. Memoirs of the Geological Survey of Great Britain.

PFIFFNER, O.A. 1986. Evolution of the north Alpine foreland basin in the Central Alps. *In*: ALLEN, P.A. & HOMEWOOD, P. (eds) *Foreland Basins*. International Association of Sedimentologists, Special Publications, **8**, 219–228.

PLATT, J.P. 1986. Dynamics of orogenic wedges and the uplift of high-pressure metamorphic rocks. *Geological Society of America Bulletin*, **97**, 1037–1053.

POWER, W.L., TULLIS, T.E. & WEEKS, J.D. 1988. Roughness and wear in brittle faulting. *Journal of Geophysical Research*, **93**, 15 268–15 278.

RAMSAY, J.G. 1969. The measurement of strain and displacement in orogenic belts. *In*: KENT, P.E., SATTERTHWAITE, G.E. & SPENCER, A.M. (eds) *Time and Place in Orogeny*. Geological Society, London, 43–79.

ROBERTSON, E.C. 1983. Relationship of fault displacement to gouge and breccia thickness. *Mining Engineering*, **35**, 1426–1432.

RUBIE, D.C. 1990. Mechanisms of reaction-enhanced deformability in minerals and rocks. *In*: BARBER, R.J. & MEREDITH, P.G. (eds) *Deformation Mechanisms of Ceramics, Minerals and Rocks*. Unwin Hyman, London, 262–295.

SCHMID, S.M. 1982. Microfabric studies as indicators of deformation mechanisms and flow laws operative in mountain building. *In*: HSÜ, K.J. (ed.) *Mountain Building Processes*. Academic Press, New York, 95–110.

SCHMID, S.M. & HANDY, M.R. 1991. Towards a genetic classification of fault rocks: geological usage and tectonophysical implications. *In*: MULLER, D.W., MCKENZIE, J.A. & WEISSERT, H. (eds) *Controversies in Modern Geology*. Academic Press, London, 339–361.

SCHMID, S.M. & KISSLING, E. 2000. The arc of the western Alps in the light of geophysical data on deep crustal structure. *Tectonics*, **19**, 62–85.

SCHOLZ, C.H. 1987. Wear and gouge formation in brittle faulting. *Geology*, **15**, 493–495.

SEGALL, P. & POLLARD, D.D. 1983. Nucleation and growth of strike-slip faults in granite. *Journal of Geophysical Research*, **88**, 555–568.

SEGALL, P. & SIMPSON, C. 1986. Nucleation of ductile shear zones on dilatant fractures. *Geology*, **14**, 56–59.

SOPER, N.J. & HUTTON, D.H.W. 1984. Late Caledonian sinistral displacements in Britain: Implications for a three plate collision model. *Tectonics*, **3**, 781–794.

SOPER, N.J. & ENGLAND, R.W. 1995. Vendean and Riphean rifting in NW Scotland. *Journal of the Geological Society, London*, **152**, 11–14.

STAHL, S.D. 1992. Thick-skinned overstep tectonics in the Jurassic Winnemucca fold-and-thrust belt, north-central Nevada, USA: Evidence from the Sonoma Range. *In*: BARTHOLOMEW, M.J., HYNDMAN, D.W., MOGK, D.W. & MASON, R. (eds) *Basement Tectonics 8: Characterisation and Comparison of Ancient and Mesozoic Continental Margins.* Kluwer, Dordrect, 249–261.

STEL, H. 1986. The effect of cyclic operation of brittle and ductile deformation on the metamorphic assemblage in cataclasites and mylonites. *Pure and Applied Geophysics*, **124**, 289–307.

STÜNITZ, H. & FITZGERALD, J.D. 1993. Deformation of granitoids at low metamorphic grade. II: Granular flow in albite-rich mylonites. *Tectonophysics*, **221**, 299–324.

TCHALENKO, J.S. 1970. Similarities between shear zones of difference magnitudes. *Bulletin of the Geological Society of America*, **81**, 1625–1640.

TRICART, P. 1984. From passive margin to continental collision: A tectonic scenario for the Western Alps. *American Journal of Science*, **284**, 97–120.

TULLIS, T.E. & WEEKS, J.D. 1986. Constitutive behaviour and stability of frictional sliding of granite. *Pure and Applied Geophysics*, **124**, 383–414.

WALLACE, R.E. & MORRIS, H.T. 1986. Characteristics of faults and shear zones in deep mines. *Pure and Applied Geophysics*, **124**, 107–125.

WHEELER, J. 1992. Importance of pressure solution and coble creep in the deformation of polymineralic rocks. *Journal of Geophysical Research*, **97**, 4579–4586.

WHITE, J.C. & WHITE, S.H. 1983. Semi-brittle deformation within the Alpine fault zone, New Zealand. *Journal of Structural Geology*, **5**, 579–589.

WHITE, S.H., EVANS, D.J. & ZHONG, D.-L. 1982. Fault rocks of the Moine Thrust Zone: microstructures and textures of selected mylonites. *Textures and Microstructures*, **5**, 33–61.

WHITE, S.H. & KNIPE, R.J. 1978. Transformation and reaction enhanced ductility in rocks. *Journal of the Geological Society, London*, **135**, 513–516.

WIBBERLEY, C.A.J. 1995. *Basement involvement and deformation in foreland thrust belts.* PhD thesis, Leeds University.

WIBBERLEY, C.A.J. 1997. A mechanical model for the reactivation of compartmental faults in basement thrust sheets, Muzelle region, Western Alps. *Journal of the Geological Society, London*, **154**, 123–128.

WIBBERLEY, C.A.J. 1999. Are feldspar-to-mica reactions necessarily reaction-softening processes in fault zones? *Journal of Structural Geology*, **21**, 1219–1227.

WIBBERLEY, C.A.J. & McCAIG, A.M. 2000. Quantifying orthoclase and albite muscovitisation sequences in fault zones. *Chemical Geology*, **165**, 181–196.

WIBBERLEY, C.A.J., PETIT, J.-P. & RIVES, T. 2000. Micromechanics of shear rupture and the control of normal stress. *Journal of Structural Geology*, **22**, 411–427.

WIBBERLEY, C.A.J. & SHIMAMOTO, T. 2003. Internal structure and permeability of major strike-slip fault zones: the Median Tectonic Line in W. Mie Prefecture, S.W. Japan. *Journal of Structural Geology*, **25**, 59–78.

WINTSCH, R.P. & KNIPE, R.J. 1985. The possible effects of deformation on chemical processes in metamorphic fault zones. *In*: THOMPSON, A.B. & RUBIE, D.C. (eds) *Advances in Physical Geochemistry*, Volume 4. Springer, Berlin, 251–268.

WINTSCH, R.P., CHRISTOFFERSEN, R. & KRONENBERG, A.K. 1995. Fluid–rock reaction weakening of fault zones. *Journal of Geophysical Research*, **100**, 13 021–13 032.

WOJTAL, S. & MITRA, G. 1986. Strain hardening and strain softening in fault zones from foreland thrust belts. *Bulletin of the Geological Society of America*, **97**, 674–687.

YONKEE, W.A. & MITRA, G. 1993. Comparison of basement deformation styles in parts of the Rocky Mountain foreland, Wyoming, and the Sevier Orogenic Belt, Northern Utah. *In*: SCHMIDT, C.J., CHASE, R.B. & ERSLEV, E.A. (eds) *Laramide Basement Deformation in the Rocky Mountain Foreland of the Western United States.* Geology Society of America, Special Publications, **280**, 197–228.

YOSHIOKA, N. 1986. Fracture energy and the variation of gouge and surface roughness during the frictional sliding of rocks. *Journal of the Physics of the Earth*, **34**, 335–355.

Geochemical variations and element transfer during shear-zone development and related episyenites at middle crust depths: insights from the Mont Blanc granite (French–Italian Alps)

M. ROSSI[1], Y. ROLLAND[2], O. VIDAL[1] & S. F. COX[3]

[1]*Laboratoire de Géodynamique des Chaînes Alpines, OSUG, BP53, Université J. Fourier, 38041 Grenoble Cédex, France (e-mail: mrossi@ujf-grenoble.fr)*

[2]*Laboratoire de Géochronologie, Géosciences Azur, 28 Avenue de Valrose, BP 2135, 06103 Nice, France*

[3]*Department of Earth and Marine Sciences, the Australian National University, Canberra, ACT 0200, Australia*

Abstract: This paper highlights the relationships between the formation of shear zones, associated quartz-rich veins and their quartz-depleted alteration haloes ('episyenites') that have formed in the Mont Blanc Massif during the Alpine orogeny. The shear zones are steeply dipping and formed late (18–13 Ma) during collisional orogeny, at mid-crustal depths (5 ± 1 kbar, 400 ± 50 °C) during uplift of the Mont Blanc Massif. Between the shear zones, nearly undeformed granite contains widely dispersed, subhorizontal veins with a quartz-dominant quartz + albite + chlorite + adularia assemblage. They do not intersect the shear zones and are surrounded by quartz-depleted alteration haloes up to several metres wide. The compositions of the shear zones and the vein-alteration haloes (episyenites) show substantial departures from the bulk composition of the host rock. Shear zones are characterized by greenschist facies assemblages (epidote-, chlorite- or K-white-mica-bearing assemblages). Each shear zone type is featured by a specific chemical change: depletions in K_2O, and enrichments in Fe_2O_3 and CaO (epidote-); with depletions in CaO, Na_2O, K_2O and slight SiO_2 enrichments (white mica–chlorite-); with depletions in SiO_2, CaO, Na_2O, K_2O and enrichments in MgO (phlogopite–chlorite shear zones). Episyenites are characterized by chemically induced porosity enhancement due to dissolution of magmatic quartz and biotite, with subsequent partial infilling of pore spaces by quartz, chlorite, albite and adularia. The vein arrays have accommodated minor vertical stretching in the Mont Blanc Massif, probably at the same time as the adjacent shear zones were accommodating more substantial vertical stretching in the massif. Coupled quartz dissolution in the wallrock alteration haloes and quartz precipitation in veins could be interpreted to reflect local mass transfer between wallrock and veins during essentially closed-system behaviour in the relatively undeformed granite domains between shear zones. In contrast, shear zones probably develop in opened systems due to their kilometric length.

Studies of fluid–rock interaction processes during deformation in metamorphic rock provide insights about fluid circulation in the middle–lower continental crust. It also provides constraints to estimate the magnitude of mass transfer during orogenic events. Mass balance calculations during fluid–rock interaction are easier when based on rocks of homogeneous composition at regional scale. This is rather common in many granitic massifs and therefore granitic rocks are good candidates for such studies.

In metamorphosed granites, fluid–rock interactions are mainly localized along shear zones,

which are narrow corridors through which fluid may migrate. Intense brittle–ductile deformation in shear zones in the middle crust is commonly associated with feldspar-to-mica breakdown in fluid-present conditions and may result in reaction weakening (e.g. Janecke & Evans 1988; Guermani & Pennacchioni 1998; Wibberley 1999). Numerous studies of shear zones in different kinds of rocks have attempted to determine the nature of the circulating fluid, and to estimate the time-integrated fluid flux using mass balance calculations (e.g. Sinha *et al.* 1986; Glazner & Bartley 1991; Dipple & Ferry 1992; Streit & Cox 1998). However, estimates

From: BRUHN, D. & BURLINI, L. (eds) 2005. *High-Strain Zones: Structure and Physical Properties*.
Geological Society, London, Special Publications, **245**, 373–396.
0305-8719/05/$15.00 © The Geological Society of London 2005.

of time-integrated fluid flow range from 10^2 to 10^6 m^3 m^{-2} (Etheridge & Cooper 1981; Marquer 1989; Jamtveit *et al.* 1990; Dipple & Ferry 1992; Streit & Cox 1998) and the structural driving mechanism controlling fluid flow (equilibrium, disequilibrium) is still debated (Marquer 1989; Ferry & Gerdes 1998; Streit & Cox 1998). Moreover, fluid transfer in granitic rocks is not restricted to advection through shear zones. Indeed, pervasive alteration of granitic rocks resulting in transport by diffusion (e.g. Cesare 1994; Watson & Wark 1997; Vidal & Durin 1999; Widmer & Thompson 2001) and pervasive fluid–rock interactions evidenced by the growth of metamorphic minerals are often observed in metagranite. This alteration pattern can result from the diffusion of elements through a static fluid at grains boundaries. Experiments demonstrated that diffusive mass transport can be a very effective for short-range (e.g. a few metres) mass transfer in the crust (Watson & Wark 1997). Observations from metapelitic rocks indicate that the formation of some mineralized veins results from diffusive mass transport during regional metamorphism. In order to understand fluid fluxes in the crust, it is primarily necessary to identify the dominant process of mass transfer, and the relationships between shear zones and more pervasive alteration features at different scales.

The present study focuses on the Mont Blanc granitic massif (NW Alps). The southern part of the massif corresponds to the deepest rocks of the batholith, while the northern part corresponds to the top of the batholith. This specific geometry allows sampling of alteration features along a vertical profile of the exhumed crustal segment. In this massif, fluid–rock interaction during Alpine deformation is associated to two distinct structural features: (1) metre- to kilometre-scale subvertical shear zones; and (2) zones of pervasive alteration (so called 'episyenites': Poty 1969; Leutwein *et al.* 1974) surrounding horizontal veins. The episyenitic alteration is characterized by the dissolution of quartz and biotite from the protolith. Such features have been described in several Hercynian massifs, in some cases in association with shear zones (e.g. Cathelineau 1986, 1987; Hecht *et al.* 1999). However, in other cases (e.g. Hålenius & Smellie 1983) there is no clear association with shear zones. The episyenites have been interpreted as resulting from the circulation of late magmatic fluids. In the Mont Blanc Massif the episyenites do not result from the circulation of late magmatic and meteoritic fluids, as it is the case for most episyenites described in the literature for Variscan European massifs

(e.g. Cathelineau 1986; Recio *et al.* 1997; Hecht *et al.* 1999). Both shear zones and veins/ episyenites developed at mid-crustal depths during the Alpine collisional event, in a relatively well-constrained tectonic setting (Bussy 1990; Rolland *et al.* 2003).

The aim of the present study is to discuss the process of episyenitization in mid-crustal conditions and its relationships with shear-zone development. We first describe the mineralogical changes in both episyenites and shear zones, and compare the geochemical variations from the protolith composition. These observations are used to discuss the processes responsible for the observed alteration pattern and related mass transfer.

Geological setting and Alpine deformation pattern

The Mont Blanc Massif (MBM) belongs to the Hercynian External Crystalline massifs of the NW Alps (Fig. 1a). It is a tectonic unit 35×10 km in size, elongated NE–SW, composed of gneisses and a granitic intrusion. Gneisses form the western rim and the southern part of the MBM (Bellière 1988) (Fig. 1b). The granites have a calc-alkaline composition (Marro 1986; Bussy 1990), and have been dated at 300 ± 3 Ma by U–Pb on zircon (Bussy & von Raumer 1994). The MBM was deformed during the Alpine orogenesis, and this phase is recorded by Rb–Sr and Ar ages of deformed and undeformed granite (Baggio 1958; Leutewin *et al.* 1974; Marshall *et al.* 1997). Pressure–Temperature ($P–T$) conditions of 300–400 °C and 1.5–4 kbar have been estimated from fluid-inclusion analysis of Alpine vein minerals (Poty *et al.* 1974; Fabre *et al.* 2002), from isochore reconstruction for quartz–chlorite–adularia–albite veins and from feldspar Na/K thermometry. Rolland *et al.* (2003) estimated peak metamorphic conditions at 450 °C and 4–6 kbar from the mineralogy of K-white mica shear zones using a multi-*equilibria* approach (Vidal & Parra 2000; Vidal *et al.* 2001). These $P–T$ conditions were reached about 13–18 Ma ago (Baggio *et al.* 1967; Leutwein *et al.* 1974). Within the Mont Blanc granites, the Alpine deformation and fluid–rock interactions resulted in an association of synkinematic cataclastic to ductile shear zones, episyenites and veins (Guermani & Pennacchioni 1998). The shear zones are all transpressive and form a complex network of anastomosing NNE–SSW (N40–60°E) and N–S (N160–20°E) components (Fig. 2) with almost vertical stretching lineations. In the nearly

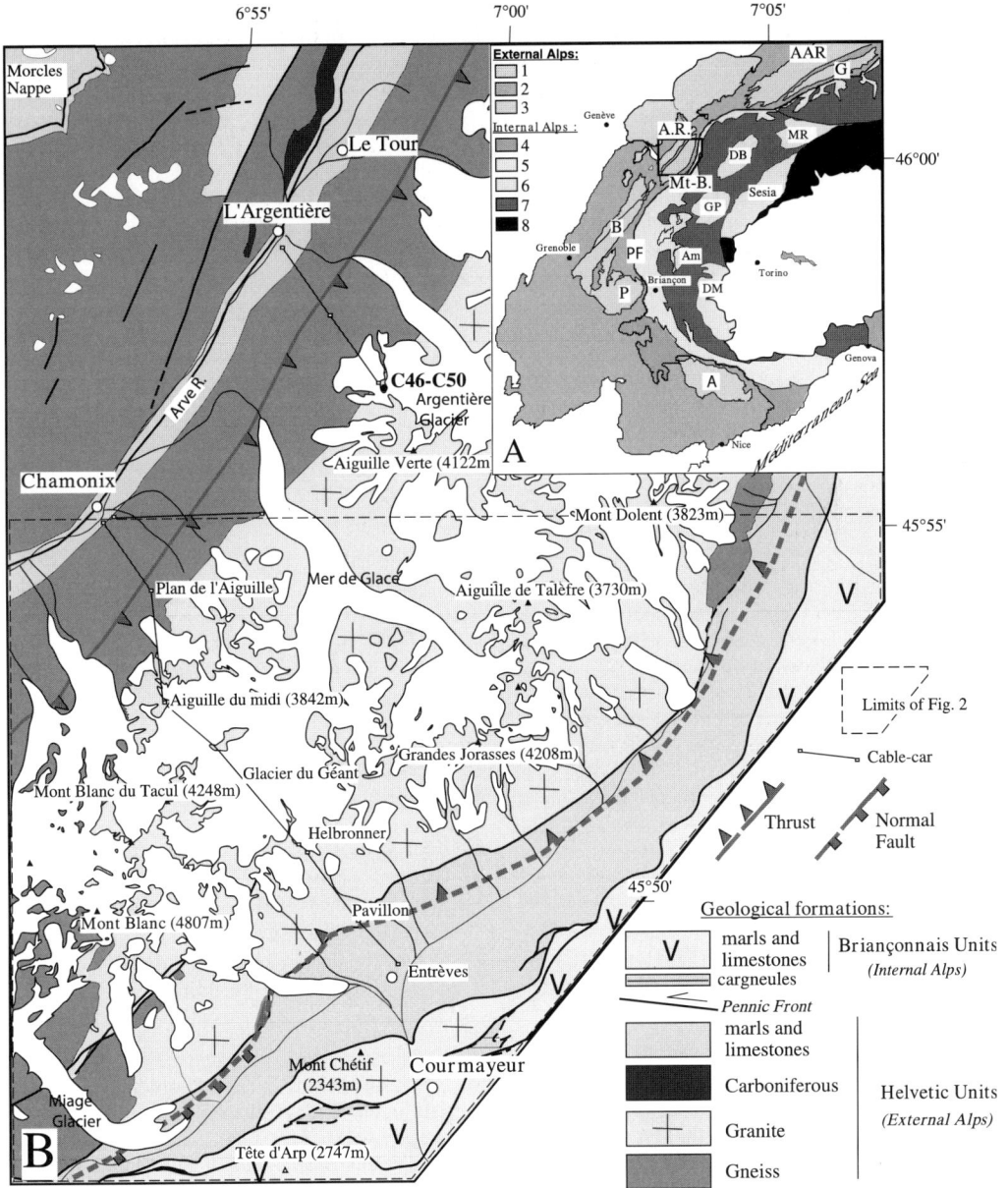

Fig. 1. (**a**) Geological map of the Western Alps. 1, pre-Alpine nappes; 2, Dauphinois/Helvetic zones; 3, external crystalline massifs; 4, exotic flysch; 5, Briançonnais zone; 6, internal crystalline massifs; 7, Schistes Lustrés complex; 8, Austro-Alpine units of the Western Alps. External crystalline massifs: Arg., Argentera; A.R., Aiguilles Rouges; Bel., Belledonne; G., Gothard; Mt-B., Mont Blanc; P., Pelvoux. Internal crystalline massifs: Am., Ambin; DB., Dent Blanche; GP, Gran Paradiso; MR, Monte Rosa. (**b**) Simplified geological map of the Mont Blanc Massif. Modified from Baggio (1958) and Bussy (1990).

undeformed domains between shear zones, sub-horizontal veins are surrounded by the episyenites haloes. The granite has a weak subvertical greens-chist facies foliation of average N40°E strike. The structural relationships of veins, shear zones and foliation indicate that Alpine deformation of the MBM involved horizontal NW–SE shortening and vertical stretching (Fig. 3).

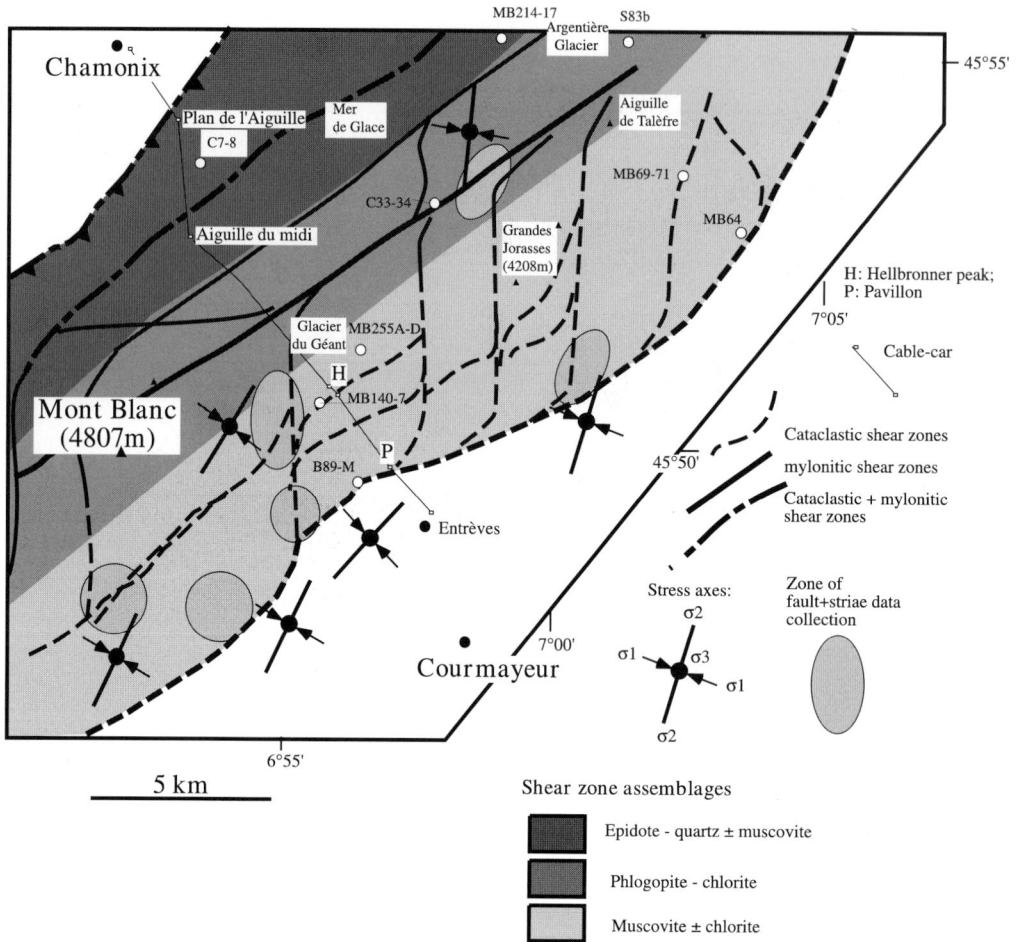

Fig. 2. Metamorphic and structural map of the Mont Blanc Massif, with location of samples discussed in the geochemical part. Bt, biotite; Chl, chlorite; Ms, muscovite; Phl, phlogopite; Qtz, quartz; Tit, titanite.

At the regional scale, in a NW–SE section of the MBM, the shear zones oriented N40°E are arranged in a fan-like geometry (Fig. 3), which define a pop-up structure with divergent NW- and SE-directed thrusting of the crystalline basement over its Helvetic cover (Antoine *et al.* 1975; Bertini *et al.* 1985; Butler 1985; Bellière 1988).

The *shear zones* consist of a relatively narrow (1–50 m) zones of intense vertical flattening surrounding large lozenge-shaped to tabular regions (100–500 m) of less deformed granite (Figs 2 and 4a) that form the Mont Blanc Massif spires. The shear zones form the depressions between the less deformed domains (Fig. 4a). Within shear zones, cataclasites, mylonites and ultramylonites are developed. Cataclasites are distinguished by the brittle deformation of quartz and feldspar, which form a fine-grained crushed matrix. In mylonites, quartz porphyroclasts show subgrain development and was recrystallized by processes of grain-boundary migration (e.g. Passchier & Trouw 1998), while plagioclase and K-feldspar show lattice bending. C–S bands develop by the elongation of K-feldspar clasts and are shown by the sigma shape of quartz aggregates. In ultramylonites, these processes give rise to a very-fine-grained matrix, whereas in mylonites K-feldspars are fractured and form domino-like structures. In shear zones, the schistosity is defined by grain shape preferred orientations of elongate metamorphic minerals (muscovite, epidote, chlorite), which crystallized as a result of breakdown of plagioclase, biotite and K-feldspar (Fig. 5b). Based on the alteration mineralogy of the shear zones, three main

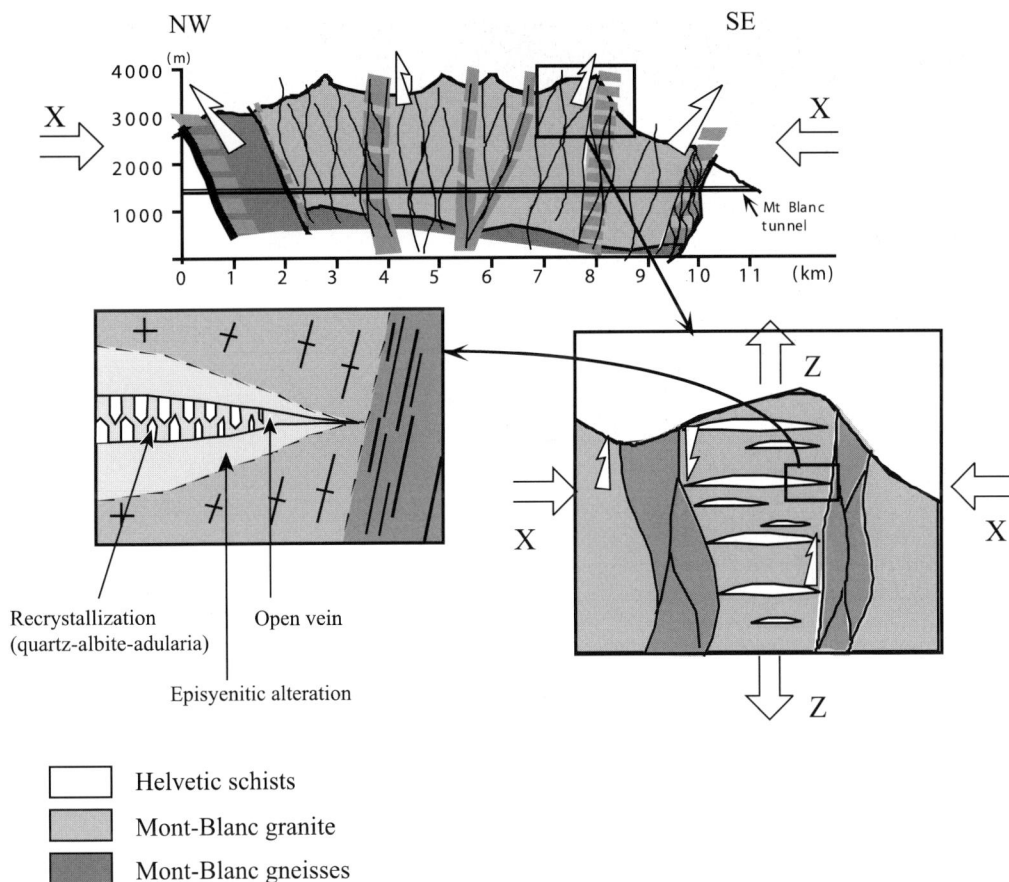

Fig. 3. Structural relationships between shear zones and episyenites in the Mont Blanc Massif.

groups of shear zones have been distinguished from NW to SE across the MBM: epidote-, chlorite- and muscovite-bearing zones (Rolland *et al.* 2003) (Fig. 2).

The *episyenites* occur in relatively unde-formed granitic lenses, mostly in the central and SE parts of the MBM. They are zones of strongly altered (almost total quartz removal and biotite-leached) granite surrounding sub-horizontal veins filled mainly by quartz, albite, adularia and chlorite (Fig. 4c & d). These veins open with the progressive attenuation of the flat-tening on shear-zones boundaries, shown by a perfectly preserved structure in the granite. Further, veins are never found cross-cutting shear zones. So it appears that veins, episyenites and shear zones are structurally related (Figs 3 and 4d). Nevertheless, episyenites developed in the undeformed granite and never cross-cut shear zones. Episyenite alteration haloes reach up to 1 m in thickness. They are highly porous

rocks produced by the dissolution of quartz and biotite, but which otherwise still preserve the original igneous texture (Fig. 5c). Secondary authigenic quartz crystals, albite, chlorite and adu-laria, as well as calcite and minor fluorite, partly fill the porosity of episyenites and open veins (Fig. 6). The transition from the protolith to the episyenite is gradational, with the strongest altera-tion (strongest loss of quartz) close to the veins.

Structural context of shear-zone deformation

Detailed analyses of shear-zone outcrops in the SE part of the MBM indicate a progressive tran-sition from relatively undeformed granite to cat-aclasites, mylonites and ultramylonites (Fig. 2). In the SE part of the massif (muscovite-bearing shear zones) we observed an evolution in both space and time from cataclasites to mylonites,

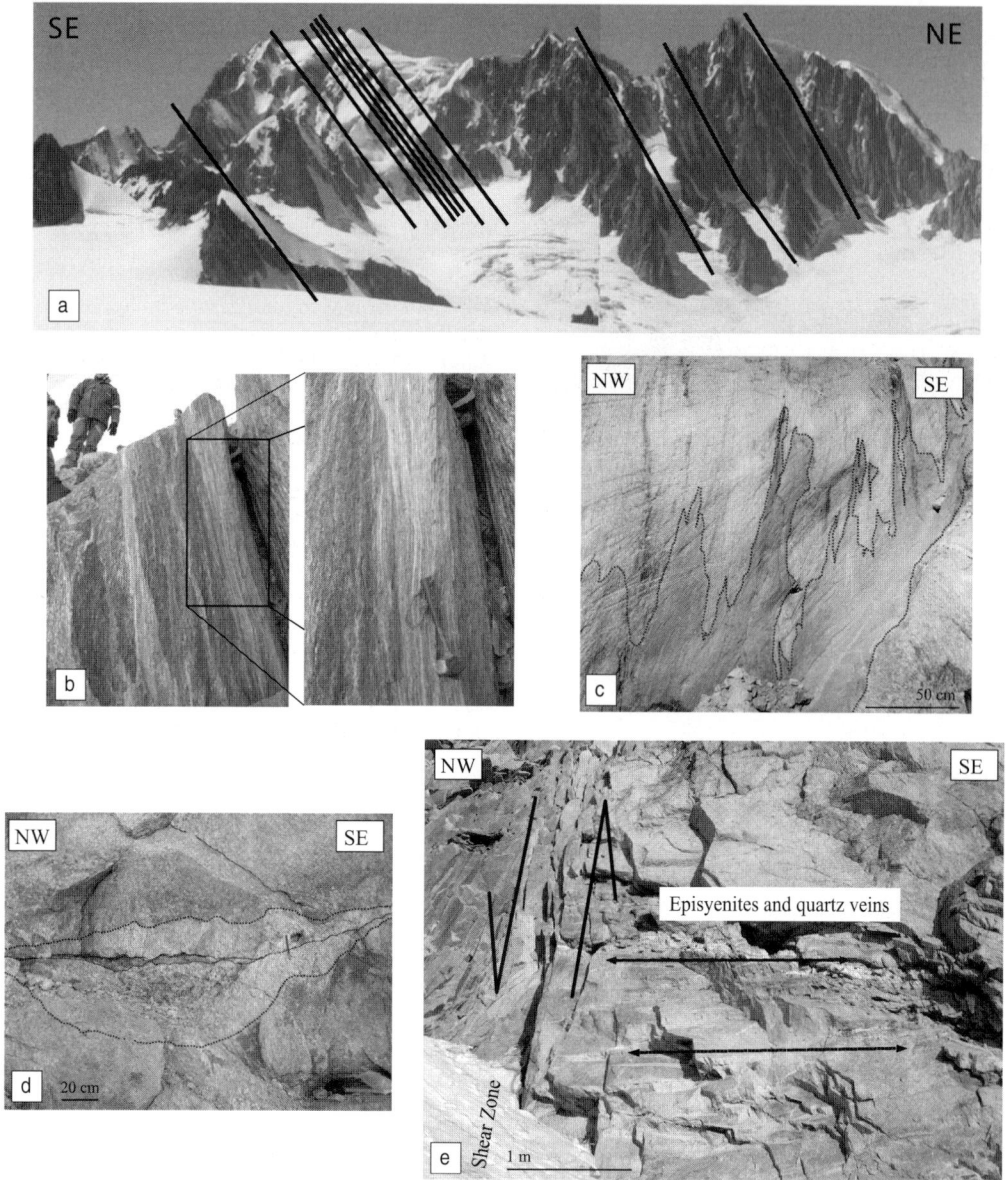

Fig. 4. Aspects of shear zones and episyenites in the Mont Blanc Massif. (**a**) Distribution of shear zones on the Mont Blanc summit. (**b**) Transition from cataclasite to mylonite and ultramylonite, Glacier du Tour. (**c**) Front of Mg-metasomatism close to a shear zone, Glacier de Leschaux. (**d**) Episyenite halo on the border of Glacier d'Argentière. (**e**) Structural relationship between 'en echelon' episyenites and the shear zone, Glacier des Améthystes.

with features demonstrating a progressive transition from brittle to subsequent ductile deformation during reaction weakening associated with the transformation of feldspars into muscovite. In the central part of the massif no cataclasites are observed, and the deformation is predominantly mylonitic, although feldspars tend to deform in a brittle manner in a ductile quartz-rich matrix. There, the transition from undeformed granite to shear zones is more progressive (order of tens of metres) than on the SE side of the MBM. Production of chlorite in shear zones could also be associated with reaction weakening. Phlogopite-bearing shear zones

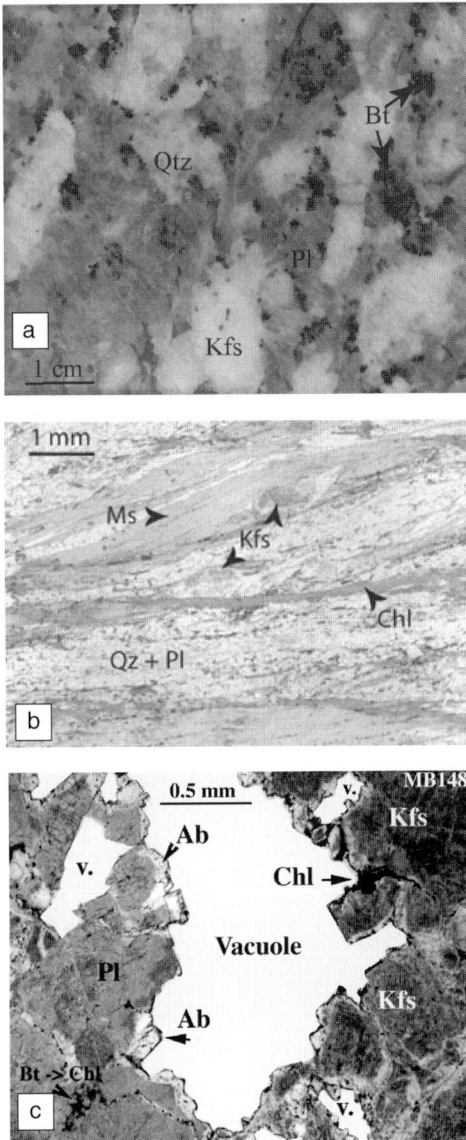

Fig. 5. Textures of the Mont Blanc granite. (a) Undeformed and unaltered granite; (b) high porosity of episyenites; (c) fine-grained matrix and C–S planes highlight by metamorphic phases. Bt, biotite; Chl, chlorite; Kfs, alkali feldspar; Pl, plagioclase; Ms, muscovite; Qtz, quartz; v, void.

NW side of the MBM, mylonites are absent and only cataclasites were observed. Here, shear zones are epidote-bearing and the limited development of phyllosilicates minerals has hindered ductile behaviour of the shear zones.

Therefore, the transformation of the granite protolith into cataclasites or mylonites is strongly influenced by the occurrence of phyllosilicate-producing reactions. Where phyllosilicate was not stable during deformation (epidote zone) the deformation remains mainly cataclastic, while it becomes mylonitic where phyllosilicates were formed due to reaction weakening associated with the transformation of feldspars into phyllosilicates (Guermani & Pennacchioni 1998).

The shear-zone network of the MBM has two main components at regional scale: one trending N40–N60°E with a partly oblique-slip dextral strike-slip sense; and the other N–S with a partly oblique-slip sinistral sense of shear (Fig. 2). Structural data for shear zones, plotted in stereoplot A of Figure 7 show that the N40°E shear-zone trend component, dipping either to the SE or to the NW, is the main shear-zone component. The NE–SW Alpine schistosity is present throughout the massif, as shown by the N40°E fabric indicated in stereo-plot B of Figure 7. The stretching lineation is defined by the elongation of feldspars, enclaves, and the growth of fibres of muscovite, calcite and chlorite on the shear-zone foliation and slip surfaces. It consistently has a plunge direction NE–SW, with a steep average plunge value of 66° to the NE (Fig. 7c). Subvertical shear zones are closely associated with horizontal veins across the massif. Veins cut across the less deformed granitic domains between shear zones, but do not cross-cut the shear zones. Furthermore, vein thickness tends to increase towards the core of the undeformed domains. They are thus interpreted to be subsynchronous to shear-zone deformation. The structural relationships of the schistosity, shear zones and veins is featured in Figure 3. From these relationships, i.e. the average N40°E subvertical strike of the foliation, the almost vertical stretching lineation average lineation value and the horizontal veins, the tectonic context of the Mont Blanc is clearly featured by horizontal NW–SE shortening and vertical extension. The plot of reverse (lineations plunge >45° on the shear plane), dextral and sinistral shear zones (lineations plunge <45° on the shear plane) is shown in Figure 7d. Similar conclusions are derived from the study of shear sense on shear zones. Shear sense was deduced macroscopically in the field from stepped striae and slickenfibres, and was checked on thin sections. On a Wulff stereoplot

are also found in the centre of the massif. A sharp, but extremely irregular, finger-like alteration front separates a phlogopite-rich part of a shear zone from a chlorite-rich part of a shear zone (Fig. 4c). This relationship suggests metasomatism with a pervasive circulation of fluid from the shear zones into the granite. On the

Fig. 6. Scanning electron microscope images of the porosity and metamorphic recrystallization. (**a**) Pore with recrystallizations of albite (on the right) and quartz (on the left); (**b**) recrystallization of adularia with traces of dissolution; (**c**) secondary bipyramidal quartz grains (in the centre) and neo-formed albite; (**d**) pore partly filled with bipyramidal quartz and vermicular chlorite; (**e**) porosity pattern of K-feldpsar; and (**f**) alteration pattern of plagioclase.

(Fig. 7d) the shear-zone data plotted appear to form two quadrants (NE and SW) where deformation is preferentially reverse + sinistral, and the other two (NW and SE) where deformation is preferentially reverse + dextral. These data clearly show that deformation observed at the massif scale is compatible with N135°E shortening and the pop-up exhumation of the MBM. This is confirmed by the inversion of fault +

striae data with the method of Angelier (1990). The results are indicated in Figure 2, showing SE–NW horizontal σ_1 and vertical σ_3 axes.

Petrography and mineralogy

Samples from undeformed granite, shear zones and episyenites were selected for mineralogical and geochemical analyses. The compositions of

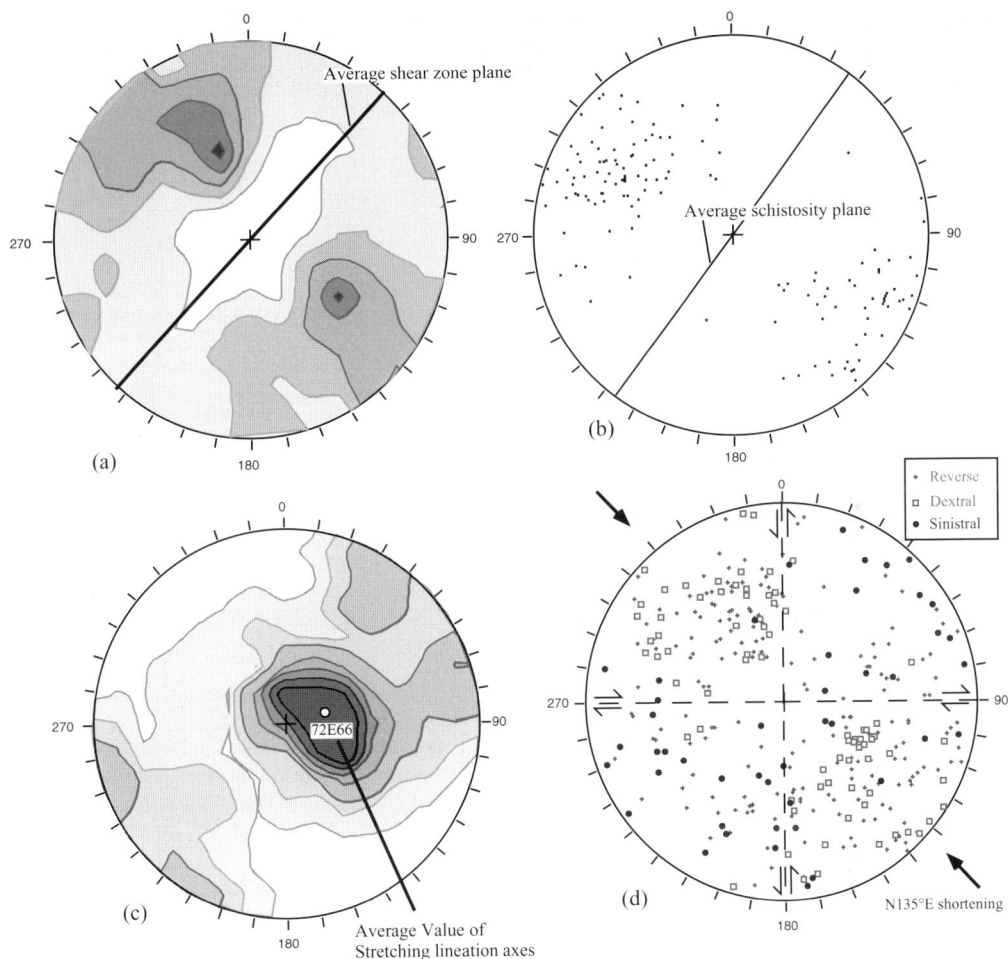

Fig. 7. Structural data of: (**a**) shear-zone pole densities; (**b**) schistosity pole densities; (**c**) lineations; and (**d**) poles of shear zones with related sense of shear. Wulff stereonets in lower hemisphere.

minerals (Tables 1 and 2) were determined by electron microprobe (CAMECA SX 100 and CAMECA Cabemax) analysis at the Laboratoire Magmas et Volcans in Clermont-Ferrand (France), and at the Laboratoire de la Lithosphère in Lyon (France). Synthetic standards were used. Working conditions were 15 kV accelerating voltage and 15 nA beam current.

Undeformed granite

Magmatic minerals. Even in the less deformed meta-granite samples, quartz grains (a few millimetres long) are slightly deformed by the Alpine event (undulatory extinction and initiation of subgrains). K-feldspars (Or_{95-98}) occur as large (few millimetres to several centimetres long)

perthitic porphyrocrystals with a local weak microcline twinning. Plagioclase is 0.1–3 mm in size. It is strongly sericitized, saussuritized and fractured on its rim. The analyses in the present study always show an albitic composition. However, Poty (1969) and Bussy (1990) described plagioclases with normal compositional zoning (An_{32-12}) from core to rim, which suggests that albitization of plagioclase at the massif's scale occurs during pervasive alteration of the granite (see below). The magmatic biotite has an homogeneous composition over the whole massif (76% SiO_2, 19% Al_2O_3 and about 5% dioctahedral component ($X_{Fe} = 0.56$). Various accessory minerals occur in the Mont Blanc granite (see Rolland *et al.* 2003 for more details).

Table 1. *Representative analyses of some minerals from the Bassin d'Argentière area*

Mineral	Biotites				Chlorites				White micas		
Samples	MB02.14	MB02.17	MB02.15		MB02.14	MB02.17	MB02.15		MB02.14	MB02.17	MB02.15
Major oxides (wt%)											
SiO_2	35.89	36.11	38.97		25.32	27.22	27.18		48.65	46.37	48.93
TiO_2	1.34	1.10	1.13		0.01	0.00	0.04		0.27	0.25	0.28
Al_2O_3	16.02	15.82	16.35		18.79	19.23	19.50		27.91	27.89	28.87
FeO	21.24	20.25	15.00		30.66	25.41	23.23		4.90	4.34	3.22
MnO	0.27	0.34	0.27		0.94	0.61	0.13		0.04	0.05	0.03
MgO	9.22	11.25	13.91		11.81	15.17	18.45		2.67	3.68	3.39
CaO	0.05	0.00	0.00		0.06	0.06	0.03		0.08	0.02	0.01
Na_2O	0.03	0.04	0.08		0.02	0.03	0.01		0.09	0.12	0.09
K_2O	10.12	10.01	9.73		0.05	0.39	0.04		10.62	10.01	10.70
Total	94.18	94.90	95.45		87.67	88.12	88.62		95.23	92.72	95.51
X-Fe	0.56	0.50	0.38		0.57	0.46	0.39		0.51	0.40	0.35
%vac	4.76	2.46	4.45	%Al	25.93	16.06	22.20	%Si	7.05	11.23	8.05
%Si	76.43	76.81	84.54	%Si	66.13	70.70	72.36	%Al	64.51	71.23	67.10
%Al	18.81	20.73	11.01	%diO	7.94	13.25	5.45	%Cel	28.45	17.54	24.86

Biotites: %vac, proportion of di-octaerdic biotites; %Si, proportion of the silica-rich biotites (phlogopite–annite); %Al, proportion of aluminium-rich biotites (siderophyllite–eastonite).
White micas: %Si, proportion of silica-rich muscovite; %Al, proportion of aluminium-rich muscovite; %Cel, proportion of celadonite.
Chlorites: %Al, proportion of aluminium-rich chlorites; %Si, proportion of silica-rich chlorites; %diO, proportion di-octaerdic chlorites.
MB02.14, unaltered and slightly deformed granite; MB02.17, mylonite; MB02.15, ultramylonite.

Metamorphic minerals. Two types of Alpine biotites crystallized at the expense of the magmatic one: (1) large green biotite crystals with cloudy patches of accessory minerals such as apatite, zircon and titanite that are remnants of original inclusions of the magmatic biotites; and (2) small-size crystals of green biotite with no accessory mineral inclusions.

In the relatively undeformed granite, epidote, chlorite and K-white mica are minor phases and rarely occur together in equilibrium contact at the grain scale, or even in the same sample.

Table 2. *Representative analyses of some minerals from the Helbronner area*

Mineral	Biotites			Chlorites				White micas
Samples	MB140	MB02.55D		MB140	MB02.55Aa verm.	MB02.55D		MB140
Major oxides (wt%)								
SiO_2	37.33	34.88		26.26	23.61	24.53		48.14
TiO_2	0.87	1.01		0.09	0.02	0.04		0.09
Al_2O_3	15.29	16.63		19.81	18.47	19.19		26.98
FeO	17.35	23.75		23.09	29.04	32.84		3.98
MnO	0.49	0.50		1.21	0.74	1.13		–
MgO	11.84	8.06		16.82	10.97	10.89		2.83
CaO	0.55	0.00		–	0.02	0.07		0.29
Na_2O	–	0.06		–	0.00	0.02		–
K_2O	8.24	9.42		0.15	0.02	0.01		10.73
Total	91.96	94.30		87.43	82.90	88.72		93.04
X-Fe	0.45	0.62		0.41	0.58	0.60		0.44
%vac	3.67	2.50	%Al	25.95	28.65	34.14	%Si	3.22
%Si	86.69	73.41	%Si	67.39	67.77	62.23	%Al	61.77
%Al	9.64	24.08	%diO	6.67	3.58	3.63	%Cel	35.01

Biotites: %vac, proportion of di-octaerdic biotites; %Si, proportion of the silica-rich biotites (phlogopite–annite); %Al, proportion of aluminium-rich biotites (siderophyllites and eastonite).
Chlorites: %Al, proportion of aluminium-rich chlorites; %Si, proportion of silica-rich chlorites; %diO, proportion di-octaerdic chlorites.
White micas: %Si, proportion of silica-rich muscovite; %Al, proportion of aluminium-rich muscovite; %Cel, proportion of celadonite.
MB140, mylonite; MB147, ultramylonite; MB02.55Aa, episyenite; MB02.55D, fresh granite verm., vermicular chlorite.

Their occurrence in the relatively undeformed granite is compatible with the mineral assemblages observed in shear zones, which are detailed below. The growth of K-white mica, chlorite or epidote is associated with the widespread albitization of plagioclase, which suggests pervasive alteration of the granites and the presence of an intergranular water-rich fluid.

Shear zones

Epidotes (pistacite) form at the expense of plagioclases or precipitate in cataclastic shear zones in the NW part of the MBM. Magmatic biotite is almost entirely replaced by aggregates of new green–brown metamorphic biotite (biotite II) defining S and C planes. Biotite II can be partly transformed into chlorite and/or muscovite depending on the location in the massif. As in meta-granite, the biotite II in shear zones has a composition closer to the phlogopite end member, with higher Si content and generally lower X_{Fe} than the magmatic biotite (Table 1). A general observation is that the observed X_{Fe} variation in biotite is related to the intensity of deformation. X_{Fe} decreases from about 0.56 in the less deformed granite to 0.50 in mylonites and about 0.38 in ultramylonites of a same outcrop (Table 1, Fig. 8). Similar evolutions are observed for chlorites and white micas (Table 1, Fig. 8): $(X_{Fe})_{musc}$ decreases from 0.51 in meta-granite to 0.40–0.60 in mylonite, and down to 0.35–0.40 in ultramylonite, and $(X_{Fe})_{chl}$ decreases from 0.57 in meta-granites to 0.44–0.50 in mylonites and 0.37–0.40 in ultra-mylonites (Table 1, Fig. 8).

In phlogopite-rich shear zones present in the massif's core, X_{Fe} is very low compared to meta-granites and other shear zones. X_{Fe} can be as low as 0.12 in biotite, 0.13 in chlorite and 0.18 in white mica. When plotting the X_{Fe} evolution from the edge to the core of the massif, a strong and local drop of X_{Fe} of phyllosilicates is observed, from X_{Fe} up to 0.90 in the massif's rims to 0.12 in phlogopite-bearing shear zones in the massif's core (Fig. 8).

As the initial chemical composition of the Mont Blanc granite is relatively homogeneous at regional scale (see the 'Geochemistry' section later), the observed variation of X_{Fe} across a NW–SE profile across the massif (Fig. 8) does not result from initial magmatic heterogeneities, but indicates a control of coupled deformation, fluid–rock reaction and pressure–temperature re-equilibration. Marked variations of X_{Fe} (e.g. samples PK4550 and PK4660) observed between areas less than a few hundred metres apart in the central part of the massif is indicative of very local fluid–rock interactions. It is consistent with the macroscopic observations, which suggest that a Mg-metasomatism was associated with the development of the phlogopite-bearing shear zones.

Episyenites

In episyenites, there is a progressive increase of quartz and biotite dissolution towards the associated vein and an increasing amount of secondary minerals (quartz, K-feldspar, albite, chlorite, ± fluorite and calcite; Figs 5c and 6) within the voids that were produced by dissolution and removal of magmatic quartz and biotite. Despite quartz dissolution and the resulting increase of porosity, the primary magmatic texture is well preserved and the pores have not collapsed during quartz dissolution. As both quartz and biotite are leached out, feldspars (microcline and albite) are the main minerals in episyenites. Magmatic plagioclases are strongly sericitized and fully albitized (An_{0-10}), and K-feldspars are slightly illitized and highly porous (Fig. 6e). Rare earth element (REE)-bearing minerals (allanite), titanite and, to a lesser extent, zircon are progressively dissolved during the episyenitization process. Euhedral adularia and high-albite crystallize in the pores (Fig. 8a & b). Authigenic hydrothermal quartz has also grown into pore spaces as small bipyramidal grains (Fig. 6c). Finally, vermicular chlorites grow in the remaining spaces

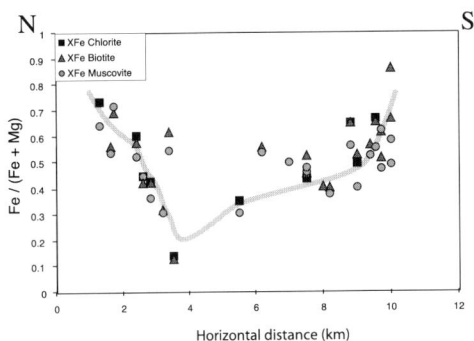

Fig. 8. Mineralogical variation of X_{Fe} (Fe/(Fe + Mg)) values across the Mont Blanc Massif shear zones. Values represent X_{Fe} in each mineral type v. distance. Asterisks are values of X_{Fe} of several whole-rock shear-zone compositions.

Table 3. *Major-element analyses of the studied samples from the Helbronner area*

Type	Mylonite	Episyenite				
Samples	MB140 Act-Ms	MB0255.Aa	MB0255.Ab	MB0255.B	MB0255.C	MB0255.D
Distance from vein (cm)	5	10	20	30	40	
Major oxides (wt%)						
SiO_2	69.64	67.35	71.99	73.38	75.84	76.80
TiO_2	0.33	0.24	0.22	0.22	0.17	0.19
Al_2O_3	14.69	17.42	15.72	14.42	12.80	12.22
Fe_2O_3	3.20	1.38	0.82	1.32	1.60	1.72
MnO	0.09	0.04	0.02	0.03	0.04	0.04
MgO	2.07	0.45	0.29	0.44	0.51	0.56
CaO	1.01	0.71	0.74	0.63	0.79	0.67
Na_2O	4.86	5.36	5.20	4.03	3.40	3.45
K_2O	2.60	5.87	4.53	4.94	4.35	3.93
P_2O_5	0.10	0.07	0.06	0.07	0.05	0.06
SO_3	0.05	–	–	–	–	–
LOI	1.36	0.74	0.62	0.78	0.83	0.77
H_2O	–	0.16	0.11	0.14	0.13	0.13
Total	100.00	99.79	100.32	100.40	100.51	100.54

LOI, loss on ignition.

$(X_{Fe} = 0.50–0.60$; Fig. 6d, Table 2). Minor fluorite and calcite are also found locally.

Geochemistry

Whole-rock analyses (Table 3) were conducted at Geoscience Australia (GA) in Canberra. Major- and trace-element analyses were performed by X-ray fluorescence spectrometry using a Philips PW2404 4 kW sequential spectrometer with a Rh tube. Detection limits can be found in Hoatson & Blake (2000).

Chemical composition of the Mont Blanc granite

In order to compare the chemical compositions of episyenites and shear zones with that of the granitic protolith, we have compared new chemical data of unaltered granite from this study with those reported by Marro (1986) and Bussy (1990) (Table 4). Bussy (1990) distinguished a central porphyritic granite (46 samples) and a fine-grained marginal facies (10 samples). The fine-grained granite is slightly enriched in SiO_2 and

Table 4. *Major-element composition of the Mont Blanc granite*

Facies	Porphyritic granite		Fine-grained granite		Porphyritic/fine-grained granite	
Major oxides (wt%)	(Marro 1986; Bussy 1990) (average, $n = 46$)	$\pm 2\sigma$	(Marro 1986; Bussy 1990) (average, $n = 10$)	$\pm 2\sigma$	(this study) (average, $n = 5$)	$\pm 2\sigma$
SiO_2	72.45	0.98	74.22	0.98	72.27	0.93
TiO_2	0.27	0.04	0.22	0.03	0.26	0.03
Al_2O_3	13.83	0.42	12.99	0.42	13.86	0.50
Fe_2O_3	1.86	0.69	1.71	0.43	2.16	0.08
FeO	0.39	0.20	0.22	0.09	–	–
MnO	0.05	0.01	0.05	0.01	0.05	0.02
MgO	0.48	0.11	0.45	0.10	0.65	0.22
CaO	1.31	0.23	1.23	0.21	1.08	0.22
Na_2O	3.60	0.17	3.52	0.20	3.73	0.13
K_2O	4.77	0.22	4.40	0.46	4.32	0.53
P_2O_5	0.09	0.01	0.10	0.14	0.07	0.02
H_2O	0.59		0.52		0.98	0.30
Total	99.73		99.63		99.59	0.23

depleted in other oxides relative to the porphyritic granite. The compositional variations between both types of granite are tight and partially superimposed (Marro 1986; Bussy 1990; Rolland *et al.* 2003) (Table 4). The standard deviations of average analyses are less than 5 wt% for SiO_2, Al_2O_3, Na_2O and K_2O (Table 4). The variations observed between the two granites are interpreted as differences in the degree of fractional crystallization (Bussy 1990). As discussed in the next section (see also Tables 3 and 4), this variation is much less important than that observed in shear zones and episyenites. This allows robust comparison of compositional differences between the initial meta-granite host rock, the hydrothermally altered shear zones and the episyenite alteration haloes around veins.

Major-element geochemistry of episyenites and shear zones

Episyenites. Chemical analyses of episyenites sampled in the SE part of the MBM (Fig. 2) and reported by Poty (1969) have been plotted in Harker diagrams (Fig. 9). On the same figure, we have also plotted the compositional variations across a single episyenite halo, from the vein (MB02.55Aa) towards the meta-granite (MB02.55D, Table 3). This profile (50 cm long) has been sampled close to Helbronner on the SE side of the massif. Figure 9 indicates a systematic decrease in SiO_2, MgO and Fe_2O_3 for all episyenites compared to the composition of meta-granites. With increasing the degree of episyenitization, episyenites become more and more depleted in SiO_2, MgO and Fe_2O_3, and enriched in Al_2O_3, K_2O and Na_2O.

The Al_2O_3 and K_2O content of episyenites plot along the trends (indicated by the dashed arrow in Fig. 9) corresponding to the relative variations of composition resulting from SiO_2 removal from the meta-granite (calculated trend indicated by the solid and dashed lines in Fig. 9). These observations suggest that the increase in Al_2O_3 and K_2O content of episynenites is mainly due to silica reduction, which induces an apparent enrichment in other elements according to mass balance calculations.

A more detailed analysis of compositional variations associated with episyenitization must take into account the changes of rock volume and density. Following Gresens' method (1967) (see also Potdevin & Marquer 1987), the absolute mass transfer of element *i* (dm_i) is given by the following equation:

$$dm_i = F_v \frac{d}{d_0} C_i - C_i^0 \qquad (1)$$

where F_v is the volume factor as defined by Gresens (1967), d is the density of the altered/deformed rock, d_0 the density of the protolith, C_i is the concentration of element *i* in the protolith (C_i^0) and in the altered or deformed rock (C_i). An example is shown in Figure 10a–d for samples MB02.55Aa–MB02.55C along the episyenite profile from Helbronner, using the composition of MB02.55D as reference. These figures show the absolute transfers of elements as a function of ($F_v \cdot F_d$) (with F_d the density factor: $F_d = d/d_0$). The convergence of the Al_2O_3, TiO_2, $\pm Na_2O$, $\pm K_2O$, P_2O_5 and MnO lines at $dm = 0$ suggests that these elements are immobile (in the range of magmatic compositions), and constrains the values of the volume and density factor ($F_v \cdot F_d$) for the different samples along the episyenite profile. With increasing degree of episyenitization ($F_v \cdot F_d$) decreases from 0.95 in sample MB02.55C to about 0.71 in sample MB02.55Aa. Preliminar measurements of the macroporosity of episyenites by filling pores with water lead to values of between 10 and 26 vol%, which are consistent with those deduced from mass balance calculation. Indeed, the decrease in ($F_v \cdot F_d$) associated with episyenitization yields an increase of porosity up to 30% ($F_v = 0.7$, $F_d = 1$). The compositional changes depend on the degree of episyenitization and the associated mineralogy changes. Quartz and biotite leaching account for the depletion in SiO_2, Fe_2O_3 and MgO at constant Al_2O_3 and K_2O, according to the simple Al and K conservative reaction:

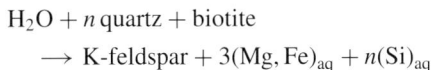

$$H_2O + n\,\text{quartz} + \text{biotite}$$
$$\rightarrow \text{K-feldspar} + 3(Mg, Fe)_{aq} + n(Si)_{aq}$$

where the subscript 'aq' stands for dissolved elements. This reaction is consistent with the observation of newly formed feldspar and eventually Fe-oxide and chlorite (if aluminium produced by the concomittent alteration of plagioclase is available) in the porosity of episyenite close to the vein (highest alteration) and the dissolution of quartz and biotite from the meta-granite (lowest alteration; Figs 5c and 6).

Shear zones. The compositions of shear zones (30 samples) from different locations of the MBM are plotted in Figure 11. Shear zones

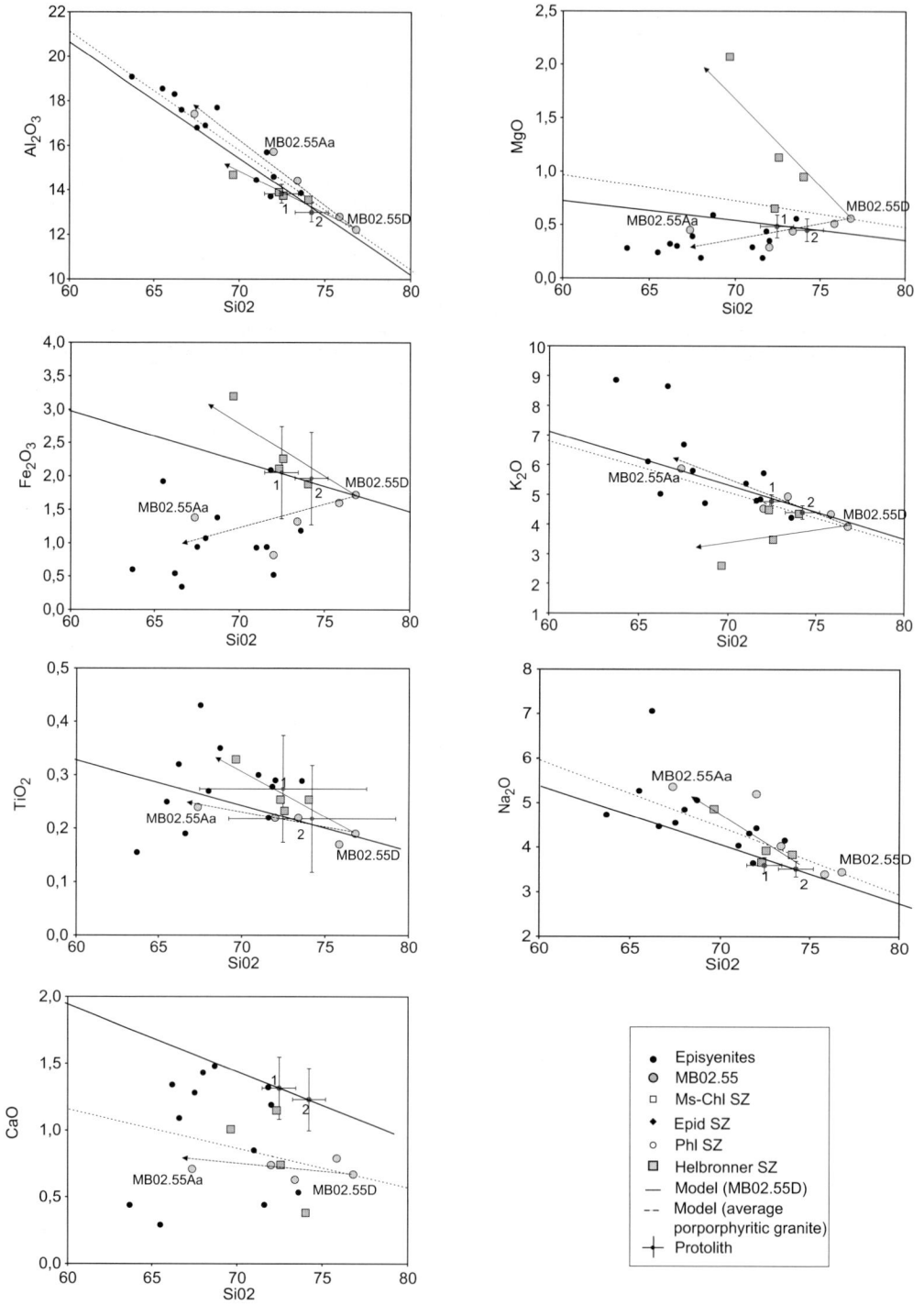

display a much larger range of SiO_2 content (37–85 wt%) than the undeformed granites (70–77 wt% SiO_2). For the other elements, various evolutions from the meta-granite composition are identified, which reflect differences in the mineralogy of the studied shear zones.

- The compositions of muscovite $\pm$ chlorite-bearing shear zones (e.g. MB140; Fig. 11 and Table 3) show SiO_2 content mostly within the range defined by the two reference compositions (1 and 2, Fig. 11), with slight enrichments observed overall. Except for CaO that is always depleted when compared with the reference compositions, the other oxide contents generally plot close to the line depicting the relative variations due to the absolute variations of SiO_2 (Fig. 11). However, when considering the whole set of shear-zone analyses, including those showing lower SiO_2 contents, we see that Fe_2O_3, MgO and, to a lesser extent, TiO_2 contents plot along trends crossing the 'SiO_2 line' with a higher slope. Therefore, the higher concentrations of these elements of the most SiO_2-depleted shear zones cannot be interpreted as a residual enrichment due to the removal of SiO_2 only. A more detailed examination of Figure 11 shows that most MgO content plots above the 'SiO_2 line', whereas most K_2O, Na_2O, CaO contents plot below, and TiO_2 and Fe_2O_3 plot sometimes below and sometimes above. These observations suggest that at MBM scale, the composition of the muscovite and chlorite shear zones is significantly different to that of the less deformed granite, with neat MgO enrichments, and K_2O, Na_2O and CaO depletions. Such differences can be interpreted as the result of removal or precipitation of SiO_2 from the protolith composition associated with removal of some oxides, leached by the fluid (especially K_2O, Na_2O and CaO) and precipitation of MgO.

- Phlogopite–chlorite-bearing shear zones are systematically strongly depleted in silica compared to the meta-granites and the other shear zones. They are furthermore depleted in K_2O, Al_2O_3 and Na_2O, and very significantly enriched in MgO with respect to the undeformed protolith. A large scatter is observed for the other elements. As evidenced in Figure 11, all these compositional variations from the meta-granite cannot be explained as the result of silica loss only.

- Epidote-bearing shear zones are mostly silica-rich compared to the protolith. They are strongly depleted in K_2O, and enriched in CaO and Fe_2O_3 (epidote crystallization). The variations of the other elements do not show any systematic enrichment and plot close to the 'SiO_2 line'.

A more quantitative estimate of the geochemical variations occurring in shear zones was made using the Gresens' approach with two representative shear zones: a muscovite- and chlorite-bearing shear zone from Helbronner (MB140), and a phlogopite- and chlorite-bearing shear zone from the massif's core (PK4660; Fig. 10e & f). As shown in Figures 9 and 11, MB140 is depleted in K_2O compared to the protolith composition, which contrasts with most chlorite- and muscovite-bearing shear zones. However, it was chosen because it is located in the same area as the episyenite profile (MB02.55), so that the comparison between both structures is easier. In both Gresens' diagrams drawn for samples MB140 and PK4660, the lack of clear convergence between the different mobility lines precludes unambiguous determination of the $(F_v \cdot F_d)$ factor (Fig. 10e & f). However, except for Na_2O and K_2O, the same observation has been made for most analysed shear zones, which always show the same relative position of the different mobility lines, but different absolute positions and slope. We have seen above (Fig. 11) that the apparent enrichment of

Fig. 9. Harker diagrams representing the major oxides compositions v. SiO_2 for episyenites and shear zones from Helbronner. Concentrations (in wt%) are normalized to 100%. The plots of sample MB02.55 represents a transect through an episyenite from core (MB02.55Aa) to rim (MB02.55D). The other episyenites data are from Poty (1969) and this study. Shear-zone data are from Rolland *et al.* (2003) and this study. The black crosses represent the average of undeformed granite (data from Marro 1986; Bussy 1990): 1 is the mean composition of the porphyritic granites and 2 is the mean composition of the fine-grained granite. The model dashed line represents enrichment/depletion due to mass balance silica addition/leaching from the porphyritic granite. The solid black line represents enrichment/depletion due to mass balance from sample MB02.55D used as a reference rock for Helbronner samples. The dashed arrow represents the trend of the transect through the episyenite MB02.55, and the black arrow the trend of shear zones from Helbronner.

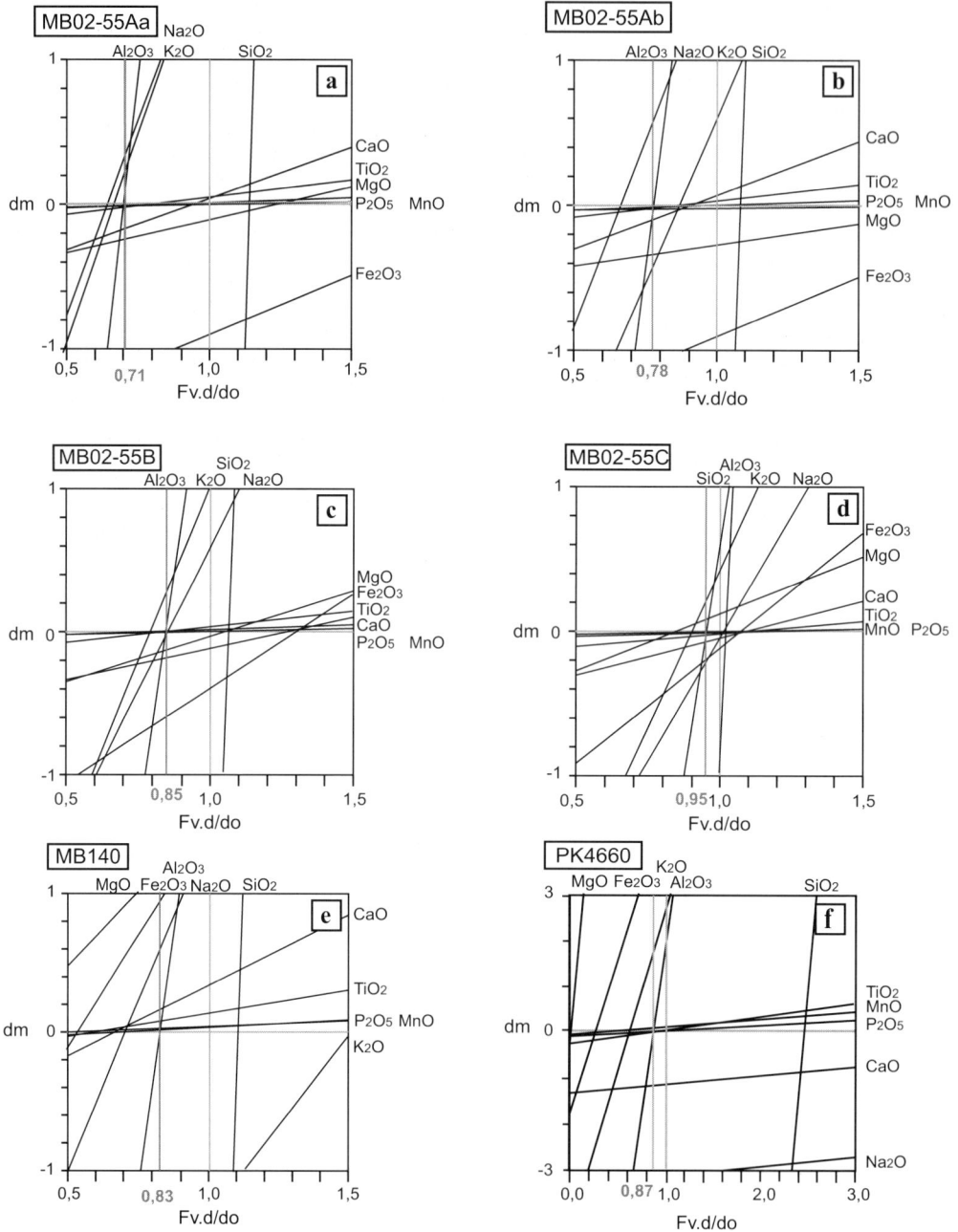

Fig. 10. Gresens' diagrams computed following the method of Gresens (1967). Mass changes (*dm*) are represented v. the density and volume factor F_vF_d. The grey line indicates the determined F_vF_d factor for each sample. Oxides that cross this line in *dm* > 0 are enriched, those in *dm* < 0 are depleted. Sample MB02.55D is used as reference rock for calculation. (a)–(d) are several samples across an episyenite transect in the Helbronner area: (**a**) MB02.55Aa; (**b**) MB0255.Ab; (**c**) MB02.55B; (**d**) MB02.55C; (**e**) and (**f**) are representative shear-zones samples: (**e**) muscovite-rich shear zone MB140; and (**f**) phlogopite-rich shear zone PK4660.

Al_2O_3 in MB140 can be explained by the loss of silica. It is therefore reasonable to assume that Al_2O_3 is immobile during the formation of this shear zone. The Al_2O_3 content of PK4660 is not explained by silica removal from the granitic protolith, but since the relative position of the mobility lines is the same we assume that Al_2O_3 was immobile in this sample too. Following these assumptions, almost the same product ($F_v \cdot F_d$) is estimated for shear zones MB140 and PK4660: ($F_v \cdot F_d$) = 0.83 for MB140 and ($F_v \cdot F_d$) = 0.87 for PK4660. For these values of ($F_v \cdot F_d$) SiO_2 is depleted in both samples MB140 and PK4660 (-25 and -57%, respectively, compared to the reference), K_2O is strongly depleted in MB140 and enriched in PK4660, and Na_2O and CaO are moderately enriched in MB140 and lost in PK4660. All the other elements show significant mass gains, with important variations being observed for MgO ($+4585\%$ in PK4660 and $+207\%$ in MB140) and Fe_2O_3 ($+216\%$ in PK4660 and $+54\%$ in MB140). These samples illustrate the great compositional variability observed for the phlogopite- or muscovite- and chlorite-bearing shear zones, which present a similar ($F_v \cdot F_d$). It is clear that the development of the phlogopite-bearing and, to a lesser extent, the muscovite- and chlorite-bearing shear zones is associated with a strong gain in MgO and loss in SiO_2. At least for the phlogopite-bearing shear zones, this gain in MgO results from strong metasomatic fluid–rock interactions, which is consistent with the macroscopic observations (see the section on 'Geological setting and Alpine deformation pattern').

Pressure–temperature–composition–mineralogy estimates

Additional data are required for a better understanding of processes leading to the mineralogical and geochemical data observed in shear zones and episyenites, and to estimate the magnitude and processes of the associated mass transfer. In particular, we need to better constrain the pressure (P) and temperature (T) conditions at which episyenites and shear zones were formed, and the relationships between these P–T conditions and the stable mineralogy for the observed different bulk-rock compositions.

P–T conditions of 300–500 °C and 1.5–6 kbar have been estimated for the Alpine metamorphism in the MBM. This wide range of P–T data reflect the different methods and rock types used for thermobarometric estimates, but also the evolution of the P–T conditions during the MBM

exhumation. The conditions derived for quartz–chlorite–adularia–albite veins from isochore reconstruction and feldspar Na/K thermometry (Cathelineau & Nieva 1985) yielded 300–400 °C and 1.5–4 kbar. The multiequilibria approach of Rolland et al. (2003) led to P–T estimates of peak metamorphic conditions of 450 °C and 4–6 kbar in muscovite ($\pm$chlorite) shear zones. In the present study, P–T conditions are calculated from the mineralogy and the composition of shear zones (MB140), undeformed granite (MB02.55D) and the most episyenitized granite (MB02.55Aa) close to a horizontal vein; these were all sampled in the same location (Helbronner area). For each bulk rock, a P–T diagram was calculated by energy minimization. This was performed with the Domino software of de Capitani & Brown (1987) using the Jun92.rgb mineral database extended with the chlorite solid solution model of Vidal et al. (2001). The results of calculation (Fig. 12) indicate that, in the case excess pore water, the topology of the petrogenetic grid remains almost the same for episyenite (MB02.55Ab), undeformed granite (MB02.55D) and shear-zone (MB140) compositions. In each case, four different assemblages are calculated to be stable in the 250–550 °C, 1–6 kbar domain: muscovite + K-feldspar + albite + quartz + H_2O are always stable with: (i) epidote + chlorite at low temperature (LT) and low pressure (LP); (ii) epidote + biotite at LT and high pressure (HP); (iii) biotite + muscovite at HT; and (iv) chlorite + muscovite at medium temperature (MT) and LP. The location of the limits between these stability fields is almost the same whatever the bulk system composition. The only significant differences are the proportions of the different minerals involved in the same paragenesis, and the composition of biotite and/or chlorite. For a fixed bulk rock composition, the calculated X_{Mg} of chlorite and biotite are similar, which is in good agreement with the natural data, and they are almost constant with varying P–T. However, the calculated X_{Mg} in biotite and chlorite is lower when using the unaltered granite composition ((X_{Mg})$_{chl}$ = 0.6 at 400 °C, 3 kbar and (X_{Mg})$_{bt}$ = 0.57 at 500 °C, 5 kbar) than when using the shear-zone composition ((X_{Mg})$_{chl}$ = 0.75 and (X_{Mg})$_{bt}$ = 0.73). This trend is also in good agreement with the observed naturally deformed assemblages.

In the *undeformed granite*, the magmatic biotite was transformed first into a metamorphic biotite, which is stable at $T > 450$ °C (point 1 in Fig. 12).

Shear zones from the Helbronner area are muscovite-rich, and also contain chlorite and chloritized biotite, but no epidote. They formed in the biotite stability field, at $T > 400\ °C$ (point 2 in Fig. 12). These conditions are comparable to those estimated by Rolland *et al.* (2003) at 4 kbar. The occurrence of chloritized biotite indicates that the Helbronner shear zones were

fluid-active below the biotite–chlorite reaction curve, possibly until point 4 (Fig. 12). In contrast, the epidote-bearing shear zones in the NW of the MBM suggest lower temperature conditions ($<400\ °C$) and possibly lower P (<3 kbar)

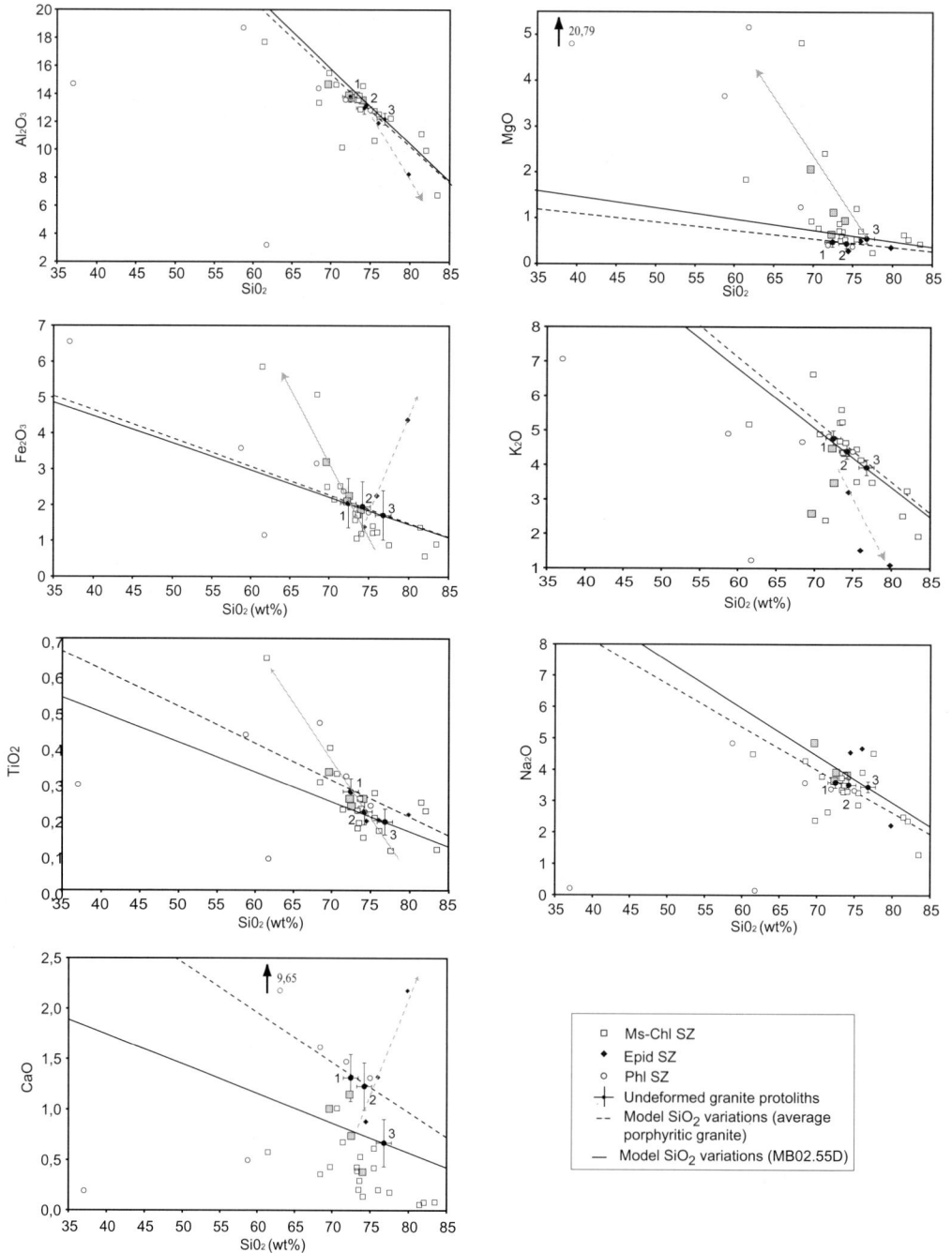

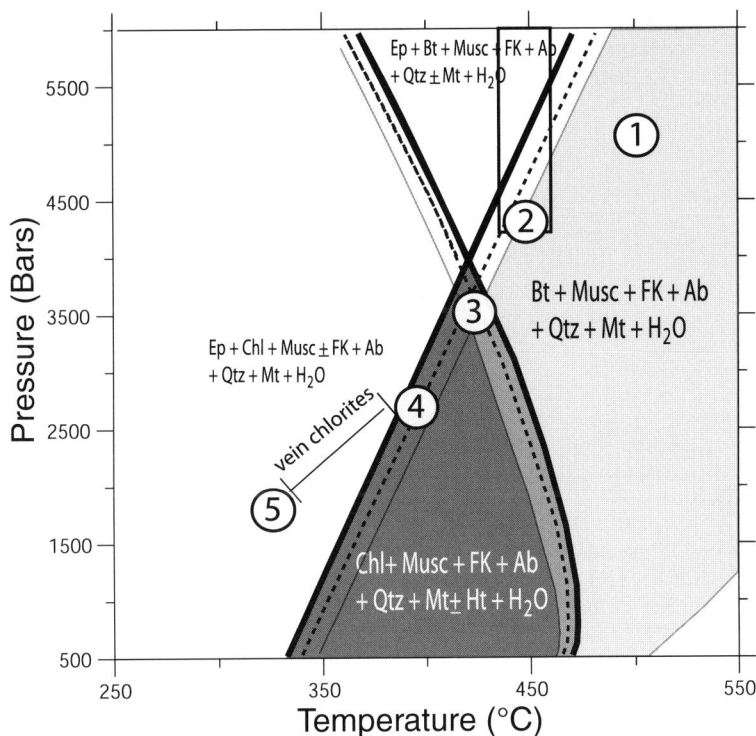

Fig. 12. $P-T$ phase diagram calculated by minimization of energy using the bulk compositions listed in Table 5. Thick lines, unaltered granite MB0255D; thin continuous lines, episyenite MB0255AB; discontinuous line, muscovite-bearing shear zone MB140. Grey area, stability field of the stable paragenesis observed in the fresh granite. Dark area, stability field of the stable paragenesis observed in the episyenites. Oblique line, range of temperatures estimated from the location of the chlorite + quartz + H_2O equilibrium (see the text) for chlorites crystallizing in the horizontal veins associated with episyenites. Box, $P-T$ conditions estimated by Rolland *et al.* (2003) for an epidote + muscovite + chlorite shear zone. Circles, possible evolution of the $P-T$ conditions in the MBM during Alpine exhumation.

conditions (points 3–5 in Fig. 12) that might be reached along a unique $P-T$ path.

In the *episyenites* from the Helbronner area (SE part of the massif) biotite is almost entirely dissolved, but some traces remain in the less altered parts of episyenites. Here, biotite is strongly chloritized. This suggests that episyenites formed below the biotite–chlorite reaction curve, at $T < 450\,°C$ (point 3 in Fig. 12). Moreover, the absence of epidote suggests that episyenites developed at $T > 350\,°C$ (point 4 in Fig. 12).

The lowest temperature for shear zones and episyenites can be estimated from the location

---

Fig. 11. Harker diagrams representing the major oxides compositions v. SiO_2 for the shear zones of the Mont Blanc Massif. Concentrations (in wt%) are normalized to 100%. Shear-zone data are from Rolland *et al.* (2003) and this study. The black crosses represent the average concentrations of the undeformed granite (data from Marro 1986; Bussy 1990): 1 is the mean composition of the porphyritic granite, 2 is the mean composition of the fine-grained granite, and 3 represents a reference sample (MB02.55D) on the edge of the Helbronner episyenite representative profile. The model dashed line represents model variations ('SiO_2' line) from the porphyritic granite (1), due to quartz dissolution or precipitation. The black line represents model variations from the reference sample MB02.55D (3), due to quartz dissolution or precipitation. The field between the two lines thus represents the possible compositional range of compositions that can be explained by quartz dissolution or precipitation. The grey solid arrow represents the trend of muscovite-bearing shear zones, and the dashed grey arrows the trend of epidote-bearing shear zones (linear correlations of concentration values).

Table 5. *Chemical analysis of chlorites and calculated temperature from selected samples from Helbronner*

Sample	MB02.55Aa episyenite protolith	MB02.55Aa episyenite porosity	MB02.55Aa episyenite porosity	MB02.55D granite	MB140 shear zone
SiO_2	23.61	22.61	23.01	24.89	26.26
TiO_2	0.02	0.00	0.02	0.05	0.09
Al_2O_3	18.47	18.15	18.70	19.05	19.81
FeO	29.04	31.21	29.41	31.92	23.09
MnO	0.74	0.73	1.03	1.19	1.21
MgO	10.97	8.49	10.37	10.66	16.82
CaO	0.02	0.02	0.06	0.08	
Na_2O	0.00	0.03	0.01	0.02	
K_2O	0.02	0.05	0.03	0.04	0.15
Total	82.90	81.29	82.64	87.91	87.43
Temperature (°C)	**373**	**352**	**418**	**362**	**392**
T °C (Cathelineau & Nieva 1985)	296	301	306	294	289
T °C (Hillier & Velde 1992)	329	340	352	324	313

of the equilibrium (clinochlore + sudoite)$_{chl}$ = (Mg-amesite)$_{chl}$ + quartz + H_2O, using the chlorite solid–solution model and thermodynamic data from Vidal *et al.* (2001), assuming that the activity of water equals unity. Chlorites filling the porosity of episyenites yield temperatures in the range 350–420 °C, while chlorites replacing biotite yield temperatures of about 370 °C. With the same approach, we calculate temperatures of about 400 °C for the chlorite from MB140, which is in good agreement with the conditions estimated above (point 3 in Fig. 12, Table 5). In the case of episyenite, chlorites filling the central vein indicate 300–370 °C, and chlorite filling the porosity indicate 350–420 °C.

All these data indicate that the *P–T* conditions corresponding to deformation and recrystallization in the shear zones (400–450 °C, 3–5 kbar), the formation of episyenite (350–420 °C) and the late crystallization of chlorites in the associated veins (300–375 °C) are not very different (Table 5). It is therefore likely that the formation of episyenite and deformation in the shear zone were contemporaneous, which is compatible with the structural observations.

Discussion

The Alpine deformation associated to the exhumation of the MBM led to the development of subvertical shear zones and the alteration of the undeformed granite around horizontal veins (episyenitization). The present work complements the data of Bertini *et al.* (1985), confirming that the shear-zone network was formed in a NW–SE horizontal compression and vertical extension context. We show that this

deformation is mainly accommodated by a complex network of NE–SW partly dextral and N–S partly sinistral shear zones, the NE–SW component being the dominant one. The N135°E shortening not only resulted in the formation of brittle–ductile faults, but also in the formation of a more pervasive NE–SW foliation, and was synchronous to the formation of horizontal veins and episyenites. Two different types of shear zones are evidenced in the massif. (1) In the central part of the MBM, phlogopite-bearing shear zones are present in association with a phlogopite-rich zone showing a sharp transition with the granitic protolith (Fig. 4c). These phlogopite-bearing shear zones clearly result from metasomatism of the granite by a Mg-rich fluid, which is also evidenced by a large drop of X_{Fe} in phyllosilicates in comparison to the X_{Fe} of phyllosilicates in the granite protolith. The circulation of this Mg-rich fluid is restricted to the central part of the massif. (2) In contrast, the mineralogy of MBM rim shear zones, the epidote-bearing shear zones and white mica-chlorite-bearing shear zones is mainly related to regional metamorphism. The granitic protolith around these shear zones shows the same metamorphic assemblages. It contains small amounts of epidote in the NW part of the massif, where epidote-bearing shear zones occur, and small amounts of chlorite and muscovite in the central and SE parts of the massif, where chlorite-mica-bearing shear zones occur. The crystallization of these metamorphic minerals, as well as the albitization of plagioclase, the breakdown of biotite and the alteration of K-feldspar, result from the variation in *P–T* conditions at regional scale. The extent of these transformations is proportional to deformation intensity, but, in contrast to the phlogopite shear zone, the formation of the mica–chlorite- and epidote-bearing shear zones does not require large amounts of fluid circulation.

Field observations show that episyenites are always found in the vicinity of the muscovite- and chlorite-bearing shear zones to which they are connected by an open vein. Both structures formed under the same tectonic regime, i.e. during NW–SE shortening and vertical extension and similar $P-T$ conditions. The development of episyenites in contact with ductile or brittle structures has already been noticed (Hålenius & Smellie 1983; Hecht et al. 1999), and is assumed to participate in the widening of these structures (Petersson & Eliasson 1997). The episyenitization process is described in the literature to result from the circulation of late magmatic fluids (e.g. Recio et al. 1997). This is not the case of the MBM episyenites, which are Alpine structures, whereas the Mont Blanc granite is Hercynian (300 ± 3 Ma; Bussy & von Raumer 1994). The dissolution of quartz and biotite from Hercynian episyenites has been explained to result from the incoming of a silica-undersaturated fluid with a higher temperature than the granite (e.g. Cathelineau 1986; Turpin et al. 1990; Recio et al. 1997). Such fluids would circulate upwards from the massif bottom in the vertical shear zones before entering the horizontal veins. This is consistent with the composition of shear zones that are most silica-depleted, such as phlogopite $\pm$ chlorite-bearing shear zones, which are clearly related to the circulation of a Mg-rich and Si-undersaturated fluid. It is also consistent to a lesser extent with the composition of chlorite–muscovite shear zones from Helbronner. However, it is not consistent with the composition of other chlorite–mica-bearing shear zones that are enriched in silica compared to the granite protolith. Moreover, all episyenites are systematically more depleted in quartz than the shear zones to which they are connected. In the hypothesis that quartz is dissolved by the circulation of a Si-undersaturated fluid flowing from the shear zones, it is hard to understand how quartz dissolution could be more important in the episyenites than in the shear zones. Moreover, quartz dissolution kinetics at 400 °C should be fast enough compared to the rate of fluid circulation in the shear zone. As all the chlorite–muscovite shear zones do contain quartz, the concentration of dissolved silica in the fluid circulating in these shear zones should rapidly increase and reach saturation. Most chlorite–muscovite shear zones are silica-enriched compared with the granite protolith.

From Figure 9 opposite trends are observed between episyenites and shear zones for MgO and Fe_2O_3, which might suggest that these elements are leached out from episyenites and transported towards the connected shear zone where they are incorporated in new minerals (mainly muscovite and chlorite). Nevertheless, the other oxides, especially K_2O that is needed to form K-white micas, do not present such systematic anti-correlation. Furthermore, in the Helbronner area, both shear zones and episyenites are depleted in silica compared to the protolith (MB02.55D, Fig. 9). Geochemical analyses of episyenites and shear zones thus indicate no systematic mass balance. On the other hand, a systematic exchange of matter is likely between episyenites and horizontal veins. Indeed, most elements lost due to quartz and biotite dissolution in the episyenites can be used for the formation of secondary quartz, albite and adularia in the porosity and in the open vein at the core of the episyenites. The secondary chlorite observed in the episyenites porosity and the veins might also have crystallized as a breakdown product of biotite dissolved in the episyenites, although chlorite always grows on all the other secondary phases, including quartz in the veins. It is therefore possible that the crystallization of chlorite post-dates the formation of episyenites. Nevertheless, whatever the exact timing of chlorite crystallization, it is likely that the elements produced during the breakdown of quartz and biotite (episyenitization) diffuse from the edge of the episyenites to the core where they are incorporated to form the secondary phases. The synchronism of quartz dissolution in the episyenites and of quartz precipitation in the veins formed at episyenites cores is suggested by the following observations: (1) quartz never seals the porosity of episyenites, which would be expected if it precipitated from an advective fluid circulating in the vein and surrounding episyenite *after* the formation of episyenite; and (2) the porosity of episyenites resulting from the dissolution of quartz reaches 30%, which is compatible with the thickness of the horizontal quartz veins (about one third that of episyenites). From these observations, it seems that the formation of episyenite does not depend on the magnitude and composition of fluid circulation in the connected shear zone. Coupled quartz dissolution in wallrock alteration haloes and quartz precipitation in veins could be interpreted to reflect local mass transfer between wallrock and veins during essentially closed-system behaviour in the relatively undeformed granite domains between shear zones. In this case, the complete removal of quartz from episyenites and its crystallization in the horizontal veins must result from a diffusive transport in response to concentration gradients at the decimetre–metre scale. Such gradients are not driven by compositional (homogeneous protolith composition) or temperature gradients. Therefore, they should be driven by pressure gradients. The origin of these pressure gradients, as well as the mechanism of episyenite formation, remains enigmatic, but it should be related to fluctuations of fluid pressure during incremental vein opening.

Conclusion

Two features are evidenced in the Mont Blanc Massif during the Alpine deformation: (1) the development of ductile shear zones; and (2) episyenite formation around horizontal veins. The development of shear zones is associated with the formation of the main weak foliation of the granite. In shear zones Al_2O_3 and, to a lesser extent, K_2O are immobile, while Fe_2O_3 and MgO can be highly mobile. Phlogopite/chlorite-bearing shear zones show very high bulk rock $Mg/(Mg + Fe)$ values, which evidence the circulation of high fluid/rock ratios of a Mg-rich fluid percolating upwards in the centre of the massif. The crystallization of epidote-, chlorite-and white-mica, as well as the albitization of plagioclase, breakdown of biotite and alteration of K-feldspar in shear zones and in the granite protolith, result from the variation in $P–T$ conditions at the regional scale. The extent of these transformations is proportional to deformation intensity. But, in contrast to the phlogopite shear zone, the formation of the mica–chlorite-and epidote-bearing shear zones does not result from intensive metasomatism. The formation of veins surrounded by an episyenitic alteration halo is contemporaneous to that of Alpine shear zones (chlorite- and white-mica-bearing shear zones) to which they are always connected. However, despite the structural relationships between shear zones and episyenites, there is no systematic mass transfer between the two features. Geochemical variations are similar within all the episyenites at the regional scale while shear-zone alteration varies largely. The episyenitization process is thus unrelated to the nature of shear-zone fluids and occur in an almost closed system at the metre–decametre scale. At the present stage, the mechanism of episyenite and opened vein formation remains enigmatic.

This work was financially supported by the CNRS-INSU (IT), Emergence and ARC programs. The manuscript was substantially improved by the reviews of D. Marquer, G. Pennacchioni and J. Evans. Thanks are due to A.-M. Boullier, P.-H. Leloup, N. Mancktelow and J.-E. Martelat for stimulating discussions.

References

ANGELIER, J. 1990. Inversion of field data in fault tectonics to obtain the regional stress. III: a new rapid inversion method by analytical means. *Geophysical Journal International*, **103**, 363–376.

ANTOINE, P., PAIRIS, J.L. & PAIRIS, B. 1975. Quelques observations nouvelles sur la structure de la couverture sédimentaire interne du massif du Mont-Blanc, entre le Col du Ferret (frontière italo-suisse) et la Tête des Fours (Savoie, France). *Géologie Alpine*, **51**, 5–23.

BAGGIO, P. 1958. *Il granito del Monte Bianco e le sue mineralizzazioni uranifere*. Studi e ricerche della divisione geomineraria CNRN, Roma, **1**, 1–130.

BAGGIO, P., FERRARA, G. & MALADORA, R. 1967. Results of some Rb/Sr ages of the rock from the Mont-Blanc tunnel. *Bollettino della Societa geologica italiana*, **86**, 193–212.

BELLIÈRE, J. 1988. On the age of mylonites within the Mont-blanc massif. *Geodinamica Acta*, **2**, 13–16.

BERTINI, G., MARCUCCI, M., NEVINI, R., PASSERINI P. & SGUAZZONI, G. 1985. Patterns of faulting in the Mont-Blanc granite. *Tectonophysics*, **111**, 65–106.

BUSSY, F. 1990. *Pétrogenèse des enclaves microgrenues sombres associées aux granitoïdes calco-alcalins: exemple des massifs varisque du Mont-Blanc (Alpes occidentales) et miocène du Monte Capanne (Ile d'Elbe, Italie)*. PhD thesis, Mémoires de Géologie (Lausanne).

BUSSY, F. & VON RAUMER, J.F. 1994. U–Pb geochronology of Palezoic magmatic events in the Mont-Blanc crystalline massif, Western Alps. *Schweizerische Mineralogische und Petrographische Mitteilungen*, **74**, 514–515.

BUTLER, R.W.H. 1985. The restoration of thrust systems and displacement continuity around the Mont-Blanc massif, NW external Alpine thrust belt. *Journal of Structural Geology*, **7**, 569–582.

CATHELINEAU, M. 1986. The hydrothermal alkali metasomatism effects on granitic rocks: quartz dissolution and related subsolidus changes. *Journal of Petrology*, **27**, 945–965.

CATHELINEAU, M. 1987. U–Th–REE mobility during albitization and quartz dissolution in granitoids: evidence from south-east Frenc Massif Central. *Bulletin de Minéralogie*, **110**, 249–259.

CATHELINEAU, M. & NIEVA, D. 1985. A chlorite solid solution geothermometer: The Los Azures (Mexico) geothermal system. *Contributions to Mineralogy and Petrology*, **91**, 235–244.

CESARE, B. 1994. Synmetamorphic veining: origin of andalusite-bearing veins in the Vedrette di Ries contact auraole, Eastern Italy. *Journal of Metamorphic Geology*, **12**, 643–653.

DE CAPITANI, C. & BROWN, T.H. 1987. The computation of chemical equilibrium in complex systems containing non-ideal solutions. *Geochimica Cosmochimica Acta*, **51**, 2639–2652.

DIPPLE, G.M. & FERRY, J.M. 1992. Metasomatism and fluid flow in ductile fault zones. *Contributions to Mineralogy and Geology*, **112**, 149–164.

ETHERIDGE, M.A. & COOPER, J.A. 1981. Rb/Sr isotopic and geochemical evolution of a recrystallized shear (mylonite) zone at Broken Hill. *Contributions to Mineralogy and Petrology*, **78**, 74–84.

FABRE, C., BOIRON, M.C., DUBESSY, J., CATHELINEAU, M. & BANKS, D.A. 2002. Paleofluid chemistry of a single fluid event: a bulk and in-situ multi-technique analysis (LIBS, Raman Spectroscopy) of an Alpine fluid (Mont-Blanc). *Chemical Geology*, **182**, 249–264.

FERRY, J.M. & GERDES, M.L. 1998. Chemically reactive fluid flow during metamorphism. *Annual Reviews of Earth and Planetary Sciences*, **26**, 255–287.

GLAZNER, A.F. & BARTLEY, J.M. 1991. Volume loss and state of strain extensional mylonites from the central Mojave Desert, California. *Journal of Structural Geology*, **13**, 584–587.

GRESENS, R.L. 1967. Composition-volume relationships of metasomatism. *Chemical Geology*, **2**, 47–65.

GUERMANI, A. & PENNACCHIONI, G. 1998. Brittle precursors of plastic deformation in a granite: an example from the Mont-Blanc massif (Helvetic, western Alps). *Journal of Structural Geology*, **20**, 135–148.

HECHT, L., THURO, K., PLINNINGER, R. & CUNEY, M. 1999. Mineralogical and geochemical characteristics of hydrothermal alteration and episyenitization in the Köningshain granites, northern Bohemian Massif, Germany. *International Journal of Earth Sciences*, **88**, 236–252.

HILLIER, S. & VELDE, B. 1992. Chlorite interstratified with a 7 Å mineral: an example from offshore norway and possible implications for the interpretation of the composition of diagenetic chlorites. *Clay Minerals*, **27**, 475–486.

HOATSON, D.M. & BLAKE, D.H. 2000. Geology and economic potential of the Palaeoproterozoic layered mafic–ultramafic intrusions in the East Kimberley, Western Australia: some new insights. *AGSO Research Newsletter*, **34**, 29–33.

HÅLENIUS, U. & SMELLIE, J.A.T. 1983. Mineralizations of the Arjeplog-Arvidsjaur-Sorsele uranium province: mineralogical studies of selected uranium occurences. *Neues Jahrbuch für Mineralogie, Abhandlungen*, **147**, 220–252.

JAMTVEIT, B., BUCHER-NURMINEN, K. & AUSTRHEIM, H. 1990. Fluid controlled eclogitization of granulites in deep crustal shear zones, Bergen arcs, Norway. *Contributions to Mineralogy and Petrology*, **104**, 184–193.

JANECKE, S.U. & EVANS, J.P. 1988. Feldspar-influenced rock rheologies. *Geology*, **16**, 1064–1067.

LEUTWEIN, F., POTY, B. SONET, J. & ZIMERMAN, J.L. 1974. Age des cavités à cristaux du granite du Mont-Blanc. *Comptes Rendus de l'Académie des Sciences de Paris*, **271**, 156–158.

MARQUER, D. 1989. Transferts de matières et déformation des granitoïdes. Aspects méthodologiques.

Schweizrische Mineralogische und Petrographische Mitteilungen, **69**, 13–33.

MARRO, C. 1986. *Les granitoïdes du Mont-Blanc en Suisse*. Thèse de doctorat, Fribourg University.

MARSHALL, D., KIRSCHNER, D. & BUSSY, F. 1997. A Variscan pressure–temperature–time path for the N-E Mont-Blanc massif. *Contributions to Mineralogy and Petrology*, **126**, 416–428.

PASSCHIER, C.W. & TROUW, R.A.J. 1998. *Microtectonics*. Springer, Berlin.

PETERSSON, J. & ELIASSON, T. 1997. Mineral evolution and element mobility during episyenitization (dequartzification) and albitization in the postkinematic Bohus granite, southwest Sweden. *Lithos*, **42**, 123–146.

POTDEVIN, J.L. & MARQUER, D. 1987. Méthodes de quantification des transferts de matière par les fluides dans les roches métamorphiques déformées. *Geodinamica Acta*, **1**, 193–206.

POTY, B. 1969. *La croissance des cristaux de quartz dans les filons sur l'exemple du filon de la Gardette (Bourg d'Oisans) et des filons du massif du Mont-Blanc*. Thèse de Doctorat, Université de Nancy.

POTY, B., STADLER, H.A. & WEISBROD, A.M. 1974. Fluid inclusions studies in quartz from fissures of the Western and Central Alps. *Schweizerische Mineralogische und Petrographische Mitteilungen*, **54**, 717–752.

RECIO, C., FALLICK, A.E., UGIDOS, J.M. & STEPHENS, W.E. 1997. Characterization of multiple fluid-granite interaction processes in the episyenites of Avila-Béjar, Central Iberian Massif, Spain. *Chemical Geology*, **143**, 127–144.

ROLLAND, Y., COX, S.F., BOULLIER, A.M., PENNACCHIONI, G. & MANCKTELOW, N. 2003. Rare Earth and trace element mobility and fractionation in mid-crustal shear zones: insights from the Mont-Blanc Massif (Western Alps). *Earth and Planetary Sciences Letters*, **214**, 203–219.

SINHA, A.K., HEWITT, D.A. & RIMSTIDT, J.D. 1986. Fluid interaction and element mobility in the development of ultramylonites. *Geology*, **14**, 883–886.

STREIT, J.E. & COX, S.F. 1998. Fluid infiltration and volume change during mid-crustal mylonitization of Proterozoic granite, King Island, Tasmania. *Journal of Metamorphic Geology*, **16**, 197–212.

TURPIN, L., LEROY, J. & SHEPPARD, S.M.F. 1990. Isotopic systematics (O, H, C, Sr, Nd) of superimposed barren and U-bearing hydrothermal systems in a Hercynian granite, Massif Central, France. *Chemical Geology*, **88**, 85–98.

VIDAL, O. & DURIN, L. 1999. Aluminium mass transfer and diffusion in water at 400–550 °C, 2 kbar in the $K_2O–Al_2O_3–SiO_2–H_2O$ system driven by a thermal gradient or by a variation of temperature with time. *Mineralogical Magazine*, **63**, 633–647.

VIDAL, O. & PARRA, T. 2000. Exhumation paths of high pressure metapelites obtained from local equilibria for chlorite–phengite assemblages. *Geological Journal*, **35**, 139–161.

VIDAL, O., PARRA, T. & TROTET, F. 2001. A thermodynamic model for Fe–Mg aluminous chlorite using data from phase equilibrium experiments and natural pelitic assemblages in the 100–600 °C, 1–25 kbar P–T range. *American Journal of Sciences*, **301**, 557–592.

WATSON, E.B & WARK, D.A. 1997. Diffusion of dissolved SiO_2 in H_2O at 1 GPa, with implications for mass transport in the crust and upper mantle. *Contributions to Mineralogy and Petrology*, **130**, 66–80.

WIBBERLEY, C. 1999. Are feldspar-to-mica reactions necessarily reaction-softening processes in fault zones? *Journal of Structural Geology*, **21**, 1219–1227.

WIDMER, T. & THOMPSON, A.B. 2001. Local origin of high pressure vein material in eclogite facies rocks of the Zermatt–Saas zone, Switzerland. *American Journal of Sciences*, **301**, 627–656.

Water distribution in dynamically recrystallized quartz grains: cathodoluminescence and micro-infrared spectroscopic mapping

JUN MUTO[1], HIROYUKI NAGAHAMA[1] & TETSUO HASHIMOTO[2]

[1]Department of Geoenvironmental Sciences, Graduate School of Science, Tohoku University, Aoba-ku, Sendai 980-8578, Japan (e-mail: mutoh@mail.tains.tohoku.ac.jp)
[2]Department of Chemistry, Faculty of Science, Niigata University, Ikarashi-Ninocho, Niigata 950-2181, Japan

Abstract: The distribution of water in dynamically recrystallized quartz aggregates in granitic mylonites was investigated by cathodoluminescence (CL) observations and micro-infrared (IR) spectroscopic mapping. It is clear from CL observations and micro-IR spectroscopic mappings that the CL intensity and contrast depend on the density of defects (e.g. the content of molecular H_2O and absorbed OH species). Dynamic recrystallization, as a recovery process of intracrystalline strain, decreases the densities of excess carriers originated from defects. The decrease in the densities of excess carriers decreases the CL intensity and contrast within each grain. To investigate the effect of water weakening in the rheological properties of the crust, CL spectroscopy and micro-IR spectroscopic mapping are essential in the determination of concentration and distribution of water in natural high-strain zones.

It has long been known that quartz deformed under high pressure and temperature is greatly weakened by H_2O and OH residing on the grain boundaries, at structural defects and within the crystalline interiors of quartz crystals (e.g. Kronenberg 1994). These phenomena lead to pronounced reductions in plastic yield strength and have been referred to as 'water weakening' or 'H_2O weakening' of quartz (McLaren et al. 1989). Previous studies on water weakening have been carried out on the species and concentration of H_2O and OH using an infrared (IR) spectrometer, as H_2O and OH species are good absorbers in the IR region of light (e.g. Katz 1962; Aines & Rossman 1984). Using IR spectroscopy, Kronenberg et al. (1990) revealed that an increase in water facilitates shear localization in a natural shear zone. However, because previous IR spectrometers measured only a single point on the specimen, we could not compare the IR spectra with deformation textures obtained by other micro-analysis. Recently, a micro-IR spectroscopic mapping method has been developed and enables us to obtain two-dimensional IR absorption images of natural quartz (Ito & Nakashima 2002; Hashimoto et al. 2003).

In addition to IR spectroscopy, cathodoluminescence (CL) observations can reveal the water content of minerals and are used to infer the fluid flow during crystal growth in natural rocks. Holness & Watt (2001) showed that the bright and dark alternating CL bands associated with grain growth could be related to a periodic and episodic infiltration of H_2O in Appin Quartzite. Furthermore, CL spectra and colour patterns have been used to infer the distribution and species of various defects in quartz, such as point defects (i.e. centres associated with vacant oxygen or silicon) (Stevens Kalceff & Phillips 1995), dislocations and chemical impurities (e.g. Al, Ti, Fe, K, Na and OH) (Waychunas 1988; Müller et al. 2000, 2002; Götze et al. 2001; Ruffini et al. 2002; Van den Kerkhof et al. 2004). From the CL observations of low-temperature mylonites, Shimamoto et al. (1991) concluded that dynamic recrystallization decreases the dislocation (and/or point defect) density and the emitted CL intensity and contrast. In semi-conductor engineering, the CL intensity and contrast have been used to infer the distribution and concentration of defects in semi-conductors by a relationship between the CL intensity and excess carrier density by defects (Löhnert & Kubalek 1984; Jakubowicz 1986; Pey et al. 1995). However, the exact relations among CL intensity, contrast and defect species in naturally deformed quartz are still poorly constrained.

From: BRUHN, D. & BURLINI, L. (eds) 2005. *High-Strain Zones: Structure and Physical Properties.*
Geological Society, London, Special Publications, **245**, 397–407.
0305-8719/05/$15.00   © The Geological Society of London 2005.

Here we describe a new method for determining the distribution of water in dynamically recrystallized quartz aggregates in mylonites from the Hatakawa shear zone (NE, Japan) using CL observations and micro-IR spectroscopic mapping methods in combination. Based on these analyses, we demonstrate that H_2O and OH species are inhomogeneously distributed in dynamically recrystallized quartz aggregates. Finally, we document that dynamic recrystallization, as a recovery process of intracrystalline strain, decreases defect densities and CL intensity and contrast according to a relationship between the CL intensity and excess carrier density by defects.

Geological setting

Mid-Cretaceous granitoids are widely distributed in the Abukuma Mountains in NE Japan. The NNW–SSE-trending major faults are exposed at the eastern margin of the Abukuma Mountains. The western and eastern faults are called the Hatakawa and Futaba shear zones, respectively (Fig. 1).

In the study area, the Hatakawa shear zone is characterized by a NNW–SSE-trending cataclasite zone approximately 50 m in width. Various granitoids are widely distributed to the west of the cataclasite zone (Fig. 1) (see also Kubo *et al.* 1990; Nakamura & Nagahama 2002). To the west of the cataclasite zone, a sinistral ductile shear zone (mylonite zone) with a thickness of 1 km is widely developed in the study area. In this mylonite zone, the mylonitic foliation strikes NNE and dips vertically, and the lineation plunges to slightly horizontal. The age of sinistral ductile deformation along this shear zone is estimated to be 106–86 Ma (Otsuki & Ehiro 1978). Sample 82 was selected for CL observation in this shear zone.

Small-scale shear zones, millimetres to several centimetres wide, are sporadically developed in the non-deformed granodiorites about 1 km west from the cataclasite zone (sample UMY). These small-scale shear zones contain well-foliated and fine-grained dark bands that are in sharp contact with undeformed massive granite. These dark bands are composed of ultramylonites (see also Takagi *et al.* 2000).

Sample UMY was selected for CL and IR measurements in these small-scale shear zones and is a non-deformed granodiorite with an ultramylonite band 1 cm in thickness (Fig. 2). This ultramylonite band is dominantly composed of very-fine-grained minerals. The fine-grained matrix is mainly composed of quartz, plagioclase, K-feldspar, chlorite, epidote and opaque

minerals (Muto & Nagahama 2004). The grain sizes of fine-grained quartz aggregates are, on average, approximately 50 μm. The quartz grains show polygonal and equant grain shapes, and no undulatory extinction. These microstructures indicate that the fine-grained quartz aggregates were formed by a dynamic recrystallization process: nucleation and growth of dislocation-free new grains from deformed grains due to the difference in dislocation density between these grains (e.g. Poirier 1985). Sample UMY-2 was selected from such a fine-grained aggregate (Fig. 2). The surrounding granodiorite shows weakly deformed textures. Quartz aggregates exhibit undulatory extinction, occasional deformation lamellae and many subgrains. The grain sizes of the weakly deformed quartz aggregates are, on average, approximately 1 mm to several millimetres with sutured grain boundaries. Sample UMY-4 was selected from the weakly deformed aggregate (Fig. 2).

Cathodoluminescence observations of mylonites

CL intensity and contrast formation by excess carriers from defects

The CL is the luminescent emission from a material irradiated with electrons. The three physical bases for the CL are the generation, the recombination and the motion of excess carriers (i.e. electrons and holes) by electron irradiation (e.g. Yacobi & Holt 1986). In the indirect-band gap materials, e.g. α-quartz (Chelikowsky & Schlüter 1977), intrinsic luminescence is known to be relatively weak, and extrinsic luminescence originated from a defect is the main luminescence phenomenon (e.g. Yacobi & Holt 1986). Based on a relationship between CL intensity and excess carrier density of defects, the CL intensity and contrast can reveal the densities and distributions of electrically active defects in semi-conductors (Löhnert & Kubalek 1984; Jakubowicz 1986; Pey *et al.* 1995). Here, we briefly introduce the theory on the formation of CL intensity and contrast by excess carriers of defects according to previous studies of CL contrast formation of defects (Löhnert & Kubalek 1984; Jakubowicz 1986; Pey *et al.* 1995).

The CL intensity far away from any defect, $I_{CL(\infty)}$, as a function of the electron beam position r is given by

$$I_{CL(\infty)} = \int_{V_s} \frac{AR\, p(r)}{\tau}\, dr^3 \qquad (1)$$

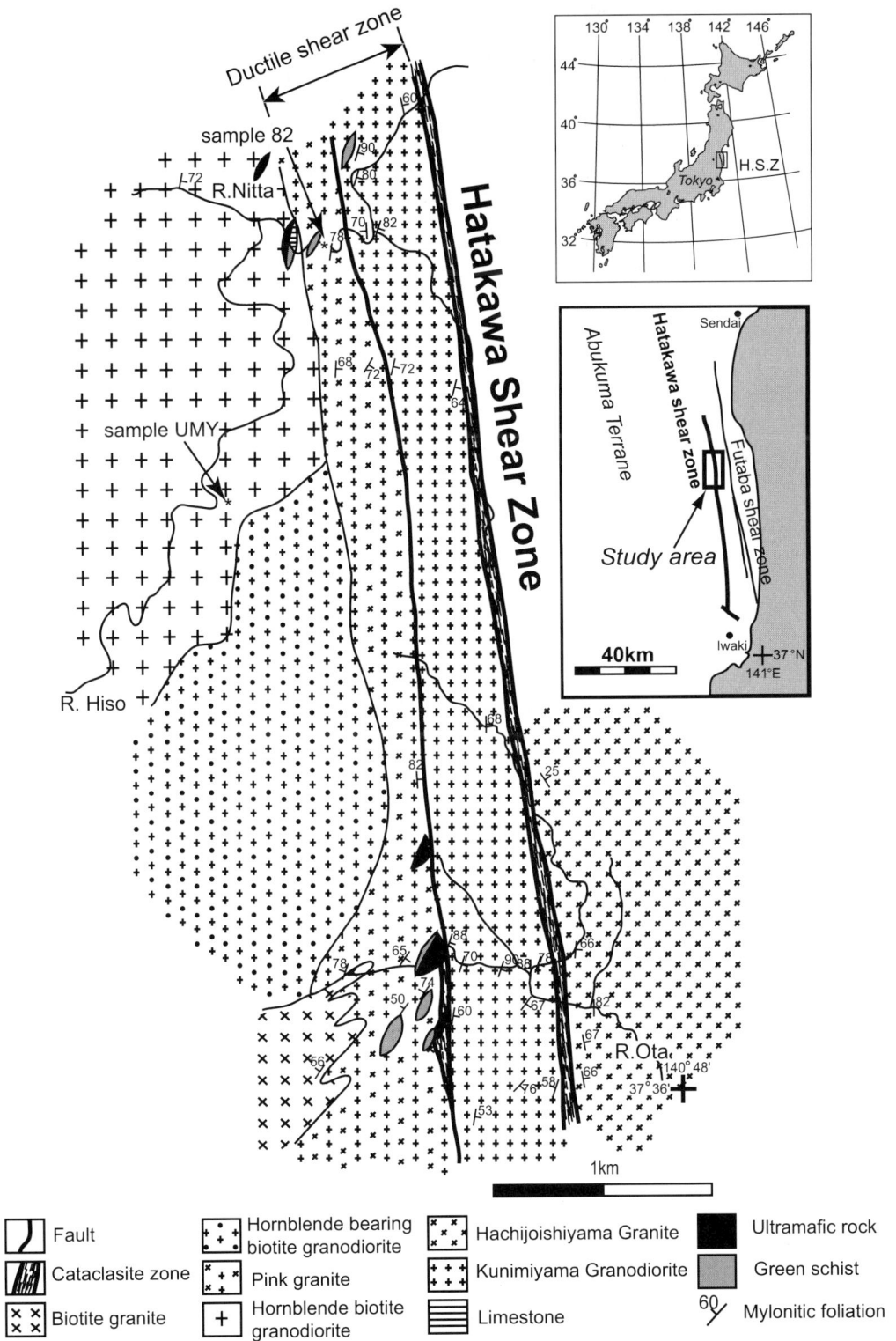

Fig. 1. Geological map of the Hatakawa shear zone and index maps of the Abukuma belts, NE Japan.

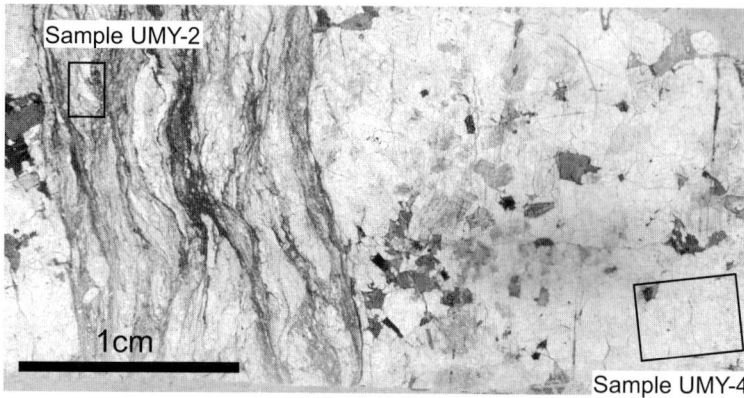

Fig. 2. Photomicrograph of an ultramylonite band under plane polarized light. Samples UMY-2 and UMY-4 represent measured fields by cathodoluminescence and micro-IR spectroscopic mappings. Sample UMY-2 corresponds to fine-grained quartz aggregate (see Fig. 5). Sample UMY-4 corresponds to weakly deformed quartz aggregate (see Fig. 3).

where V_s is the whole volume of the defect-free region, $p(r)$ and τ are the excess carrier density and its lifetime in the defect free region, and A and R are correction factors for self-absorption inside a medium and reflection losses at the surface, respectively. Although the correction factors A and R also depend on the position r, this can be neglected in most cases for simplicity (Löhnert & Kubalek 1984). The CL intensity in the presence of defects is

$$I_{CL(\tau')} = \int_{V_s} \frac{AR\,p(r)}{\tau}\mathrm{d}r^3 + \int_{V_d}\left(\frac{1}{\tau'}-\frac{1}{\tau}\right)AR\,p'(r)\,\mathrm{d}r^3$$

(2)

where V_d is the whole volume of defects, $p'(r)$ and τ' are the excess carrier density and its lifetime in the presence of defects. Here, the CL contrast $C(\tau')$ is defined as the ratio of the CL intensity of a homogeneous medium without a defect-containing $I_{CL(\infty)}$ to that of a defect-containing CL intensity $I_{CL(\tau')}$ as follows:

$$C(\tau') = \frac{I_{CL(\tau')} - I_{CL(\infty)}}{I_{CL(\infty)}}$$

$$= \frac{\int_{V_d}\left(\frac{1}{\tau'}-\frac{1}{\tau}\right)AR\,p'(r)\,\mathrm{d}r^3}{\int_{V_s}\frac{AR\,p(r)}{\tau}\mathrm{d}r^3}$$

$$= \frac{1}{I_{CL(\infty)}}\int_{V_d}\left(\frac{1}{\tau'}-\frac{1}{\tau}\right)AR\,p'(r)\,\mathrm{d}r^3 \quad (3)$$

Equation (3) shows that in the case where the CL intensity, $I_{CL(\infty)}$, is constant in a defect-free

region, V_s, the CL contrast, $C(\tau')$, depends on the excess carrier density, $p'(r)$, and its lifetime, τ', in the presence of defects. From equations (2) and (3), the density of excess carrier originated from defects (i.e. defect density) characterizes CL intensity and contrast. From this point of view there are various types of defects inducing excess carriers by electron irradiations, such as point defects, dislocations and chemical impurities (e.g. Al, Ti, Fe, K, Na and OH).

Methods and sample preparations

The analysed samples were selected from Hata-kawa mylonites (Samples 82 and UMY). Conventional petrographic thin sections were polished using 3 μm diamond paste and subsequently coated with a 20–40 nm-thick film of carbon. The sections were then examined using a scanning electron microscopy (SEM) operating at an accelerating voltage of 15 keV. The CL observations were made by Oxford MiniCL (Oxford instruments). Panchromatic CL images (unresolved wavelength CL emission) were detected by an optical fibre attached to a photo-multiplier tube (PMT). The PMT output signal was amplified to form bright and dark CL images on the SEM screen. As monochromatic CL images were obtained in these measurements, we kept CL contrast and brightness constant during measurements.

Cathodoluminescence observations of deformed quartz aggregates

Typical CL microstructures of quartz aggregates are shown in Figures 3–5, together with

(a)

(b)

(c)

(d)

55.0

45.0

35.0

150μm

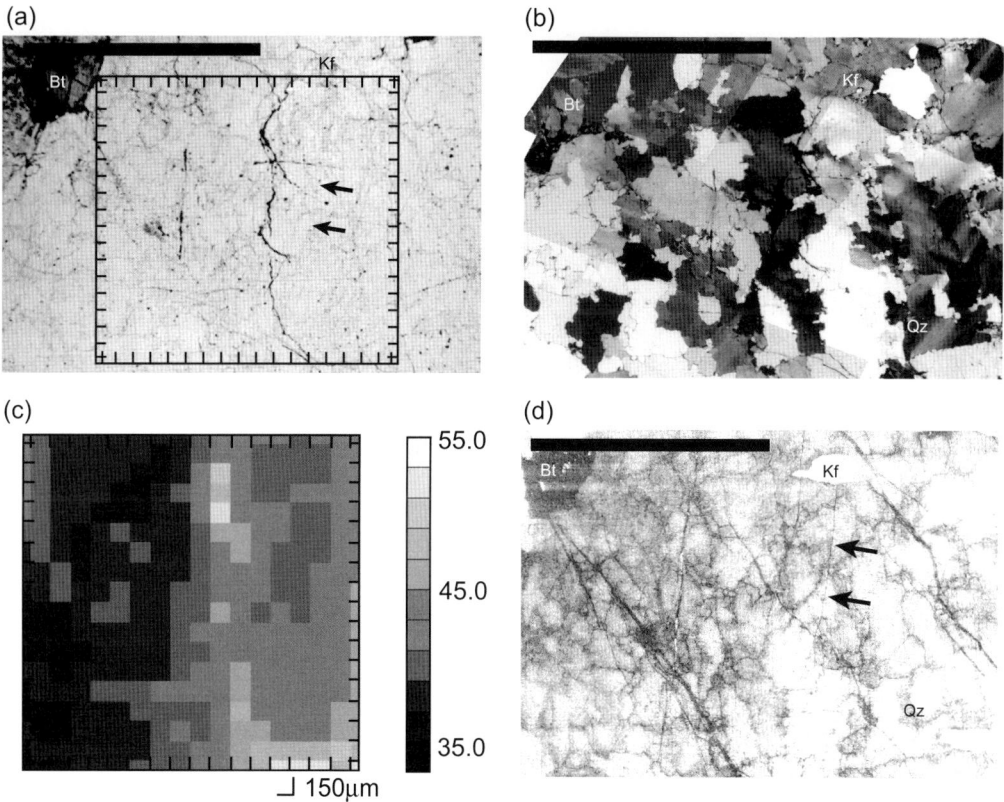

Fig. 3. Microstructures of weakly deformed quartz aggregate (sample UMY-4). (**a**) Photomicrograph of weakly deformed quartz aggregate under plane polarized light. Square area is the mapped region using a micro-IR spectrometer. (**b**) Photomicrograph of the same field as (a) between crossed nicols. (**c**) Two-dimensional integrated infrared (IR) absorbance (arbitrary units = a.u.) mapping of the square area in (a) ranging from 3300 to 3600 cm^{-1} with a 150×150 μm aperture. Note that the right-hand side of the square area shows high absorbance. Moreover, the optically visible crack shows high absorbance. (**d**) The cathodoluminescence (CL) image of the same field as (a). From the comparison (a) with (d), narrow dark bands indicated by arrows correspond to the optically visible, transgranular fluid inclusions or linear arrays of bubbles. Moreover, the region of inhomogeneous bright luminescence (right-hand side of d) shows strong undulatory extinction. Only conspicuous grains are identified on these photographs. Abbreviations: Qz, quartz; Bt, biotite; Kf, potassium feldspar. The scale bars in (a), (b) and (d) represent 2 mm.

corresponding optical micrographs. These quartz aggregates have been derived from a granitic magma and are not secondary quartz healing fractured or pull-apart grains by solution-precipitation processes. Although we conducted CL observations in various scales, there are apparent differences in CL microstructures between weakly deformed quartz aggregates and fine-grained ones. Thus, we can classify the observed CL microstructures of quartz into two groups according to the CL intensity and contrast (scale bars are 2 mm in Fig. 3a and 500 μm in Fig. 5c): (i) inhomogeneously and brightly luminescence scanning weakly deformed aggregates; and (ii) homogenously and weakly luminescence scanning fine-grained aggregates.

In the CL observations of weakly deformed quartz aggregates, the grains (grain sizes approximately 1 mm) with an inhomogeneous and bright luminescence have a strong undulatory extinction, serrated grain boundaries and deformation lamellae (the right-hand sides of Figs 3d and 4c). In some grains, the grain boundaries can be recognized more clearly in CL images than in optical microscope images (Fig. 4). These grains have high-angle boundaries that have fairly small misorientation angles of the c-axis and are identified by low luminescence. Moreover, we can often recognize narrow dark bands or dots in CL images (Fig. 3d). Narrow dark bands or dots correspond to optically visible, isolated fluid inclusions or

(a)

(b)

(c)

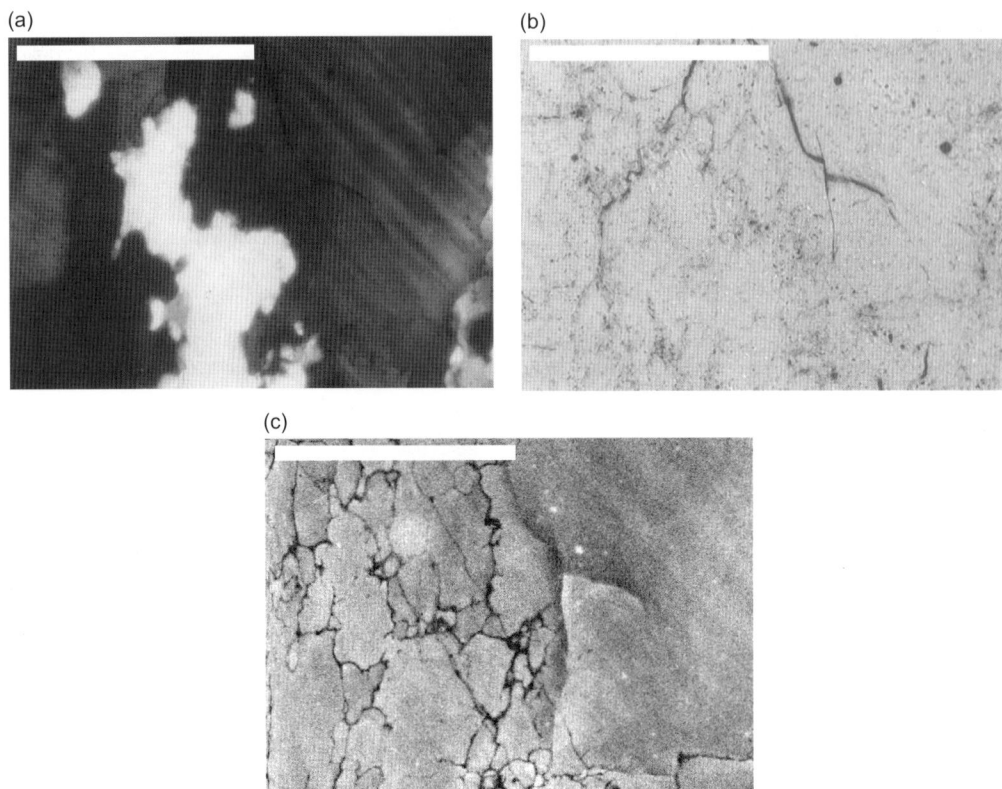

Fig. 4. Microstructures of deformed quartz aggregate in weakly deformed mylonite (sample 82). (**a**) Photomicrograph of the deformed quartz aggregate between crossed nicols. The development of the deformation lamellae is observed in some grains (right-hand side). (**b**) Photomicrograph of the same field as (a) under plane polarized light. (**c**) The CL image of the same field as (a). Note that the grains with banded luminescence correspond to the grains with deformation lamellae. Moreover, the grain boundaries of some quartz grains are found more clearly in this CL image than in the optical image (a). The scale bars represent 200 μm.

semi-planar bubble arrays (Fig. 3a & d) (see also Kanaori 1986). Most narrow bands were intragranular or transgranular. However, some narrow bands observed in the CL images were not clearly distinguishable optically.

On the other hand, fine-grained quartz aggregates (grain sizes approximately 50 μm) are generally non-luminescent or homogeneously and weakly luminescent (Fig. 5c & d). Moreover, intragranular or transgranular narrow dark bands are hardly recognized. Grain boundaries in the fine-grained quartz aggregate are less recognizable in CL image than those of host grains in the weakly deformed samples. These homogeneously and weakly luminescing grains show highly recrystallized polygonal grain shapes and no undulatory extinction, and have hardly any arrays of fluid inclusions. The highly

deformed quartz is recrystallized more pervasively as a recovery process of intracrystalline strain and is less luminescent than the quartz found in the other weakly deformed zones.

From these CL observations of deformed quartz aggregates, in spite of the differences in the observed scales, it emerges that inhomogeneously and brightly luminescent regions correspond to weakly deformed regions, and CL images clearly show undulatory extinction and deformation lamellae. On the other hand, homogeneously and weakly luminescent regions correspond to fine-grained regions. This decrease in CL contrast and intensity (brightness) caused by dynamic recrystallization as a recovery process of intracrystalline strain is in agreement with the CL observations of Shimamato *et al.* (1991) on Ho-oh mylonites.

(a)

(b)

(c)

(d)

(e)

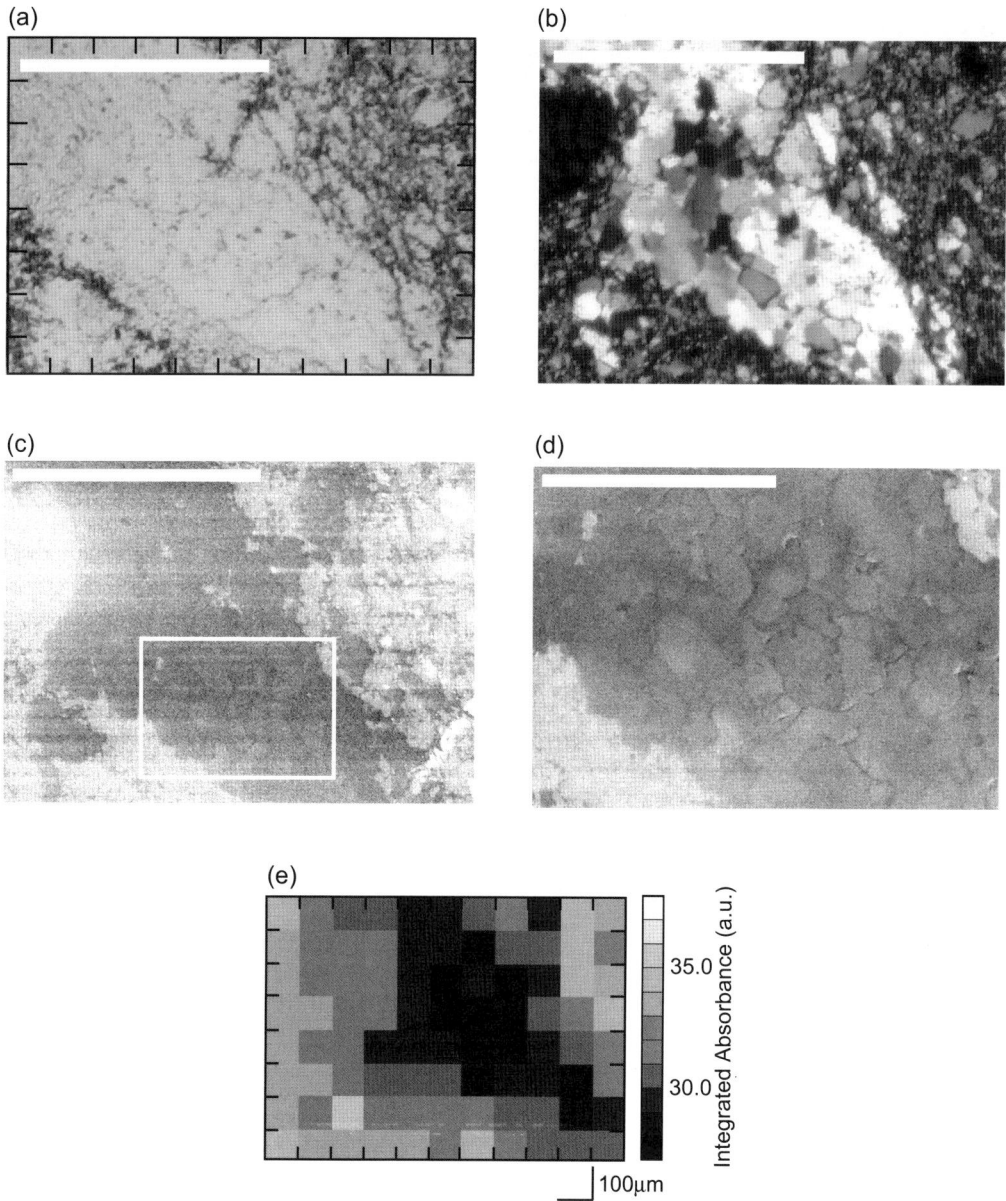

Fig. 5. Microstructures of fine-grained quartz aggregate (sample UMY-2). (**a**) Photomicrograph of the fine-grained quartz aggregate under plane polarized light. (**b**) Photomicrograph of the same quartz aggregate as (a) between crossed nicols. (**c**) The CL image of the same field as (a). (**d**) The CL image of quartz aggregate squared by the small rectangle in (c). Fine-grained quartz aggregate is non-luminescent or homogeneously and weakly luminescent. The intragranular or transgranular narrow dark bands are hardly recognized. Grain boundaries in the fine-grained quartz aggregate are less recognizable than those of weakly deformed one (see Figs 3 and 4). The scale bars represent 500 μm in (a)–(c), 50 μm in (d). (**e**) Two-dimensional integrated IR absorbance (a.u.) mapping of the same area as (a) ranging from 3300 to 3600 cm^{-1} using a 100 × 100 μm aperture. Note that integrated absorption of the fine-grained quartz aggregate is lower than that of weakly deformed one (Fig. 3c).

Micro-infrared spectroscopy of mylonites

Methods

When a molecule is irradiated with IR light, it can absorb the light if the light has the same energy as the vibrational or rotational transitions of the molecule (e.g. Brittain *et al.* 1970). As H_2O and OH species are efficient absorbers of IR light, IR spectroscopy can determine the species and the concentration of H_2O and OH groups (e.g. Katz 1962; Aines & Rossman 1984).

The distribution of water in deformed quartz aggregates were measured with a microscopic Fourier Transform IR (FT-IR) spectrometer (Perkin Elmer, Spectrum GX FT-IR system and AutoIMAGE System) using conventional petrographic thin sections with a thickness of 30 μm. The microscope stage is driven automatically by a computer control and a two-dimensional mapping image of the IR absorption spectrum can be generated. The scanning IR spectroscopy was carried out over square regions of about 2.5×1.3 mm in sample UMY-2 and about 2.6×2.6 mm in sample UMY-4 using 100×100 and 150×150 μm apertures, respectively. At each position, the IR-transmission spectrum was measured five times within about 2.5 s. The scanning intervals were 200 μm and the measured range of wavelengths was 2900–4000 cm^{-1} in a wavenumber. The peak determinations were made under a 4 cm^{-1} resolution in a wavenumber. The IR absorption spectra were recorded at room temperature. All obtained spectral data were stored in the computer memory and two-dimensional mappings of IR absorption ranging from 3300 to 3600 cm^{-1} were constructed.

Water content in a specimen can be calculated from the measured absorbance value and the sample thickness according to Lambert–Beer's law (e.g. Brode 1949), which shows a linear relation among the content of a chemical species, the sample thickness and the absorbance. From this point of view, the relative concentration of H_2O and OH can be compared in the same specimen under the assumption that the thickness is constant in the same thin sample. However, we have not obtained the correct values of H_2O and OH content in quartz aggregates yet because of the necessity of accurate determinations for sample thicknesses in the order of micrometres.

Micro-infrared spectroscopic mapping of deformed quartz aggregates

The IR spectra for deformed quartz are shown in Figure 6. In the measured spectral range from

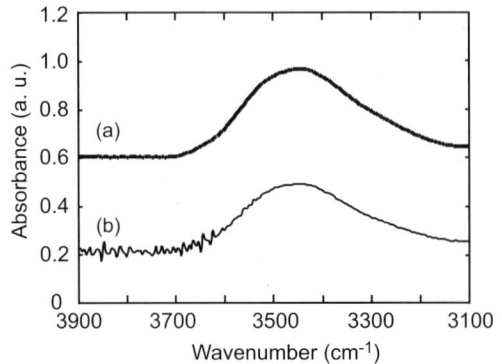

Fig. 6. Representative IR absorption spectra (a.u.) for deformed quartz grains. (**a**) Spectrum of weakly deformed quartz grain (sample UMY-4, bold spectrum). (**b**) Spectrum of fine-grained quartz grain (sample UMY-2, thin spectrum). Both spectra show absorption peaks that are centred at 3450 cm^{-1}, broadened from 3300 to 3600 cm^{-1} and are weakly asymmetric.

2900 to 4000 cm^{-1}, all of the IR spectra show broad-band absorptions around 3450 cm^{-1}. These broad peaks are centred at 3450 cm^{-1}, broadened from 3300 to 3600 cm^{-1} and are weakly asymmetric.

On two-dimensional mapping (Figs 3c and 5e) we can recognize high absorbance along the optically visible crack in Figure 3c. Moreover, in the weakly deformed quartz aggregate, we observe that the right-hand side shows a higher absorbance value (around 50 arbitrary units = a.u.) than the left-hand side (around 40 a.u.). On the other hand, the fine-grained quartz aggregate (sample UMY-2; Fig. 5e) shows low absorbance (around 30 a.u.) compared with the weakly deformed one (sample UMY-4; Fig. 3c). Furthermore, we see homogeneous absorption in the fine-grained quartz aggregate (sample UMY-2).

Discussion

From CL observations of weakly deformed quartz aggregates (Figs 3d and 4c), the CL images clearly reflect the undulatory extinction and deformation lamellae. Moreover, dynamic recrystallization, as a recovery process of intracrystalline strain, decreases CL contrast and intensity (Fig. 5c & d). From deformation experiments of quartz, deformation lamellae are usually composed of arrays of dislocations (Christie *et al.* 1964; Drury 1993). When the dislocations are related to the CL intensity and contrast, the lifetime and density of excess carriers originated from dislocations depend on dislocation densities (see also Hamilton *et al.* 1983).

Although there are no data available for dislocation density in this paper, from equations (2) and (3) we infer that increases in dislocation densities correspond to increases in CL intensity and contrast. During deformation, dynamic recrystallization is the softening process that competes with a strain-hardening process, and causes a decrease in dislocation density as a recovery process of intracrystalline strain (e.g. Nicolas & Poirier 1976). Hence, CL contrast and intensity within each grain decrease with grain-size refinement by dynamic recrystallization (Fig. 5c & d).

In the present study, we mention the relationships between the presence of broad IR absorption peaks around $3300-3600 \text{ cm}^{-1}$ and the species of H_2O and OH. From the IR mapping of the weakly deformed quartz aggregate (sample UMY-4; Fig. 3c), the absorbance values of $3300-3600 \text{ cm}^{-1}$ broad peaks increase from approximately 40 a.u. on the left-hand side to 50 a.u. on the right-hand side. This increase tendency corresponds to the increase in CL intensity and contrast from the left-hand side to the right-hand side in the CL image (Fig. 3d). On the other hand, the fine-grained quartz aggregate shows an homogenously and weakly luminescent image (Fig. 5c & d), and also shows homogeneous and low absorption values of approximately 30 a.u. (Fig. 5e). From the previous studies of IR spectroscopy on H_2O or OH species in quartz, the obtained broad peaks around $3300-3600 \text{ cm}^{-1}$ indicate the presence of free H_2O in fluid inclusions, bubbles (e.g. Aines & Rossman 1984) and at grain boundaries (Ito & Nakashima 2002), protonated water molecules $H^+ \cdot 6H_2O$ (Kim et al. 2002) or adsorbed OH defects as dangling bonds (Muster et al. 2001). These molecular H_2O and absorbed OH species are referred to as water in this paper. From these IR images, water is inhomogeneously distributed in the weakly deformed quartz aggregate (Fig. 3c) and its content decreases as a recovery process of intracrystalline strain (Fig. 5e). Moreover, the correspondence between CL images and IR mapping shows that CL intensity and contrast depend on the water content.

Equation (3) is satisfied with the correspondences among the CL intensity, contrast and the water content. In hydrous SiO_2, the radiolysis of the OH leads to the formation of non-bridging oxygen hole centres (NBOHC) and the production of hydrogen radicals (i.e. $\equiv Si-O-H \rightarrow \equiv Si-O \cdot \cdot H$) (Griscom 1985; Stevens Kalceff & Phillips 1995). The NBOHC and hydrogen radicals from residual H_2O and/or absorbed OH cause the CL emission (Koyama 1980; Stevens Kalceff & Phillips 1995). The recombinations of NBOHC and hydrogen radicals are activated by the increase in water content. Therefore, the increase in excess carriers originated from H_2O and OH increases CL intensity $I_{CL(\tau')}$ in equation (2), and increases CL contrast $C(\tau')$ in equation (3). On the other hand, dynamic recrystallization decreases the water content within each grain as a recovery process of intracrystalline strain, and reduces the CL intensity and contrast within each grain of fine-grained quartz aggregate (Fig. 5).

From equations (2) and (3), the increases in CL intensity and contrast can also be caused from increases in other types of defect inducing excess carriers by electron irradiations, such as point defects and chemical impurities (e.g. Al, Ti, Fe, K and Na). Although we do not have any evidence of fluid circulation during deformation and also did not estimate the actual water content, the deformed rock we observed is of magmatic origin and generally shows a low content of chemical impurities (e.g. Al, Ti: $100-300$ ppm; Müller et al. 2000, 2002; Van den Kerkhof et al. 2004) compared with its water content (e.g. $4000-10 000$ ppm; Kronenberg & Wolf 1990; Kronenberg et al. 1990). So, only detailed chemical analysis can rule out the possibility of CL emission from chemical impurities. Therefore, CL emission from the chemical impurities is relatively weak and the main source of CL is given by the water content (i.e. molecular H_2O and absorbed OH species).

Conclusions

Based on the IR mapping, we have presented observations which show that water (molecular H_2O and absorbed OH species) is inhomogeneously distributed in weakly deformed quartz aggregates. Furthermore, the comparison of CL observations with IR mapping indicates that CL contrast and the intensity of deformed quartz aggregates depend on a content of defects (e.g. water content). On the other hand, dynamic recrystallization, as a recovery process of intracrystalline strain, decreases defect density and CL contrast and intensity within each grain. The CL spectroscopies and micro-IR spectroscopic mapping can determine the concentrations and distributions of water in natural high-strain zones, which control the rheological properties of lithosphere.

The authors would like to thank T. Yamaguchi for his help in IR measurements. The authors also acknowledge T. Ouchi for guidance through the CL operations. We also express our sincere thanks to L. Burlini, D. Bruhn, B. den Brok and an anonymous reviewer for their insightful review comments which improved the manuscript. We also acknowledge B. Cramer for improving the English of our manuscript. The first author (J. Muto) was financially supported by the 21st Century Centre-Of-Excellence programme, 'Advanced Science and Technology Centre for the Dynamic Earth', of Tohoku University.

References

AINES, R.D. & ROSSMAN, G.R. 1984. Water in minerals? A peak in the infrared. *Journal of Geophysical Research*, **89**, 4059–4071.

BRITTAIN, E.F.H., GEORGE, W.O. & WELLS, C.H.J. 1970. *Introduction to Molecular Spectroscopy: Theory and Experiment.* Academic Press, London.

BRODE, W.R. 1949. The presentation of absorption spectra data. *Journal of the Optical Society of America*, **39**, 1022–1031.

CHELIKOWSKY, J.R. & SCHLÜTER, M. 1977. Electron states in α-quartz: A self-consistent pseudopotential calculation. *Physical Review B*, **15**, 4020–4029.

CHRISTIE, J.M., GRIGGS, D.T. & CARTER, N.L. 1964. Experimental evidence of basal slip in quartz. *Journal of Geology*, **72**, 734–756.

DRURY, M.R. 1993. Deformation lamellae in metals and minerals. *In:* BOLAND, J.N. & GERALD, J.D.F. (eds) *Defects and Processes in the Solid State: Geoscience Applications.* Developments in Petrology, **14**, 195–212.

GÖTZE, J., PLÖTZE, M. & HABERMANN, D. 2001. Origin, spectral characteristics and practical applications of the cathodoluminescence (CL) of quartz – a review. *Mineralogy and Petrology*, **71**, 225–250.

GRISCOM, D.L. 1985. Defect structure of glasses. *Journal of Non-crystalline Solids*, **73**, 51–77.

HAMILTON, B., PEAKER, A.R. & WIGHT, D.R. 1983. Luminescence and deep level studies of line dislocations in gallium phosphide. *Journal de Physique III*, **C4-44**, 233–241.

HASHIMOTO, T., YAMAGUCHI, T., FUJITA, H. & YANAGAWA, Y. 2003. Comparison of infrared spectrometric characteristics of Al–OH impurities and thermoluminescence patterns in natural quartz slices at temperature below 0 °C. *Radiation Measurements*, **37**, 479–485.

HOLNESS, M.B. & WATT, G.R. 2001. Quartz recrystallization and fluid flow during contact metamorphism: a cathodoluminescence study. *Geofluids*, **1**, 215–228.

ITO, Y. & NAKASHIMA, S. 2002. Water distribution in low-grade siliceous metamorphic rocks by micro-FTIR and its relation to grain size: a case from the Kanto Mountain region, Japan. *Chemical Geology*, **189**, 1–18.

JAKUBOWICZ, A. 1986. Theory of cathodoluminescence contrast from localized defects in semiconductors. *Journal of Applied Physics*, **59**, 2205–2209.

KANAORI, Y. 1986. A SEM cathodoluminescence study of quartz in mildly deformed granite from the region of the Atotsugawa fault, central Japan. *Tectonophysics*, **131**, 133–146.

KATZ, A. 1962. Hydrogen in alpha-quartz. *Philips Research Reports*, **17**, 133–195, 201–279.

KIM, J., SCHMITT, U.W., GRUETZMACHER, J.A., VOTH, G.A. & SCHERER, N.E. 2002. The vibrational spectrum of the hydrated proton: Comparison of experiment, simulation, and normal mode analysis. *Journal of Chemical Physics*, **116**, 737–746.

KOYAMA, H. 1980. Cathodoluminescence study of SiO_2. *Journal of Applied Physics*, **51**, 2228–2235.

KRONENBERG, A.K. 1994. Hydrogen speciation and chemical weakening of quartz. *Reviews in Mineralogy*, **29**, 123–176.

KRONENBERG, A.K. & WOLF, G.H. 1990. Fourier transform infrared spectroscopy determinations of intragranular water content in quartz-bearing rocks: implications for hydrolytic weakening in the laboratory and within the earth. *Tectonophysics*, **172**, 255–271.

KRONENBERG, A.K., SEGALL, P. & WOLF, G.H. 1990. Hydrolytic weakening and penetrative deformation within a natural shear zone. *In:* DUBA, A.G., DURHAM, W.B., HANDIN, J.W. & WANG, H.F. (eds) *The Brittle–Ductile Transition in Rocks.* American Geophysical Union, Geophysical Monographs, **56**, 21–36.

KUBO, K., YANAGISAWA, Y., YOSHIOKA, T., YAMAMOTO, T. & TAKIZAWA, F. 1990. *Geology of the Haramachi and Ōmika District.* With geological sheet map at 1:50 000, Geological Survey of Japan, Tsukuba (in Japanese with English abstract).

LÖHNERT, K. & KUBALEK, E. 1984. The cathodoluminescence contrast formation of localized nonradiative defects in semiconductors. *Physica Status Solidi A*, **83**, 307–314.

MCLAREN, A.C., GERALD, J.D.F. & GERRETSEN, J. 1989. Dislocation nucleation and multiplication in synthetic quartz – relevance to water weakening. *Physics and Chemistry of Minerals*, **16**, 465–482.

MÜLLER, A., SELTMANN, R. & BEHR, H.-J. 2000. Application of cathodoluminescence to magmatic quartz in a tin granite – case study from the Schellerhau Granite Complex, Earstern Erzgebirge, Germany. *Mineralium Deposita*, **35**, 169–189.

MÜLLER, A., LENNOX, P. & TRZEBSKI, R. 2002. Cathodoluminescence and micro-structural evidence for crystallisation and deformation processes of granites in the Eastern Lachlan Fold Belt (SE Australia). *Contributions to Mineralogy and Petrology*, **143**, 510–524.

MUSTER, T.H., PRESTIDGE, C.A. & HAYES, R.A. 2001. Water absorption kinetics and contact angles of silica particles. *Colloids and Surface A: Physicochemical and Engineering Aspects*, **176**, 253–266.

MUTO, J. & NAGAHAMA, H. 2004. Dielectric anisotropy and deformation of crustal rocks: physical interaction theory and dielectric mylonites. *Physics of the Earth and Planetary Interiors*, **141**, 27–35.

NAKAMURA, N. & NAGAHAMA, H. 2002. Tribocheimical wearing in S–C mylonites and its implication to lithosphere stress level. *Earth Planets Space*, **54**, 1103–1108.

NICOLAS, A. & POIRIER, J.P. 1976. *Crystalline Plasticity and Solid State Flow in Metamorphic Rocks.* Wiley, London.

OTSUKI, K. & EHIRO, M. 1978. Major strike-slip faults and their bearing on spreading in the Japan Sea. *Journal of Physics of the Earth*, **26**, Supplement, S537–S555.

PEY, K.L., CHAN, D.S.H. & PHANG, J.C.H. 1995. Cathodoluminescence contrast of localized defects part I. Numerical model for simulation. *Scanning Microscopy*, **9**, 355–366.

POIRIER, J.P. 1985. *Creep of Crystals.* Cambridge University Press, Cambridge.

RUFFINI, R., BORGHI, A., COSSIO, R., OLMI, F. & VAGGELLI, G. 2002. Volcanic quartz growth zoning identified by cathodoluminescence and EPMA studies. *Mikrochimica Acta*, **139**, 151–158.

SHIMAMOTO, T., KANAORI, Y. & ASAI, K. 1991. Cathodoluminescence observations on low-temperature mylonites: potential for detection of solution-precipitation microstructures. *Journal of Structural Geology*, **13**, 967–973.

STEVENS KALCEFF, M.A. & PHILLIPS, M.R. 1995. Cathodoluminescence microcharacterization of the defect structure of quartz. *Physical Review B*, **52**, 3122–3134.

TAKAGI, H., GOTO, K. & SHIGEMATSU, N. 2000. Ultramylonite bands derived from cataclasite and pseudotachylyte in granites, northeast Japan. *Journal of Structural Geology*, **22**, 1325–1339.

VAN DEN KERKHOF, A.M., KRONZ, A., SIMON, K. & SCHERER, T. 2004. Fluid-controlled quartz recovery in granulite as revealed by cathodoluminescence and trace element analysis (Bamble sector, Norway). *Contributions to Mineralogy and Petrology*, **146**, 637–652.

WAYCHUNAS, G.A. 1988. Luminescence, X-ray emission and new spectroscopies. *Reviews in Mineralogy*, **18**, 639–698.

YACOBI, B.G. & HOLT, D.B. 1986. Cathodoluminescence scanning electron microscopy of semiconductors. *Journal of Applied Physics*, **59**, R1–R24.

Application of Geographical Information Systems to shape-fabric analysis

F. J. FERNÁNDEZ[1], R. MENÉNDEZ-DUARTE[1,2], J. ALLER[1] & F. BASTIDA[1]

[1]*Departamento de Geología, Universidad de Oviedo, C/Jesús Arias de Velasco s/n, 33005 Oviedo, Spain (e-mail: brojos@geol.uniovi.es)*
[2]*INDUROT, Universidad de Oviedo, Campus de Mieres, 33600 Mieres, Spain*

Abstract: A Geographical Information System (GIS) has been applied to shape-fabric analysis and strain measurement. It works in either raster or vectorial format, allows determination of many geometrical features of grains and stores additional descriptive information linked to all these graphic elements. Some new applications for shape characterization and strain measurement have been developed, e.g. the shape parameter *S*, the Fry method installed as a variant that uses only contiguous grains (TFry method) and a new method (ASPAS method), based on the intersection lengths of the lines of a radial network with the grain boundaries. A compilation of finite-strain estimates on a tectonite sample has been used to compare the results obtained by the new methods. As the shape fabric does not register the full strain history, the strain ratios obtained from the shape-fabric analysis are lower in most cases than those obtained using pretectonic markers.

Shape-fabric analysis was a time-consuming task before digital image processing, when grain-boundary maps were produced by manual tracing. Despite this limitation, Sander (1950) designed the axial distribution analysis (AVA) method, which graphically establishes the relations between crystallographic preferred orientation (CPO) and the shape fabric. Three decades after this work, the first computer applications to strain analysis based on centre-to-centre techniques were developed (Fry 1979; Erslev 1988; Erslev & Ge 1990). The first program to analyse shape fabric (Panozzo 1983) was based on the projections of lines or ellipses, which represent the best-fit representation of grains, on a reference axis (*x*-axis). If a set of lines or ellipses displays a preferred orientation, the average length of projection of the lines or ellipses on the *x*-axis changes with direction in the co-ordinate plane. Nowadays, the most popular image-processing software for shape-fabric analysis is the NIH image (for Macintosh, NIH 1999) or its equivalent, Scion Image (for Windows, Scion Corporation 1999). This software can be used to measure the area, centroid, perimeter and so on of grains in boundary maps. It also performs automated particle analysis, and provides tools for measuring path lengths, angles, sizes and length measurements. Moreover, automatic grain-boundary detection and grain size analysis are possible using orientated images from petrographic thin sections. This method has been programmed as a set of macros for the NIH image (Heilbronner 2000). In combination with this technique, the *c*-axis preferred orientation of uniaxial mineral aggregates can be also determined automatically by computer-aided microscopy (Panozzo-Heilbronner & Pauli 1993).

However, the handling of very large data sets, as well as the possibility to easily link the grain-shape analysis to other information about the grains gives the use of the geographical information system (GIS) large advantages (Bonham-Carter 1994). GIS is commonly used in Earth science, and it allows the same geometric parameters as NIH to be obtained from the shape fabric. In addition, GIS can work in both raster and vectorial formats, and it defines the topology, which is the spatial relationships between connecting or adjacent arcs, nodes, polygons and points. In a vector model GIS, each element represented on a map corresponds to a geometric shape that is defined by a series of co-ordinate pairs (e.g. points, lines and polygons defined by groups of interconnected lines). Each grain of the shape fabric is targeted as a polygon with a variety of descriptive information (i.e. shape-grain parameters, mineralogy or

From: BRUHN, D. & BURLINI, L. (eds) 2005. *High-Strain Zones: Structure and Physical Properties.*
Geological Society, London, Special Publications, **245**, 409–420.
0305-8719/05/$15.00 © The Geological Society of London 2005.

crystallographic orientation). The conversion of grains to discrete polygons has some clear and rapid applications in shape-fabric analysis. In a section, the components of a rock are closed polygons of a certain size, shape, relative position and relationship with neighbouring polygons. The GIS has the advantage that it can comprise the possibility of executing complex spatial operations and managing a large amount of graphical data. Nevertheless, other microstructural analysis methods such as the autocorrelation function (Heilbronner 1992) or the intercept method (Launeau & Robin 1996) may be more suitable in some very fine-grained rocks in which it is difficult to derive the outline map.

The main aim of this paper is to describe a number of GIS applications to shape-fabric analysis. To do this, shape-fabric parameters have been revised and some of them redefined. Three samples of experimental and natural tectonites from different strain conditions have been used to illustrate the applicability of GIS for shape-fabric analysis, as well as to compare the suitability of the different methods applied. In addition, four subroutines applying GIS to strain analysis are placed in the public domain http://www.indurot.uniovi.es/docdisp/doc.html.

Sample preparation and image processing

The GIS used in this application was ArcInfo v.7.1.2, in which the topological model is Arc-Node type. The basic logical entity is the arc, a series of points that start and end at a node. A polygon is a list of the arcs that determines a closed area (ESRI 1992). The macros of strain analysis have been programmed in AML, the ARC Macro Language (ESRI 1994), which includes a set of directives and in-line functions that can be used interactively or in AML programs. In addition to Arcinfo commands, the program also includes commands from the operating system (UNIX HP-UX 10.20).

Shape-fabric analysis was conducted on photomicrographs or, in some cases, on SEM images to a scale large enough to image perfectly the grain boundaries. Nowadays automatic grain-boundary detection is possible using multispectral image methods (e.g. Swan & Garrat 1995; Heilbronner 2000) but not using GIS. Despite their evident advantages, all automatic methods still have problems in samples where the colour contrast between grains is low, or where grain boundaries produce interlocking irregular contacts of variable sharpness. Finally, the map of grain-boundary outlines is transformed into a raster image file either by scanning the hand-drawn sketch or directly if working with automatic grain-boundary images.

Before opening the raster file with a GIS, it is convenient to vectorize using a single central line of high resolution and constant thickness in order to have polygons closely resembling the grains. If we convert the file using the GIS, the raster image is converted to a vector-based format using the 'gridpoly' command of ArcInfo. The fabric map in a vector-based format is a coverage that contains polygons with label points and a polygon attribute table. In GIS the coverage consists of topologically linked geometric features and their stored associated descriptive data. The graphical co-ordinates are also registered, maintaining the real dimensions of the represented elements.

The computation of the area, perimeter and centroid of a polygon using Arcinfo is obtained directly from its co-ordinates (x, z) or (y, z), i.e. the centroid is given by the average (x, z) co-ordinates of the outlines that define each polygon. If we assign attributes to the polygons from other available data, such as their mineralogy or other features, we can obtain quantitative data of the sample, such as the modal composition, the mean, the maximum and minimum grain sizes, the geometry of the grains or other additional features (crystallographic preferred orientation, cracking data sets, etc).

Shape-fabric parameters

Three main geometric features of the shape fabric have been analysed in this work: grain size, grain shape and grain orientation. These are recorded using vector-based and raster files of the shape-fabric map. The size of single grains is defined by their area. The grain size of a sample has been characterized using two parameters: the percentage area of the different size ranges and the size frequency (Fig. 1).

In order to describe the shape of the grains we propose two complementary parameters: grain ellipticity and area/square perimeter ratio. Grain ellipticity is defined by the aspect ratio of the largest ellipse that can be inscribed inside each grain (a/b ratio, where a is the longest and b the shortest ellipses semi-axis). As ellipticity is sensitive to the bulk shape of the grains, but not to the detailed shape of their boundaries, the ellipticity parameter is complemented with the shape parameter (S). S is calculated by:

$$S = \frac{4\pi A_i}{P_i^2} \qquad (1)$$

where A_i is the area and P_i the perimeter of each grain. S is the inverse of the traditional shape

PROGRESSIVE SHEAR STRAIN (0.8T$_M$)

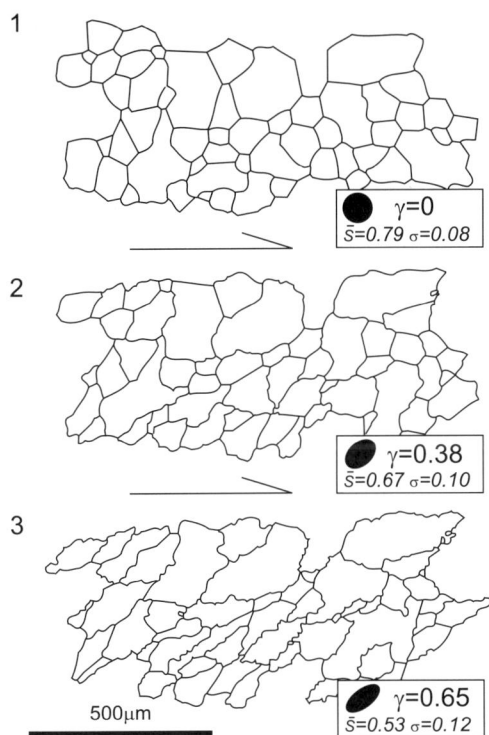

Fig. 1. Scanned grain-boundary outlines of experimental deformed octachloropropane described by Means & Dong (1998). The shear strain γ is calculated from the passive markers in the experiment. $\bar{S}$ is the arithmetic average of the shape parameter S and σ is the standard deviation. T_M is the temperature of melting.

factor 1 (Underwood 1970, p. 228). Note that in this way, S is grain-size independent and $S = 1$ for a perfect circle, but it decreases for increasing deviations from the circle. The arithmetic average, $\bar{S}$, together with its standard deviation (σ) can be used as a number representative of the grain-boundaries map. This parameter might be also useful as a geothermometer (Kruhl & Nega 1996) or as a palaeopiezometer (Takahashi *et al.* 1998) if the degree of serration of the grain boundaries is related to the deformation conditions.

The angle, ϕ, between the major axis of the greatest inscribed ellipse and a reference line defines the grain orientation. This angle and the ellipticity are the parameters of the R_f/ϕ strain analysis method (Ramsay 1967; Dunnet 1969; Ramsay & Huber 1983; Lisle 1985), which are also used in the PAROR method (Panozzo 1983). In Arcinfo, the ellipticity and the angle ϕ are obtained using the grid function 'Zonal Geometry'.

The shape-fabric maps of Figure 1 show the effect of shearing octachloropropane (Means & Dong 1998). Changes in shape fabric are characterized using the grain size and orientation histograms, and the diagram of ellipticity v. S (Fig. 2) where the curve E corresponding to perfectly elliptic shapes is also plotted. The distance from a point to E in Figure 2c gives a measure of the grain-boundary irregularity. General features observed during progressive shear strain are: (i) the three grain size histograms are similar; only the most strained shape fabric developed a new range of fine grain size, characteristic of the dynamically recrystallized grains; (ii) the orientation histograms show a decrease in the standard deviation, and the appearance of a maximum at $35°$–$40°$ to the shear direction; and (iii) $\bar{S}$ decreases and ellipticity increases progressively as the fabric changes from a 'foam structure' composed of polygonal grains to a deformed shape fabric with irregular boundary grains or when grain boundaries become more serrated (Figs 1 and 2).

Strain analysis using GIS

GIS can be used to apply the R_f/ϕ method (Ramsay 1967; Dunnet 1969; Ramsay & Huber 1983; Lisle 1985), which is grain-shape dependent. Two applications based on the Fry method (Fry 1979) have also been developed and a new method based on the linear intercept method (Exner 1972; Panozzo 1984, 1987; Launeau & Robin 1996), the 'ASPAS method'. These applications are now available in three new macros loaded in the public domain http://www.indurot.uniovi.es/docdisp/doc.html.

The TFry method

The shape methods, such as R_f/ϕ and ASPAS, are sensitive to the orientation of the grains, but not to their position in the sample sections. This latter aspect can be analysed with centre-to-centre techniques (e.g. Ramsay 1967; Fry 1979; Lisle 1979), which use the anisotropy of an initially uniform anti-clustered spatial distribution of centre points as a measure of strain. These methods are therefore complementary to the shape methods. We have installed the Fry method using GIS commands and programmed it as a macro called 'TFry.aml'. An important concept in the TFry method is the definition of contiguity between grains. Contiguity is the topological concept that allows the vector data model to determine adjacency. The segments have direction, so that the left and right side of the grains can be determined. In TFry analysis, all the grains that have an arc in common with the grain considered in each iteration can be selected. The centroids of

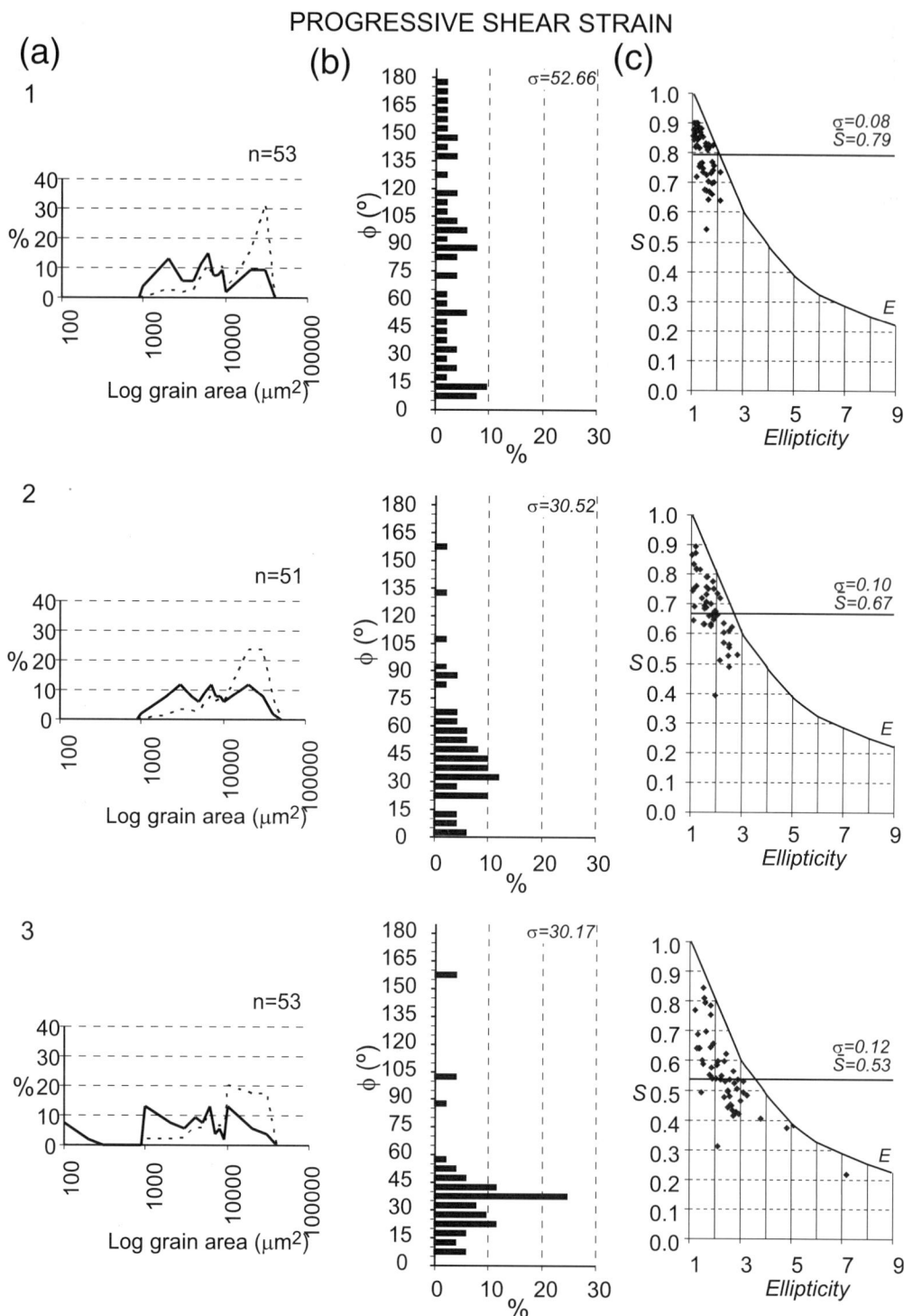

Fig. 2. Shape-fabric analysis of samples 1, 2 and 3 in Figure 1. (**a**) Grain size diagrams showing the size frequency (black line) and the area per cent (dashed line) of the different size ranges. (**b**) Histograms showing the frequency of the orientation (ϕ) of the major ellipse inscribed in each grain and σ is the standard deviation. (**c**) Shape parameter (S) v. ellipticity diagrams; $\bar{S}$ is the arithmetic mean of S (horizontal black line).

the adjacent polygons are displaced to an arbitrary fixed point (Fig. 3b), and this operation is repeated for each grain of the shape-fabric map (Fig. 3a). For homogeneously deformed rocks, the initial distances between grain centroids become modified in proportion to the value of the longitudinal strain along that direction (Ramsay 1967, p. 197). The graphic result is a cloud of points spread around the arbitrary fixed point (Fig. 3c). Note that the selection of a subset of nearest grains is defined by the contiguity property automatically when using GIS, whereas the exact definition of what constitutes a nearest grain leads to some

subjectivity in the measurements of previous programs based on the Fry method (Erslev & Ge 1990; McNaught 1994). An attempt to solve this problem by the application of Delaunay triangulation to the nearest-grains method of strain analysis has been developed by Mulchrone (2002).

The classical Fry method may be also useful in fabrics where grains are not in contact and the GIS contiguity-command cannot be applied, such as the fabrics of some conglomerates, or the fabrics of polycrystalline aggregates composed of different minerals and porphyroaphanitic textures. The macro 'Fry.aml' is

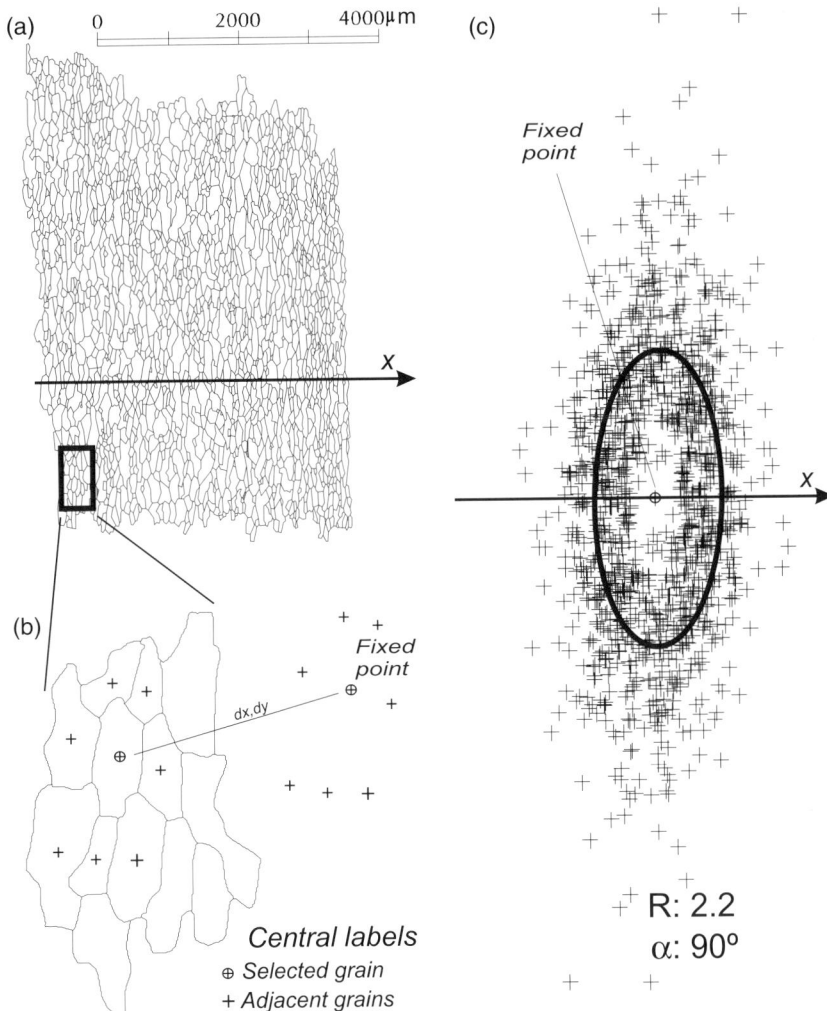

Fig. 3. Explanation of the TFry method procedure used in this work. (**a**) Grain-boundaries map. (**b**) For each grain, the centre of the adjacent grains is selected. All these points are displaced a distance (dx, dy) equal to the distance between the centre of the first grain and a predetermined fixed point. (**c**) The repetition of this operation for all the grains of the boundaries map yields a cloud of points that defines the strain ellipse.

(a)

$$\Sigma p_i = p_0 + p_1 \dots + p_8.$$
$$\Delta \varphi : 2°$$

(b)

30
34
38
42
46
50

p_7 p_8 p_1

p_6 p_0 p_2 X

p_5 p_4 p_3

(c)

X

0   2000   4000μm

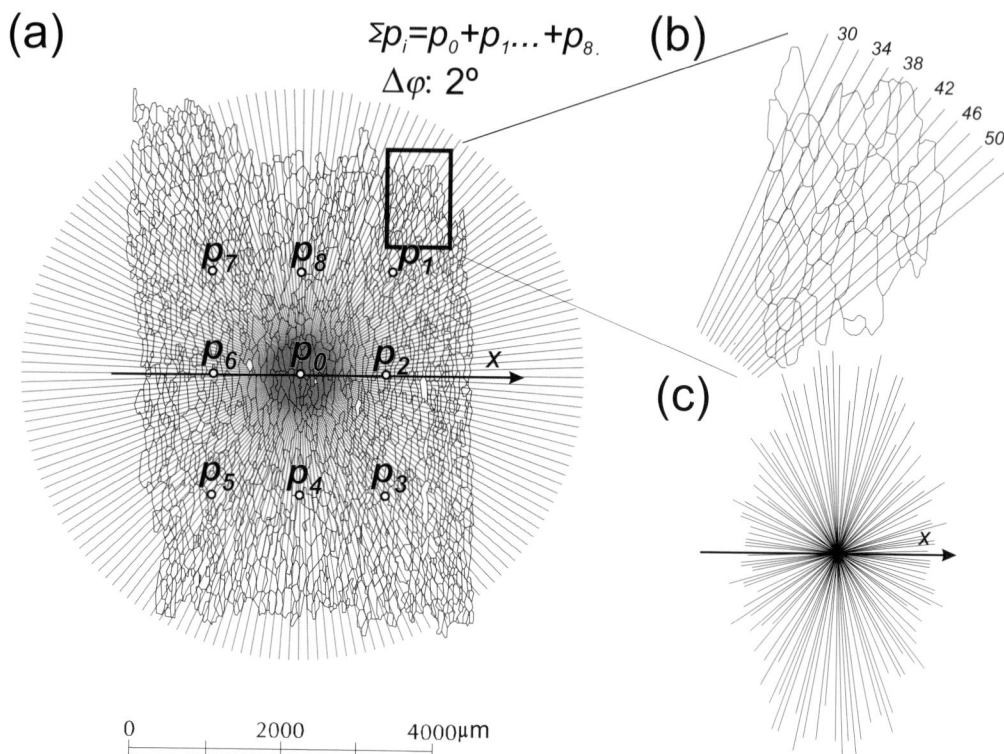

Fig. 4. ASPAS method procedure. The ASPAS network with a rotation angle $\Delta \varphi = 2°$ intersects with the grain boundaries (**a**), dividing each line into a variable number of segments (**b**). The average length of these segments in each direction defines the individual ASPAS diagram (**c**). This procedure repeated for the nine points ($p_0 - p_8$) shown in (a) allows the representative shape to be obtained (see the text for details).

programmed with the GIS commands necessary to use the classical Fry method.

The ASPAS method

This method is based on the linear intercept method (Exner 1972) and allows the determination of the representative shape of the fabric. It has been programmed as a macro called 'Aspas.aml'.

The method involves the superposition of the outlines map over a network of radiating lines (ASPAS) centred on a reference point (p_i). This network resembles the cross-piece of a sail ('aspas' in Spanish). The ASPAS is developed

as an arc coverage with a constant angular interval ($\Delta\varphi$) and it is large enough to cover the analysed area shape. In the superposition of the coverage, the ASPAS intersects with grain boundaries, dividing each initial line of the ASPAS in a variable number of segments (Fig. 4a & b). The mean intersection length along every line ($\bar{d}$) is a function of its orientation (φ). Then, we can construct a rose diagram to represent the function $\bar{d}(\varphi)$ (Fig. 4c). As this rose diagram is dependent on the p_i chosen and grain size 'layering', several rose diagrams placed on different reference points of the same fabric (Figs 4a, 5a and 6a) are used to minimize

Fig. 5. Shape-fabric analysis of the Caurel quartzite. (**a**) Grain-boundaries map. Different grey tones represent quartz grains A and B discriminated from the grain size diagram. (**b**) Grain size diagram showing the size frequency (black line) and the area per cent (dashed line) of the different size ranges. (**c**) Histograms showing the frequency of the orientation (ϕ) of the major ellipse inscribed in each grain for grains A and B, and diagrams of the shape parameter (S) v. ellipticity for grains A and B; σ is the standard deviation, $\bar{S}$ is the arithmetic mean of S (horizontal black line). (**d**) Individual ASPAS diagrams calculated for the nine centres p_i. (**e**) Representative shape and total ASPAS diagrams. (**f**) R_f/ϕ diagram. Dashed lines represent the R_f maximum and minimum calculated following the same procedure that Ramsay & Huber (1985, p. 83); contour intervals 1% per area. (**g**) TFry analysis plot.

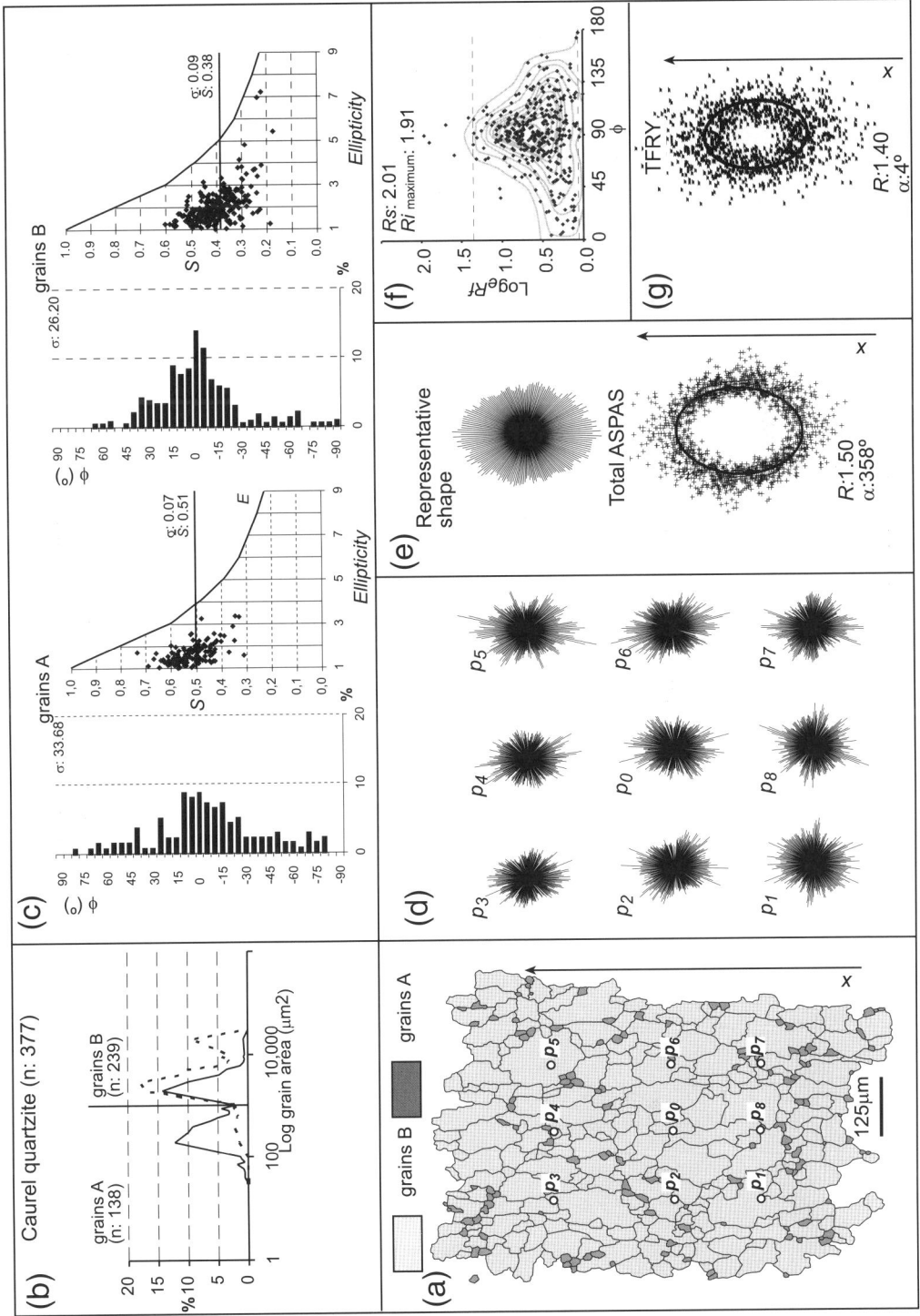

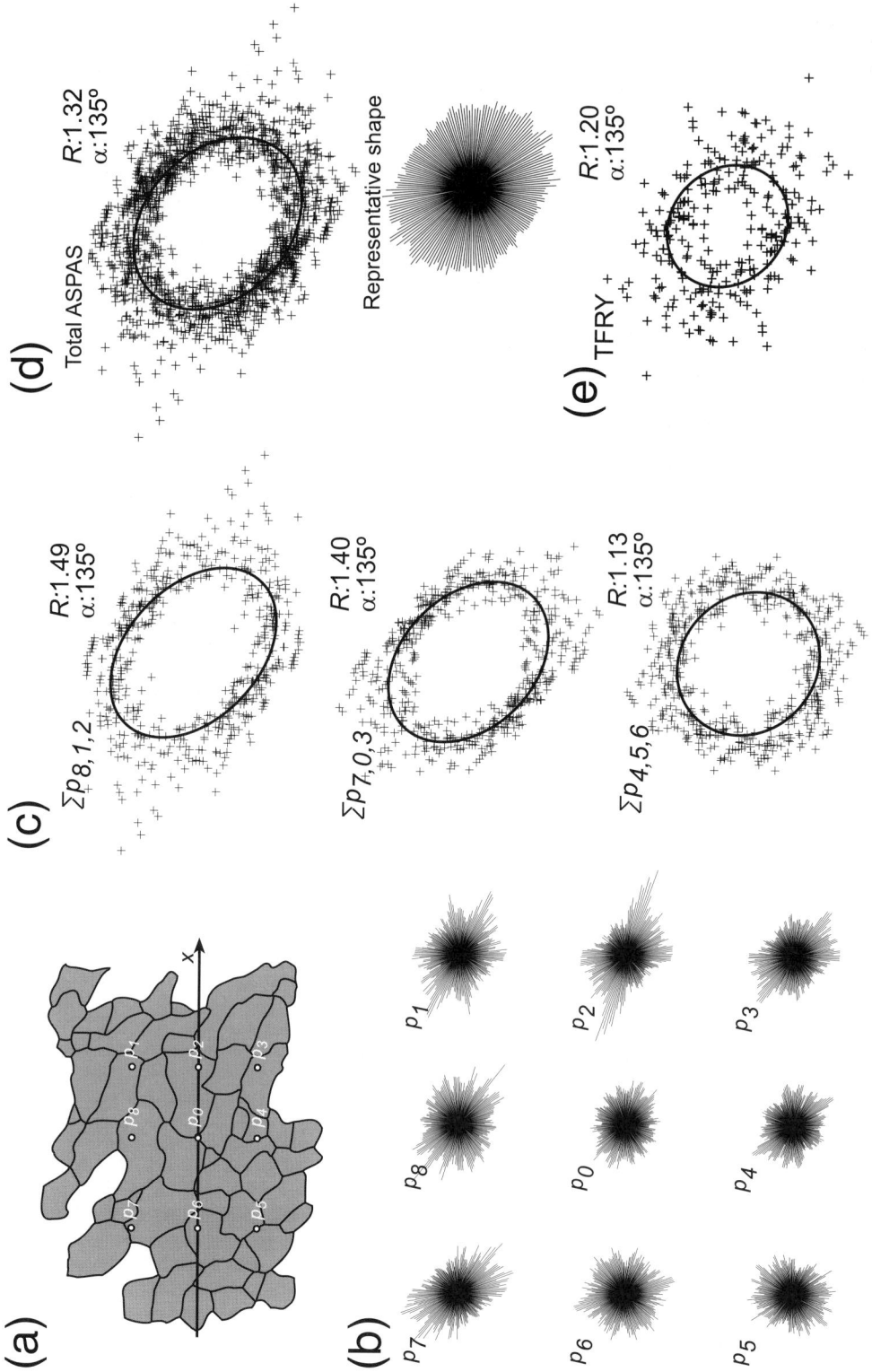

(a)

(b) p_1 p_2 p_3 p_8 p_0 p_4 p_7 p_6 p_5

(c) $\Sigma p_{8,1,2}$ *R*:1.49 α:135° $\Sigma p_{7,0,3}$ *R*:1.40 α:135° $\Sigma p_{4,5,6}$ *R*:1.13 α:135°

(d) Total ASPAS *R*:1.32 α:135° Representative shape

(e) TFRY *R*:1.20 α:135°

this local effect. Based on our experience, we recommend following the standard procedure with $\Delta\varphi = 2°$ and constructing nine rose diagrams, corresponding to the vertices of a rectangular grid (Fig. 4a) with p_0 placed approximately on the centre of the shape fabric analysed.

Comparison of the different rose diagrams can be used to detect heterogeneities of the shape fabric (Figs 5d and 6b). Calculating the weighted average of the segment lengths for each direction in all the rose diagrams, we can construct a new rose diagram named the ASPAS-representative shape (Figs 5e and 6d). In addition, we use the end points of the segments of all the rose diagrams to fit an ellipse by the least-squares method (Figs 5e and 6d); we refer to this as the total ASPAS-ellipse. Assuming a homogeneous original fabric and strain, the total ASPAS-ellipse approaches the shape and orientation of the strain ellipse. Similar methods have been developed by Panozzo (1987) and Launeau & Robin (1996). The advantage of the ASPAS method is that it allows the representative shape to be obtained directly, and small heterogeneities of the shape fabric can be detected. The confidence of this fit measures to what extent the characteristic shape represents the strain ellipse.

Least-squares approximation to the finite-strain ellipse

TFry and ASPAS methods yield a graphic with a cloud of points and a network of segments of variable length, respectively. As these figures are digital files in vector-based format, we automatically have pairs of co-ordinates for the points at the ends of the segments. From these pairs of co-ordinates it is possible to complete the graphical or mathematical estimation of a best-fit strain ellipse. The standard least-squares method (Bartlett 1915) was used in this analysis because the solution of the least-squares equations gives a unique, unbiased result. With the results obtained from the TFry, ASPAS and PAROR methods, we have calculated the coefficients A, B and C of an ellipse equation of the form:

$$Ax^2 + Bxy + Cy^2 - 1 = 0 \qquad (2)$$

using the appendix of Erslev & Ge (1990). To obtain the semi-axes and orientation of the ellipse we must apply an axes rotation defined by an angle α given by:

$$\tan 2\alpha = B/(A - C) \qquad (3)$$

where α is the angle between the long semi-axis of ellipse, measured from the positive side of the x-axis.

With the rotation of axes, equation (2) is transformed into:

$$A'(x')^2 + C'(y')^2 - 1 = 0. \qquad (4)$$

Using the invariants (e.g. Thomas 1967, p. 543), we obtain:

$$B^2 - 4AC = -4A'C' \qquad (5)$$

$$C' = (A + C) - A'. \qquad (6)$$

Solving this equation system, we can obtain A' and C', where the long and short semi-axes of the ellipse are:

$$a = (A')^{-1/2}$$

$$b = (C')^{-1/2}. \qquad (7)$$

The macro 'Ajuste.aml' is programmed with the GIS commands necessary to apply this approximation method to the data base of the TFry and ASPAS analysis.

Application to naturally deformed quartzites

Two tectonites have been chosen to illustrate the applicability of GIS to shape-fabric analysis and to discuss the constraints of some methods of strain analysis. The first sample has been used for a detailed analysis of shape fabric and strain in a tectonite with incipient recrystallization. A general comparison of the ASPAS method with other methods of strain analysis was made on the second sample.

Tectonites with incipient recrystallization

The fabric analysed (Fig. 5a) is from a quartzite sample collected in a inverted limb of an isoclinal kilometric-scale recumbent fold with

Fig. 6. Detailed ASPAS analysis of a quartzite (after fig. 7.16, p. 118, of Ramsay & Hubber 1983). (**a**) Fabric outlines showing the nine centres p_i. (**b**) Individual ASPAS diagrams calculated for the nine centres p_i. (**c**) Total ASPAS diagrams obtained adding the data of three individual ASPAS diagrams: p_1, p_2 and p_8 (above); p_0, p_3 and p_7 (centre); and p_4, p_5 and p_6 (below). (**d**) ASPAS-representative shape and total ASPAS diagrams. (**e**) TFry analysis plot. The numerical results of all these analyses are given in Table 1.

associated cleavage, located in the Caurel region (NW of the Iberian Massif). This tectonite was formed during the development of the fold at low-temperature conditions (epizone after Sarmiento *et al.* 1999). The tectonite is formed by heterometric and elongated grains with irregular boundaries. The grain size histograms (Fig. 5b) indicate two main grain populations. Coarsest grains 'A grains' have usually undulose or patchy extinction and elongated shapes ($R > 2$) with shape parameter $S = 0.38$; they define a good shape orientation (Fig. 5c). Small grains 'B grains' do not have intracrystalline deformation microstructures; they have weakly elongated shapes ($R < 2$), $S = 0.51$, a rough shape orientation (Fig. 5c) and have been interpreted as a result of dynamic recrystallization.

The ASPAS method has been applied to grains labelled A, following the standard procedure (Fig. 5a & d). Some small differences can be observed between the nine corresponding ASPAS diagrams (Fig. 5d) suggesting local heterogeneities of the shape fabric. The characteristic shape (Fig. 5e) gives a more even form than the individual ASPAS diagrams, suggesting that the local irregularities have been balanced. The cloud of points of the total ASPAS diagram defines an ellipse with a R of 1.5 (Fig. 5e).

In order to check the results of the ASPAS method, the R_f/ϕ and the TFry methods have been applied (Fig. 5f & g, respectively) to grains A of the quartzite sample. The TFry method gives a slightly lower aspect ratio ($R = 1.4$) than ASPAS, and R_f/ϕ, applied following the same procedure as that of Ramsay & Huber (1983, p. 83), gives a higher value ($R = 2.01$), but the scatter of data in the R_f/ϕ

diagram (Fig. 5f) suggests that this method, based on the assumption of an original elliptical shape, is not suitable in this case.

Tectonites without dynamical recrystallized grains

As a second example we analyse a section parallel to the bedding of a quartzite sample from the South Mountain anticline (NW Scotland) that has been previously studied by several authors (Ramsay & Huber 1983, pp. 73, 117 and 124; Panozzo 1987; Launeau & Robin 1996); it will allow us the comparison of results (Table 1). This quartzite was deformed predominantly by intracrystalline plasticity at relatively low-pressure and temperature conditions. A bulk strain ratio of 1.6 is derived from the elliptical sections of the *Scolithus* tubes on the bedding surface of this rock by Ramsay & Huber (1983, p. 81). The bulk strain can be compared with the aspect ratio and orientation of long axis of the strain ellipse obtained by the ASPAS, TFry and R_f/ϕ methods, and also with the previous results of other authors applying different methods (Fig. 6 and Table 1).

The fabric map has only 57 grains. A visual inspection of this fabric reveals the existence of a SW domain with near isometric grains, whereas the grains are elongated in the NW–SE direction on the rest of the fabric map. Consequently, at this scale, the shape fabric and the strain seem to be heterogeneous. Therefore, the analysis of this sample will allow us to evaluate the sensitivity of the different methods to these heterogeneities.

Table 1. *Comparison between different measurements made on a quartzite from Ramsay & Huber (1983, p. 118): data from this work*, from Ramsay & Huber (1983)[†], from Panozzo (1987)[‡] and from Launeau & Robin (1996)[§]. R is the axial ratio of the fitted ellipse and α is the angle between the long semi-axis of the ellipse and the x-axis measured from the positive side of the x-axis*

Method	Analysed fabric element	R	α ($°$)
Fry[†]	Centre points	1.7	142
Fry[‡]	Centre points	1.59	152
TFRY*	Centroids	1.20[a]	135
SURFOR[‡]	Outlines	1.28	145
Inverse SURFOR[‡]	Outlines	1.27	150
Inverse SURFOR[‡]	Tie lines	1.37	150
Inverse SURFOR[§]	Tie lines	1.29	153
Mean intercept length[§]	Outlines	1.29	143
Average inertia tensor[§]	Inertia tensor of grains	1.40	144
Cosine directions[§]	Inertia tensor long axes	1.59	145
R_f/ϕ*	R_{grain} and ϕ_{grain}	1.21	135
Total ASPAS diagram*	Outlines	1.32	135

Differences between the individual ASPAS diagrams confirm the existence of local heterogeneities in the shape fabric (Fig. 6b). As in the preceding sample, the characteristic shape gives a more even form than the individual ASPAS diagrams. The total ASPAS ellipse has a R value of 1.32 (Fig. 6d). In addition, three total ASPAS diagrams for different domains of the sample are presented in Figure 6c. They show a heterogeneous strain, with a minimum R for the SW domain of 1.13, and R values of 1.40 and 1.49 for the other domains. In addition, the TFry method has also been used to measure strain in this sample (Fig. 6e). It gives a poorly defined ellipse with $R = 1.20$.

The comparison of the above results with those obtained by other methods is shown in Table 1. In this table the R values range between 1.7 and 1.2. It can be seen that the cosine directions and the two values obtained by the classical Fry method give the higher values, whereas the TFry and the R_f/ϕ give the lower values. The rest of the methods give values ranging between 1.27 and 1.40. In this case, the small amount of grains in the fabric analysed gives rise to a poorly defined strain ellipse in the centre-to-centre methods (TFry and classical Fry). This is probably the cause of the extreme values given by these methods.

R values of all the intersection methods (Table 1) are similar; nevertheless they are always lower than the bulk strain measured by Ramsay & Huber (1983). This is probably due to the methods based on the shape fabric not registering the full strain history.

Discussion

The grain-boundaries shape can be quantitatively defined by different shape factors such as the parameter S and the PARIS factor (Panozzo & Hürlimann 1983) or by the fractal dimensions of grain boundaries (e.g. Kruhl & Nega 1996; Takahashi *et al.* 1998). Moreover, the above indices of shape are a function of the aspect ratio of the grains. Therefore, we propose the use of these grain-boundaries shape parameters always together with the ellipticity, in order to provide the necessary characterization of the grain shape for the purpose of establishing the possible relations with the rock-forming conditions. In addition to the characterization of grain shape, we have developed some new applications for strain measurement: a new variant of the Fry method that uses only contiguous polygons (TFry method) and the ASPAS method, based on the intersection lengths of the lines of a radial network with the grains.

The application of intersection methods, including results of previous studies, to a tectonite sample gives similar results. Nevertheless, the resulting strain ratios are lower than those obtained using pretectonic markers, suggesting that the shape fabric does not register the full strain history. Consequently, it is possible that the strain-measurement methods based on the shape-fabric analysis give in general only a minimum strain ratio.

The results of the analyses of the naturally deformed quartzites suggest that it is very useful to carry out a grain size and shape-fabric study before the strain analysis in order to determine whether, and to what extent, the shape fabric is affected by dynamical recrystallization. The TFry method might give better results for tectonites without recrystallization where the contiguity concept is appropriate; whereas ASPAS results lie between TFry and R_f/ϕ values in quartzites with incipient recrystallization, suggesting a better performance of the ASPAS in this case.

Conclusions

In this paper we show for the first time that GIS has an important application in the field of fabric analysis. The capabilities of GIS to execute complex spatial operations and to manage a large amount of graphical data have been used in this work to carry out detailed shape-fabric analysis. Particular qualities of GIS for shape-fabric analysis are: (i) it permits determination of a wide variety of grain characteristics, e.g. centroids, centroid-to-centroid distance, grain area and perimeter, ellipticity and orientation of the largest inscribed ellipse, adjacency relationships, etc.; (ii) it can work in both raster and vectorial format, allowing in each case the format more suitable for each task involved in the analysis; and (iii) it stores descriptive information linked to all these graphic elements, such as chemical and mineralogical composition, crystallographic orientation, fracture data, etc.

The new applications developed in this paper, e.g. the shape parameter S, the modified Fry method and the ASPAS method, are better suited than the existing methods to the analyses of fabrics with incipient dynamical recrystallization or where local heterogeneities are present in the fabrics.

Thorough and constructive reviews by R. Panozzo-Heilbronner and an anonymous referee significantly improved the first version of this paper. The revised version has also benefited from a critical review by R.J. Lisle and from the patience and 'cosmetic changes' by L. Burlini. This contribution was carried out with the financial support of the DGICYT projects BPE 2002-00187 and DGE99-PB98-1545. GIS was supplied by the INDUROT (University of Oviedo).

References

BARTLETT, D.P. 1915. *General Principles of the Method of Least Squares with Applications.* Rumford Press, Concord, NH.

BONHAM-CARTER, G.F. 1994. *Geographic Information Systems for Geoscientists.* Pergamon, Tarrytown, NY.

DUNNET, D. 1969. A technique of finite strain analysis using elliptical particles. *Tectonophysics*, **7**, 117–136.

ERSLEV, E.A. 1988. Normalized centre-to-centre strain analysis of packed aggregates. *Journal of Structural Geology*, **10**, 201–209.

ERSLEV, E.A. & GE, H. 1990, Least-squares centre-to-centre and mean object ellipse fabric analysis. *Journal of Structural Geology*, **12**, 1047–1059.

ESRI. 1992. *Understanding GIS. The ARC/INFO Method.* Enviromental Systems Research Institute, Redlands, CA.

ESRI. 1994. *ARC Macro Language. Developing ARC/INFO Menus and Macros with AML.* Enviromental Systems Research Institute, Redlands, CA.

EXNER, H.E. 1972. Analysis of grain- and particle-size distribution in metallic materials. *Institute of Physical Metalurgy and Metal Review*, **159**, 25–42.

FRY, N. 1979. Random point distribution and strain measurement in rocks. *Tectonophysics*, **60**, 89–105.

HEILBRONNER, R. 1992. The autocorrelation function: an image processing tool for fabric analysis. *Tectonophysics*, **212**, 351–370.

HEILBRONNER, R. 2000. Automatic grain boundary detection and grain size analysis using polarization micrographs or orientation images. *Journal of Structural Geology*, **22**, 969–981.

KRUHL, J.H. & NEGA, M. 1996. The fractal shape of sutured quartz grain boundaries: application as a geothermometer. *Geologische Rundschau*, **85**, 38–43.

LAUNEAU, P. & ROBIN, P.-Y.F. 1996. Fabric analysis using the intercept method. *Tectonophysics*, **267**, 91–119.

LISLE, R.J. 1979. Strain analysis using deformed pebbles: the influence of initial pebble shape. *Tectonophysics*, **60**, 263–277.

LISLE, R.J. 1985. *Geological Strain Analysis. A Manual for the R_f/ϕ Technique.* Pergamon Press, Oxford.

MEANS, W.D. & DONG, H.G. 1998. Post-deformational recovery of microstructure in sheared octaclorpropane (C3Cl8). *In*: SNOKE, A.W., TULLIS, J. & TODD, V.R. (eds) *Fault-related*

Rocks: A Photographic Atlas. Princeton University Press, New Jersey, 450–453.

MCNAUGHT, M. 1994. Modifying the normalized Fry method for aggregates of non-elliptical grains. *Journal of Structural Geology*, **16**, 493–503.

MULCHRONE, K.F. 2002. Application of Delaunay triangulation to the nearest neighbour method of strain analysis. *Journal of Structural Geology*, **25**, 689–702.

NIH. 1999. National Institute of Health (NIH) image 1.62, public domain image analysis software. http://rsb.info.nih.gov/nih-image/download.html.

PANOZZO, R. 1983. Two-dimensional analysis of shape fabric using projections of lines in a plane. *Tectonophysics*, **95**, 279–294.

PANOZZO, R. 1984. Two-dimensional strain from the orientation of lines in a plane. *Journal of Structural Geology*, **6**, 215–222.

PANOZZO, R. 1987. Two-dimensional strain by the inverse SURFOR wheel. *Journal of Structural Geology*, **9**, 115–119.

PANOZZO, R. & HÜRLIMANN, H. 1983. A simple method for the quantitative discrimination of convex and convex–concave lines. *Microscopica Acta*, **87**, 169–176.

PANOZZO-HEILBRONNER, R. & PAULI, C. 1993. Integrated spatial and orientation analysis of quartz c-axes by computer-aided microscopy. *Journal of Structural Geology*, **15**, 369–382.

RAMSAY, J.G. 1967. *Folding and Fracturing of Rocks.* McGraw-Hill, New York.

RAMSAY, J.G. & HUBER, M.I. 1983. *The Techniques of Modern Structural Geology, Volume 1: Strain Analysis.* Academic Press, London.

SANDER, B. 1950. *Einführung in die Gefügekunde der geologischen Körper, zweiter Teil: Die Korngefüge.* Springer, Wien.

SARMIENTO, G.N., GARCÍA-LÓPEZ, S. & BASTIDA, F. 1999. Conodont colour alteration indices (CAI) of Upper Ordovician limestones from the Iberian Peninsula. *Geologie en Mijnbouw*, **77**, 77–91.

SCION CORPORATION. 1999. Scientific image acquisition and analysis software for Windows. http://www.scioncorp.com/.

SWAN, A.R.H. & GARRAT, J.A. 1995. Image analysis of petrographics textures and fabrics using semi-variance. *Mineralogical Magazine*, **59**, 189–196.

TAKAHASHI, M., NAGAHAMA, H., MASUDA, T. & FUJIMURA, A. 1998. Fractal analysis of experimentally, dynamically recrystallized quartz grains and its possible application as a strain rate meter. *Journal of Structural Geology*, **20**, 269–275.

THOMAS, G.B. 1967. *Elements of Calculus and Analytical Geometry.* Addison-Wesley, Menlo Park, CA.

UNDERWOOD, E.E. 1970. *Quatitative Stereology.* Addison-Wesley, London.

Rigid polygons in shear

DANIEL W. SCHMID

*Physics of Geological Processes, University of Oslo, Pb 1048 Blindern,
0316 Oslo, Norway (e-mail: schmid@fys.uio.no)*

Abstract: Clasts, inclusions and intrusions in shear are potential recorders of strain, stress, rheology and metamorphism. In order to extract the recorded information, it is essential to have analytical and numerical theories that describe the deformation mechanics of such bodies. To overcome the simplifications of the commonly employed ellipsoid-based shape approximation, a combination of Muskhelishvili-type analytical solutions and finite-element method calculations is used to study the behaviour of (quasi) rigid polygons in shear. The results confirm that the polygon rotation and the pressure perturbation outside rigid polygonal clasts are well approximated by ellipse-based theories. However, this observation does not hold for the inside of these polygons, which show strongly varying values of pressure perturbation and maximum shear stress. For example, pressure perturbations inside the polygons are usually the opposite of the neighbouring matrix values across the polygon–matrix interface. This complex behaviour is summarized in the ellipse decomposition rule that allows for a qualitative understanding of the pressure perturbation in and around a wide range of polygons in shear. Other quantities studied include maximum values of overpressure relative to the shortening stress, and the area that undergoes overpressure with respect to the clast size. The results demonstrate that overpressure can be twice as large as the rock strength.

Particles, ranging in size from clasts to plutons, that have been subjected to shear potentially record important information about the geological past and may be used to decipher the kinematic history, metamorphosis and the mechanical behaviour of a certain outcrop or region. To achieve this, a sound understanding of the mechanics of particles in pure and simple shear is required. The available analytical theories (Bilby *et al.* 1975; Eshelby 1957, 1959; Ghosh & Ramberg 1976; Jeffery 1922; Schmid & Podladchikov 2003) approximate the geometry of particles with ellipsoidal shapes. Arbaret *et al.* (2001) have shown by means of analogue modelling that the ellipsoidal-shape simplification is justified in regard to kinematics. Yet, a theoretical foundation for this observation is lacking (cf. Treagus & Lan 2003) and, even more importantly, it remains unclear to what extent the ellipsoidal-shape simplification holds for dynamic key parameters, such as pressure and maximum shear stress, that drive metamorphic reactions and determine the deformation mechanism.

The consequences of imperfect or non-ellipsoidal geometry can be significant. For example, rhomboidal particle shapes may enhance the development of shape preferred orientations (SPOs) (Ceriani *et al.* 2003), which renders the existing analytical theories unsuitable and, consequently, the far-field flow conditions cannot be reconstructed based on the theory derived by Ghosh & Ramberg (1976). Other examples include the understanding of mineral growth in rocks (Fletcher & Merino 2001), kinetics of phase transitions (Perrillat *et al.* 2003) and strength of minerals (Mosenfelder *et al.* 2000), all of which employ ellipsoidal-shape simplifications for cases where local perturbations caused by imperfect geometries could shift the system to a different equilibrium. In particular, non-ellipsoidal shapes drive local pressure perturbations around particles that are likely to deviate even more strongly from lithostatic values than previously established (Kenkmann & Dresen 1998; Tenczer *et al.* 2001; Schmid & Podladchikov 2003). Hence, the barometric interpretation of a mineral assemblage requires an analytical theory for the dynamics in and around non-ellipsoidal particles.

While the ellipsoidal-shape simplification may hold for the dynamic parameters in the vicinity of a clast, it is almost certainly an oversimplification for the inside of the clast. Perfectly ellipsoidal clasts have the exceptional property that all stress and strain-rate components inside the

From: Bruhn, D. & Burlini, L. (eds) 2005. *High-Strain Zones: Structure and Physical Properties.*
Geological Society, London, Special Publications, **245**, 421–431.
0305-8719/05/$15.00 © The Geological Society of London 2005.

clast are constant under homogenous far-field boundary conditions such as pure or simple shear. This is unlikely to be the case for general polygonal shapes.

A combination of two different approaches is employed in this paper to illustrate the complexities that arise with natural, non-ellipsoidal shapes: analytical solutions and finite-element models. Both have distinct advantages. Analytical solutions are extremely valuable as they allow for large parameter studies and generally help to understand a problem, but they can only be derived for a small range of problems. On the other hand, finite-element models can deal with a large range of problems, but must be re-run for every parameter change and therefore allow for the understanding of individual cases but are less useful for capturing the general character of the problem. In the context of this paper, an analytical solution based on Muskhelishvili's (1953) method is presented that is valid outside a special class of rigid triangles and squares. The inside of these triangles and squares is calculated with finite-element models, which are also applied to deal with clasts of more complex shape. Both analytics and numerics are two dimensional and assume plane strain. The material behaviour is linear viscous (Newtonian) and incompressible. The term 'rigid' employed in this paper designates the situation where the viscosity contrasts between clast and matrix is

so high that a further increase in viscosity contrast does not change the stress fields. The restriction to rigid (or quasi-rigid) clasts does not cause much loss of generality. As shown by Schmid & Podladchikov (2003), the infinite viscosity contrast limit can be taken as representative for clast–matrix viscosity contrasts as low as 10:1.

Methods

Analytical solution

The most flexible method for obtaining analytical solutions to complex geometries in two-dimensional elastic or incompressible viscous problems was developed by Muskhelishvili (1953). Co-ordinate transformations are a basic part of this method and allow for solution finding in geometrically simpler image domains. For example, a square in the physical domain can be transformed into a circle in the image domain, facilitating the derivation of the analytical solution for the square (Fig. 1).

In this paper, the hypotrochoid (cf. mathworld.wolfram.com\Hypotrochoid.html) transformation is used:

$$z = R\left(\frac{1}{\zeta} + m\zeta^n\right). \qquad (1)$$

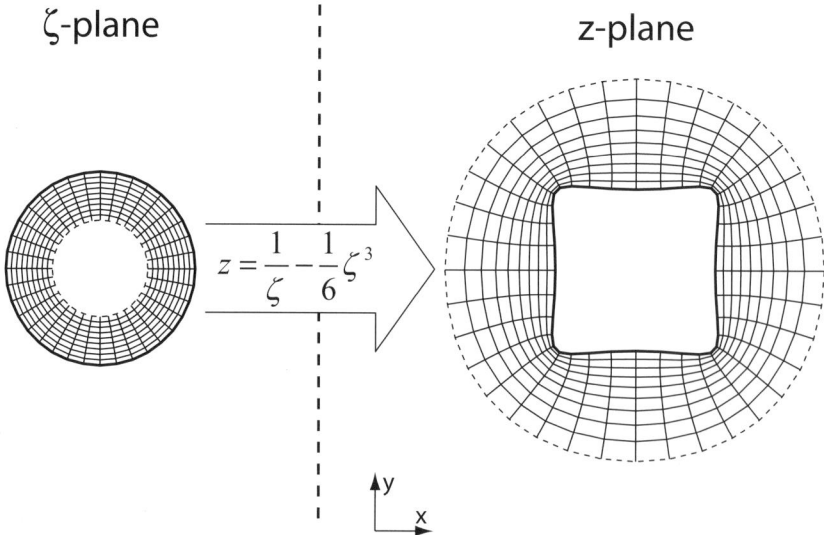

Fig. 1. Hypotrochoid mapping example. Setting $R = 1$, $m = -1/6$, $n = 3$ maps the inside of the unit circle in the ζ-plane to the outside of an approximated square in the z-plane. Note the correspondence between the bold solid lines and the dashed lines in the two planes. Since $z \to \infty$ for $\zeta \to 0$, the inside of the unit circle in the ζ-plane corresponds to the entire plane outside of the hypotrochoid in the physical plane z.

Here z is the complex co-ordinate in the physical plane, ζ the complex co-ordinate in the image plane, if m is non-zero $n + 1$ is the number of vertexes (n is a positive, non-zero integer) and R is used to scale the size of the hypotrochoid. An example of hypotrochoid mapping is given in Figure 1.

The hypotrochoid transform does not only map the inside of the unit circle in ζ to the outside of the hypotrochoid in z, but also maps the outside of the unit circle in ζ to the outside of the hypotrochoid in z. Therefore, possible solutions are restricted to the outside of the hypotrochoid in the physical domain and must implement the behaviour of the interior of the hypotrochoid through the interface boundary condition. In this paper, the hypotrochoids are modelled as rigid objects and, hence, the possible velocities at the interface are restricted to rigid-body rotation with a rotation rate that is determined through the condition that the resultant moment acting on the rigid hypotrochoid must be zero (Muskhelishvili 1953, p. 349). Savin (1961, pp. 281–293) gives the solution for rigid triangles and squares in uniaxial tension for an elastic matrix. As the instantaneous, incompressible elastic and viscous problem are mathematically identical and it is admissible to perform linear solution superposition, it is straightforward to translate Savin's (1961) solutions to rigid triangles and squares embedded in a viscous matrix and subjected to pure shear conditions (general shear is discussed in the subsection on 'Kinematics'). The resulting expressions are:

$$\frac{\phi_{tri}}{2\mu\dot{\varepsilon}R} = -\zeta e^{2i\alpha} \tag{2}$$

$$\frac{\psi_{tri}}{2\mu\dot{\varepsilon}R} = -\zeta^{-1}e^{-2i\alpha} + \frac{\zeta^3 + m}{2m\zeta^3 - 1}e^{2i\alpha} \tag{3}$$

$$\frac{\phi_{squ}}{2\mu\dot{\varepsilon}R} = -\zeta\left(\frac{me^{-2i\alpha} - e^{2i\alpha}}{m^2 - 1}\right) \tag{4}$$

$$\frac{\psi_{squ}}{2\mu\dot{\varepsilon}R} = -\zeta^{-1}e^{-2i\alpha}$$
$$+ \frac{(3m^2 + 1)\zeta^3}{(3m\zeta^4 - 1)}\left(\frac{me^{-2i\alpha} - e^{2i\alpha}}{m^2 - 1}\right) \tag{5}$$

where μ is the viscosity of the matrix, $\dot{\varepsilon}$ the pure shear strain rate, $i = \sqrt{-1}$, α is the inclination of the clast with respect to the far-field flow, and the subscripts tri and squ denote solutions for the rigid hypotrochoid triangle and square, respectively. The Muskhelishvili solution of biharmonic

problems, such as the one solved here, consists of two different complex potentials, ϕ and ψ. From these, all stress, strain rates and velocities can be deduced with a set of rules that can be found in various works, including Muskhelishvili (1953), Savin (1961), Jaeger & Cook (1979) and Schmid & Podladchikov (2003). For example, the expression for the pressure perturbation p (for the given boundary conditions, pressure can only be determined up to a lithostatic constant) is:

$$p = -2\Re\left(\frac{\partial\phi}{\partial\zeta}\left(\frac{\partial z}{\partial\zeta}\right)^{-1}\right) \tag{6}$$

where $\Re$ means the real part and the minus sign is due to the convention that compressive pressure is positive. Substituting ϕ_{tri} and ϕ_{squ} into equation (6) we obtain:

$$\frac{p_{tri}}{2\mu\dot{\varepsilon}} = -2\Re\left(\frac{\zeta^2 e^{2i\alpha}}{1 - 2m\zeta^3}\right) \tag{7}$$

$$\frac{p_{squ}}{2\mu\dot{\varepsilon}} = -2\Re\left(\frac{\zeta^2(2me^{-2i\alpha} - 2e^{2i\alpha})}{(1 - m^2)(3m\zeta^4 - 1)}\right). \tag{8}$$

These relatively simple expressions determine the entire pressure perturbation field outside isolated rigid triangles and squares, respectively. By substituting $|\zeta| = 1$ into equations (7) and (8) the pressure at the clast interface is obtained, setting $m = 0$ the hypotrochoid transform yields a circle and the corresponding pressure perturbations are found.

The second key parameter of interest for the understanding of clasts in shear zones is the effective or maximum shear stress, τ, which is calculated as (e.g. Ranalli 1995):

$$\tau = \sqrt{\left(\frac{\sigma_{xx} - \sigma_{yy}}{2}\right)^2 + \sigma_{xy}^2} \tag{9}$$

where σ_{xx}, σ_{yy} and σ_{xy} are the components of the total stress tensor. The maximum shear stress is important because it indicates proximity to failure (von Mises criterion) and generally influences which deformation mechanism is activated. The resulting expressions for the maximum shear stress around the rigid triangles and squares are too lengthy to be reproduced here. However, upon request, these and all other analytical expressions used in this work are available from the author in the form of MATLAB scripts used to produce the figures in this paper.

Finite-element model

The numerical model used is a personally developed, two-dimensional finite-element method (FEM) code solving the Stokes equations for slow flow of incompressible, viscous materials. The type of element employed is a seven-node Crouzeix–Raviart triangle (Crouzeix & Raviart 1973) featuring continuous (bubble-node enriched), quadratic basis functions for the velocities and discontinuous linear basis functions for pressure. The incompressibility constraint is taken care of by Uzawa iterations (Cuvelier et al. 1986). For triangular and square objects, the FEM is only required to provide the solution for the inside of the clasts. While it is theoretically possible to combine the analytical solution and the FEM through the clast interface, it is more practical to calculate the entire domain with the FEM (with the pure shear velocity-boundary conditions chosen at a suitable distance as not to perturb the clast) and then combine the analytically derived clast-outside and the numerically obtained clast-inside for the plots. This procedure also allows for cross-checking between the two different approaches outside the clast. The observed differences, due to the numerical discretization of the FEM, were always smaller than 1%. As both the matrix and the clast are treated as linear viscous materials in the FEM model the clast can only be quasi-rigid, which is achieved by setting the viscosity contrast between clast and matrix suitably high, in the presented cases at 10 000:1.

Results and discussion

Kinematics

The possible velocities of a rigid clast are rotation and/or translation, the latter being irrelevant for the studied case. The rigid-body rotation inherent to equations (2)–(5) is zero and therefore the hypotrochoid approximated triangle and square do not have a preferred orientation in pure shear. In order to translate the obtained solutions to general, plane strain, shear, one must at every instant determine the orientation and magnitude of the principal flow axis (Ramsay & Huber 1983), then orient and match the pure shear solutions given here accordingly. In addition, the rigid-body rotation component of the general shear has to be determined and added to the velocity field. Hence, in simple shear the rotation rate of rigid triangles and squares will be half the applied shear strain rate, which is simply the rigid-body rotation component of the applied simple shear

(cf. Casey et al. 1983). This result was numerically observed by Treagus & Lan (2003) and is identical to the rotation rate, which Jeffery (1922) derived for the rigid circle in simple shear. The result is also a theoretical confirmation of the experimentally established hypothesis that a wide range of particles can be modelled with sufficient accuracy using the analytical theory derived for elliptical shapes (e.g. Arbaret et al. 2001). One would expect that small perturbations to the perfect equilateral, equiangular shape will not significantly influence the rotation in simple shear. However, in pure shear, small shape imperfections will cause a rotational component and consequently the polygon is likely to orient itself as would a rigid ellipse (cf. Ghosh & Ramberg 1976).

Stresses

As outlined in the 'Methods' section, pressure and maximum shear stress may be regarded as the key components of the stress tensor. The sign conventions used imply that positive stresses designate tension and positive pressure means compression. All analysed stress components are normalized by the characteristic background stress, which is $|2\mu\dot{\varepsilon}|$. This is the absolute value of the far-field stress imposed by the applied pure shear flow.

Inclination. Despite the fact that equilateral, equiangular triangles and squares do not show a preferred orientation concerning the rotation rate, all dynamic parameters such as pressure indeed depend on the inclination of the clast to the applied pure shear flow. Figure 2 illustrates this by considering the rigid square in horizontal pure shear shortening as an example. Analysing the pressure perturbation in the matrix, we observe that the parts of the matrix that are pressed against the clast experience high pressures (relative compression) and the parts that are pulled away from the clast show pressure lows (relative tension) of identical magnitude. In contrast, the inside of the clast shows less intuitive results. In the horizontal position (Fig. 2a) the inside of the clast experiences pressure perturbations that are opposite to the corresponding matrix values across the interface. Therefore, the sides of the square that face the shortening component of the applied pure shear actually experience pressure lows (tension). The second end-member inclination is displayed in Figure 2c, where the vertexes face the applied pure shear. Here the pressure perturbation inside the clast is synchronized with the outside. The intermediate inclination (Fig. 2b) results in a

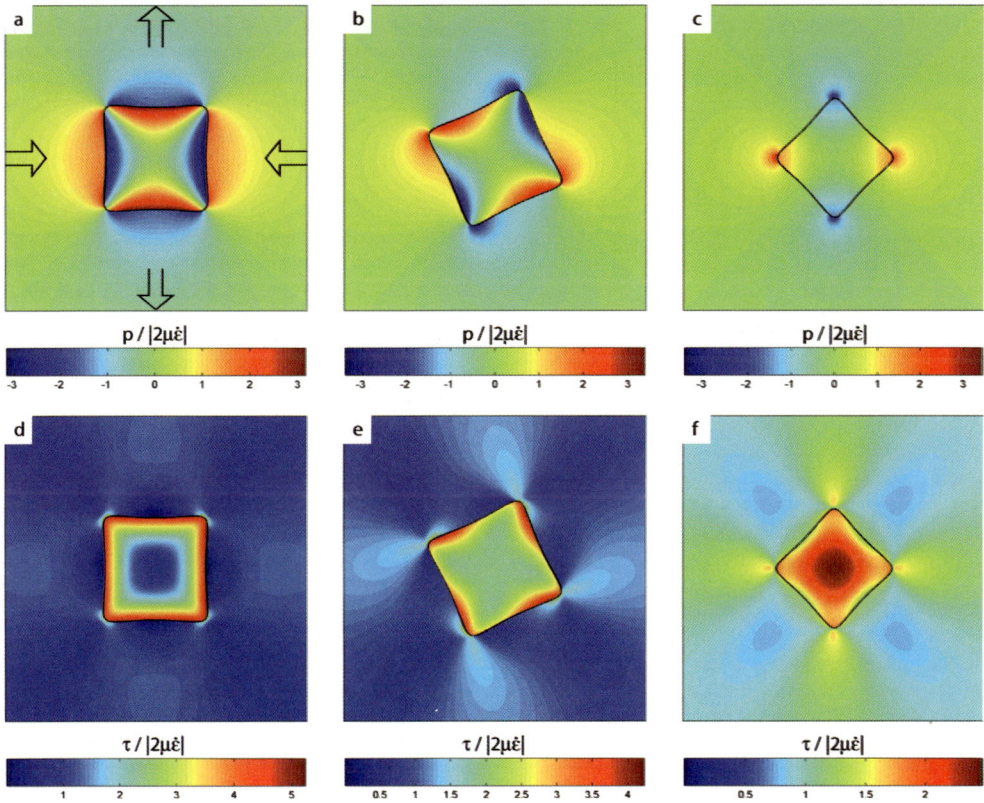

Fig. 2. Rigid square in horizontal shortening pure shear. Influence of clast inclination with respect to the far-field flow on pressure perturbation (**a**)–(**c**) and maximum shear stress (**d**)–(**f**). Columns correspond to $0°$, $25°$ and $45°$ inclination, respectively (left to right). Remark: the outside of the square is based on the analytical solution given in the text; the inside of the clast is calculated with a finite-element model, which approximates the rigid-clast behaviour by setting the viscosity of the clast at 10 000 times higher than that of the matrix. This is the case for all figures.

situation similar to Figure 2a. It may be useful to point out the following two points. (i) Pressure, opposite to normal and shear stresses, does not have to be continuous through interfaces and material boundaries. (ii) The magnitude of the pressure perturbation and all other stress tensor components is not a function of the object size (cf. absence of R in equations 7 and 8).

The maximum shear stress distribution differs significantly from the pressure perturbation. As the clast is competent and acts as an obstacle in the pure shear flow it must sustain the largest maximum shear stress values in the entire domain, with the maxima occurring inside the clast–matrix interface. The exception is the $45°$ inclination position, where the maximum shear stress is found in the clast centre and has a magnitude that is approximately half of the $0°$ inclination maxima. One may view the $45°$

position as exceptionally 'aerodynamic to the pure shear wind' with no clast faces directly exposed to applied flow.

A study of how pressure perturbation and maximum shear stress in the matrix vary with inclination and radius of curvature is displayed in Figure 3. It is intriguing that the pressure perturbation maximum (Fig. 3a) for small tip curvature radii ($<10^{-1}$) does not occur when the clast vertexes are in line with the pure shear, but at an intermediate inclination of $22.5°$. The plot of maximum shear stress (Fig. 3b) confirms the observation that in the $45°$ position the square clast presents the least obstacle for the pure shear flow and hence the smallest τ values occur in this position.

In order to obtain dimensional quantities from the normalized pressure perturbations and maximum shear stresses discussed here, it is

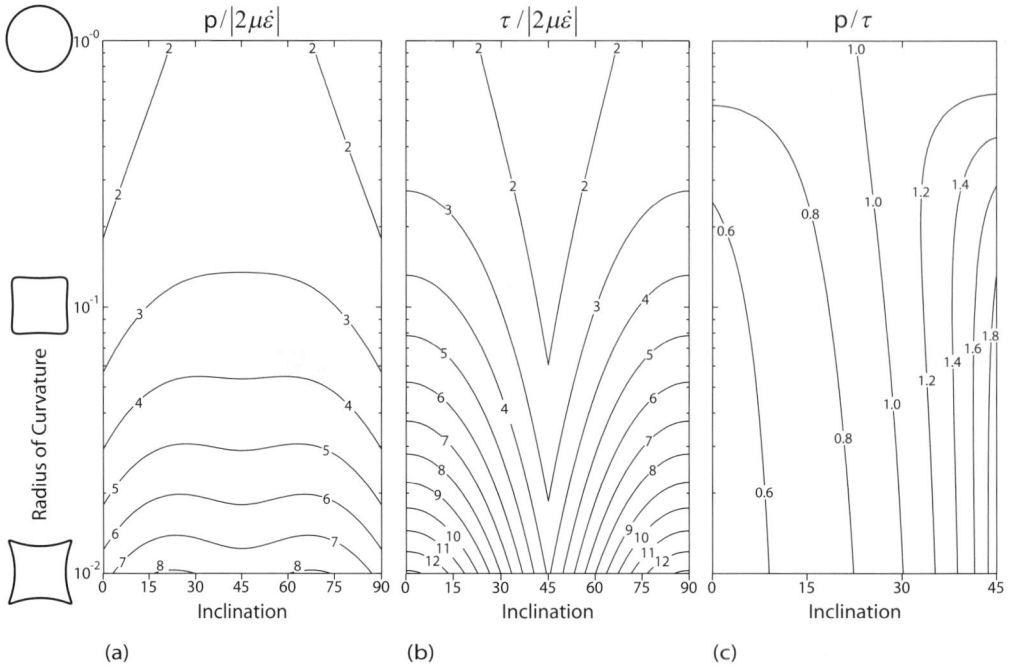

Fig. 3. Maxima in the matrix (i.e. at the clast–matrix interface) caused by the presence of a rigid square of variable inclination and tip curvature in pure shear. The tip curvature is given in terms of the analytically calculated tip curvature radius divided by half the maximum clast diameter going through the clast centre. The vertical axes in (a)–(c) are identical. (**a**) Pressure perturbation; (**b**) maximum shear stress; and (**c**) ratio between pressure and maximum shear stress.

sufficient to choose values for μ and $\dot{\varepsilon}$ that, depending on how realistic the chosen values are, will yield a wide range of numbers for possible pressure perturbations. However, it may be argued that these over- and underpressures can only occur if the strength of the rock is not exceeded; the latter being largely determined by how much shear stress the rock can sustain. Hence, if the maximum shear stress is limited it is essential to know how the pressure perturbation relates to this, as displayed in Figure 3c. It follows that the pressure perturbation can be up to approximately twice the maximum shear stress if the square is oriented in the 45° position. Given the vertical nature of the contour lines this relationship is valid for a wide range of squares with different vertex sharpness.

Clast shape and vertex curvature. Elliptical-shape simplifications are usually employed based on the assumption that they are a good approximation for more complex shapes. Figures 4 and 5 show pressure perturbation and

maximum shear stress comparisons between rigid elliptical clasts (solution taken from Schmid & Podladchikov 2003), triangular and square clasts. Columns show the different types of shapes with identical vertex curvature and rows show the influence of increasing the vertex sharpness. As ellipsoidal clasts have the previously mentioned property of constant values inside, it is evident from the above discussion that the ellipse approximation indeed oversimplifies the characteristics of the interior of general polygonal shapes.

The ellipse-based predictions of pressure perturbation in the matrix are quite accurate as demonstrated by the pressure maxima within columns of Figure 4. These maxima are a function of the vertex radius; they go from twofold overpressures (relative to $|2\mu\dot{\varepsilon}|$) in the case of extremely blunt vertexes (circle) to ever-increasing values as the vertex becomes more acute. Again, these seemingly unbounded pressure perturbation values are limited by the strength of the rock and therefore one must

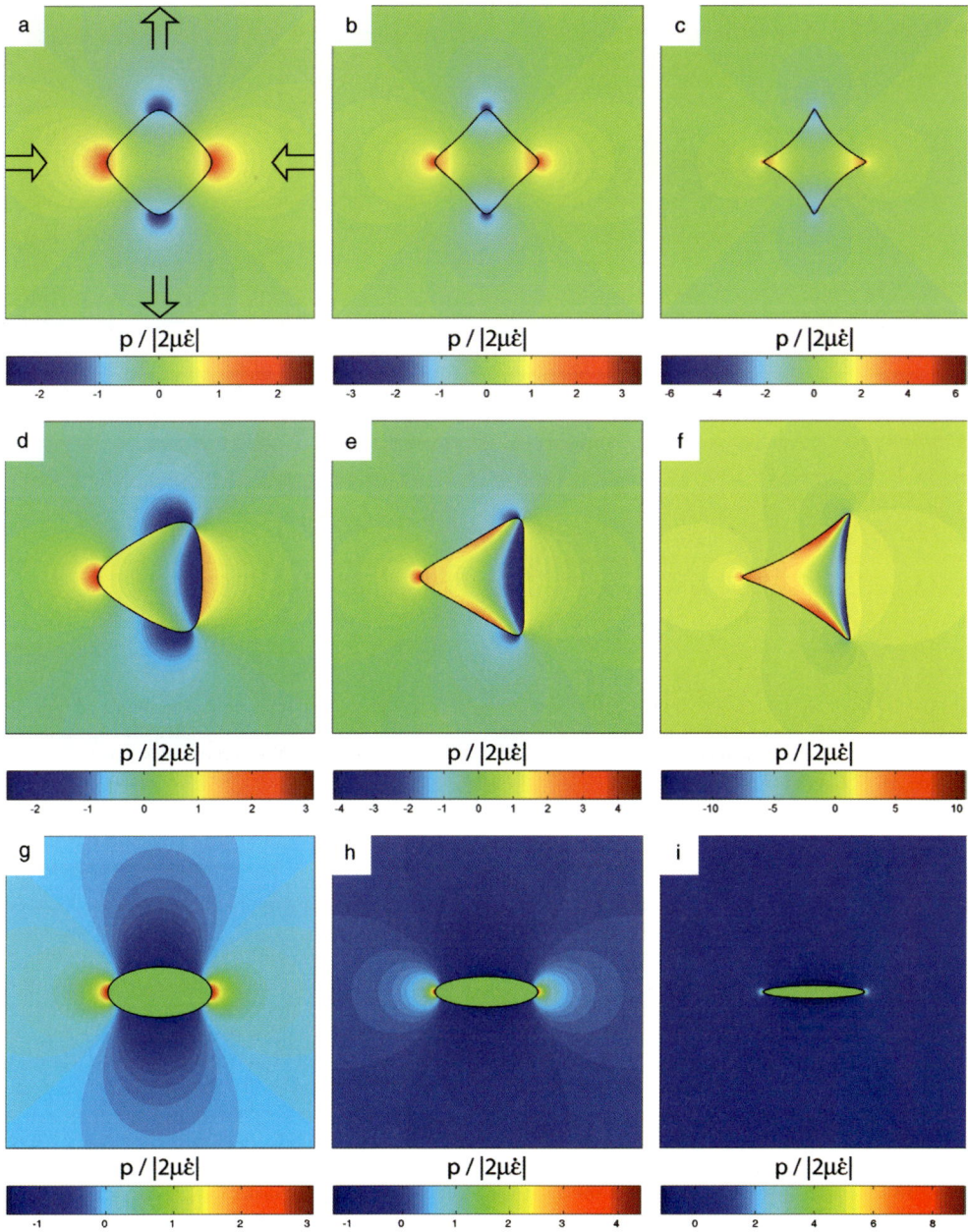

Fig. 4. Comparison of the pressure perturbation generated by different objects oriented in homologous positions with respect to the horizontal shortening pure shear. The tip curvatures are constant in the individual columns. As the complete analytical solution (including the inside) exists for elliptical clasts, no finite-element calculation was required for the ellipse row.

analyse the pressure perturbation in relation to the maximum shear stress (Fig. 6). In the case of the rigid elliptical clast, the possible overpressure is also (cf. square) twice the maximum shear strength. The rigid triangle, however, is basically incapable of generating a pressure perturbation that significantly exceeds τ. Another interesting quantity is the size of the area outside a rigid

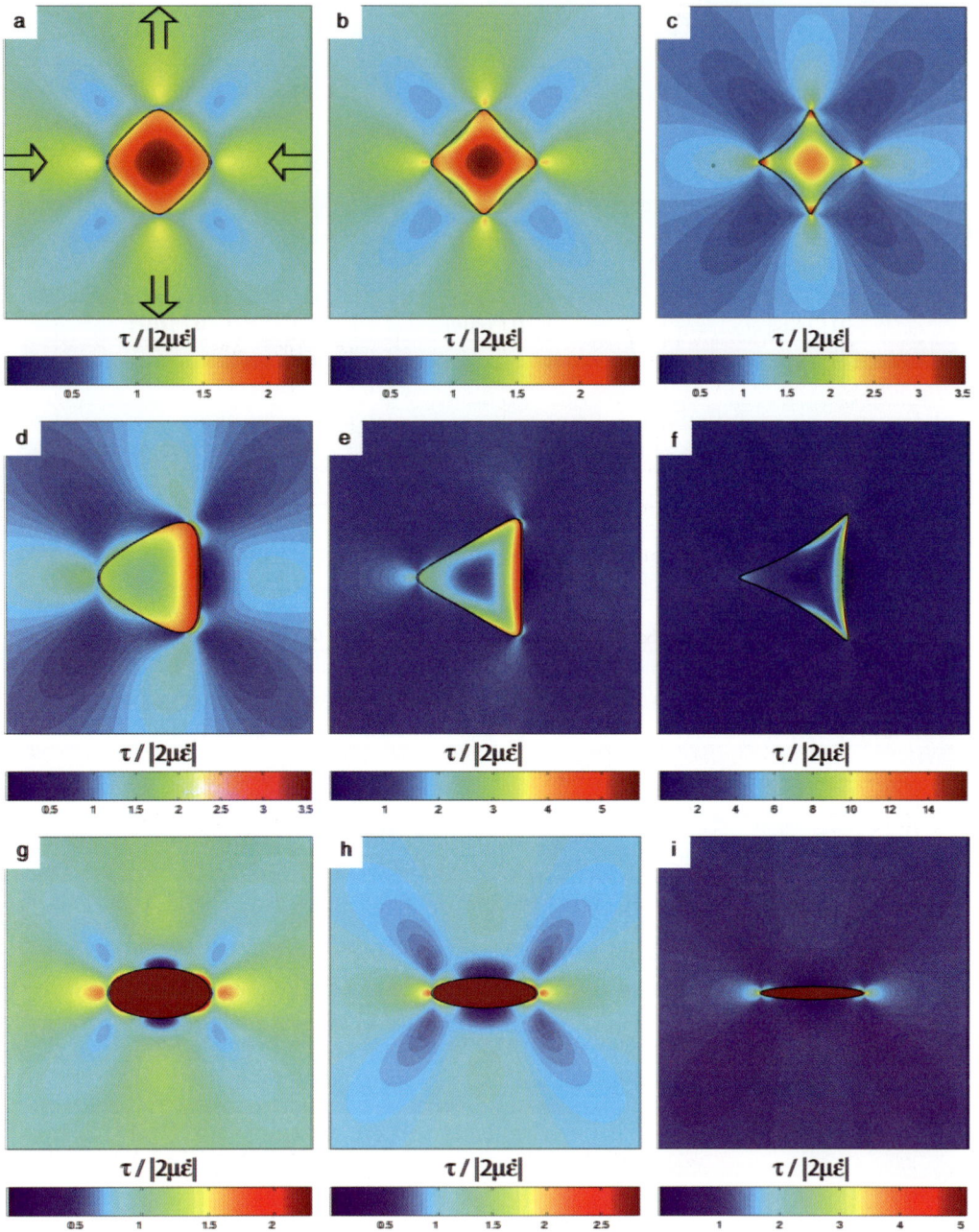

Fig. 5. Maximum shear stress equivalent to Figure 4. Comparison of the maximum shear stress generated by different objects oriented in homologous positions with respect to the horizontal shortening pure shear.

clast that experiences a certain pressure perturbation. Figure 7 illustrates this in terms of area size that experiences twofold or higher overpressure created by a single vertex. The investigated range of shapes does not yield areas that exceed 1–2% of the clast size and it is clear that areas experiencing higher pressure are even smaller (Fig. 4). Yet, even small areas around clasts

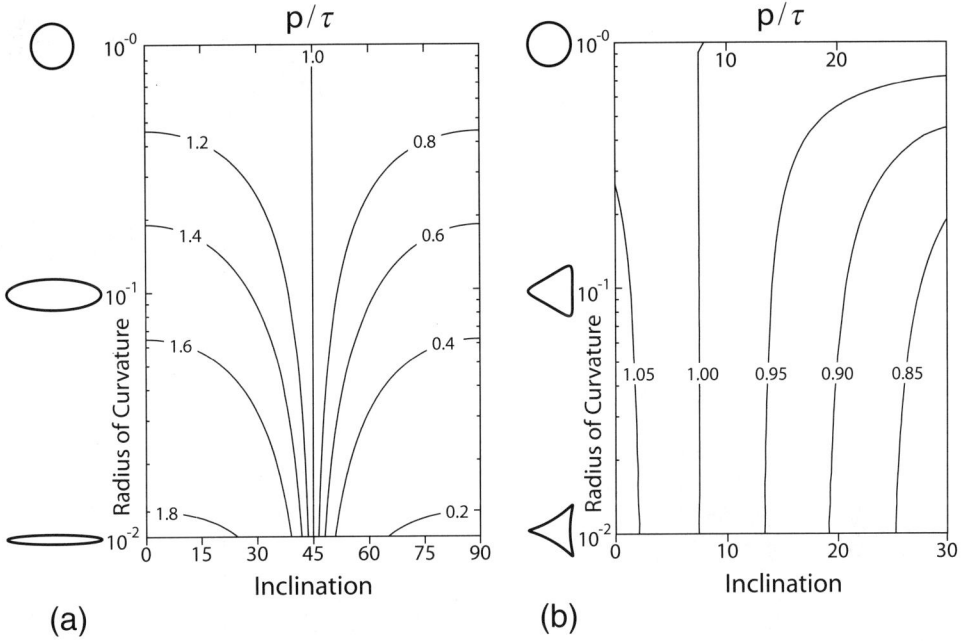

Fig. 6. Ratio between pressure perturbation and maximum shear stress around (**a**) rigid ellipses and (**b**) rigid triangles.

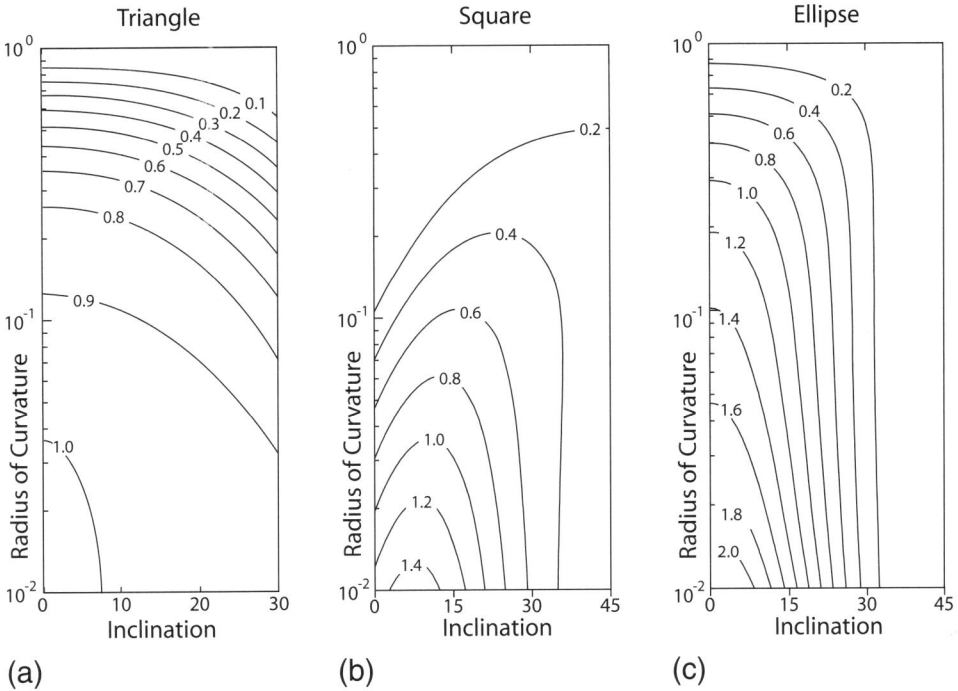

Fig. 7. Contour plots of the area size outside a rigid clast that shows twofold or larger overpressures relative to $|2\mu\dot{\varepsilon}|$. The area is given in percent of the object size and represents the perturbation created by a single vertex.

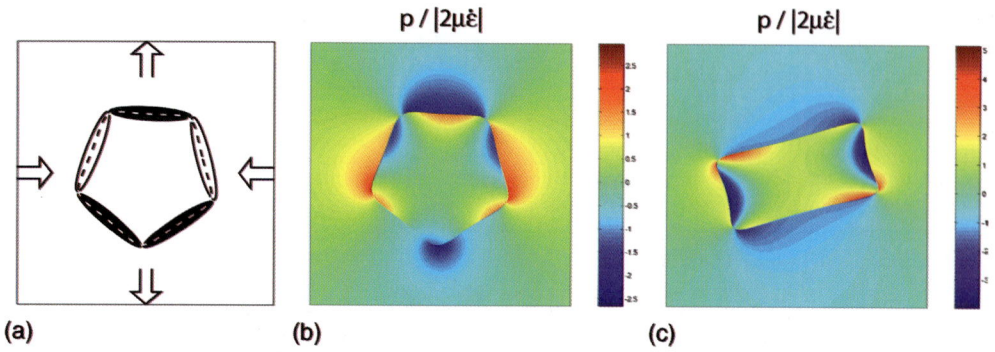

Fig. 8. Illustration of the ellipse decomposition rule. (**a**) shows a rigid pentagon (dashed line) in horizontal pure shear. According to the ellipse-decomposition rule, the edges of the pentagon are approximated by competent ellipses and, depending on their orientation relative to the applied pure shear flow, these ellipses indicate pressure-perturbation lows (empty ellipses) and highs (filled ellipses). This simple rule is confirmed in (**b**), which is the FEM calculated equivalent to (**a**). The rectangle in (**c**) demonstrates that the rule also holds for objects that are not of the equilateral type.

may suffice to drive a phase change or to start a metamorphic reaction. Furthermore, the given analytical solutions are not a function of size and may be applicable to objects as large as pluton intrusions.

The pressure perturbations inside the different types of clasts indicate that vertices directly aligned with the applied pure shear flow synchronize the pressure perturbation outside and inside clast interfaces. On the other hand, clast edges facing the shortening or extension of the pure shear show pressure perturbations that are opposite to what one would deduce from the far-field flow. This is clearly displayed in the case of the triangles (Fig. 4d–f), but cannot be observed for the square in Figure 4 because the 45° inclination of the edges with respect to the applied pure shear does not allow for a certain preference.

Aside from being a measure of proximity to failure, the maximum shear stress figure (Fig. 5) illustrates two points. First, maximum shear stress in and around rigid clasts increases as vertexes become more acute. Second, the inside of edges that are suitably aligned with the pure shear flow exhibit much lower maximum shear stress than edges that are facing directly a shortening or extension direction, as is illustrated by the triangle (Fig. 5d & f).

Ellipse decomposition rule

The observed characteristics of the pressure-perturbation field inside and outside rigid clasts can be encompassed by the ellipse-decomposition rule. This rule postulates that, in order to estimate the pressure perturbations in and around competent objects in shear, it suffices

to decompose the outline of the object into ellipses and base the pressure estimates on these ellipses (cf. Fig. 8). Treating these ellipses as isolated rigid clasts, one obtains the pressure perturbation for the inside of the rigid object: ellipses whose long axis lies in the extension direction undergo relative underpressure, whereas ellipses whose long axes lie in the shortening direction experience overpressure, and ellipses that lie in intermediate positions, i.e. at 45° inclination to the pure shear, remain at background (lithostatic) pressure. Once the inside pressure perturbations are qualitatively known, the outside values can be obtained by flipping the sign of the pressure perturbation inside the ellipses. The exception to the rule is the case of blunt vertexes that directly face one of the principal directions of the applied shear. In such cases, the pressure perturbation from the nearby matrix can penetrate into the clast and overprint the ellipse rule (cf. bottom pentagon vertex Fig. 8b).

Conclusions

Employing a combination of Muskhelishvili-based analytical solutions and finite-element method calculations, this paper illustrates the behaviour of rigid equilateral, equiangular polygons in shear and therefore represents a step towards understanding natural, non-ellipsoidal particles in shear. The following intriguing results were obtained. (i) Despite their vertexes, equilateral, equiangular triangles and squares do not rotate in pure shear. (ii) Pressure perturbations around rigid clast vertexes can be predicted with sufficient accuracy based on rigid ellipses with identical radii of curvature. (iii) Pressure perturbations and all other dynamic parameters, such as maximum shear stress inside polygonal clasts, do not show constant values and therefore ellipse-based predictions are invalid. (iv) Pressure perturbations inside polygonal clasts show, with the exception of blunt vertexes that lie in the principal flow direction, values that are opposite to what one would expect based on the understanding of pressure perturbation outside rigid clasts. (v) The complex behaviour of the pressure-perturbation field can be qualitatively approximated with the ellipse-decomposition rule. (vi) Based on the vertex curvature, matrix viscosity and far-field shortening rate, a large range of overpressure values is possible. However, if the rock strength is limited by the maximum shear stress, then the maximum overpressure perturbation is restricted to approximately twice the rock strength. (vii) The area of matrix material that experiences twofold or higher overpressure relative to the shortening stress is less than 2% of the object size (per polygon vertex). (viii) The obtained results are not only valid for pure shear, but can be applied to general shear by identifying the principal flow axes and adding the rigid-body rotation component of the flow. (ix) The given solutions do not have size limitations per se and are thought to be applicable to objects varying from small clasts in shear zones up to (with restrictions) pluton intrusions.

I would like to thank Y. Podladchikov for his everlasting enthusiasm for science in general and for his input to this paper in particular. I thank D. Clamond for his valuable advice concerning complex analysis, and H. Paul, N. Onderdonk and S. Schmalholz for help in finishing this paper. Finally, I acknowledge the reviews by M. Casey, T. Masuda and V. Tenczer.

References

ARBARET, L., MANCKTELOW, N.S. & BURG, J.P. 2001. Effect of shape and orientation on rigid particle rotation and matrix deformation in simple shear flow. *Journal of Structural Geology*, **23**, 113–125.

BILBY, B.A., ESHELBY, J.D. & KUNDU, A.K. 1975. The change of shape of a viscous ellipsoidal region embedded in a slowly deforming matrix having a different viscosity. *Tectonophysics*, **28**, 265–274.

CASEY, M., DIETRICH, D. & RAMSAY, J.G. 1983. Methods for determining deformation history for chocolate tablet boudinage with fibrous crystals. *Tectonophysics*, **92**, 211–239.

CERIANI, S., MANCKTELOW, N.S. & PENNACCHIONI, G. 2003. Analogue modelling of the influence of shape and particle/matrix interface lubrication on the rotational behaviour of rigid particles in simple

shear. *Journal of Structural Geology*, **25**, 2005–2021.

CROUZEIX, M. & RAVIART, P.A. 1973. Conforming and nonconforming finite-element methods for solving stationary Stokes equations. *Revue Francaise d' Automatique Informatique Recherche Operationnelle*, **7**, 33–75.

CUVELIER, C., SEGAL, A. & STEENHOVEN, A.A. 1986. *Finite Element Methods and Navier–Stokes Equations*. Reidel, Dordrecht.

ESHELBY, J.D. 1957. The determination of the elastic field of an ellipsoidal inclusion, and related problems. *Proceedings of the Royal Society of London*, **A241**, 376–396.

ESHELBY, J.D. 1959. The elastic field outside an ellipsoidal inclusion. *Proceedings of the Royal Society of London*, **A252**, 561–569.

FLETCHER, R.C. & MERINO, E. 2001. Mineral growth in rocks: Kinetic–rheological models of replacement, vein formation, and syntectonic crystallization. *Geochimica et Cosmochimica Acta*, **65**, 3733–3748.

GHOSH, S.K. & RAMBERG, H. 1976. Reorientation of inclusions by combination of pure shear and simple shear. *Tectonophysics*, **34**, 1–70.

JAEGER, J.C. & COOK, N.G.W. 1979. *Fundamentals of Rock Mechanics*. Chapman & Hall, London.

JEFFERY, G.B. 1922. The motion of ellipsoidal particles immersed in a viscous fluid. *Proceedings of the Royal Society of London*, **A102**, 161–179.

KENKMANN, T. & DRESEN, G. 1998. Stress gradients around porphyroclasts: palaeopiezometric estimates and numerical modelling. *Journal of Structural Geology*, **20**, 163–173.

MOSENFELDER, J.L., CONNOLLY, J.A.D., RUBIE, D.C. & LIU, M. 2000. Strength of (Mg,Fe)(2)SiO₄ wadsleyite determined by relaxation of transformation stress. *Physics of the Earth and Planetary Interiors*, **120**, 63–78.

MUSKHELISHVILI, N.I. 1953. *Some Basic Problems of the Mathematical Theory of Elasticity*. Noordhoff, Groningen.

PERRILLAT, J.P., DANIEL, I., LARDEAUX, J.M. & CARDON, H. 2003. Kinetics of the coesite–quartz transition: Application to the exhumation of ultrahigh-pressure rocks. *Journal of Petrology*, **44**, 773–788.

RAMSAY, J.G. & HUBER, M.I. 1983. *Strain Analysis*. Academic Press, London.

RANALLI, G. 1995. *Rheology of the Earth*. Chapman & Hall, London.

SAVIN, G.N. 1961. *Stress Concentration Around Holes*. Pergamon Press, New York.

SCHMID, D.W. & PODLADCHIKOV, Y.Y. 2003. Analytical solutions for deformable elliptical inclusions in general shear. *Geophysical Journal International*, **155**, 269–288.

TENCZER, V., STÜWE, K. & BARR, T.D. 2001. Pressure anomalies around cylindrical objects in simple shear. *Journal of Structural Geology*, **23**, 777–788.

TREAGUS, S.H. & LAN, L. 2003. Simple shear of deformable square objects. *Journal of Structural Geology*, **25**, 1993–2003.

Strain localization conditions in porous rock using a two-yield surface constitutive model

KATHLEEN A. ISSEN[1] & VENNELA CHALLA[2]

[1]*Mechanical and Aeronautical Engineering, 207 CAMP, Box 5725,*
8 Clarkson Avenue, Clarkson University, Potsdam, NY 13699-5725,
USA (e-mail: issenka@clarkson.edu)
[2]*Mechanical and Aeronautical Engineering, Box 5727, 8 Clarkson Avenue,*
Clarkson University, Potsdam, NY 13699-5727, USA

Abstract: This work examines theoretical conditions for localized deformation in porous rock, with emphasis on two recently identified deformation structures: compaction bands and dilation bands. Field and laboratory observations report that compaction/dilation bands consist of pure compressional/dilational deformation, which form perpendicular to maximum/minimum compression. A bifurcation approach is employed, with a two-yield surface constitutive model, to develop localization conditions under axisymmetric stress states for different stress paths. The first yield surface corresponds to a dilatant, frictional-damage mechanism (brittle regime), while the yield surface cap corresponds to a compactant mechanism (ductile regime). In the transitional regime, where both mechanisms are active, this model successfully predicts compaction bands and shear bands observed in axisymmetric compression tests. Due to discontinuities in the predicted band angle for probable material parameter values, this model may explain the lack of low angle compacting shear band observations in experiments. The two-yield surface model may also be applicable for a non-traditional axisymmetric extension stress path: increasing confining pressure with constant axial compression. Conditions for dilation band formation for this stress path are significantly less restrictive than corresponding compaction band conditions, suggesting that dilation bands could be a common deformation mode for high porosity sandstone.

Strain localization is a common deformation mode in geological materials, occurring at scales ranging from intragranular to global. Shear bands (also referred to as shear fractures or deformation band faults) have long been observed in rocks in both field and laboratory settings. These planar structures form at an angle to the direction of maximum compression and consist of shear deformation, often accompanied by either dilatant or compactant deformation normal to the band. Two other planar deformation structures have recently been identified in high porosity rock in the field and in laboratory specimens. The first are compaction bands, oriented perpendicular to maximum compression, consisting of pure compressive deformation. The second are dilation bands, oriented perpendicular to minimum compression (maximum extension), consisting of pure dilational deformation. This work examines theoretical localization conditions for high porosity rock, sandstone in particular, with emphasis on compaction band or dilation band formation. The

effects of stress state, stress path and constitutive model are also considered. Relevant field and experimental observations and data are cited to provide the basis for model development and assessment.

Background

Field and laboratory observations of localized deformation

Mollema & Antonellini (1996) first identified compaction bands in aeolian Navajo sandstone of 20–25% porosity (East Kaibab monocline, south central Utah). They described these bands as thin tabular zones of pure compressional deformation (no shear offset) that formed perpendicular to the direction of maximum compression in the compression quadrants at the tips of shear bands. Microstructural investigations revealed grain crushing within the band, and the band porosity was found to be a 'few per cent or less'. In laboratory drilling experiments

From: BRUHN, D. & BURLINI, L. (eds) 2005. *High-Strain Zones: Structure and Physical Properties.*
Geological Society, London, Special Publications, **245**, 433–452.
0305-8719/05/$15.00 © The Geological Society of London 2005.

on Berea sandstone (22–25% porosity), Haimson & Song (1998) and Haimson (2001) observed compacted zones of crushed grains, where the crush zone propagated perpendicular to the drilling axis. The loose, crushed grains were then washed out by drilling fluid, leaving behind long tabular slot-like borehole breakouts oriented perpendicular to maximum compression. These 'anti-dilatant' structures, which could be interpreted as compaction bands, contrast with the short V-shaped breakouts typically observed in low porosity rocks.

Olsson (1999) tested Castlegate sandstone (28% porosity) under axisymmetric compression (ASC). Specimens exhibited either compaction bands and/or shear bands, with observed band angles of $0°$ (compaction bands) to $43°$. (In the present paper, the band angle is defined as the angle between the *band normal* and the direction of maximum compression, in contrast to the experimental rock mechanics convention that defines it as the angle between the *band* and the direction of maximum compression.) Olsson & Holcomb (2000) used acoustic emissions to track the onset and propagation of compaction bands in Castlegate sandstone. They found that compaction bands initiated at the sample ends and propagated to the centre of the specimen with continued loading, such that the entire specimen was eventually transformed to the lower porosity compacted material. Microstructural observations by DiGiovanni et al. (2000) found that porosity reduction within the bands resulted from grain crushing and subsequent rotation of grain fragments.

Bésuelle et al. (2000) and Bésuelle (2001) examined localization in Vosges sandstone (22% porosity). Under ASC at lower confining pressures, specimens exhibited either dilating shear bands or axial splitting. It was suggested that axial splitting could be interpreted as 'extension bands' (i.e. dilation bands, $90°$ band angle). At increased confining pressures, shear bands occurred. Observed shear band angles decreased with increasing confining pressure from $54°$ to $37°$ and the volume strain evolved from dilatant to compactant. Specimens tested in axisymmetric extension (ASE) exhibited either dilating shear bands or dilation bands. Reported band angles were $69°–90°$.

Wong and colleagues (Wu et al. 2000; Klein et al. 2001; Wong et al. 2001) conducted ASC tests on Bentheim, Berea and Darley Dale sandstones (porosities of 23, 21 and 13%, respectively) at various confining pressures, including the transition from brittle faulting to cataclastic flow. At lower confining pressures, in the brittle regime, shear bands were observed. However,

at intermediate confining pressures (transitional regime), low angle (less than $45°$) shear bands, conjugate shear bands and/or compaction bands occurred near the sample ends. At higher confining pressures within the transitional regime, only compaction bands were observed. Multiple discrete bands formed near the specimen ends and the compaction band zones thickened with increasing axial strain until the entire specimen exhibited compaction bands. At high confining pressures, uniform compaction (cataclastic flow) with no evidence of compaction localization was observed (Wong et al. 1992, 1997). Microstructural observations by Menéndez et al. (1996) and Wu et al. (2000) found that at low confining pressures (brittle regime), prior to shear-band formation, axial intragranular cracking and shear-induced debonding occur. At high confining pressures (cataclastic flow) grain crushing and pore collapse are observed. Wong et al. (2001) indicated that both damage processes are active at intermediate confining pressures (transitional regime), where compaction bands form.

Du Bernard et al. (2002) recently reported field observations of dilation bands. These bands were identified in poorly consolidated sand (Savage Creek marine terrace near McKinleyville, northern California). The porosity of the undeformed sand at the time of dilation band formation was estimated to be 22–38%. The dilation bands formed in the extensional quadrant associated with conjugate sets of shear bands, typically originating from the shear band tips. Approximately 7% fewer detrital grains occur inside the dilation bands, compared with the surrounding material. They interpret this 7% as an increase in porosity or dilatancy occurring with band formation. The bands exhibit no macroscopic shear offset, and no evidence of grain breakage. Further, they indicate that the dilation bands, consisting of a zone of increased porosity, are a different failure mode than jointing or planar opening-mode fractures, which consist of two discrete surfaces (Du Bernard et al. 2002, p. 29-2 fig. 1).

To our knowledge, field and laboratory observations of compaction bands and dilation bands in geomaterials are limited to high porosity sandstone and sand. However, the theoretical framework employed here (and discussed in the next sections) is sufficiently general to enable application to other porous geomaterials. This implies that compaction band formation may be possible in other porous geomaterials that exhibit a brittle–ductile transitional regime similar to porous sandstone. In addition, theoretical results suggest that dilation band formation is possible not only in high porosity materials, similar to those cited above, but for low porosity materials that exhibit sufficiently dilatant behaviour.

The existence or formation of compaction bands and/or dilation bands in rock could significantly affect applications such as drilling, fluid extraction, injection and sequestration. Haimson (2001) suggested that compaction band-like borehole breakouts, where the drilling fluid washes out the compacted material, could result in substantial sand production. Holcomb & Olsson (2003) determined that the permeability dropped two orders of magnitude due to compaction localization, which could affect fluid flow within reservoirs as well as fluid injection and extraction. In contrast, it is possible that dilation bands could act as fluid pathways, due to increased porosity and potentially increased permeability. Wawersik et al. (2001) indicated that greenhouse gases, carbon dioxide in particular, could be captured and sequestered terrestrially, possibly in oil or gas reservoirs or aquifers. However, the presence of compaction bands or dilation bands, which could alter permeability, could impact containment effectiveness. Furthermore, as strain localization occurs at many scales in the Earth's crust, understanding the conditions that lead to localized deformation is fundamentally important to the tectonics and geophysics fields. Thus, the constitutive models and methods developed for predicting strain localization in high porosity rock have potentially broad impacts across several disciplines.

Bands of localized compaction have also been observed in high porosity cellular materials loaded in uniaxial compression. Papka & Kyriakides (1998a, b) found that the initially uniform compaction in polycarbonate honeycombs and aluminium honeycombs leads to collapse of a single row of cells perpendicular to the axial direction, similar in appearance to compaction bands in sandstone. Under continued loading, additional rows collapse into the first row such that the entire specimen is transformed to the crushed configuration. Several researchers (Andrews et al. 1999; Gradinger & Rammerstorfer 1998; Bart-Smith et al. 2000; Bastawros et al. 2000) found similar behaviour in aluminium foam: strain localizes in one or more bands of collapsed cells, before or near peak stress. These bands consist of mostly compressional deformation, oriented approximately perpendicular to the axial direction (Bart-Smith et al. 2000; Bastawros et al. 2000). Bands of localized compaction have also been observed in steel foams (Park & Nutt 2001). Thus, results regarding band formation in high porosity

sandstone will contribute towards the development of an overall understanding of strain localization conditions in other porous materials.

Constitutive relations

Olsson (1999) suggested that the localization theory of Rudnicki & Rice (1975), originally developed for shear localization in low porosity rock, could be used to predict compaction band formation as well. Recent re-examinations of Rudnicki & Rice (1975) determined compaction band conditions and highlighted the theoretical possibility of dilation band formation (Issen & Rudnicki 2000, 2001; Bésuelle 2001). As will be discussed later, localization conditions determined using this approach are strongly influenced by details of the constitutive relations. Material parameter values reported by Olsson (1999) and Wong et al. (2001), for specimens that exhibited compaction bands, did not satisfy the compaction band conditions determined by Issen & Rudnicki (2000, 2001) using a single-yield surface model. These findings, along with microstructural observations discussed above, motivated development of a two-yield surface constitutive model for high porosity sandstone (Issen 2002), shown in Figure 1. The first surface is a 'shear' yield surface with positive pressure dependence, representing a dilatant, frictional mechanism, such as the axial microcracking observed in the brittle regime. The second is a yield surface cap, with negative pressure dependence, corresponding to

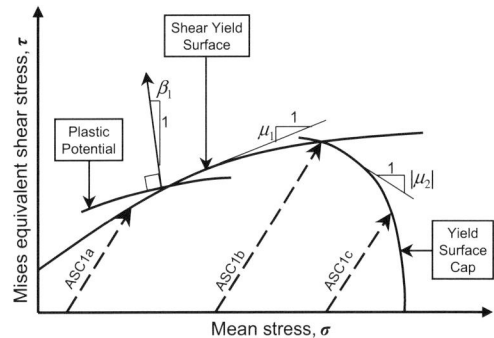

Fig. 1. Two-yield surface model for high porosity rock. The shear yield surface slope, μ_1, is positive, while the slope of the cap, μ_2, is negative. Axisymmetric compression loading at a low confining pressure intersects the shear yield surface (path ASC1a), but at a higher confining pressure intersects the cap (path ASC1c). At an intermediate confining pressure it is possible to intersect the region where the yield surfaces meet (path ASC1b), thus activating both yield surfaces.

a compactant mechanism: grain crushing and pore collapse during cataclastic flow. Wong *et al.* (1992) confirmed the existence of a yield surface cap for high porosity sandstone, similar to that suggested by DiMaggio & Sandler (1971) for soils. Some key details of the derivation of the two-yield surface model are provided below.

It is assumed that the material behaviour can be decomposed into a recoverable elastic response plus an irrecoverable inelastic response. For simplicity, isotropic elasticity is assumed, such that increments of elastic strain are given by $d\varepsilon_{ij}^e = (1/2G)[\delta_{ik}\delta_{jl} - (\nu/1+\nu)\delta_{ij}\delta_{kl}]d\sigma_{kl}$, where ν is Poisson's ratio and G is the elastic shear modulus. The Kronecker delta is $\delta_{ij}(=1$ if $i = j$, and $= 0$ if $i \neq j$), with repeated subscripts implying summation. Classical plasticity methods are used to determine the expression for the increments of inelastic strain. This requires specification of a yield surface or surfaces in stress space, a 'yield envelope', that encloses the stress states corresponding to elastic response. Inelastic response occurs for stress states falling on the yield surface(s). For increasing inelastic loading the yield surface evolves (e.g. expands) such that the stress state remains on the yield surface.

The yield surfaces are assumed to depend on the first invariant of stress through the mean stress (positive in compression), $\sigma = -\sigma_{kk}/3$, and the second invariant of deviatoric stress, $\tau = \sqrt{s_{ij}s_{ij}/2}$, where the deviatoric stress is $s_{ij} = \sigma_{ij} - (1/3)\sigma_{kk}\delta_{ij}$. For the traditional ASC test (constant confining pressure with increasingly compressive axial load) the intermediate principle stress is non-unique. In this case $\sigma = (\sigma_a + 2\sigma_c)/3$, where σ_a and σ_c are the axial stress and confining pressures, respectively (positive in compression), and τ is related to the differential stress: i.e. $\tau = (\sigma_a - \sigma_c)/\sqrt{3}$. The two-yield surface model is depicted in Figure 1, in τ–σ stress space, with three traditional ASC paths shown. The lower confining pressure path (ASC1a) intersects the shear yield surface, the higher confining pressure path (ASC1c) intersects the cap. For intermediate confining pressures, the loading path (ASC1b) intersects the region where the two-yield surfaces meet, thus activating both yield surfaces and representing the occurrence of both damage processes.

Note that, in its current formulation, the two-yield surface model is strictly only applicable at the point where the two-yield surfaces intersect. This point could be reached if the loading path intersected one surface, which, under continued loading was pushed outward until the

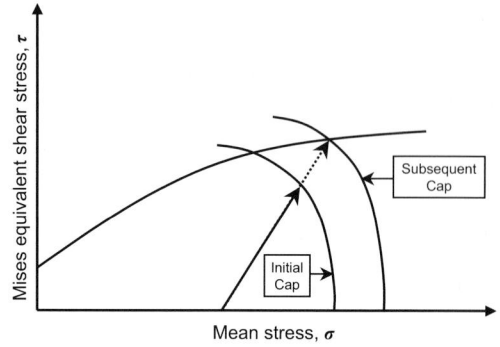

Fig. 2. After the cap is intersected, continued loading pushes the cap outwards until the shear yield surface is also activated. The hardening modulus, h, is infinite when the shear yield surface is first activated.

other surface was also intersected, thus activating both yield surfaces (see Fig. 2). In addition, while a schematic model from Wong *et al.* (1992, p. 289, fig. 9), which shows the cap yield surface expansion under hydrostatic consolidation, implies a related outwards shift in the brittle failure surface, we are unaware of any systematic studies to determine the influence of one activated yield surface on the other. Finally, the fitting of two-yield surfaces through experimental data is a somewhat subjective process, typically resulting in a region of possible intersection points that fall within the observed brittle–ductile transitional regime. Thus, we refer to a 'region' where the two surfaces intersect, rather than an intersection point.

The yield surface expressions (yield functions) are defined by $F_1 = \tau - f_1(\sigma, \gamma^p) = 0$ for the shear yield surface and $F_2 = \sigma - f_2(\tau, \varepsilon^p) = 0$ for the cap. The accumulated inelastic shear strain, γ^p, is used to track inelastic deformation history for the first surface, required in order to specify evolution of the yield surface with continued inelastic loading. The initial and subsequent yield surfaces, therefore, are a family of contours of constant inelastic shear strain. The inelastic shear strain increment is defined as $d\gamma^p = \sqrt{2de_{ij}^p de_{ij}^p}$, where $e_{ij} = \varepsilon_{ij} - (1/3)\varepsilon_{kk}\delta_{ij}$ is the deviatoric strain. Examination of mechanical data for Bentheim sandstone (courtesy of P. Baud and T.-f. Wong, from Wong *et al.* 2001) indicates that the first yield surface does expand with increasing inelastic shear strain, while the cap expands with increasing inelastic volume strain. Similarly, Wong *et al.* (1992) found that the cap expands with porosity reduction, which is related to accumulation of inelastic volume strain. Therefore, for the cap, accumulated inelastic volume strain, ε^p, was selected to track inelastic deformation history. The inelastic volume strain (positive in compression) is $\varepsilon^p = -\varepsilon_{kk}^p$. For the traditional ASC test, $\gamma^p = 2(\varepsilon_a^p - \varepsilon_\ell^p)/\sqrt{3}$ and $\varepsilon^p = \varepsilon_a^p + 2\varepsilon_\ell^p$, where ε_a and ε_ℓ are the axial and lateral strains (positive in compression), respectively. The local slope of the yield surface, also known as the 'friction coefficient', is given by $\mu_\alpha = -(\partial F_\alpha/\partial\sigma)/(\partial F_\alpha/\partial\tau)$, where $\alpha = 1$ refers to the shear-yield surface and 2 refers to the cap-yield surface (no summation on α). Typically, $\mu_1 > 0$ and $\mu_2 < 0$ are assumed, as shown in Figure 1.

For metal plasticity, the condition of 'normality' is applicable, as the inelastic strain increment vector is typically normal to the yield surface. This is also referred to as 'associated flow', as the direction of the inelastic strain increment is associated with the yield surface. Typically, normality is not true for geomaterials: the inelastic strain increments are not perpendicular to the yield surface (a condition known as non-associated flow), and are instead perpendicular to another surface called a plastic potential (see Fig. 1). For the two-yield surface model, non-associated flow is assumed for both surfaces, with plastic potentials given by $\Gamma_1 = \tau - g_1(\sigma, \gamma^p)$ and $\Gamma_2 = \sigma - g_2(\tau, \varepsilon^p)$. The local slope of a plastic potential is $\beta_\alpha = -(\partial\Gamma_\alpha/\partial\sigma)/(\partial\Gamma_\alpha/\partial\tau)$, also know as the dilation coefficient, where $\beta > 0$ for dilatation, and $\beta < 0$ for volume compaction (typically $\beta_1 > 0$ and $\beta_2 < 0$ for the shear yield surface and cap, respectively). Note that normality is often assumed to hold true for the cap. This condition is recovered by taking $\beta_2 = \mu_2$.

While the above expressions for the yield surfaces and plastic potentials are general, for a particular material, the functions f_1, f_2, g_1 and g_2 could be specified using experimentally determined mechanical data. In their study of several porous sandstones, Wong et al. (1997) suggested that the brittle failure envelope is parabolic and the cap is elliptical. Although these envelopes are not strictly yield surfaces, the shapes of the actual yield surfaces are expected to be similar. As will be shown later the surface slopes (μ_1, μ_2, β_1 and μ_2) strongly influence theoretical localization conditions, and, in some cases, small changes in these parameters can produce significant changes in the predicted band orientation.

The inelastic strain increment for a single yield surface is determined by the expression $d\varepsilon_{ij}^p = d\lambda(\partial\Gamma/\partial\sigma_{ij})$, where $d\lambda$ is a non-negative scalar that specifies the magnitude of the inelastic strain increment. The direction is specified by

$\partial\Gamma/\partial\sigma_{ij}$. The $d\lambda$ is determined from the consistency condition, $dF = 0$, which requires the stress state to remain on the yield surface during inelastic loading. Using a single yield surface, F_1 and plastic potential, Γ_1, given above, Holcomb & Rudnicki (2003) derived the Rudnicki & Rice (1975) expression for inelastic strain increments for a pressure-dependent material. The magnitude and direction were found to be $d\lambda = (s_{kl}/2\tau + \mu_1\delta_{ij}/3)/h$ and $\partial\Gamma/\partial\sigma_{ij} = s_{ij}/2\tau + \beta_1\delta_{ij}/3$, respectively. The resulting expression is:

$$(d\varepsilon_{ij}^p)_1 = \frac{1}{h}\left(\frac{s_{ij}}{2\tau} + \beta_1\frac{1}{3}\delta_{ij}\right)\left(\frac{s_{kl}}{2\tau} + \mu_1\frac{1}{3}\delta_{kl}\right)d\sigma_{kl}$$

(1)

where the hardening modulus, h, is the slope of the shear stress–inelastic shear strain curve at constant mean stress.

Koiter (1960) suggested that the total plastic strain increment due to the activation of multiple yield surfaces is determined by summing the increments from all active yield surfaces. Therefore, for the two-yield surface model, $d\varepsilon_{ij}^p = (d\varepsilon_{ij}^p)_1 + (d\varepsilon_{ij}^p)_2$, where $(d\varepsilon_{ij}^p)_\alpha = d\lambda_\alpha\partial\Gamma_\alpha/\partial\sigma_{ij}$, and the $d\lambda_\alpha$ are determined from the consistency conditions, $dF_\alpha = 0$. (A detailed derivation of this model is given in Issen 2002.) The final expression for the increments of inelastic strain for the two-yield surface model is:

$$d\varepsilon_{ij}^p = \frac{1}{hk\beta_2\mu_2}\left[\left(a\frac{s_{ij}}{2\tau} + b\frac{1}{3}\delta_{ij}\right)\frac{s_{kl}}{2\tau}\right.$$
$$\left. + \left(c\frac{s_{ij}}{2\tau} + d\frac{1}{3}\delta_{ij}\right)\frac{1}{3}\delta_{kl}\right]d\sigma_{kl} \quad (2)$$

where $a = h + k\beta_2\mu_2$, $b = \beta_2(h + k\beta_1\mu_2)$, $c = \mu_2(h + k\beta_2\mu_1)$ and $d = \beta_2\mu_2(h + k\beta_1\mu_1)$. The 'bulk' hardening modulus, k, is related to the cap yield surface and is defined as the slope of the mean stress–inelastic volume strain curve at constant shear stress. The total strain increment is determined by adding the elastic and inelastic strain increments: $d\varepsilon_{ij} = d\varepsilon_{ij}^e + d\varepsilon_{ij}^p = R_{ijkl}d\sigma_{kl}$. In the current formulation, the material parameters (h, k, β_1, μ_1, β_2, μ_2) are determined experimentally for a given specimen loaded on a given stress path. However, using mechanical data from a suite of tests for a single material, the initial and subsequent yield surfaces and plastic potentials could be determined, enabling development of specific expressions for these material parameters. These expressions could be incorporated into the constitutive relations (1) or (2), which would then be applicable for loading on any stress path.

Deviatoric stress states and stress paths

Testing of geomaterials is typically conducted using cylindrical specimens under axisymmetric stress conditions, which can be classified using the deviatoric stress state as either axisymmetric compression (ASC) or axisymmetric extension (ASE), determined by the relationship between the principal stresses. Using engineering conventions, the principal stresses are ordered $\sigma_I \geq \sigma_{II} \geq \sigma_{III}$ (positive in tension), while using geomechanics conventions $\sigma_1 \geq \sigma_2 \geq \sigma_3$ (positive in compression), such that $\sigma_1 = -\sigma_{III}$, $\sigma_2 = -\sigma_{II}$ and $\sigma_3 = -\sigma_I$. ASC is defined as $\sigma_I = \sigma_{II} > \sigma_{III}$ (equivalently, $\sigma_1 > \sigma_2 = \sigma_3$), while ASE is defined as $\sigma_I > \sigma_{II} = \sigma_{III}$ (equivalently, $\sigma_1 = \sigma_2 > \sigma_3$). The current work focuses on these two stress states for two reasons. First, the experimental observations used to evaluate theoretical predictions were obtained via either ASC or ASE tests. Second, it will be seen that ASC is the most favourable stress state for compaction band formation, while ASE favours dilation band formation. Clearly, these stress states represent the extremes, and field conditions will often lie in between ($\sigma_1 > \sigma_2 > \sigma_3$). However, if σ_2 is close to σ_1 (or to σ_3), then conditions for compaction band (or dilation band) formation may still be favourable.

For the ASC stress state, the traditional ASC stress path (ASC1 in Fig. 3) consists of a constant confining pressure with an increasingly compressive axial load, ($\sigma_1 > \sigma_2 = \sigma_3 = $ constant). However, many other stress paths are possible within the ASC stress state. For example, an alternative path (ASC2 in Fig. 3) consists of a constant axial stress with decreasing confining pressure ($\sigma_1 = $ constant $> \sigma_2 = \sigma_3$). A constant mean stress ASC path is obtained by decreasing the confining pressure while increasing the axial stress, such that both $\sigma_1 > \sigma_2 = \sigma_3$ and $\sigma = (\sigma_1 + \sigma_2 + \sigma_3)/3 = $ constant remain true. Similarly, multiple stress paths are possible within the ASE stress state. The traditional ASE path (ASE1 on Fig. 3) is achieved by reducing the compressive axial stress at constant confining pressure ($\sigma_1 = \sigma_2 = $ constant $> \sigma_3$), while an alternate path (ASE2 in Fig. 3) consists of maintaining constant axial stress while increasing the confining pressure ($\sigma_1 = \sigma_2 > \sigma_3 = $ constant).

The stress path and the shape of the envelope formed by the yield surfaces will determine what portion of the yield surface envelope is

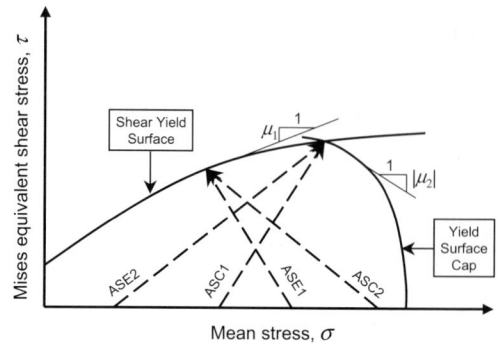

Fig. 3. Two-yield surface model showing axisymmetric extension (ASE1, ASE2) and axisymmetric compression (ASC1, ASC2) loading paths. Traditional paths are ASC1 (constant confining pressure with increasing axial compression) and ASE1 (constant confining pressure with decreasing axial compression). Path ASE2 (constant axial compression with increasing confining pressure) may intersect the region where the surfaces meet, thus activating both yield surfaces.

accessible. For example, by selecting the proper constant confining pressure, path ASE1 (Fig. 3) will intersect only the shear yield surface, where a single-yield surface model is applicable. Similarly, by selecting a high confining pressure for path ASC1 (Fig. 1, path ASC1c), the load path will intersect only the cap, again implying a single-yield surface model. If an intermediate confining pressure is selected for ASC1 (see Fig. 3, or path ASC1b in Fig. 1), the load path will intersect the region where the two-yield surface model may be appropriate. Similarly, the constant axial stress for ASE2 (Fig. 3) can be selected such that the load path also intersects the region where a two-yield surface model may be applicable. While microstructural and mechanical data imply that two-yield surfaces may be warranted, the manner in which these two surfaces meet is still unclear. If the cap meets the shear yield surface at a vertex then the two-yield surface model seems appropriate, while if the surfaces meet smoothly a single-yield surface constitutive model is suggested. Therefore, while the focus of this paper is on two-yield surface results, single-yield surface results will also be presented for comparison purposes.

Localization conditions

Localization theory

In their investigation of conditions for localized deformation in pressure-sensitive dilatant materials, Rudnicki & Rice (1975) suggested that formation of a planar band of localized strain could be viewed as a bifurcation from homogeneous deformation. Combining the conditions of kinematic compatibility and stress equilibrium with the constitutive relation, they determined that a planar band, with band normal components n_i, is possible when $\det|n_i L_{ijkl} n_l| = 0$ is satisfied. The modulus tensor, L_{ijkl}, is defined by $d\sigma_{ij} = L_{ijkl} d\varepsilon_{kl}$. For the two-yield surface model, Issen (2002) provides the expression for L_{ijkl} (too lengthy to repeat here) determined by inverting the incremental strain–stress relation, $d\varepsilon_{ij} = R_{ijkl} d\sigma_{kl}$, described above. The expression for the critical value of the hardening modulus, $h_{2\,ys}$, at the inception of planar-band formation is found by substituting L_{ijkl} into the localization condition and solving for h:

$$h_{2\,ys} = \frac{k\beta_2\mu_2\eta_1 + C}{k\beta_2\mu_2 - \eta_2} \tag{3}$$

where

$$
\eta_\alpha = G\left\{ \frac{1+v}{9(1-v)}(\beta_\alpha - \mu_\alpha)^2 \right.
$$
$$
- \frac{1+v}{1-v}\left[\frac{1}{2}N_{22} - \frac{1}{3}(\beta_\alpha + \mu_\alpha)\right]^2
$$
$$
\left. + \frac{3}{4}N_{22}^2 + M \right\} \tag{4}
$$

$$N_{22} = \frac{s_{22}}{\tau} = n_\mathrm{I}^2 N_\mathrm{I} + n_\mathrm{II}^2 N_\mathrm{II} + n_\mathrm{III}^2 N_\mathrm{III} \tag{5}$$

$$M = n_\mathrm{I}^2 N_\mathrm{I}^2 + n_\mathrm{II}^2 N_\mathrm{II}^2 + n_\mathrm{III}^2 N_\mathrm{III}^2 - N_{22}^2 - 1 \tag{6}$$

$$C = \frac{4}{9}\left(\frac{1+v}{1-v}\right)G^2(\beta_1 - \beta_2)$$
$$\times (\mu_1 - \mu_2)\left(\frac{3}{4}N_{22}^2 + M\right) \tag{7}$$

where $K = $ I, II, III are the principal stress directions ($\sigma_\mathrm{I} \geq \sigma_\mathrm{II} \geq \sigma_\mathrm{III}$) and n_K are the components of the band normal. The subscript α appearing on the η, β and μ terms is equal to 1 for the shear yield surface and 2 for the cap. The parameters $N_K = s_K/\tau$ represent the deviatoric stress state, where s_K are the principal deviatoric stresses (τ is the Mises equivalent shear stress defined earlier). Thus, the above expressions are applicable for any stress state. For ASC $N_\mathrm{I} = N_\mathrm{II} = 1/\sqrt{3}$ and $N_\mathrm{III} = -2/\sqrt{3}$, while for ASE $N_\mathrm{I} = 2/\sqrt{3}$ and $N_\mathrm{II} = N_\mathrm{III} = -1/\sqrt{3}$. Although this work focuses on ASC and ASE, localization conditions can be developed for non-axisymmetric

stress states ($\sigma_1 > \sigma_2 > \sigma_3$) as well, possibly corresponding to known or assumed field conditions.

For a single-yield surface model, the inelastic constitutive relation of (1) is combined with isotropic elasticity to determine the expression for the tensor modulus, L_{ijkl} (see Rudnicki & Rice 1975). Applying the localization condition results in the following expression for the critical hardening modulus for a single active yield surface:

$$h_{1\,ys} = \eta \tag{8}$$

where η is given by equation (4), dropping the α subscript. As the single-yield surface constitutive formulation of Rudnicki & Rice (1975) uses a single hardening modulus, h, the critical hardening modulus is $h_{1\,ys}$. However, the two-yield surface model employs two hardening moduli, h and k. To facilitate comparison with the single-yield surface model, Issen (2002) elected to solve $\det|n_i L_{ijkl} n_l| = 0$ for h as well. However, if only the cap is active, then solving $\det|n_i L_{ijkl} n_l| = 0$ leads to a critical hardening modulus, $k_{1\,ys} = \eta/\beta_2\mu_2$. In this case, assuming β_2 and μ_2 are both negative, Issen (2002) showed that the band orientation predictions are identical to the single-yield surface results ($h_{1\,ys} = \eta$), which are briefly recapped below. In addition, Rudnicki (2004) examined the transition from shear localization to compaction band formation in terms of confining pressure, for ASC paths that intersect an elliptical cap.

From the above discussion, it is apparent that the value of the critical hardening modulus, h_{ys}, varies with: (1) stress path (which determines the proper constitutive model); (2) stress state (represented by the N_K); (3) band orientation (represented by n_K); and (4) material parameters (i.e. values of β_α, μ_α, v, k, G). Since h is typically assumed to decrease with continued inelastic loading, the band orientation resulting in the largest value of the critical hardening modulus, for a chosen stress path, stress state and set of material parameter values, is the band orientation predicted to occur first. These band orientation predictions are presented graphically in Figures 4–11, and are discussed in detail in the following sections.

While mechanical loading obviously influences localization conditions, the development of compaction, dilation or shear bands in field settings will also be affected by fluid and chemical effects. For example, in their examination of Berea, Boise, Darley Dale and Gosford sandstones (porosities of 11–35%), Baud et al. (2000) found that the size and location of elliptical yield envelopes under wet conditions were significantly different than those for dry conditions. Rice (1975) determined that, for fluid-saturated rock, dilatant hardening occurs during undrained deformation, although the stability of the homogenous response is limited by the underlying drained behaviour. However, similar results for high porosity rock will be more complex, as material response may be dilatant or compactant. While important, investigation of these effects is beyond the scope of the present work.

Localization conditions for axisymmetric stress states

For axisymmetric stress states (ASC or ASE) the direction of the intermediate principal stress is not uniquely defined, and $n_{II}^2 = 0$ can be assumed without loss of generality. The relationship between the remaining components of the band normal is $n_I^2 + n_{III}^2 = 1$. Alternatively, this relationship can be written in terms of the axial and lateral directions as $n_a^2 + n_\ell^2 = 1$. For ASC, the axial and lateral directions are the III and I directions (alternatively, the 1 and 3 directions), respectively, such that $n_{III}^2 = n_1^2 = n_a^2$ and $n_I^2 = n_3^2 = n_\ell^2$. For ASE, the axial and lateral

directions are the I and III directions (alternatively, the 3 and 1 directions), respectively, such that $n_I^2 = n_3^2 = n_a^2$ and $n_{III}^2 = n_1^2 = n_\ell^2$. After substituting the appropriate values of the N_K for ASC and for ASE into equations (4)–(7), the resulting equations for C and η_α can be rearranged to reveal certain symmetries between the expressions for ASC and ASE. These symmetries allow us to write the following expressions for both axisymmetric stress states:

$$\eta_\alpha = -\frac{4G(1+v)}{9(1-v)}\left[\frac{27n_\ell^4}{8(1+v)}\right.$$

$$+ \delta\frac{3\sqrt{3}}{4}(\beta_\alpha + \mu_\alpha - \delta\sqrt{3})n_\ell^2$$

$$\left. + \left(\beta_\alpha - \delta\frac{\sqrt{3}}{2}\right)\left(\mu_\alpha - \delta\frac{\sqrt{3}}{2}\right)\right] \quad (9)$$

$$C = -\frac{(1+v)}{3(1-v)}G^2(\beta_1 - \beta_2)(\mu_1 - \mu_2)n_\ell^4 \quad (10)$$

where δ is defined as -1 for ASC and $+1$ for ASE. The term C represents the severity of the vertex, $(\beta_1 - \beta_2)(\mu_1 - \mu_2)$, and is always nonpositive. In equations (9) and (10), the band-normal component, n_ℓ^2, may vary smoothly

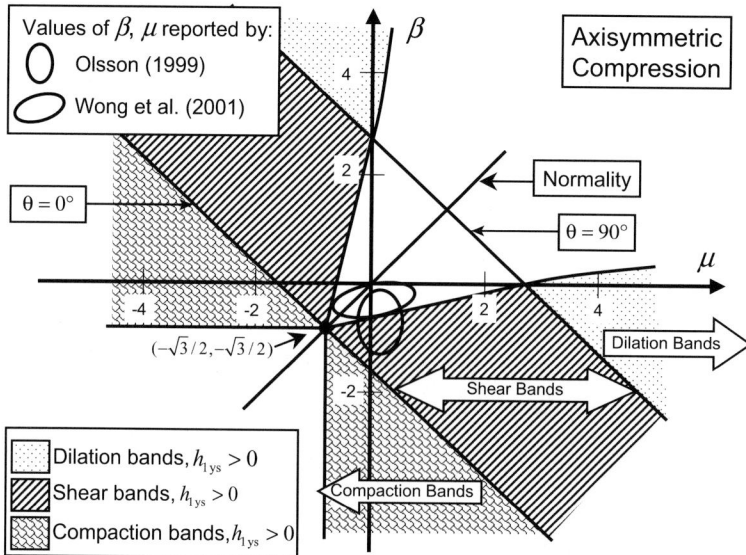

Fig. 4. Predicted band orientation as a function of β and μ for a single active yield surface for ASC. The shaded regions represent positive values of the critical hardening modulus. Compaction bands are predicted below the $\theta = 0°$ line: $\beta + \mu \leq -\sqrt{3}$. Dilation bands are predicted above the $\theta = 90°$ line: $\beta + \mu \geq \sqrt{3}(2-v)/(1+v)$. Shear bands, with band orientations varying from $0°$ to $90°$, are predicted between these lines. The band orientation, θ, is the angle between the band normal and the axial direction. Only shear bands are predicted for reported values of β and μ, while compaction bands and/or shear bands were observed.

from 0 and 1, corresponding to a spectrum of band angles from $0°$ (compaction bands) to $90°$ (dilation bands). In addition, as discussed in detail by Bésuelle (2001), the predicted strain type inside the band at the inception of localization varies with the band orientation. Thus, compaction bands, shear bands and dilation bands are not three distinct band types, but are members of a spectrum of possible bands, ranging from pure compaction bands (perpendicular to maximum compression), through compacting shear bands, pure shear bands, dilating shear bands, to pure dilation bands (parallel to maximum compression).

Under ASC or ASE, $0 < n_\ell^2 < 1$ corresponds to shear bands. For ASC, $n_\ell^2 = 0$ and $n_\ell^2 = 1$ correspond to compaction bands and dilation bands, respectively, while for ASE, $n_\ell^2 = 0$ for dilation bands and $n_\ell^2 = 1$ for compaction bands. Issen & Rudnicki (2000) determined that ASC is the most favourable stress state for compaction band formation for a single-yield surface model. Similarly, it can be shown that ASE is the most favourable stress state for dilation band formation. While not rigorously proven, it will be apparent from the band orientation predictions discussed later, that the most favourable stress states for compaction bands and dilation bands are ASC and ASE,

respectively, for the two-yield surface model as well. Therefore, for either axisymmetric stress state, the band of primary interest is the band oriented perpendicular to the axial direction, denoted with superscript, $\perp$. Thus, equations (9) and (10), can be written as:

$$\eta_\alpha^\perp = -\frac{4G(1+v)}{9(1-v)}\left(\beta_\alpha - \delta\frac{\sqrt{3}}{2}\right)\left(\mu_\alpha - \delta\frac{\sqrt{3}}{2}\right)$$

(11)

$$C^\perp \equiv 0 \qquad (12)$$

where the $\perp$ represents compaction bands for ASC and dilation bands for ASE.

Single-yield surface model results

To facilitate comparison with the two-yield surface model, results from a single-yield surface model are summarized briefly. This model is applicable for loading paths that intersect only one yield surface (e.g. path ASE1 and ASC2 in Fig. 3, and path ASC1a and ASC1c in Fig. 1). Note that some form of the following results have been presented by several authors (Rudnicki & Rice 1975; Ottosen & Runesson

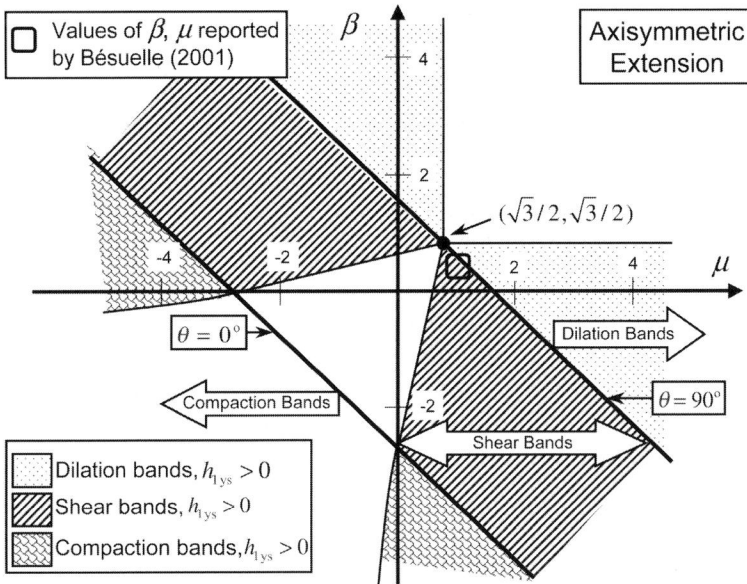

Fig. 5. Predicted band orientation as a function of β and μ for a single active yield surface for ASE. The shaded regions represent positive values of the critical hardening modulus. Compaction bands are predicted below the $\theta = 0°$ line: $\beta + \mu \leq -\sqrt{3}(2-v)/(1+v)$. Dilation bands are predicted above the $\theta = 90°$ line: $\beta + \mu \geq \sqrt{3}$. Shear bands, with band orientations varying from $0°$ to $90°$, are predicted between these lines. The band orientation, θ, is the angle between the band normal and the axial direction.

1991; Perrin & LeBlond 1993; Rudnicki & Olsson 1998; Issen & Rudnicki 2000, 2001; Bésuelle 2001). For any general stress state, the expression for the critical hardening modulus is $h_{1\,ys} = \eta$, where η is given by either equation (4), (9) or (11) and the subscript, α, is dropped. Shear bands are predicted when:

$$-\frac{3(N_{\mathrm{I}} + \nu N_{\mathrm{II}})}{1 + \nu} \leq \beta + \mu \leq -\frac{3(N_{\mathrm{III}} + \nu N_{\mathrm{II}})}{1 + \nu}.$$
(13)

Compaction bands are predicted when the left-side of equation (13) is violated, and dilation bands when the right-side is violated. For ASC, equation (13) becomes $-\sqrt{3} \leq \beta + \mu \leq \sqrt{3}(2 - \nu)/(1 + \nu)$, while for ASE it becomes $-\sqrt{3}(2 - \nu)/(1 + \nu) \leq \beta + \mu \leq \sqrt{3}$. These results are depicted graphically in the $\beta - \mu$ plane in Figures 4 and 5 for ASC and ASE, respectively (for $\nu = 1/5$). Shear bands are predicted for values of β and μ between the two bold diagonal lines, while compaction bands are predicted below the lower line and dilation bands above the upper line. The shaded regions represent values of β and μ, where $h_{1\,ys} > 0$. Notice that deviations from normality promote localization, as the critical hardening modulus is negative when $\beta = \mu$.

For ASC (Fig. 4) the ovals indicate the range within which the experimental data falls for high porosity sandstone loaded at intermediate confining pressures (corresponding to the region where the two-yield surface model may be applicable). Note that the theory predicts only shear bands, while compaction bands and/or shear bands were observed (Olsson 1999; Wong et al. 2001). Alternatively, for samples loaded at higher confining pressures, the loading path intersects only the cap, near the mean stress axis, where μ is quite negative, and symmetry considerations require normality. In this case $h_{1\,ys}$ is strongly negative, such that localization is prohibited, as confirmed experimentally, where only uniform compaction is observed. For ASC at very low confining pressures (intersecting only the shear yield surface), Bésuelle (2001) observed axial splitting in one specimen, and reported values of $\beta = 1.55$, $\mu = 0.97$ and $\nu = 0.4$, which satisfy the ASC condition for dilation band formation. Both the predicted and observed values of the critical hardening modulus were negative.

Bésuelle et al. (2000) and Bésuelle (2001) observed dilation bands and/or dilating shear bands for ASE tests (path ASE1). The reported values of $0.37 < \beta < 0.59$ and $0.97 < \mu <$

1.15, are represented by the small square in Figure 5. The largest of these values satisfy the dilation band condition, $\beta + \mu \geq \sqrt{3}$. Thus, reported values of β and μ predict the experimentally observed dilation bands or dilating shear bands. As expected for the shear yield surface, normality is not applicable: all values of β are less than $\sqrt{3}/2$, while all values of μ are greater than $\sqrt{3}/2$. These values satisfy the ASE requirement for a positive critical hardening modulus for dilation bands, $(\beta - \sqrt{3}/2) \times (\mu - \sqrt{3}/2) \leq 0$, implying localization prior to peak stress, which corresponds well to their observations of localization at or prior to peak stress.

From the above results, bifurcation theory using a single-yield surface model successfully predicts localization conditions when the loading path clearly intersects only one yield surface. However, in the transitional regime where two deformation processes are active, the single-yield surface model does not predict the observed compaction bands. Therefore, localization conditions for the two-yield surface model are considered next.

Two-yield surface model results for $k = 0$

If localization occurs either at the stress peak or along a horizontal plateau in the mean stress–volume strain curve, then $k \approx 0$. For the two-yield surface model, from equation (3), when $k = 0$, then $h_{2\,ys} = -C/\eta_2$ and therefore, $h_{2\,ys}^{\perp} \equiv 0$ (where $h = 0$ corresponds to a peak or constant stress plateau in the shear stress–shear strain curve). Since C is always non-positive, the sign of $h_{2\,ys}$ is determined only by η_2. Bands perpendicular to the axial direction are predicted when the critical hardening moduli for the other band orientations (denoted by λ) are negative: $h_{2\,ys}^{\lambda} \equiv 0 > h_{2\,ys}$. This condition reduces to $0 > n_\ell^4/\eta_2^\lambda$, which no longer depends on the parameters from the first yield surface, β_1 and μ_1. Therefore, band orientation predictions can be depicted graphically in the $\beta_2 - \mu_2$ plane for a particular stress state, as shown in Figure 6 for ASC, and Figure 7 for ASE. In both figures, compaction bands are predicted in the fish-scale-patterned regions, shear bands in the unshaded areas and dilation bands in the dotted regions. For values corresponding to normality ($\beta_2 = \mu_2$, which is often assumed for the cap) and many values representing mild deviations from normality, compaction bands are predicted for ASC and dilation bands are predicted for ASE. Thus, similar to the single-yield surface findings, ASC favours compaction band

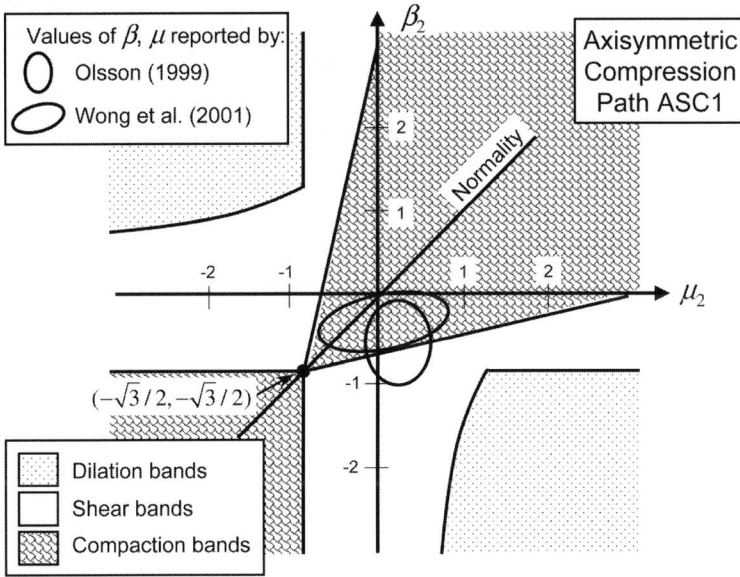

Fig. 6. Band orientation predictions as a function of β_2 and μ_2 for the two-yield surface model when $k = 0$ for path ASC1. Compaction bands are predicted in the fish-scale-patterned regions, with h_2 $_{ys} = 0$. Shear bands are predicted in the unshaded areas, while dilation bands are predicted in the dotted regions. The ovals represent values of β and μ when compaction bands and/or shear bands were observed.

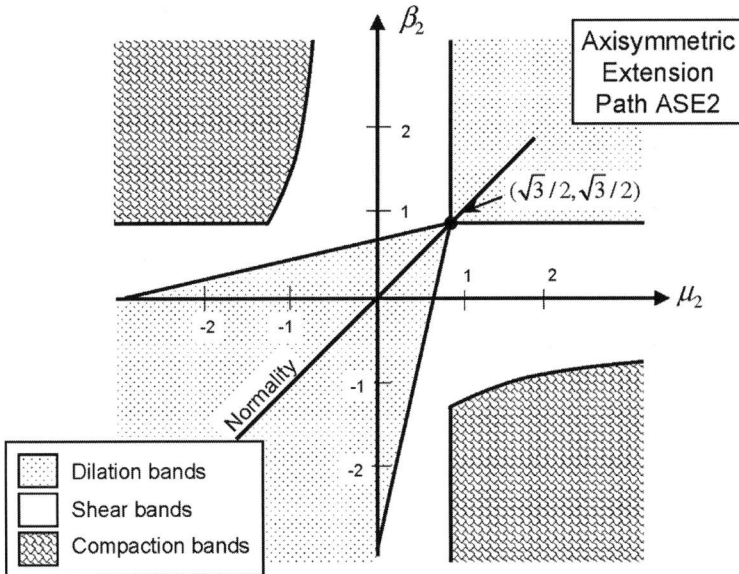

Fig. 7. Band orientation predictions as a function of β_2 and μ_2 for the two-yield surface model when $k = 0$ for path ASE2. Dilation bands are predicted in the dotted regions, with h_2 $_{ys} = 0$. Shear bands are predicted in the unshaded areas, while compaction bands are predicted in the dotted regions.

formation while ASE favours dilation band formation. If β_2 and μ_2 are negative (another common assumption), then, for ASE, dilation bands are predicted exclusively, except for improbably large deviations from normality. However, for ASC, compaction bands or shear bands are possible for negative values of β_2 and μ_2.

For both ASC and ASE, shear bands require some deviation from cap normality, and the critical hardening modulus, given by $h_{2\,\mathrm{ys}} = -C/\eta_2$, is found to be infinitely large (Challa & Issen 2004). During inelastic loading, an infinitely large hardening modulus could correspond to initial activation of the shear yield surface as the cap is pushed outward (see Fig. 2). The predicted band angle, as shown in Figure 8 for ASC, is determined by setting the denominator, given by equation (9), equal to zero and solving for the band normal component, n_ℓ^2, for specific values of β_2 and μ_2 within the shear band zone. This shear band zone for ASC is discussed in more detail next, as it falls within regions of β_2 and μ_2 values that are of interest. While the following discussion could also be applied to the ASE plot, it would be of little practical interest, since the shear band zone falls primarily in regions corresponding to improbable values of β_2 and μ_2.

As discussed by Challa & Issen (2004) for ASC (see Fig. 8), for values of β_2 and μ_2 in the shear band zone, but below $\beta_2 = -\sqrt{3}/2$ the equation $\eta_2 = 0$ has only one positive root, n_ℓ^2. However, above $\beta_2 = -\sqrt{3}/2$ two positive roots exist, such that two distinct band angles are possible for a given point (β_2, μ_2). For example, at point A in Figure 8, the roots are $n_\ell^2 = 0$ and $n_\ell^2 = 0.5$, corresponding to band angles of $0°$ (compaction bands) and $\pm 45°$, respectively. At point B, roots are $n_\ell^2 = 0.25$ and $n_\ell^2 = 0.375$, corresponding to band angles of $\pm 30°$ and $\pm 37.7°$, respectively. Notice that small changes in the values of β_2 and μ_2 could shift point B into the compaction band region with a band angle of $0°$. In addition, each root, n_ℓ^2, represents two band normals, $\pm n_\ell$, possibly corresponding to conjugate shear bands.

The values of β and μ reported by Olsson (1999) and Wong et al. (2001) for ASC are shown in the small ovals in Figures 6 and 8. Unlike the single-yield surface model (see Fig. 4), both the observed compaction bands and shear bands are predicted by the two-yield surface model. In addition, the critical hardening modulus for compaction bands is zero, which agrees with experimental observations of compaction band formation near peak stress. Some of Olsson's data fall within the shear

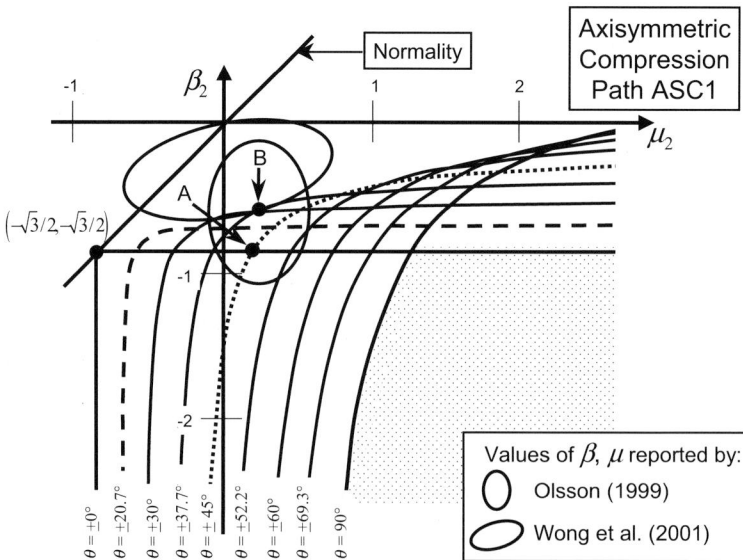

Fig. 8. Predicted shear band angles, as a function of β_2 and μ_2, for $k = 0$ for the two-yield surface model. In this region, $h_{2\,\mathrm{ys}} \to \infty$, possibly corresponding to initial activation of the shear yield surface as the cap is pushed out (see Fig. 2). Below $\beta_2 = -\sqrt{3}/2$ only one band orientation is predicted, while above two distinct band angles are possible for the same (β_2, μ_2). At point A, band angles of either $0°$ or $\pm 45°$ are theoretically possible, while band angles of either $\pm 30°$ or $\pm 37.7°$ are possible at point B.

band region where the predicted band angles are $0°-55°$, which correlate well with observed band angles of $0°$ and $14°-65°$. Band angles of $0°$ and $35°-65°$ were observed in several other porous sandstones (Baud et al. 2004). Although the results of Wong et al. (2001) fall almost entirely in the compaction band region, some of the data are very close to predicted shear band angles of $45°$ or less, such that shear bands would be predicted for small changes in reported values of β and μ. Note that the two-yield surface model requires values for four parameters: β_1, β_2, μ_1 and μ_2, while only two values, β and μ, were reported. While not rigorously proven, $\beta_1 < \beta < \beta_2$ and $\mu_1 < \mu < \mu_2$ are reasonable assumptions. Therefore, if values for β_2 and μ_2 were extrapolated from the mechanical data, it is probable that β_2 and μ_2 would be more negative than reported values of β and μ, shifting the ovals in Figures 6 and 8 downwards and to the left. If this shift were not too extreme, both the observed compaction bands and shear bands would still be predicted.

While compaction bands and/or shear bands have been observed in ASC specimens, reports of very low angle shear bands are uncommon. For example, Olsson (1999) observed no shear-band angles below $14°$, while Baud et al. (2004) report no band angles below $35°$. This scarcity of very low angle shear bands is difficult to explain using a single-yield surface model, where predicted band angles increase smoothly with increasing values of $\beta + \mu$. However, Rudnicki (2004) notes that the band angle increases quickly from $0°$ as values of $\beta + \mu$ increase. Even so, the transition is still smooth, unlike the two-yield surface model, where there are discontinuities in the predicted band angle. As discussed above, small perturbations in the β_2 and μ_2 values for the two-yield surface model can result in significant changes in the predicted band angle. For example, two similar specimens subjected to the same test would probably have similar, but not identical, values of β_2 and μ_2, such that compaction bands could be predicted for one specimen, while a $35°$ shear band could be predicted for the other. Therefore, the two-yield surface model provides a possible explanation for the lack of very low angle shear bands in experiments on high porosity sandstone loaded in the brittle–ductile transition regime.

Results for ASC for k > 0

Challa & Issen (2004) examined the influence of the bulk hardening modulus, k, on localization conditions. Key aspects of their results are summarized here for two cases: $k = G/100$ and $k = G/10$. These values were selected based upon the examination of mechanical data for Bentheim sandstone (courtesy of P. Baud and T.-F. Wong, from Wong et al. 2001), in which compaction bands were observed to first form near peak stress and propagate along a jagged, sometimes sloping, plateau in the mean stress–volume strain curve (see, for example, Wong et al. 2001, p. 2521, fig. 1). From the overall plateau slope, we determined that k is approximately zero for the 120 MPa confining-pressure test, and increases with increasing confining pressure to approximately $k = G/10$ for the 350 MPa test. These small positive values of k could also correspond to localization just prior to the initial peak stress or near the small peaks observed along the jagged plateau.

The band orientation predictions for $k = G/100$ and $k = G/10$ are shown graphically in Figures 9 and 10, respectively. Note that, although predicted band angles depend on β_1 and μ_1 when $k \neq 0$, the results do not vary significantly for small changes within the range of probable values. Therefore, $\beta_1 = \sqrt{3}/4$ and $\mu_1 = 3\sqrt{3}/4$ were assumed, and results are again plotted in the $\beta_2 - \mu_2$ plane. In general, compaction band conditions for slightly positive values of k are less favourable than those for $k = 0$. Compaction bands (fish-scale-patterned regions) are predicted in a more restricted region of β_2 and μ_2 (see Figs 9 and 10) and Challa & Issen (2004) found that the critical hardening modulus is negative, implying localization later in the loading program.

In addition, for some negative values of β_2 and μ_2 below line A–C (wave-patterned areas in Figs 9 and 10), including some values at/near normality, low angle shear bands are now predicted, where compaction bands would be expected if $k = 0$. This low angle shear band region expands for increasing k, until compaction bands are completely prohibited for most negative values of β_2 and μ_2 when $k = G/10$ (Fig. 10). However, the band orientation predictions for experimentally reported values of β and μ for $0 < k < G/10$ are similar to those for $k = 0$, and correlate well with the observed band orientations represented by the small ovals. While dilation bands appear to be predicted for values of β_2 and μ_2 near $-\sqrt{3}/2$ (checkered region), Challa & Issen (2004) found the corresponding values of the critical hardening modulus to be improbably negative, thus prohibiting localization for values of β_2 and μ_2 in this region. These results contrast with those for $k = 0$, where localization is possible for all values of β_2 and μ_2, and

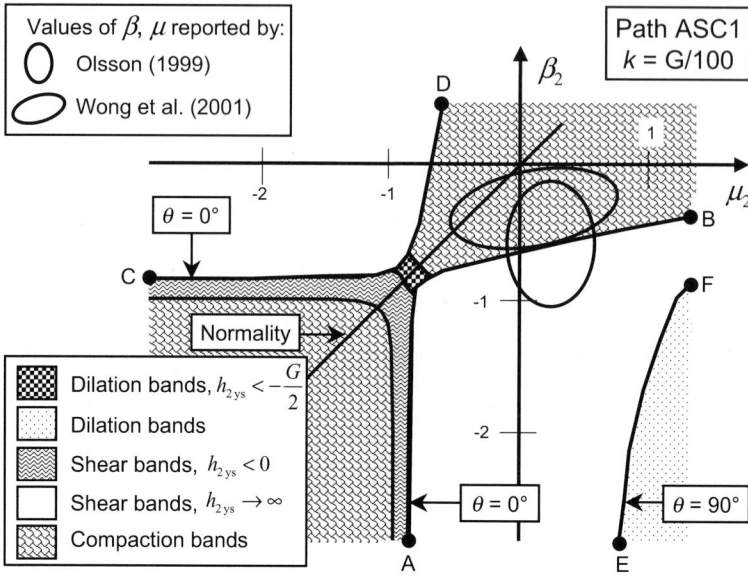

Fig. 9. Band orientation predictions as a function of β_2 and μ_2 for $k = G/100$, using the two-yield surface model ($\beta_1 = \sqrt{3}/4$, $\mu_1 = 3\sqrt{3}/4$). Compaction bands are predicted in the regions with the fish-scale pattern, with $h_{2\,ys} < 0$. Shear bands are expected in the unshaded regions ($h_{2\,ys} \to \infty$) and the wave-patterned regions ($h_{2\,ys} < 0$). Dilation bands are predicted in the dotted regions ($h_{2\,ys} < 0$). While dilation bands are also predicted in the checkered region, the required critical hardening modulus value, $h_{2\,ys} < -G/2$, is improbable.

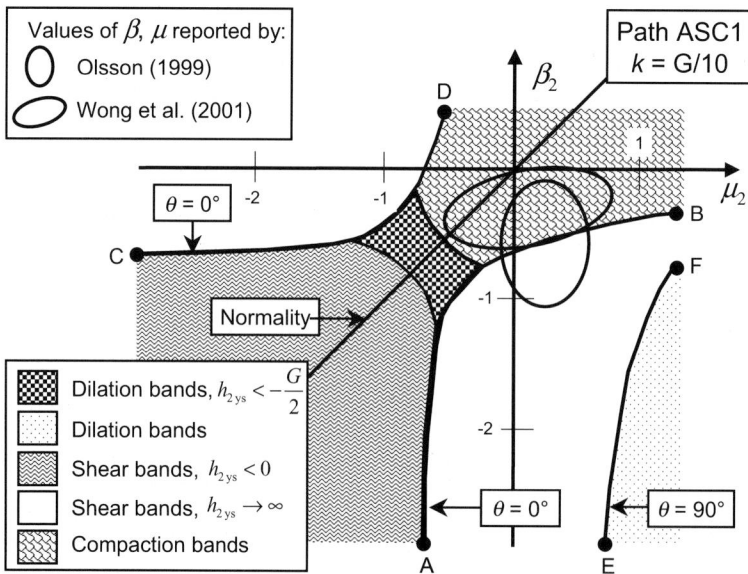

Fig. 10. Band orientation predictions as a function of β_2 and μ_2 for $k = G/10$, using the two-yield surface model ($\beta_1 = \sqrt{3}/4$, $\mu_1 = 3\sqrt{3}/4$). Compaction bands are predicted only above line B–D in the fish-scale-patterned areas, with $h_{2\,ys} < 0$. Shear bands are expected in the unshaded regions ($h_{2\,ys} \to \infty$) and the wave-patterned regions ($h_{2\,ys} < 0$). Dilation bands are predicted in the dotted regions ($h_{2\,ys} < 0$). While dilation bands are predicted in the checkered region, the required critical hardening modulus value, $h_{2\,ys} < -G/2$, is improbable.

the critical hardening modulus is always non-negative.

Results for ASE for $k > 0$

As no experimental results regarding dilation-band formation in high-porosity materials loaded on path ASE2 are available to guide theoretical analyses, for simplicity we choose to examine values similar to the ASC case: $k = 0$ and $0 < k < G/5$. These rough assumptions serve as a framework with which to begin a discussion of dilation band conditions for the two-yield surface model. Future experimental work will provide improved bounds for more detailed analyses. The case of $k = 0$ was discussed above, and it was determined that dilation bands are predicted, with $h_{2\,ys} = 0$ (localization at peak stress), for all negative values of β_2 and μ_2 except those representing improbable deviations from cap normality. As, for ASC tests, k is observed to decrease with ongoing inelastic loading, but typically remains positive, it is assumed that behaviour under ASE will be similar, and only non-negative values of k are examined. The analysis also focuses on results for negative values of β_2 and μ_2, which are reasonable assumptions for the cap, which is typically depicted as negative-sloping and where inelastic volume strain is compactive.

In the ASC analysis, predicted band angles were found to vary significantly with β_2 and μ_2, but were less sensitive to changes in β_1 and μ_1. However, for ASE, band orientation predictions are more sensitive to changes in β_1 and μ_1. Therefore, results are presented in the β_1–μ_1 plane for selected values of k, β_2 and μ_2. Only positive values of β_1 and μ_1 are considered here, corresponding to a positively sloping shear yield surface with dilatant behaviour. These assumptions are consistent with observed high porosity sandstone behaviour for loading paths that clearly intersect only the shear yield surface. For example, for tests on path ASE1, Bésuelle (2001) reports values of $0.37 < \beta < 0.59$ and $0.97 < \mu < 1.15$ for Vosges sandstone.

The hardening modulus, h, is again assumed to decrease with continued inelastic loading; therefore, the predicted band orientation is that which has the largest value of the critical hardening modulus, for a given set of material parameter values. Predicted band angles are shown graphically in the β_1–μ_1 plane, assuming $\beta_2 = \mu_2 = -\sqrt{3}/4$ for $k = G/10$ (Fig. 11a) and $k = G/5$ (Fig. 11b). Lines of constant band angle are labeled. Recall that the band angle is defined as the angle between the band normal and the direction maximum compression (for ASE, this is also

the angle between the band and the axial direction). Near the axes, if either β_1 or μ_1 is small (approaching zero), shear bands are predicted, with band angles as low as 77° for $k = G/10$ and 67° for $k = G/5$. Dilation bands are predicted for larger positive values of β_1 and μ_1. While normality was assumed on the cap, deviations from normality that are not improbably large provide results similar to those shown in Figure 11. In addition, for values of β_2 and μ_2 less negative than $-\sqrt{3}/4$, the dilation band line moves closer to the origin, so dilation bands are possible for smaller values of β_1 and μ_1. For values of β_2 and μ_2 more negative than $-\sqrt{3}/4$, the dilation band line moves away from the origin, so dilation band formation requires larger values of β_1 and μ_1.

Unfortunately, as we know of no ASE tests conducted using a load path where both yield surfaces could be active, experimental data are not available for comparison. The ASE1 load path and confining pressures used by Bésuelle (2001) seem likely to intersect the shear yield surface only. However, for ASE2 paths that meet the shear yield surface near the intersection with the cap, it is likely that the material response due to the shear yield surface will be less, not more, dilatant, and that the slope of the yield surface will be less, not more, positive. Therefore, it is hypothesized that the values of β_1 and μ_1 near the cap intersection will be similar to or less positive than those reported by Bésuelle (2001). These values of β_1 and μ_1 are represented in Figure 11 by the small square. For $k = G/10$ only dilation bands are predicted, while for $k = G/5$ dilation bands or high angle shear bands (band angle greater than 77°) are predicted. If actual values of β_1 and μ_1 are smaller than those reported by Bésuelle (2001), but still positive, then dilation bands could still be predicted, particularly for smaller values of k. While these results are promising, data from ASE experiments using a stress path that intersects the region where the two surfaces meet is clearly required to fully evaluate theoretical findings.

The effect of $k > 0$ on the critical hardening modulus, $h_{1\,ys}$, is examined next. When β_2 and μ_2 are negative, then $\eta_2^\perp < 0$, and the denominator of equation (3) is always positive (for ASE, the $\perp$ superscript refers to dilation bands). Recalling that $C^\perp \equiv 0$, the sign of $h_{2\,ys}^\perp$ depends only on the sign of $\eta_1^\perp$, which is positive when $(\beta_1 - \sqrt{3}/2)(\mu_1 - \sqrt{3}/2) < 0$, a condition easily satisfied by some appropriate deviation from normality. For example, the values of β_1 and μ_1 reported by Bésuelle (2001) for ASE satisfy this condition. Therefore, it is

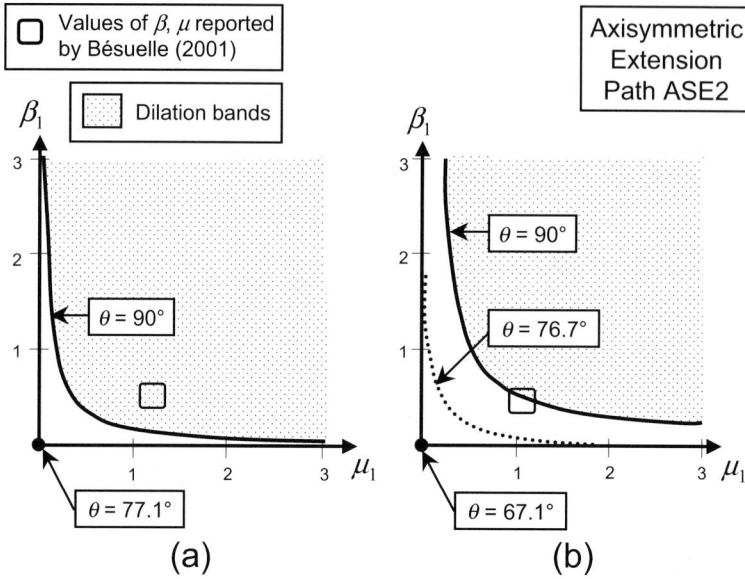

Fig. 11. Band orientation predictions, as a function of β_1 and μ_1 for axisymmetric extension loading (path ASE2) for the two-yield surface model ($\beta_2 = \mu_2 = -\sqrt{3}/4$), when (**a**) $k = G/10$ and, (**b**) $k = G/5$. Dilation bands are predicted in the dotted regions, while high angle shear bands are predicted in the unshaded regions. The square represents values of β and μ when dilation bands or high angle dilating shear bands were observed for path ASE1 tests.

possible, perhaps probable, that $h^{\perp}_{2\,\mathrm{ys}}$ will be positive for $k > 0$, while $k = 0$ requires $h^{\perp}_{2\,\mathrm{ys}} \equiv 0$. Thus, $k > 0$ promotes dilation-band formation in a loading program where h decreases with ongoing inelastic deformation. Note that this contrasts with the ASC results, where $k > 0$ results in a negative critical hardening modulus for compaction bands, inhibiting localization.

Summary and conclusions

Compaction bands, zones of pure compressional deformation oriented perpendicular to maximum compression, and dilation bands, zones of pure dilational deformation oriented perpendicular to minimum compression (maximum extension), were recently observed in high porosity sandstone in both field and laboratory settings. These deformation bands represent the end members in a spectrum of possible shear band orientations with corresponding band strain (e.g. shear strain accompanied by either compactional or dilatational deformation normal to the band). This work focused on examining theoretical conditions for compaction band and dilation band formation in porous rock, including the influence of stress state, stress path and constitutive model. Results were found to compare

favourably with available experimental observations.

The Rudnicki & Rice (1975) bifurcation theory was used to determine localization conditions. As this formulation suggests that band formation is possible due to a constitutive instability, the resulting localization conditions are sensitive to the choice of constitutive model. At low confining pressures, high porosity sandstone experiences dilatant microcracking; at high confining pressures, grain crushing and pore collapse predominate; while at intermediate confining pressures, both processes are active. This suggests use of a two-yield surface constitutive model, which macroscopically represents the two microstructural damage processes. This model consists of a positively sloping shear yield surface ($\mu_1 > 0$) corresponding to dilatant deformation ($\beta_1 > 0$) at lower mean stresses, and a negatively sloping cap ($\mu_2 < 0$) corresponding to compactant behaviour ($\beta_2 < 0$) at higher mean stresses.

Localization conditions also vary with stress state. Axisymmetric compression (ASC), where the axial stress is more compressive than the confining pressure, is the most favourable stress state for compaction band formation. However, axisymmetric extension (ASE), where the confining pressure is more compressive than the axial stress, is the most favourable stress state for dilation band formation. For a given stress state, multiple stress paths are possible. The chosen stress path determines which portion of the yield envelope is accessible and, therefore, which constitutive model is appropriate.

For stress paths that clearly intersect only one yield surface, the localization conditions determined using a single-yield surface model correlate well with experimental observations. Bésuelle (2001) found good agreement between observed and predicted band orientations/strain types (compacting shear bands, dilating shear bands and dilation bands) for ASC and ASE tests on Vosges sandstone, where loading paths clearly intersected only the shear yield surface. Wong et al. (1992, 1997) observed uniform compaction for high confining pressure ASC tests where the load path intersects only the cap. This is in accordance with theoretical predictions: for loading paths intersecting the cap near the mean stress axis, normality is appropriate, such that the value of the critical hardening modulus required for localization is prohibitively negative.

Compaction bands, and sometimes shear bands or conjugate shear bands, are observed under ASC at intermediate confining pressures, corresponding to the brittle–ductile transition (Olsson 1999; Wong et al. 2001). In this transitional regime, as both damage processes are active, the two-yield surface model may be applicable. The need for a different constitutive model in this regime is also evident when examining theoretical localization conditions: the single-yield surface model does not predict the observed compaction bands for reported values of key material parameters. Using the two-yield surface model, Challa & Issen (2004) determined that compaction band formation is favoured at peak stress, when the bulk hardening modulus, k (slope of the mean stress–inelastic volume strain curve), and hardening modulus, h (slope of the shear stress–inelastic shear strain curve), are both zero. When $k = 0$, both the observed compaction bands and shear bands are predicted for reported values of β and μ. Shear bands require some deviation from cap normality, and the critical hardening modulus for shear localization is infinitely large, possibly corresponding to activation of the shear yield surface by pushing out a previously activated cap.

Compaction band conditions were found to be less favourable when $0 < k \leq G/10$, corresponding to localization either prior to the peak or along the mildly sloping stress plateau often observed in high porosity sandstone. As the value of k increases compaction bands are

predicted for a reduced range of material parameter values, and if $k = G/10$ compaction bands are no longer predicted for strongly negative values of β_2 and μ_2. The critical value of the hardening modulus, h, at the onset of localization is negative for $k > 0$, implying localization later in the inelastic loading program.

While compaction bands, shear bands and/or conjugate shear bands are observed at intermediate confining pressures, very low angle shear bands are rarely reported. It is difficult to explain the lack of low angle shear band observations using the single-yield surface model, where a continuum of increasing band angles is predicted for increasing values of $\beta + \mu$. However, using the two-yield surface model, a discontinuity exists in the predicted band angle in the region of reported β and μ values. Therefore, if the β_2 and μ_2 values corresponding to a compaction band were perturbed only slightly, moderate angle shear bands would then be predicted, while very low angle shear bands require a larger perturbation. In addition, in a relatively small region of β_2 and μ_2 values, which include the reported β and μ values, two distinct shear band angles are theoretically possible for a single set of β_2 and μ_2 values. This feature enables the model to predict the wide range of shear band angles observed for ASC loading of high porosity sandstone at intermediate confining pressures.

The traditional ASE loading path (constant confining pressure with decreasing axial compression/increasing extension) will typically intersect the shear yield surface only. Dilation bands are possible for dilatant behaviour, where $\beta_1 + \mu_1 \geq \sqrt{3}$. The critical hardening modulus is positive for sufficient deviations from normality: $(\beta_1 - \sqrt{3}/2)(\mu_1 - \sqrt{3}/2) \leq 0$. Although fairly restrictive, these conditions have been satisfied experimentally (Bésuelle 2001). However, using an alternate ASE loading path (e.g. increasing confining pressure with constant axial compression), it is possible to intersect the yield envelope in the region where the two-yield surface model is applicable for high porosity materials. In this case, the dilation band conditions were found to be significantly less restrictive than the single-yield surface model. If localization occurs at peak stress, $k = 0$ and $h = 0$, then dilation bands are predicted exclusively for all negative values of β_2 and μ_2 (except those representing improbable deviations from normality), and any values of β_1 and μ_1 that satisfy the convex corner requirement: $(\beta_1 - \beta_2)(\mu_1 - \mu_2) > 0$. If $k > 0$, the most favourable conditions for dilation-band formation occur when k is small, β_2 and μ_2 are small magnitude negative numbers, and β_1 and μ_1 are positive, but not too small. Larger values of β_1 and μ_1 are required to form dilation bands as k gets larger or as β_2 and μ_2 become more negative. Although experimental data for this stress path are not available, using the values β_1 and μ_1 reported by Bésuelle (2001) for traditional ASE stress paths, dilation bands and dilating shear bands would be predicted for moderately negative values of β_2 and μ_2 when $0 < k \leq G/5$. In addition, when $k > 0$, the critical hardening modulus is positive, suggesting possible localization prior to peak stress.

Although theoretical results suggest experimental conditions favourable for dilation band formation, it may be difficult to unequivocally identify dilation bands. Recall that dilation bands are defined as a zone of increased porosity, in contrast with joints or planar opening-mode fractures, which consist of two discrete surfaces. Furthermore, the space between the two discrete surfaces of a joint or opening-mode fracture is empty, while the dilation band is several grain diameters wide and contains grains with increased spacing (thus the increased porosity) when compared with grain spacing in the surrounding non-band material. The photomicrograph of a dilation band observed by Du Bernard et al. (2002, p. 29-3, fig. 3) clearly shows such a zone of increased porosity. However, capturing the formation of such a zone may be difficult in some materials and/or settings. For example, if a dilation band were to form in a traditional laboratory ASE test on sandstone, continued loading would probably break enough grain bonds to separate the specimen into two portions along the dilation band. This process could leave little or no evidence that the dilation band ever existed, as the resulting two discrete surfaces may be indistinguishable from a through-going crack.

Results presented in this paper suggest that the localization theory of Rudnicki & Rice (1975), originally developed for shear localization in low porosity rock, is useful in predicting localized band formation in high porosity sandstone, with the following caveat. Localization conditions determined using this approach are sensitive to details in the constitutive relation. Therefore, care must be taken in selecting a constitutive model that adequately represents the material behaviour for the stress path of interest. While a single-yield surface model was found to be sufficient for loading in either the brittle or the ductile regimes, a two-yield surface model better represented the observed localization conditions in the brittle–ductile transition.

After the proper constitutive model is selected or developed, localization theory can be used to predict possible formation of yet-unobserved band types. For example, the basis for theoretical compaction band and dilation band conditions was identified by Rudnicki & Rice (1975), and subsequently refined by others, years before the recent field and laboratory observations of these band types. Similarly, in this paper, localization theory was employed to predict deformation modes for infrequently explored stress states and stress paths. The compaction bands predicted using the two-yield surface model are now observed frequently in axisymmetric compression experiments. Using this same model, the conditions for dilation band formation under axisymmetric extension are significantly less restrictive than those for the aforementioned compaction bands. This suggests that dilation band formation is at least possible, if not likely, in high porosity rock loaded in axisymmetric extension in the brittle–ductile transition regime. This intriguing theoretical result clearly suggests that further experimental examination is warranted.

The work presented here is of broad interest to several disciplines. The presence or formation of compaction bands or dilation bands could significantly affect applications such as drilling, fluid extraction, injection and sequestration, as these bands could act as barriers to, or conduits for, fluid flow. Strain localization, which occurs at many scales in the Earth's crust, is of fundamental importance to the tectonics and geophysics fields. In addition, localized deformation modes, similar to compaction bands, have also been observed in other porous materials such as metal foam and honeycombs. Therefore, these results may also be applied towards improving the understanding of the mechanics of all porous materials.

Financial support for this research was provided by the National Science Foundation, Directorate for Geosciences, Division of Earth Sciences via grants EAR-0106932 and EAR-0310085 to Clarkson University, in collaboration with T.-F. Wong of SUNY Stony Brook. The authors thank P. Baud and T.-F. Wong for providing the mechanical data for Bentheim sandstone.

References

ANDREWS, E., SANDERS, W. & GIBSON, L.J. 1999. Compressive and tensile behavior of aluminum foams. Materials Science and Engineering, **A270**, 113–124.

BART-SMITH, H., BASTAWROS, A.F., MUMM, D.R., EVANS, A.G., SYPECK, D.J. & WADLEY, H.N.G. 2000. Compressive deformation and yielding mechanisms in cellular Al alloys determined using X-ray tomography and surface strain mapping. *Acta Materialia*, **46**, 3583–3592.

BASTAWROS, A., BART-SMITH, H. & EVANS, A.G. 2000. Experimental analysis of deformation mechanisms in a closed cell aluminum foam. *Journal of the Mechanics and Physics of Solids*, **48**, 301–322.

BAUD, P., ZHU, W. & WONG, T.-F. 2000. Failure mode and weakening effect of water on sandstone. *Journal of Geophysical Research*, **105**, 16 371–16 389.

BAUD, P., KLEIN, E. & WONG, T.-F. 2004. Compaction localization in porous sandstones: spatial evolution of damage and acoustic emission activity. *Journal of Structural Geology*, **26**, 603–624.

BÉSUELLE, P. 2001. Compacting and dilating shear bands in porous rock: Theoretical and experimental conditions. *Journal of Geophysical Research*, **106**, 13 435–13 442.

BÉSUELLE, P., DESRUES, J. & RAYNAUD, S. 2000. Experimental characterization of the localization phenomenon inside a Vosges sandstone in a triaxial cell. *International Journal of Rock Mechanics and Mining Sciences*, **37**, 1223–1237.

CHALLA, V. & ISSEN, K.A. 2004. Conditions for compaction band formation in porous rock using a two yield surface model. *Journal of Engineering Mechanics*, **130**, 1089–1097.

DiGIOVANNI, A.A., FREDRICH, J.T., HOLCOMB, D.J. & OLSSON, W.A. 2000. Micromechanics of compaction in an analogue reservoir sandstone. *In*: GIRARD, J., LIEBMAN, M., BREEDS, C. & DOE, T. (eds) *Proceedings of the North American Rock Mechanics Symposium*. A.A. Balkema, Amsterdam. 1153–1158.

DiMAGGIO, F.L. & SANDLER, I.S. 1971. Material model for granular soils. *Journal of the Engineering Mechanics Division, American Society of Civil Engineers*, **97**, 935–949.

DU BERNARD, X., EICHHUBL, P. & AYDIN, A. 2002. Dilation bands: A new form of localized failure in granular media. *Geophysical Research Letters*, **29**, 2176, doi:10.1029/2002GL015966.

GRADINGER, R. & RAMMERSTORFER, F.G. 1998. On the influence of meso-inhomogeneities on the crush worthiness of metal foams. *Acta Materialia*, **47**, 143–148.

HAIMSON, B.C. 2001. Fracture-like borehole breakouts in high porosity sandstone: Are they caused by compaction bands? *Physics and Chemistry of the Earth (A)*, **26**, 15–20.

HAIMSON, B.C. & SONG, I. 1998. Borehole breakouts in Berea sandstone: two porosity-dependent distinct shapes and mechanisms of formation. *Rock Mechanics in Petroleum Engineering, Society of Petroleum Engineers*, **1**, 229–238.

HOLCOMB, D.J. & OLSSON, W.A. 2003. Compaction localization and fluid flow. *Journal of Geophysical Research*, **108**(B6), 2290.

HOLCOMB, D.J. & RUDNICKI, J.W. 2001. Inelastic constitutive properties and shear localization in Tennessee marble. *International Journal of Numerical and Analytical Methods in Geomechanics*, **25**, 109–129.

ISSEN, K.A. 2002. The influence of constitutive models on localization conditions for porous rock. *Engineering Fracture Mechanics*, **69**, 1891–1906.

ISSEN, K.A. & RUDNICKI, J.W. 2000. Conditions for compaction bands in porous rock. *Journal of Geophysical Research*, **105**, 21 529–21 536.

ISSEN, K.A. & RUDNICKI, J.W. 2001. Theory of compaction bands in porous rock. *Physics and Chemistry of the Earth (A)*, **26**, 95–100.

KLEIN, E., BAUD, P., REUSCHLÉ, T. & WONG, T.-F. 2001. Mechanical behavior and failure mode of Bentheim sandstone under triaxial compression. *Physics and Chemistry of the Earth (A)*, **26**, 21–25.

KOITER, W.T. 1960. General theorems for elastic–plastic solids. *In*: SNEDDON I.N. & HILL, R. (eds) *Progress in Solid Mechanics*. North-Holland, Amsterdam, 165–221.

MENÉNDEZ, B., ZHU, W. & WONG, T.-F. 1996. Micromechanics of brittle faulting and cataclastic flow in Berea sandstone. *Journal of Structural Geology*, **18**, 1–16.

MOLLEMA, P.N. & ANTONELLINI, M.A. 1996. Compaction bands: a structural analog for anti-mode I cracks in Aeolian sandstone. *Tectonophysics*, **267**, 209–228.

OLSSON, W.A. 1999. Theoretical and experimental investigation of compaction bands. *Journal of Geophysical Research*, **104**, 7219–7228.

OLSSON, W.A. & HOLCOMB, D.J. 2000. Compaction localization in porous rock. *Geophysical Research Letters*, **27**, 3537–3540.

OTTOSEN, N.S. & RUNESSON, K. 1991. Properties of discontinuous bifurcation solutions in elastoplasticity. *International Journal of Solids and Structures*, **27**, 401–421.

PAPKA, S.D. & KYRIAKIDES, S. 1998a. In-plane crushing of a polycarbonate honeycomb. *International Journal of Solids and Structures*, **35**, 239–267.

PAPKA, S.D. & KYRIAKIDES, S. 1998b. Experiments and full-scale numerical simulations of in-plane crushing of a honeycomb. *Acta Materialia*, **46**, 2765–2776.

PARK, C. & NUTT, S.R. 2001. Anisotropy and strain localization in steel foam. *Material Science and Engineering*, **A299**, 68–74.

PERRIN, G. & LEBLOND, L.B. 1993. Rudnicki and Rice's analysis of strain localization revisited. *Journal of Applied Mechanics*, **60**, 842–846.

RICE, J.R. 1975. On the stability of dilatant hardening for saturated rock masses. *Journal of Geophysical Research*, **80**, 1531–1536.

RUDNICKI, J.W. 2004. Shear and compaction band formation on an elliptic yield cap. *Journal of Geophysical Research – Solid Earth*, **109**, B03402, doi: 10.1029/2003JB002633.

RUDNICKI, J.W. & RICE, J.R. 1975. Conditions for the localization of deformation in pressure-sensitive dilatant materials. *Journal of Mechanics and Physics of Solids*, **23**, 371–394.

RUDNICKI, J.W & OLSSON, W.A. 1998. Reexamination of fault angles predicted by shear localization theory. *International Journal of Rock Mechanics and Mining Sciences*, **35**, 4–5.

WAWERSIK, W.R., ORR F.M., JR. *ET AL.* 2001. Terrestrial sequestration of CO_2: An assessment of research needs. *Advances in Geophysics*, **43**, 97–177.

WONG, T.-F., SZETO, H. & ZHANG, J. 1992. Effect of loading path and porosity on the failure mode of porous rocks. *Applied Mechanical Review*, **45**, 281–293.

WONG, T.-F., DAVID, C. & ZHU, W. 1997. The transition from brittle faulting to cataclastic flow in porous sandstones: mechanical deformation. *Journal of Geophysical Research*, **102**, 3009–3025.

WONG, T.-F., BAUD, P. & KLEIN, E. 2001. Localized failure modes in compactant porous rock. *Geophysical Research Letters*, **28**, 2521–2524.

WU, X.Y., BAUD, P. & WONG, T.-F. 2000. Micromechanics of compressive failure and spatial evolution of anisotropic damage in Darley Dale sandstone. *International Journal of Rock Mechanics and Mining Sciences*, **37**, 143–160.

Index

Note: Page numbers in italic, e.g. *375*, refer to figures. Page numbers in bold, e.g. **176**, signify entries in tables.